SMART LIGHT-RESPONSIVE MATERIALS

SMART LIGHT-RESPONSIVE MATERIALS

Azobenzene-Containing Polymers and Liquid Crystals

EDITED BY

Yue Zhao
University of Sherbrooke
Sherbrooke, Quebec, Canada

Tomiki Ikeda
Tokyo Institute of Technology
Tokyo, Japan

A JOHN WILEY & SONS, INC., PUBLICATION

Published by John Wiley & Sons, Inc., Hoboken, New Jersey
Published simultaneously in Canada

Library of Congress Cataloging-in-Publication Data:

Zhao, Yue, 1961-
Smart light-responsive materials : azobenzene-containing polymers and liquid crystals/Yue Zhao and Tomiki Ikeda.
p. cm.
Includes index.
ISBN 978-0-470-17578-1 (cloth)
1. Smart materials. 2. Polymers–Optical properties. 3. Azo compounds. 4. Liquid crystals. I. Ikeda, Tomiki. II. Title.
TA418.9.S62.Z53 2009
620.1′1295–dc22

2008035493

Printed in the United States of America

10 9 8 7 6 5 4 3 2 1

CONTENTS

PREFACE

Azobenzene and its derivatives are fascinating molecules that display the reversible photoisomerization between the more stable trans and the less stable cis isomers. Although photoisomerization can result in important changes of properties for azobenzene molecules on their own, such as a large change in molecular shape and dipole moment, changes that can be imparted to polymers and liquid crystals when azobenzene is part of their structures or is associated with them are even more interesting. Indeed, reversible photoisomerization in azobenzene-containing polymers and liquid crystals enables the use of light as a powerful external stimulus to control or trigger the change of the properties of these two important classes of soft materials. For this reason, over the past two decades or so, there has been considerable worldwide research dedicated to azo–polymers and liquid crystals, ranging from fundamental studies to exploitation of applications. A number of important discoveries made in the 1990s have had a pivotal impact on this field. These include the surface relief grating that can readily be inscribed on azobenzene polymers using an interference pattern as a result of photoinduced mass transport and the isothermal photochemical liquid crystalline (LC)-to-isotropic (order-disorder) phase transitions because of the perturbation effect arising from the trans-cis photoisomerization. Today, the research field of azo-functional materials remains extremely active.

Although most research in the 1990s dealt mainly with the physical and optical properties of azo–polymers and liquid crystals important for optical information storage and switching, the field has witnessed important new developments and directions over the past 6 to 7 years. Amazing new phenomena continue to be discovered, such as the light-controllable bending of cross-linked LC azo–polymers, which, by showing how drastic the effect of azo–photoisomerization can have on a material, and offer new appealing opportunities. In recent years, there have been increasing efforts toward the development of diverse functional materials through rational molecular and materials designs that make use of established knowledge, and newer applications other than information storage and switching have emerged. Convinced that ongoing and future research on azobenzene-based light-enabled smart materials will have great potential and impact on both fundamental and applied research, we think it is time to edit a book that, by reviewing recent developments and showing perspectives, provides a forum for discussion and exchange of new ideas. We would like to thank all the contributors for their great effort in helping us put together a book that should not only benefit researchers who work on azo–polymers and liquid crystals but should also be of interest to those who develop light-responsive materials without using azobenzene, as many of the discussed strategies and ideas about azobenzene could be adapted to other chromophores.

The vitality and sustained interest of this field can easily be noticed from the many research papers on azobenzene-based materials that continue to appear. Obviously, this book cannot cover all new, post-2000 developments. As editors, we have tried to ensure that all chapters are relevant to the theme of the book, with regard to research works that promise development of light-enabled smart materials based mainly on azo–polymers and liquid crystals. Despite the apparent diversity of the topics covered in this book, the cohesion of all chapters and the link between the different chapters are solid. Chapter 1 (Yager and Barrett) introduces basic azobenzene photochemistry, photophysics, and the wide variety of azo–materials. In Chapter 2 (Stumpe et al.), which reviews and discusses the photoinduced phenomena in supramolecular azo–materials, the basic background is set to help the general readership understand the fundamental aspects involved in and the ideas and interests behind the various types of smart azo–materials discussed in this book. This is followed by three chapters on photoinduced motion and the photomechanical effect of LC azo–polymers; (Chapter 3, Yu and Ikeda), amorphous azo–polymers (Chapter 4, Yager and Barrett), and colloidal particles (Chapter 5, Wang). The conversion of photoenergy into mechanical energy is certainly a major new direction in the field. In contrast to the colloidal particles self-assembled by amphiphilic random copolymers, micellar aggregates are the subjects of Chapter 6 (Zhao) and Chapter 7 (Tribet) and address the self-assembly of amphiphilic block copolymers and hydrophobically modified polymers, respectively. Solution self-assembled light-responsive micro- and nanostructures of azo–polymers and their potential applications as discussed in these chapters represent another exciting new research direction. Likewise, the research works presented in Chapter 8 (Seki) and Chapter 9 (Watanabe) explore azo–polymers in two dimensions and on surface. The next two chapters, Chapter 10 (Kurihara) and Chapter 11 (Zhao), mainly concern smart light-sensitive materials of small-molecule LCs. The last two chapters, Chapter 12 (Yu and Ikeda) and Chapter 13 (Liu and Brinker), provide excellent examples of new azo–materials and architectures, with a focus on azo-block copolymers in the solid state and azo-hydride silica materials, respectively.

Research on azobenzene-based smart materials is dynamically progressing. We hope this book gives a critical review of the new developments and shows new directions. However, what we want most for this book to accomplish is to generate interest among graduate students and young researchers in this exciting field and help spark their imaginations with ideas for creative research. This is essential to ensure further research and development, and to help maintain the excitement in this field for many years to come.

YUE ZHAO
TOMIKI IKEDA

Sherbrooke, Quebec, Canada
Tokyo, Japan
January 2009

CONTRIBUTORS

PROFESSOR CHRISTOPHER J. BARRETT Department of Chemistry, McGill University, Montreal, Canada

PROFESSOR C. JEFFREY BRINKER Advanced Materials Laboratory, University of New Mexico and Sandia National Laboratories, Albuquerque, New Mexico

LEONID M. GOLDENBERG Fraunhofer Institute for Applied Polymer Research, Science Campus Golm, Potsdam, Germany

PROFESSOR TOMIKI IKEDA Chemical Resources Laboratory, Tokyo Institute of Technology, Yokohama, Japan

OLGA KULIKOVSKA Fraunhofer Institute for Applied Polymer Research, Science Campus Golm, Potsdam, Germany

PROFESSOR SEIJI KURIHARA Graduate School of Science and Technology, Kumamoto University, Kumamoto, Japan

DR. NANGUO LIU Specialty Chemicals Business, Dow Corning Corporation, Midland, Michigan

PROFESSOR TAKAHIRO SEKI Department of Molecular Design and Engineering, Graduate School of Engineering, Nagoya University, Chikusa, Nagoya, Japan

DR. JOACHIM STUMPE Fraunhofer Institute for Applied Polymer Research, Science Campus Golm, Potsdam, Germany

PROFESSOR CHRISTOPHE TRIBET Macromolecular Laboratory of Physics and Chemistry, University of Paris, Paris, France

PROFESSOR XIAOGONG WANG Department of Chemical Engineering, Tsinghua University, Beijing, China

DR. OSAMU WATANABE Toyota Central Research and Development Laboratories, Inc., Nagakute, Aichi, Japan

DR. KEVIN G. YAGER Polymers Division, National Institute of Standards and Technology, Gaithersburg, Maryland

PROFESSOR HAIFENG YU Department of Chemistry, Nagaoka University of Technology, Nagaoka, Japan

PROFESSOR YANLEI YU Department of Materials Science, Fudan University, Shanghai, China

YURIY ZAKREVSKYY Fraunhofer Institute for Applied Polymer Research, Science Campus Golm, Potsdam, Germany

PROFESSOR YUE ZHAO Polymer and Liquid Crystal Laboratory, Department of Chemistry, University of Sherbrooke, Québec, Canada

1

AZOBENZENE POLYMERS FOR PHOTONIC APPLICATIONS

Kevin G. Yager and Christopher J. Barrett

1.1. INTRODUCTION TO AZOBENZENE

Azobenzene, with two phenyl rings separated by an azo (–N=N–) bond, serves as the parent molecule for a broad class of aromatic azo compounds. These chromophores are versatile molecules, and have received much attention in research areas both fundamental and applied. The strong electronic absorption maximum can be tailored by ring substitution to fall anywhere from the ultraviolet (UV) to visible red regions, allowing chemical fine-tuning of color. This, combined with the fact that these azo groups are relatively robust and chemically stable, has prompted extensive study of azobenzene-based structures as dyes and colorants. The rigid mesogenic shape of the molecule is well suited to spontaneous organization into liquid crystalline (LC) phases, and hence polymers doped or functionalized with azobenzene-based chromophores (azo polymers) are common as LC media. With appropriate electron-donor–acceptor ring substitution, the π electron delocalization of the extended aromatic structure can yield high optical nonlinearity, and zo chromophores have seen extensive study for nonlinear optical applications as well. One of the most interesting properties of these chromophores however, and the main subject of this review, is the readily induced and reversible isomerization about the azo bond between the trans and cis geometric isomers and the geometric changes that result when azo chromophores are incorporated into polymers and other materials. This light-induced interconversion allows systems incorporating azobenzenes to be used as photoswitches, effecting rapid and reversible control over a variety of chemical, mechanical, electronic, and optical properties.

Smart Light-Responsive Materials. Edited by Yue Zhao and Tomiki Ikeda

Perhaps of a range as wide as the interesting phenomena displayed by azo aromatic compounds is the variety of molecular systems into which these chromophores can be incorporated. In addition to LC media and amorphous glasses, azobenzenes can be incorporated into self-assembled monolayers and superlattices, sol–gel silica glasses, and various biomaterials. The photochromic or photoswitchable nature of azobenzenes can also be used to control the properties of novel small molecules, using an attached aromatic azo group. A review will be presented here of the photochemical and photophysical nature of chromophores in host polymers, the geometric and orientational consequences of this isomerization, and some of the interesting ways in which these phenomena have been expolited recently to exert control over solution and biochemical properties using light. This photoisomerization can be exploited as a photoswitch to orient the chromophore (which induces birefringence), or even to perform all-optical surface topography patterning. These photomotions enable many interesting applications, ranging from optical components and lithography to sensors and smart materials.

1.1.1. Azobenzene Chromophores

In this text, as in most on the subject, we use "azobenzene" and "azo" in a general way: to refer to the class of compounds that exhibit the core azobenzene structure, with different ring substitution patterns (even though, strictly, these compounds should be referred to as "diazenes"). There are many properties common to nearly all azobenzene molecules. The most obvious is the strong electronic absorption of the conjugated π system. The absorption spectrum can be tailored, via the ring substitution pattern, to lie anywhere from the UV to the visible red region. It is not surprising that azobenzenes were originally used as dyes and colorants, and up to 70% of the world's commercial dyes are still azobenzene-based (Zollinger, 1987, 1961). The geometrically rigid structure and large aspect ratio of azobenzene molecules make them ideal mesogens: azobenzene small molecules and polymers functionalized with azobenzene can exhibit LC phases (Möhlmann and van der Vorst, 1989; Kwolek et al., 1985). The most startling and intriguing characteristic of the azobenzenes is their highly efficient and fully reversible photoisomerization. Azobenzenes have two stable isomeric states, a thermally stable trans configuration and a metastable cis form. Remarkably, the azo chromophore can interconvert between these isomers upon absorption of a photon. For most azobenzenes, the molecule can be optically isomerized from trans to cis with light anywhere within the broad absorption band, and the molecule will subsequently thermally relax back to the trans state on a timescale dictated by the substitution pattern. This clean photochemistry is central to azobenzene's potential use as a tool for nanopatterning.

Azobenzenes can be separated into three spectroscopic classes, well described by Rau (1990): azobenzene-type molecules, aminoazobenzene-type molecules, and pseudo-stilbenes (refer to Fig. 1.1 for examples). The particulars of their absorption spectra (shown in Fig. 1.2) give rise to their prominent colors: yellow, orange, and red, respectively. Many azos exhibit absorption characteristics similar to the unsubstituted azobenzene archetype. These molecules exhibit

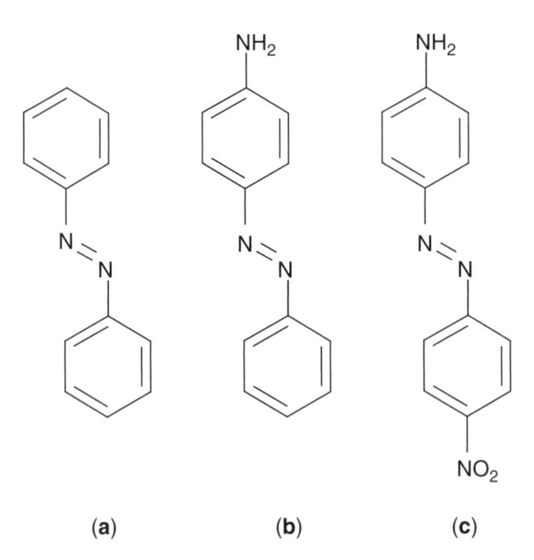

Figure 1.1. Examples of azomolecules classified as (**a**) azobenzenes, (**b**) aminoazobenzenes, and (**c**) pseudo-stilbenes.

a low intensity n→π* band in the visible region and a much stronger π→π* band in the UV. Although the n→π* is symmetry-forbidden for *trans*-azobenzene (C_{2h}), vibrational coupling and some extent of nonplanarity nevertheless make it observable (Rau, 1968).

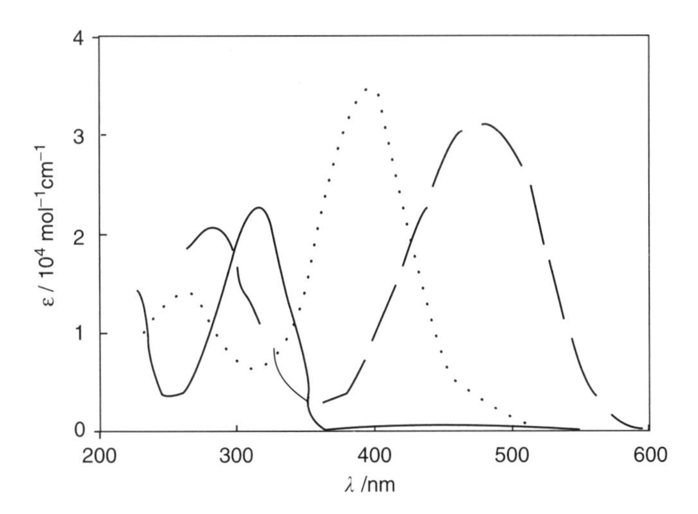

Figure 1.2. Schematic of typical absorbance spectra for *trans*-azobenzenes. The azobenzene-type molecules (*solid line*) have a strong absorption in the UV, and a low intensity band in the visible (barely visible in the graph). The aminoazobenzenes (*dotted line*) and pseudo-stilbenes (*dashed line*) typically have strong overlapped absorptions in the visible region.

Adding substituents to the azobenzene rings may lead to minor or major changes in spectroscopic character. Of particular interest is ortho- or para-substitution with an electron-donating group (usually an amino, $-NH_2$), which results in a new class of compounds. In these aminoazobenzenes, the $n \rightarrow \pi^*$ and $\pi \rightarrow \pi^*$ bands are much closer. In fact, the $n \rightarrow \pi^*$ may be completely buried beneath the intense $\pi \rightarrow \pi^*$. Although azobenzenes are fairly insensitive to solvent polarity, aminoazobenzene absorption bands shift to higher energy in nonpolar solvents and shift to lower energy in polar solvents. Substituting azobenzene at the 4 and 4′ positions with an electron-donor and an electron-acceptor (such as an amino and a nitro, $-NO_2$, group) leads to a strongly asymmetric electron distribution (often referred to as a "push–pull" substitution pattern). This shifts the $\pi \rightarrow \pi^*$ absorption to lower energy, toward the red and past the $n \rightarrow \pi^*$. This reversed ordering of the absorption bands defines the third spectroscopic class, the pseudo-stilbenes (in analogy to stilbene, phenyl–C=C–phenyl). The pseudo-stilbenes are very sensitive to local environment, which can be useful in some applications.

Especially in condensed phases, the azos are also sensitive to packing and aggregation. The π–π stacking gives rise to shifts of the absorption spectrum. If the azo dipoles have a parallel (head-to-head) alignment, they are called J-aggregates, and give rise to a redshift of the spectrum (bathochromic) as compared with the isolated chromophore. If the dipoles are antiparallel (head-to-tail), they are called H-aggregates and lead to a blueshift (hypsochromic). Fluorescence is seen in some aminoazobenzenes and many pseudo-stilbenes but not in azobenzenes, whereas phosphorescence is absent in all the three classes. By altering the electron density, the substitution pattern necessarily affects the dipole moment, and in fact all the higher order multipole moments. This becomes significant in many nonlinear optical (NLO) studies. For instance, the chromophore's dipole moment can be used to orient with an applied electric field (poling), and the higher order moments of course define the molecule's nonlinear response (Delaire and Nakatani, 2000). In particular, the strongly asymmetric distribution of the delocalized electrons that results from push–pull substitution results in an excellent NLO chromophore.

1.1.2. Azobenzene Photochemistry

Key to some of the most intriguing results and interesting applications of azobenzenes is the facile and reversible photoisomerization about the azo bond, converting between the trans (E) and cis (Z) geometric isomers (Fig. 1.3). This photoisomerization is completely reversible and free from side reactions, prompting Rau to characterize it as "one of the cleanest photoreactions known."(Rau, 1990) The trans isomer is more stable by $\sim 50\,kJ\,mol^{-1}$ (Mita et al., 1989; Schulze et al., 1977), and the energy barrier to the photoexcited state (barrier to isomerization) is on the order of $200\,kJ\,mol^{-1}$ (Monti et al., 1982). Thus, in the dark, most azobenzene molecules will be found in the trans form. On absorption of a photon (with a wavelength in the trans absorption band), the azobenzene will convert, with high efficiency, into the cis isomer. A second wavelength of light (corresponding to the cis absorption band) can cause the back-conversion.

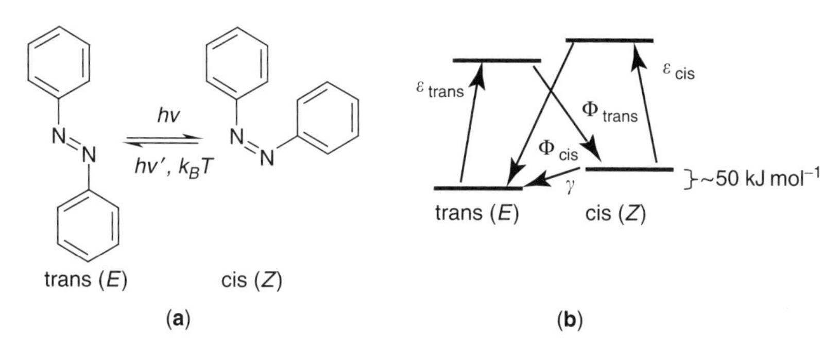

Figure 1.3. **(a)** Azobenzene can convert between trans and cis states photochemically and relaxes to the more stable trans state thermally. **(b)** Simplified state model for azobenzenes. The trans and cis extinction coefficients are denoted by ε_{trans} and ε_{cis}. The Φ refer to quantum yields of photoisomerization, and γ is the thermal relaxation rate constant.

These sphotoisomerizations usually have picosecond timescales (Lednev et al., 1996; Kobayashi et al., 1979). Alternately, azos will thermally reconvert from the cis into trans state, with a timescale ranging from milliseconds to hours, depending on the substitution pattern and local environment. More specifically, the lifetimes for azobenzenes, aminoazobenzenes, and pseudo-stilbenes are usually on the order of hours, minutes, and seconds, respectively. The energy barrier for thermal isomerization is on the order of 90 kJ mol^{-1} (Brown and Granneman, 1975; Haberfield et al., 1975). Considerable work has gone into elongating the cis lifetime, with the goal of creating truly bistable photoswitchable systems. Bulky ring substituents can be used to hinder the thermal back reaction. For instance, a polyurethane main-chain azo exhibited a lifetime of 4 days (thermal rate constant of $k = 2.8 \times 10^{-6}\,s^{-1}$, at 3°C) (Lamarre and Sung, 1983), and an azobenzene parasubstituted with bulky pendants had a lifetime of 60 days ($k < 2 \times 10^{-7}\,s^{-1}$, at room temperature) (Shirota et al., 1998). The conformational strain of macrocylic azo compounds can also be used to lock the cis state, where lifetimes of 20 days ($k = 5.9 \times 10^{-7}\,s^{-1}$) (Norikane et al., 2003), 1 year (half-life 400 days, $k = 2 \times 10^{-8}\,s^{-1}$) (Rottger and Rau, 1996; Rau and Roettger, 1994), or even 6 years ($k = 4.9 \times 10^{-9}\,s^{-1}$) (Nagamani et al., 2005) were observed. Similarly, using the hydrogen bonding of a peptide segment to generate a cyclic structure, a cis lifetime of ~40 days ($k = 2.9 \times 10^{-7}\,s^{-1}$) was demonstrated (Vollmer et al., 1999). Of course, one can also generate a system that starts in the cis state and where isomerization (in either direction) is completely hindered. For instance, attachment to a surface (Kerzhner et al., 1983), direct synthesis of ringlike azo molecules (Funke and Gruetzmacher, 1987), and crystallization of the cis form (Hartley, 1938, 1937) can be used to maintain one state, but such systems are obviously not bistable photoswitches.

A bulk azo sample or solution under illumination will achieve a photostationary state, with a steady-state trans–cis composition based on the competing effects of photoisomerization into the cis state, thermal relaxation back to the trans state, and possibly cis reconversion upon light absorption. The steady-state composition is unique to each system, as it depends on the quantum yields for the two processes (Φ_{trans} and Φ_{cis}) and the thermal relaxation rate constant. The composition also depends on irradiation intensity, wavelength, temperature, and the matrix (gas phase, solution, liquid crystal, sol–gel, monolayer, polymer matrix, etc.). Azos are photochromic (their color changes on illumination), since the effective absorption spectrum (a combination of the trans and cis spectra) changes with light intensity. Thus absorption spectroscopy can be conveniently used to measure the cis fraction in the steady state (Rau et al., 1990; Fischer, 1967), and the subsequent thermal relaxation to an all-trans state (Beltrame et al., 1993; Hair et al., 1990; Eisenbach, 1980a; Gabor and Fischer, 1971). Nuclear magnetic resonance (NMR) spectroscopy can also be used (Magennis et al., 2005). Under moderate irradiation, the composition of the photostationary state is predominantly cis for azobenzenes, mixed for aminoazobenzenes, and predominantly trans for pseudo-stilbenes. In the dark, the cis fraction is below most detection limits, and the sample can be considered to be in an all-trans state. Isomerization is induced by irradiating with a wavelength within the azo's absorption spectrum, preferably close to λ_{max}. Modern experiments typically use laser excitation with polarization control, delivering on the order of 1–100 mW cm^{-2} of power to the sample. Various lasers cover the spectral range of interest, from the UV (Ar^+ line at 350 nm) through blue (Ar^+ at 488 nm), green (Ar^+ at 514 nm, YAG at 532 nm, HeNe at 545 nm), and into the red (HeNe at 633 nm, GaAs at 675 nm).

The ring substitution pattern affects both the trans and the cis absorption spectra, and for certain patterns, the absorption spectra of the two isomers overlap significantly (notably for the pseudo-stilbenes). In these cases, a single wavelength of light effectuates both the forward and reverse reaction, leading to a mixed stationary state and continual interconversion of the molecules. For some interesting azobenzene photomotions, this rapid and efficient cycling of chromophores is advantageous, whereas in cases where the azo chromophore is used as a switch, it is clearly undesirable.

The mechanism of isomerization has undergone considerable debate. Isomerization takes place either through a rotation about the N–N bond, with rupture of the π bond, or through inversion, with a semilinear and hybrizidized transition state, where the π bond remains intact (refer to Fig. 1.4). The thermal back-relaxation is agreed to be via rotation, whereas for the photochemical isomerization, both mechanisms appear viable (Xie et al., 1993). Historically, the rotation mechanism (as necessarily occurs in stilbene) was favored for photoisomerization, with some early hints that inversion may be contributing (Gegiou et al., 1968). More recent experiments, based on matrix or molecular constraints to the azobenzene isomerization, strongly support inversion (Altomare et al., 1997; Liu et al., 1992; Naito et al., 1991; Rau and Lueddecke, 1982). Studies using picosecond Raman and femtosecond fluorescence show a double bond (N=N) in the excited state, confirming the inversion mechanism (Fujino et al., 2001; Fujino

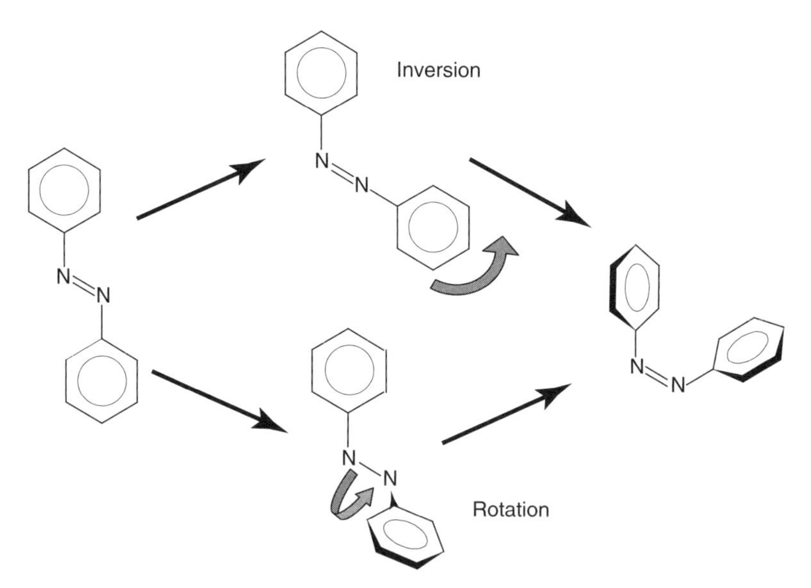

Figure 1.4. The mechanism of azobenzene isomerization proceeds either via rotation or inversion. The cis state has the phenyl rings tilted at 90° with respect to the CNNC plane.

and Tahara, 2000). In contrast, Ho et al. (2001) found evidence that the pathway is compound-specific: a nitro-substituted azobenzene photoisomerized via the rotation pathway. Furthermore, ab initio and density functional theory calculations indicate that both pathways are energetically accessible, although inversion is preferred (Angeli et al., 1996; Jursic, 1996). Thus, both mechanisms may be competing, with a different one dominating depending on the particular chromophore and environment. The emerging consensus nevertheless appears to be that inversion is the dominant pathway for most azobenzenes (Ikeda and Tsutsumi, 1995). The availability of the inversion mechanism explains how azos are able to isomerize easily even in rigid matrices, such as glassy polymers, since the inversion mechanism has a much smaller free volume requirement than rotation.

The thermal back-relaxation is generally first order, although a glassy polymer matrix can lead to anomalously fast decay components (Barrett et al., 1995, 1994; Paik and Morawetz, 1972; Priest and Sifain, 1971), attributed to a distribution of chromophores in highly strained configurations. Higher matrix crystallinity increases the rate of decay (Sarkar et al., 2001). The decay rate can act as a probe of local environment and molecular conformation (Tanaka et al., 2004; Norman and Barrett, 2002). The back-relaxation of azobenzene is acid catalyzed (Rau et al., 1981), although strongly acidic conditions will lead to side reactions (Hartley, 1938). For the parent azobenzene molecule, quantum yields (which can be indirectly measured spectroscopically (Shen and Rau, 1991; Priest and Sifain, 1971; Malkin and Fischer, 1962) are on the order of 0.6 for the trans→cis photoconversion, and 0.25 for the back photoreaction. Solvent has a small effect, increasing the trans→cis and decreasing the cis→trans yield as polarity increases (Bortolus and Monti, 1979). Aminoazobenzenes and pseudo-stilbenes isomerize very quickly and can have quantum yields as high as 0.7–0.8.

1.1.3. Classes of Azobenzene Systems

Azobenzenes are robust and versatile moieties, and have been extensively investigated as small molecules, pendants on other molecular structures, or incorporated (doped or covalently bound) into a wide variety of amorphous, crystalline, or LC polymeric systems. Noteworthy examples include self-assembled monolayers and superlattices (Yitzchaik and Marks, 1996), sol–gel silica glasses (Levy and Esquivias, 1995), and biomaterials (Gallot et al., 1996; Willner and Rubin, 1996; Sisido et al., 1991a). A number of small molecules incorporating azobenzene have been synthesized, including crown ethers (Shinkai et al., 1983), cyclodextrins (Jung et al., 1996; Yamamura et al., 1996), proteins such as bacteriorhodopsin (Singh et al., 1996), and three-dimensional (3-D) polycyclics such as cubane (Chen et al., 1997b) and adamantane (Chen et al., 1995). Typically, azo chromophores are embedded in a solid matrix for studies and devices. As a result, matrix effects are inescapable: the behavior of the chromophore is altered due to the matrix, and in turn, the chromophore alters the matrix (Ichimura, 2000). Although either could be viewed as a nuisance, both are in fact useful: the chromophore can be used as a probe of the matrix (free volume, polarizability, mobility, etc.), and when the matrix couples to chromophore motion, molecular motions can be translated to larger length scales. Thus, the incorporation strategy is critical to exploiting azobenzene's unique behavior.

1.1.3.1. Amorphous Polymer Thin Films. Doping azobenzenes into polymer matrices is a convenient inclusion technique (Birabassov et al., 1998; Labarthet et al., 1998). These "guest–host" systems can be cast or spin-coated from solution mixtures of polymer and azo small molecules, where the azo content in the thin film is easily adjusted via concentration. Although doping leaves the azo chromophores free to undergo photoinduced motion unhindered, it has been found that many interesting photomechanical effects do not couple to the matrix in these systems. Furthermore, the azo mobility often leads to instabilities, such as phase separation or microcrystallization. Thus, one of the most attractive methodologies for incorporating azobenzene into functional materials is by covalent attachment to polymers. The resulting materials benefit from the inherent stability, rigidity, and processability of polymers, in addition to the unusual photoresponsive behavior of the azo moieties. Both side-chain and main-chain azobenzene polymers have been prepared (Viswanathan et al., 1999) (Fig. 1.5). Reported synthetic strategies involve either polymerizing azobenzene-functionalized monomers (Ho et al., 1996; Natansohn et al., 1992) or postfunctionalizing

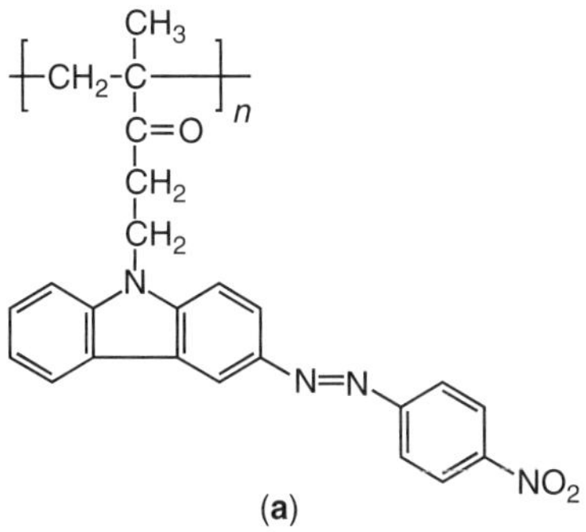

Figure 1.5. Examples of azo polymer structures, showing that both (a) side-chain and (b) main-chain architectures are possible.

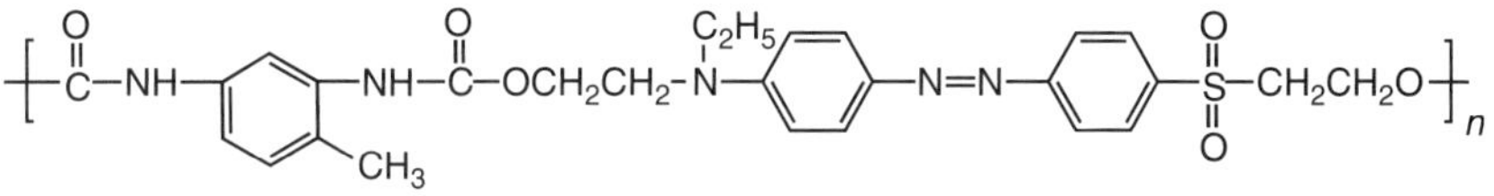

a polymer that has an appropriate pendant group (usually a phenyl) (Wang et al., 1997a,b,c). The first method is preferred for its simplicity and control of sequence distribution. The second takes advantage of commonly available starting materials, but may require more reaction steps. Many different backbones have been used as scaffolds for azo moieties, including imides (Agolini and Gay, 1970), esters (Anderle et al., 1989), urethanes (Furukawa et al., 1967), ethers (Bignozzi et al., 1999), organometallic ferrocene polymers (Liu et al., 1997), dendrimers (Junge and McGrath, 1997; Mekelburger et al., 1993), and even conjugated polydiacetylenes (Sukwattanasinitt et al., 1998), polyacetylenes (Teraguchi and Masuda, 2000), and main-chain azobenzenes (Izumi et al., 2000a,b). The most common azo polymers are acrylates (Morino et al., 1998), methacrylates (Altomare et al., 2001), and isocyanates (Tsutsumi et al., 1996). Thin films are usually prepared by spin-coating (Han and Ichimura, 2001; Blinov et al., 1998; Weh et al., 1998; Ichimura et al., 1996), although there are also many examples of using solvent evaporation, the Langmuir–Blodgett technique (Silva et al., 2002; Razna et al., 1999; Jianhua et al., 1998; Seki et al., 1993), and self-assembled monolayers (Evans et al., 1998). Spin-cast films are typically annealed above the polymer glass transition temperature (T_g) to remove residual solvent and erase any hydrodynamically induced anisotropy. Recently molecular glasses have been investigated as alternatives to amorphous polymer systems (Mallia and Tamaoki, 2003). These monodisperse systems appear to maintain the desirable photomotions and photoswitching properties, while allowing precise control of molecular architecture and thus material properties (Naito and Miura, 1993).

1.1.3.2. Liquid Crystals. Azobenzenes are anisotropic, rigid molecules and as such are ideal candidates to act as mesogens: molecules that form LC mesophases. Many examples of small-molecule azobenzene liquid crystals have been studied. Some azo polymers also form LC phases (refer to Fig. 1.6 for a typical structure). For side-chain azobenzenes, a certain amount of mobility is required for LC phases to be present; as a rule, if the tether between the chromophore and the backbone is less than 6 alkyl units long, the polymer will exhibit an amorphous and isotropic solid-state phase, whereas if the spacer is longer, LC phases typically form. The photoisomerization of azobenzene leads to modification of the phase and alignment (director) in LC systems (Shibaev et al., 2003; Ichimura, 2000). The director of a liquid crystal phase can be modified by orienting chromophores doped into the phase (Sun et al., 1992; Anderle et al., 1991) by using an azobenzene-modified "command surface" (Chen and Brady, 1993; Ichimura et al., 1993; Gibbons et al., 1991), using azo copolymers (Wiesner et al., 1991), and, of course, in pure azobenzene LC phases (Hvilsted et al., 1995; Stumpe et al., 1991). One can force the LC phase to adopt an in-plane order (director parallel to surface), homeotropic alignment (director perpendicular to surface), tilted or even biaxial orientation (Yaroschuk et al., 2001). These changes are fast and reversible. Although the *trans*-azobenzenes are excellent mesogens, the cis-azos typically are not. If even a small number of azomolecules are

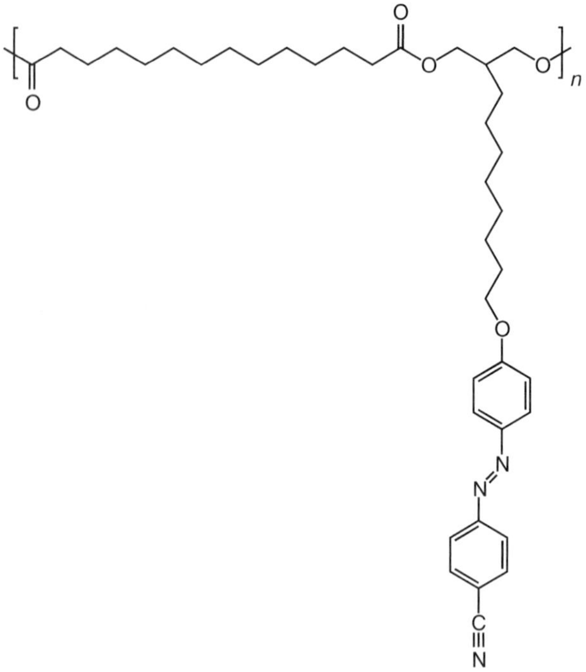

Figure 1.6. A typical liquid-crystalline side-chain azobenzene polymer.

distributed in an LC phase, trans→cis isomerization can destabilize the phase by lowering the nematic-to-isotropic phase transition temperature (Eich and Wendorff, 1990). This enables fast isothermal photocontrol of phase transitions (Kato et al., 1996; Hayashi et al., 1995; Ikeda and Tsutsumi, 1995; Ikeda et al., 1990). Since these modulations are photoinitiated, it is straightforward to create patterns (Shannon et al., 1994). These LC photoswitching effects are obviously attractive in many applications, such as for display devices, optical memories (Gibbons et al., 1991), electro-optics (Luk and Abbott, 2003), and modulating the polarization of ferroelectric liquid crystals (Fischer et al., 1997; Ikeda et al., 1993).

1.1.3.3. Dendrimers. Dendrimers have been investigated as unique structures to exploit and harness azobenzene's photochemistry (Momotake and Arai, 2004a,b; Villavicencio and McGrath, 2002). Dendritic and branched molecular architectures can have better solubility properties and can be used to control undesired aggregation, resulting in higher quality films for optical applications (Campbell et al., 2006; Ma et al., 2002). Dendrimers with strongly absorbing pendants can act as antenna, harvesting light and making it available, via intramolecular energy transfer, to the dendrimer core. In dendrimers with azo cores, this allows for the activation of isomerization using a wavelength outside of the azo-absorption band (since the dendrimer arms absorb and transfer energy to the core) (Aida et al., 1998; Jiang and Aida, 1997). Furthermore, the configurational change that results from the core isomerization will translate into a larger scale geometric change. For instance, in a dendrimer with three azobenzene arms (Fig. 1.7), the various isomerization combinations (EEE, EEZ, EZZ, and ZZZ) could all be separated by thin-layer chromatography because of their different physical properties (Junge and McGrath, 1999). The conformational change associated with isomerization modifies (typically reduces) the hydrodynamic volume, with the specific extent of conformational change depending strongly on where the azo units are incorporated (Li and McGrath, 2000).

1.1.3.4. Polyelectrolyte Multilayers. A new facile and versatile film preparation technique, layer-by-layer electrostatic self-assembly, has become the subject of intensive research since its introduction by Decher (Decher, 1997; Decher and Schmitt, 1992; Decher and Hong, 1991; Decher et al., 1991). In this technique, a charged or hydrophilic substrate is immersed in a solution of charged polymers (polyelectrolytes), which adsorb irreversibly onto the substrate. After rinsing, the substrate is then immersed in a solution containing a polyelectrolyte of opposite charge, which adsorbs electrostatically to the charged polymer monolayer. Because each layer of adsorbed polymer reverses the surface charge, one can build up an arbitrary number of alternating polycation–polyanion layers. These polyelectrolyte multilayers (PEMs) are easy to prepare, use benign (all-aqueous) chemistry, and are inherently tunable (Decher et al., 1998; Hammond, 1999; Knoll, 1996). Specifically, by adjusting the ionic strength (Steitz et al., 2000; Linford et al., 1998; Lösche et al., 1998; Sukhorukov et al., 1996) or pH (Burke and Barrett, 2003a; Chung and Rubner, 2002; Wang et al., 2002; Shiratori and

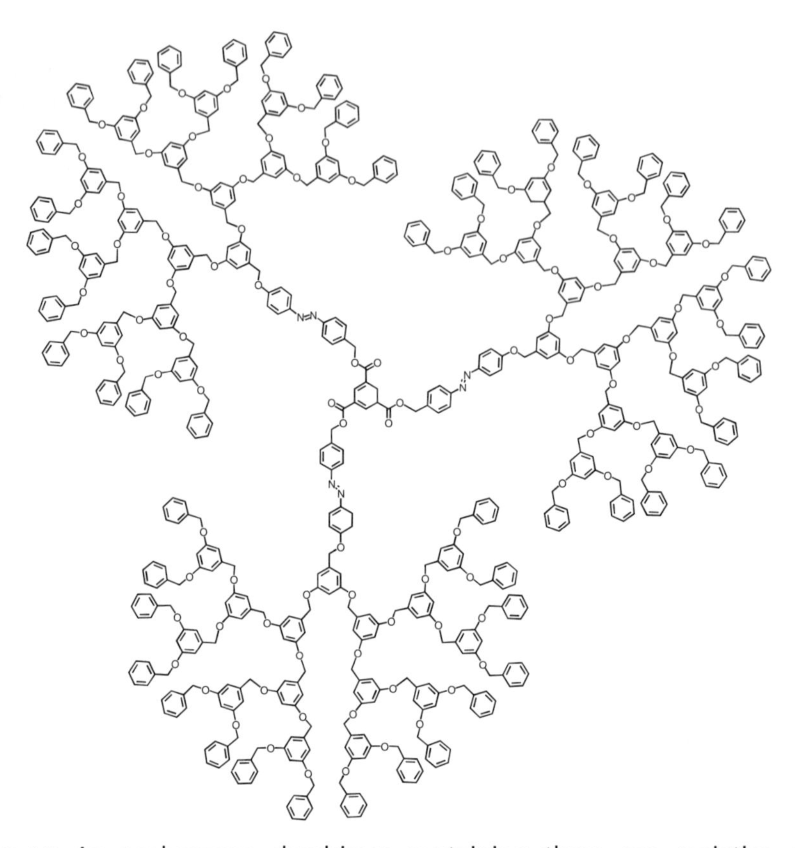

Figure 1.7. An azobenzene dendrimer containing three azo moieties. Each chromophore has two isomerization states (trans and cis), leading to four distinct photoisomers for the dendrimer molecule. All four isomers have different physical properties.

Rubner, 2000) (in the case of "weak" polyelectrolytes) of the assembly solution, the polyelectrolyte chain conformation is modified, and hence the resulting film architecture is tuned. For instance, one can control thickness (Lösche et al., 1998; Dubas and Schlenoff, 1999), permeability (Rmaile and Schlenoff, 2003), morphology (Antipov et al., 2003; McAloney et al., 2003; Mendelsohn et al., 2000), and density (Dragan et al., 2003). Recently the technique has been modified to assemble the alternate layers using a spin-coater, which reduces the assembly times and adsorption solution volumes considerably (Chiarelli et al., 2001; Cho et al., 2001; Lee et al., 2003, 2001).

As a film preparation technique, this method has numerous advantages. The adsorption of the polymers is quasithermodynamic, with the chains adsorbing into a local minimum, which makes the films stable against many defects (dewetting, pinhole formation, etc.). Importantly, the technique is not limited to flat surfaces: any geometry that can be immersed in solution (or have solution flowed through) is suitable. Colloids have been efficiently coated with PEMs (Caruso, 2001; Sukhorukov et al., 1998b), and by dissolving the core one can also form hollow PEM microcapsules (Sukhorukov et al., 1998a). Multilayers can be formed on nearly any material (glass, quartz, silicon, most metals, etc.) and are robust against thermal and solvent treatment (Mermut and Barrett, 2001). One of the main interests in PEMs is due to their inherent biocompatibility (Richert et al., 2002): multilayers have been formed on enzyme microcrystals (Jin et al., 2001), used to encapsulate living cells (Diaspro et al., 2002) and coat arterial walls (Thierry et al., 2003). Perhaps the most useful feature of the multilayering technique is its ability to incorporate secondary functional groups into the thin film structure. The location of these functional units (which may be small molecules, pendants on the polyelectrolyte chains, or particles) within the multilayer stack can be controlled with subnanometer precision. A wide variety of functionalities have been demonstrated, including organic molecules (He et al., 2000a), synthetic polymers (Balasubramanian et al., 1998), biopolymers (Burke and Barrett, 2003b), natural proteins (Caruso and Möhwald, 1999), colloids (Lvov et al., 1997), inorganic nanoparticles (Kotov et al., 1995), clay platelets (Kleinfeld and Ferguson, 1994) (used as a nacre biomimic [Tang et al., 2003]), dendrimers (Watanabe and Regen, 1994), electrochemically active species (Knoll, 1996), functionalized C_{60} (Mattoussi et al., 2000), and even, counterintuitively, uncharged and nonpolar polymer chains (Rouse and Ferguson, 2002).

Many research groups have investigated the possibility of incorporating optically responsive azobenzene chromophores into the versatile PEM structures (examples presented in Fig. 1.8), including Advincula (Advincula et al., 2001, 2003; Advincula, 2002; Ishikawa et al., 2002), Kumar and Tripathy (Lee et al., 2000; Balasubramanian et al., 1998), Tieke (Ziegler et al., 2002; Toutianoush et al., 1999; Toutianoush and Tieke, 1998; ; Saremi and Tieke, 1998),

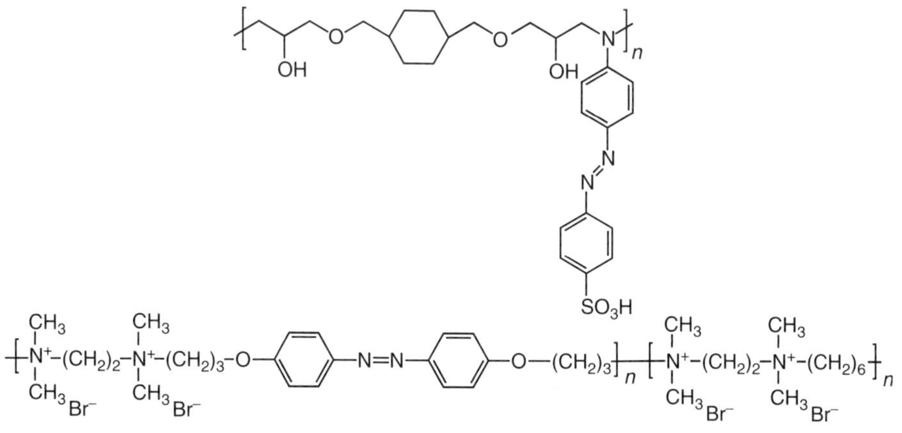

Figure 1.8. Examples of water-soluble azo polyelectrolytes, which can be used in the preparation of photoactive polyelectrolyte multilayers.

Heflin (Van Cott et al., 2002), and Barrett (Mermut and Barrett, 2003; Mermut et al., 2003). In some cases, copolymers are synthesized, where some of the repeat units are charged groups and some are azo chromophores (Suzuki et al., 2003; Wu et al., 2001a). These materials may, however, have solubility issues, as the azo chromophore is typically not water-soluble. Efforts have therefore gone into synthesizing azo ionomers (Jung et al., 2002; Hong et al., 2000), or polymers where the charge appears on the azobenzene unit (Wang et al., 1998, 2004; Wu et al., 2001a). The azobenzene chromophore may also be created by postfunctionalization of an assembled PEM (Lee et al., 2000). Azobenzene-functionalized PEMs have demonstrated all of the unique photophysics associated with the chromophore, including induced birefringence (Ishikawa et al., 2002; Park and Advincula, 2002) and surface mass transport (Wang et al., 1998) (which is described in more detail in Section 1.3). It should be noted, however, that in general the quality of the patterning is lower (Wang et al., 2004), presumably because of the constraints to chain motion that the ionic "cross-links" engender. There are many examples of performing the multilayering with a polyelectrolyte and a small molecule azobenzene ionic dye (Dragan et al., 2003). In contrast to conventional doped systems, the chromophores in these systems do not suffer from aggregation instabilities (Advincula et al., 2001), and the azo photomotions do couple to the matrix, as evidenced by birefringence (dos Santos Jr. et al., 2003; Bian et al., 2000) and surface patterning (He et al., 2000a,b). These effects can again be attributed to the fact that the ionic attachment points act as cross-links in a dry PEM sample. The aggregation and photochemical behavior of the azo chromophore (absorbance spectrum, isomerization rate, etc.) vary depending on the nature of the counterpolymer (Dante et al., 1999) (and of course, is affected by any ionic ring substituent). These may be viewed as undesirable matrix effects, or as a way to tune the chromophore response. The multilayering technique does not offer the precision and reproducibility of conventional inorganic film preparation techniques. It is, however, simple, versatile, and offers the possibility of combining unique structures and functionalities (for instance, it has been used to create superhydrophobic surfaces [Zhai et al., 2004], to make azo photochromic hollow shells [Jung et al., 2002], and is amenable to patterning [Nyamjav and Ivanisevic, 2004]). Although it is unlikely to replace established techniques for high performance devices, it may find applications in certain niches (coatings, disposable electronics, biomedical devices, etc.).

1.2. PHOTOINDUCED MOTIONS AND MODULATIONS

Irradiation with light produces molecular changes in azobenzenes, and under appropriate conditions, these changes can translate into larger scale motions and even modulation of material properties. Following Natansohn and Rochon (2002), we will describe motions roughly in order of increasing size scale. However, since the motion on any size scale invariably affects (and is affected by) other scales, clear divisions are not possible. In all cases, some of the implicated applications, photoswitching, and photomodulations will be outlined.

1.2.1. Molecular Motion

The fundamental molecular photomotion in azobenzenes is the geometrical change that occurs on absorption of light. In *cis*-azobenzene, the phenyl rings are twisted at 90° relative to the C–N = N–C plane (Naito et al., 1991; Uznanski et al., 1991). Isomerization reduces the distance between the 4 and 4′ positions from 0.99 nm in the trans state to 0.55 nm in the cis state (Brown, 1966; Hampson and Robertson, 1941; de Lange et al., 1939). This geometric change increases the dipole moment: whereas the trans form has no dipole moment, the cis form has a dipole moment of 3.1 D (Hartley, 1937). The free volume requirement of the cis is larger than that of the trans (Naito et al., 1993), and it has been estimated that the minimum free volume pocket required to allow isomerization to proceed via the inversion pathway (Naito et al., 1991; Paik and Morawetz, 1972) is 0.12 nm^3, and $\sim$0.38 nm^3 via the rotation pathway (Lamarre and Sung, 1983). The effects of matrix free volume constraints on photochemical reactions in general have been considered (Weiss et al., 1993). The geometrical changes in azobenzene are very large, by molecular standards, and it is thus no surprise that isomerization modifies a wide host of material properties.

This molecular displacement generates a nanoscale force, which has been measured in single-molecule force spectroscopy experiments (Holland et al., 2003; Hugel et al., 2002) and compared with theory (Neuert et al., 2005). In these experiments, illumination causes contraction of an azobenzene polymer, showing that each chromophore can exert pN molecular forces on demand. A pseudo-rotaxane that can be reversibly threaded–dethreaded using light has been called an "artificial molecular-level machine"(Balzani et al., 2001; Asakawa et al., 1999). The ability to activate and power molecular-level devices using light is of course attractive since it circumvents the limitations inherent to diffusion or wiring. The fast response and lack of waste products in azo isomerization are also advantageous. Coupling these molecular-scale motions to do useful work is of course the next challenging step. Progress in this direction is evident from a wide variety of molecular switches that have been synthesized. For example, an azo linking two porphyrin rings enabled photocontrol of electron transfer (Tsuchiya, 1999). In another example, dramatically different hydrogen-bonding networks (intermolecular and intramolecular) can be favored on the basis of the isomeric state of the azo group linking two cyclic peptides (Steinem et al., 1999; Vollmer et al., 1999).

1.2.2. Photobiological Experiments

The molecular conformation change of the azo chromophore can be used to switch the conformation and hence properties of larger molecular systems to which it is attached. This is particularly interesting in the case of inclusion within molecular-scale biological systems. The bridging of biology and physical chemistry is an ever-expanding research domain. It is no surprise that the clean and unique azo photochemistry has been applied to switching biological systems (Willner and Rubin, 1996). One of the earliest investigations of azobenzene in a biological context involved embedding azobenzene molecules into a model membrane system (Balasubramanian et al., 1975). On isomerization, the lamellae were disrupted and rearranged, which also changed the enzymatic activity of membrane-bound proteins. The catalytic activity of a cyclodextrin with a histidine and azobenzene pendant was photocontrollable because the trans version of the azo pendant can bind inside the cyclodextrin pocket, whereas the cis version liberated the catalytic site (Lee and Ueno, 2001). Photoregulation of polypeptide structure has been an active area of research (Ciardelli and Pieroni, 2001), with the azobenzenes making significant contributions. Azo-modified poly (L-alanine) (Sisido et al., 1991a,b), poly(L-glutamic acid) (Houben et al., 1983; Pieroni et al., 1980), and poly(L-lysine) (Malcolm and Pieroni, 1990), among others, have been prepared. Depending on the system, photoisomerization may cause no change (Houben et al., 1983) or can induce a substantial conformational change, including transitions from ordered chiral helix to disordered achiral chain (Fissi et al., 1996; Yamamoto and Nishida, 1991; Montagnoli et al., 1983), changes in the α-helix content, or even reversible α-helix to β-sheet conversions (Fissi et al., 1987). Also, owing to the change in local electrostatic environment, the pK_a of the polypeptides can be controlled in these systems.

Covalent attachment of azobenzene units to enzymes can modify protein activity by distorting the protein structure with isomerization. This was used to control the enzyme activity of papain (Willner and Rubin, 1993; Willner et al., 1991a) and the catalytic efficiency of lysozyme (Inada et al., 2005). A different methodology is to immobilize the protein of interest inside a photoisomerizable copolymer matrix, which was used to control α-chymotrypsin (Willner and Rubin, 1993; Willner et al., 1991b, 1993). The azobenzene need not be directly incorporated into an enzyme of interest. In one case, the activity of tyrosinase could be modified by isomerization of small-molecule azo inhibitors (Komori et al., 2004). The photoselective binding of short peptide fragments into enzymes can be used to inhibit, thus control, activity (Harvey and Abell, 2000, 2001). Similarly, the binding of an azopeptide with a monoclonal antibody was found to be photoreversible (Harada et al., 1991). The photoresponse of azobenzene can thus be used to control the availability of key biomolecules. In one case, NAD^+ was modified with an azobenzene group, and introduced into a mixture with an antibody that binds to the trans form (Hohsaka et al., 1994). This binding makes NAD^+ unavailable, whereas irradiation of the solution with UV light induces the trans to cis isomerization, and thereby liberates NAD^+.

Bioengineering has more recently been broadened by expanding the natural protein alphabet with artificial amino acids. This enables novel and nonnatural protein sequences to be created, while still exploiting the highly efficient natural synthesis machinery. Chiral azobenzene amino acids have been synthesized and incorporated into protein sequences (Wang and Schultz, 2004). The introduction of artificial photoactive residues opens the possibility of photocontrol of biological processes. For instance, *Escherichia coli* variants were selectively evolved that would incorporate azobenzene amino acids into proteins, which enabled photocontrol of protein binding in that organism (Bose et al., 2006). For instance, photocontrol of the binding affinity of a transcription factor to its promoter, allowed for, in essence, light control of gene expression in the organism. In another case, a (negatively charged) hydrophilic azobenzene amino acid was incorporated into a restriction enzyme, and enabled control of activity with light (Nakayama et al., 2004, 2005). Specifically, the *trans*-azo residue was positioned at the dimer interface, and disrupted association, whereas in the cis state, the proteins could aggregate and exhibit normal biological activity. It has also been suggested that the rapid switching of azobenzene could be used as a "molecular shuttle" for electron transduction in enzyme systems (Voinova and Jonson, 2004). In effect, this would mean that light could be efficiently used to alter behavior in yet another class of enzymes. Incorporation of azobenzene into DNA is another interesting way to control biological systems. In one case, the duplex of modified DNA could be reversibly switched (Asanuma et al., 2001), since the *trans*-azobenzene intercalates between base pairs and helps bind the two strands of the double helix together, whereas the *cis*-azobenzene disrupted the duplex (Liang et al., 2003). By incorporating an azobenzene unit into the promoter region of an otherwise normal DNA sequence, it was possible to photocontrol gene expression (Liu et al., 2005). In this case, the trans versus cis states of the azo unit have different interactions with the polymerase enzyme.

These experiments suggest an overall strategy to control biological systems using light. A complex biochemical pathway can be controlled by photoregulating the activity or availability of a key biomolecule. This allows one to turn a biological process on and off at will using light. The use inside living organisms is obviously more complicated, but one can reasonably easily apply these principles to control biological processes in industrially relevant settings. The ability to quickly and cleanly switch biological activity using a short light pulse may find application in new microfluidic devices, which need to be able to address specific device regions and may rely on natural molecular machinery to carry out certain tasks. Azobenzenes present unique opportunities in the biological sciences for studying complex biological systems, in addition to controlling them. A bacteriorhodopsin analog with a central azobenzene molecule, rather than the retinal, was prepared as a model system for studying rhodopsin (Singh et al., 1996). As expected, the azobenzene molecule did not interact as favorably with the protein host as strongly as the natural retinal. Despite this, the azo molecules could be coupled into the protein (in the absence of retinal) and led to significant shifts in the physicochemical properties of the complex. Moreover, the azo molecule could be used as a probe of the inner protein domain (sensing pH, for instance). A particularly elegant experiment involved using azobenzenes to monitor protein folding (Bredenbeck et al., 2003; Spörlein et al., 2002). Femtosecond two-dimensional infrared (2-D IR) spectroscopy was used as a gauge of the distances between carbonyl groups in the peptide. An azobenzene chromophore, incorporated inside the polypeptide chain, acted as the photoswitch, initiating a conformational change, hence initiating protein folding, on demand. Simultaneous time-resolved measurements of the azo spectra allowed determination of the folding dynamics. This unique measurement of protein-folding behavior was possible because of the phototriggering nature of the azo unit. Ultrafast laser pulse experiments are used to study a large number of chemical reactions, providing detail not possible before. This technique is, however, obviously limited to systems where the chemical events can be phototriggered. By incorporating azobenzene units into new systems, one can generate a phototriggerable system from an otherwise photoinactive one. This strategy can thus be applied to a wide range of problems in chemical dynamics, with biological systems being obvious targets.

1.2.3. Photoorientation

Azobenzene chromophores can be oriented using polarized light (Yu and Ikeda, 2004; Ichimura, 2000) via a statistical selection process, described schematically in Fig. 1.9. Azobenzenes preferentially absorb light polarized along their transition dipole axis (long axis of the azomolecule). The probability of absorption varies as $\cos^2\phi$, where ϕ is the angle between the light polarization and the azo dipole axis.

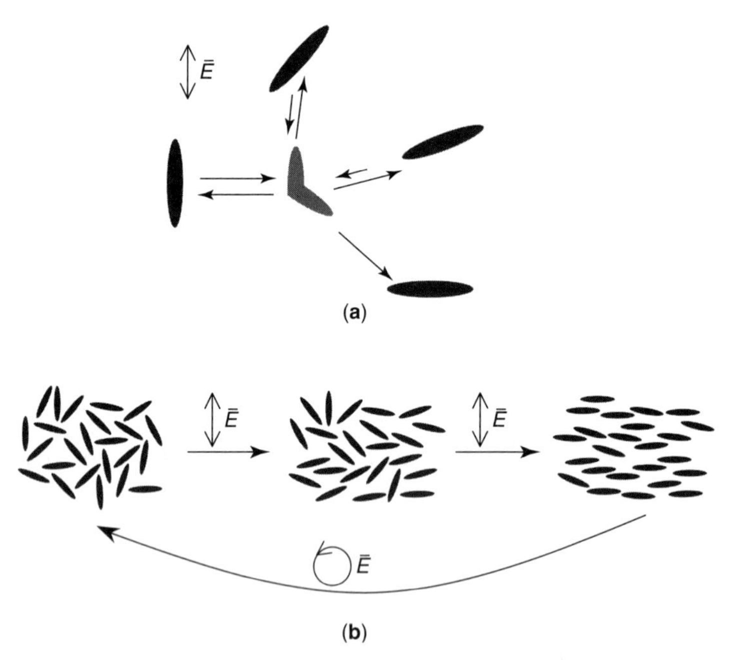

Figure 1.9. Statistical photoorientation of azomolecules. (**a**) The molecules aligned along the polarization direction of the incident light absorb, isomerize, and reorient. Those aligned perpendicular cannot absorb and remain fixed. (**b**) Irradiation of an isotropic samples leads to accumulation of chromophores in the perpendicular direction. Circularly polarized light restores isotropy.

Thus, azos oriented along the polarization of the light will absorb, whereas those oriented against the light polarization will not. For a given initial angular distribution of chromophores, many will absorb, convert into the cis form and then revert to the trans form with a new random direction. Those chromophores that fall perpendicular to the light polarization will no longer isomerize and reorient; hence, there is a net depletion of chromophores aligned with the light polarization, with a concomitant increase in the population of chromophores aligned perpendicular (i.e., orientation hole burning). This statistical reorientation is fast and gives rise to strong birefringence (anisotropy in refractive index) and dichroism (anisotropy in absorption spectrum) because of the large anisotropy of the azo electronic system. The process is especially efficient because of the mesogenlike cooperative motion that the azobenzene groups facilitate even in amorphous samples below T_g (Wiesner et al., 1991). Since the process requires cycling of the chromophores between the trans and cis states, the pseudo-stilbenes have the fastest response.

The orientation due to polarized light is reversible. The direction can be modified by using a new polarization angle for the irradiating light. Circularly polarized light will randomize the chromophore orientations. It must be emphasized, however, that there is another preferential alignment direction during irradiation: along the axis of the incoming light. It is unavoidable that chromophores will efficiently build up aligned along the irradiation axis, but this is often ignored in the literature, or characterized as "photobleaching" when in fact it is a reversible photoalignment (albeit one that reduces the absorbance as viewed by any photoprobe). Because unpolarized light can photoorient (along the axis of illumination) (Han and Ichimura, 2001), even sunlight is suitable. The motion of the sun through the sky over the course of a day can cause orientation at different tilt angles (Ichimura and Han, 2000). This causes chromophores at different depths to be oriented in different directions, which produces a net chiral helical ordering in the film of a particular handedness (on the basis of the hemisphere in which the experiment is performed). The implications of such results to the origin of absolute chirality in biological systems are intriguing.

1.2.3.1. Birefringence. Irradiation with light polarized in the y-direction will lead to net alignment of chromophores in the x-direction. As a result, the refractive index probed in the x-direction, n_x, will measure the azo long axis, and will be larger than n_y. Birefringence is the anisotropy in refractive index: $\Delta n = n_x - n_y$. Photoalignment in azobenzene systems can achieve extremely high values of Δn, up to 0.3–0.5 at $\sim$633 nm (Hagen and Bieringer, 2001; Natansohn et al., 1994). Importantly, very high birefringence values can be obtained far outside of the azo-absorption band, which means that the birefringence can be utilized/measured without disturbing the chromophores. An in-plane isotropic state ($n_x = n_y$) can be restored by irradiation with circularly polarized light, and a fully isotropic state can be obtained by heating above the T_g.

The exact nature of the orientation can be rigorously quantified using optical techniques. Using surface plasmon resonance spectroscopy or waveguide spectroscopy, the three orthogonal refractive indices in an oriented sample can be measured (Tawa and Knoll, 2002). Stokes polarimetry can be used to fully characterize the optical anisotropy, separating linear and circular components (Hore et al., 2002). The anisotropy of the cis population during irradiation can also be measured in some systems (Buffeteau et al., 2001; Sekkat et al., 1995b), where it is found that, as with trans, there is an enrichment perpendicular to the irradiation polarization. In some LC systems, however, it may occur that the cis population preferentially aligns with the irradiating polarization (which may be attributed to an optical Fréedericksz transition) (Hore et al., 2003).

The birefringence can be written and erased hundreds of thousands of times, which is important technologically (Holme et al., 1996). Amorphous polymer systems with relatively high T_g exhibit good temporal stability of any induced orientation. Upon heating, some order will be lost, with full isotropy restored after heating past T_g. A short spacer between the chromophore and the polymer backbone slows the growth of birefringence yet promotes stability, owing to hindered motion. Surprisingly, main-chain azos can achieve high levels of birefringence, indicating relatively high polymer mobility (Wu et al., 2001b; Xu et al., 2000; Lee et al., 1998). As might be expected, (nanosecond) pulsed experiments lead to thermal effects, which enhance chromophore motion and thereby induced greater birefringence at the same net dose compared with continuous-wave (cw) experiments (Cimrová et al., 2002; Hildebrandt et al., 1998). At very high pulsed fluence, the thermal effects were too great and erased the induced birefringence.

The easily inscribed and erased birefringence has a number of unique applications. Most readily, it can be used to create wave plates (Shi et al., 1991a) and polarization filters, which can be used to separate right-handed from left-handed circularly polarized light (Natansohn and Rochon, 1999). The strong refractive index contrast, if patterned into a line, can serve as a channel waveguide (Watanabe et al., 1996; Shi et al., 1991b). This offers the unique possibility of optical devices that can be patterned, erased, and reused. In principle, these photonic circuits could be altered during device operation, enabling optical routing of optical signals (i.e., optical computing). The switching of orientational order can thus be used as an all-optical switch (Shishido et al., 1997). By illuminating an azo sample with a spatially varying light pattern, birefringence gratings can also be formed (Eichler et al., 2001; Nikolova et al., 1996; Couture and Lessard, 1988). These are phase gratings, as opposed to amplitude gratings, and diffract light on the basis of spatial variation of the refractive index. This is the essence of holography: two interfering coherent beams generate a spatially varying light pattern, which is encoded into the material. Under illumination of the material with one of the beams, the diffraction reproduces the other encoded beam. In the case of liquid crystal samples, light induces a spatial pattern of nematic and isotropic zones (which have different refractive indices). These holographic phase grating can be rapidly formed, erased, and switched (Yamamoto et al., 2001).

1.2.3.2. Nonlinear Optics. The requirement for NLO response in any material is an asymmetric (strictly, anharmonic) response of the electronic system. Pseudo-stilbenes, which have push–pull substituents, have a strongly asymmetric electron distribution, which makes them ideal NLO molecules (see, for instance, Fig. 1.10). For a bulk NLO response, one requires an overall noncentrosymmetric material. This requirement is achieved in many inorganic crystals. In organic systems, the broken symmetry is typically obtained by applying an electric field at a temperature sufficient to allow for the molecular dipoles to align with the field. This process is called electric field poling and is accomplished using interdigitated or flat electrodes or a sharp charged needle (or grid) held above a grounded sample (called corona poling). The NLO response is typically quantified using second-harmonic generation (SHG; the emission of light at double the frequency of the incident beam), the electrooptic effect (change of refractive index on application of an electric field), or wave-mixing experiments (where various frequencies of light can be synthesized or enhanced). These also constitute the main applications of NLO materials: they can be used to synthesize new frequencies of light, to electrically switch a beam, or to allow two beams of light to interact and couple (which can form the basis of an all-optical switch) (Eaton, 1991).

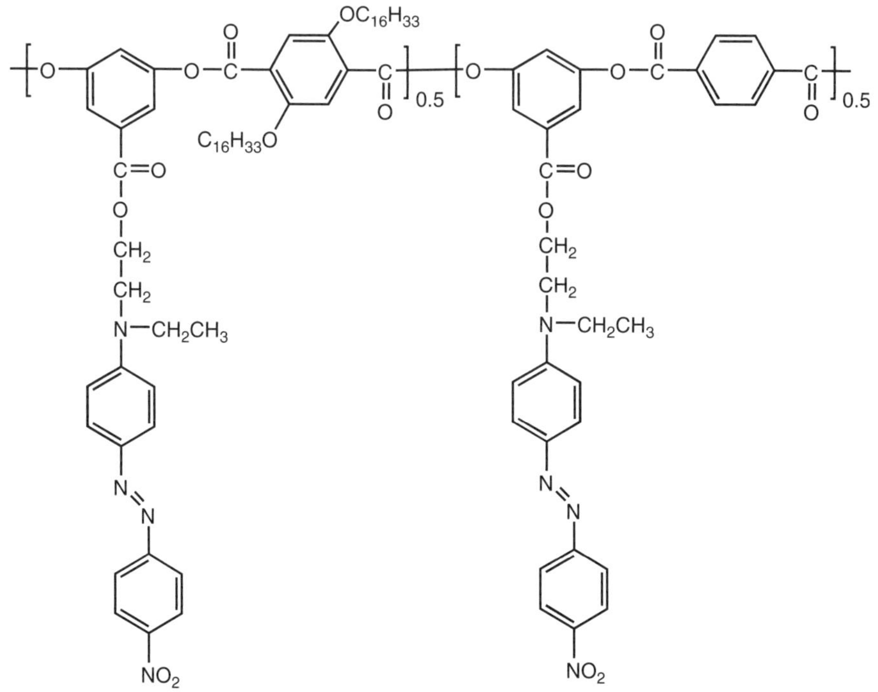

$OC_{16}H_{33}$ O O O C O O C C 0.5 O O C O C C 0.5 $C_{16}H_{33}O$ O C=O C=O O O CH_2 CH_2 CH_2 CH_2 N—CH_2CH_3 N—CH_2CH_3 N N N N NO_2 NO_2

Figure 1.10. Example of a nonlinear optic azo polymer used for photoassisted poling.

Azo polymers have been shown to be excellent NLO materials (Yesodha et al., 2004; Delaire and Nakatani, 2000; Dalton et al., 1995; Burland et al., 1994). In azo systems, one has the additional advantage of using light to affect the chromophores. Although photoalignment orients the chromophore axis, it does not select out a preferred direction for the molecular dipole (thus, an equal number of dipoles point "left" and "right"). In fact, evidence suggests that dipoles in these systems tend to orient antiparallel (H-aggregates) (Meng et al., 1996; Brown et al., 1995), thereby canceling polar order. Nevertheless, the photoalignment can be used to facilitate the electric poling, enabling it to be performed at room temperature and with a small DC field (Jiang et al., 1996; Sekkat et al., 1995a; Blanchard and Mitchell, 1993a,b). Furthermore, by using polarized light and its harmonic, a net noncentrosymmetry can be obtained in an all-optical process (Zhong et al., 2001; Nunzi et al., 1998). This occurs because the mixture of a primary beam and its second harmonic creates a directional electric field in the material.

Another interesting approach for NLO uses dendrons ("half-dendrimers") with azo functionalities (Yokoyama et al., 2000). The dendritic architecture forces all the chromophores within the dendron to align, which strongly enhances the NLO response. The dendron had a first-order molecular hyperpolarizability 20 times larger than the monomer. With regard to applications, the azos have been shown to function as electro-optical switches (Yamane et al., 1999) and exhibit photorefraction (Iftime et al., 2002; Steenwinckel et al., 2001; Barrett et al., 1998; Ho et al., 1996), an NLO effect where photoconductivity permits light to establish a space charge grating, whose associated index grating refracts a probe light beam.

1.2.4. Domain Motion

The orientation and reorientation of LC domains has already been outlined. The azo chromophores act as mesogens and their photoalignment becomes transferred to the LC host. A very small azo content (a few mol% [Ikeda et al., 1990]) can lead to orientational control of LC domains. This is an excellent example of amplification of the azomolecular motion. The phase of a liquid crystal can also be switched with light. Irradiation produces cis isomers, which are poor mesogens and destabilize the nematic phase, thereby inducing a phase transition to the isotropic state. There are comparatively few examples of phototriggered increases in LC ordering. In one case, a nanoscale phase separation of the cis isomers led to a net increase in the order parameter of the LC phase (Prasad and Nair, 2001). In another system, a chiral azo was found to induce a cholesteric phase when it was in the cis state (Ruslim and Ichimura, 2001).

With LC (Nikolova et al., 1997) or preoriented amorphous samples (Ivanov et al., 2000), one can photoinduce a chiral domain structure. Incident circularly polarized light becomes elliptically polarized because of the first oriented layer. This ellipse subsequently reorients deeper chromophores, which in turn modify the ellipticity of the light. This reorientation continues throughout the film depth. Overall, a chiral ordering of the chromophore domains is established (Nikolova et al., 2000). Remarkably, one can switch between a right- and left-handed supramolecular helix at will, by changing the incident light handedness. There are many other examples of photocontrol of supramolecular order. The pitch of a cholesteric LC can be modified by isomerization (Sackmann, 1971). Biomacromolecular variants abound (Willner and Rubin, 1996). Azo-modified polypeptides can be photoswitched between ordered states (α-helix or β-sheet) and a random coil (Everlof and Jaycox, 2000; Fissi et al., 1996; Yamamoto and Nishida, 1991). The duplex of modified DNA can be reversibly switched (Asanuma et al., 2001), and the catalytic activity of histidine can be controlled (Lee and Ueno, 2001).

Photoisomerization can also affect self-assembly behavior at the domain level. On irradiation, one can induce a phase change (Aoki et al., 2000), a solubility change (Arai and Kawabata, 1995; Yamamoto et al., 1990), crystallization (Ebralidze and Mumladze, 1990), or even reversal of phase separation (Effing and Kwak, 1995). The critical micelle concentration (cmc) and surface activity can also be modified (Yang et al., 1995). In an amphiphilic polypeptide system, self-assembled micelles were formed in the dark and could be disaggregated with light (Higuchi et al., 1994). When allowed to assemble as a transmembrane structure, the aggregate could be reversibly formed and destroyed using light, which allowed for reversible photoswitching of ion transport (Higuchi et al., 1995a). Related experiments on methacrylates (Chen et al., 1997a; Angiolini et al., 1995) and polypeptides (Sisido et al., 1991b) showed that a polymer's chiral helix could be reversibly suppressed on irradiation. In a series of polyisocyanate polymers, it could be selected whether irradiation would suppress or increase chirality (Müller and Zentel, 1996; Maxein and Zentel, 1995).

1.2.5. Macroscopic Motion

It is interesting to study whether the azobenzene molecular conformational rearrangements can result in changes to bulk phenomena, or even to macroscopic motion. The first consideration is whether the material expands to an appreciable extent. In monolayers, it is well established that the larger molecular size of the cis isomer leads to a corresponding lateral expansion (Higuchi et al., 1995b), which can modify other bulk properties. For instance, this allows photomodulation of a monolayer's water contact angle (Siewierski et al., 1996) or surface potential (Stiller et al., 1999). Using fluorinated azo polymer, good photocontrol (Feng et al., 2001) and photopatterning (Moller et al., 1998) of wettability has been demonstrated. A monolayer of azo-modified calixarene, when irradiated with a light gradient, produced a gradient in surface energy sufficient to move a macroscopic oil droplet (Ichimura et al., 2000), suggesting possible applications in microfluidics. Modest photoinduced contact angle changes for thin polymer films have also been reported (Sarkar et al., 2001). Recently an azobenzene copolymer assembled into polyelectrolyte multilayer showed a modest 2° change in contact angle with UV light irradiation. However, when the same copolymer was assembled onto a patterned substrate, the change in contact angle upon irradiation was enhanced to 70° (Jiang et al., 2005). It is well established that surface roughness plays a role in contact angle and many systems can be optimized to give rise to a large change in surface properties.

In layered inorganic systems with intercalated azobenzenes, reversible photo-changes in the basal spacing (on the order of 4%) can be achieved (Fujita et al., 1998, 2001). In polymer films, there is some evidence that the film thickness increases, as measured by ellipsometry (Shi et al., 1991b) (the refractive index certainly changes [Ivanov et al., 1995], but this is not an unambiguous demonstration of expansion–contraction). Experiments that show that external applied pressure tends to hinder photoisomerization (Kleideiter et al., 2000) are related. Photocontraction for semicrystalline main-chain azos has been measured (Eisenbach, 1980b; Agolini and Gay, 1970). This photomechanical response presumably occurs because of the shortening of the polymer chains upon trans→cis conversion. However, photoexpansion would seem to be contradicted by positron lifetime experiments that suggest no change in microscopic free volume cavity size during irradiation (Algers et al., 2004). More conclusive experiments are in order.

The most convincing demonstration of macroscopic motion due to azo isomerization is the mechanical bending and unbending of a free-standing polymer film (Ikeda et al., 2003; Yu et al., 2003). The macroscopic bending direction may be selected either with polarized light or by aligning the chromophores with rubbing. Bending occurs in these relatively thick films because the free surface (which absorbs light) contracts, whereas the interior of the film (which is not irradiated owing to the strong absorption of the upper part of the film) does not contract. Because the direction of bending can be controlled with polarized light, the materials enable full directional photomechanical control (Yu et al., 2005). This photomechanical deformation has also been used to drive macroscopic motion of a floating film (Camacho-Lopez et al., 2004). That these materials contract (rather than expand) appears again to be related to the main-chain azo groups and may also be related to the LC nature of the cross-linked gels. For a thin film floating on a water surface, a contraction in the direction of polarized light was seen for LC materials, whereas an expansion was seen for amorphous materials (Bublitz et al., 2000). A related amplification of azo motion to macroscopic motion is the photoinduced bending of a microcantilever coated with an azobenzene monolayer (Ji et al., 2004). One can also invert the coupling of mechanical and optical effects: by stretching an elastomeric azo film containing a grating, one can affect its wavelength-selection properties and orient chromophores (Bai and Zhao, 2001).

1.2.6. Other Applications of Azobenzenes

1.2.6.1. Photoswitches. As already pointed out, the azo isomerization can be used to photoswitch a wide variety of other properties (at numerous size scales). In addition to the optical changes already described, it is worth noting that the transient change in material refractive index (owing to the different n of cis and trans) can itself act as a photoswitch (Barley et al., 1991). The azo photochromism has even been suggested as a possible optical neural network element (Sumaru et al., 1999). Binding and transport properties can also be photoswitched (Weh et al., 1998; Anzai and Osa, 1994). In some systems the redox potential and ionic conductivity can be switched with light (Willner and Willner, 1997). Crown ethers (Zawisza et al., 1999; Tokuhisa et al., 1994) and calixarenes (Reynier et al., 1998) functionalized with azobenzene can be used as reversible ion-binding systems. Thus, ion transport can be photoregulated. In other cases, the transport properties can be photocontrolled not via binding but on the basis of changes in pore sizes (Sata et al., 2000; Abraham and Purushothaman, 1998; Kano et al., 1980). In a particularly elegant example, the size of nanochannels could be modified by irradiating azo ligands that decorate the channel walls (Liu et al., 2004). Azo-derivatized gramicidin ion channels represent a unique case where ion transport can be photocontrolled by the optical manipulation of a biomolecule (Lien et al., 1996). In addition to obvious applications in controlled transport, this offers the possibility of studying cells by controlling the timing of ion exchange processes. Photoinduced catalysis is also possible, for instance, using molecules where only the cis form is catalytically active (Wuerthner and Rebek, 1995). Extension of the molecular imprinting technique to azo polymers allows for photoswitching of binding activity with respect to the imprinted molecule (Minoura et al., 2004).

1.2.6.2. Photoprobes. The properties of an azo chromophore (spectrum, isomerization kinetics, etc.) depend strongly on the local environment. This enables the possibility of using the chromophore as a molecular sensing element: a photoprobe. For instance, it has been found that many azo properties depend on local H^+ concentration, to the extent that the azo can in fact be used as a pH meter (Uznanski and Pecherz, 2002; Mermut and Barrett, 2001). As mentioned earlier, the isomerization kinetics can also be used as a probe of free volume (Naito et al., 1993; Lamarre and Sung, 1983), local aggregation (Norman and Barrett, 2002), or phase transitions. The azo molecule is small and exhibits clean photochemistry, which makes it more versatile and robust than many other photoprobes. The rate of isomerization is also remarkably insensitive to temperature (Yamamoto, 1986), yet sensitive to local solvent conditions (Li et al., 2006; Norman and Barrett, 2002). This is an area of research that deserves considerably more attention.

In a more sophisticated example, azo chromophores were used to monitor protein folding (Bredenbeck et al., 2003; Spörlein et al., 2002). Specifically, femtosecond 2-D IR spectroscopy was used to monitor the distances between carbonyl groups in the peptide. An azo chromophore, incorporated inside the polypeptide chain, was used as a photoswitch to initiate a conformational change, hence initiate protein folding, on demand. Combined with time-resolved monitoring of the azo spectrum, this allows the deconvolution of folding dynamics. Pump-probe ultrafast laser pulse experiments are being used to study many different chemical reactions, but are obviously limited to reactions that can be triggered by light. Incorporating azobenzene into the experiment allows a wider range of reactions to be phototriggered.

1.2.6.3. Optical Data Storage. The azos have been investigated as optical storage media for some time. Early proofs of principle were on Langmuir–Blodgett films, using photochromism (Liu et al., 1990) or birefringence (Dhanabalan et al., 1999). Increasingly, amorphous polymer systems are being recognized as promising materials. In these easily processed systems, the birefringence is strong, stable, and switchable, making them ideal for optical memories. A single domain could encode one bit by either being isotropic or birefringent, a difference that is easily probed optically. The Δn values are large enough, in fact, that a gray-level algorithm could be used, where each domain stores more than one bit of data. On the negative side, the photoalignment generated in the direction of the read–write beam leads to an effective loss of material performance with time. Full anisotropy could be restored with heat, however (which can be local and photoinduced, with appropriate device setup). The feasibility of storing $\sim$30 GB of data on a single layer of a removable disk using this gray-level approach has been demonstrated (Hagen and Bieringer, 2001).

Even the fastest photoinduced birefringence in azo systems requires milliseconds and is slow compared with most computer timescales. However, optical data storage is amenable to gray-level read–write (Sabi et al., 2001) and to storing–retrieving full 2-D "pages" of data at a time. In principle, azo systems could achieve high data storage and retrieval speeds. The full 3-D volume of a material can be used by encoding many layers of 2-D data (pages) one on top of the other (Kawata and Kawata, 2000; Ishikawa et al., 1998). This is accomplished by moving the optical focal plane through the material.

An intriguing possibility for high density storage is to use angular multiplexing (Hagen and Bieringer, 2001). By storing multiple superimposed holograms in a single material, the data density is increased dramatically, and the whole 3-D volume of the material is exploited (Ramanujam et al., 2001). Volume-phase holograms in azo systems can have diffraction efficiencies greater than 90% (Zilker et al., 1998), making data readout robust. The hologram is encoded by interfering a reference beam and a writing beam inside the sample volume, at a particular angle. The write beam, having passed through a spatial light modulator (SLM), has a pattern corresponding to the data, which is then holographically encoded in the sample. The entire page of data is written at once. By selecting different angles, new pages of data can be written. To readout a page, the azo sample is set at the correct angle and illuminated with the reference beam. The resulting diffraction pattern is imaged on a charge-coupled device (CCD) array, which measures the encoded beam pattern (data). The volume of data and transmission rate is clearly large: projections of $\sim$1000 GB in a single disk have been made. Since the entire hologram image is stored throughout the material, the technique is fairly insensitive to dust, scratches, and pinpoint defects.

The use of azo-substituted peptide oligomers appears to enable control of the order, hence optimization for holographic applications (Berg et al., 1996). Optical memories would be considerably enhanced by using two-photon processes. This allows the addressable volume to be smaller and better defined, while reducing cross-talk between encoded pages. Some azo chromophores exhibit "biphotonic" phenomena, which could be employed to enhance optical data storage.

1.2.6.4. Surface Mass Transport. In 1995, a surprising and unprecedented optical effect was discovered in polymer thin films containing the azo chromophore Disperse Red 1 (DR1). The Natansohn–Rochon (Rochon et al., 1995) research team and the Tripathy–Kumar collaboration (Kim et al., 1995) simultaneously and independently discovered a large-scale surface mass transport when the films were irradiated with a light interference pattern. In a typical experiment, two coherent laser beams, with a wavelength in the azo-absorption band, are intersected at the sample surface. The sample usually consists of a thin spin-cast film (10–1000 nm) of an amorphous azo polymer on a transparent substrate. The sinusoidal light interference pattern at the sample surface leads to a sinusoidal surface patterning, that is, a surface relief grating (SRG). These gratings were found to be extremely large, up to hundreds of nanometers, as confirmed by atomic force microscopy (AFM). The SRGs diffract very efficiently, and in retrospect, it is clear that many reports of large diffraction efficiency before 1995, attributed to birefringence, were in fact due to surface gratings. The process occurs readily at room temperature (well below the T_g of the amorphous polymers used) with moderate irradiation ($1–100\,mW\,cm^{-2}$) over seconds to minutes. The phenomenon is a reversible mass transport, not irreversible material ablation, since a flat film with the original thickness is recovered upon heating above T_g. Critically, it requires the presence and isomerization of azobenzene chromophores. Other absorbing but nonisomerizing chromophores do not produce SRGs. Many other systems can exhibit optical surface patterning (Yamaki et al., 2000), but the amplitude of the modification is much smaller, does not involve mass transport, and usually requires additional processing steps. The all-optical patterning unique to azobenzenes has been studied intensively since its discovery, yet there remains controversy regarding the mechanism. The competing interpretations are evaluated in Chapter 4, where they are discussed at length. Many reviews of the remarkable body of experimental results are available (Natansohn and Rochon, 2002; Delaire and Nakatani, 2000; Yager and Barrett, 2001; Viswanathan et al., 1999).

ACKNOWLEDGMENT

This work is dedicated to Professors Almeria Natansohn and Sukant Tripathy, teachers and pioneers in the field of azo polymers, who were unable to see the completion of this book.

REFERENCES

Abraham G, Purushothaman E. 1998. Synthesis and photostimulated dilation changes of polymers with azobenzene cross-links. Indian J Chem Technol 5(4):213–216.

Advincula R, Park M-K, Baba A, Kaneko F. 2003. Photoalignment in ultrathin films of a layer-by-layer deposited water-soluble azobenzene dye. Langmuir, 19(3):654–665.

Advincula RC. 2002. Polyelectrolyte layer-by-layer self-assembled multilayers containing azobenzene dyes. In: Tripathy SK, Kumar J, Nalwa HS, editors. Handbook of Polyelectrolytes and Their Applications. Stevenson Ranch (CA): American Scientific Publishers, p. 65–97.

Advincula RC, Fells E, Park M-K. 2001. Molecularly ordered low molecular weight azobenzene dyes and polycation alternate multilayer films: aggregation, layer order, and photoalignment. Chem Mater 13(9):2870–2878.

Agolini F, Gay FP. 1970. Synthesis and properties of azoaromatic polymers. Macromolecules 3(3):349–351.

Aida T, Jiang D-L, Yashima E, Okamoto Y. 1998. A new approach to light-harvesting with dendritic antenna. Thin Solid Films 331(1–2):254–258.

Algers J, Sperr P, Egger W, Liszkay L, Kogel G, Baerdemaeker J, Maurer FHJ. 2004. Free volume determination of azobenzene-PMMA copolymer by a pulsed low-energy positron lifetime beam with in-situ UV illumination. Macromolecules 37(21):8035–8042.

Altomare A, et al. 2001. Synthesis and polymerization of amphiphilic methacrylates containing permanent dipole azobenzene chromophores. J Polym Sci, Part A 39(17):2957–2977.

Altomare A, Ciardelli F, Tirelli N, Solaro R. 1997. 4-Vinylazobenzene: polymerizability and photochromic properties of its polymers. Macromolecules 30(5):1298–1303.

Anderle K, Birenheide R, Eich M, Wendorff JH. 1989. Laser-induced reorientation of the optical axis in liquid-crystalline side chain polymers. Makromol Chem, Rapid Commun 10(9):477–483.

Anderle K, Birenheide R, Werner MJA, Wendorff JH. 1991. Molecular addressing? Studies on light-induced reorientation in liquid-crystalline side chain polymers. Liq Cryst 9(5):691–699.

Angeli C, Cimiraglia R, Hofmann H-J. 1996. On the competition between the inversion and rotation mechanisms in the cis-trans thermal isomerization of diazene. Chem Phys Lett 259(3–4):276–282.

Angiolini L, Caretti D, Carlini C, Salatelli E. 1995. Optically active polymers bearing side-chain photochromic moieties: synthesis and chiroptical properties of methacrylic and acrylic homopolymers with pendant L-lactic acid or L-alanine residues connected to trans-4-aminoazobenzene. Macromol Chem Phys 196(9):2737–2750.

Antipov AA, Sukhorukov GB, Möhwald H. 2003. Influence of the ionic strength on the polyelectrolyte multilayers' permeability. Langmuir 19(6):2444–2448.

Anzai J-I, Osa T. 1994. Photosensitive artificial membranes based on azobenzene and spirobenzopyran derivatives. Tetrahedron 50(14):4039–4070.

Aoki K, Nakagawa M, Ichimura K. 2000. Self-assembly of amphoteric azopyridine carboxylic acids: organized structures and macroscopic organized morphology influenced by heat, pH change, and light. J Am Chem Soc 122(44):10997–11004.

Arai K, Kawabata Y. 1995. Changes in the sol-gel transformation behavior of azobenzene moiety-containing methyl cellulose irradiated with UV light. Macromol Rapid Commun 16(12):875–880.

Asakawa M, et al. 1999. Photoactive azobenzene-containing supramolecular complexes and related interlocked molecular compounds. Chem Eur J 5(3):860–875.

Asanuma H, Liang X, Yoshida T, Komiyama M. 2001. Photocontrol of DNA duplex formation by using azobenzene-bearing oligonucleotides. ChemBioChem 2(1):39–44.

Bai S, Zhao Y. 2001. Azobenzene-containing thermoplastic elastomers: coupling mechanical and optical effects. Macromolecules 34(26):9032–9038.

Balasubramanian D, Subramani S, Kumar C. 1975. Modification of a model membrane structure by embedded photochrome. Nature 254(5497):252–254.

Balasubramanian S, et al. 1998. Azo chromophore-functionalized polyelectrolytes. 2. Acentric self-assembly through a layer-by-layer deposition process. Chem Mater 10(6):1554–1560.

Balzani V, Credi A, Marchioni F, Stoddart JF. 2001. Artificial molecular-level machines. Dethreading-rethreading of a pseudorotaxane powered exclusively by light energy. Chem Commun 18:1860–1861.

Barley SH, Gilbert A, Mitchell GR. 1991. Photoinduced reversible refractive-index changes in tailored siloxane-based polymers. J Mater Chem 1(3):481–482.

Barrett C, Natansohn A, Rochon P. 1994. Thermal cis-trans isomerization rates of azobenzenes bound in the side chain of some copolymers and blends. Macromolecules 27(17):4781–4786.

Barrett C, Natansohn A, Rochon P. 1995. Cis-trans thermal isomerization rates of bound and doped azobenzenes in a series of polymers. Chem Mater 7(5):899–903.

Barrett C, Choudhury B, Natansohn A, Rochon P. 1998. Azocarbazole polymethacrylates as single-component electrooptic materials. Macromolecules 31(15):4845–4851.

Beltrame PL, et al. 1993. Thermal cis-trans-isomerization of azo dyes in poly(methyl methacrylate) matrix—a kinetic-study. J Appl Polym Sci 49(12):2235–2239.

Berg RH, Hvilsted S, Ramanujam PS. 1996. Peptide oligomers for holographic data storage. Nature 383:505–508.

Bian S, He J-A, Li L, Kumar J, Tripathy SK. 2000. Large photoinduced birefringence in Azo dye/polyion films assembled by electrostatic sequential adsorption. Adv Mater 12(16):1202–1205.

Bignozzi MC, et al. 1999. Liquid crystal poly(glycidyl ether)s by anionic polymerization and polymer-analogous reaction. Polymer J (Tokyo) 31(11-1):913–919.

Birabassov R, et al. 1998. Thick dye-doped poly(methyl methacrylate) films for real-time holography. Appl Opt 37(35):8264–8269.

Blanchard PM, Mitchell GR. 1993a. A comparison of photoinduced poling and thermal poling of azo-dye-doped polymer films for second order nonlinear optical applications. Appl Phys Lett 63(15):2038–2040.

Blanchard PM, Mitchell GR. 1993b. Localized room temperature photo-induced poling of azo-dye-doped polymer films for second-order nonlinear optical phenomena. J Phys D 26(3):500–503.

Blinov LM, Kozlovsky MV, Ozaki M, Skarp K, Yoshino K. 1998. Photoinduced dichroism and optical anisotropy in a liquid-crystalline azobenzene side chain polymer caused by anisotropic angular distribution of trans and cis isomers. J Appl Phys 84(7):3860–3866.

Bortolus P, Monti S. 1979. Cis-trans photoisomerization of azobenzene. Solvent and triplet donors effects. J Phys Chem 83(6):648–652.

Bose M, Groff D, Xie J, Brustad E, Schultz PG. 2006. The incorporation of a photoisomerizable amino acid into proteins in E. coli. J Am Chem Soc 128(2):388–389.

Bredenbeck J, et al. 2003. Transient 2D-IR spectroscopy: snapshots of the nonequilibrium ensemble during the picosecond conformational transition of a small peptide. J Phys Chem B 107(33):8654–8660.

Brown CJ. 1966. A refinement of the crystal structure of azobenzene. Acta Cryst 21(1): 146–152.

Brown D, Natansohn A, Rochon P. 1995. Azo polymers for reversible optical storage. 5. Orientation and dipolar interactions of azobenzene side groups in copolymers and blends containing methyl methacrylate structural units. Macromolecules 28(18):6116–6123.

Brown EV, Granneman GR. 1975. Cis-trans isomerism in pyridyl analogs of azobenzene—kinetic and molecular-orbital analysis. J Am Chem Soc 97(3):621–627.

Bublitz D, et al. 2000. Photoinduced deformation of azobenzene polyester films. Appl Phys B: Lasers Opt 70(6):863–865.

Buffeteau T, Labarthet FL, Pézolet M, Sourisseau C. 2001. Dynamics of photoinduced orientation of nonpolar azobenzene groups in polymer films. Characterization of the cis isomers by visible and FTIR spectroscopies. Macromolecules 34(21):7514–7521.

Burke SE, Barrett CJ. 2003a. Acid-base equilibria of weak polyelectrolytes in multilayer thin films. Langmuir 19(8):3297–3303.

Burke SE, Barrett CJ. 2003b. pH-Responsive properties of multilayered poly(L-lysine)/hyaluronic acid surfaces. Biomacromolecules 4(6):1773–1783.

Burland DM, Miller RD, Walsh CA. 1994. Second-order nonlinearity in poled-polymer systems. Chem Rev 94(1):31–75.

Camacho-Lopez M, Finkelmann H, Palffy-Muhoray P, Shelley M. 2004. Fast liquid-crystal elastomer swims into the dark. Nat Mate 3(5):307–310.

Campbell VE, et al. 2006. Chromophore orientation dynamics, phase stability, and photorefractive effects in branched azobenzene chromophores. Macromolecules 39(3):957–961.

Caruso F. 2001. Nanoengineering of particle surfaces. Adv Mater 13(1):11–22.

Caruso F, Möhwald H. 1999. Protein multilayer formation on colloids through a stepwise self-assembly technique. J Am Chem Soc 1212(25):6039–6046.

Chen AG, Brady DJ. 1993. Two-wavelength reversible holograms in azo-dye doped nematic liquid crystals. Appl Phys Lett 62(23):2920–2922.

Chen JP, Gao JP, Wang ZY. 1997a. Preparation and photochemical study of soluble optically active block copolymethacrylates and azo-containing random copolymethacrylates. J Polym Sci, Part A 35(1):9–16.

Chen SH, et al. 1995. Novel glass-forming organic materials. 1. Adamantane with pendant cholesteryl, disperse red 1, and nematogenic groups. Macromolecules 28(23):7775–7778.

Chen SH, Mastrangelo JC, Shi H, Blanton TN, Bashir-Hashemi A. 1997b. Novel glass-forming organic materials. 3. Cubane with pendant nematogens, carbazole, and disperse red 1. Macromolecules 30(1):93–97.

Chiarelli PA, et al. 2001. Controlled fabrication of polyelectrolyte multilayer thin films using spin-assembly. Adv Mater 13(15):1167–1171.

Cho J, Char K, Hong J-D, Lee K-B. 2001. Fabrication of highly ordered multilayer films using a spin self-assembly method. Adv Mater 13(14):1076–1078.

Chung AJ, Rubner MF. 2002. Methods of loading and releasing low molecular weight cationic molecules in weak polyelectrolyte multilayer films. Langmuir 18(4):1176–1183.

Ciardelli F, Pieroni O. 2001. Photoswitchable polypeptides. In: Feringa BL, editor Molecular Switches, Hoboken, NJ: Wiley, p. 399–441.

Cimrová V, et al. 2002. Comparison of the birefringence in an azobenzene-side-chain copolymer induced by pulsed and continuous-wave irradiation. Appl Phys Lett 81(7):1228–1230.

Couture JJA, Lessard RA. 1988. Modulation transfer function measurements for thin layers of azo dyes in PVA matrix used as an optical encoding material. Appl Opt 27(16):3368–3374.

Dalton LR, et al. 1995. Synthesis and processing of improved organic second-order nonlinear optical materials for applications in photonics. Chem Mater 7(6):1060–1081.

Dante S, Advincula RC, Frank CW, Stroeve P. 1999. Photoisomerization of polyionic layer-by-layer films containing azobenzene. Langmuir 15(1):193–201.

de Lange JJ, Robertson JM, Woodward I. 1939. X-ray crystal analysis of trans-azobenzene. Proc Roy Soc (London) A171:398–410.

Decher G. 1997. Fuzzy nanoassemblies: toward layered polymeric multicomposites. Science 277(5330):1232–1237.

Decher G, Hong JD. 1991. Buildup of ultrathin multilayer films by a self-assembly process: II. Consecutive adsorption of anionic and cationic bipolar amphiphiles and polyelectrolytes on charged surfaces. Berichte der Bunsen-Gesellschaft 95(11):1430–1434.

Decher G, Schmitt J. 1992. Fine-tuning of the film thickness of ultrathin multilayer films composed of consecutively alternating layers of anionic and cationic polyelectrolytes. Prog Colloid Polym Sci 89(Trends Colloid Interface Sci VI):160–164.

Decher G, Hong JD, Schmitt J. 1991. Buildup of ultrathin multilayer films by a self-assembly process: III. Consecutively alternating adsorption of anionic and cationic polyelectrolytes on charged surfaces. Thin Solid Films 210–211(Part 2):831–835.

Decher G, Eckle M, Schmitt J, Struth B. 1998. Layer-by-layer assembled multicomposite films. Curr Opin Colloid Interface Sci 3(1):32–39.

Delaire JA, Nakatani K. 2000. Linear and nonlinear optical properties of photochromic molecules and materials. Chem Rev 100(5):1817–1846.

Dhanabalan A, et al. 1999. Optical storage in mixed langmuir-blodgett (LB) films of disperse Red-19 Isophorone polyurethane and cadmium stearate. Langmuir 15(13):4560–4564.

Diaspro A, Silvano D, Krol S, Cavalleri O, Gliozzi A. 2002. Single living cell encapsulation in nano-organized polyelectrolyte shells. Langmuir 18(13):5047–5050.

dos Santos DS Jr., et al. 2003. Light-induced storage in layer-by-layer films of chitosan and an azo dye. Biomacromolecules 4(6):1502–1505.

Dragan S, Schwarz S, Eichhorn K-J, Lunkwitz K. 2003. Electrostatic self-assembled nanoarchitectures between polycations of integral type and azo dyes. Colloids Surf A 195(1–3):243–251.

Dubas ST, Schlenoff JB. 1999. Factors controlling the growth of polyelectrolyte multilayers. Macromolecules 32(24):8153–8160.

Eaton DF. 1991. Nonlinear optical materials. Science 253(5017):281–287.

Ebralidze TD, Mumladze AN. 1990. Light-induced anisotropy in azo-dye-colored materials. Appl Opt 29(4):446–447.

Effing JJ, Kwak JCT. 1995. Photoswitchable phase separation in hydrophobically modified polyacrylamide/surfactant systems. Angew Chem Int Ed Engl 34(1):88–90.

Eich M, Wendorff J. 1990. Laser-induced gratings and spectroscopy in monodomains of liquid-crystalline polymers. J Opt Soc Am B 7(8):1428.

Eichler HJ, Orlic S, Schulz R, Rübner J. 2001. Holographic reflection gratings in azobenzene polymers. Opt Lett 26(9):581–583.

Eisenbach CD. 1980a. Cis-trans isomerization of aromatic azo chromophores, incorporated in the hard segments of poly(ester urethane)s. Macromol Rapid Commun 1(5):287–292.

Eisenbach CD. 1980b. Isomerization of aromatic azo chromophores in poly(ethyl acrylate) networks and photomechanical effect. Polymer 21(10):1175–1179.

Evans SD, Johnson SR, Ringsdorf H, Williams LM, Wolf H. 1998. Photoswitching of azobenzene derivatives formed on planar and colloidal gold surfaces. Langmuir 14(22):6436–6440.

Everlof GJ, Jaycox GD. 2000. Stimuli-responsive polymers. 4. Photo- and thermo-regulated chiroptical behavior in azobenzene-modified polymers fitted with main chain spirobiindane turns and chiral binaphthyl bends. Polymer 41(17):6527–6536.

Feng CL, et al. 2001. Reversible wettability of photoresponsive fluorine-containing azobenzene polymer in langmuir-blodgett films. Langmuir 17(15):4593–4597.

Fischer B, et al. 1997. The packing of azobenzene dye moieties and mesogens in polysiloxane copolymers and its impact on the opto-dielectric effect. Liq Cryst 22(1):65–74.

Fischer E. 1967. Calculation of photostationary states in systems A-B when only A is known. J Phys Chem 71(11):3704–3706.

Fissi A, Pieroni O, Ciardelli F. 1987. Photoresponsive polymers—azobenzene-containing poly(L-lysine). Biopolymers 26(12):1993–2007.

Fissi A, Pieroni O, Balestreri E, Amato C. 1996. Photoresponsive polypeptides. Photomodulation of the macromolecular structure in poly(N((phenylazophenyl)sulfonyl)- L-lysine). Macromolecules 29(13):4680–4685.

Fujino T, Tahara T. 2000. Picosecond time-resolved raman study of trans-azobenzene. J Phys Chem A 104(18):4203–4210.

Fujino T, Arzhantsev SY, Tahara T. 2001. Femtosecond time-resolved fluorescence study of photoisomerization of trans-azobenzene. J Phys Chem A 105(35):8123–8129.

Fujita T, Iyi N, Klapyta Z. 1998. Preparation of azobenzene-mica complex and its photoresponse to ultraviolet irradiation. Mater Res Bull 33(11):1693–1701.

Fujita T, Iyi N, Klapyta Z. 2001. Optimum conditions for photoresponse of azobenzene-organophilic tetrasilicic mica complexes. Mater Res Bull 36(3–4):557–571.

Funke U, Gruetzmacher HF. 1987. Dithiadiaza[n.2]paracyclophenes. Tetrahedron 43(16):3787–3795.

Furukawa J, Takamori S, Yamashita S. 1967. Preparation of block copolymers with a macro-azo-nitrile as an initiator. Angew Makromol Chem 1(1):92–104.

Gabor G, Fischer E. 1971. Spectra and cis-trans isomerism in highly dipolar derivatives of azobenzene. J Phys Chem 75(4):581–583.

Gallot B, Fafiotte M, Fissi A, Pieroni O. 1996. Liquid-crystalline structure of poly(L-lysine) containing azobenzene units in the side chain. Macromol Rapid Commun 17(8):493–501.

Gegiou D, Muszkat KA, Fischer E. 1968. Temperature dependence of photoisomerization. V. Effect of substituents on the photoisomerization of stilbenes and azobenzenes. J Am Chem Soc 90(15):3907–3918.

Gibbons WM, Shannon PJ, Sun S-T, Swetlin BJ. 1991. Surface-mediated alignment of nematic liquid crystals with polarized laser light. Nature 351(6321):49–50.

Haberfield P, Block PM, Lux MS. 1975. Enthalpies of solvent transfer of transition-states in cis-trans isomerization of azo-compounds—rotation vs nitrogen inversion mechanism. J Am Chem Soc 97(20):5804–5806.

Hagen R, Bieringer T. 2001. Photoaddressable polymers for optical data storage. Adv Mater 13(23):1805–1810.

Hair SR, Taylor GA, Schultz LW. 1990. An easily implemented flash-photolysis experiment for the physical-chemistry laboratory—the isomerization of 4-Anilino-4′-Nitroazobenzene. J Chem Educ 67(8):709–712.

Hammond PT. 1999. Recent explorations in electrostatic multilayer thin film assembly. Curr Opin Colloid Interface Sci 4(6):430–442.

Hampson GC, Robertson JM. 1941. Bond length and resonance in the cis-azobenzene molecule. J Chem Soc Abstracts:409–413.

Han M, Ichimura K. 2001. Tilt orientation of p-methoxyazobenzene side chains in liquid crystalline polymer films by irradiation with nonpolarized light. Macromolecules 34(1):82–89.

Harada M, Sisido M, Hirose J, Nakanishi M. 1991. Photoreversible antigen-antibody reactions. FEBS Lett 286(1–2):6–8.

Hartley GS. 1937. Cis form of azobenzene. Nature 140:281.

Hartley GS. 1938. Cis form of azobenzene and the velocity of the thermal cis-trans conversion of azobenzene and some derivatives. J Chem Soc Abstracts:633–642.

Harvey AJ, Abell AD. 2000. Azobenzene-containing, peptidyl a-ketoesters as photobiological switches of a-chymotrypsin. Tetrahedron 56(50):9763–9771.

Harvey AJ, Abell AD. 2001. a-Ketoester-based photobiological switches: synthesis, peptide chain extension and assay against a-chymotrypsin. Bioorg Med Chem Lett 11(18):2441–2444.

Hayashi T, et al. 1995. Photo-induced phase transition of side chain liquid crystalline copolymers with photochromic group. Eur Polym J 31(1):23–28.

He J-A, et al. 2000a. Photochemical behavior and formation of surface relief grating on self-assembled polyion/dye composite film. J Phys Chem B 104(45):10513–10521.

He J-A, et al. 2000b. Surface relief gratings from electrostatically layered azo dye films. Appl Phys Lett 76(22):3233–3235.

Higuchi M, Minoura N, Kinoshita T. 1994. Photocontrol of micellar structure of an azobenzene containing amphiphilic sequential polypeptide. Chem Lett (2):227–230.

Higuchi M, Minoura N, Kinoshita T. 1995a. Photoinduced structural and functional changes of an azobenzene containing amphiphilic sequential polypeptide. Macromolecules 28(14):4981–4985.

Higuchi M, Minoura N, Kinoshita T. 1995b. Photo-responsive behavior of a monolayer composed of an azobenzene containing polypeptide in the main-chain. Colloid Polym Sci 273(11):1022–1027.

Hildebrandt R, et al. 1998. Time-resolved investigation of photoinduced birefringence in azobenzene side-chain polyester films. Phys Rev Lett 81:5548–5551.

Ho C-H, Yang K-N, Lee S-N. 2001. Mechanistic study of trans-cis isomerization of the substituted azobenzene moiety bound on a liquid-crystalline polymer. J Polym Sci, Part A 39(13):2296–2307.

Ho MS, et al. 1996. Synthesis and optical properties of poly{(4-nitrophenyl)-[3-[N-[2-(methacryloyloxy)ethyl]carbazolyl]]diazene}. Macromolecules 29(13):4613–4618.

Hohsaka T, Kawashima K, Sisido M. 1994. Photoswitching of Nad+-mediated enzyme reaction through photoreversible antigen-antibody reaction. J Am Chem Soc 116(1):413–414.

Holland NB, et al. 2003. Single molecule force spectroscopy of azobenzene polymers: switching elasticity of single photochromic macromolecules. Macromolecules 36(6):2015–2023.

Holme NCR, Ramanujam PS, Hvilsted S. 1996. 10,000 optical write, read, and erase cycles in an azobenzene sidechain liquid-crystalline polyester. Opt Lett 21(12):902–904.

Hong J-D, Jung B-D, Kim CH, Kim K. 2000. Effects of spacer chain lengths on layered nanostructures assembled with main-chain azobenzene ionenes and polyelectrolytes. Macromolecules 33(21):7905–7911.

Hore DK, Natansohn A, Rochon PL. 2002. Optical anisotropy as a probe of structural order by stokes polarimetry. J Phys Chem B 106(35):9004–9012.

Hore DK, Natansohn AL, Rochon PL. 2003. Anomalous cis isomer orientation in a liquid crystalline azo polymer on irradiation with linearly polarized light. J Phys Chem B 107(10):2197–2204.

Houben JL, et al. 1983. Azobenzene-containing poly(L-glutamates)—photochromism and conformation in solution. Int J Biol Macromol 5(2):94–100.

Hugel T, et al. 2002. Single-molecule optomechanical cycle. Science 296(5570):1103–1106.

Hvilsted S, Andruzzi F, Kulinna C, Siesler HW, Ramanujam PS. 1995. Novel side-chain liquid crystalline polyester architecture for reversible optical storage. Macromolecules 28(7):2172–2183.

Ichimura K. 2000. Photoalignment of liquid-crystal systems. Chem Rev 100(5):1847–1873.

Ichimura K, Han M. 2000. Molecular reorientation induced by sunshine suggesting the generation of absolute chirality. Chem Lett 29(3):286–287.

Ichimura K, Hayashi Y, Akiyama H, Ikeda T, Ishizuki N. 1993. Photo-optical liquid crystal cells driven by molecular rotors. Appl Phys Lett 63(4):449–451.

Ichimura K, Momose M, Kudo K, Akiyama H, Ishizuki N. 1996. Surface-assisted formation of anisotropic dye molecular films. Thin Solid Films 284–285:557–560.

Ichimura K, Oh S-K, Nakagawa M. 2000. Light-driven motion of liquids on a photo-responsive surface. Science 288(5471):1624–1626.

Iftime G, Labarthet FL, Natansohn A, Rochon P, Murti K. 2002. Main chain-containing azo-tetraphenyldiaminobiphenyl photorefractive polymers. Macromolecules 14(1):168–174.

Ikeda T, Tsutsumi O. 1995. Optical switching and image storage by means of azobenzene liquid-crystal films. Science 268(5219):1873–1875.

Ikeda T, Horiuchi S, Karanjit DB, Kurihara S, Tazuke S. 1990. Photochemically induced isothermal phase transition in polymer liquid crystals with mesogenic phenyl benzoate side chains. 2. Photochemically induced isothermal phase transition behaviors. Macromolecules 23(1):42–48.

Ikeda T, Sasaki T, Ichimura K. 1993. Photochemical switching of polarization in ferroelectric liquid-crystal films. Nature 361:428–430.

Ikeda T, Nakano M, Yu Y, Tsutsumi O, Kanazawa A. 2003. Anisotropic bending and unbending behavior of azobenzene liquid-crystalline gels by light exposure. Adv Mater 15(3):201–205.

Inada T, Terabayashi T, Yamaguchi Y, Kato K, Kikuchi K. 2005. Modulation of the catalytic mechanism of hen egg white lysozyme (HEWL) by photochromism of azobenzene. J Photochem Photobiol A 175(2–3):100–107.

Ishikawa J, et al. 2002. Photo-induced in-plane alignment of LC molecules on layer-by-layer self-assembled films containing azo dyes evaluated by attenuated total reflection measurements. Colloids Surf A 198–200:917–922.

Ishikawa M, et al. 1998. Reflection-typeconfocal readout for multilayered optical memory. Opt Lett 23(22):1781–1783.

Ivanov M, Todorov T, Nikolova L, Tomova N, Dragostinova V. 1995. Photoinduced changes in the refractive index of azo-dye/polymer systems. Appl Phys Lett 66(17):2174–2176.

Ivanov M, et al. 2000. Light-induced optical activity in optically ordered amorphous side-chain azobenzene containing polymer. J Mod Opt 47(5):861–867.

Izumi A, Teraguchi M, Nomura R, Masuda T. 2000a. Synthesis of conjugated polymers with azobenzene moieties in the main chain. J Polym Sci, Part A 38(7):1057–1063.

Izumi A, Teraguchi M, Nomura R, Masuda T. 2000b. Synthesis of poly(p-phenylene)-based photoresponsive conjugated polymers having azobenzene units in the main chain. Macromolecules 33(15):5347–5352.

Ji HF, et al. 2004. Photon-driven nanomechanical cyclic motion. Chem Commun (22):2532–2533.

Jiang D-L, Aida T. 1997. Photoisomerization in dendrimers by harvesting of low-energy photons. Nature 388(6641):454–456.

Jiang WH, et al. 2005. Photo-switched wettability on an electrostatic self-assembly azobenzene monolayer. Chem Commun (Cambridge, UK) (28):3550–3552.

Jiang XL, Li L, Kumar J, Tripathy SK. 1996. Photoassisted poling induced second harmonic generation with in-plane anisotropy in azobenzene containing polymer films. Appl Phys Lett 69(24):3629–3631.

Jianhua G, et al. 1998. Surface plasmon resonance research on photoinduced switch properties of liquid crystalline azobenzene polymer Langmuir-Blodgett films. Supramol Sci 5(5–6):675–678.

Jin W, Shi X, Caruso F. 2001. High activity enzyme microcrystal multilayer films. J Am Chem Soc 123(33):8121–8122.

Jung B-D, et al. 2002. Photochromic hollow shells: photoisomerization of azobenzene polyionene in solution, in multilayer assemblies on planar and spherical surfaces. Colloids Surf A 198–200:483–489.

Jung JH, Takehisa C, Sakata Y, Kaneda T. 1996. p-(4-Nitrophenylazo)phenol dye-bridged permethylated a-cyclodextrin dimer: synthesis and self-aggregation in dilute aqueous solution. Chem Lett (2):147–148.

Junge DM, McGrath DV. 1997. Photoresponsive dendrimers. Chem Commun (9):857–858.

Junge DM, McGrath DV. 1999. Photoresponsive azobenzene-containing dendrimers with multiple discrete states. J Am Chem Soc 121(20):4912–4913.

Jursic BS. 1996. Ab initio and density functional theory study of the diazene isomerization. Chem Phys Lett 261(1–2):13–17.

Kano K, et al. 1980. Photoresponsive membranes. Regulation of membrane properties by photoreversible cis-trans isomerization of azobenzenes. Chem Lett (4):421–424.

Kato T, Hirota N, Fujishima A, Frechet JMJ. 1996. Supramolecular hydrogen-bonded liquid-crystalline polymer complexes. Design of side-chain polymers and a host-guest system by noncovalent interaction. J Polym Sci, Part A 34(1):57–62.

Kawata S, Kawata Y. 2000. Three-dimensional optical data storage using photochromic materials. Chem Rev 100(5):1777–1788.

Kerzhner BK, Kofanov VI, Vrubel TL. 1983. Photoisomerization of aromatic azo compounds adsorbed on a hydroxylated surface. Zhurnal Obshchei Khimii 53(10):2303–2306.

Kim DY, Tripathy SK, Li L, Kumar J. 1995. Laser-induced holographic surface relief gratings on nonlinear optical polymer films. Appl Phys Lett 66(10):1166–1168.

Kleideiter G, Sekkat Z, Kreiter M, Lechner MD, Knoll W. 2000. Photoisomerization of disperse red one in films of poly(methyl-methacrylate) at high pressure. J Mol Struct 521(1–3):167–178.

Kleinfeld ER, Ferguson GS. 1994. Stepwise formation of multilayered nanostructural films from macromolecular precursors. Science 265(5170):370–373.

Knoll W. 1996. Self-assembled microstructures at interfaces. Curr Opin Colloid Interface Sci 1(1):137–143.

Kobayashi T, Degenkolb EO, Rentzepis PM. 1979. Picosecond spectroscopy of 1-phenylazo-2-hydroxynaphthalene. J Phys Chem 83(19):2431–2434.

Komori K, Yatagai K, Tatsuma T. 2004. Activity regulation of tyrosinase by using photoisomerizable inhibitors. J Biotechnol 108(1):11–16.

Kotov NA, Dekany I, Fendler JH. 1995. Layer-by-layer self-assembly of polyelectrolyte-semiconductor nanoparticle composite films. J Phys Chem 99(35):13065–13069.

Kwolek S, Morgan P, Schaefgen J. 1985. Encyclopedia of Polymer Science and Engineering, 9. New York: John-Wiley.

Labarthet FL, Buffeteau T, Sourisseau C. 1998. Analyses of the diffraction efficiencies, birefringence, and surface relief gratings on azobenzene-containing polymer films. J Phys Chem B 102(15):2654–2662.

Lamarre L, Sung CSP. 1983. Studies of physical aging and molecular motion by azochromophoric labels attached to the main chains of amorphous polymers. Macromolecules 16(11):1729–1736.

Lednev IK, Ye T-Q, Hester RE, Moore JN. 1996. Femtosecond time-resolved UV-visible absorption spectroscopy of trans-azobenzene in solution. J Phys Chem 100(32): 13338–13341.

Lee S-H, et al. 2000. Azo polymer multilayer films by electrostatic self-assembly and layer-by-layer post azo functionalization. Macromolecules 33(17):6534–6540.

Lee S-S, et al. 2001. Layer-by-layer deposited multilayer assemblies of ionene-type polyelectrolytes based on the spin-coating method. Macromolecules 34(16):5358–5360.

Lee S-S, Lee K-B, Hong J-D. 2003. Evidence for spin coating electrostatic self-assembly of polyelectrolytes. Langmuir 19(18):7592–7596.

Lee TS, et al. 1998. Photoinduced surface relief gratings in high-Tg main-chain azoaromatic polymer films. J Polym Sci, Part A 36(2):283–289.

Lee W-S, Ueno A. 2001. Photocontrol of the catalytic activity of a beta-cyclodextrin bearing azobenzene and histidine moieties as a pendant group. Macromol Rapid Commun 22(6):448–450.

Levy D, Esquivias L. 1995. Sol-gel processing of optical and electrooptical materials. Adv Mater 7(2):120–129.

Li S, McGrath DV. 2000. Effect of macromolecular isomerism on the photomodulation of dendrimer properties. J Am Chem Soc 122(28):6795–6796.

Li Y, Deng Y, Tong X, Wang X. 2006. Formation of photoresponsive uniform colloidal spheres from an amphiphilic azobenzene-containing random copolymer. Macromolecules 39(3):1108–1115.

Liang X, et al. 2003. NMR Study on the photoresponsive DNA tethering an azobenzene. Assignment of the absolute configuration of two diastereomers and structure determination of their duplexes in the *trans*-form. J Am Chem Soc 125(52):16408–16415.

Lien L, Jaikaran DCJ, Zhang Z, Woolley GA. 1996. Photomodulated blocking of gramicidin ion channels. J Am Chem Soc 118(48):12222–12223.

Linford MR, Auch M, Möhwald H. 1998. Nonmonotonic effect of ionic strength on surface dye extraction during dye-polyelectrolyte multilayer formation. J Am Chem Soc 120(1):178–182.

Liu M, Asanuma H, Komiyama M. 2005. Azobenzene-tethered T7 promoter for efficient photoregulation of transcription. J Am Chem Soc 128(3):1009–1015.

Liu N, et al. 2004. Photoregulation of mass transport through a photoresponsive azobenzene-modified nanoporous membrane. Nano Lett 4(4):551–554.

Liu X-H, Bruce DW, Manners I. 1997. Novel calamitic side-chain metallomesogenic polymers with ferrocene in the backbone: synthesis and properties of thermotropic liquid-crystalline poly(ferrocenylsilanes). Chem Commun (3):289–290.

Liu ZF, Hashimoto K, Fujishima A. 1990. Photoelectrochemical information storage using an azobenzene derivative. Nature 347:658–660.

Liu ZF, Morigaki K, Enomoto T, Hashimoto K, Fujishima A. 1992. Kinetic studies on the thermal cis-trans isomerization of an azo compound in the assembled monolayer film. J Phys Chem 96(4):1875–1880.

Lösche M, Schmitt J, Decher G, Bouwman WG, Kjaer K. 1998. Detailed structure of molecularly thin polyelectrolyte multilayer films on solid substrates as revealed by neutron reflectometry. Macromolecules 31(25):8893–8906.

Luk Y-Y, Abbott NL. 2003. Surface-driven switching of liquid crystals using redox-active groups on electrodes. Science 301(5633):623–626.

Lvov Y, Ariga K, Onda M, Ichinose I, Kunitake T. 1997. Alternate assembly of ordered multilayers of SiO_2 and other nanoparticles and polyions. Langmuir 13(23):6195–6203.

Ma H, et al. 2002. Highly efficient and thermally stable electro-optical dendrimers for photonics. Adv Funct Mater 12(9):565–574.

Magennis SW, Mackay FS, Jones AC, Tait KM, Sadler PJ. 2005. Two-photon-induced photoisomerization of an azo dye. Chem Mater 17(8):2059–2062.

Malcolm BR, Pieroni O. 1990. The photoresponse of an azobenzene-containing poly(L-lysine) in the monolayer state. Biopolymers 29(6–7):1121–1123.

Malkin S, Fischer E. 1962. Temperature dependence of photoisomerization. Part II.1 Quantum yields of cis-trans isomerizations in azo-compounds. J Phys Chem 66(12):2482–2486.

Mallia VA, Tamaoki N. 2003. Photoresponsive vitrifiable chiral dimesogens: photo-thermal modulation of microscopic disordering in helical superstructure and glass-forming properties. J Mater Chem 13(2):219–224.

Mattoussi H, Rubner MF, Zhou F, Kumar J, Tripathy SK. 2000. Photovoltaic heterostructure devices made of sequentially adsorbed poly(phenylene vinylene) and functionalized C60. Appl Phys Lett 77(10):1540–1542.

Maxein G, Zentel R. 1995. Photochemical inversion of the helical twist sense in chiral polyisocyanates. Macromolecules 28(24):8438–8440.

McAloney RA, Dudnik V, Goh MC. 2003. Kinetics of salt-induced annealing of a polyelectrolyte multilayer film morphology. Langmuir 19(9):3947–3952.

Mekelburger HB, Rissanen K, Voegtle F. 1993. Repetitive synthesis of bulky dendrimers—a reversibly photoactive dendrimer with six azobenzene side chains. Chem Ber 126(5):1161–1169.

Mendelsohn JD, et al. 2000. Fabrication of microporous thin films from polyelectrolyte multilayers. Langmuir 16(11):5017–5023.

Meng X, Natansohn A, Barrett C, Rochon P. 1996. Azo polymers for reversible optical storage. 10. Cooperative motion of polar side groups in amorphous polymers. Macromolecules 29(3):946–952.

Mermut O, Barrett CJ. 2001. Stable sensor layers self-assembled onto surfaces using azobenzene-containing polyelectrolytes. Analyst (Cambridge, UK) 126(11):1861–1865.

Mermut O, Barrett CJ. 2003. Effects of charge density and counterions on the assembly of polyelectrolyte multilayers. J Phys Chem B 107(11):2525–2530.

Mermut O, Lefebvre J, Gray DG, Barrett CJ. 2003. Structural and mechanical properties of polyelectrolyte multilayer films studied by AFM. Macromolecules 36(23):8819–8824.

Minoura N, et al. 2004. Preparation of azobenzene-containing polymer membranes that function in photoregulated molecular recognition. Macromolecules 37(25):9571–9576.

Mita I, Horie K, Hirao K. 1989. Photochemistry in polymer solids. 9. Photoisomerization of azobenzene in a polycarbonate film. Macromolecules 22(2):558–563.

Möhlmann G, van der Vorst C. 1989. Side Chain Liquid Crystal Polymers. Glasgow: Plenum and Hall.

Moller G, Harke M, Motschmann H, Prescher D. 1998. Controlling microdroplet formation by light. Langmuir 14(18):4955–4957.

Momotake A, Arai T. 2004a. Photochemistry and photophysics of stilbene dendrimers and related compounds. J Photochem Photobiol C 5(1):1–25.

Momotake A, Arai T. 2004b. Synthesis, excited state properties, and dynamic structural change of photoresponsive dendrimers. Polymer 45(16):5369–5390.

Montagnoli G, Pieroni O, Suzuki S. 1983. Control of peptide chain conformation by photoisomerising chromophores: Enzymes and model compounds. Polymer Photochem 3(4):279–294.

Monti S, Orlandi G, Palmieri P. 1982. Features of the photochemically active state surfaces of azobenzene. Chem Phys 71(1):87–99.

Morino S, Kaiho A, Ichimura K. 1998. Photogeneration and modification of birefringence in crosslinked films of liquid crystal/polymer composites. Appl Phys Lett 73(10):1317–1319.

Müller M, Zentel R. 1996. Interplay of chiral side chains and helical main chains in polyisocyanates. Macromolecules 29(5):1609–1617.

Nagamani SA, Norikane Y, Tamaoki N. 2005. Photoinduced hinge-like molecular motion: studies on xanthene-based cyclic azobenzene dimers. J Org Chem 70(23):9304–9313.

Naito K, Miura A. 1993. Molecular design for nonpolymeric organic dye glasses with thermal stability: relations between thermodynamic parameters and amorphous properties. J Phys Chem 97(23):6240–6248.

Naito T, Horie K, Mita I. 1991. Photochemistry in polymer solids. 11. The effects of the size of reaction groups and the mode of photoisomerization on photochromic reactions in polycarbonate film. Macromolecules 24(10):2907–2911.

Naito T, Horie K, Mita I. 1993. Photochemistry in polymer solids: 12. Effects of main-chain structures and formation of hydrogen bonds on photoisomerization of azobenzene in various polymer films. Polymer 34(19):4140–4145.

Nakayama K, Endo M, Majima T. 2004. Photochemical regulation of the activity of an endonuclease BamHI using an azobenzene moiety incorporated site-selectively into the dimer interface. Chem Commun (Cambridge, UK) (21):2386–2387.

Nakayama K, Endo M, Majima T. 2005. A hydrophilic azobenzene-bearing amino acid for photochemical control of a restriction enzyme BamHI. Bioconjug Chem 16(6):1360–1366.

Natansohn A, Rochon P. 1999. Photoinduced motions in azobenzene-based amorphous polymers: possible photonic devices. Adv Mater 11(6):1387–1391.

Natansohn A, Rochon P. 2002. Photoinduced motions in azo-containing polymers. Chem Rev 102(11):4139–4176.

Natansohn A, Rochon P, Gosselin J, Xie S. 1992. Azo polymers for reversible optical storage. 1. Poly[4′-[[2-(acryloyloxy)ethyl]ethylamino]-4-nitroazobenzene]. Macromolecules 25(8):2268–2273.

Natansohn A, et al. 1994. Azo polymers for reversible optical storage. 4. Cooperative motion of rigid groups in semicrystalline polymers. Macromolecules 27(9):2580–2585.

Nikolova L, et al. 2000. Self-induced light polarization rotation in azobenzene-containing polymers. Appl Phys Lett 77(5):657–659.

Nikolova L, et al. 1996. Polarization holographic gratings in side-chain azobenzene polyesters with linear and circular photoanisotropy. Appl Opt 35(20):3835–3840.

Nikolova L, et al. 1997. Photoinduced circular anisotropy in side-chain azobenzene polyesters. Opt Mater 8(4):255–258.

Neuert G, Hugel T, Netz RR, Gaub HE. 2005. Elasticity of poly(azobenzene-peptides). Macromolecules 39(2):789–797.

Norikane Y, Kitamoto K, Tamaoki N. 2003. Novel crystal structure, cis-trans isomerization, and host property of meta-substituted macrocyclic azobenzenes with the shortest linkers. J Org Chem 68(22):8291–8304.

Norman LL, Barrett CJ. 2002. Solution properties of self-assembled amphiphilic copolymers determined by isomerization spectroscopy. J Phys Chem B 106(34):8499–8503.

Nunzi J-M, Fiorini C, Etilé A-C, Kajzar F. 1998. All-optical poling in polymers: dynamical aspects and perspectives. Pure Appl Opt 7(2):141–150.

Nyamjav D, Ivanisevic A. 2004. Properties of polyelectrolyte templates generated by dip-pen nanolithography and microcontact printing. Chem Mater 16(25):5216–5219.

Paik CS, Morawetz H. 1972. Photochemical and thermal isomerization of azoaromatic residues in the side chains and the backbone of polymers in bulk. Macromolecules 5(2):171–177.

Park M-K, Advincula RC. 2002. In-plane photoalignment of liquid crystals by azobenzene-polyelectrolyte layer-by-layer ultrathin films. Langmuir 18(11):4532–4535.

Pieroni O, Houben JL, Fissi A, Costantino P. 1980. Reversible conformational-changes induced by light in poly(L-glutamic acid) with photochromic side-chains. J Am Chem Soc 102(18):5913–5915.

Prasad SK, Nair GG. 2001. Effects of photo-controlled nanophase segregation in a re-entrant nematic liquid crystal. Adv Mater 13(1):40–43.

Priest WJ, Sifain MM. 1971. Photochemical and thermal isomerization in polymer matrixes. Azo compounds in polystyrene. J Polym Sci, Part A 9(11):3161–3168.

Ramanujam PS, et al. 2001. Physics and technology of optical storage in polymer thin films. Synth Met 124(1):145–150.

Rau H. 1968. Radiationless deactivation of azo compounds and light fastness of azo dyes. Berichte der Bunsen-Gesellschaft 72(3):408–414.

Rau H. 1990. Photoisomerization of Azobenzenes. In: Rebek J, editor. Photochemistry and Photophysics. Boca Raton (FL): CRC Press, p. 119–141.

Rau H, Lueddecke E. 1982. On the rotation-inversion controversy on photoisomerization of azobenzenes. Experimental proof of inversion. J Am Chem Soc 104(6):1616–1620.

Rau H, Roettger D. 1994. Photochromic azobenzenes which are stable in the trans and cis forms. Mol Cryst Liq Cryst Sci Technol, A 246:143–146.

Rau H, Crosby AD, Schaufler A, Frank R. 1981. Triplet-sensitized photoreaction of azobenzene in sulfuric-acid. Z Naturforsch, A: Phys Sci 36(11):1180–1186.

Rau H, Greiner G, Gauglitz G, Meier H. 1990. Photochemical quantum yields in the A-B system when only the spectrum of A is known. J Phys Chem 94(17):6523–6524.

Razna J, Hodge P, West D, Kucharski S. 1999. NLO properties of polymeric Langmuir-Blodgett films of sulfonamide-substituted azobenzenes. J Mater Chem 9(8):1693–1698.

Reynier N, Dozol J-F, Saadioui M, Asfari Z, Vicens J. 1998. Complexation properties of a new photosensitive calix[4]arene crown ether containing azo unit in the lower rim towards alkali cations. Tetrahedron Lett 39(36):6461–6464.

Richert L, et al. 2002. Cell interactions with polyelectrolyte multilayer films. Biomacromolecules 3(6):1170–1178.

Rmaile HH, Schlenoff JB. 2003. Optically active polyelectrolyte multilayers as membranes for chiral separations. J Am Chem Soc 125(22):6602–6603.

Rochon P, Batalla E, Natansohn A. 1995. Optically induced surface gratings on azoaromatic polymer films. Appl Phys Lett 66(2):136–138.

Rottger D, Rau H. 1996. Photochemistry of azobenzenophanes with three-membered bridges. J Photochem Photobiol A Chem 101(2–3):205–214.

Rouse JH, Ferguson GS. 2002. Stepwise incorporation of nonpolar polymers within polyelectrolyte multilayers. Langmuir 18(20):7635–7640.

Ruslim C, Ichimura K. 2001. Conformation-assisted amplification of chirality transfer of chiral Z-azobenzenes. Adv Mater 13(1):37–40.

Sabi Y, et al. 2001. Photoaddressable polymers for rewritable optical disk systems. Jpn J Appl Phys, Part 1 40(3B):1613–1618.

Sackmann E. 1971. Photochemically induced reversible color changes in cholesteric liquid crystals. J Am Chem Soc 93(25):7088–7090.

Saremi F, Tieke B. 1998. Photoinduced switching in self-assembled multilayers of an azobenzene bolaamphiphile and polyelectrolytes. Adv Mater 10(5):389–391.

Sarkar N, Sarkar A, Sivaram S. 2001. Isomerization behavior of aromatic azo chromophores bound to semicrystalline polymer films. J Appl Polym Sci 81(12):2923–2928.

Sata T, Shimokawa Y, Matsusaki K. 2000. Preparation of ion-permeable membranes having an azobenzene moiety and their transport properties in electrodialysis. J Membr Sci 171(1):31–43.

Schulze FW, Petrick HJ, Cammenga HK, Klinge H. 1977. Thermodynamic properties of the structural analogs benzo[c]cinnoline, trans-azobenzene, and cis-azobenzene. Z Phys Chem (Muenchen, Ger) 107(1):1–19.

Seki T, et al. 1993. "Command surfaces" of Langmuir-Blodgett films. Photoregulations of liquid crystal alignment by molecularly tailored surface azobenzene layers. Langmuir 9(1):211–218.

Sekkat Z, Kang C-S, Aust EF, Wegner G, Knoll W. 1995a. Room-temperature photoinduced poling and thermal poling of a rigid main-chain polymer with polar azo dyes in the side chain. Chem Mater 7(1):142–147.

Sekkat Z, Wood J, Knoll W. 1995b. Reorientation mechanism of azobenzenes within the trans-cis photoisomerization. J Phys Chem 99(47):17226–17234.

Shannon PJ, Gibbons WM, Sun ST. 1994. Patterned optical properties in photopolymerized surface-aligned liquid-crystal films. Nature 368(6471):532–533.

Shen YQ, Rau H. 1991. The environmentally controlled photoisomerization of probe molecules containing azobenzene moieties in solid poly(methyl methacrylate). Macromol Chem Phys 192(4):945–957.

Shi Y, Steier WH, Yu L, Chen M, Dalton LR. 1991a. Large photoinduced birefringence in an optically nonlinear polyester polymer. Appl Phys Lett 59(23):2935–2937.

Shi Y, Steier WH, Yu L, Chen M, Dalton LR. 1991b. Large stable photoinduced refractive index change in a nonlinear optical polyester polymer with disperse red side groups. Appl Phys Lett 58(11):1131–1133.

Shibaev V, Bobrovsky A, Boiko N. 2003. Photoactive liquid crystalline polymer systems with light-controllable structure and optical properties. Prog Polym Sci 28:729–836.

Shinkai S, Minami T, Kusano Y, Manabe O. 1983. Photoresponsive crown ethers. 8. Azobenzenophane-type switched-on crown ethers which exhibit an all-or-nothing change in ion-binding ability. J Am Chem Soc 105(7):1851–1856.

Shiratori SS, Rubner MF. 2000. pH-Dependent thickness behavior of sequentially adsorbed layers of weak polyelectrolytes. Macromolecules 33(11):4213–4219.

Shirota Y, Moriwaki K, Yoshikawa S, Ujike T, Nakano H. 1998. 4-[Di(biphenyl-4-yl)amino]azobenzene and 4,4′-bis[bis(4′-*tert*-butylbiphenyl-4-yl)amino]azobenzene as a novel family of photochromic amorphous molecular materials. J Mater Chem 8(12):2579–2581.

Shishido A, et al. 1997. Rapid optical switching by means of photoinduced change in refractive index of azobenzene liquid crystals detected by reflection-mode analysis. J Am Chem Soc 119(33):7791–7796.

Siewierski LM, Brittain WJ, Petrash S, Foster MD. 1996. Photoresponsive monolayers containing in-chain azobenzene. Langmuir 12(24):5838–5844.

Silva JR, Dall'Agnol FF, Oliveira ON Jr., Giacometti JA. 2002. Temperature dependence of photoinduced birefringence in mixed Langmuir–Blodgett (LB) films of azobenzene-containing polymers. Polymer 43(13):3753–3757.

Singh AK, Das J, Majumdar N. 1996. Novel bacteriorhodopsin analogs based on azo chromophores. J Am Chem Soc 118(26):6185–6191.

Sisido M, Ishikawa Y, Harada M, Itoh K. 1991a. Helically arranged azobenzene chromophores along a polypeptide chain. 2. Prediction of conformations and calculation of theoretical circular dichroism. Macromolecules 24(14):3999–4003.

Sisido M, Ishikawa Y, Itoh K, Tazuke S. 1991b. Helically arranged azobenzene chromophores along a polypeptide chain. 1. Synthesis and circular dichroism. Macromolecules 24(14):3993–3998.

Spörlein S, et al. 2002. Ultrafast spectroscopy reveals subnanosecond peptide conformational dynamics and validates molecular dynamics simulation. Proc Natl Acad Sci USA 99(12):7998–8002.

Steenwinckel DV, Hendrickx E, Persoons A. 2001. Large dynamic ranges in photorefractive NLO polymers and NLO-polymer-dispersed liquid crystals using a bifunctional chromophore as a charge transporter. Chem Mater 14(4):1230–1237.

Steinem C, Janshoff A, Vollmer MS, Ghadiri MR. 1999. Reversible photoisomerization of self-organized cylindrical peptide assemblies at air-water and solid interfaces. Langmuir 15(11):3956–3964.

Steitz R, Leiner V, Siebrecht R, Klitzing Rv. 2000. Influence of the ionic strength on the structure of polyelectrolyte films at the solid/liquid interface. Colloids Surf A 163(1):63–70.

Stiller B, et al. 1999. Self-assembled monolayers of novel azobenzenes for optically induced switching. Mater Sci Eng C 8–9:385–389.

Stumpe J, et al. 1991. Photoreaction in mesogenic media. 5. Photoinduced optical anisotropy of liquid-crystalline side-chain polymers with azochromophores by linearly polarized light of low intensity. Makromol Chem, Rapid Commun 12(2):81–87.

Sukhorukov GB, Schmitt J, Decher G. 1996. Reversible swelling of polyanion/polycation multilayer films in solutions of different ionic strength. Berichte der Bunsen-Gesellschaft 100(6):948–953.

Sukhorukov GB, et al. 1998a. Stepwise polyelectrolyte assembly on particle surfaces: a novel approach to colloid design. Polym Adv Technol 9(10–11):759–767.

Sukhorukov GB, et al. 1998b. Layer-by-layer self assembly of polyelectrolytes on colloidal particles. Colloids Surf A 137(1–3):253–266.

Sukwattanasinitt M, et al. 1998. Functionalizable self-assembling polydiacetylenes and their optical properties. Chem Mater 10(1):27–29.

Sumaru K, Inui S, Yamanaka T. 1999. Preliminary investigation of self-organization pattern mapping system based on photochromism. Opt Eng 38(2):274–283.

Sun ST, Gibbons WM, Shannon PJ. 1992. Alignment of guest-host liquid crystals with polarized laser light. Liq Cryst 12(5):869–874.

Suzuki I, Sato K, Koga M, Chen Q, Anzai J-I. 2003. Polyelectrolyte layered assemblies containing azobenzene-modified polymer and anionic cyclodextrins. Mater Sci Eng C 23(5):579–583.

Tanaka K, Tateishi Y, Nagamura T. 2004. Photoisomerization of azobenzene probes tagged to polystyrene in thin films. Macromolecules 37(22):8188–8190.

Tang Z, Kotov NA, Magonov S, Ozturk B. 2003. Nanostructured artificial nacre. Nat Mater 2:413–418.

Tawa K, Knoll W. 2002. Out-of-plane photoreorientation of azo dyes in polymer thin films studied by surface plasmon resonance spectroscopy. Macromolecules 35(18):7018–7023.

Teraguchi M, Masuda T. 2000. Synthesis and properties of polyacetylenes having azobenzene pendant groups. Macromolecules 33(2):240–242.

Thierry B, Winnik FM, Merhi Y, Tabrizian M. 2003. Nanocoatings onto arteries via layer-by-layer deposition: toward the in vivo repair of damaged blood vessels. J Am Chem Soc 125(25):7494–7495.

Tokuhisa H, Yokoyama M, Kimura K. 1994. Photoresponsive ion-conducting behavior of polysiloxanes carrying a crowned azobenzene moiety at the side chain. Macromolecules 27(7):1842–1846.

Toutianoush A, Tieke B. 1998. Photoinduced switching in self-assembled multilayers of azobenzene-containing ionene polycations and anionic polyelectrolytes. Macromol Rapid Commun 19(11):591–595.

Toutianoush A, Saremi F, Tieke B. 1999. Photoinduced switching in self-assembled multilayers of ionenes and bolaamphiphiles containing azobenzene units. Mater Sci Eng C C8–C9:343–352.

Tsuchiya S. 1999. Intramolecular electron transfer of diporphyrins comprised of electron-deficient porphyrin and electron-rich porphyrin with photocontrolled isomerization. J Am Chem Soc 121(1):48–53.

Tsutsumi N, Yoshizaki S, Sakai W, Kiyotsukuri T. 1996. Thermally stable nonlinear optical polymers. MCLC S&T, Section B: Nonlinear Optics 15(1–4):387–390.

Uznanski P, Pecherz J. 2002. Surface plasmon resonance of azobenzene-incorporated polyelectrolyte thin films as an H+ indicator. J Appl Polym Sci 86(6):1459–1464.

Uznanski P, Kryszewski M, Thulstrup EW. 1991. Linear dichroism and trans-cis photo-isomerization studies of azobenzene molecules in oriented polyethylene matrix. Eur Polym J 27(1):41–43.

Van Cott KE, et al. 2002. Layer-By-layer deposition and ordering of low-molecular-weight dye molecules for second-order nonlinear optics. Angew Chem Int Ed 41(17):3236–3238.

Villavicencio O, McGrath DV. 2002. Azobenzene-containing dendrimers. Advances in Dendritic Macromolecules 5:1–44.

Viswanathan NK, et al. 1999. Surface relief structures on azo polymer films. J Mater Chem 9(9):1941–1955.

Voinova MV, Jonson M. 2004. Electronic transduction in model enzyme sensors assisted by a photoisomerizable azo-polymer. Biosens Bioelectron 20(6):1106–1110.

Vollmer MS, Clark TD, Steinem C, Ghadiri MR. 1999. Photoswitchable hydrogen-bonding in self-organized cylindrical peptide systems. Angew Chem Int Ed 38(11):1598–1601.

Wang H, He Y, Tuo X, Wang X. 2004. Sequentially adsorbed electrostatic multilayers of branched side-chain polyelectrolytes bearing donor-acceptor type azo chromophores. Macromolecules 37(1):135–146.

Wang L, Schultz PG. 2004. Expanding the genetic code. Angew Chem Int Ed 44(1):34–66.

Wang TC, Rubner MF, Cohen RE. 2002. Polyelectrolyte multilayer nanoreactors for preparing silver nanoparticle composites: controlling metal concentration and nanoparticle size. Langmuir 18(8):3370–3375.

Wang X, et al. 1997a. Self-assembled second order nonlinear optical multilayer azo polymer. Macromol Rapid Commun 18(6):451–459.

Wang X, et al. 1997b. Epoxy-based nonlinear optical polymers functionalized with tricyanovinyl chromophores. Chem Mater 9(1):45–50.

Wang X, et al. 1997c. Epoxy-based nonlinear optical polymers from post azo coupling reaction. Macromolecules 30(2):219–225.

Wang X, Balasubramanian S, Kumar J, Tripathy SK, Li L. 1998. Azo chromophore-functionalized polyelectrolytes. 1. Synthesis, characterization, and photoprocessing. Chem Mater 10(6):1546–1553.

Watanabe O, Tsuchimori M, Okada A. 1996. Two-step refractive index changes by photoisomerization and photobleaching processes in the films of nonlinear optical polyurethanes and a urethane-urea copolymer. J Mater Chem 6(9):1487–1492.

Watanabe S, Regen SL. 1994. Dendrimers as building blocks for multilayer construction. J Am Chem Soc 116(19):8855–8856.

Weh K, et al. 1998. Modification of the transport properties of a polymethacrylate-azobenzene membrane by photochemical switching. Chem Eng Technol 21(5):408–412.

Weiss RG, Ramamurthy V, Hammond GS. 1993. Photochemistry in organized and confining media: a model. Acc Chem Res 25(10):530–536.

Wiesner U, Reynolds N, Boeffel C, Spiess HW. 1991. Photoinduced reorientation in liquid-crystalline polymers below the glass transition temperature studied by time-dependent infrared spectroscopy. Makromol Chem, Rapid Commun 12(8):457–464.

Willner I, Rubin S. 1993. Reversible photoregulation of the activities of proteins. Reactive Polymers 21(3):177–186.

Willner I, Rubin S. 1996. Control of the structure and functions of biomaterials by light. Angew Chem Int Ed Engl 35(4):367–385.

Willner I, Willner B. 1997. Molecular optoelectronic systems. Adv Mater 9(4):351–355.

Willner I, Rubin S, Riklin A. 1991a. Photoregulation of papain activity through anchoring photochromic azo groups to the enzyme backbone. J Am Chem Soc 113(9):3321–3325.

Willner I, Rubin S, Zor T. 1991b. Photoregulation of alpha-chymotrypsin by its immobilization in a photochromic azobenzene copolymer. J Am Chem Soc 113(10):4013–4014.

Willner I, Rubin S, Shatzmiller R, Zor T. 1993. Reversible light-stimulated activation and deactivation of alpha-chymotrypsin by its immobilization in photoisomerizable copolymers. J Am Chem Soc 115(19):8690–8694.

Wu L, Tuo X, Cheng H, Chen Z, Wang X. 2001a. Synthesis, photoresponsive behavior, and self-assembly of poly(acrylic acid)-based azo polyelectrolytes. Macromolecules 34(23):8005–8013.

Wu Y, Natansohn A, Rochon P. 2001b. Photoinduced birefringence and surface relief gratings in novel polyurethanes with azobenzene groups in the main chain. Macromolecules 34(22):7822–7828.

Wuerthner F, Rebek J, Jr. 1995. Light-switchable catalysis in synthetic receptors. Angew Chem Int Ed Engl 34(4):446–450.

Xie S, Natansohn A, Rochon P. 1993. Recent developments in aromatic azo polymers research. Chem Mater 5(4):403–411.

Xu Z-S, Drnoyan V, Natansohn A, Rochon P. 2000. Novel polyesters with amino-sulfone azobenzene chromophores in the main chain. J Polym Sci, Part A 38(12):2245–2253.

Yager KG, Barrett CJ. 2001. All-optical patterning of azo polymer films. Curr Opin Solid State Mater Sci 5(6):487–494.

Yamaki S, Nakagawa M, Morino S, Ichimura K. 2000. Surface relief gratings generated by a photocrosslinkable polymer with styrylpyridine side chains. Appl Phys Lett 76(18):2520–2522.

Yamamoto H. 1986. Synthesis and reversible photochromism of azo aromatic poly(L-lysine). Macromolecules 19(10):2472–2476.

Yamamoto H, Nishida A. 1991. Photoresponsive peptide and polypeptide systems. Part 9. Synthesis and reversible photochromism of azo aromatic poly(L-a,g-diaminobutyric acid). Polym Int 24(3):145–148.

Yamamoto H, Nishida A, Takimoto T, Nagai A. 1990. Photoresponsive peptide and polypeptide systems. VIII. Synthesis and reversible photochromism of azo aromatic poly(L-ornithine). J Polym Sci, Part A: Polym Chem 28(1):67–74.

Yamamoto T, et al. 2001. Phase-type gratings formed by photochemical phase transition of polymer azobenzene liquid crystal. 2. Rapid switching of diffraction beams in thin films. J Phys Chem B 105(12):2308–2313.

Yamamura H, et al. 1996. A cyclodextrin derivative with cation carrying ability: heptakis(3,6-anhydro)-b-cyclodextrin 2-O-p-phenylazobenzoate. Chem Lett(9):799–800.

Yamane H, Kikuchi H, Kajiyama T. 1999. Laser-addressing rewritable optical information storage of (liquid crystalline side chain copolymer/liquid crystals/photo-responsive molecule) ternary composite systems. Polymer 40(17):4777–4785.

Yang L, Takisawa N, Hayashita T, Shirahama K. 1995. Colloid chemical characterization of the photosurfactant 4-Ethylazobenzene 4′-(Oxyethyl)trimethylammonium Bromide. J Phys Chem 99(21):8799–8803.

Yaroschuk O, et al. 2001. Light induced structures in liquid crystalline side-chain polymers with azobenzene functional groups. J Chem Phys 114(12):5330–5337.

Yesodha SK, Pillai CKS, Tsutsumi N. 2004. Stable polymeric materials for nonlinear optics: a review based on azobenzene systems. Prog Polym Sci 29:45–74.

Yitzchaik S, Marks TJ. 1996. Chromophoric self-assembled superlattices. Acc Chem Res 29(4):197–202.

Yokoyama S, Nakahama T, Otomo A, Mashiko S. 2000. Intermolecular coupling enhancement of the molecular hyperpolarizability in multichromophoric dipolar dendrons. J Am Chem Soc 122(13):3174–3181.

Yu Y, Nakano M, Ikeda T. 2003. Photomechanics: Directed bending of a polymer film by light. Nature 425:145.

Yu YL, Ikeda T. 2004. Alignment modulation of azobenzene-containing liquid crystal systems by photochemical reactions. J Photochem Photobiol C: Photochem Rev 5(3):247–265.

Yu YL, Nakano M, Maeda T, Kondo M, Ikeda T. 2005. Precisely direction-controllable bending of cross-linked liquid-crystalline polymer films by light. Mol Cryst Liq Cryst 436:1235–1244.

Zawisza I, Bilewicz R, Luboch E, Biernat JF. 1999. Properties of Z and E isomers of azocrown ethers in monolayer assemblies at the air–water interface. Thin Solid Films 348(1–2):173–179.

Zhai L, Cebeci FÇ, Cohen RE, Rubner MF. 2004. Stable superhydrophobic coatings from polyelectrolyte multilayers. Nano Lett 4(7):1349–1353.

Zhong X, et al. 2001. Identification of the alignment of azobenzene molecules induced by all-optical poling in polymer films. Opt Commun 190(1–4):333–337.

Ziegler A, Stumpe J, Toutianoush A, Tieke B. 2002. Photoorientation of azobenzene moieties in self-assembled polyelectrolyte multilayers. Colloids Surf A 198–200:777–784.

Zilker SJ, et al. 1998. Holographic data storage in amorphous polymers. Adv Mater 10(11):855–859.

Zollinger H. 1961. Azo and Diazo Chemistry. New York: Interscience.

Zollinger H. 1987. Colour Chemistry, Synthesis, Properties, and Applications of Organic Dyes. Weinheim: VCH.

2

PHOTO-INDUCED PHENOMENA IN SUPRAMOLECULAR AZOBENZENE MATERIALS

Joachim Stumpe, Olga Kulikovska, Leonid M. Goldenberg, and Yuriy Zakrevskyy

2.1. INTRODUCTION

Probably, there is no other molecule as azobenzene that has been so well known for a long time and, nevertheless, still brings off surprising phenomena. Though azobenzene once started off as a simple dye—a molecule just absorbing light—it has been recognized as photochrome, a molecule changed by light and has since become indispensable in optics as a molecule controlling light. Recently, it has revealed a completely different characteristic—a molecule moved by light. The combination of all these properties in one molecule makes it favored for a wide range of optical applications. The application of azobenzenes is further supported by the availability of various commercial dyes or by simplicity of their synthesis. Different material aspects and optical applications of azobenzene may be found in some recent reviews (Seki, 2007; Shibaev et al., 2003, 1996; Natansohn and Rochon, 2002; Yager and Barrett, 2001; Viswanathan et al., 1999; Fuhrmann and Wendorff, 1997; Ichimura, 1996; Ichimura et al., 1994).

Most azobenzene derivatives can be isomerized reversibly between the E and Z isomeric forms by light or heat. Some derivatives may show phototautomerism, photodimerization, photoreduction, and fading. The effects discussed in this chapter are based on E–Z photoisomerization of aromatic azobenzenes (Rau, 1990) also known in the literature as trans–cis photoisomerization.

Photoisomerization reaction of azobenzene allows the reversible change of molecular properties such as absorbance, dipole moment, polarity, geometric shape, and others. These molecular changes may be used to control macroscopic properties of materials. For example, the change of the geometric form of

Smart Light-Responsive Materials. Edited by Yue Zhao and Tomiki Ikeda

molecules can affect the formation of molecular aggregates (Stumpe et al., 1996; Shimomura, 1993), induce phase transitions in the material (Hasegawa et al., 1999), or change the enzyme activity (Willner et al., 1991). Recently, the photoinduced change of the geometric shape of macroscopic objects has been demonstrated (Ikeda et al., 2007; Bublitz et al., 2000; Seki et al., 1996).

If the irradiating light activates both E–Z and back Z–E transitions, the steady state of photoreaction is established, whereas the molecules undergo repeated isomerization cycles. In many cases, it is not an equilibrium concentration of both isomeric forms that is important, it is the light-induced dynamics, that is, the continuing movement of the chromophore. In other words, the isomerizing azobenzene molecule works as a light-driven molecular "motor," whose power and force direction may be controlled by the parameters of the irradiating light, such as wavelength, polarization, or intensity distribution. This statement is kept in mind when discussing the photoinduced processes in this chapter.

The irradiation with polarized light results in the preferential orientation of azobenzene molecules perpendicularly to the light electric field vector—orientational order is induced, the system becomes anisotropic. The photoinduction of molecular order causes plenty of macroscopic effects, including optical anisotropy, which is one of the subjects in this chapter. After the actinic light is switched off, the photoproduct reacts back to the thermally stable E isomeric form and the initial state is restored at the molecular level. It may be changed thermally or by irradiation with actinic light. The uniformity of the photoreaction and rebuilding of the initial state at the molecular level guarantee the complete reversibility of the processes. Note that an important advantage of azobenzenes is that both photoisomerization processes take place with high quantum yields and without further side reactions. Therefore, in contrast to most other chromophores, a number of isomerization cycles can be carried out without material destruction. All these effects have been widely addressed in the literature. Since the pioneer work on photoinduction of optical anisotropy in photosensitive organic materials (Weigert, 1929), the dynamic photoorientation in azobenzene-containing viscous solutions (Makushenko et al., 1971) and dye-polymer blends (Gibbson et al., 1991; Todorov et al., 1984), and the stable anisotropy in functionalized glassy polymers (Eich et al., 1987) have been studied. Typical optical applications of the photoinduced order include polarization elements (Petrova et al., 2003), diffraction optical elements (Martinez-Ponce et al., 2004), waveguides (Karimi-Alavijeh et al., 2008; Luo et al., 2007), optical storage (Nedelchev et al., 2003; Hagen and Bieringer, 2001; Sabi et al., 2001), all-optical switching (Luo et al., 2005), multiplexing (Ilieva et al., 2005), alignment of liquid crystals (Ichimura, 1996; Ichimura et al., 1994) and bulk alignment of liquid crystalline polymers (Zen et al., 2002; Rosenhauer et al., 2001; Han et al., 2000; Fischer et al., 1997; Stumpe et al., 1991), and orientation of further functional groups in liquid crystalline (LC) polymers (Rosenhauer et al., 2005).

Repeated E–Z isomerization may also result in a translation motion of the molecules. This motion becomes manifested on spatially inhomogeneous irradiation. The phenomenon has been known as a light-driven mass transport or, with regard to optical applications, a light-induced formation of surface relief gratings (SRG) (Kim et al., 1995; Rochon et al., 1995). It implies an undulation of a film surface upon irradiation with the spatially modulated light. Such all-optical deformation is a quite unexpected phenomenon as it takes place in the solid films below glass transition temperature. The induced deformation is thermally and light reversible. Another peculiarity of the process is its polarization selectivity: the mass transport is effective only in the direction in which the light electrical field vector is modulated. Currently, it is accepted in the literature that the repeated E–Z isomerization is a necessary condition for this process. The nature of a driving force is still not understood. Several models have been proposed to describe the mechanism responsible for the surface relief formation but none can explain all observed effects. The conclusion is difficult, as the phenomenon has been reported in a variety of materials where not only different chromophores are used but also the materials properties differ drastically. The question of the interinfluence of molecular orientation and their translation motion remains unclear too. More detailed explanations on the suggested mechanisms may be found in the original works (Saphiannikova and Neher, 2005; Baldus and Zilker, 2001; Bublitz, 2001; Fukuda et al., 2001b; Kumar et al., 1998; Lefin et al., 1998; Pedersen et al., 1998; Barrett et al., 1996) and numerous reviews (Oliveira et al., 2005; Natansohn and Rochon, 2002; Yager and Barrett, 2001; Viswanathan et al., 1999). Formation of SRG is also addressed in Chapter 4 of this book and in Section 2.3 of this chapter. All-optical formation of SRG has attracted much attention because of its potential for applications. Typical applications include diffractive optical elements (Goldenberg et al., 2008a,b; Kulikovska et al., 2008; Natansohn and Rochon, 2002; Viswanathan et al., 1999), waveguide coupler (Choi et al., 2003; Kang et al., 2003; Barrett et al., 1997), lasing (Ye et al., 2007), and optical storage (Oliveira et al., 2005; Egami et al., 2000). Applications of nano- and microstructured surfaces as aligning layers (Kim et al., 2002; Dantsker et al., 2001), analysis of optical near field (Fukuda et al., 2001a), templates (Kim et al., 2006a), nanodevice processing (Ohdaira et al., 2006), and biology (Barille et al., 2006; Baac et al., 2004) also attract more and more interest.

All applications discussed earlier just give a general idea about the photo-induced processes of azobenzene related to the topics of this chapter. For the sake of simplicity, the authors have kept away from the material questions. Nevertheless, all discussed processes depend not only on the chromophore but are strongly affected by its surroundings and the macroscopic properties of the medium. Especially, the stability of the induced changes is determined mostly by the embedding matrix. Vice versa, the photoinduced motions of chromophore can modify its surroundings. This interplay of light, light-sensitive moiety, and embedding matrix is the key for the understanding of the photoinduced orientation and mass transport. It is also the key for design of any applicable material. It is kept in mind as a leading idea by reviewing the results in this chapter.

Following consideration can help to relate different photoreactions of chromophore, influence of the matrix and the observed photoinduced processes. The authors' view is outlined in Fig. 2.1. Depending on the irradiation conditions,

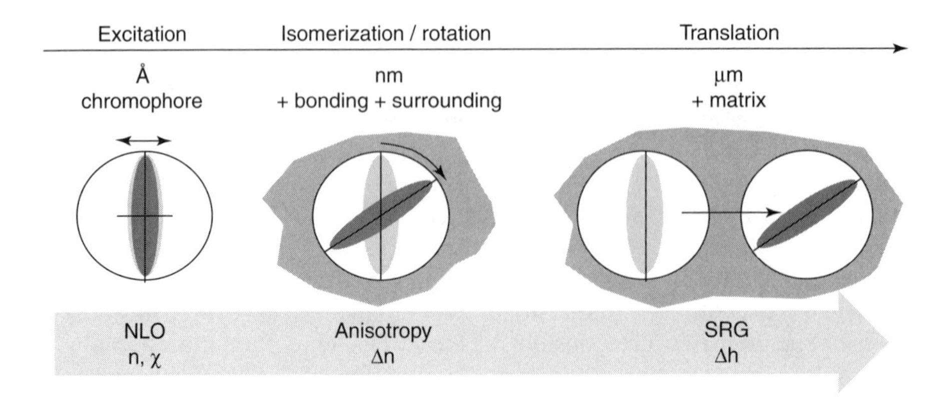

Figure 2.1. Photoinduced processes in azobenzene and their length scales.

the molecular photoreaction of azobenzene induces motions on a quite different length scale. The photoexcitation involves the chromophore alone and correspondingly may be related to the scale of angstroms. It may be used, for example, in nonlinear optics (NLO). The isomerization additionally involves the local surroundings of the chromophore on the nanometer scale and depends strongly, for example, on the free volume and intermolecular interactions. Also, the orientation is a more or less local process on nanometer scale, which, in turn, can affect the orientational order of the neighboring molecular segments. It is important if the induction of birefringence and, especially, its stability is considered. On the contrary, the mass transport implies the translation of chromophore on a micrometer scale that necessarily produces a movement of the neighboring molecular segments or even of the whole molecules. The intermolecular interactions are essential for this process. But additionally, the matrix properties and morphology become of particular importance if the formation of SRG is considered. Moreover, the stability of SRG is provided almost exclusively by the matrix.

Now, the azobenzene materials undergoing both effects should be discussed in more detail. Photoorientation of azobenzenes takes place in solutions, liquid crystals, or glassy polymers; amount and stability of the induced anisotropy strongly depend on the dynamics of the matrix. In contrast to photoorientation, light-induced mass transport was observed only in glassy systems. With respect to this, both effects can take place in a variety of quite different photochromic matrices, which are schematically depicted in Fig. 2.2. So far the main interest was focused on glassy polymers. One should distinguish between azobenzene-doped polymers (Fig. 2.2a) and azobenzene-functionalized polymers (Fig. 2.2b), in which the photochromic group is covalently attached to the polymer backbone either as part of the polymer backbone (main-chain polymers) or as side group (side-chain polymers, Fig. 2.2b). Both the effects were observed in amorphous or liquid crystalline polymers, whereas liquid crystallinity can cause specific effects. In principle, guest–host systems consisting of azobenzene-doped polymers would

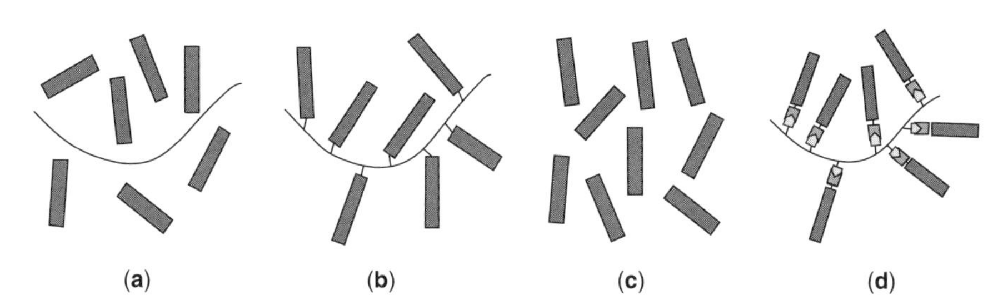

Figure 2.2. Types of azobenzene-containing materials used for photoinduced generation of anisotropy and mass transport: (**a**) guest–host system, (**b**) azobenzene-containing side-chain polymer, (**c**) film-forming low-molecular-weight azobenzene derivatives, and (**d**) supramolecular (H- or ionic bound) azobenzene complexes.

present a cost-effective solution, but the restricted azobenzene solubility, phase separation, and molecular aggregation due to $\pi\pi$ stacking limit the loading level and the efficiency, respectively. With regard to mass transport, some authors explain poor SRG-forming properties of guest–host materials with the argument that nontethered azobenzenes cannot transfer the translational motion to the matrix, and in this way, a migration of the polymer chain could not take place or is strongly restricted. The widely accepted conclusion is that covalent bonding between the photoactive group and the inactive matrix polymer is required for efficient induction process and stable SRG.

Therefore, a large variety of azobenzene-functionalized polymers with different chemical structure, molecular weight, glass transition temperature, and azobenzene content have been investigated (see Natansohn and Rochon, 2002; Chapter 4 of this book). Almost all of them undergo the photoorientation and light-driven mass transport; the efficiency of the processes is being strongly affected by the above-mentioned parameters of the polymers. Especially side-chain azobenzene-containing polymers have been shown as very effective materials for photoorientation and SRG formation.

Multistep syntheses are required to manufacture such functionalized polymers from commercially available chemicals, and consequently, they are expensive. The exact control of polarity, molecular weight, glass transition temperature, and viscosity are rather difficult in such polymers, and reproducibility is a problem. The availability and variety of specifically functionalized structures is strongly limited. Moreover, the purification and the solubility of the polymers are difficult problems. Restricted solubility requires using environmentally nonfriendly organic solvents. Therefore, a number of activities to develop alternatives to glassy functionalized polymers were pursued.

There are only few examples for the induction of permanent anisotropy in bulk films of neat azobenzenes (Tanino et al., 2007; Ishow et al., 2006; Presnyakov et al., 2006; Shirota, 2005; Ando et al., 2003; Chigrinov et al., 2003, 2002; Nakano et al., 2002; Perschke and Fuhrmann, 2002; Stracke et al., 2000; Utsumi et al., 2000; Shirota et al., 1998; Klima and Waniek, 1957). If the crystallization can be suppressed, the induction of anisotropy is possible in films of low-molecular-weight azobenzene derivatives (Fig. 2.2). Surprisingly, SRG formation in a single crystal of 4-(dimethylamino)azobenzene was reported (Nakano et al., 2002).

Low-molecular-weight glass-forming systems are an alternative to functionalized polymers (Tanino et al., 2007; Ishow et al., 2006; Presnyakov et al., 2006; Shirota, 2005; Ando et al., 2003; Chigrinov et al., 2002, 2003; Nakano et al., 2002; Perschke and Fuhrmann, 2002; Stracke et al., 2000; Utsumi et al., 2000; Shirota et al., 1998; Klima and Waniek, 1957). Sterically hindered bulky azobenzene-containing molecules are a new family of functionalized materials that form amorphous films by spin coating and vacuum deposition without any dilution of the system. Such materials are free from polymer chains and any restricting entanglement of the backbone. The results indicate that the existence of a polymer chain is not a necessary condition for induction of anisotropy and SRG formation. The bulky structure of the molecules is a necessary condition to prevent the crystallization and ensure film-forming properties. Moreover, the glass transition of the materials is rather low, leading to low stability of the photoinduced structures. As in the case of polymers, the alternative use of low-molecular-weight glasses is also restricted by their specific design and complicated synthesis. The bulky structure restricts the inducible anisotropy. In addition to amorphous glasses, there is one example for low-molecular-weight liquid crystalline glass-forming material (Stracke et al., 2000). Interestingly, that anisotropy and SRG, photoinduced in the glassy state of spin-coated films at room temperature, were amplified by annealing in the mesophase.

Industrial application of photoinduced anisotropy or light-induced SRG formation requires fulfilling of a number of additional parameters. At first the photochromic material has to be transferred in films of good optical quality. This can be achieved using different materials as discussed above and by different methods of film preparations, such as spin coating, doctor blading, Langmuir–Blodgett, or layer-by-layer (LbL) techniques. In addition to the basic effects, a number of further requirements become important, such as the easy processing and handling of the films. Especially film-forming properties, use of good and environmental solvents, restriction of crystallization and aggregation, good optical quality of the films before and after induction, stability of the photoinduced effects, and, of course, the price of the materials are very important requirements for any application. Moreover, the matrix can introduce specific properties such as liquid crystallinity or further functionalization. Therefore, there is still a challenge to find low cost systems capable of effective light-induced orientation and SRG formation.

The supramolecular chemistry with its variety of intermolecular interactions can be seen as exactly right strategy in design of functional materials. According to the broad definition by Lehn (1995), supramolecular chemistry is chemistry beyond the molecule. It is the chemistry of intermolecular bonds forming supramolecular complexes or phases. It may be divided into broad, partially overlapping areas that deal first with supramolecules, more or less defined species resulting from specific intermolecular association of at least two complementary components (molecular recognition) and, second, with molecular assemblies, polymolecular systems formed by the spontaneous association (self-assembly) of a large number of components to specific phases characterized by defined microscopic organization. In both cases, chemical species are held together and organized by means of noncovalent-binding interactions. Both types of supramolecular systems are characterized by new quality of properties compared with those of the single components. Moreover, such organization of molecular structural and dynamic features allows fine-tuning of the rigidity and flexibility, of order and properties dynamics. These approaches offer easy ways of functionalization.

Supramolecular chemistry typically involves the complexes based on molecular recognition, which include crown ethers, catenates, rotaxanes, etc. Azobenzene derivatives are often used as structural elements in such supramolecular assembly. Already in 1980 (Shinkai et al., 1980), azobenzene moiety had been introduced into crown ether. Other recent examples include azobenzene–cyclodextrin complexes (Callari et al., 2006), azobenzene-attached nanotube–cyclodextrin complexes (Descalzo et al., 2006), and molecular recognition complexes with DNA (Haruta et al., 2008). Numerous examples of photoswitchable azobenzene supramolecular systems in solutions can be found (Yagai et al., 2005; Balzani et al., 2002). But there are no investigations in these systems concerning photoorientation and mass transport.

There are a large number of publications on photoinduced anisotropy in supramolecular-ordered azobenzene-containing systems, such as liquid crystals (Kreuzer et al., 2000; Szabados et al., 1998), liquid crystalline polymers (Rosenhauer et al., 2001; Han et al., 2000), LbL (Geue et al., 1997; Stumpe et al., 1996), self-assembled monolayers (Sekkat et al., 1995), and adsorbed azobenezene molecules (West et al., 2001). But mass transport resulting in SRG was only reported till recently in glassy systems.

The considerations here will be limited to azobenzene supramolecular materials for photoorientation and photoinduced mass transport. These materials are based on H-bonds and ionic interactions of opposite-charged species. Two types of azobenzene-containing supramolecular complexes either with low-molecular-weight surfactants or related macromolecules, which means polyelectrolytes or polymers able to form H-bonding, will be reported that are schematically depicted in Fig. 2.2d. The use of such supramolecular building blocks' principle would offer new perspectives for the design of functionalized materials. In addition, this approach allows to use environment-friendly solvents for material preparation and to considerably decrease the price of the materials by applying of commercially available components as building blocks.

The photoinduced orientation is reviewed in Section 2.2. Most of the chemical structures of the compounds are also presented in this chapter, and the formation of the corresponding complexes is discussed. These complexes include LbL multilayer systems formed by polyelectrolytes and either azobenzenes (or other low-molecular-weight azobenzene derivatives) or main-chain and side-chain azobenzene-containing polyelectrolytes. The results on photoinduced orientation in supramolecular complexes based on either H-bonds or ionic interactions will also be discussed. The ionic complexes are formed by either polyelectrolytes or surfactants and azobenzenes (similar to ones used for LbL process). Complexes between azobenzenes and surfactants formed by ionic self-assembly (ISA) will be concentrated on, as they have shown the best results for photoorientation.

Section 2.3 is dedicated to the light-induced formation of relief structures in supramolecular materials. To the best of the authors' knowledge, not a lot of results have been published up to date. The authors have tried to include all results in their review, starting with LbL films and continuing with ionic and H-bond complexes, both amorphous and liquid crystalline ones. Wherever possible, they have commented on different mechanisms of the relief formation in these materials. In particular, they have emphasized the role of the matrix for the light-driven mass transport. They have also tried to combine these very recent results with those well known from the literature on complex formation, materials properties, and light-driven mass transport, if they found it useful for understanding.

In summary, the authors derive the advantages and drawbacks of the new supramolecular materials for photoinduced orientation and SRG formation. Also, the questions that were aroused only on the application of the supramolecular systems for orientation and SRG formation are briefly discussed.

2.2. PHOTOORIENTATION

Under illumination with polarized light, azobenzene moieties undergo angle-selective E–Z photoisomerization cycles depending on the angle between the transition dipole moment of the azobenzene unit and the electric field vector of the exiting light. The linear polarized irradiation tends to orient the chromophores perpendicular to the electric field vector of the exiting light, generating an oblate orientation distribution. This photoorientation process takes place via a number of angular-dependent photoselection events in the steady state of the photoisomerization, and it is connected with the rotational diffusion of the chromophores. This effect occurs in all media. However, stability of the induced anisotropy depends on the matrix properties. In some materials (e.g, glassy polymers), this effect may be cooperative, which means causing an orientation of nonphotochromic side groups and other polymer segments toward the same direction and often to a comparable degree of anisotropy (Shibaev et al., 1996; McArdle, 1992). The photoinduced order is stable in the glassy state of the matrix, and the anisotropy disappears above the glass transition temperature.

Considerable research efforts are currently focused on the interplay between molecular architecture, intermolecular interactions, order, and macroscopic properties (Lehn, 1995; Ikkala and ten Brinke, 2002; Moore, 1966). The construction of materials with different functionalities, easy processing, high anisotropy and stability through self-assembly, and self-organization processes, in which molecules associate spontaneously into ordered aggregates as a result of noncovalent interactions or entropic factors, is becoming one of the primary frontiers of materials research. To achieve this goal, several strategies such as H-bonding (Macdonald and Whitesides, 1994), metal coordination (Lehn, 1990), charge-assisted H-bonding (Hosseini, 2003), and, more recently, ISA (Faul and Antonetti, 2003) have been investigated. A rather large number of these activities are directed to generate mesophases, where the mesogenic units are formed by intermolecular interactions.

The mutual order is encoded not only in the shape and chemical functionality of the objects involved but also in the strength and directionality of the secondary interactions used. In classical supramolecular chemistry, these interactions are usually hydrogen bridges or coordinative metal binding. However, molecular recognition and amphiphilic association should also be considered. Table 2.1 summarizes the most important interactions playing the main role in organization of organic supramolecular materials. Manipulation of structural and macroscopic order in films and bulk solids, however, remains a major challenge for all of these approaches.

The approach of deriving new photosensitive materials using secondary noncovalent interactions like H-bonding and ionic interactions between different units such as polymers and low-molecular-weight functionalized units has gained a great interest in the past decade. This approach allows the development of a new generation of smart materials and their application in optoelectronics, holography, data recording, and imaging systems. New materials may advantageously combine the properties of polymers (film forming) and low-molecular-weight components (photochromism, high optical activity, easy processing, etc.).

The simplest way to use secondary interactions in design of photosensitive material would be to introduce them between the polymer backbone and functional photosensitive groups. This approach would allow direct comparison of photoorientation properties with already well-studied materials. A general application of this idea between different high-molecular-weight and

TABLE 2.1. Interactions and Some of Their Properties

Interaction	Strength (kJ/mol)	Range	Character
van der Waals	~50	Short	Nonselective, nondirectional
H-bonding	5–65	Short	Selective, directional
Coordination binding	50–200	Short	directional
Fit-interactions	10–100	Short	Very selective
Amphiphilic	5–50	Short	Nonselective
Ionic	50–250[a]	Long	Nonselective
Covalent	350	Short	irreversible

[a]Data are for organic media, dependent on solvent and ion solution.

Source: Faul and Antonetti 2003. Reproduced with permission from Wiley-VCH.

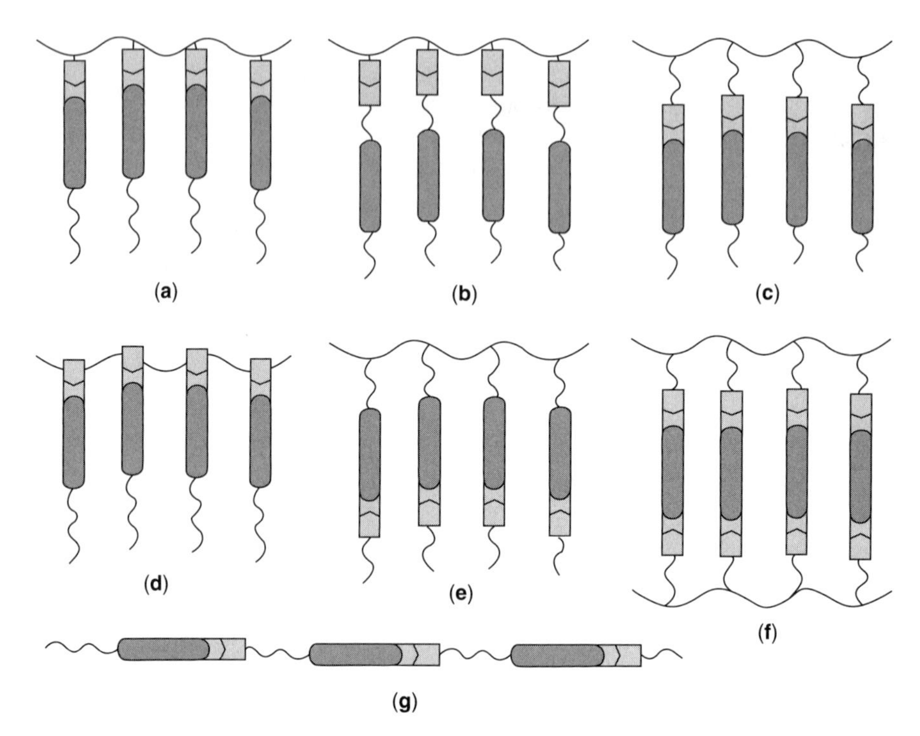

Figure 2.3. Principles of molecular construction of LC polymers by self-assembly induced by hydrogen or ionic bonding.

low-molecular-weight tectonic unit is shown in Fig. 2.3. For the first time, this approach was realized for photosensitive polymer–azobenzene unit complexes using ionic interaction (Fig. 2.2d) (Meyer et al., 1991) and H-bonding (Fig. 2.4a) (Kato et al., 1996), where nonlinear optical properties and influence of UV-vis irradiation on a reversible isothermal nematic–isotropic phase transition have been investigated, respectively.

The development of H-bonded complexes is now considered. An influence of UV irradiation on optical properties of low-molecular-weight azobenzene-containing material (Fig. 2.4b) has been investigated (Aoki et al., 2000) on the basis of such interactions. The first observation of photoinduced optical anisotropy in H-bonded complexes of azobenzene dyes and copolymers (Fig. 2.4b) has been recently demonstrated (Medvedev et al., 2005). In this case, the induced anisotropy was stable, and the maximum dichroic ratio of 2 has been observed. A kinetics of the induction of birefringence (maximum value of ca. 0.01) in one of these complexes is shown in Fig. 2.5. An influence of H-bonding on the mesomorphic and photoorientation properties was recently demonstrated (Cui and Zhao, 2004). In this approach, the amorphous azopyridine side-chain polymer was converted into liquid crystalline polymers through self-assembly with a series of commercially available, aliphatic, and aromatic carboxylic acids (Fig. 2.4d).

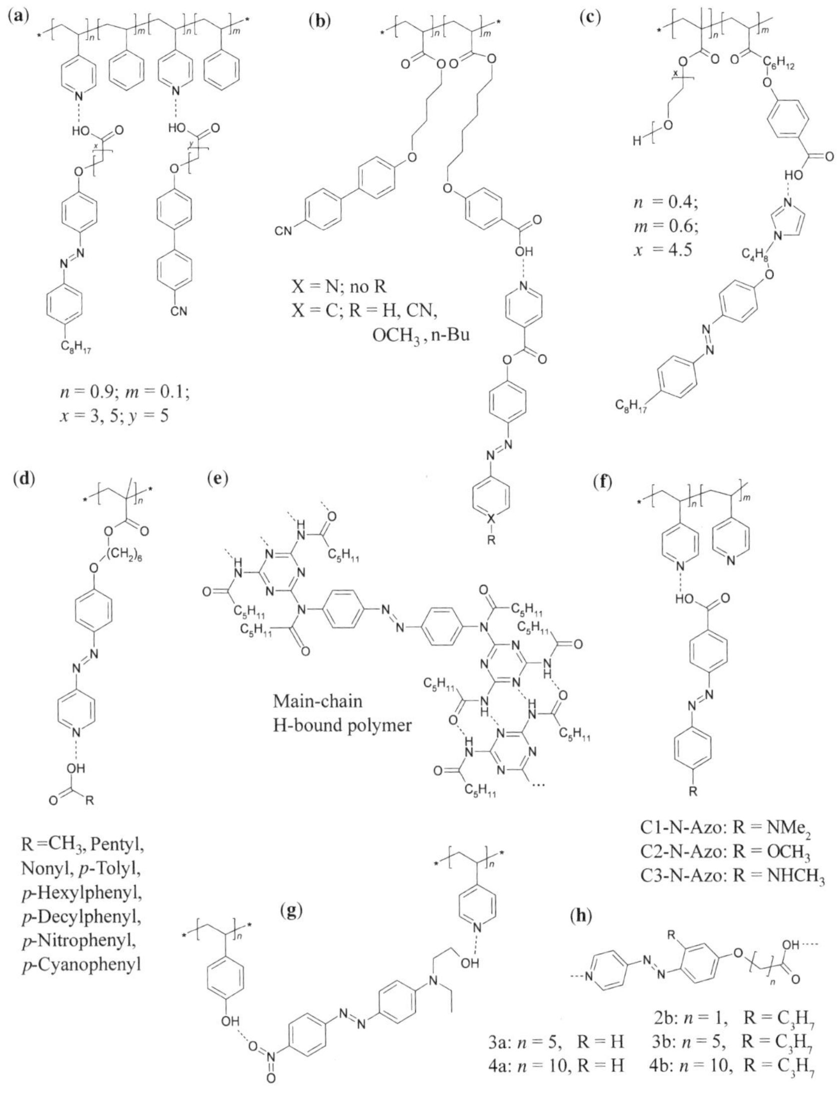

Figure 2.4. Chemical structures of complexes with hydrogen bonds.

The measurement of photoinduced birefringence revealed very different anisotropy. Considering the numerous compounds that could readily be complexed with this type of polymer, including chiral acids, phenols, and metals, the approach of using side-chain azopyridine polymers offers the possibility to produce a large number of new photoactive liquid crystalline materials without exhaustive synthetic efforts and with new properties to exploit. And really the approach has been exploited in recent works. The materials similar to that

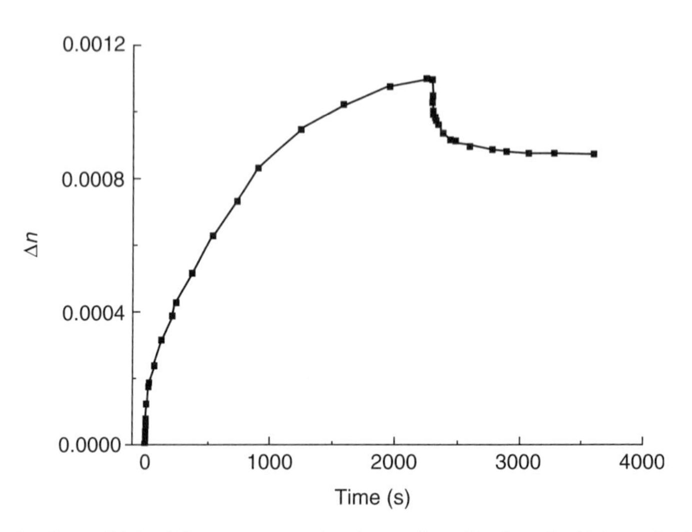

Figure 2.5. A rise of birefringence under laser irradiation in H-bonded azobenzene polymer (Fig. 2.4: m = 39%, X = C, R = H) homeotropic film at room temperature. The dashed lines indicate the moment when the laser irradiation was switched off (P = 0.15 W/cm^2). *Source*: Medvedev et al., 2005. Reprinted with permission from American Chemical Society.

depicted in Fig. 2.4a have been reported (Gao et al., 2007b). However, the films of this H-bonded azobenzene polymer (Fig. 2.4f) exhibited dichroic ratio of only 1.2. When in another investigation (Priimagi et al., 2007), commercially available azodye (Disperse Red 1) forming relatively week H-bond with the same polyvinylpyridine (Fig. 2.4h) has been used, relatively high birefringence of 0.035 has been attained. Even higher value of birefringence (0.04, highest value for H-bonded complexes) has been obtained when the same azobenzene derivative was used in combination with polyvinylphenol using the opportunity of another H-bonding (Fig. 2.4h). All this improvement in photoinduced orientation (guest–host system of the same dyes and polystyrene (PS) exhibited only birefringence of 0.005) was attributed to the prevention of dye aggregation due to formation of H-bonding [Priimagi et al., 2007). Rather different approach was realized earlier (Gao et al., 2007a), and it is based on intermolecular multiple H-bonding in azobenzene derivative forming an H-bonded main-chain azobenzene polymer (Fig. 2.4e). The latter material showed rather higher dichroic ratio of 5.5. On the contrary, mass transport properties of these last two polymers (discussed later) were vice versa. All these data cannot be analyzed, as different values (birefringence and dichrosim) have been measured; however, it shows that by a variation of H-bond strength, structure of the azobenzene and polymer, significant modification of optical properties is possible.

In contrast to the H-bonded complexes, complexes based on ionic interactions have gained a much greater interest. Such complexes emerged as a multilayer assembly produced by an electrostatic LbL technique. LbL deposition was introduced in the early 1990s (Lvov et al., 1993) and has since gained a great popularity in the fabrication of thin films of different materials from simple ionic species to colloidal particles and biological macromolecules (Hammond, 2000; Decher et al., 1998). These solid films consist of adjacent layers of oppositely charged polyelectrolytes and different other low molecular materials, nanoparticles, colloids, etc. In this case, ionic interaction is not only used to form a molecular complex but also results in layered structure at a molecular level. The films are manufactured by subsequent dipping of a substrate in solutions of these oppositely charged species (Fig. 2.6).

According to the authors, LbL films, which are self-assembled layered structures, could be considered as solid-state supramolecular complexes formed by ionic interactions. Azobenzene-containing materials also found their place with this popular technique and could be divided from chemical point of view to azobenzene-containing polyelectrolytes (Aldea et al., 2007; He et al., 2004; Wang et al., 2004a,b, 1998; Zucolotto et al., 2004b, 2002; Wu et al., 2001; Lee et al., 2000; Balasubramanian et al., 1998; Lvov et al., 1997), with azobenzene moieties in main- or side chain, and low-molecular-weight azobenzene derivatives (Zucolotto et al., 2003; dos Santos et al., 2002; Shinbo et al., 2002; Ziegler et al., 2002; Dragan et al., 2001; He et al., 2000; Advincula et al., 1999; Saremi and Tieke, 1998); some of them are conventional ionic azodyes.

The requirement and restriction to azodye in this method is a presence of at least two charged groups to be able to bind to oppositely charged sites in the polyelectolyte chains from two different adjacent layers (Fig. 2.6). As second nonactive component, usual polyelectrolytes such as PDADMAC, PAH, PEI,

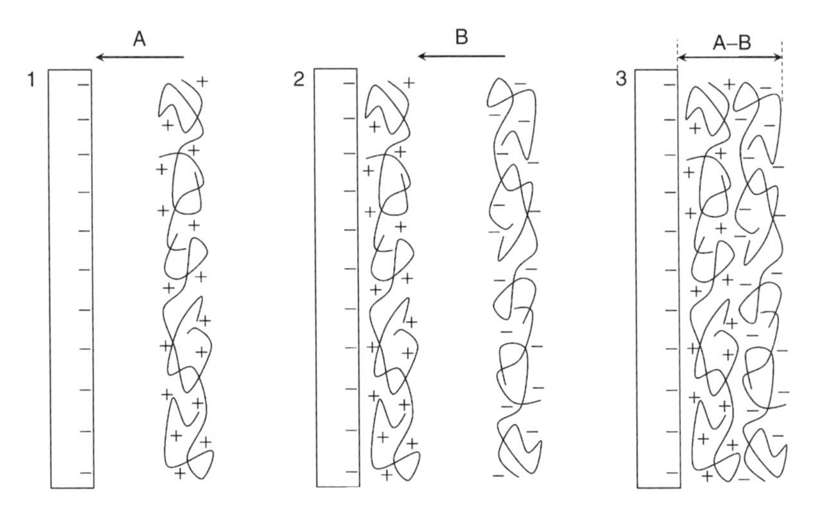

Figure 2.6. Schematic illustration of the preparation of alternate polycation–polyanion multilayer assemblies onto a negatively charged substrate.

PVS, or PSS or rather exotic ones like chitosan, carragenan, cyclodextrine, or dendritic polyelectrolytes have been employed. LbL layers with low-molecular-weight azobenzene species could be considered as analogue of typical materials used for photoorientation and photoinduced mass transport, namely, azobenzene side-chain polymers. In this case, covalent bonding of azobenzene moieties and polymer is replaced with ionic interactions (weak or strong depending on the type of polyelectolyte) (compare Fig. 2.2b and d). The above-mentioned restriction to the structure of azodye is also of crucial importance to resulting photochemical properties. Namely this causes a restriction on the freedom of movement of the azobenzene species, which seems to be of particular importance for the photoinduced mass transport. Depending on the number of charged groups and polyelectrolyte type, these restrictions would be more or less pronounced.

Photoisomerization has been investigated for some of these solid-state complexes. As generally LbL procedure is rather tedious and time consuming (even using robotic technique available nowadays), from the practical viewpoint it would be attractive only in case of rather thin films. Figure 2.7 displays low molecular azobenzene derivatives, some of them commercially available, which were used to study photoinduced orientation. Both negatively and positively charged dyes have been employed.

Relatively high dichroic ratio up to 2.3 due to geometry of the molecule (high aspect ratio of Direct Red 80) has been achieved (Advincula et al., 2003). To manufacture these films with PDADMAC of ca. 100-nm thickness, 100 layers were necessary. Less effective photoorientation of spin-coated and LbL films with other polyelectrolyte (PAH, PEI) was attributed to specific formation of J-aggregates in LbL films. In the LbL films produced from chitosan and Sunset Yellow (dos Santos et al., 2002), spontaneous birefringence of 0.04 for the film of 300-nm thickness was observed; however, the value was not affected by the light. This fact underlines the importance of selection of dye and polyelectrolyte.

Some indications of the influence of ionic azobenzene derivative structure (Fig. 2.7) and polyelectrolyte can be deduced from a number of azobenzene derivative studied (Ziegler et al., 2002). All studied LbL films (24 layers) exhibited only out-of-plane orientation. Only for anionic-derivative Z1 (indicated in Fig. 2.7) and only with PDADMAC, dichroic ratio of 2 was observed. In the layers of Z1, but with PEI as polyelectrolyte, photoorientation and photoisomerization are totally restricted by strong aggregation. The high value of dichroism of PDADMAC/Z1 layers should be caused by the hexamethylene spacer of Z1, which results in higher mobility necessary for the orientation process. Cationic Z2–Z5 were all manufactured into LbL films using the same polyelectrolyte PSS, and maximum value of 1.4 (dichroic ratio) was found.

Instead of low-molecular-weight azobenzene derivatives, some azobenzene-containing polyelectrolytes were also employed (Aldea et al., 2007; He et al., 2004; Wang et al., 2004a,b, 1998; Zucolotto et al., 2004b, 2002,; Wu et al., 2001; Lee et al., 2000; ; Balasubramanian et al., 1998; Lvov et al., 1997). By chemical structure, azobenzene-containing polyelectrolytes can be divided into a group with azobenzene moiety situated in the main chain and a group with side-chain azobenzene

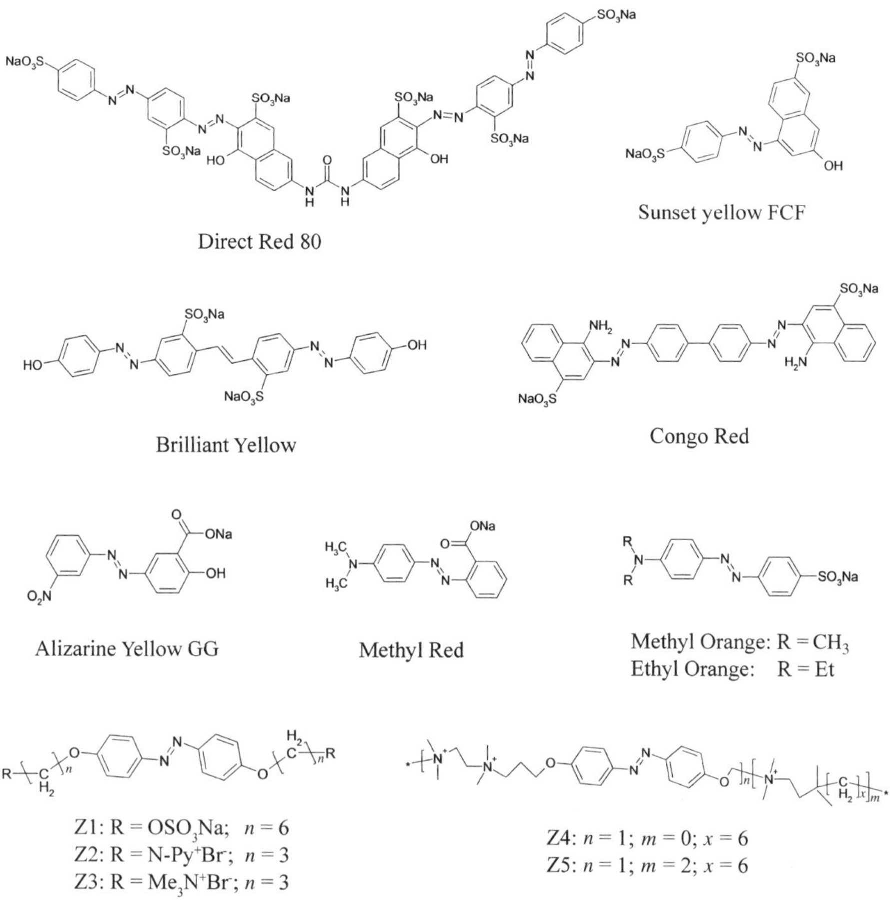

Direct Red 80

Sunset yellow FCF

Brilliant Yellow

Congo Red

Alizarine Yellow GG

Methyl Red

Methyl Orange: $R = CH_3$
Ethyl Orange: $R = Et$

Z1: $R = OSO_3Na$; $n = 6$
Z2: $R = N\text{-}Py^+Br^-$; $n = 3$
Z3: $R = Me_3N^+Br^-$; $n = 3$

Z4: $n = 1$; $m = 0$; $x = 6$
Z5: $n = 1$; $m = 2$; $x = 6$

Figure 2.7. Chemical formula of azodyes used for the preparation of LbL multilayers and supramolecular complexes.

(Fig. 2.8). Side-chain polymers are more typical, and commercially available polymer PAZO (Aldrich) was already used for LbL assembly in 1997 (Lvov et al., 1997).

Both PAZO and PS-119 (Fig. 2.8) could be considered as strong polyelectrolytes being in salt form. It is important that charged group be connected to azobenzene moiety, and therefore one can expect the ionic interactions responsible for the formation of LbL structure to restrict azobenzene motion. Birefringence value of 0.09 for the 200-nm film (dos Santos et al., 2006) was achieved for structures of PS-119 with cationic dendrimers. Interestingly, for higher dendrimer generation (with higher charge density) stronger absorption of PS-119 has been observed. That fact led to the lower values of birefringence, which is related to restriction of chromophore molecular motion. For PS-119/PAH architecture

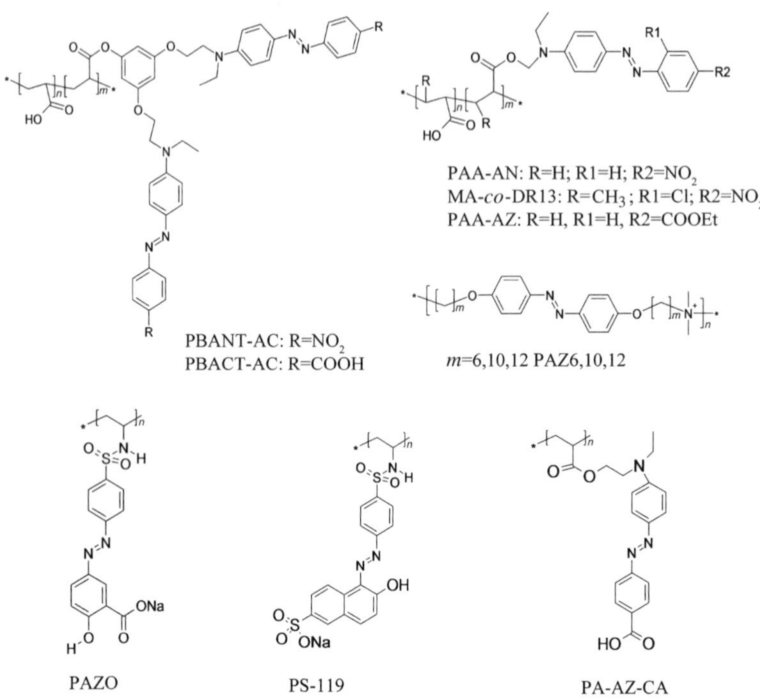

PBANT-AC: R=NO_2
PBACT-AC: R=COOH

PAA-AN: R=H; R1=H; R2=NO_2
MA-*co*-DR13: R=CH_3; R1=Cl; R2=NO_2
PAA-AZ: R=H, R1=H, R2=COOEt

m=6,10,12 PAZ6,10,12

PAZO PS-119 PA-AZ-CA

Figure 2.8. Chemical formula of azobenzene-containing polyelectrolytes used for the preparation of LbL multilayers and supramolecular complexes.

(Zucolotto et al., 2004b), the structure and charge density of cationic polyelectrolyte PAH can be governed by pH during LbL deposition, and it changes the LbL films properties. By increasing the pH from acidic to basic, it was possible to increase the bilayer thickness by the factor of 2.4 because of the loose conformation of PAH in the LbL structure. The authors compared the films with similar amount of azobenzene (the same optical absorption) prepared at different pH and came to the conclusion that an increase in the pH led to a slowing of kinetics of the induction of photoorientation by the factor of 4. For another similar polyelectrolyte PAZO (Fig. 2.8), the data on photoinduced orientation in LbL structures are not available. In contrast to LbL structures, the results on ionic complex of PAZO with PEI (Stumpe et al., 2006a) showed that dichroism was lower in this case compared with the neat PAZO films (Stumpe et al., 2006b; Goldenberg et al., 2005), approximately proportional to a decrease in azobenzene loading. It seems that there is no restriction of azobenzene motion by ionic interactions with oppositely charged polyelectrolyte in this complex. This demonstrates that the layered structure of the LbL films should be responsible for this restriction.

Two other azobenzene-containing polyelectrolytes (Fig. 2.8) are copolymers. The important difference is that ionizable group (COOH) is attached not to the azobenzene but to another unit on the polymer chain. Therefore, one could expect that ionic interactions, necessary for the LbL organization, will have lower impact on the properties of azobenzene moiety. However, some of these copolymers (PBANT-AC, PBACT-AC, PAA-AZ) exhibited only low dichotic ratio (up to 1.3) (Wang et al., 2004a,b). Other LbL structures with PAA-AN, MA-*co*-DR13 showed birefringence value up to 0.068 (Zucolotto et al., 2002; Lee et al., 2000). It should be noted that spin-coated films (Lee et al., 2000) showed the same results, and it means that there is no advantage in tedious LbL assembly in this case. Also, the authors (Zucolotto et al., 2002) mentioned much longer time necessary to induce maximum value of orientation in LbL systems, and this was attributed to the influence of ionic and H-bonding interactions. In addition, the drier was the LbL film, the more sluggish was the kinetics of induction of photoorientation. In addition to all the above-mentioned side-chain polyelectrolytes, some azobenzene-containing polyelectrolytes with chromophore in main chain have been investigated (Fig. 2.8) (Jung et al., 2003; Hong et al., 1999). Unfortunately, only small value of dichroic ratio of about 1.4 was achieved in this case.

In contrast to LbL structures, ionic supramolecular complexes (schematically depicted in Fig. 2.2d) can be first formed in solution and then build into films. The complex formation can be proved by FT-IR and NMR spectra (Zhang et al., 2008; Kulikovska et al., 2007; Xiao et al., 2007; Lin et al., 2003, 2002). In this case, the azobenzene units, although bound with ionic interactions to polymer chains, are devoid of confinement of layered structure. The first observation of photoorientation was recently demonstrated in such polyelectrolyte–azobenzene complex of azodye and PEI (Kulikovska et al., 2007; Stumpe et al., 2006a). In the investigated materials, only temporally stable photoinduced anisotropy was found contrary to the glassy polymers. Figure 2.9 displays the induction and relaxation of optical anisotropy in a film of the material on the basis of PEI and the dye Alizarin Yellow GG (see Fig. 2.7). Irradiation of the film with 488-nm light gives rise to the orthogonal polarization component of the transmitted probe beam (633 nm) absent in the incident beam, indicating the induction of optical anisotropy. After the switch-off of the actinic light, the induced anisotropy relaxes, whereas the relaxation is well described by a two-exponential decay function. Alternation of the polarization of the actinic light allows switching off of the induced molecular orientation, as demonstrated in the Fig. 2.9b. The observed instability of the light-induced orientation of chromophores is caused by the ionic nature of the azobenzene–polyelectrolyte materials. Unlike as in the functionalized polymers (Fukuda et al., 2006; Kulikovska et al., 2002), in this case polymer chains provide for a spatial distribution of charges that counteracts the orientation effect of polarized light, returning the ionic chromophores to their initial positions.

Stable light-induced anisotropy in ionic complexes of poly(vinyl-*N*-alkylpyridinium) with Methyl Orange has been recently obtained (Zhang et al., 2008; Xiao et al., 2007). Both complexes have chemical structure (see Fig. 2.10a) similar to the complexes reported earlier (Lin et al., 2003, 2002). However, supposedly due to different synthetic procedure, they form smectic A (SmA) liquid crystalline phases. Induction of birefringence in one of these complexes is shown in Fig. 2.11.

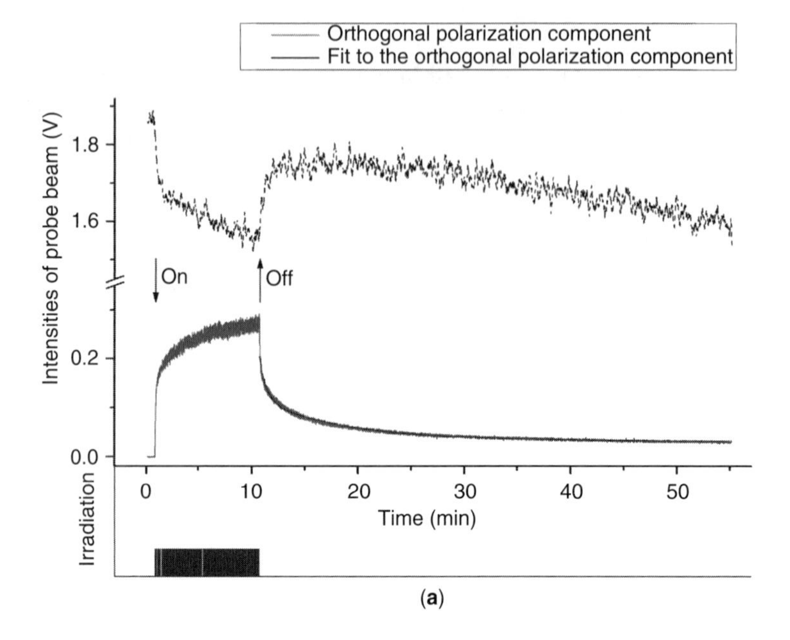

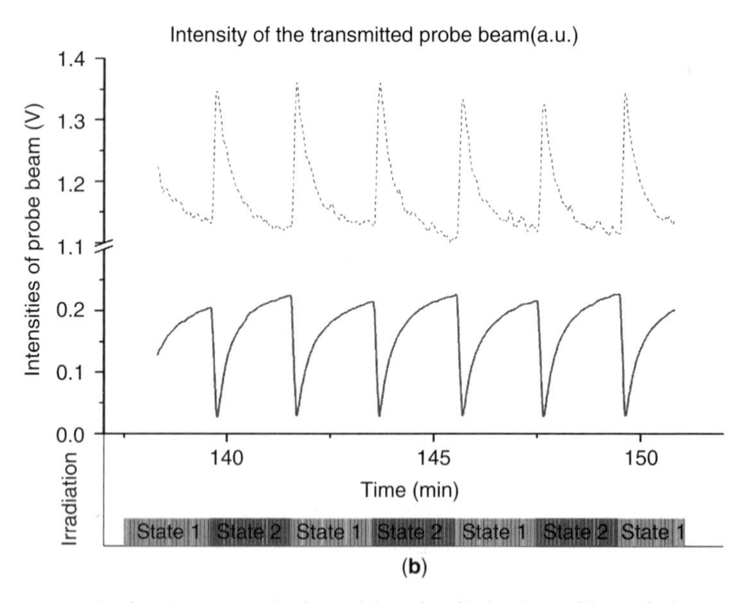

Figure 2.9. Optical anisotropy induced by the light in a film of the compound Alizarin Yellow GG-PEI: kinetics of the parallel (*dashed black line*) and orthogonal (*solid gray line*) polarization components of the transmitted probe beam. (**a**) Induction and relaxation, solid black line presents two-exponential decay fit to the data, and (**b**) switching by alternation of polarization of the irradiating beam. *Source*: Kulikovska et al., 2007. Reproduced with permission from American Chemical Society.

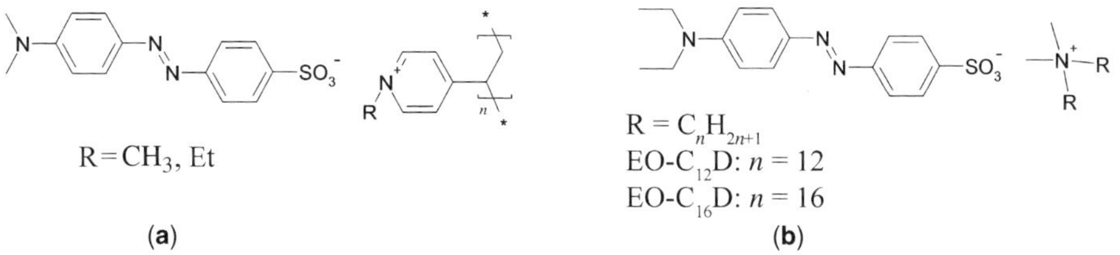

Figure 2.10. Chemical structure of complexes of poly(vinyl-*N*-alkylpyridinium) with Methyl Orange and the Ethyl Orange (EO) with $C_{12}D$ and $C_{16}D$ surfactants.

The maximum value of photoinduced birefringence achieved were 0.067 and 0.31 for ethyl- and methyl-quaternized polyvinylpyridine (see Fig. 2.10a), respectively (Zhang et al., 2008; Xiao et al., 2007). The authors believe that such a great difference in photoorientation properties is hardly attributed to structural difference consisting of one methylene group, but more probably to the difference in synthesis, purification, film preparation, etc.

A rather simple approach for design of low-molecular-weight ionic supramolecular complexes has been recently proposed (Zakrevskyy et al., 2005, 2004; Guan et al., 2003). These new materials are formed by introduction of oppositely charged ions between different low molecular tectonic units and are called ISA complexes. In contrast to the discussed LbL and polyelectrolyte-based ionic complexes, the polymer unit is usually replaced by low-molecular-weight surfactant. Application of the ISA concept to produce photosensitive low-molecular-weight materials with good film-forming properties, which are capable of undergoing light-induced generation of optical anisotropy, has been proved to be very successful (Zakrevskyy et al., 2006a). The photosensitive ISA complexes consist of charged azobenzene-containing photosensitive units complexed with oppositely charged surfactants. The chemical structure of the most detailed investigated complexes is shown in Fig. 2.10b. Other examples of the photosensitive complexes can be found in a study by Zakrevskyy et al. (2006b). It should be noted that proper design of ISA complexes, based on functionalized tectonic units and surfactants, results in a new strategy for creation of functionalized liquid crystals. This approach can lead to specific optical functions of the material.

Induction of optical anisotropy in spin-coated film of the ISA complexes under irradiation with polarized light of an Ar^+ laser was very effective (Zakrevskyy et al., 2007). Dichroic ratios of 50 for the EO–$C_{12}D$ complex and of 20 for the EO–$C_{16}D$ complex have been obtained (Fig. 2.12a and b). To the best of the authors' knowledge, this is the highest value of photoinduced anisotropy. It was shown that both complexes exist in highly ordered columnar LC phases at room temperature, and their "photobehavior" is strongly affected by their liquid crystallinity. Spin-coated films of the complexes are built from LC domains. The photoalignment of the complexes depends strongly on the size of the domains. In the case of small domains, the photoalignment is the most effective. Photoreorientation in the complexes is not local like in azobenzene-containing polymers, but is a cooperative process and connected to mechanical rotation of domains.

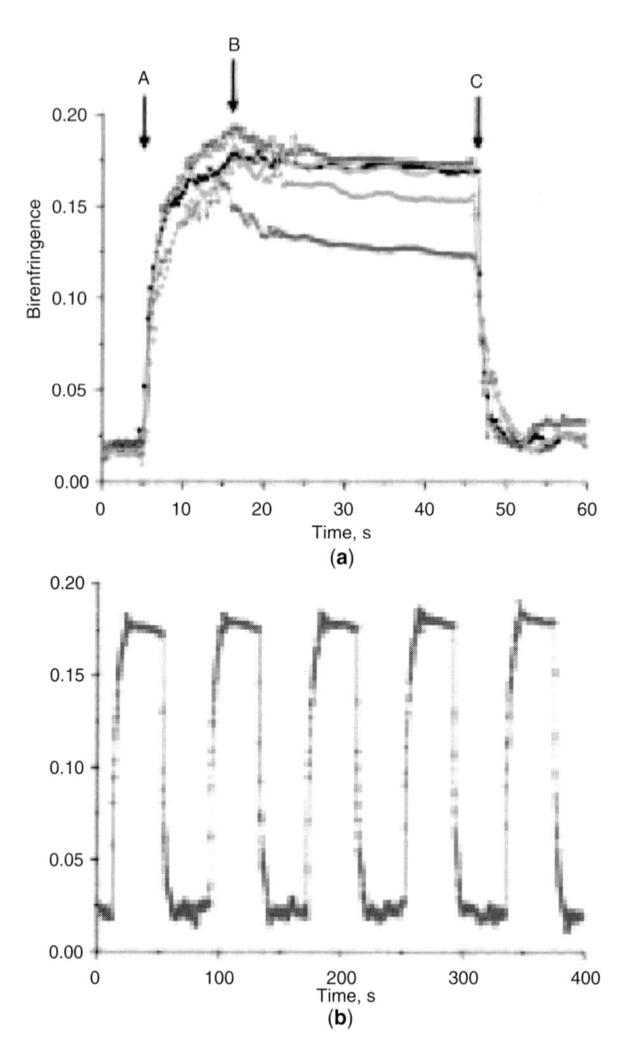

Figure 2.11. Photoinduced birefringence (PIB) in a spin-coated film (200-nm thick) of the MO/poly(vinyl-*N*-methylpyridinium) complex. PIB curves (a) (A, write ON 10 s; B, write OFF 30 s; C, erasure ON) taken at the same spot after different thermal treatment. (b) Multiple write-erase cycles (25°C; write ON 10 s, write OFF 30 s, erase ON 30 s, erase OFF 10 s; only the first five of eight identical cycles are shown. *Source*: Zhang et al., 2008. Reprinted with permission from American Chemical Society.

Domains in the initial film are aligned in such a way that columns of the complex are parallel to the substrate with isotropic orientational distribution parallel to the film plane. During irradiation, there is an in-plane reorientation of domains (columns).

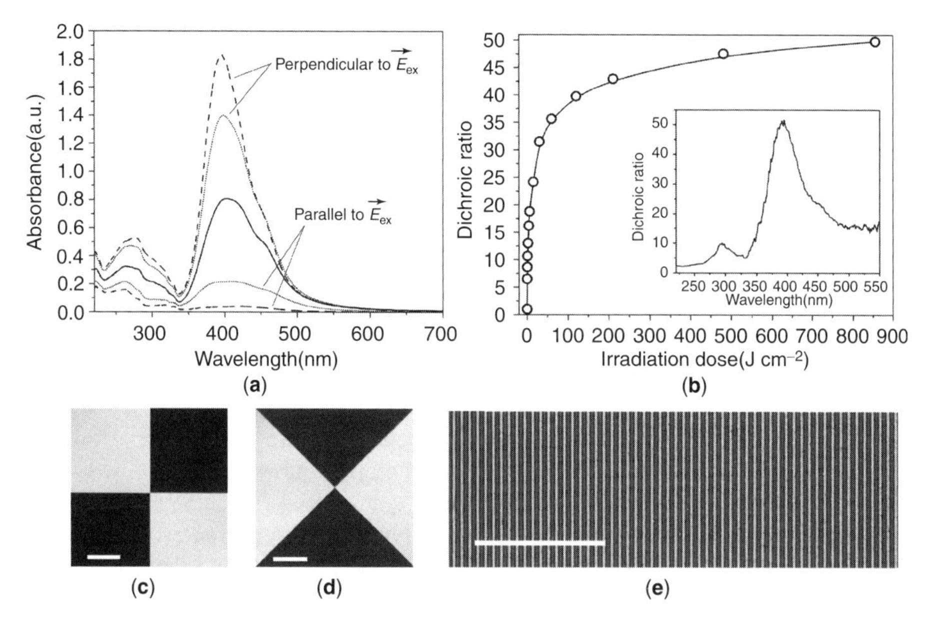

Figure 2.12. (**a**) Changes of polarized absorbance spectra in a film of the EO–C_{12}D complex under irradiation with linearly polarized light E_{ex} of an Ar^+ laser (λ_{ex} = 488 nm, P = 50 mW/cm^2): solid line, initial spectrum; dotted line, after irradiation dose of 0.25 J/cm^2; dashed line, after irradiation dose of 850 J/cm^2. (**b**) Kinetics of induced dichroic ratio in a film of the complex calculated at λ_{test} = 400 nm; inset graph: spectral dependence of induced dichroic ratio after an irradiation dose of 850 J/cm^2. (**c,d**) Photograph (under crossed polarizers) of a film of the EO–C_{16}D complex when the second irradiation was prepared through a mask with polarized light of an Ar^+ laser (λ_{ex} = 488 nm, P = 50 mW/cm^2) at an angle 45° to the first one; crossed polarizers parallel to the sides of the pictures; scale bar, 100 μm. (**e**) Picture of the pattern in the film of the EO–C_{16}D complex after irradiation with two circularly polarized interfering beams of an Ar^+ laser (λ_{ex} = 488 nm, exposure dose = 1 kJ/cm^2); scale bar, 25 μm. *Source*: Zakrevskyy et al., 2006a. Reproduced with permission from Wiley-VCH Verlag GmbH & Co. KGaA.

Thus, potential application of the ISA complexes for optical data storage has also been demonstrated. Photoreorientation of the induced optical structures in the complexes has been successfully achieved. Well-defined optical patterns were obtained when a second irradiation was applied through a standard photomask (Fig. 2.12c,d). Induction of optical phase gratings in films of the azobenzene-containing ISA complexes was also observed. A diffraction efficiency (DE) of approximately 7% was obtained on 100-nm thick films (calculated value of Δn that corresponds to this efficiency is ~0.54) of the EO + C_{16}D complex after irradiation with two circularly polarized interfering beams of an Ar^+ laser.

The resulting optical pattern is shown in Fig. 2.12e. However, no formation of SRGs was observed for the investigated liquid crystalline complexes.

The last examples clearly demonstrate that photosensitive low molecular film-forming ISA complexes are viable alternatives to photosensitive polymers. The ISA complexes have several advantages in comparison to functionalized polymers: (i) low production cost due to cheap starting materials; (ii) ability to induce higher values of anisotropy; (iii) thermal stability of induced anisotropy; and (iv) solubility of ISA complexes in environment-friendly solvents such as ethanol.

2.3. SURFACE RELIEF GRATINGS

In the introduction to this chapter, the process of the SRG formation, the suggested mechanisms, and the mostly used materials have been described. Here the authors would like to underline that most results concerning SRG generation have been obtained using functionalized polymers, that is, materials with the azobenzene units covalently attached to the polymer chains. The importance of these results can scarcely be exaggerated. The investigations brought out a new process, namely, the light-driven mass transport. The process is peculiar to azobenzene-containing materials. Following investigations led to understanding of the process and brought a variety of photosensitive materials and new applications of them. The latter, in turn, gave rise to a demand for an effective, good optical quality, easy processable but low cost material. Besides, the flexibility in design is of high importance for applications, as the materials may be tuned to match specific requirements, such as wavelength, combination with anisotropy, reversibility, long time, thermal and chemical stability, wetting/dewetting, and adhesion. The existing functionalized polymers could not meet all these requirements, since the complicated synthesis is necessary for each material variation that limited the uptake of SRG for applications. The use of the supramolecular approach to design azobenzene-containing materials capable of light-induced mass transport can both improve the understanding of the process and revolutionize its applications.

The photoinduction of optical anisotropy has been studied in a number of the supramolecular systems, as discussed in the previous part of the chapter. On the contrary, the choice of the supramolecular systems, in which the formation of SRG has been reported, is rather restricted. One of the reasons can be a prevailing opinion that the covalent bonding of azobenzene to polymer chain is a necessary condition for the effective light-induced mass transport. First attempts to inscribe SRG in the supramolecular system were performed on LbL films (see description of the LbL system in the previous part of this chapter), but the obtained gratings were not very deep. Although the applicability of the approach has been principally demonstrated, the obtained results rather supported the current opinion that the mass transport may be effective only in the covalently bound materials. That is why the very effective inscription of SRG in the spin-coated films of the ionic complexes of azodyes and polyelectrolyte first demonstrated in 2006 (Stumpe et al., 2006a) was rather unexpected. Although only a few examples of SRG formation in the noncovalent bonded systems other than LbL films have been reported in the literature so far, the authors believe that the supramolecular approach presents a very prospective alternative to the functionalized polymers.

Further, the authors will review to the best of their knowledge the results on the SRG formation in supramolecular materials, starting with LbL films and continuing with ionic and H-bonded complexes. They will also comment on the complex formation, materials properties, and light-driven mass transport—well known from the literature—if they think it relevant. They hope that combining these results will bring them closer to understanding this unexpected phenomenon and highlight the advantages and perspectives of these materials for optical applications.

The inscription of SRG in LbL systems has been reported in some publications. The researchers used either charged azodyes or azobenzene-containing polyelectrolytes to prepare films by electrostatic LbL deposition technique (see the discussion of these systems concerning photo-orientation process in previous part of the chapter). In LbL films, multilayers of azobenzene-containing polyelectrolytes (He et al., 2004; Wang et al., 2004a; Camilo et al., 2003; Zucolotto et al., 2003; Lee et al., 2000) or azodyes (Zucolotto et al., 2004a, 2003; Constantino et al., 2002; He et al., 2000) are alternated with nonphotoactive polyelectrolytes. The influence of the confinement on the isomerization behavior and orientation dynamics of azobenzene chromophores has been investigated. The review of these results can be found in Oliveira et al. (2005). Photoorientation concerns close-order motion of azobenzene chromophore (rotation), whereas the formation of SRG involves long-distance movement (mass transport) of chromophores and, possibly, of nonphotoactive polymer chains. According to Zucolotto et al. (2003), the electrostatic interactions between active and inert layers serve as a mean for the SRG formation but, on the contrary, tend to hinder the photoisomerization of azobenzenes. This inherent conflict together with restriction of molecules freedom by layered structure explains rather ineffective SRG inscription in LbL architectures.

The materials used for recording of SRG were similar and are depicted in the Figs. 2.7 and 2.8. Between them, three azodyes, Direct Red 80, Congo Red, and Brilliant Yellow (Fig. 2.7), were employed. For PDADMAC/(Congo Red) structures (He et al., 2000), the dependence of SRG efficiency on the number of layers was investigated. Maximum relief height of 140 nm at corresponding first-order DE of 17% has been achieved for 2-μm thick film consisting of 640 layers. This is actually the highest value reported so far for LbL layers. Dependence of the achieved relief height on the number of bilayers is shown in Fig. 2.13. As the film thickness is of comparable value, used generally for polymer materials, low efficiency cannot be attributed to layer thickness. Also, writing kinetics for thick layers was rather slow—at the irradiation intensity of 150 mW/cm^2, the maximum of first-order DE was achieved in 75 min.

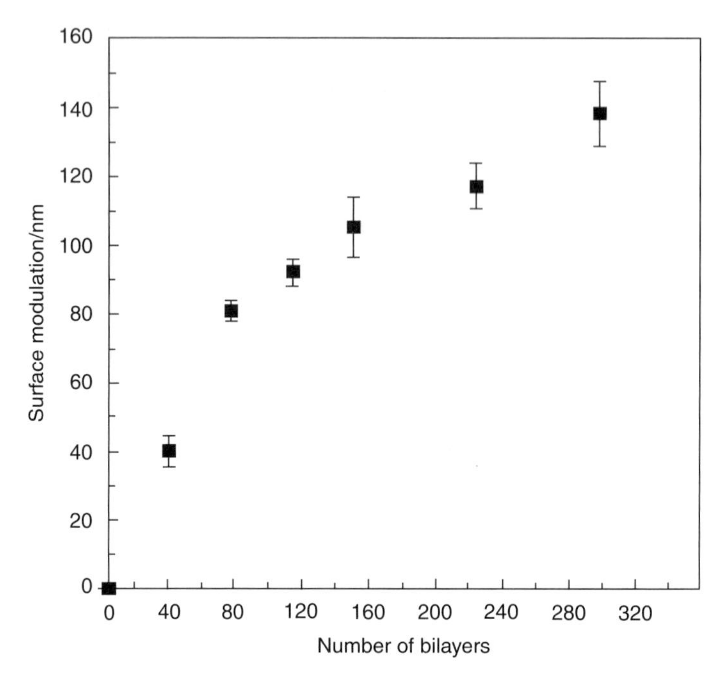

Figure 2.13. Dependence of surface modulation amplitude (according to AFM measurements) on the number of deposited bilayers in PDADMAC/(Congo Red) LbL films. *Source*: He et al., 2000. Reprinted with permission from American Chemical Society.

The authors mentioned two differences in behavior in contrast to azobenzene-containing polymers, namely, bleaching during SRG inscription (azodyes, in particular Congo Red, were photochemically not so stable as azobenzene-containing polymers) and recording efficiency independence on polarization configuration of the writing beams. The authors mention here well-known (Brown, 1971) property of azobenzene derivatives with OH or NH_2 group in ortho-position. The photochemistry of such derivatives is dominated by phototautomerization through proton transfer, leading to bleaching.

Very thin LbL films of (Direct Red 80)/PDADMAC consisting of 20 layers exhibited only 5-nm deep SRG and that required 3-h exposure (Kaneko et al., 2002; Shinbo et al., 2002). Generally one cannot expect much higher modulation depth in the films of 15-nm thickness.

Otherwise, when thick films consisting of 600 layers of Brilliant Yellow alternated with either PDAMAC or PAH were used (Zucolotto et al., 2003), higher amplitude of 120 nm has been achieved after 1-h exposure. It may be adverted here for comparison to the formation of the ionic complex using the same azodye (Brilliant Yellow), discussed later in more detail, where SRGs with DE of 15% have been inscribed at the comparable conditions (Stumpe et al.,

2006a). In contrast to the results with Congo Red (He et al., 2000), strong dependence on the polarization configuration of the writing beams was observed for the layers with Brilliant Yellow (Zucolotto et al., 2003). It evidences different mechanisms of SRG formation. Also, according to the authors (Zucolotto et al., 2003), bleaching is less important in the case of Brilliant Yellow in comparison with Congo Red, as already discussed.

The results with azobenzene-containing polyelectrolyte-based LbL structures were even less encouraging. Strong ionic bonds between the layers were presumed to be responsible for working against the light-driven mass transport process, which is initiated at the sample surface (Lee et al., 2000). Using postdeposition owing to azo coupling, it was possible to obtain LbL layers of PAA-AN (Fig. 2.8), where rather shallow SRG with pick-to-valley height of 25 nm could be written. At the same time, SRG could be readily inscribed onto the spin-coated layers of the same polymer (Lee et al., 2000).

Some light-induced surface modulation was also observed for PBACT-AC and PBANT-AC polyelectrolytes (Fig. 2.8) multilayers films. The SRG amplitude was really small, which corresponds quite well with the spin-coated film of PBANT-AC that also yielded only 85-nm deep SRG (Wang et al., 2004b). Similar results (16-nm amplitude after ca. 40-min writing) were obtained for PA–AZ–CA (Fig. 2.8) layers (He et al., 2004). For PAA-AZ LbL multilayers, any surface modulations have been observed (Wang et al., 2004a). Also, in this case, one cannot exclude that holographic writing conditions were not optimized.

For polyelectrolyte PS-119 (Fig. 2.8), a dependence of recording rate on pH upon film preparation similar to that discussed earlier for photoorientation has been found (Zucolotto et al., 2004b). Gratings of 26- and 12-nm depth were obtained after 2-h writing in the films produced at pH 4 and pH 10, respectively. It was associated with higher free volume available in films produced at low pH. When the same azobenzene-containing polyelectrolyte (PS-119) was alternated by charged dendrimers (dos Santos, 2006), higher efficiency for low generation dendrimer was obtained (SRG amplitude of 30 nm). However, all these values were rather small, which bring the question about what might have been the reason for the low efficiency of the grating formation in LbL systems. In the authors' opinion, it is closely related to the mechanisms of grating formation in these systems. Is it the light-driven motion of chromophores dragging the inert polymer chains by means of ionic interactions in adjacent layers (Zucolotto et al., 2004a), whereas the latter suppresses the reversible isomerization of azobenzene and, at the same time, the confinement to the layers hinders any translation motion? Or are there other mechanisms, such as photodegradation (bleaching) (He et al., 2000), that may dominate the formation of grating? Nevertheless, the applied approach, demonstrating the inscription of SRG in materials with ionic interactions, presents a new insight into physics of light-induced mass transport in azobenzene-based materials. The importance of this approach is particularly enhanced by the possibility of the unique, molecular-level control over the material properties (Oliveira et al., 2005; Halabieh et al., 2004).

A different supramolecular approach has been successfully used to design azobenzene-containing materials capable of very effective SRG formation (Kulikovska et al., 2007; Stumpe et al., 2006a). The materials consist of ionic complexes of polyelectrolytes and oppositely charged azobenzene derivatives as counter ions. Fig. 2.2 explains schematically the material structures in comparison to the conventional azobenzene-containing materials reported for SRG inscription. For the first time, a very effective light-induced mass transport—even superior to that in the functionalized polymers—has been demonstrated in the systems, where the chromophores are attached to the passive polymer chains by the noncovalent interactions. SRGs with the surface modulation depth up to 1.65 μm have been obtained using these materials. Formation of such deep relief structures evidences that alternatively to covalent bonding, the ionic interactions keep the chains of passive polymer linked to the chromophores, strongly enough to be involved into the light-induced motion of chromophores.

Since the interactions between azobenzene moiety and a polymer matrix became of particular importance in this latter case of the light-induced mass transport, a few studies on noncovalent intermolecular interactions that have been widely exploited in supramolecular chemistry are discussed. They were used, for example, to prepare self-assembled materials (Meier and Schubert, 2005; Pollino and Weck, 2005; Faul and Antonetti, 2003; Ikkala and ten Brinke, 2002). A number of polyelectrolyte–surfactant complexes that form films with low surface energies have been reported (Thünemann et al., 2000, 1999; Antonietti et al., 1994). The majority of investigations have dealt with dye–dye intermolecular interactions, but less effort has been directed toward the study of noncovalent interactions between the active molecules and host material. The effect of electronic properties of the host polymer and specific chromophore–host interactions on the electro-optic coefficient of doped polymer films has been investigated in Banach et al. (1999). Recently, the influence of noncovalent interactions between dye molecules and host polymer on the aggregation tendency of the guest (dye) molecules has been studied (Priimagi et al., 2007, 2006, 2005). The effect of polyelectrolyte matrix on the film properties has been investigated as well (Meyer et al., 1991; Simmrock et al., 1989).

The new materials under discussion have polymerlike properties and thus can be processed from solution to form films with excellent optical quality and transmission in the visible spectral range. Generally, the polyelectrolyte as a polymer matrix determines the mechanical, thermal, and morphological properties of the system. For example, the presence of flexible chain segments or a certain amount of irregularity along the polymeric structure reduces recrystallization tendency and improves the stability of the glassy state (Meyer et al., 1991; Simmrock et al., 1989). By these means the thermal behavior and processability of the materials can be influenced. The authors demonstrated variation of the glass transition temperature from −22°C to 191°C just by change of organic moieties in either main- or side chains of polyelectrolyte (Simmrock et al., 1989). Also, the charge density and the pH value can affect the structures and the surface energies as it has been shown for the polyelectrolyte–surfactant complexes (Faul and

Antonetti, 2003; Thünemann, 2002; Thünemann et al., 2000, 1999; Antonietti et al., 1994).

In turn, optical properties are mainly determined by azobenzene chromophore. The refractive index is substantially influenced by the polarizability of azobenzene moiety, and the absorption of the chromophores plays a decisive role in the spectral dispersion. Different chromophores have been studied, especially for nonlinear optical applications (Meyer et al., 1991). Azobenzene chromophores are also responsible for the photoinduced birefringence and dichroism resulting from repeated E–Z isomerization. As these effects complain spatial orientation of chromophores, they may be influenced by the type of bonding and by the chromophores surroundings, that is, by the polyelectrolyte matrix in the case of the materials under discussion. For example, it has been postulated that the stacking structure of the LbL films of the PDADMAC/(Congo Red) produces a confined geometry to enhance the stability of the Z-isomer of Congo Red (He et al., 2000). It was shown that covalent bonding of chromophores to the polymer matrix limits the degree of freedom and thus stabilizes their light- and electrical field–induced orientation (Marino et al., 2000; Boilot et al., 1998). But the main effect achieved by covalent bonding the chromophores to the polymer matrix was the suppression of their aggregation and thus the higher dye loading, which was impossible in "guest–host" materials. Understandably, high concentrations of chromophores are desirable for any optical effect in these materials, including nonlinearity, anisotropy, and SRG. Concentration dependence of SRG formation has been investigated on a series of functionalized polymers (Börger et al., 2005). The higher the azobenzene concentration, the more efficient the grating formation, unless it becomes suppressed by increasing aggregation of chromophores, which is not excluded even in covalently bound systems. The prevention of dye aggregation is one of the key challenges of functionalized polymer systems. Intermolecular interactions have been demonstrated to be convenient means to reduce the tendency of the molecules to aggregate (Priimagi et al., 2007, 2006, 2005). Using ionic interaction between an ionic dye and a polyelectrolyte, a 1:1 molar ratio of dye molecules can be attached to the polymer simply by mixing the constituents (Priimagi et al., 2006; Stumpe et al., 2006a).

Ionic complexes of polyelectrolytes and charged azobenzenes have optical response of the azobenzene, although influenced by the polymer matrix, as discussed earlier. The repeated E–Z isomerization is considered as a primer condition for the light-induced mass transport. Peculiar to the latter is the fact that the process is not more a local light-induced event but implies a translational motion of chromophores (Fig. 2.1). Moreover, the mass is transported only if the chromophores involve the passive polymer chains in their translation. This explains the high efficiency of SRG formation in functionalized polymers (Natansohn and Rochon, 2002; Viswanathan et al., 1999), why the covalent bonding has been considered as a necessary condition for the effective light-induced mass transport (Oliveira et al., 2005; Zucolotto et al., 2003; Fiorini et al., 2000; Viswanathan et al., 1999;), and why the extremely effective SRG formation in the materials with ionic-bounded azobenzene (Kulikovska et al., 2007; Stumpe et al., 2006a) was rather unexpected. It is also why in the latter case the process is determined not only by chromophore itself but to a large extent also by the intermolecular interactions and by the passive matrix (Kulikovska et al., 2007, 2008a). Because of its nonlocal nature, the light-driven diffusion presents a very sensible and elegant tool to study the intermolecular interactions.

The applicability of the supramolecular approach for SRG recording has been demonstrated by combination of different charged azobenzene derivatives with different polyelectrolytes. Such azobenzene derivatives as Brilliant Yellow, Alizarin Yellow GG, and Methyl Red, all commercially available, have been used as active components. The chemical structures of the compounds are shown in Fig. 2.7. Note that the absorption may be tuned by a proper selection of the dye component that can match specific applications where given laser wavelengths must be used.

Generally, a number of possibilities may be explored—in terms of the chromophores, matrix, and processing—in order to optimize material properties. A highly branched polyelectrolyte, PEI, with high charge density and a polyelectrolyte with quite stiff backbone, PDADMAC, were used as polymer counterparts. Different functional groups and charge densities provide for the different strengths of ionic interactions. Both the interaction strength and the molecular conformation are expected to influence film-forming properties and photochemical processes, including light-driven mass transport. The material formulations described earlier tap in no case the full potential of the approach but serve a purpose to demonstrate its versatility and to encourage investigations of further material combinations. A few other examples reported in the literature will be discussed later.

A distinguishing feature of the materials is that they were produced by the mixing of water–alcohol solutions of azobenzene derivatives and the polyelectrolytes. The materials were readily processed into films by conventional film preparation techniques, such as spin coating and casting. If the complex was precipitated, the product was separated and dissolved in another solvent. Otherwise the complexes were sufficiently soluble in water–alcohol mixture to form films of good optical quality with thicknesses in the range from 100 nm to a few micrometers directly from the solutions.

As it is typically for holographic inscription of SRG in azobenzene-containing materials, the films were irradiated with an interference pattern of two coherent orthogonally polarized beams at wavelength of 488 nm within the absorption band of the materials. For most azobenzene-containing materials, the orthogonal polarization configurations—either linear $\pm 45^{\circ}$ or circular—have been found to be more effective. The interference pattern from the two orthogonally polarized laser beams creates no bright and dark regions but regions with different light polarization states across the material. Note that generally the polarization selectivity of SRG inscription may be seen as an evidence of the light-driven mass transport mechanism. On the contrary, higher efficiency of the intensity pattern recording may point the presence of mechanism other than mass transport. For example, no polarization dependence was observed in LbL films of low-molecular-weight azobenzene compound (He et al., 2000), where the contribution of dye bleaching has been suggested. Also, in the sol–gel materials, in which some cross-linking of polymer takes place, the efficiency of polarization recording was minimal (Darracq et al., 1998).

The result of holographic irradiation is a modulation of film thickness that replicates the interference pattern created by the laser beams. This enables the hologram to be encoded as a relief pattern. A weak probe 633-nm laser has been typically used to diffract from this phase hologram. The DE of nth diffracted order may be calculated as $\eta_n = I_n/\Sigma I_n$, where I_n are the intensities of the nth-order diffracted beams. In the literature, typically the DE of the first diffraction order has been characterized. Note that it is correct until the recorded gratings are not very deep (that is often the case) and the grating profile remains sinusoidal. Deep relief gratings have many diffraction orders in Raman–Nath regime of diffraction, and their profiles deviate often from sinusoidal form. In this case, the characterization of grating by the total DEs proved to be an advantage. Total DE is determined as the sum of DEs of all nonzero orders, $\eta_{tot} = \eta_{\pm 1} + \eta_{\pm 2} + \cdots$.

The best-reported example of recording kinetics in ionic complexes is shown in Fig. 2.14 (Kulikovska et al., 2007). The grating was recorded in the material (Methyl Red)/PDADMAC with moderate irradiation intensities of 250 mW/cm^2. After 20 min of irradiation, 98% of the probe light was diffracted by the recorded grating. Maximum diffraction of 45% in first diffraction order was reached after 17 min, whereas maximum of 35% of light diffracted into second diffraction

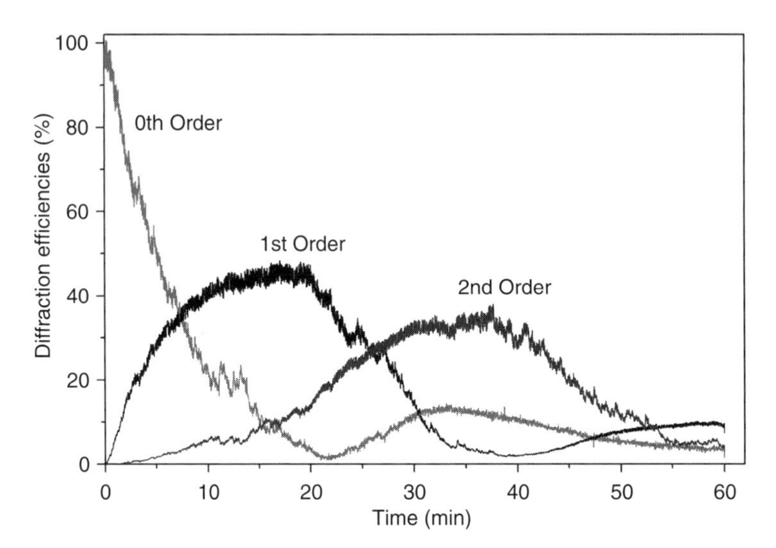

Figure 2.14. Kinetics of the diffraction efficiencies of the 0th, 1st, and 2nd diffraction orders upon holographic irradiation in (Methyl Red)/PDADMAC film (Fig. 2.7). *Source*: Kulikovska et al., 2007. Reprinted with permission from American Chemical Society.

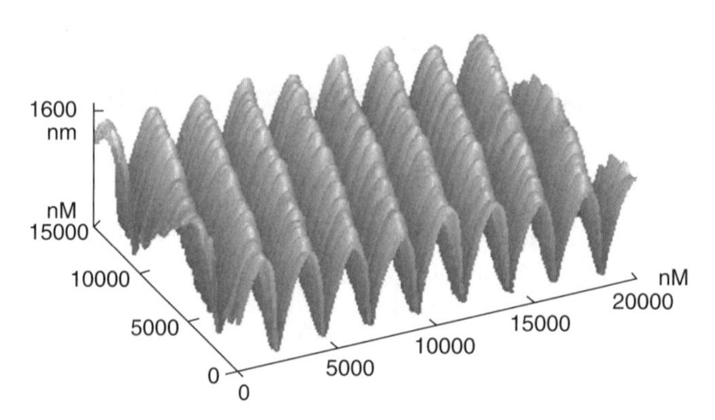

Figure 2.15. AFM image of surface relief grating inscribed in (Methyl Red)/PDADMAC film (Fig. 2.7). *Source*: Kulikovska et al., 2007. Reprinted with permission from American Chemical Society.

orders after 35 min of irradiation. The diffraction behavior reveals a very deep grating. An atomic force microscopy (AFM) image of the corresponding grating with a peak-to-valley height of ca. 1.65 μm and a period of 2.3 μm is presented in Fig. 2.15. One may estimate the recording rate to be as high as 10%/min for this grating. Very high modulation depths of relief gratings achieved in these materials are the highest values among the azobenzene-containing materials reported so far.

In other respects, the inscription of SRG in ionic complex materials was comparable to the azobenzene-functionalized polymers. The gratings could be erased, if were irradiated with spatially homogeneous unpolarized UV light, and then recorded again into the same film. Also, the successive recording was demonstrated as a proof of the process reversibility. In this way, two-dimensional (2-D) cross-gratings consisting of up to 13 linear gratings have been successively inscribed into the film. Figure 2.16 displays an example of the 2-D grating adopted from Kulikovska et al. (2008b).

Remarkably, because the induced SRGs are thermally very stable, only heating the samples above 150°C can erase them. The thermal stability of the light-induced surface structures in the ionic complex materials is worthy to discuss here in comparison to the stability of gratings in the azobenzene-functionalized polymers. It is well established now that SRGs in polymers are stable below the glass transition temperature of the polymers and can be erased if heated above T_g. It has been noted that the higher the glass transition temperature, the lower is the SRG formation efficiency. Only shallow, though thermally stable, gratings could be induced in polymers with high T_g of about 150°C (Che et al., 2002; Lee et al., 1998) and in polymer with rigid backbone, for which no T_g was found (Sukwattanasinitt et al., 1998). Very effective SRGs, stable up to T_g of 147°C, were inscribed in azobenzene-based amorphous molecular material with a spirobifluorene core (Chun et al., 2003). The high efficiency was attributed to the high concentration of azobenzene groups per molecule and the good mobility of

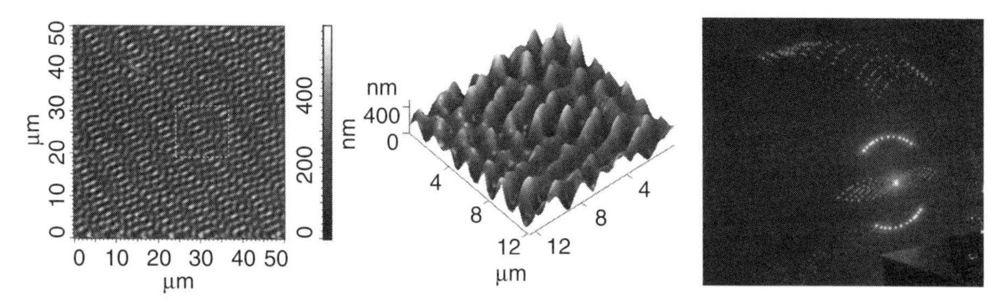

Figure 2.16. Relief structure recorded in film of the (Methyl Red)/PDADMAC material (Fig. 2.7). Topology of the grating is presented on the left, cutting of the grating in 3-D view in the middle, and photograph of diffraction by the grating on the right. *Source*: Kulikovska et al., 2008b. Reprinted with permission from SPIE. See color insert.

molecules due to the absence of long polymer backbones and the lack of complicated entanglements (Chun et al., 2003). Generally, glassy state of polymers may be considered as a guarantor for the stability of induced structures. Since no glass transition temperature was found for the azobenzene–polyelectrolyte complexes reported in Kulikovska et al. (2007), the high thermal stability of SRG in these materials was related to their ionic nature.

Another material with highly thermally stable SRG and photoorientation (polarization gratings) is azobenzene-containing polyelectrolyte PAZO (Fig. 2.8). It was already discussed in connection with photoorientation in LbL system in Section 2.2. SRGs in this material were stable till degradation temperature of 225°C (Goldenberg et al., 2005) and birefringence grating till at least 100°C (Stumpe et al., 2006b). This was also explained by the ionic nature of this polyelectrolyte (Goldenberg et al., 2005).

Subsequently, photoinduced birefringence and surface relief structures have been reported in a photosensitive supramolecular material obtained by the ISA of poly(1-ethylvinylpyridinium bromide) and of commercially available azobenzene dye Methyl Orange (Xiao et al., 2007), as shown in Fig. 2.10. However, a different mechanism of relief formation reported by the authors has to be stressed here. The complex has a rigid molecular structure because of the absence of a spacer between the main chain and azobenzene unit and exhibits a lamellar organization with a periodicity of 2.7 nm. Within this lamellar self-assembly, the disordered polyelectrolyte soft segments were alternated by the stacking of the Methyl Orange rigid side chain. Laser-induced periodic surface structures (LIPSS) have been formed upon irradiation with an s-polarized beam of a pulsed UV laser operating at the wavelength of 355 nm if the intensity was high enough. The period of the relief structure was strongly dependent on the incident angle of the laser beam. At a period of about 400 nm, a relief depth of 100 nm was attained. Formation of LIPSS was considered to be a result of thermal-induced macroscopic mass movement of molecules. The high thermal stability of the induced relief structures up to 130°C is noticeable. The complex itself was stable at least up to 275°C. The surfaces with LIPSS were used to align liquid crystals. It should be commented here that probably the mechanism of LIPSS formation is different from the light-induced mass transport discussed in this chapter. The authors investigated neither the induction of relief structures upon modulated irradiation nor the light reversibility of LIPSS, which makes it difficult to compare these results.

Recently, the inscription of SRG in a similar dye–polyelectrolyte spacer-free ionic complex shown in Fig. 2.10 has been reported (Zhang et al., 2008). Although there is no spacer between the azobenzene unit and polymer chain, the material exhibits liquid crystalline SmA order. SRGs with a grating spacing of 1 μm were induced in a 5- to 10-μm thick film using rather low irradiation intensities of 65 mW/cm^2 (Zhang et al., 2008; supporting information). The surface modulation depth approached 200 nm after 2.25-h irradiation. The photoinduced birefringence in this material has been reported to be both temporally and thermally stable, but the authors did not comment on the thermal stability of the induced SRG. The polarization dependence of the grating inscription, especially interesting in the case of liquid crystal material, was not commented on either.

The supramolecular approach would not be complete without the materials based on hydrogen bonds. Among intermolecular interactions, the latter is characterized by its directionality and moderate strength (Table 2.1). It is very interesting to clarify whether the light-driven mass transport demonstrated for ionic complexes of azobenzene is as well effective in the H-bonded supramolecular materials.

Recently, a main-chain supramolecular azobenzene polymer based on self-complementary quadruple hydrogen bonds has been developed (Fig. 2.4e) (Gao et al., 2007b). It consists of azobenzene-containing building blocks connected by H-bonds forming a main-chain polymer as depicted in Fig. 2.4e. The films prepared by casting exhibited 15-nm deep SRG with grating spacing of 780 nm upon 40-min holographic irradiation using p-polarized beams of moderate intensity of 100 mW/cm^2. The authors report that the induced structures are thermally stable up to 100°C. Moreover, the mechanism of thermal erasing of the grating has been investigated by in situ temperature-dependent FT–IR spectroscopy. It has been shown that the amide groups change from Z into E conformation by heating to 150°C, which breaks quadruple hydrogen bonds. Such disassembly of supramolecular polymer leads to the disappearance of the surface relief structure.

As the gratings inscription in the H-bonded main-chain supramolecular polymer was very ineffective, the authors developed an azobenzene-containing supramolecular side-chain polymer, as it is shown in Fig. 2.4f (Gao et al., 2007a). A 300-nm deep grating was inscribed in this material under similar irradiation conditions. Note that at the same time the photoorientation ability of the azobenzene was a few times smaller than in the supramolecular main-chain polymer. The induced SRGs are thermally stable up to 120°C. A series of materials was obtained by using azobenzene derivatives with different types of the electron-withdrawing groups in the para-position of the azobenzene units. It has been shown that the SRG formation is more effective in materials having strong electron-donating substituents in azobenzene moiety.

Two published works reviewed in the following text demonstrated additional possibilities given by supramolecular approach, and with these works the review of reported results on the surface relief structures in supramolecular azobenzene-containing materials is completed.

The supramolecular approach based on the ionic interactions (Kulikovska et al., 2007; Stumpe et al., 2006a) was additionally extended to polyelectrolytes formed in situ by sol–gel reaction of alkoxysilanes (Kulikovska et al., 2008a; Stumpe et al., 2006a). The complex is shown in Fig. 2.17. Similarly, different commercially available ionic azobenzene derivatives were used as active components (Stumpe et al., 2006a). The aminosubstituted alkoxysilane (3-aminopropyl)-triethoxysilane (APTES) was used as an example of a passive component. The materials were produced by mixing of alcoholic solutions of the azobenzene derivatives and APTES followed by sol–gel reaction catalyzed by HCl. No glass transition has been observed for the investigated materials (Kulikovska et al., 2008a). In addition to the ionic azobenzene–polyelectrolyte complexes here, the charged chromophores form the ionic complexes with amino groups of polymer chains. But at the same time the polysiloxane network is formed by polycondensation. This additional irreversible cross-linking of polymer backbones differ the materials from all other supramolecular materials used for SRG formation. The degree of condensation and the degree of cross-linking could be tuned by the choice and amount of catalyst, by the choice of the solvent, and by the thermal treatment or aging of films. Up to 1.2-μm deep SRGs were inscribed into these materials. Figure 2.18 presents an example of 2-D relief structure induced by successive recording of linear gratings. The reported polarization dependence of

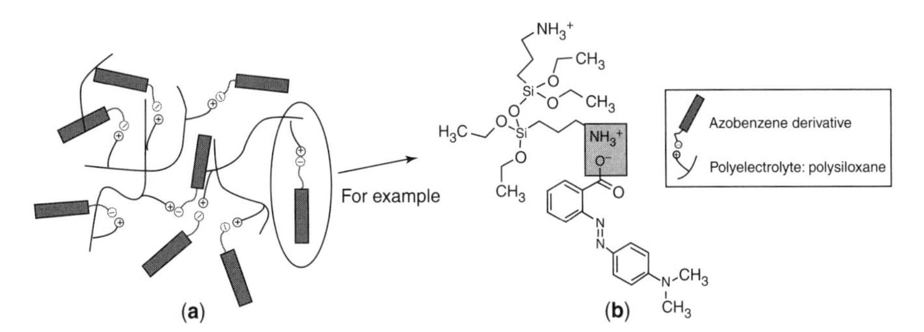

Figure 2.17. Schematic presentation of supramolecular Methyl Red sol–gel material. *Source*: Kulikovska, 2008a. Reprinted with permission from American Chemical Society.

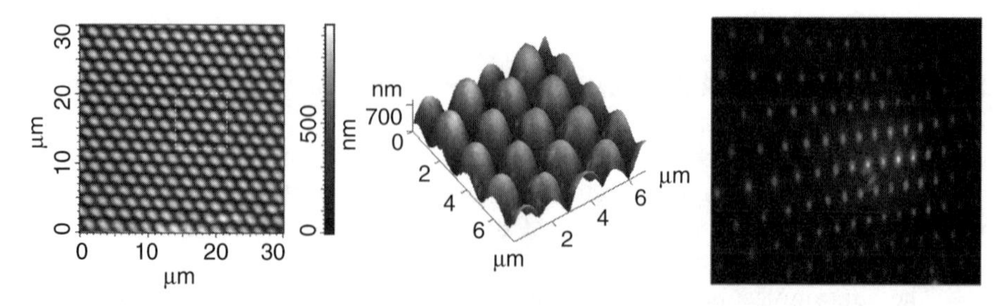

Figure 2.18. Relief structure recorded in film of the Methyl Red sol–gel material (Fig. 2.17). Topology of the grating is presented on the left, cutting of the grating in 3-D view in the middle, and photograph of diffraction by the grating on the right. *Source*: Kulikovska et al., 2008b. Reprinted with permission from 2008 SPIE. See color insert.

grating formation and the possibility of multiple successive recordings evidence the mechanism of grating formation as light-induced mass transport, although affected by the presence of irreversible cross-linking.

Note that the sol–gel materials based on polysiloxane have been known for their durability and good optical quality. If functionalized with azobenzene moieties, they exhibit nonlinear optical properties and photoinduced anisotropy. There are a few examples of SRG inscribed onto materials (Kim et al., 2006b; Chaput et al., 2000; Frey et al., 2000; Darracq et al., 1998). The gratings could be inscribed only in freshly prepared films, whereas in aged films, the cross-linking of the polymer completely hindered SRG formation.

The ionic sol–gel materials (Kulikovska et al., 2008a; Kulikovska et al., 2009) differ by the ionic bonding between the matrix and chromophore. Obviously, the presence of ionic groups and interactions affect the film properties. In these materials, the aging of films is also accompanied by chemical cross-linking of the matrix, but is much slower. The films had to be either 1-month-old at room conditions or heated at 150°C for 30 min before the SRG formation was suppressed completely. Possibly, the freshly prepared films contain only short polysiloxane oligomers that are sufficient for the formation of good quality films and effective inscription of SRG. The light-induced mass transport is not hindered until a critical degree of cross-linking is accumulated. Moreover, it becomes optimal at a certain degree of cross-linking. The formation of grating was essentially sped-up by the thermal pretreatment of spin-coated films: heating the films at 70°C for 30 min led to an increase in the recording rate from 0.18 to 0.28/min compared with the freshly prepared films. The depth of gratings in pretreated films was almost three times larger (700 nm) than that in the freshly prepared films (250 nm).

All these results lead to the conclusion that both the rate of grating formation and the achieved relief modulation strongly depend on the film stiffness and the molecular mobility of the polymer chains. If certain mobility is required to transport the material by light, certain stiffness is necessary to prevent any relaxation of the induced relief structure. The compromise between these opposite effects has to be found for effective grating formation. Thermal treatment is only one of the possible ways to shift this equilibrium.

As expected, the gratings recorded in ionic materials are thermally very stable. After heating the sample to 100°C, they remained stable with $\eta_{norm} \approx 90\%$, where $\eta_{norm} = \eta_{heated}/\eta_{recorded}$. Heating to 150°C erased the grating to $\eta_{norm} \approx 75\%$, but even after heating the sample to 200°C, $\eta_{norm} \approx 40\%$ retained. The grating could not be written again onto the thermally treated film that may be explained by further sol–gel condensation, which increases the rigidity of the matrix. Thus the latter approach combines a very high efficiency of SRG formation with a high stability of the induced surface structures provided by the simultaneously formed polymer network.

Another example of the additional functionalization of the supramolecular material presents the H-bonded system, where the azobenzene moieties can be detached from the polymer chains and even removed from the film (Zettsu et al., 2008). The removal of the strongly colored azobenzene derivative is of particular advantage for the applications such as liquid crystal alignment layer. Chemical structure of the H-bonded complex is shown in Fig. 2.4c. The complex exhibited a single glass transition at −9°C and complicated subsequent mesophase at temperatures up to 109°C. In the temperature region of liquid crystalline phase, a smectic phase was formed. Spin-coated films of typical thickness, 50 nm, were prepared. Very characteristic for the material was that the E–Z photoisomerization caused the phase transition from the liquid crystalline phase to isotropic melt in the film. It has been verified that the photoisomerization and the associating phase transition did not break the hydrogen bonding. When the formation of intermolecular H-bonds was insufficient, the efficiency of SRG formation reduced drastically. Most effectively the SRGs were inscribed in the smectic mesophase. Remarkably, the relief height of 110 nm was obtained, starting from an originally 50-nm thick film. Another characteristic feature of the SRG formation in this material is its polarization independence. The authors classified the mechanism of mass transfer in the H-bonded material as "phototriggered" surface relief formation rather than a "photoinduced" one, similar to the functionalized polymer reported earlier (Zettsu et al., 2007). This phototriggered process has been observed only for azobenzene polymers exhibiting a liquid crystal phase at the ambient temperature, independently of the polymer architectures (Zettsu and Seki, 2004). The selective extraction of azobenzene units from the film was performed after the chemical postfixation of the SRG structure. The strategy of this process is indicated in Fig. 2.19. The extraction of azobenzene derivative was followed by a drastic reduction in the height of SRG from 250 to 50 nm, approximately corresponding to the loss of mass. The authors reported that their attempts to recover the volume contraction by adapting hydrogen accepting guest molecules were not successful. Anyway, the extraction of azobenzene from the SRG has demonstrated the additional functionality granted by the supramolecular character of the material.

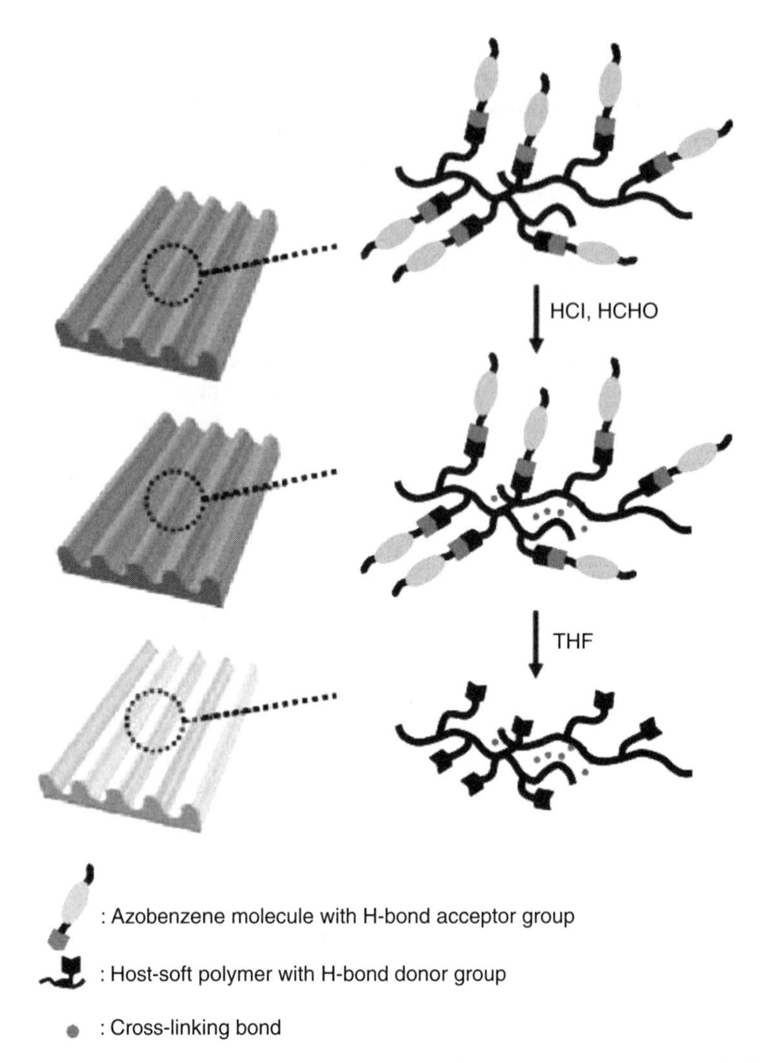

Figure 2.19. Schematic illustration of the SRG formation in an H-bonded supramolecular liquid crystalline polymer film and the strategy for the selective detachment of azobenzene unit. *Source*: Zettsu et al., 2008. Reproduced with permission from Wiley-VCH.

2.4. CONCLUSION AND OUTLOOK

In this chapter, the authors have reviewed the very recent results on photoinduction of anisotropy and SRGs in azobenzene-based supramolecular materials. They have tried to discuss these results starting from and in the framework of other materials well known for these phenomena, like low-weight molecular compounds and doped and functionalized polymers, emphasizing the fine distinctions brought by the supramolecular character of the new materials.

The new materials are based on H-bond bridges or electrostatic interactions of oppositely charged ionic species. The advantages provided by the application of the basic building block principle of supramolecular chemistry are beyond any doubt: (i) a large variety of ionic and H-bonding azobenzene derivatives are easily accessible; (ii) very high dye concentrations with homogeneous distribution can be obtained; (iii) a large variety of ionic and H-bonding polymer matrixes are known and readily available; (iv) practically any material property can be optimized by combination of appropriate passive polymer and active dye components: flexibility in design is of high importance for applications, as the materials may be tuned to match specific requirements, such as wavelength, combination with anisotropy, reversibility, stability, wetting/dewetting, adhesion; (v) materials can be processed from environment friendly water–alcohol solutions, which on the contrary causes their humidity sensitivity; and (vi) owing to their solubility in polar solvents, the materials can be combined with conventional polymers for fabrication of stacked layers. But the reviewed research answered the decisive questions about the possibility of the photoinduction of molecular orientation and mass transport and their efficiencies and stabilities of the induced structures in the new materials. Just to highlight the achievements, the remarkably high photoinduced molecular orientation in ISA complexes is noticed here, resulting in the highest dichroic ratio of 1:50 reported up to now. The most effective light-induced mass transport was observed in the ionic complexes of azobenzene and polyelectrolyte, where the SRGs as deep as 1.65 μm were recorded. These results are comparable if not superior to the best functionalized polymers. They open new perspectives for applications of azobenzene-containing materials.

Such a high efficiency of photoinduced processes in supramolecular materials has not been completely explained so far. Moreover it gave rise to the fundamental questions about the role of the intermolecular interactions and especially of the polymer matrix for photoinduced orientation and mass transport. The effects remained relatively unnoticed as long as the high efficiency of the processes was attributed to the covalent bonding of azobenzene to polymer chains. With demonstration of very effective orientation and mass transport in materials with noncovalent interactions they become very important. In this respect, a comparison of a series of supramolecular materials would be quite interesting. Further it would be useful to analyze the photoinduced processes in relation to their length scales under consideration of the effects of intermolecular interactions and matrix properties, such as elasticity, free volume, order, and morphology.

One more problem may be reconsidered in relation to the photoinduced processes in supramolecular materials—thermal stability of the induced structures and how is it affected by the ionic interactions. First studies demonstrated relatively high thermal stability of induced anisotropy and SRG in both ionic and H-bond material systems. The ionic interactions have been suggested to increase the stability. Precise explanation of the effect requires further investigations.

The reviewed results show that since the first demonstration of the photo-induced anisotropy in supramolecular complexes in 2005 and SRGs in 2006 the field has attracted much attention. The appeared works clearly demonstrated the eligibility of the supramolecular approach for light-induced processes and outlined the possibilities of material design. The achieved values shape very positive prospects for optical application. It can be expected that systematic studies will result in further improvement of the effects.

The fundamental questions emerged from this approach are of particular importance not only for supramolecular materials but also for understanding of light-induced orientation and mass transport. The solutions may give new insights into physics of the processes. This combination of both application and fundamental aspects makes the approach so interesting for further investigations.

REFERENCES

Advincula R, Park MK, Baba A, Kaneko F. 2003. Photoalignment in ultrathin films of a layer-by-layer deposited water-soluble azobenzene dye. Langmuir 19:684–696.

Advincula RC, Baba A, Kaneko F. 1999. Polymer ultrathin films via alternate self-assembly adsorption of polyelectrolyte and azo-dye molecules: photo-induced alignment and LC display properties. MRS Proc Spring Symp D Liquid Crystal Materials and Devices, San Francisco, 559.

Aldea G, Gutierrez H, Nunzi JM, Chitanu GC, Sylla M, Simionescu BC. 2007. Second harmonic generation diagnostic of layer-by-layer deposition from Disperse Red 1 – functionalized maleic anhydride copolymer. Opt Mater 29:1640–1646.

Ando H, Takahashi T, Nakano H, Shirota Y. 2003. Comparative studies of the formation of surface relief grating. Amorphous molecular material vs vinyl polymer. Chem Lett 32:710–711.

Antonietti M, Conrad J, Thünemann A. 1994. Polyelectrolyte-surfactant complexes: a new type of solid, mesomorphous material. Macromolecules 27: 6007–6011.

Aoki K, Nakagawa M, Ichimura K. 2000. Self-Assembly of amphoteric azopyridine carboxylic acids: organized structures and macroscopic organized morphology influenced by heat, pH change, and light. J Am Chem Soc 122:10997–11004.

Baac H, Lee JH, Seo JM, Park TH, Chung H, Lee SD, Kim SJ. 2004. Submicron-scale topographical control of cell growth using holographic surface relief grating. Mater Sci Eng C 24:209–212.

Balasubramanian S, Wang X, Wang HC, Yang K, Kumar J, Tripathy SK. 1998. Azo chromophore-functionalized polyelectrolytes. 2. Acentric self-assembly through a layer-by-layer deposition process. Chem Mater 10:1554–1560.

Baldus O, Zilker SJ. 2001. Surface relief grating in photoaddressable polymers generated by cw holography. Appl Phys B 72:425–427.

Balzani V, Credi A, Venturi M. 2002. Molecular Devices and Machines—A Journey into the Nano World. New York, London, Sydney, Toronto: Wiley-Interscience.

Banach MJ, Alexander MD Jr, Caracci S, Vaia RA. 1999. Enhancement of electrooptic coefficient of doped films through optimization of chromophore environment. Chem Mater 11:2554–2561.

Barille R, Kandjani SA, Dabos-Seignon S, Nunzi J-M, Letournel F, Ortyl E, Kucharski S. 2006. Neuron growth engineering on a photoinduced surface relief grating: a tool for plastic neuroelectronics. Proc SPIE 6191:61911Q–61918Q.

Barrett C, Natansohn A, Rochon P. 1996. Mechanism of optically inscribed high-efficiency diffraction gratings in azo polymer films. J Phys Chem 100:8836–8842.

Barrett C, Natansohn A, Rochon P. 1997. Photoinscription of channel waveguides and grating couplers in azobenzene polymer thin films. Proc SPIE 3006:441–449.

Boilot JP, Biteau J, Chaput F, Gacoin T, Brun A, Darracq B, Georges P, Levy Y. 1998. Organic-inorganic solids by sol-gel processing: optical applications. Pure Appl Opt 7:169–177.

Börger V, Kulikovska O, G-Hubmann K, Stumpe J, Huber M, Menzel H. 2005. Novel polymers to study the influence of the azobenzene content on the photo-induced surface relief grating formation. Macromol Chem Phys 206:1488–1496.

Brown GH, editor, 1971. Photochromism. Techniques of Chemistry. Vol. III. New York, London, Sydney, Toronto: Wiley-Interscience. p. 510.

Bublitz D, Fleek B, Wenke L. 2001. A model for surface-relief formation in azobenzene polymers. Appl Phys B 72:931–936.

Bublitz D, Helgert M, Fleck B, Wenke L, Hvilsted S, Ramanujam PS. 2000. Photoinduced deformation of azobenzene polyester film. Appl Phys B 70:863–865.

Callari F, Petraia S, Sortino S. 2006. Highly photoresponsive monolayer-protected gold clusters by self-assembly of a cyclodextrin–azobenzene-derived supramolecular complex. Chem Commun 1009–1011.

Camilo CS, dos Santos DSJ, Rodrigues JJJ, Vega ML, Campana Filho SP, Oliveira ONJ, Mendonça CR. 2003. Surface-relief gratings and photoinduced birefringence in layer-by-layer films of chitosan and an azopolymer. Biomacromolecules 4:1583–1588.

Chaput F, Lahlil K, Biteau J, Boilot J-P, Darracq B, Levy Y, Peretti J, Safarov VI, Lehn J-M, Fernandez-Acebes A. 2000. Design of optical components and optical data storage in photochromic sol-gel films containing dithienylethene or azobenzene derivatives. Proc SPIE 3943:32–37.

Che Y, Sugihara O, Fujimura H, Okamoto N, Egami C, Kawata Y, Tsuchimori M, Watanabe O. 2002. Stable surface relief grating with second-order nonlinearity on urethane-urea copolymer film. Opt Mater 21:79–82.

Chigrinov V, Prudnikova E, Kozenkov V, Kwok H. 2002. Synthesis and properties of azo dye aligning layer for liquid crystal cells. Liq Cryst 29:1321–1327.

Chigrinov V, Muravski A, Kwok HS. 2003. Anchoring properties of photoaligned azo-dye materials. Phys Rev E 68:061702–061705.

Choi S, Kim KR, Oh K, Chun CM, Kim MJ, Yoo SJ, Kim DY. 2003. Interferometric inscription of surface relief gratings on optical fiber using azo polymer film. Appl Phys Lett 83:1080–1082.

Chun C, Kim M-J, Vak D, Kim DY. 2003. A novel azobenzene-based amorphous molecular material with a spiro linked bifluorene. J Mater Chem 13:2904–2909.

Constantino CJL, Aroca RF, He J-A, Zucolotto V, Li L, Oliveira ON, Kumar J, Tripathy SK. 2002. Raman microscopy and mapping as a probe for photodegradation in surface relief gratings recorded on layer-by-layer films of congo red/polyelectrolyte. Appl Spectrosc 56:187–191.

Cui L, Zhao Y. 2004. Azopyridine side chain polymers: an efficient way to prepare photoactive liquid crystalline materials through self-assembly. Chem Mater 16:2076–2082.

Dantsker D, Kumar J, Tripathy SK. 2001. Optical alignment of liquid crystals. J Appl Phys 89:4318–4325.

Darracq B, Chaput F, Lahlil K, Lévy Y, Boilot J-P. 1998. Photoinscription of surface relief gratings on azo-hybrid gels. Adv Mater 10:1133–1136.

Decher G, Eckle M, Schmitt J, Struth B. 1998. Layer-by-layer assembled multicomposite films. Curr Opin Coll Interface Sci 3:32–39.

Descalzo AB, Martinez-Maez R, Sancenon F, Hoffmann K, Rurack K. 2006. The supramolecular chemistry of organic–inorganic hybrid materials. Angew Chem Int Ed Engl 45:5924–5948

dos Santos DS Jr, Bassi A, Misoguti L, Ginani MF, de Oliveira ON Jr, Mendonca CR. 2002. Spontaneous birefringence in layer-by-layer films of chitosan and azo dye sunset yellow. Macromol Rapid Commun 23:975–977.

dos Santos DS Jr, Cardoso MR, Leite FL, Aroca RF, Mattoso LHC, Oliveira ON Jr, Mendonca CR. 2006. The role of azopolymer/dendrimer layer-by-layer film architecture in photoinduced birefringence and the formation of surface-relief gratings. Langmuir 22:6177–6180.

Dragan S, Schwarz S, Eichhorn KJ, Lunkwitz K. 2001. Electrostatic self-assembled nanoarchitectures between polycations of integral type and azo dyes. Coll Surf A 195:243–251.

Egami C, Kawata Y, Aoshima Y, Alasfar S, Sugihara O, Fujimura H, Okamoto N. 2000. Two-stage optical data storage in azo polymers. Jpn J Appl Phys 39:1558–1561.

Eich M, Wendorff JH, Reck B, Ringsdorf H. 1987. Reversible digital and holographic optical storage in polymeric liquid crystal. Makromol Chem Rapid Commun 8:59–63.

Faul CFJ, Antonetti M. 2003. Ionic self-assembly: facile synthesis of supramolecular materials. Adv Mater 15:673–683.

Fiorini C, Prudhomme N, de Veyrac G, Maurin I, Raimond P, Nunzi J-M. 2000. Molecular migration mechanism for laser induced surface relief grating formation. Synth Met 115:121–125.

Fischer T, Läsker L, Czapla S, Rübner J, Stumpe J. 1997. Interdependence of photo-orientation and thermotropic self-organization in photochromic liquid crystalline polymers. Mol Cryst Liq Cryst 298:213–220.

Frey L, Darracq B, Chaput F, Lahlil K, Jonathan JM, Roosen G, Boilot JP, Levy Y. 2000. Surface and volume gratings investigated by the moving grating technique in sol-gel materials. Optics Commun 173:1–16.

Fuhrmann T, Wendorff HJ. 1997. Optical data storage. In: Brostow W, Collyer A, editors. Polymer Liquid Crystals. Vol 4. London: Chapman and Hall.

Fukuda T, Sumaru K, Kimura T, Matsuda H. 2001a. Observation of optical near-field as photo-induced surface relief formation. Jpn J Appl Phys 40:L900–L902.

Fukuda T, Sumaru K, Kimura T, Matsuda H. 2001b. Photofabrication of surface relief structure – mechanism and application. J.Photochem Photobiol A 145:35–39.

Fukuda T, Kim JY, Barada D, Senzaki T, Yase KJ. 2006. Molecular design and synthesis of copolymers with large photoinduced birefringence. J Photochem Photobiol A 182:262–268.

Gao J, He Y, Xu H, Sing B, Zhang X, Wang Z, Wang X. 2007a. Azobenzene-containing supramolecular polymer films for laser-induced surface relief gratings. Chem Mater 19:14–17.

Gao J, He Y, Liu F, Zhang X, Wang Z, Wang X. 2007b. Azobenzene-containing supramolecular side-chain polymer films for laser-induced surface relief gratings. Chem Mater 19:3877–3881.

Geue T, Ziegler A, Stumpe J. 1997. Light induced orientation phenomena in LB multilayers. Macromolecules 30:5729–5738.

Gibbson WM, Shannon PJ, Sun ST, Swetlin BJ. 1991. Surface-mediated alignment of nematic liquid crystals with polarized light. Nature 358:49–50.

Goldenberg L, Gritsai Y, Kulikovska O, Stumpe J. 2008a. 3D planarized diffraction structures based on surface relief gratings in azobenzene materials. Opt Lett 33: 1309–1311.

Goldenberg L, Gritsai Y, Sahno O, Kulikovska O, Stumpe J. 2008b. All-optical fabrication of 2D, 3D and hierarchic structures using step-by-step approach and a single polymer phase mask. Proc 1st Med Photon Conf Ischia, Italy, 28–30.

Goldenberg LM, Kulikovska O, Stumpe J. 2005. Thermally stable holographic surface relief gratings and switchable optical anisotropy in films of an azobenzene-containing polyelectrolyte. Langmuir 21:4794–4796.

Guan Y, Zakrevskyy Y, Stumpe J, Antonietti M, Faul CFJ. 2003. Perylenediimide-surfactant complexes: thermotropic liquid-crystalline materials via ionic self-assembly. Chem Commun 894–895.

Hagen R, Bieringer T. 2001. Photoaddressable polymers for optical data storage. Adv Mater 13:1805–1810.

Halabieh RHE, Mermut O, Barrett CJ. 2004. Using light to control physical properties of polymers and surfaces with azobenzene chromophores. Pure Appl Chem 76:1445–1465.

Hammond PT. 2000. Recent explorations in electrostatic multilayer thin film assembly. Curr Opin Coll Interface Sci 4:430–442.

Han M, Morino S, Ichimura K. 2000. Factors affecting in-plane and out-of-plane photoorientation of azobenzene side chains attached to liquid crystalline polymers induced by irradiation with linearly polarized light. Macromolecules 33:6360–6371.

Haruta O, Matsuo Y, Hashimoto Y, Niikura K, Ijiro K. 2008. Sequence-specific control of azobenzene assemblies by molecular recognition of DNA. Langmuir 24:2618–2625.

Hasegawa M, Yamamoto T, Kanazawa A, Shinono T, Ikeda T. 1999. Photochemically induced dynamic grating by means of side chain polymer liquid crystals. Chem Mater 11:2764–2769.

He JA, Bian S, Li L, Kumar J, Tripathy SK, Samuelson LA. 2000. Photochemical behavior and formation of surface relief grating on self-assembled polyion/dye composite film. J Phys Chem B 104:10513–10521.

He Y, Wang H, Tuo X, Deng W, Wang X. 2004. Synthesis, self-assembly and photoinduced surface-relief gratings of a polyacrylate-based azo polyelectrolyte. Opt Mater 26:89–93.

Hong JD, Park ES, Park AL.1999. Effects of added salt on photochemical isomerization of azobenzene in alternate multilayer assemblies: bipolar amphiphile-polyelectrolyte. Langmuir 15:6515–6521.

Hosseini MW. 2003. Molecular tectonics: from molecular recognition of anions to molecular networks. Coord Chem Rev 240:157–166.

Ikeda T, Mamuya J, Yanleu Y. 2007. Photomechanics of liquid crystalline elastomers. Angew Chem Int Ed Engl 46:506–528.

Ikkala O, ten Brinke G. 2002. Functional materials based on self-assembly of polymeric supramolecules. Science 295:2407–2409.

Ichimura K, 1996. Photoregulation of liquid crystal alignment by photochromic molecules and polymeric thin films. In: Shibaev V, editor. Polymers as Electrooptical and Photooptical Active Media. Berlin: Springer.

Ichimura K, Seki T, Kawanishi Y, Suzuki Y, Sauragi M, Tamaki T. 1994. Photocontrols of liquid crystal alignment by command surfaces. In: Irie M, editor. Photo-reactive Materials for Ultrahigh Density Optical Memories. Amsterdam: Elsevier.

Ilieva D, Nedelchev L, Petrova T, Tomova N, Dragostinova V, Nikolova L. 2005. Holographic multiplexing using photoinduced anisotropy and surface relief in azopolymer films. J Opt A 7:35–39.

Ishow E, Lebon B, He Y, Wang X, Bouteiller L, Galmiche L, Nakatani K. 2006. Structural and photoisomerization cross studies of polar photochromic monomeric glasses forming surface relief gratings. Chem Mater 18:1261–1268.

Jung BD, Stumpe J, Hong JD. 2003. Influence of multilayer nanostructures on photoisomerization and photoorientation of azobenzene. Thin Sol Films 441:261–270.

Kaneko F, Kato T, Baba A, Shinbo K, Kato K, Advincula RC. 2002. Photo-induced fabrication of surface relief gratings in alternate self-assembled films containing azo dye and alignments of LC molecules. Coll Surf A 198–200:805–810.

Kang JW, Kim MJ, Kim JP, Yoo SJ, Lee JS, Kim DY, Kim JJ. Polymeric wavelength filters fabricated using holographic surface relief gratings on azobenzene-containing polymer films. 2003. Appl Phys Lett 80:3823–3825.

Karimi-Alavijeh H, Parsanasab G-M, Baghban M-A, Sarailou E, Gharavi A, Javadpour S, Shkunov V. 2008. Fabrication of graded index waveguides in azo polymers using a direct writing technique. Appl Phys Lett 92:041105–041103.

Kato T, Hirota N, Fujishima A, Frechet JMJ. 1996. Supramolecular hydrogen-bonded liquid-crystalline polymer complexes. Design of side-chain polymers and a host-guest system by noncovalent interaction. J Polym Sci A 3:57–62.

Kim DY, Tripathy SK, Li L, Kumar J. 1995. Laser-induced holographic surface relief gratings on nonlinear optical polymer films. Appl Phys Lett 66:1166–1168.

Kim JY, Kim TH, Kimura T, Fukuda T, Matsuda H. 2002. Surface relief grating and liquid crystal alignment on azobenzene functionalized polymers. Opt Mater 21:627–631.

Kim M-R, Choi Y-I, Park S-W, Lee J-W, Lee J-K. 2006b. Syntheses and optical properties of hybrid materials containing azobenzene groups via sol-gel process for reversible optical storage. J Appl Polym Sci 100:4811–4818.

Kim SS, Chun C, Hong JC, Kim DY. 2006a. Well-ordered TiO_2 nanostructures fabricated using surface relief gratings on polymer films. J Mater Chem 16:370–375.

Klima M, Waniek L. 1957. Über die besondere mechanische Wirkung des polarisieren Lichts. Die Naturwiss 44:370–371.

Kreuzer M, Marrucci L, Paparo D. 2000. Light-induced modification of kinetic molecuclar properties; enhancement of optical Kerr effect in absorbing liquids, photoinduced torque and molecular motors in dye-doped nematics. J Nonlin Optical Phys Mater 9:157–182.

Kulikovska O, Gharagozloo-Hubmann K, Stumpe J. 2002. Polymer surface relief structures caused by light-driven diffusion. Proc SPIE 4802:85–95.

Kulikovska O, Goldenberg LM, Stumpe J. 2007. Supramolecular azobenzene based materials for optical generation of microstructures. Chem Mater 19:3343–3348.

Kulikovska O, Goldenberg L, Kulikovsky L, Stumpe J. 2008a. Smart ionic sol-gel based azobenzene materials for optical generation of microstructures. Chem Mater 20:3528–3534.

Kulikovska O, Kulikovsky L, Goldenberg LM, Stumpe J. 2008b. Generation of microstructures in novel supramolecular ionic materials based on azobenzene. Proc SPIE 6999:69990I.

Kulikovsky L, Kulikovska O, Goldenberg L, Stumpe J. 2009. Phenomenology of photochemical processes in new ionic sol-gel based azobenzene materials. Chem Mater submitted.

Kumar J, Li L, Jiang XL, Kim DY, Lee TS, Tripathy SK. 1998. Gradient force: the mechanism for surface relief grating formation in azobenzene functionalized polymers. Appl Phys Lett 72:2096–2098.

Lee SH, Balasubramanian S, Kim DY, Viswanathan NK, Bian S, Kumar J, Tripathy SK. 2000. Azo polymer multilayer films by electrostatic self-assembly and layer-by-layer post azo functionalization. Macromolecules 33:6534–6540.

Lee TS, Kim DY, Jiang XL, Li L, Kumar J, Tripathy SK. 1998. Photoinduced surface-relief gratings in high-tg main-chain azoaromatic polymer-films. J Polym Sci A 36:283–289.

Lehn JM. 1990. Perspectives in supramolecular chemistry—from molecular recognition towards molecular information processing and self-organization. Angew Chem Int Ed Engl 29:1304–1319.

Lehn JM. 1995. Supramolecular Chemistry: Concepts and Perspectives. Weinheim: Wiley-VCH.

Lefin P, Fiorini C, Nunzi J-M. 1998. Anisotropy of the photo-induced translation diffusion of azobenzene dyes in polymer matrices. Pure Appl Opt 7:71–82.

Lin X, Zhong A, Chen D, Zhou Z, He B. 2002. Studies on self-assembly and characterization of polyelectrolytes and organic dyes. J Appl Pol Sci 85:638–644.

Lin X, Zhong A, Chen D, Zhou Z, He B. 2003. Studies on self-assembly and characterization of polyelectrolytes and organic dyes. J Appl Pol Sci 87:369–374.

Luo Y, She W, Wu S, Zeng F, Yao S. 2005. Improvement of all-optical switching effect based on azobenzene-containing polymer films. Appl Phys B 80:77–80.

Luo Y, Zhou J, Yan Q, Su W, Li Z, Zhang Q, Huang J, Wang K. 2007. Optical manipulable polymer optical fiber Bragg gratings with azopolymer as core material. Appl Phys Lett 91:071110–071113.

Lvov Y, Decher G, Möhwald H. 1993. Assembly, structural characterization, and thermal behavior of layer-by-layer deposited ultrathin films of poly(vinyl sulfate) and poly-(allylamine). Langmuir 9:481–486.

Lvov Y, Yanada S, Kunitake T. 1997. Non-linear optical effects in layer-by-layer alternate films of polycations and an azobenzene-containing polyanion. Thin Sol Films 300:107–112.

Macdonald JC, Whitesides GM. 1994. Solid-state structures of hydrogen-bonded tapes based on cyclic secondary diamides. Chem Rev 94:2383–2420.

Makushenko AM, Neporent BS, Stolbova OV. 1971. Reversible orientational photodichroism and photoisomerisation of aromatic azo compounds. I. The model of the system and II. Azobenzene and substituted azobenzene derivatives. Opt Spectr 31:295–299 and 741–748.

Marino IG, Bersani D, Lottici PP. 2000. Photo-induced birefringence in dr1-doped sol–gel silica and ormosils thin films. Opt Mater 15:175–180.

Martinez-Ponce G, Petrova T, Tomova N, Dragostinova V, Todorov T, Nikolova L. 2004. Bifocal-polarization holographic lens. Optic Lett 29:1001–1003.

McArdle CB, editor, 1992. Applied Photochromic Polymer Systems. New York: Blackie & Son Ltd.

Medvedev AV, Barmatov EB, Medvedev AS, Shibaev VP, Ivanov SA, Kozlovsky M, Stumpe J. 2005. Phase behavior and photooptical properties of liquid crystalline functionalized copolymers with low-molecular-mass dopants stabilized by hydrogen bonds. Macromolecules 38:2223–2229.

Meier MAR, Schubert US. 2005. Combinatorial evaluation of the host-guest chemistry of star-shaped block copolymers. J Comb Chem 7:356–359.

Meyer WH, Pecherz J, Maty A, Wegner G. 1991. Polyelectrolite glasses with linear and nonlinear optical properties. Adv Mater 3:153–156.

Moore JS. 1966. Molecular architecture and supramolecular chemistry. Curr Opin Solid State Mater Sci 1:777–788.

Nakano H, Takahashi T, Kadota T, Shirota Y. 2002. Formation of the surface relief grating using a novel azobenzene-based photochromic amorphous molecular material. Adv Mater 14:1157–1160.

Natansohn A, Rochon P. 2002. Photoinduced motions in azo-containing polymers. Chem Rev 102:4139–4175.

Nedelchev LL, Matharu AS, Hvilsted S, Ramanujam PS. 2003. Photoinduced anisotropy in a family of amorphous azobenzene polyesters for optical storage. Appl Opt 42:5918–5927.

Ohdaira Y, Noguchi K, Shinbo K, Kato K, Kaneko F. 2006. Nano-fabrication of surface relief gratings on azo dye films utilizing interference of evanescent waves on prism. Coll Surf A 284–285:556–560.

Oliveira ON Jr, dos Santos DS Jr, Balogh DT, Zucolotto V, Mendonca CR. 2005. Optical storage and surface-relief gratings in azobenzene-containing nanostructured films. Adv Coll Inter Sci 116:179–192.

Pedersen TG, Johansen PM, Holme NCR, Ramanujam PS, Hvilsted S. 1998. Mean-field theory of photoinduced formation of surface reliefs in side-chain azobenzene polymers. Phys Rev Lett 80:89–91.

Perschke A, Fuhrmann T. 2002. Molecular azo glasses as grating couplers and resonators for optical devices. Adv Mater 14:841–843.

Petrova TS, Mancheva I, Nacheeva E, Tomava E, Dragostinova V, Todorova T, Nikolova L. 2003. New azobenzene polymers for light-controlled optical elements. J Mat Sci Mater Electron 14:823–824.

Pollino JM, Weck M. 2005. Non-covalent side-chain polymers: design principles, functionalization strategies, and perspectives. Chem Soc Rev 34:193–207.

Presnyakov V, Asatryan K, Galstian T, Chigrinov V. 2006. Optical polarization grating induced liquid crystal micro-structure using azo-dye command layer. Opt Expr 14:10558–10564.

Priimagi A, Cattaneo S, Ras RHA, Valkama S, Ikkala O, Kauranen M. 2005. Polymer-dye complexes: a facile method for high doping level and aggregation control of dye molecules. Chem Mater 17:5798–5802.

Priimagi A, Cattaneo S, Ras RHA, Valkama S, Ikkala O, Kauranen M. 2006. Polymer-dye complexes: supramolecular route toward functional optical materials. Proc SPIE 6192:61922–61928.

Priimagi A, Kaivola M, Rodriguez FJ, Kauranen M. 2007. Enhanced photoinduced birefringence in polymer-dy complexes: hydrogen bonding makes a difference. Appl Phys Lett 90:121103–121106.

Rau H. 1990. Azo compounds. In: Dürr H, Bounas-Laurent H, editors. Photochromism, Molecules and Systems. Amsterdam: Elsevier.

Rochon P, Batalla E, Natansohn A. 1995. Optically induced surface gratings on azoaromatic polymer films. Appl Phys Lett 66:136–138.

Rosenhauer R, Fischer T, Czapla S, Stumpe J, Vinuales A, Pinol M, Serrano JL. 2001. Photo-induced alignment of LC polymers by photoorientation and thermotropic self-organisation. Mol Cryst Liq Cryst 364:295–304.

Rosenhauer R, Fischer T, Stumpe J, Vinuales A, Gimenez R, Pinol M, Serrano JL, Broer D. 2005. Light-induced orientation of liquid crystalline terpolymers containing azobenzene and dye moieties. Macromolecules 38:2213–2222.

Sabi Y, Yamamoto M, Watanabe H, Bieringer T, Haarer D, Hagen R, Kostromine SG, Berneth H. 2001. Photoaddressable polymers for rewritable optical disc systems. Jpn J Appl Phys 40:1613–1618.

Saphiannikova M, Neher D. 2005. Thermodynamic theory of light-induced material transport in amorphous azobenzene polymer films. J Phys Chem B 109:19428–19436.

Saremi F, Tieke B. 1998. Photoinduced switching in self-assembled multilayers of an azobenzene bolaamphiphile and polyelectrolytes. Adv Mater 10:388–391.

Seki T. 2007. Smart photoresponsive polymer systems organized in two dimensions. Bull Chem Soc Jpn 80:2084–2102.

Seki T, Fukuda R, Yokoi M, Tamaki T, Ichimura K. 1996. Photomechanical response of azobenzene containing monolayers on water surface. Bull Chem Soc Jpn 69:2375–2381.

Sekkat Z, Wood J, Geerts Y, Knoll W, 1995. A smart ultrathin photochromic layer. Langmuir 11:2856–2859.

Shibaev VP, Kostromin SG, Ivanov SA. 1996. Comb-shaped polymers with mesogenic side groups as electro- and photooptical active media. In: Shibaev VP, editor. Polymers as Electrooptical and Photo-optical Active Media. Berlin: Springer-Verlag.

Shibaev VP, Bobrovsky A, Boiko N. 2003. Photoactive liquid crystalline polymer systems with light controllable structure and optical properties. Prog Polym Sci 28:729–836.

Shimomura M. 1993. Preparation of ultrathin polymer films based on two-dimensional molecular ordering. Prog Polym Sci 18:295–339.

Shinbo K, Baba A, Kaneko F, Kato T, Kato K, Advincula RC, Knoll W. 2002. In situ investigations on the preparations of layer-by-layer films containing azobenzene and applications for LC display devices. Mater Sci Eng C 22:319–325.

Shinkai S, Nakaji T, Nishida Y, Ogawa T, Manabe O. 1980. Photoresposive crown ethers. 1. Cis-trans isomerism of azobenzene as a tool to enforce conforamtional change of crown ethers and polymers. J Am Chem Soc 102:5860–5865.

Shirota Y. 2005. Photo- and electroactive amorphous molecular materials-molecular design, syntheses, reactions, properties, and applications. J Mater Chem 15:75–93.

Shirota Y, Moriwaki K, Yoshikawa S, Ujike T, Nakano H. 1998. 4-[Di(biphenyl-4-ylamino]azobenzene and 4,4′-bis[bis(4′-tertbutylbiphenyl-4-yl)amino]azobenzene as a novel family of photochromic amorphous molecular materials. J Mater Chem 8:2579–2581.

Simmrock H-U, Mathy A, Dominguez L, Meyer WH, Wegner G. 1989. Polymers with a high refractive index and low optical dispersion. Ang Chem Inter Ed Eng 101:1148–1149.

Stracke A, Wendorff, JH, Goldmann D, Janietz D. 2000. Optical data storage in a smectic mesophase; thermal amplification of light-induced chromophore orientation and surface relief gratings. Liq Cryst 27:1049–1057.

Stumpe J, Müller L, Kreysig D, Hauck G., Koswig HD, Ruhmann R, Rübner J. 1991. Photoinduced optical anisotropy of liquid crystalline side-chain polymer with azochromophores by low intensity of linearly polarized light. Makromol Chem Rapid Commun 12:81–87.

Stumpe J, Fischer T, Menzel H. 1996. Langmuir-Blodgett-films of photochromic polyglutamates, relation between photochemical modification and thermotropic properties. Macromolecules 29:2831–2842.

Stumpe J, Goldenberg LM, Kulikovska O. 2006a. Film forming material and preparation of surface relief and optically anisotropic structures by irradiating a film of the said material. European Patent Application PCT/EP2005/009346, EP1632520, WO2006024500.

Stumpe J, Goldenberg LM, Kulikovska O. 2006b. Photoactive film, its preparation and use, and preparation of surface relief grating and optically anisotropic structures by irradiating said film. European Patent Application PCT/EP2005/009346, WO2006061419.

Sukwattanasinitt M, Wang X, Li L, Jiang X, Kumar J, Tripathy SK, Sandman DJ. 1998. Functionalizable self-assembling polydiacetylenes and their optical properties. Chem Mater 10:27–29.

Szabados L, Janossy I, Kosa T, 1998. Laser-induced bulk effects in nematic liquid crystals doped with azo dyes. Mol Cryst Liq Cryst 320:239–248.

Tanino T, Yoshikawa S, Ujike T, Nagahama D, Moriwaki K, Takahashi T, Kotani Y, Nakano H, Shirota Y. 2007. Creation of azobenzene-based photochromic amorphous molecular materials—synthesis, glass-forming properties, and photochromic response. J Mater Chem 17:4953–4963.

Thünemann AF. 2002. Polyelectrolyte-surfactant complexes (synthesis, structure and materials aspects). Prog Polym Sci 27:1473–1572.

Thünemann AF, Schnoeller U, Nuyken O, Voit B. 1999. Self-assembled complexes of diazosulfonate polymers with low surface energies. Macromolecules 32:7414–7421.

Thünemann AF, Schnöller U, Nuyken O, Voit B. 2000. Diazosulfonate polymer complexes: structure and wettability. Macromolecules 33:5665–5671.

Todorov T, Nikolaeva L, Tomova T. 1984. Polarisation holography 2: polarisation holographic gratings in photoanisotropic materials with and without intrinsic birefringence. Appl Opt 23:4588–4591.

Utsumi H, Nagahama D, Nakano H, Shirota Y. 2000. A novel family of photochromic amorphous molecular materials based on dithienylethene. J Mater Chem. 10:2436–2437.

Viswanathan NK, Kim DY, Bian S, Williams J, Liu W, Li L, Samuelson L, Kumar J, Tripathy SK. 1999. Surface relief structures on azo polymer films. J Mater Chem 9:1941–1955.

Wang G, He Y, Wang X, Jiang L. 2004a. Self-assembly and optical properties of poly(acrylic acid)-based azo polyelectrolyte. Thin Sol Films 458:143–148.

Wang H, He Y, Tuo X, Wang X. 2004b. Sequentially adsorbed electrostatic multilayers of branched side-chain polyelectrolytes bearing donor-acceptor type azo chromophores. Macromolecules 37:135–146.

Wang X, Balasubramanian S, Kumar J, Tripathy SK, Li L. 1998. Azo chromophore-functionalized polyelectrolytes. 1. Synthesis, characterization, and photoprocessing. Chem Mater 10:1546–1553.

Weigert F. 1929. Photodichroismus and photoanisotropie. Z Phys Chem 3:377.

West, J, Su L, Reshnikov Y. 2001. Photoalignment using adsorbed dichroic molecules. Mol Cryst Liq Cryst 364:199–210.

Willner I, Rubin S, Riklin A. 1991. Photoregulation of papin activity through anchoring photochromic azo groups to the enzyme backbone. J Am Chem Soc 113:3321–3325.

Wu L, Tuo X, Cheng H, Chen Z, Wang X. 2001. Synthesis, photoresponsive behavior, and self-assembly of poly(acrylicacid)-based azo polyelectrolytes. Macromolecules 34:8005–8013.

Xiao S, Lu X, Lu Q. 2007. Photosensitive polymer from ionic self-assembly of azobenzene dye and poly(ionic liquid) and its alignment characteristic toward liquid crystal molecules. Macromolecules 40:7944–7950.

Yagai S, Karatsu T, Kitamura A. 2005. Photocontrollable self-assembly. Chem Eur J 11:4054–4063.

Yager GY, Barrett CJ. 2001. All-optical patterming of azo polymer films. Curr Opin Sol State Mater Sci 5:457–494.

Ye C, Wong KY, He Y, Wang X. 2007. Distributed feedback sol-gel zirconia waveguide lasers based on surface relief gratings. Optic Expr 15:936–944.

Zakrevskyy Y, Faul CFJ, Guan Y, Stumpe J. 2004. Alignment of a perylene-based ionic self-assembly complex in thermotropic and lyotropic liquid-crystalline phases. Adv Funct Mater 14:835–841.

Zakrevskyy Y, Smarsly B, Stumpe J, Faul CFJ. 2005. Highly ordered monodomain ionic self-assembled liquid-crystalline materials. Phys Rev E 71:021701-01–021701-12.

Zakrevskyy Y, Stumpe J, Faul CFJ. 2006a. A supramolecular approach to optically anisotropic materials: photosensitive ionic self-assembly complexes. Adv Mater 18:2133–2136.

Zakrevskyy Y, Faul CFJ, Stumpe J. 2006b. Film forming photosensitive materials for the light induced generation of optical anisotropy. Patent Application WO2006/117403 A1.

Zakrevskyy Y, Stumpe J, Smarsly B, Faul CFJ. 2007. Photo-induction of optical anisotropy in azobenzene containing ionic self-assembly liquid-crystalline material. Phys Rev E 75:031703-01–031703-12.

Zen A, Neher D, Bauer C, Asawapirom U, Scherf U, Hagen R, Kostromine S, Mahrt RF. 2002. Polarization-sensitive photoconductivity in aligned polyfluorene layers. Appl Phys Lett 80:4699-4701.

Zettsu N, Seki T. 2004. Highly efficient photogeneration of surface relief structure and its immobilization in cross-linkable liquid crystalline azobenzene polymers. Macromolecules 37:8692–8698.

Zettsu N, Ogasawara T, Arakawa R, Nagano S, Ubukata T, Seki T. 2007. Highly photosensitive surface relief gratings formation in a liquid crystalline azobenzene polymer: New implications for the migration process. Macromolecules 40:4607–4613.

Zettsu N, Ogasawara T, Mizoshita N, Nagano S, Seki T. 2008. Photo-triggered surface relief grating formation in supramolecular liquid crystalline polymer systems with detachable azobenzene unit. Adv Mater 20:516–521.

Zhang Q, Bazuin CG, Barrett CJ. 2008. Simple spacer-free dye-polyelectrolyte ionic complex: side-chain liquid crystal order with high and stable photoinduced birefringence. Chem Mater 20:29–31.

Ziegler A, Stumpe J, Toutianoush A, Tieke B. 2002. Photoorientation of azobenzene moieties in self-assembled polyelectrolyte multilayers. Coll Surf A 198–200:777–784.

Zucolotto V, Mendonca CR, dos Santos DS Jr, Balogh DT, Zilio SC, Oliveira ON Jr, Constantino CJL, Aroca RF. 2002. The influence of electrostatic and H-bonding interactions on the optical storage of layer-by-layer films of an azopolymer. Polymer 43:4645–4650.

Zucolotto V, He JA, Constantino CJL, Neto NMB, Rodrigues JJ Jr, Mendonca CR, Zılio SC, Li L, Aroca RF, Oliveira ON Jr, Kumar J. 2003. Mechanisms of surface-relief gratings formation in layer-by-layer films from azodyes. Polymer 44:6129–6133.

Zucolotto V, Barbosa Neto NM, Rodrigues JJ Jr, Constantino CJL, Zílio SC, Mendonça CR, Aroca RF, Oliveira ON Jr. 2004a. Photoinduced phenomena in layer-by-layer films of poly(allylamine hydrochloride) and brilliant yellow azodye. J Nanosci Nanotech 4:855–860.

Zucolotto V, Strack PJ, Santos FR, Balogh DT, Constantino CJL, Mendonca CR, Oliveira ON, Jr. 2004b. Molecular engineering strategies to control photo-induced birefringence and surface-relief gratings on layer-by-layer films from an azopolymer. Thin Sol Films 453–454:110–113.

3

PHOTODEFORMABLE MATERIALS AND PHOTOMECHANICAL EFFECTS BASED ON AZOBENZENE-CONTAINING POLYMERS AND LIQUID CRYSTALS

Yanlei Yu and Tomiki Ikeda

3.1. INTRODUCTION

Living systems respond to external stimuli by adapting themselves to changing conditions. Polymer scientists have been trying to mimic this behavior for the past 20 years, creating the so-called smart polymers (SPs). The characteristic feature that actually makes them "smart" is their ability to respond to very slight changes in the surrounding environment, such as temperature, pH, light, magnetic or electric field, ionic factors, and biological molecules. The responses are manifested as changes in one or more of the following: shape, surface characteristics, solubility, formation of an intricate molecular assembly, a sol-to-gel transition, and others (Kumar et al., 2007). The uniqueness of these materials lies not only in the fast macroscopic changes occurring in their structure but also in these transitions being reversible. SPs have very promising applications in the biomedical field as delivery systems of therapeutic agents, tissue engineering scaffolds, cell culture supports, bioseparation devices, sensors, or actuator systems.

Among various SPs, the materials that can sense signals and produce a definite dynamic response in the form of a change in shape or volume are under active investigation, such as shape-memory polymers (Lendlein et al., 2005), polymer gels (Liu and Calver, 2000; Zriny et al., 2000; Osada and Gong, 1998; Suzuki et al.,

Smart Light-Responsive Materials. Edited by Yue Zhao and Tomiki Ikeda

1996; Hu et al., 1995; Yoshida et al., 1995; Zhang et al., 1995; Osada et al., 1992; Mamada et al., 1990; Suzuki and Tanaka, 1990), conjugated polymers (Bay et al., 2003; Otero and Cortés, 2003; Smela, 2003; Jager et al., 2000; Baughman, 1996), carbon nanotubes (Spinks et al., 2002; Baughman et al., 1999; Kim and Lieber, 1999; Zhang and Iijima, 1999), and dielectric elastomers (Pelrine et al., 2000). Besides, light is a particularly fascinating stimulus because it can be precisely modulated in terms of wavelength, polarization direction, and intensity, allowing noncontact control. Related technologies, such as lasers, optics, and optical fibers, are well established and available. Therefore, light-responsive polymers that can undergo a photoinduced shape or volume change, named as photodeformable polymers, have attracted more and more interest of researchers.

To be light responsive, polymers have to be equipped with photosensitive functional groups or fillers. The incorporation of such photosensitive groups or molecules into a tailored polymer surrounding in combination with a functionalization process is a well-established strategy for transferring effects from the molecular level into effects that are macroscopically visible. For instance, the most ubiquitous natural molecule for reversible shape change is the rhodopsin–retinal protein system that enables vision. The small retinal molecule embedded in a cage of rhodopsin helices isomerizes from a cis geometry to a trans geometry around a C=C double bond with the absorption of a single photon. The modest shape change of just a few angstroms is quickly amplified, however, and sets off a cascade of larger and larger shape and chemical changes, eventually culminating in an electrical signal to the brain of a vision event. Here, the energy of the input photon is amplified thousands of times in the process. Complicated biochemical pathways then revert the trans isomer back to cis and set the system backup for another cascade upon subsequent absorption.

Although we are yet unable to reconstruct the biological photoresponsive systems as they are, it is worthwhile mimicking their processes. From the viewpoint of reversibility, speed, and simplicity of incorporation, perhaps azobenzene chromophores are the best photosensitive molecules to do this artificial mimic. Trans and cis states can be switched in picoseconds even with low power light, reversibility of 10^5 and 10^6 cycles is routine before fatigue, and a wide variety of molecular architectures is available to the synthetic materials' chemist, permitting facile anchoring and compatibility, as well as chemical and physical amplification of the simple geometric change (Ikeda, 2003; Natansohn and Rochon, 2002). Therefore, this chapter describes various photodeformable polymer systems containing azobenzenes such as polymer gels, solid films, and liquid crystals (LCs), focusing attention on light-responsive LC elastomers (LCEs) that are recently developed photomechanical systems. Other photodeformable polymer systems such as colloidal particles and monolayers for which good reviews are included in the following chapters are skipped here. By using their deformations, we can convert light energy into mechanical power directly (photomechanical effects). These systems have obvious attractiveness in basic science and are potential for many applications, for example, serving as artificial muscles, sensors, microrobots, micropumps, and actuators.

3.2. PHOTODEFORMABLE MATERIALS BASED ON AZOBENZENE-CONTAINING POLYMER GELS

The most extensive investigations on SPs have been carried out on the polymer gels in which a network of long polymer molecules holds the liquid in place and so gives the gel what solidity it has. Chemically cross-linked gels can undergo a transition between a collapsed and an expanded state, that is, between a shrunk and a swollen state (Fig. 3.1). The difference between these two states could reach several orders of magnitude. This behavior has been very attractive for the applications of the gels as potential actuators, sensors, controllable membranes for separations, and modulators for drug delivery (Osada and Gong, 1998).

Studies on the light-responsive gels began as early as the 1970s. Van der Veen and Prins (1971) prepared a gel system consisting of a low-molecular-weight (LMW) chrysophenine dye (**1**) and a water-swollen gel of poly(2-hydroxyethyl methacrylate) cross-linked with ethylene glycol dimethylacylate. The polymer gel was found to contract upon irradiation of UV light, because the trans–cis isomerization causes a reduction in hydrophobicity of the dye, librating the dyes from the polymer chain to the surrounding solution. The same type of the photoinduced deformations was also observed in cross-linked poly(methacrylic acid) on which chrysophenine was absorbed (Van der Veen and Prins, 1974).

In contrast, adsorption of the positively charged dye, 4-phenylazophenyl trimethylammonium iodide (**2**), to the cross-linked poly(methacrylic acid) film led to the reverse of the photoresponsive behavior described earlier (Van der Veen et al., 1974). Swelling occurred upon irradiation of UV light, whereas contraction took place during relaxation in the dark. The aqueous polymeric acid gel combined with the positively charged trans form takes on a more hydrophobic and globular conformation. After irradiation of UV light, the more soluble cis isomer is formed, causing the dye to drop off the polymer. As a result, the polymer relaxes to an extended conformation, and the sample swells.

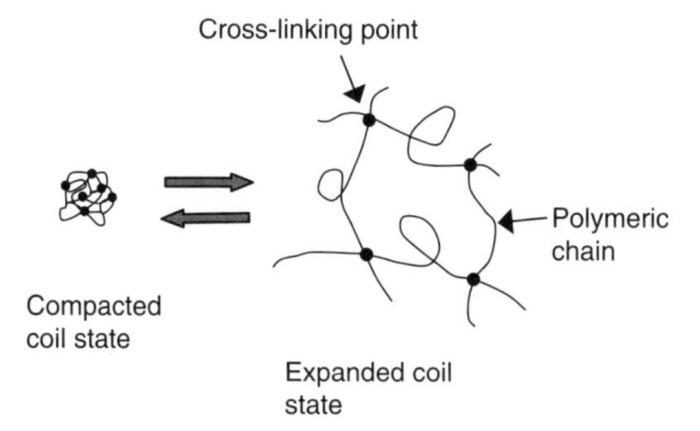

Figure 3.1. Schematic representation of a gel in its collapsed and swollen states.

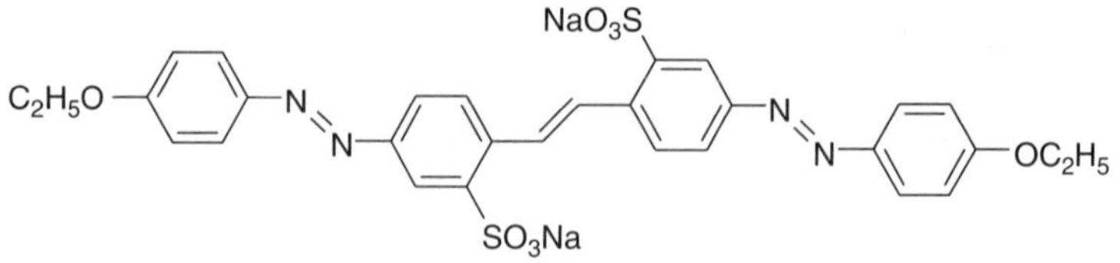

1

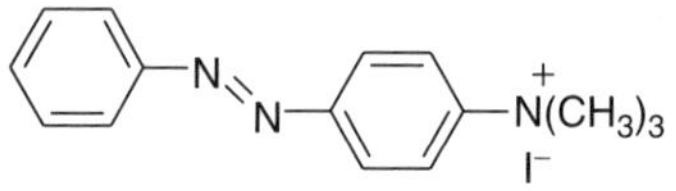

2

Compared with the gels sensitive to such stimuli as pH and heat, the studies on the light-responsive gels develop much more slowly. There have been few reports concerning the photoinduced changes in shape or volume of the polymer gels for about 30 years, except that Irie and Suzuki et al. have respectively done some excellent work on modification of the side chains of the gel networks with functional groups other than azobenzenes that either undergo ionization upon irradiation of UV light (Irie and Kungwatchakun, 1995; Ishikawa et al., 1991; Mamada et al., 1990; Irie, 1986) or cause local heating by absorbing energy from visible light (Suzuki et al., 1996; Suzuki and Tanaka, 1990).

Recently, a successful strategy to manipulate the swelling response by light has been reported on poly(*N*-isopropylacrylamide) gels wherein the cross-links contain azobenzene groups. During the deswelling, irradiation of the gels with UV light was found to increase the expulsion of water from the swollen gels and decrease the water content of the gels by as much as 20%–30%. They suggested that trans–cis isomerization of the cross-links causes a local stress field and lower entropy of the network, which in turn leads to an increase in unfavorable interaction of the network with the solvent, contributing to the increased expulsion of water from the gels. Furthermore, an increase in the cross-link density was found to reduce the magnitude of this change (Mun-Sik and Vinay, 2002).

Recently, semirigid poly(amide acid) gels containing azobenzene groups in the backbone structure were synthesized with a symmetric trifunctional cross-linker, 1,3,5-tris(4-aminophenyl) benzene (TAPB) (Fig. 3.2). Irradiation of laser light with the wavelength of 405 nm induced a local volume change in the gels owing to the deformation of the network structure, and the volume change was reversible by irradiation with visible light or heat. The viscoelastic properties of the gels before and after light irradiation were also investigated, and a twofold increase in the storage modulus of the gels was observed upon irradiation of 405-nm laser light (Hosono et al., 2007).

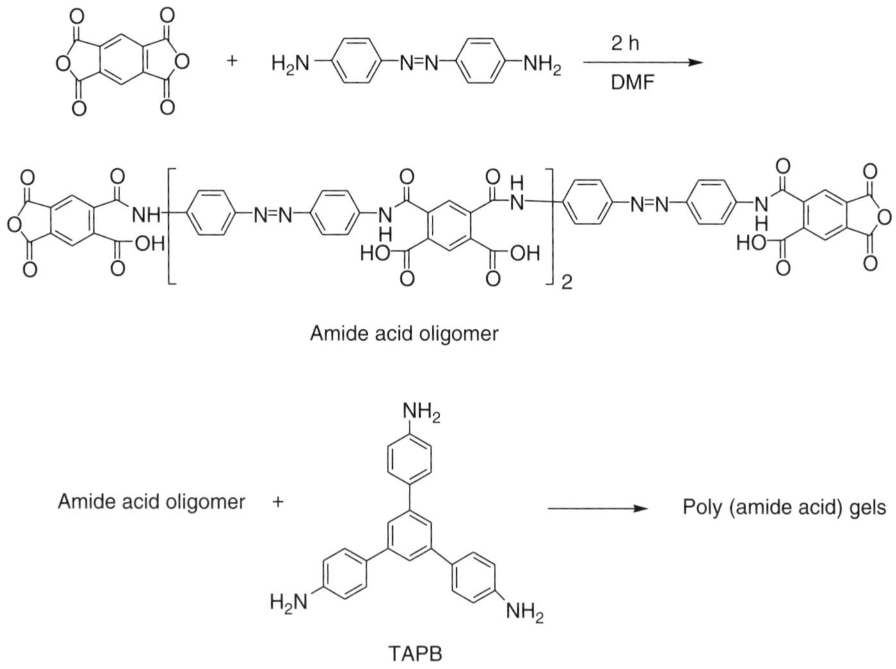

Figure 3.2. Synthesis route of poly(amide acid) gels containing azobenzene groups in their backbones and cross-linked with TAPB. *Source*: Hosono et al., 2007.

Seki and coworkers (2007) developed a new type of photoresponsive polymer gel by introducing an azobenzene moiety on the mobile cyclodextrin part as the sliding cross-linking unit. The synthesis scheme of the gel is shown in Fig. 3.3. The polyrotaxane (PR) was composed of a noncovalent complex between poly(ethylene oxide) and α-cyclodextrin (α-CD, filling ratio: 28% of full complexation) capped with an adamantanamine unit at each end. Onto the sliding ring of CD, three types of azobenzene derivatives (6Az10-, 6Az$(EO)_3$-, and CNAz$(EO)_3$-) were linked by esterification. Gelation reactions, namely, formation of figure-of-eight sliding cross-linkages of the azobenzene-linked PR, were performed with 1,1-carbonyl diimidazole in dimethyl-sulfoxide (DMSO). Figure 3.4 indicates volume changes of the gels in DMSO upon irradiation of alternate UV and visible light. For all three systems, irradiation of UV light induced expansion of the gels, and visible light the reverse. The change in D/D_0 for 6Az10-PR gel reached 25%, which corresponds to a volume change of 80%–100%. It is inferred that such an effective molecule-to-material transformation can be ascribed to the dynamic nature of the cross-linkers (Sakai et al., 2007).

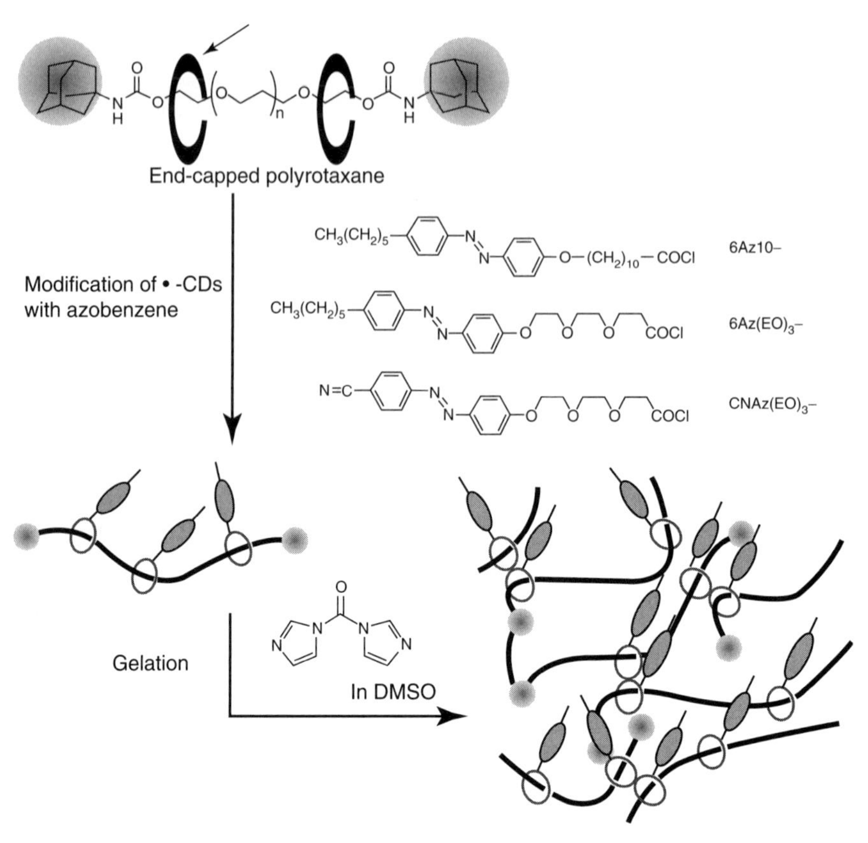

Figure 3.3. Synthesis scheme of azobenzene-containing slide-ring gels. *Source*: Sakai et al., 2007.

3.3. PHOTODEFORMABLE MATERIALS BASED ON AZOBENZENE-CONTAINING SOLID FILMS

Polymer gels have attracted special attention for possible applications in the biomedical field; however, low elastic modulus and low yield strength of the gels provide important limitations for their performance. Polymers are one of the most superior materials in view of their high processability, ability to form self-standing films with thicknesses from nanometers to centimeters, lightweight, flexibility in molecular design, and precisely controllable synthesis. Many kinds of polymers have been put to practical use in daily life and industry. From this point of view, polymer actuators capable of responding and deforming in response to external stimuli are most desirable for practical applications. Various chemical and physical stimuli have been applied to induce deformation of polymer actuators, for example, temperature changes (Liu et al., 2002), electric fields (Fukushima et al., 2005; Otero and Cortés, 2003), and solvent composition (Gao et al., 2003).

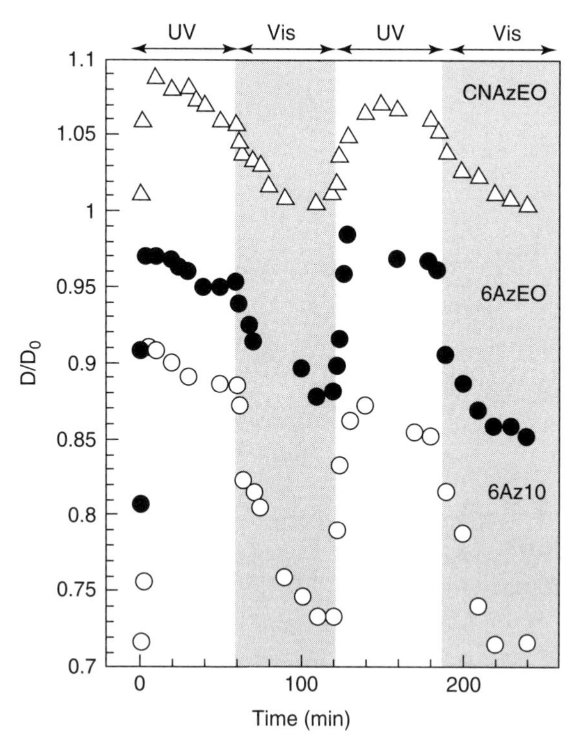

Figure 3.4. Swelling behavior of slide-ring gels in DMSO upon alternate irradiation of UV and visible light. D and D_0 are diameters of the gel and inner diameter of mold capillary, respectively. *Source*: Sakai et al., 2007.

The use of structural changes of photoisomerizable chromophores to change the size of polymers was first proposed by Merian (1966). He observed that a nylon filament fabric dyed with an azobenzene derivative shrank upon photoirradiation. This effect is ascribed to the photochemical structural change of the azobenzene moiety absorbed on the nylon fibers. However, the observed shrinkage was very small (only $\sim$0.1%), and subsequent to this work, much effort was made to find new photomechanical systems with an enhanced efficiency (Irie, 1990; Smets and De Blauwe, 1974).

Eisenbach (1980) investigated the photomechanical effect of poly(ethyl acrylate) networks (**3**) cross-linked with azobenzene moieties and observed that the polymer network contracted upon exposure to UV light (caused by the trans–cis isomerization of the azobenzene cross-links) and expanded upon irradiation of visible light (caused by cis–trans back-isomerization; Fig. 3.5). This photomechanical effect is mainly due to the conformational change of the azobenzene cross-links by the trans–cis isomerization of the azobenzene chromophore. However, the degree of deformation was small (0.2%).

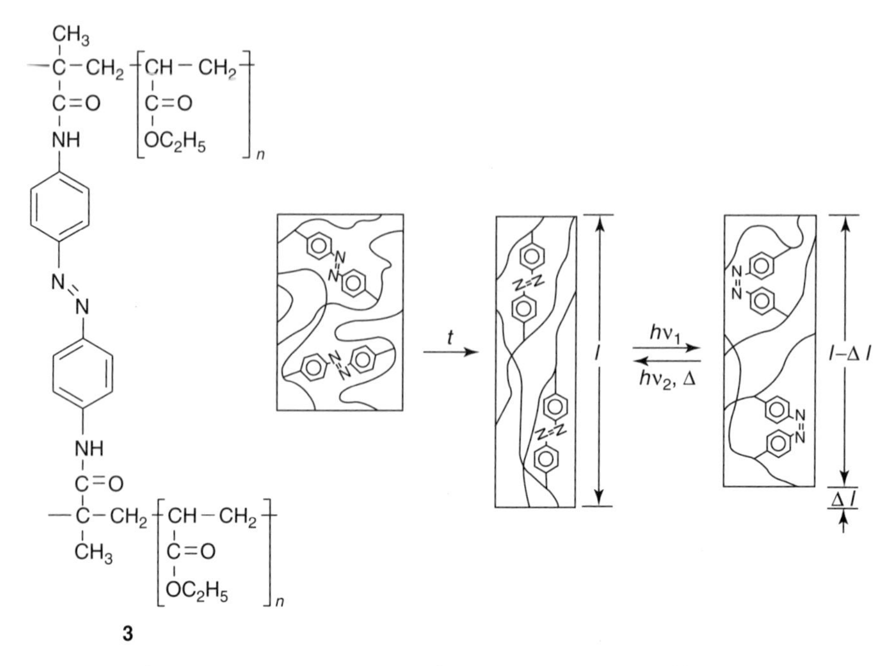

Figure 3.5. Schematic representation of photomechanical effect in a poly(ethyl acrylate) network with azobenzene cross-links upon irradiation. *f* = Force. *Source*: Eisenbach, 1980.

Matějka et al. synthesized several types of photochromic polymers based on a copolymer of maleic anhydride and styrene with azobenzene moieties both in the side chains and cross-links of the polymer network (Matějka et al., 1981; Matějka and Dusek, 1981; Matějka et al., 1979) . The photomechanical effect was enhanced by an increase in the content of photochromic groups, and the photoinduced contraction of the sample amounted to 1% for a polymer with 5.4 mol% azobenzene moieties.

The photoinduced expansion of thin films of polymers (**4**) containing azobenzene chromophores was explored in real time by single wavelength ellipsometry (Fig. 3.6; Tanchak and Barret, 2005). The initial expansion of the azobenzene polymer films of thickness ranging from 25 to 140 nm was irreversible and amounted to 1.5%–4%. Subsequently, reversible expansion was observed with repeated irradiation cycles; the relative expansion was 0.6%–1.6%.

Recent development of single-molecular force spectroscopy by atomic force microscopy (AFM) techniques has enabled quite successfully the measurement of mechanical force produced at a molecular level. Gaub and coworkers reported on a phototriggered polymer, which contains azobenzene units as part of the main polymer backbone (**5**; Adoc = 1-adamantyloxycarbonyl) (Holland et al., 2003; Hugel et al., 2002). This key study demonstrates a photomechanical cycle using

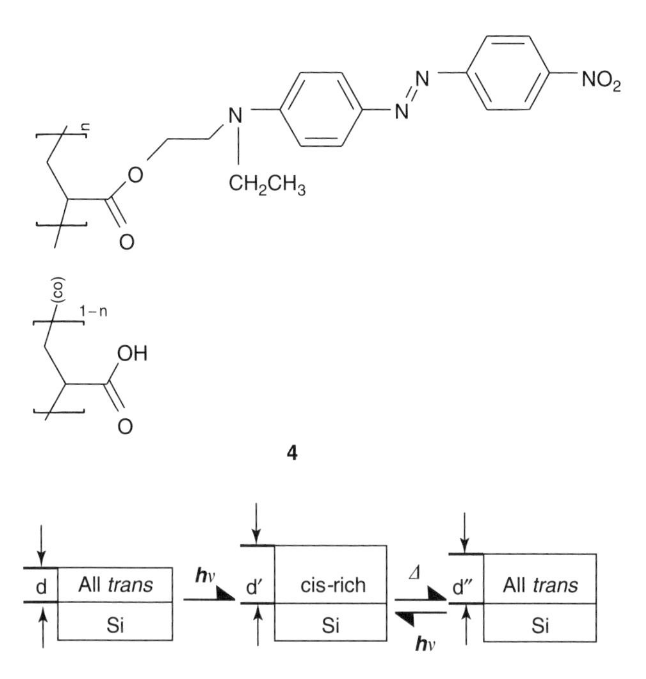

Figure 3.6. Schematic representation of photoexpansion effect in the thin films. *Source*: Tanchak and Barret, 2005.

a single azobenzene polymer molecule (Fig. 3.7), representing the first experimental work using photomechanical energy conversion in a single-molecule device, in which a polymer is covalently coupled to an AFM tip and a glass slide. The azobenzene units are reversibly switched at two distinctly different wavelengths between an extended trans and a contracted cis configuration. Applying focused UV light ($\lambda = 420$ nm) to the glass substrate having the polymer bound to it results in stretching of the polymer since its conformation is switched to the all-trans state. Subsequent irradiation with light of $\lambda_{max} = 356$ nm causes relaxation of the polymer and conversion to the all-cis conformation. The material can be repeatedly cycled between these two states, even when an external load is put on the AFM tip. The mechanical work performed by the azobenzene polymer strand by trans–cis photoisomerization was approximately 4.5×10^{-20} J. This mechanical work at the molecular level results from a macroscopic photoexcitation, and the real quantum efficiency of the photomechanical work for the given cycle in their AFM setup was only on the order of 10^{-18}. However, a maximum efficiency of the photomechanical energy conversion at a molecular level can be estimated as 0.1, if it is assumed that each switching of a single azobenzene unit is initiated by a single photon with an energy of 5.5×10^{-19} J ($\lambda = 365$ nm) (Holland et al., 2003; Hugel et al., 2002).

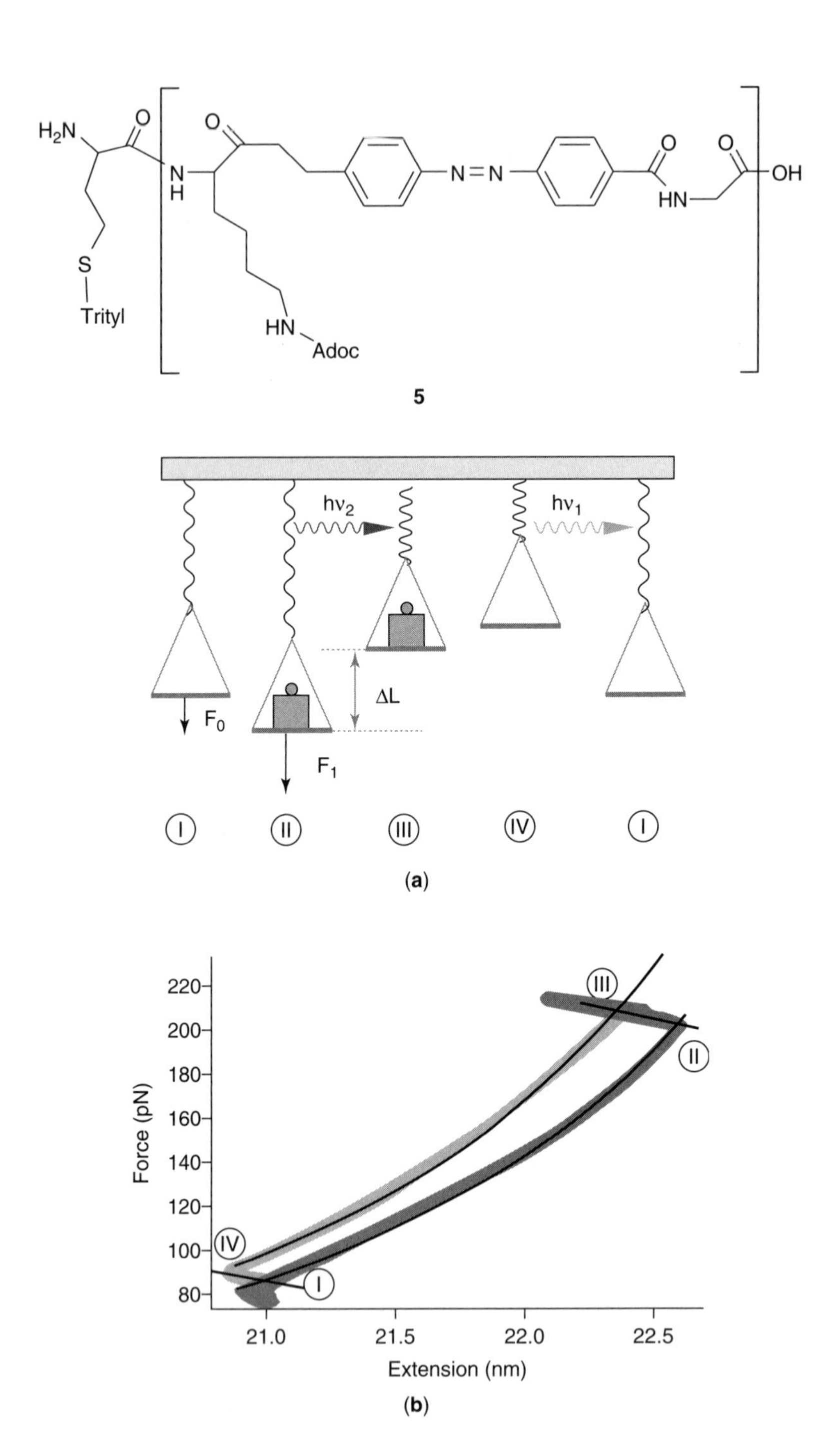

Figure 3.7. Schematics of four states (I–IV) of a work cycle based on **(a)** a single polyazopeptide molecule and **(b)** the experimental realization of the operating cycle. *Source*: Hugel et al., 2002.

Photoinduced reversible changes in elasticity of semi-interpenetrating network films bearing azobenzene moieties (Components **6a**,**b**) were achieved by irradiation with UV and visible light (Kim et al., 2005a). The network films were prepared by cationic copolymerization of azobenzene-containing vinyl ethers in a linear polycarbonate matrix. The network film showed reversible deformation by switching the UV light on and off. The photomechanical effect is attributed to a reversible change between the highly aggregated and dissociated state of the azobenzene groups (Fig. 3.8) (Kim et al., 2005a,b,c).

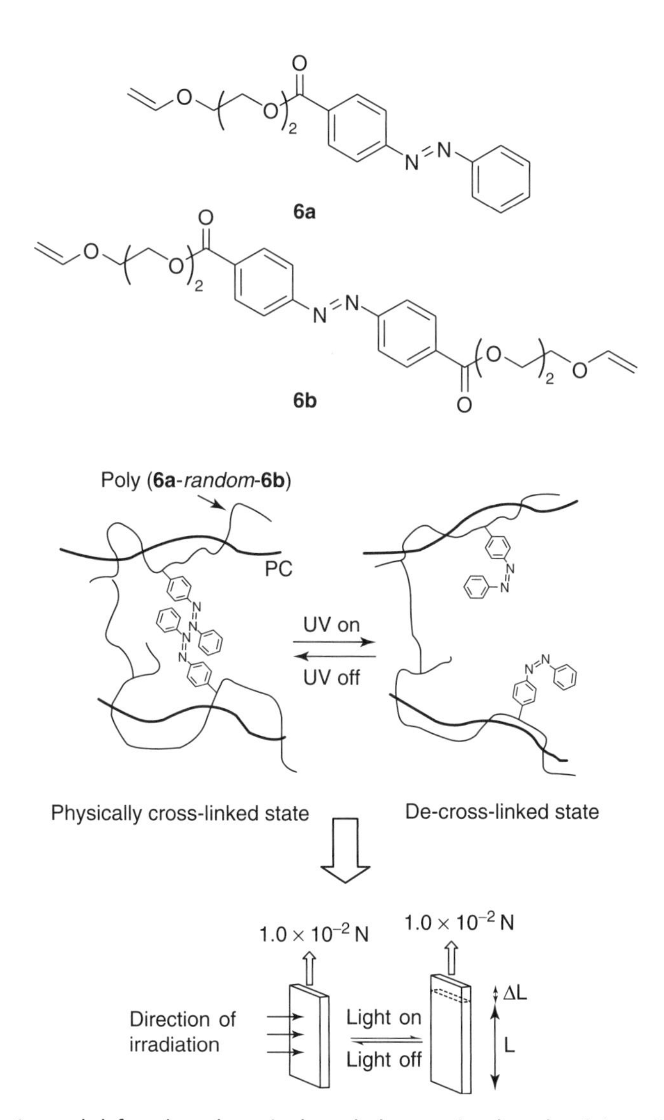

Figure 3.8. A model for the photoinduced change in the elasticity. *Source*: Kim et al., 2005a.

Azobenzene-containing gels and polymer films are especially used for applications; however, in general the gels reported have a disadvantage in that the response is slow and the degree of deformation of the polymer films is too small to be practically used. It is generally agreed now that it is crucial to develop only photomechanical systems that can undergo fast and large deformations. The main challenge in the development of such polymer systems is the conversion of photoinduced effects in some light-reversible functional groups or fillers into a significant macroscopic deformation of the host materials. The gels and polymer films used in the studies described earlier were amorphous, without microscopic or macroscopic order, and thus their deformations are isotropic. If materials with anisotropic physical properties are used, the mechanical power produced could increase significantly, and more control can be realized.

3.4. PHOTODEFORMABLE MATERIALS BASED ON AZOBENZENE-CONTAINING LCs

3.4.1. LCs and LCEs

For a long time it has been known that anisotropic molecules, the so-called mesogenic groups, can form LC phases. The LC phases are thermodynamically stable phases that exist in a temperature range between the crystalline phase and the isotropic melt. The remarkable properties of LC phases stem from their combination of long-range order of crystals with mobility of fluids, such as birefringence, fluidity, and alignment change by external fields. LC polymers are well-known polymer materials that combine the self-organization of the mesogenic groups into the ordered structure of LC phases with some typical polymer properties. For example, above the glass transition temperature (T_g), these polymers are in the LC state and can be ordered by electric or magnetic fields. Cooling down below T_g, the system vitrifies yielding anisotropic glasses. This makes the material suitable for optical applications and for storage devices.

LCEs are unique because of the combination of the anisotropic aspects of the LC phases and the rubber elasticity of polymer networks. Because of the LC properties, mesogens in LCEs show alignment, and the alignment of mesogens can be coupled with polymer network structures. This coupling gives rise to many characteristic properties of LCEs. There are a number of excellent books and review articles on the chemistry, physics, and theory of LCEs (Xie and Zhang, 2005; Warner and Terentjev, 2003; Terentjev, 1999; Brand, 1998; Cotton and Hardouin, 1997; Warner and Terentjev, 1996; Kelly, 1995; Zentel, 1989). LCEs have been a hot topic recently because large deformations of LCE materials can be induced by changing the alignment of mesogens in LCEs by external stimuli such as electric fields, changes in temperature, and light. These characteristics make LCEs extremely useful as raw materials for artificial muscles and soft actuators.

LCEs are usually lightly cross-linked networks. The cross-linking density is known to have a great influence on the macroscopic properties and the phase structures (Küpfer et al., 1994; Finkelmann and Rehage, 1984). The mobility of chain segments is decreased with an increase in the cross-linking points, and consequently the mobility of mesogens in the vicinity of a cross-link is suppressed. A cross-link is recognized as a defect in the LC structure, and an increase in the cross-linking density produces an increasing number of defects. Therefore, LC polymers with a high cross-linking density are referred to as LC thermosetting polymers (duromers), which are distinguished from LCEs.

3.4.2. General Methods of Preparation of LCEs

The concept of LCEs was first proposed by de Gennes (1975), and the first example of an LCE was prepared by Finkelmann et al. (1981). Since then, a variety of LCEs with various structures of main chains and various kinds of mesogens have been prepared. Generally, there are two methods of preparation of LCEs: one is the two-step method (Küpfer and Finkelmann, 1994; Zentel and Benalia, 1987; Zentel and Reckert, 1986), and the other is the one-step method (Broer et al., 1988). In the former method, well-defined weak networks are synthesized in the first step. These networks are then deformed with a constant load to induce network anisotropy. In the second reaction step, cross-linking reactions fix the network anisotropy. The advantage of this method is that the induced network anisotropy in the first step is reproducible, so that well-aligned elastomers are obtained (Küpfer and Finkelmann, 1994; Zentel and Benalia, 1987; Zentel and Reckert, 1986). Uniformly aligned mesogenic monomers with two reactive groups or prepolymers with reactive groups can be photochemically or thermally polymerized or cross-linked by nonmesogenic cross-linking agents to give macroscopically aligned LCEs and anisotropic LC networks with different cross-linking densities, in which the macroscopic orientation of the LC states in the solid samples is fixed (Fig. 3.9; Küpfer and Finkelmann, 1991). Broer et al. (1988) developed the one-step method to prepare highly oriented LC side-chain polymers, namely, the *in situ* photopolymerization of macroscopically aligned LC monomers. Highly ordered polymers may be obtained by polymerizing ordered LC monomers.

LCEs with various LC phases, such as nematic (Tammer et al., 2005; Kundler and Finkelmann, 1998), cholesteric (He et al., 2005; Meng et al., 2005; Finkelmann et al., 2001), smectic (Beyer and Zentel, 2005; Hiraoka et al., 2005; Zanna et al., 2002; Lehmann et al., 2001; Brodowsky et al., 1999; Nishikawa and Finkelmann, 1999; Gebhard and Zentel, 1998; Nishikawa and Finkelmann, 1997; Brehmer and Zentel, 1994) , and discotic (Disch et al., 1995) were prepared by polymerization of various LC monomers containing more than one polymerizable group. For example, a cholesteric LCE (Components **7a–d**) with a spontaneous and uniform alignment in the form of a helical structure was synthesized (Finkelmann et al., 2001). Dye-doped cholesteric LCEs can act as mirrorless lasers, in which the wavelength of the laser emission can be tuned by external mechanical deformation.

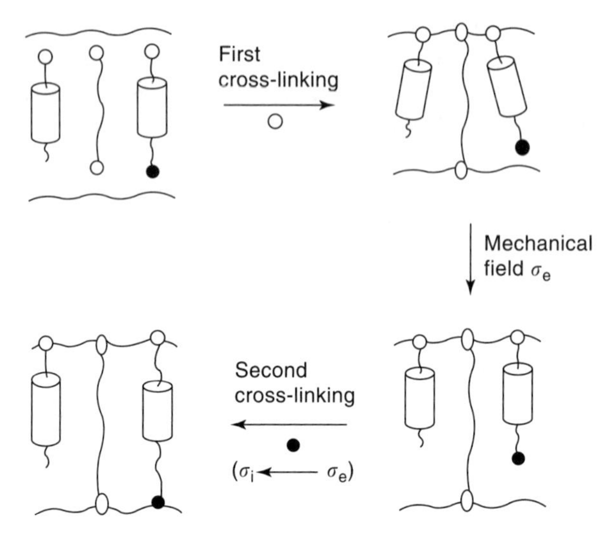

Figure 3.9. Preparation of LCEs by the two-step method. *Source*: Küpfer and Finkelmann, 1994.

A discotic LCE containing triphenylene group was also synthesized by the two-step process (Disch et al., 1995). Samples with a chemically fixed macroscopic alignment of the director (monodomains) was prepared by application of a uniaxial mechanical field during the synthesis of the LCEs.

7a: 100 mol%

7b: 10 mol%

7c: 18 mol%

7d: 72 mol%

Broer et al. (1989) investigated the alignment of a mesogen in a diacrylate monomer before and after photopolymerization. Before polymerization, the values of birefringence and the order parameter decreased with an increase in temperature. With the progress of polymerization, the polymer main chains that formed prevented close packing of mesogens, which resulted in a decrease in birefringence when the alignment of mesogens before polymerization was high; in contrast, the polymer chains enhanced the packing of mesogens when the alignment of mesogens before polymerization was low.

Serrano and coworkers studied the photopolymerization of mixtures of mono- and difunctional LC monomers to yield polymer materials with different degrees of cross-linking (Sánchez et al., 1999). Oriented films can be prepared by controlling the alignment of the monomers. The birefringence of photopolymerizable samples and polymer films, as well as its temperature dependence, was evaluated. The birefringence of the films increased as the content of the monofunctional LC monomer increased.

Furthermore, LCEs have been prepared by block copolymerization and hydrogen bonds (Cui et al., 2004; Li et al., 2004). Li et al. (2004) proposed a musclelike material with a lamellar structure based on a nematic triblock copolymer (Components **8a–c**, Fig. 3.10). The material consists of a repeated series of nematic (N) polymer blocks and conventional rubber (R) blocks. The synthesis of block copolymers with well-defined structures and narrow molecular-weight distributions is a crucial step in the production of artificial muscle based on triblock elastomers. Talroze and coworkers studied the structure and the alignment behavior of LC networks stabilized by hydrogen bonds under mechanical stress (Shandryuk et al., 2003). They synthesized poly[4-(6-acryloyloxyhexyloxy)benzoic acid], which

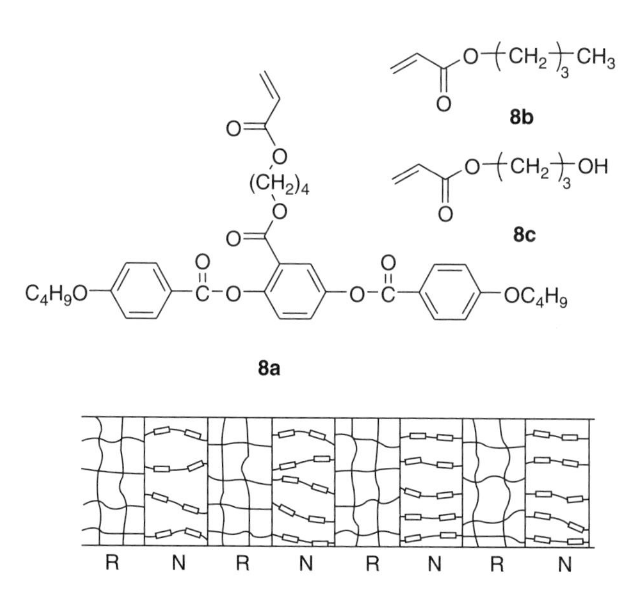

Figure 3.10. Chemical structures of a side-on nematic monomer (8a) and a striated artificial muscle with a lamellar structure based on a nematic triblock copolymer, RNR (N: nematic; R: rubber). *Source*: Li et al., 2004.

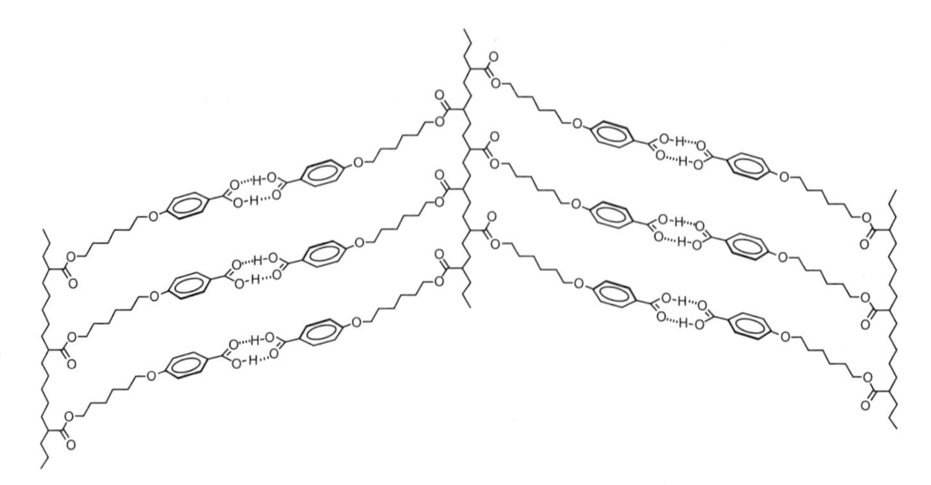

Figure 3.11. Plausible structure of polymer network based on H-bonded rodlike dimers formed by carboxyl monomer units incorporated in the polymer side chains.

exhibits a smectic LC phase over a broad temperature range. Addition of up to 10 mol% of any LMW benzoic acid and, as a result, the formation of mixed dimers do not influence the smectic C_A structure (Fig. 3.11). The amorphous azopyridine polymer can easily be converted into LC polymers through hydrogen bonding with a series of commercially available aliphatic and aromatic carboxylic acids (Cui and Zhao, 2004). The pure acids have only a crystal phase that melts at high temperatures; after mixing them with the azopyridine polymer, the complex that formed showed a new LC phase. This observation is strong evidence for the formation of hydrogen-bonded complexes, because neither the pure acids nor the azopyridine polymer shows any LC phase.

3.4.3. Temperature-/Electricity-/pH-Responsive LCEs

De Gennes (1975) proposed the possibility of using LCEs as artificial muscles, by taking advantage of their substantial uniaxial contraction in the direction of the director axis. The alignment of mesogens can be coupled with polymer network structures in LCEs; thus, if one heats nematic LCE films toward the nematic–isotropic phase transition temperature, the nematic order will decrease, and when the phase transition temperature is exceeded, one observes a disordered state of mesogens. Through this phase transition, the LCE films show a general contraction along the alignment direction of the mesogens, and if the temperature is lowered back below the phase transition temperature, the LCE films revert to their original size by expanding. This anisotropic deformation of the LCE films can be very large, and along with good mechanical properties, thus providing the LCE materials with promising properties as artificial muscles.

Anisotropic deformation of monodomain LCEs by a thermal phase transition from an LC to an isotropic state was first reported by Küpfer and Finkelmann (1994). A nematic LCE prepared by the two-step process contracted by about 26% owing to the change in order parameter as a result of the change in molecular alignment of mesogens. This anisotropic deformation behavior of LCEs has been a subject of extensive experimental and theoretical studies (Stenull and Lubensky, 2005; Mao et al., 2001; Warner et al., 1988). Warner and Terentjev established a relation between the nematic order parameter S and the effective backbone anisotropy of polymer chains forming the rubbery network. The relation is expressed by a dimensionless ratio of the principal step lengths parallel and perpendicular to the nematic director ($R_{\parallel}/R_{\perp}$) (Warner and Terentjev, 2003, 1996). In the nematic phase, this ratio is larger than unity ($\boldsymbol{R}_{\parallel}/\boldsymbol{R}_{\perp} \approx 1.3$) (Warner and Terentjev, 1996), but after a nematic–isotropic phase transition, this ratio approaches unity as a result of the formation of a random coil of polymer chains, which makes the polymer material contract along the director axis of LCEs. In the smectic A phase, the ratio $R_{\parallel}/R_{\perp}$ is in general smaller than unity because the polymer chains are likely to exist between the smectic layers (Cotton and Hardouin, 1997).

There have been a number of efforts to develop artificial musclelike materials (Buguin et al., 2006; Mol et al., 2005; Saikrasun et al., 2005; Stenull and Lubensky, 2005; Li et al., 2004; Naciri et al., 2003; Clarke et al., 2001; Mao et al., 2001; Thomsen et al., 2001). LCE films with a splayed or twisted molecular alignment display a well-controlled deformation as a function of temperature (Fig. 3.12) (Mol et al., 2005). The twisted films show a complex macroscopic deformation owing to the formation of saddlelike geometries, whereas the deformation of the splayed structure is smooth and well controlled (Mol et al., 2005). Wermter and Finkelmann (2001) reported a new type of LC coelastomers composed of LC side chains and LC main-chain polymers (Components **10a–c**) as network strands. Thanks to the direct coupling of the LC main-chain segments to the network anisotropy, the thermoelastic response was increased remarkably with an increase of the concentration of these segments. An elongation in the direction of the director by up to a factor of 4, relative to the length of the networks in an isotropic state, was observed in the nematic state (Fig. 3.13) (Wermter and Finkelmann, 2001). Ratna and coworkers prepared LCEs with laterally attached side-chain mesogens (**11**) and studied their thermoelastic properties across a nematic–isotropic phase transition (Thomsen et al., 2001). The LCEs showed a large contraction (strain change) of 35%–40% through the phase transition, and the maximum retractive force generated was 270 kPa, which is comparable to that of a skeletal muscle (Fig. 3.14). This increase in the stress during the isostrain measurements is related to the entropy change of the polymer chains and interpreted as a result of the initial work of the wormlike (prolate) to coil transition in the polymer backbones with laterally coupled mesogenic side chains.

Skeletal muscles are anisotropic, that is, they exhibit contraction and elongation along the fiber axis. Naciri et al. (2003) described a method of preparing LC fibers from a side-chain LC terpolymer containing two side-chain mesogens

9a: R = H, $x = 6$

9b: R = CH_3, $x = 6$

9c: R = CH_3, $x = 3$

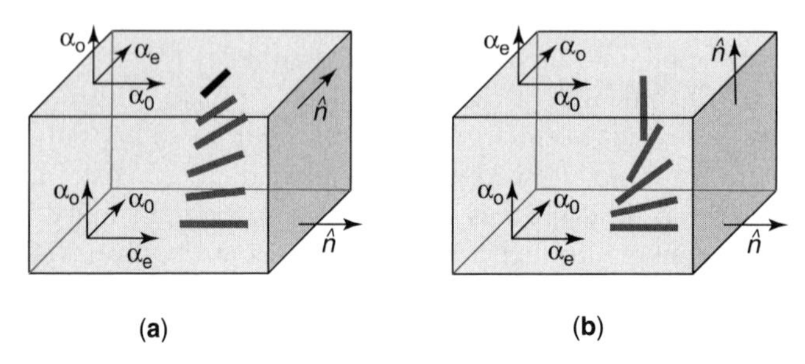

Figure 3.12. LCE films with a (**a**) splayed or (**b**) twisted molecular alignment. α_o and α_e indicate perpendicular and parallel coefficients of thermal expansion. *Source*: Mol et al., 2005.

(Components **11a,b**) and a nonmesogenic group that acts as a reactive site for cross-linking. Fibers were drawn from a melt of the polymer and a cross-linker, 4,4′-methylenebis(phenyl isocyanate) (MDI). The fibers that were formed showed high LC alignment when observed by polarizing optical microscopy. The thermo-elastic response exhibited strain changes through the nematic–isotropic phase transition of ~35%. A retractive force of nearly 300 kPa was also observed in the isotropic phase. Yusuf et al. (2004a,b, 2003) investigated the swelling behavior of the LCE films in LMWLCs. Figure 3.15 shows the shape changes of dry LCEs during heating and cooling processes of planar and homeotropic samples. The planar sample contracted parallel to the director and expanded perpendicular to the director upon heating. The homeotropic sample expanded on heating. All samples reverted to their initial shapes upon cooling. Furthermore, the swelling behavior of both polydomain and monodomain LCE films was studied. Polydomain LCEs swell equally in all three dimensions, whereas monodomain LCEs swell isotropically only in two dimensions, but not in three dimensions.

Another useful property of LCEs is that their shape changes by applied electric fields, electromechanical responses, owing to reorientation effects induced by electric field. Zentel first found tiny changes of LCEs swollen in LMWLCs under large fields in 1986 (Zentel, 1986). Subsequently, Barnes et al. (1989) observed 20% contraction of polydomain elastomers swollen in an isotropic LMWLC. Kishi et al.

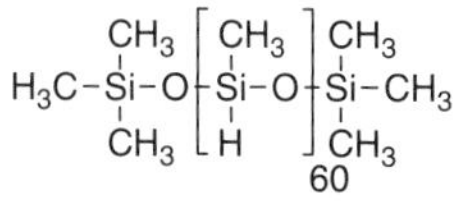

10a

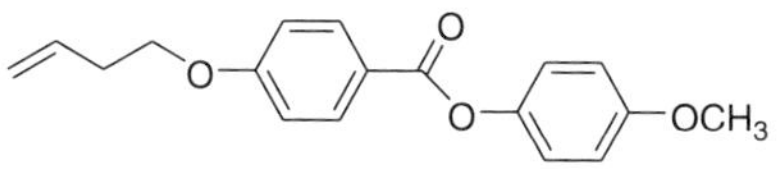

10b

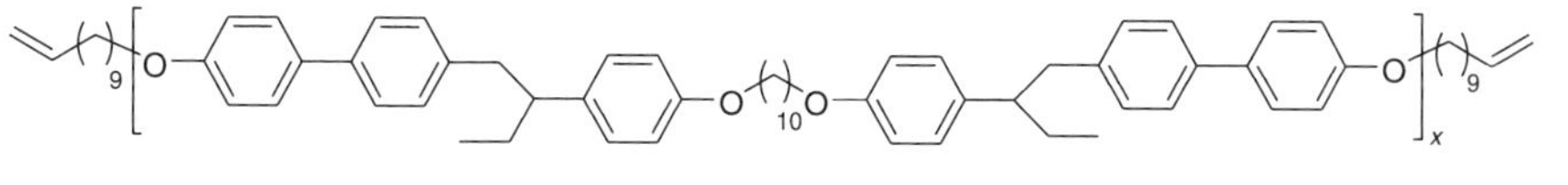

10c

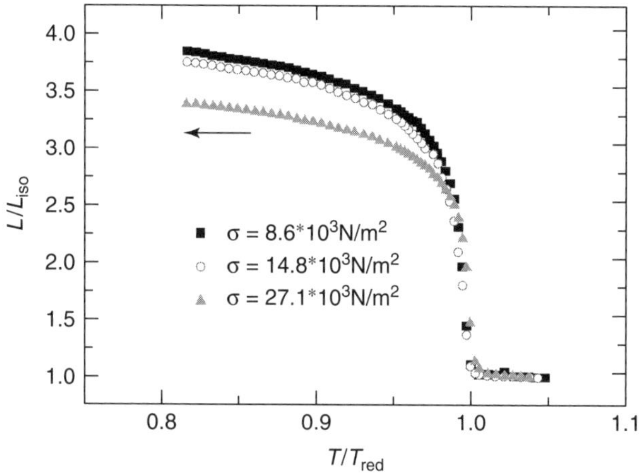

Figure 3.13. Thermoelastic response of an LCE prepared from Components 10a–c. σ: mechanical field; L: length of the network in the nematic state; L_{iso}: length of the network in the isotropic state; T_{red}: reduced temperature. *Source*: Wermter and Finkelmann, 2001.

(1994) then reported quantitative results on shape changes of swollen polydomain LCEs under a dc electric field (0.3 V/μm). Recently, Courty et al. (2003) demonstrated a fairly large electromechanical effect in an LCE embedded with carbon nanotubes, but needed a large applied field (1 V/μm). Furthermore, Yusuf et al. (2005) reported measurable shape changes (a maximum 13% contraction) in LCEs swollen with an LMWLC under small fields (0.01–1 V/μm). Kremer and coworkers developed a new material that showed high and fast strains of 4% by electrostriction under a much lower applied electric field (by two orders of magnitude) than those reported previously. It consists of ultrathin ferroelectric LCE films, which exhibited 4% strain at only 1.5 MV/m (Fig. 3.16; Lehmann et al., 2001). The thinning of the film by 4% in the ferroelectric LCE corresponds to an electrically induced tilt angle (electroclinic effect) of $\theta = 16^\circ$ in the sample cosine model.

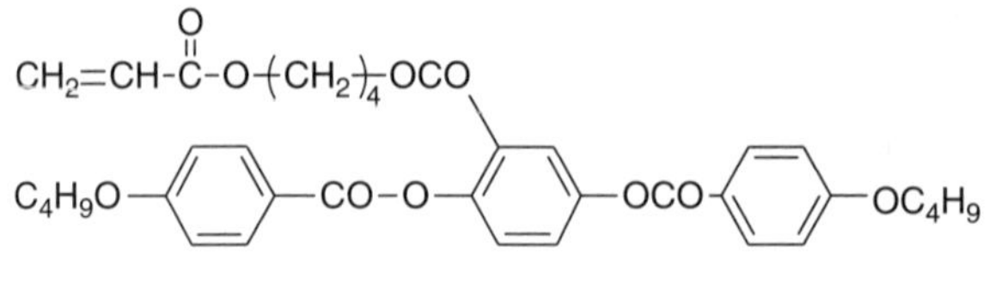

11a

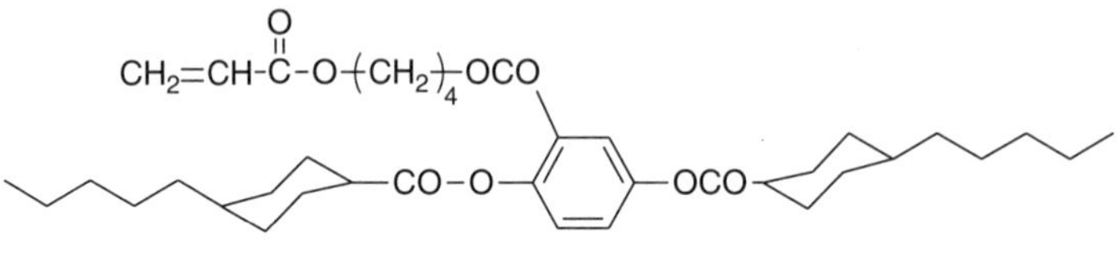

11b

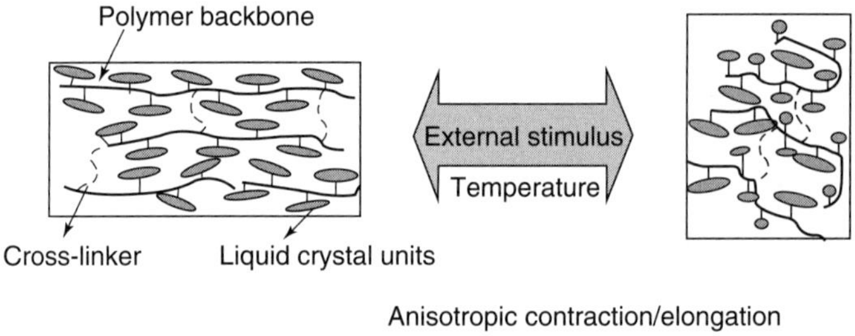

Figure 3.14. LCEs with laterally attached side-chain mesogens and their deformation behavior. *Source*: Naciri et al., 2003.

Broer and coworkers proposed pH- or water-controlled actuators based on two simple concepts: an aligned LC network, consisting of both covalent and secondary bonds, and stimulus-controlled molecular switching between acidic and neutral states (Fig. 3.17; Harris et al., 2005a). The uniaxially aligned film responded equally to water or pH in all regions of the film and only elongated by a small amount when exposed to a uniform stimulus. Differences in the pH value or humidity of the upper and lower surfaces induced a large degree of bending of the film. The twisted and splayed configurations do not require environmental gradients to produce macroscopic motion. In both cases, the preferred expansion directions on the opposite surfaces of a film were offset by 90°, and a uniform stimulus resulted in expansion gradients over the thickness of the film and bending behavior similar to thermal deformation in a metallic bilayer.

3.4.4. Photoresponsive Behavior of Chromophore-Containing LCs

3.4.4.1. Photochemical Phase Transitions of LCs. Cooperative motion of molecules in LC phases may be most advantageous in changing the molecular alignment by external stimuli. If a small portion of LC molecules change their

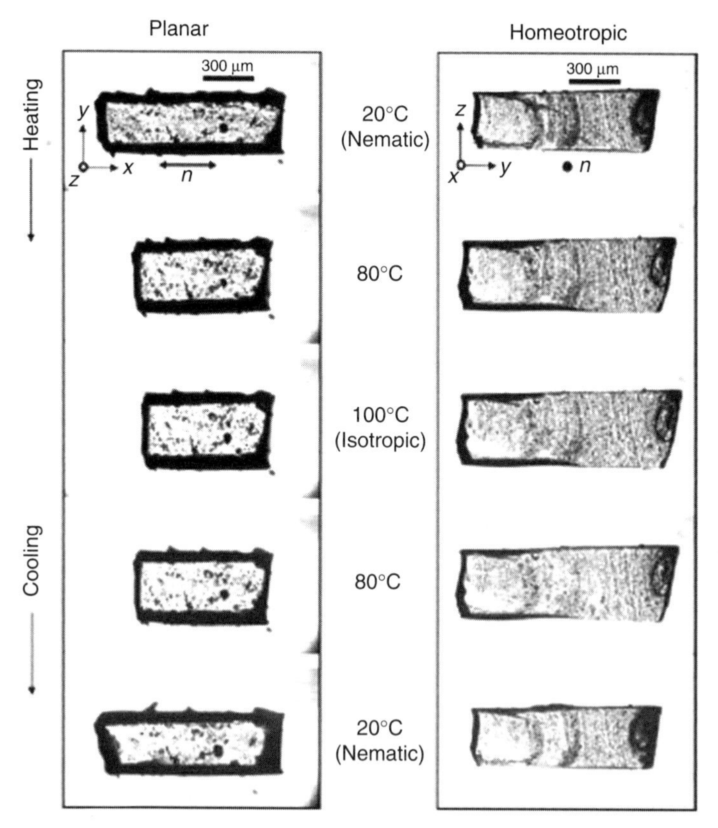

Figure 3.15. Shape change of an LCE (planar and homeotropic samples) as a function of temperature. *Source*: Yusuf et al., 2004a.

alignment in response to an external stimulus, the other LC molecules also change their alignment. The energy required to induce an alignment change of only 1 mol% of the LC molecules is enough to bring about the alignment change of the whole system. In other words, a huge amplification is possible in LC systems. When a small amount of a photochromic molecule, such as an azobenzene, stilbene, spiropyran, and fulgide derivative, is added to LCs and the resulting guest/host mixture is irradiated to cause photochemical reactions of the photochromic guest molecules, an LC to isotropic phase transition of the mixture can be induced isothermally. Tazuke et al. (1987) reported the first explicit example of a nematic–isotropic phase transition induced by trans–cis photoisomerization of an azobenzene guest molecule dispersed in a nematic LC.

The photochemically induced phase transitions ("photochemical phase transitions") are interpreted in terms of a change in the phase transition temperature of LC systems upon accumulation of one isomer of the photochromic guest molecule (Sung et al., 2002; Legge and Mitchell, 1992). The trans form of the

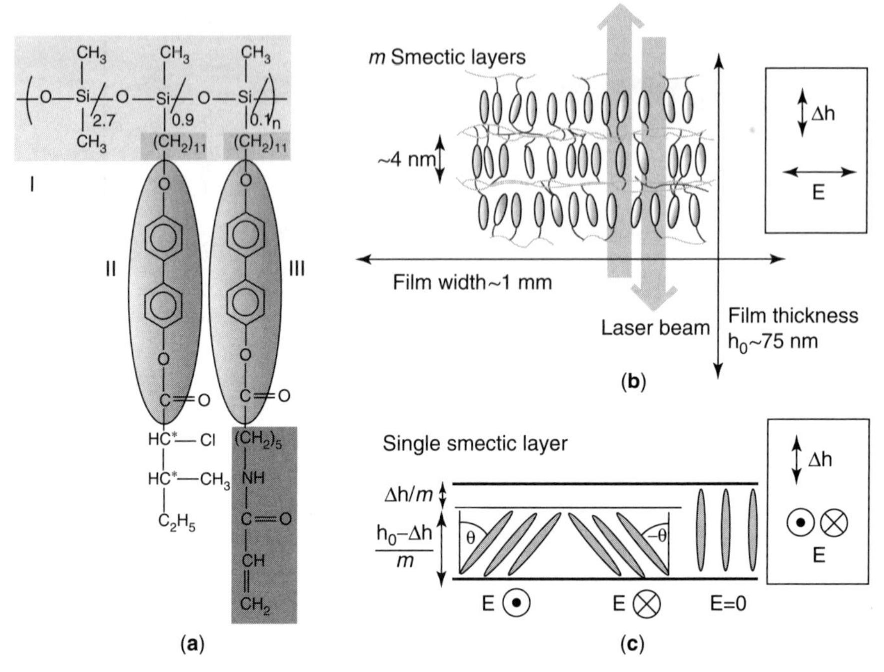

Figure 3.16. Electroclinic effect in ferroelectric LCEs: **(a)** Chemical structure of sample. **(b)** Measurement geometry: the beam in the interferometer passes twice through the film to measure the electrically induced thickness modulation. **(c)** The viewing angle is turned by 90° around the layer normal compared with that in **(b)**. *Source*: Lehmann et al., 2001.

azobenzenes, for example, has a rodlike shape, which stabilizes the phase structure of the LC phase, whereas its cis isomer is bent and tends to destabilize the phase structure of the mixture. As a result, the LC–isotropic phase transition temperature (T_c) of the mixture with the cis form (T_{cc}) is much lower than that with the trans form (T_{ct}) (Fig. 3.18). If the temperature of the sample (T) is set at a temperature between T_{ct} and T_{cc}, and the sample is irradiated to cause trans–cis photoisomerization of the azobenzene guest molecules, T_c decreases with an accumulation of the cis form. When T_c becomes lower than the irradiation temperature T, an LC–isotropic phase transition of the sample is induced. Photochromic reactions are usually reversible, so the sample reverts to the initial LC phase through cis–trans back-isomerization. Thus, phase transitions of LC systems can be induced isothermally and reversibly by photochemical reactions of photoresponsive guest molecules (Fig. 3.18).

3.4.4.2. Photochemical Flip of Polarization and Alignment Change in Ferroelectric LCs. Photochromic reactions can also strongly influence the phase structures of various LCs. Ferroelectric LCs exhibit spontaneous

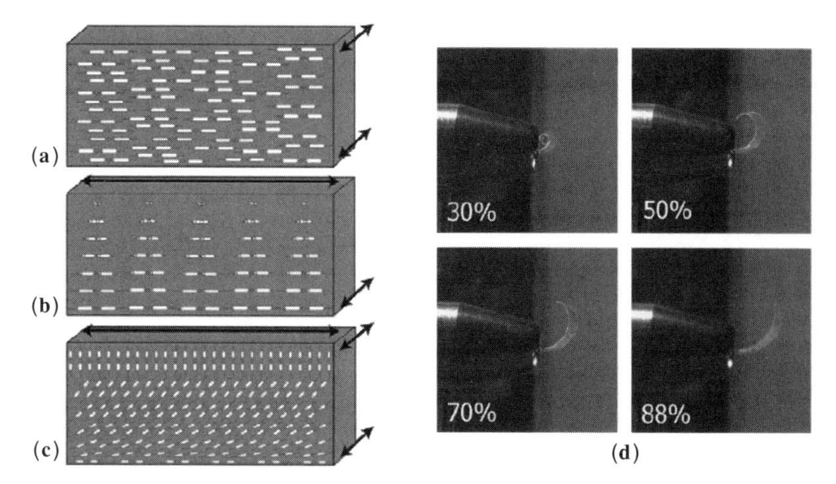

Figure 3.17. Director-orientation configurations for bending motion: (a) uniaxially, (b) twisted, and (c) splayed aligned network film. The arrows indicate the preferred expansion directions. (d) Photographs of the twisted film under the relative humidity conditions. *Source*: Harris et al., 2005a.

polarization (P_s) and show microsecond responses to changes in applied electric fields in a surface-stabilized state (flip of polarization) (Fukuda and Takezoe, 1990). When a mixture of an azobenzene (**12a**; 3 mol%) and a ferroelectric LC (**12b**; Fig. 3.19) prepared in a very thin LC cell in a surface-stabilized state was irradiated with UV light ($\lambda = 366$ nm), trans–cis photoisomerization of the azobenzene molecule led the threshold electric field for the flip of polarization (coercive force, E_c) of ferroelectric LCs to change upon photoirradiation (Ikeda et al., 1993). Ferroelectric LCs in the surface-stabilized state show a hysteresis between the applied electric field and polarization (Fukuda and Takezoe, 1990), and the hysteresis of the *trans*-azobenzene/ferroelectric LC mixture was different from that of the *cis*-azobenzene/ferroelectric LC. This effect of molecular shape on the coercive force is very similar to that described earlier for the different T_c values of the azobenzene/nematic LC mixtures. The phase structure of the chiral smectic C phase of the ferroelectric LCs is not disorganized significantly in the azobenzene/ferroelectric LC mixture with the azobenzene in the trans form, whereas the phase structure of the smectic C* phase is seriously affected when the azobenzene is in the cis form. The threshold value for the flip of polarization is reduced significantly as a result of the decrease in order. In view of these properties, a new mode of optical switching of ferroelectric LCs (photochemical flip of polarization of ferroelectric LCs) has been proposed (Fig. 3.19; Sasaki et al., 1994; Ikeda et al., 1993). A flip of polarization induced at the irradiated sites can lead to a change in alignment of the ferroelectric LCs. These changes in polarization and alignment of ferroelectric LCs produce an optical contrast between the irradiated and non-irradiated sites, which remain unchanged (memory effect).

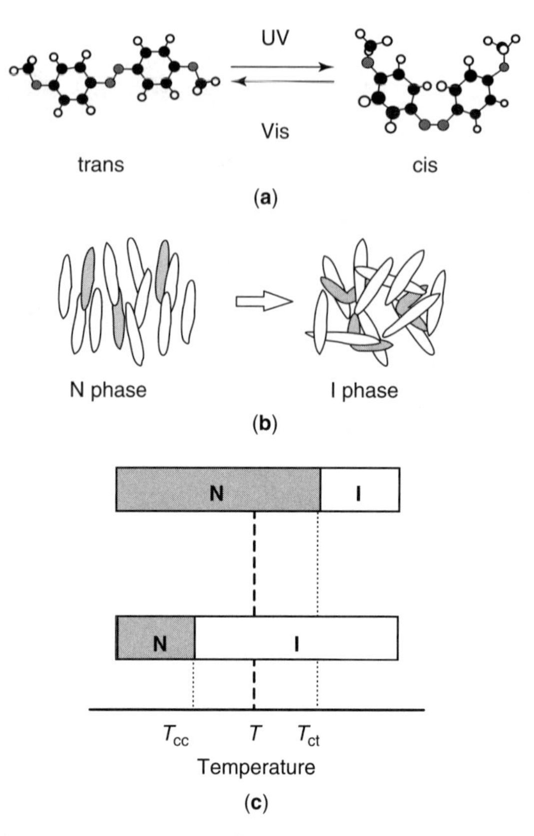

Figure 3.18. (**a**) Photoisomerization of 4,4′-disubstituted azobenzene derivative and (**b, c**) phase diagrams of the photochemical phase transition of azobenzene/LC systems (N, nematic; I, isotropic). *Source*: Ikeda, 2003.

Various effects on the flip of polarization in the photochromic guest/ferroelectric LC host systems have been investigated in detail, including the structure of ferroelectric LC hosts (Sasaki and Ikeda, 1995a, 1993), the structure of photochromic guests (Sasaki and Ikeda, 1995b), temperature (Sasaki and Ikeda, 1995c), bias voltage (Sasaki et al., 1994), and the change in spontaneous polarization (Sasaki and Ikeda, 1995c). The photochemical flip of polarization was also examined in antiferroelectric LCs. It was found that the flip of polarization can be induced similarly in the azobenzene/antiferroelectric LC mixtures upon irradiation to cause trans–cis photoisomerization of the guest molecule (Moriyama et al., 1993).

Much attention has been paid to molecular design of effective guest molecules to induce the photochemical flip of polarization in ferroelectric LC systems. An azobenzene derivative with a chiral cyclic carbonate group (**13**) was designed to induce a large value of polarization and examined as a chiral dopant to induce a

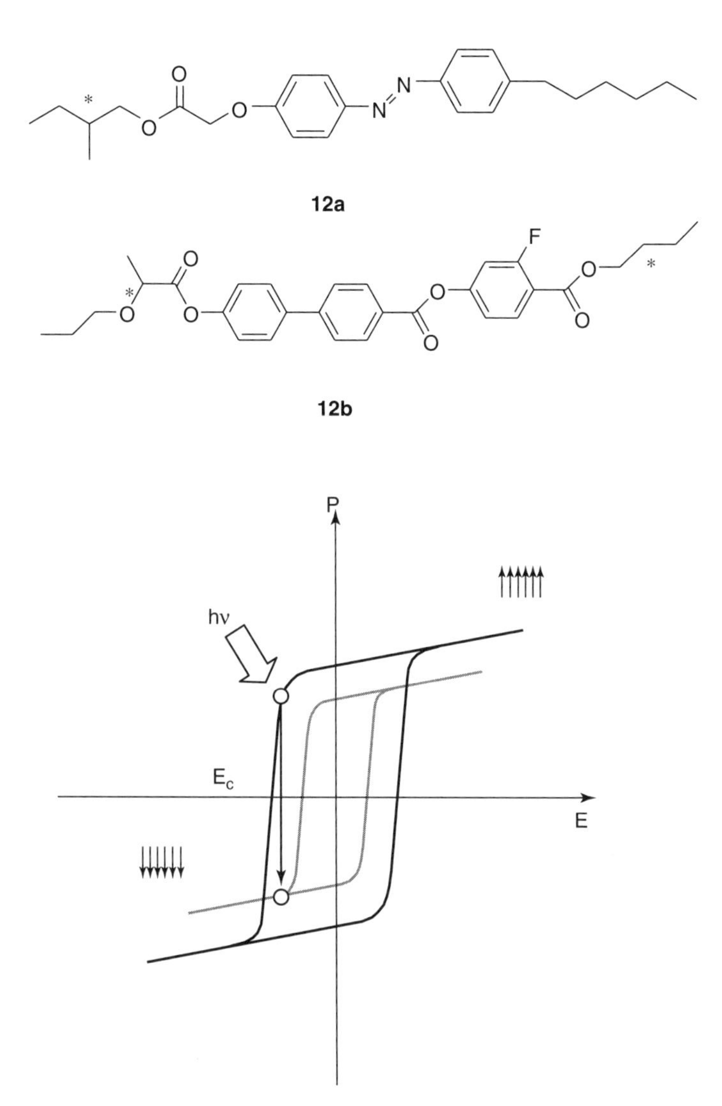

Figure 3.19. Photochemical flip of polarization in ferroelectric LCs. *Source*: Sasaki et al., 1994.

smectic C* phase (Negishi et al., 1996a). In this system, the chiral dopant also acts as a photoresponsive molecule; therefore, a change in molecular shape of the dopant is expected to affect significantly the phase structure of the smectic C* phase because the molecular shape of the dopant is crucial for the induction of the smectic C* phase. The azobenzene with a cyclic carbonate is quite effective at inducing the photochemical flip of polarization of the ferroelectric LC mixture (Negishi et al., 1996a). Furthermore, azobenzene derivatives that exhibit antiferroelectric properties were developed, and their photoresponsive behavior was examined (**14a**, **b**) (Negishi et al., 1996b). The photochemical flip of polarization in these antiferroelectric LCs is induced very effectively (Negishi et al., 1996b), and a device fabricated with these antiferroelectric LCs was explored (Shirota and Yamaguchi, 1997).

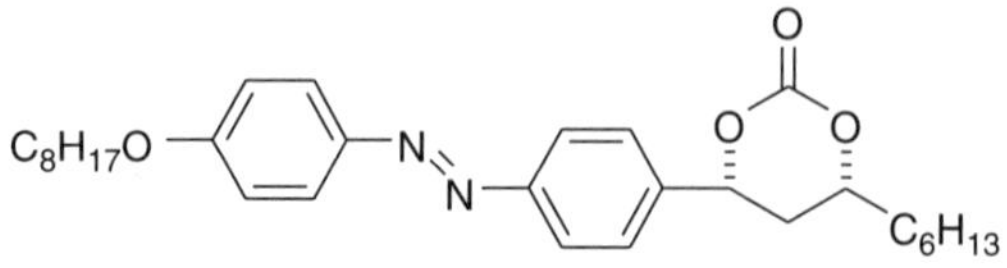

13

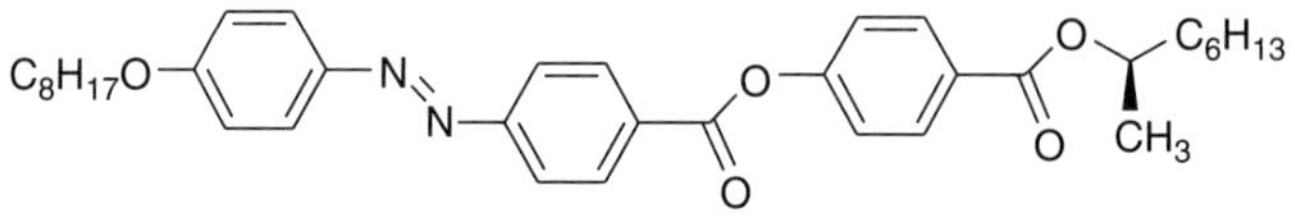

14a

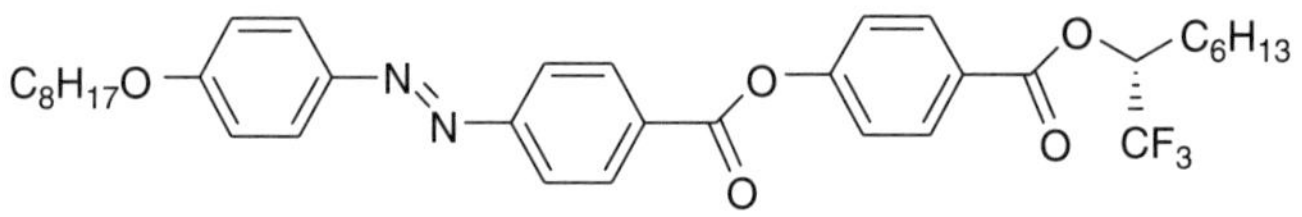

14b

3.4.4.3. Photochemical Phase Transition of Azobenzene LC Polymers. LC polymers possess properties of both polymers and LCs and currently are regarded as promising photonic materials because of their advantageous properties mentioned earlier. Wendorff and coworkers reported the first example of a photon-mode photoresponse in LC polymers, namely, holographic recording in LC polymer films containing azobenzene moieties and mesogenic groups (Eich et al., 1987; Eich and Wendorff, 1987). Ikeda et al. (1990a,b, 1988) reported the first example of the photochemical phase transition in LC polymers; they demonstrated that by irradiation of LC polymers doped with LMW azobenzene derivatives with UV light to cause trans–cis isomerization led to a nematic–isotropic phase transition; upon cis–trans back-isomerization, the LC polymers reverted to the initial nematic phase. Although doping the chromophores in a matrix is most convenient, the resultant LC polymer systems often exhibit instabilities, such as phase separation and microcrystallization. This occurs because of the mobility of the azobenzene chromophores in the matrix and the propensity of the dipolar azobenzene units to form aggregates. Higher quality LC polymer systems are obtained when the azobenzene moiety is covalently bound to the host polymer matrix, and this approach is now preferable for most applications. These materials combine the stability and processability of LC polymers, with the unusual photoresponsive behavior of the azobenzene groups. A variety of

LC copolymers were prepared, and their photochemical phase transition behavior was examined (Tsutsumi et al., 1998a; Sasaki et al., 1992; Ikeda et al., 1990b,c). One of the important factors of the photoresponsive LCs is their responses to optical stimuli. In this respect, the response time of the photochemical phase transition has been explored by time-resolved measurements (Sasaki et al., 1992; Ikeda et al., 1991; Kurihara et al., 1990).

Photochromic reactions are in general very fast, occurring on a timescale of picoseconds. If an ultrafast laser with a pulse width on the picosecond scale is used as an excitation light source to induce phase transitions of LC systems containing photochromic molecules, photochemical reactions of the photochromic molecules can be completed in picoseconds, and the T_c of the system can be decreased below the irradiation temperature on this timescale. This means that immediately after pulse irradiation, a nonequilibrium state is produced, which is thermodynamically an isotropic phase in its equilibrium state but shows an anisotropy because orientational relaxation of mesogens is not completed. The response of the whole LC systems depends strongly on this orientational relaxation of the mesogens, which in fact is the rate-determining step, especially in highly viscous LC polymers that require a relatively long time (Ikeda, 2003).

A new system has been developed in which every mesogen in the LC or LC polymer is photosensitive (Ikeda, 2001; Tsutsumi et al., 1998b; Tsutsumi et al., 1997; Ikeda and Tsutsumi, 1995). For example, the azobenzene moiety could play roles as both a mesogen and a photosensitive moiety (Fig. 3.20) in azobenzene derivatives that form LC phases. These azobenzene LCs show a stable LC phase only when the azobenzene moiety is in the trans form, but do not show an LC phase when all the azobenzene moieties are in the cis form. Examination of these azobenzene LCs revealed that a nematic–isotropic phase transition is induced in the azobenzene LC polymers within 200 ns over a wide temperature range under optimized conditions (Ikeda, 2001).

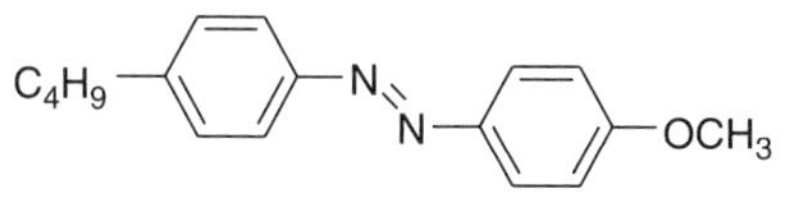

C_4H_9 N N OCH_3 **15**

K 35 N 48 I

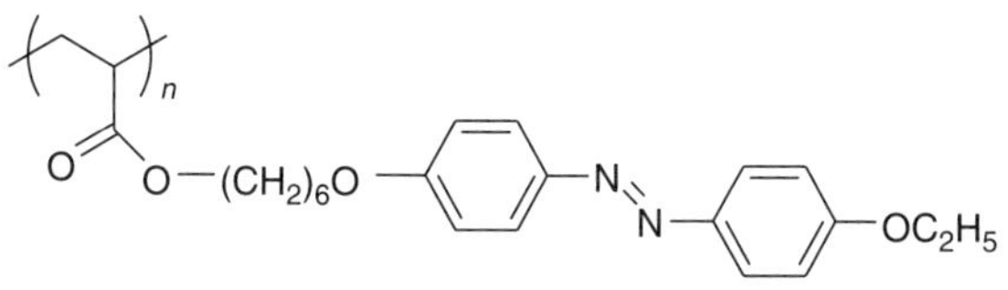

n O O $(CH_2)_6O$ N N OC_2H_5 **16**

G 45 N 150 I

Figure 3.20. Examples of azobenzene LCs.

3.4.5. Light-Responsive LCEs

3.4.5.1. Photoinduced Contraction and Expansion of LCEs. Light as a stimulus enables noncontact activation at ambient temperature. In this way, external heating, which is unfavorable for certain applications and is used to stimulate conventional deformable polymers, can be avoided; thus, decades of intensive studies as aforementioned have focused on the discovery of photoresponsive polymer systems that exhibit fast and large macroscopic shape changes.

As described in Section 3.4.3, LCEs show thermoelastic properties: upon transition from the nematic to the isotropic phase, they contract along the alignment direction of mesogens, and upon cooling below the phase transition temperature, they expand. If this property of LCEs is combined with photochemical phase transition (or photochemically induced decrease of nematic order), it is expected that deformation of LCEs can be induced by light (Cviklinski et al., 2002; Hogan et al., 2002; Finkelmann and Nishikawa, 2001). In fact, Finkelmann and coworkers succeeded in inducing a contraction by 20% in an azobenzene-containing LCE (Components **17a–f**) upon exposure to UV light to cause the trans–cis isomerization of the azobenzene moiety (Fig. 3.21) (Finkelmann and Nishikawa, 2001). They synthesized monodomain nematic LCEs with polysiloxane main chains and azobenzene chromophores at the cross-links. From the viewpoint of photomechanical effects, the subtle variation in nematic order upon trans–cis isomerization causes a significant uniaxial deformation of the LCs along the director axis when the LC molecules are strongly associated by covalent cross-linking to form a three-dimensional polymer network. Terentjev and coworkers incorporated a wide range of azobenzene derivatives into LCEs as photoresponsive moieties and examined their deformation behavior upon exposure to UV light (Cviklinski et al., 2002; Hogan et al., 2002).

Keller and coworkers synthesized monodomain nematic side-on elastomers containing azobenzenes (Components **18a,b**) by photopolymerization with a near-infrared photoinitiator (Li et al., 2003). The photopolymerization was performed with aligned nematic azobenzene monomers in conventional LC cells. Thin films of these LCEs showed fast (less than 1 min) photochemical contraction of up to 18% upon irradiation with UV light and a slow thermal back-reaction in the dark (Fig. 3.22).

3.4.5.2. Photoinduced Three-Dimensional Movements of LCEs. Two-dimensional movements of LCE films have since been demonstrated, and many three-dimensional examples have been envisaged and are under preparation. Ikeda and coworkers were the first to report photoinduced bending behavior of macroscopic LC gels (Ikeda et al., 2003) and LCEs (Mamiya et al., 2008; Yamada et al., 2008; Kondo et al., 2007; Yu et al., 2007; Kondo et al., 2006; Yu et al., 2004a,b; Ikeda et al., 2003; Yu et al., 2003) containing azobenzenes. In comparison with a two-dimensional contraction or expansion, the bending mode, a full three-dimensional movement, could be advantageous for a variety of real manipulation applications. Figure 3.23 shows the bending and unbending

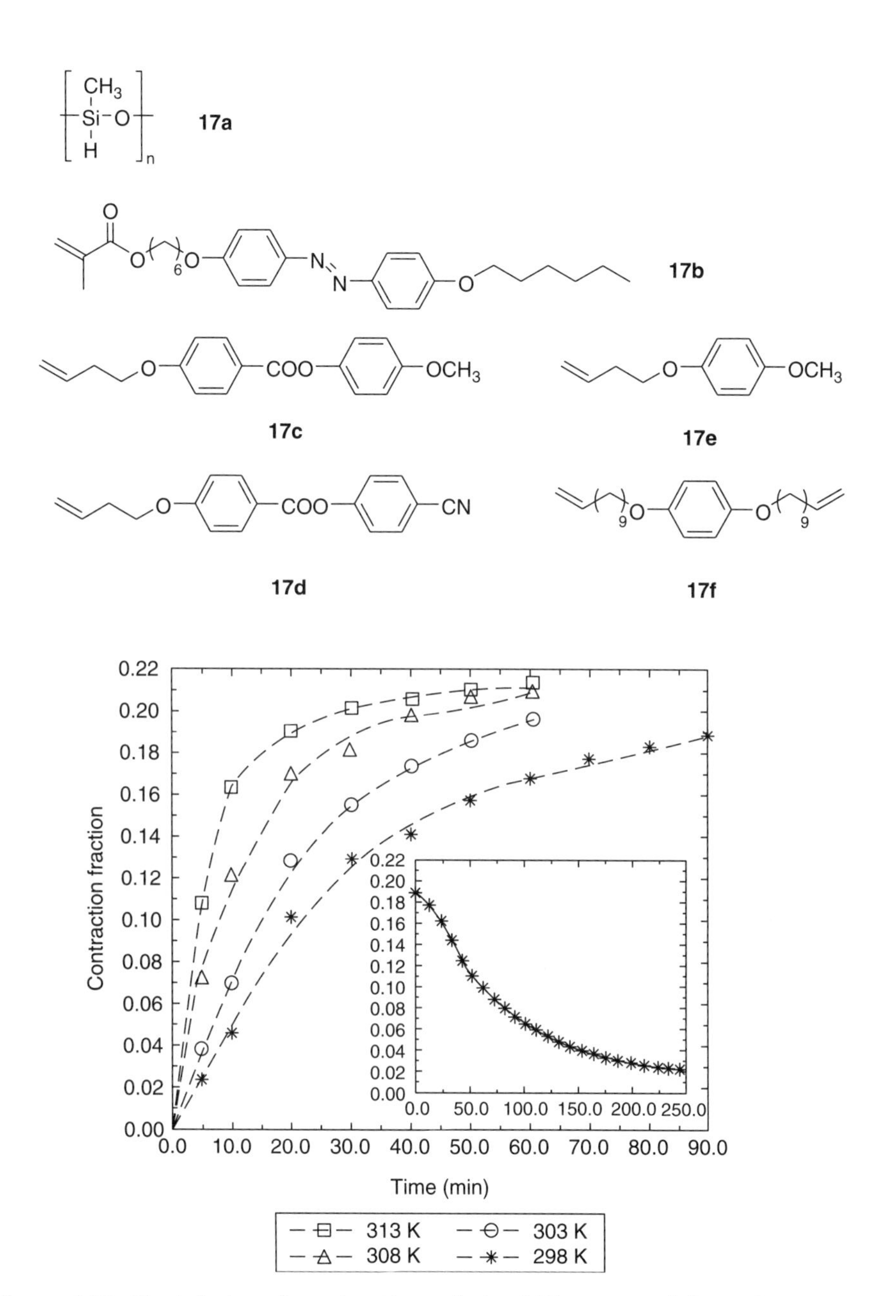

Figure 3.21. Photoinduced contraction of the LCE prepared from Components **17a–f**. (*Inset*) Recovery of the contracted LCE at 298 K after irradiation was switched off. *Source*: Finkelmann and Nishikawa, 2001.

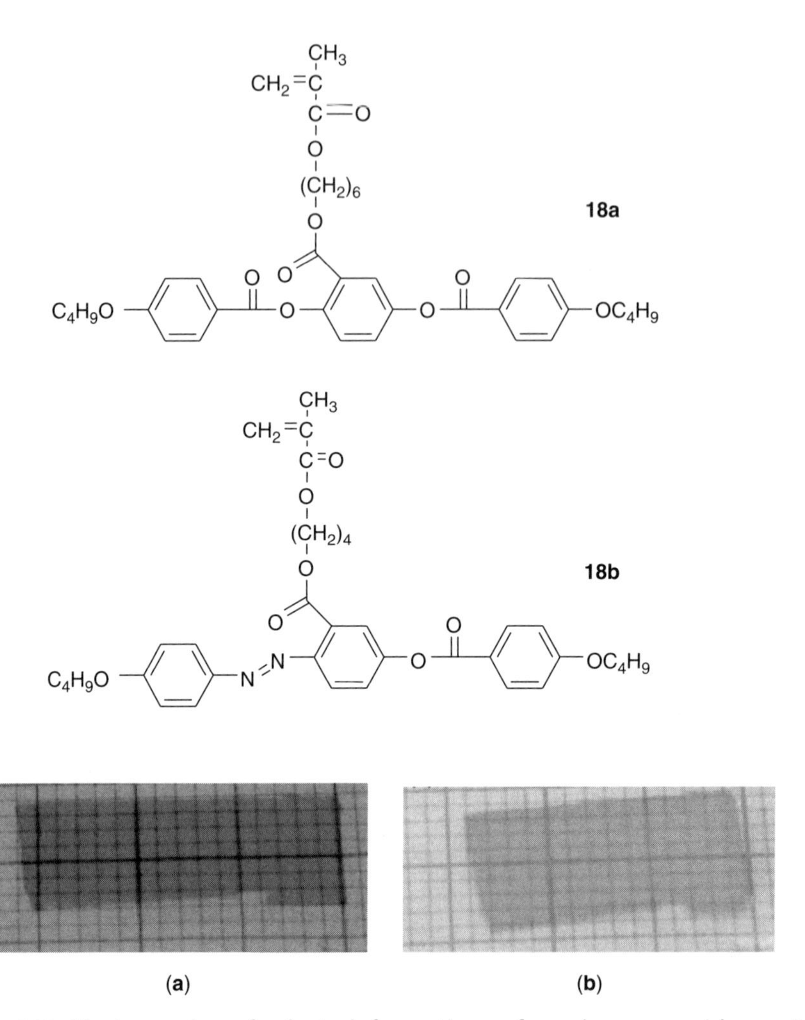

Figure 3.22. Photographs of photodeformation of azobenzene side-on LCE (**a**) before irradiation (**b**) and under irradiation with UV light. *Source*: Li et al., 2003.

processes induced by irradiation with UV and visible light, respectively. The monodomain LCE film bent along the rubbing direction toward the irradiation source, and the bent film reverted to the initial flat state upon exposure to visible light. This bending and unbending behavior was reversible and could be controlled by simply changing the wavelength of the incident light. Furthermore, when the film was rotated at 90°, the bending was again observed along the rubbing direction. These results demonstrate that the bending is anisotropically induced and occurs only along the rubbing direction of the alignment layers.

Irradiation with UV light gives rise to the trans–cis isomerization of azobenzene moieties and thus destabilization of the nematic phase (decrease in

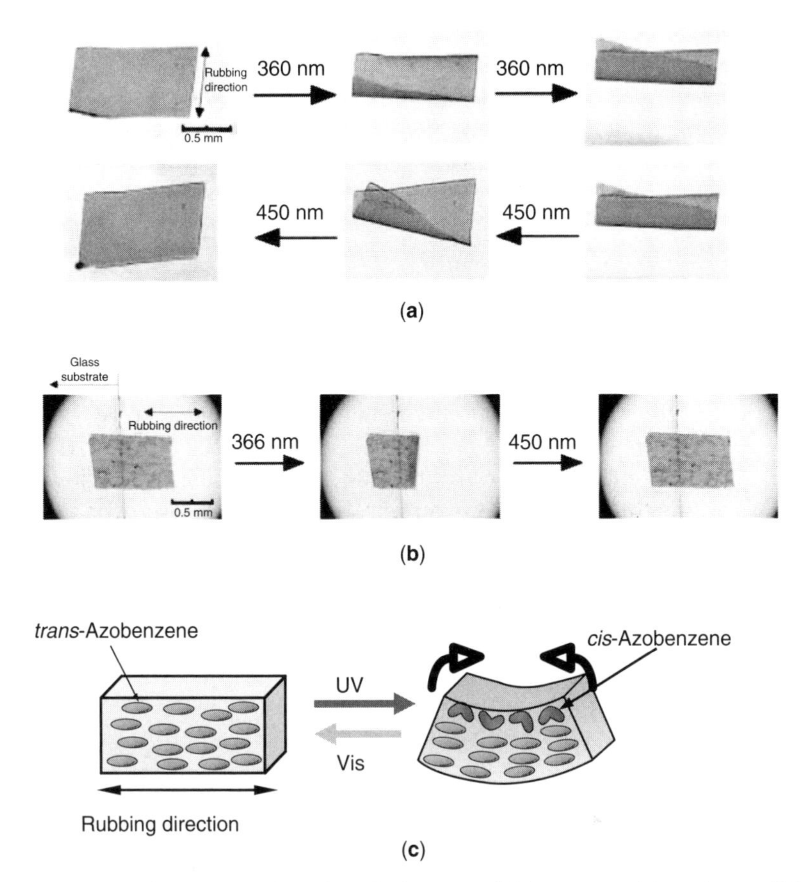

Figure 3.23. Bending and unbending behavior of (**a**) an LC gel in toluene (**b**) and an LCE film in air. (**c**) Plausible mechanism of the photoinduced bending of LCE films. *Source*: Ikeda et al., 2003.

nematic order) even as far as to produce a nematic–isotropic phase transition of the LC systems, as mentioned in Section 3.4.4. However, the extinction coefficient of the azobenzene moieties at around 360 nm is large and thus more than 99% of the incident photons are absorbed by a surface with a thickness less than 1 μm. As the thickness of the films used was 20 μm, the decrease in nematic order occurs only in the surface region facing the incident light, but in the bulk of the film, the *trans*-azobenzene moieties remain unchanged. As a result, the volume contraction is generated only in the surface layer, thus causing the film to bend toward the irradiation source (Fig. 3.23). Furthermore, the azobenzene moieties are preferentially aligned along the rubbing direction of the alignment layers, and the decrease in the alignment order of the azobenzene moieties is thus produced only along this direction, which contributes to the anisotropic bending behavior.

Monodomain LCE films with different cross-linking densities were prepared by copolymerization of Components **19a** and **19b** (Yu et al., 2004b). The films showed the same bending behavior, but the maximum extents of bending were different among the films with different cross-linking densities (Fig. 3.24). Because films with higher cross-linking densities show higher order parameter, the decrease in alignment order of the azobenzene moieties gives rise to a larger volume contraction along the rubbing direction, thus contributing to a larger extent of bending of the film along this direction.

A variety of LCE films with different alignments of the azobenzene mesogens were prepared and examined to make clear how the alignment affects the photoinduced bending behavior. By means of the selective absorption of linearly polarized light in polydomain LCE films, Ikeda and coworkers were able to control the direction of photoinduced bending so that a single polydomain LCE film could be bent repeatedly and precisely along any chosen direction (Fig. 3.25; Yu et al., 2003). The film bent toward the irradiation source in a direction parallel to the polarization of the light. Homeotropically aligned films were found to show a completely different bending behavior; upon exposure to UV light, they bent away from the actinic light source (Fig. 3.26; Kondo et al., 2006). Because the alignment direction of the azobenzene mesogens in the homeotropic films is perpendicular to the film surface; thus, exposure to UV light induces an isotropic expansion contributing to the bending behavior in a completely opposite direction. Furthermore, LCE films with a homogenous alignment on one surface and a homeotropic alignment on the opposite surface (hybrid alignment) were prepared, and their bending behavior was investigated (Kondo et al., 2007). Upon irradiation of UV light on the homogeneous surface, the film bent toward the light source along the alignment direction, whereas the film bent away from the actinic light source when the homeotropic surface was irradiated. This demonstrates that the bending direction is determined by the surface alignment treatment. Upon irradiation from both surfaces of the film, the bending speed was greatly enhanced at the same time.

It is well known that ferroelectric LCs have two advantages: one is their high degree of order of mesogens and the other is that their molecular alignment can be controlled quickly by applying an electric field due to the presence of spontaneous polarization. Therefore, ferroelectric LCE films with a high LC order and a low T_g value were prepared by in situ photopolymerization of oriented LC monomers (**20a,b**) under an electric field (Yu et al., 2007). Irradiation with UV light led the films to bend at room temperature toward the actinic light source with a tilt to the rubbing direction of the alignment layer (Fig. 3.27). The bending process was completed within 500 ms of irradiation from a laser beam. Moreover, the mechanical force generated by photoirradiation reached about 220 kPa (Fig. 3.27), which is similar to the contraction force of human muscles ($\sim$300 kPa).

As mentioned earlier, hydrogen bonds can be used for the preparation of LCEs by exploiting intermolecular hydrogen bonding. Ikeda and coworkers fabricated hydrogen-bonded LCE films using two cross-linkers (**21a,b**) capable of recognizing hydrogen bond donor molecules at the pyridyl ends (Mamiya et al., 2008). A complex of a copolymer (**21c**), which has both a carboxyl group and an

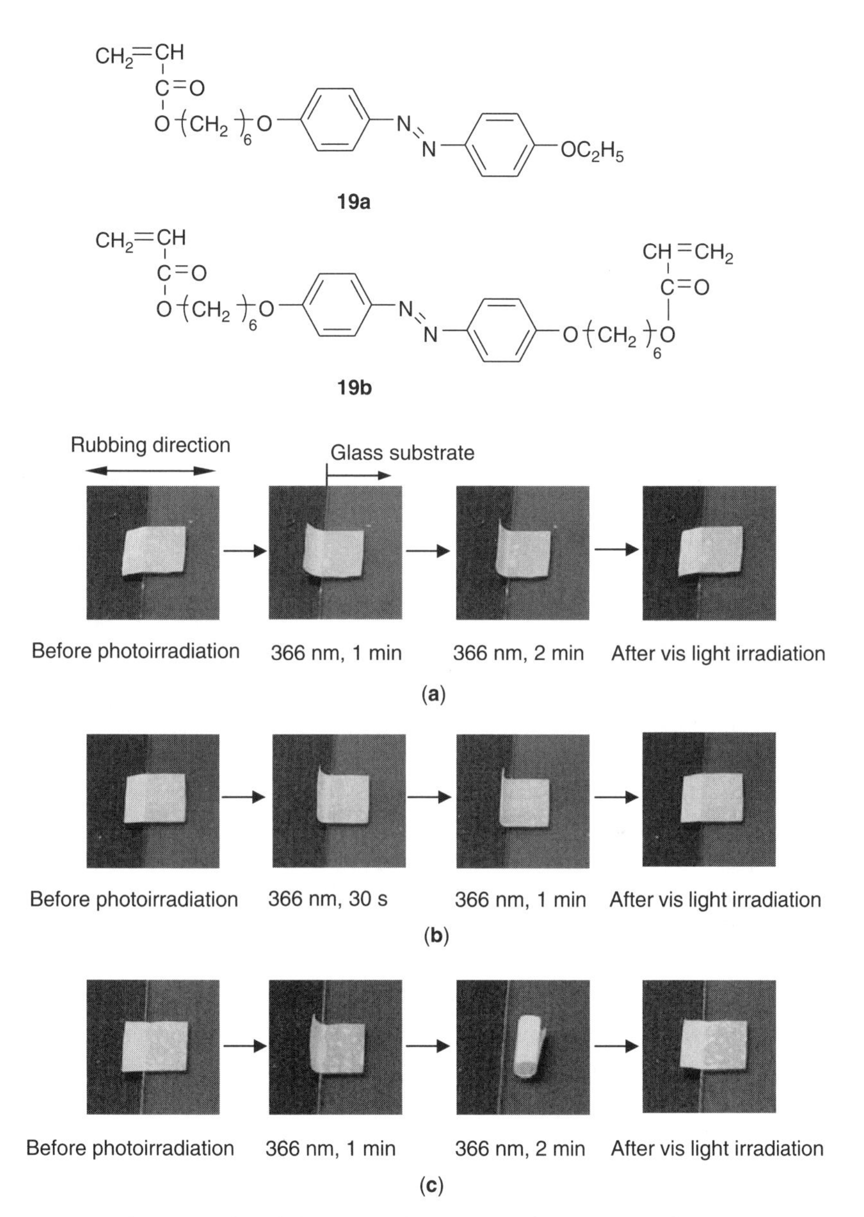

Figure 3.24. Photographs of the monodomain LCE films with different cross-linking densities exhibiting photoinduced bending and unbending behavior. Cross-linker concentration: **(a)** 5 mol%, **(b)** 10 mol%, and **(c)** 50 mol%. *Source*: Yu et al., 2004b.

azobenzene moiety, and the cross-linker was obtained from a tetrahydrofurane (THF) solution; then the melt complex was sandwiched between two sodium chloride (NaCl) plates with rubbing treatment to prepare hydrogen-bonded LCE films. It was found that the film without the azobenzene groups in the cross-links

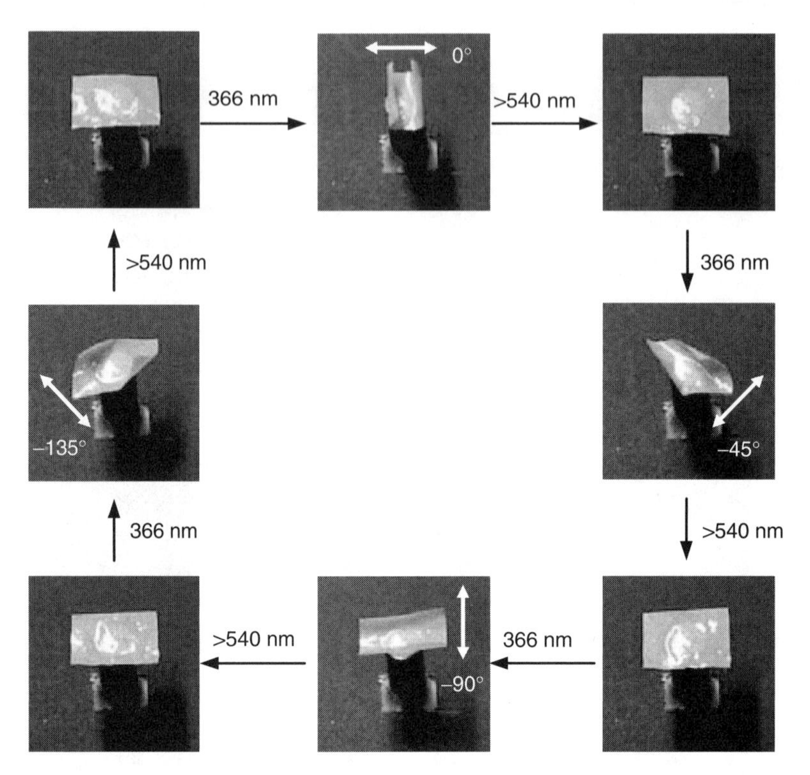

Figure 3.25. Precise control of the bending direction of a film by linearly polarized light: photographs of the polydomain film in different directions in response to irradiation by linearly polarized light at different angles of polarization (*white arrows*) at $\lambda = 366$ nm; the bent films are flattened by irradiation with visible light at $\lambda > 540$ nm. *Source*: Yu et al., 2003. See color insert.

showed no deformation on exposure to UV light, whereas the film with azobenzene cross-links bent toward the actinic light source along the alignment direction of the mesogens (Fig. 3.28). The photoinduced bending and unbending of the hydrogen-bonded films are similar to that of the chemically bonded films. These results indicate that a structural change caused by photoisomerization of the azobenzene moieties at the cross-links plays an important role in the photoinduced bending of the hydrogen-bonded LCE films. In other words, it means that the cross-links formed by noncovalent bonds can convert the motion of the mesogens into a macroscopic change of the LCE films. This kind of self-assembly LCE films could be reconstructed through cross-link and decross-link of hydrogen bonds, and the photoinduced bending of the reformed LCE films could be induced repeatedly.

Several other interesting movements of LCEs in response to light were also reported. Palffy-Muhoray and coworkers demonstrated that the mechanical deformation of an LCE sample doped with azobenzene dyes (**22a–d**) in response

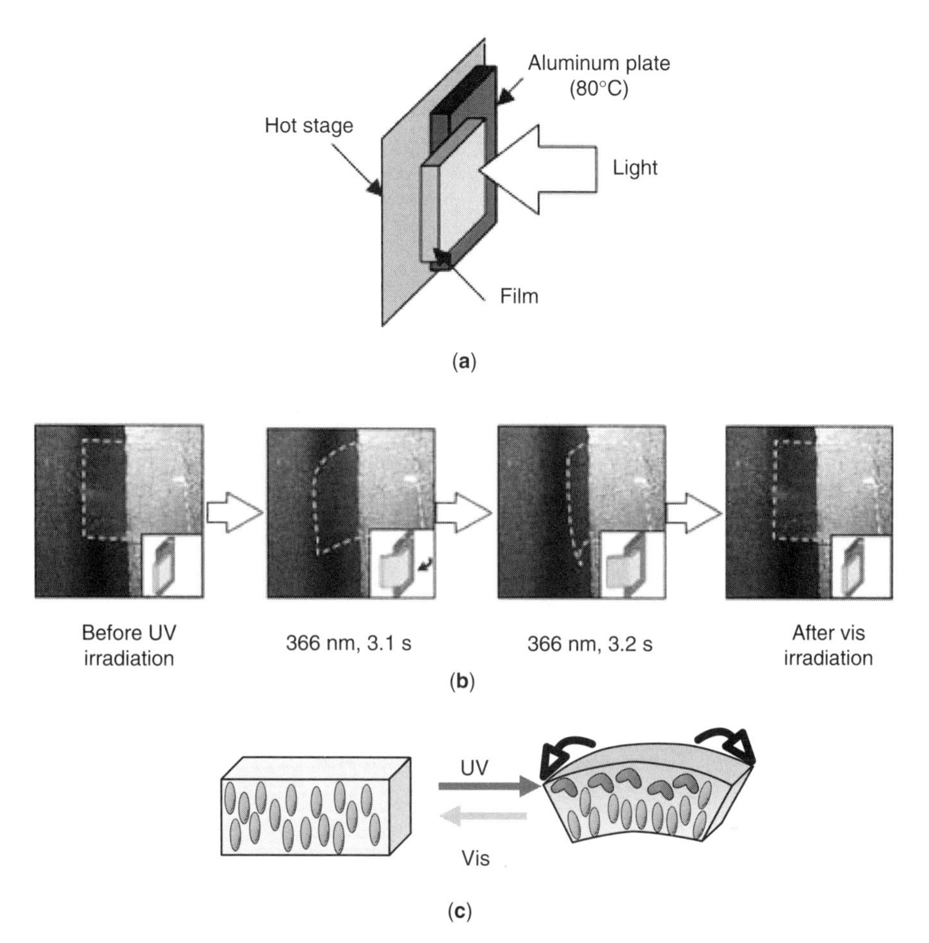

Figure 3.26. (**a**) Experimental setup and (**b**) photographs of the homeotropic film that exhibits photoinduced bending and unbending behavior. The white dash lines show the edges of the films, and the inset of each photograph is a schematic illustration of the film state. (**c**) Schematic illustration of the bending mechanism in the homeotropic film. *Source*: Kondo et al., 2006.

to nonuniform illumination by visible light becomes very large (the sample bends by more than 60°) (Camacho-Lopez et al., 2004). When laser beam is shone from above onto a dye-doped LCE sample floating on water, the LCE swims away from the laser beam, with an action resembling that of flatfish (Fig. 3.29). Tabiryan et al. (2005) developed an azobenzene LCE film with extraordinarily strong and fast mechanical response to the influence of a laser beam. The direction of the photoinduced bending or twisting of LCE can be reversed by changing the polarization of the laser beam. The phenomenon is a result of photoinduced reorientation of azobenzene moieties in the LCE (Fig. 3.30). Broer and coworkers

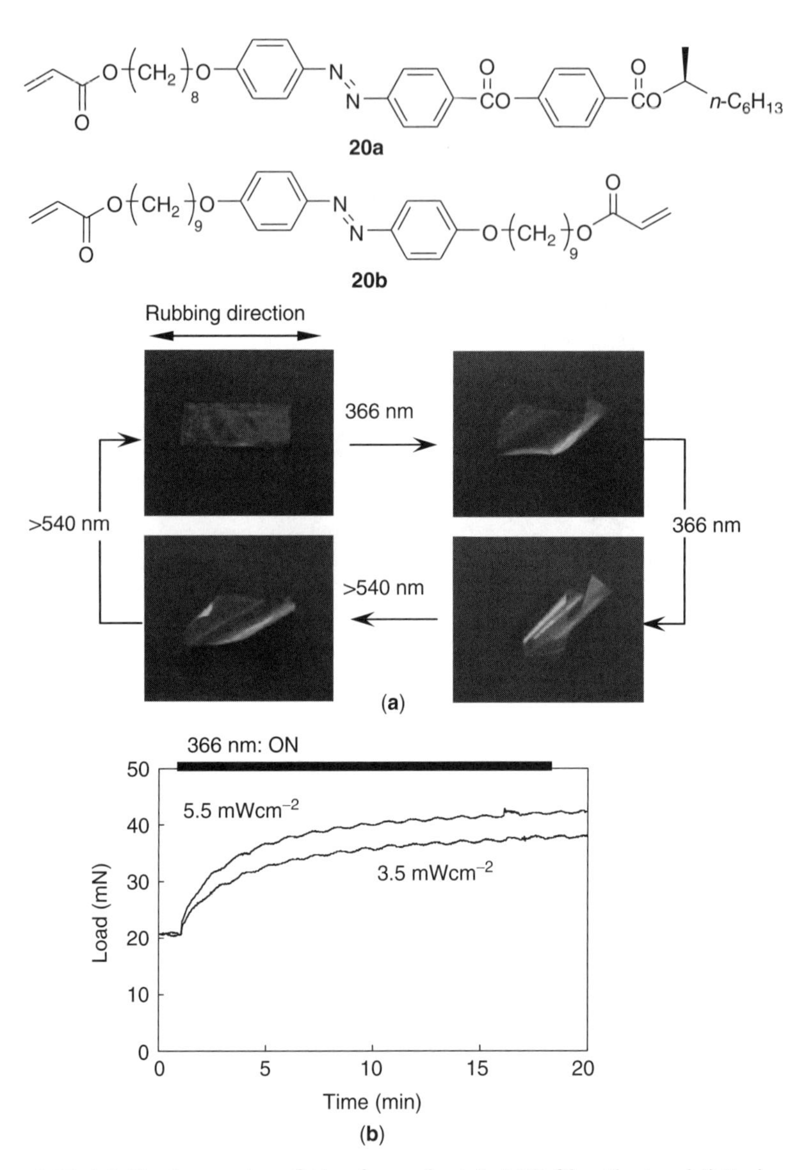

Figure 3.27. (**a**) Photographs of the ferroelectric LCE film that exhibits bending and unbending behavior upon alternative irradiation with UV and visible light at room temperature: the film bent toward the actinic light source along the alignment direction of mesogens in response to irradiation at $\lambda = 366$ nm, and being flattened again by irradiation with visible light at $\lambda > 540$ nm. (**b**) Change of the load on the ferroelectric LCE film when exposed to UV light at $\lambda = 366$ nm with different intensities at 50°C. The cross-section area of the film is 5 mm × 20 μm. An external force of 20 mN was loaded initially on the film to keep the length of the film unchanged. *Source*: Yu et al., 2007.

Hydrogen-bond acceptors

21a **21b**

Hydrogen-bond donor

21c

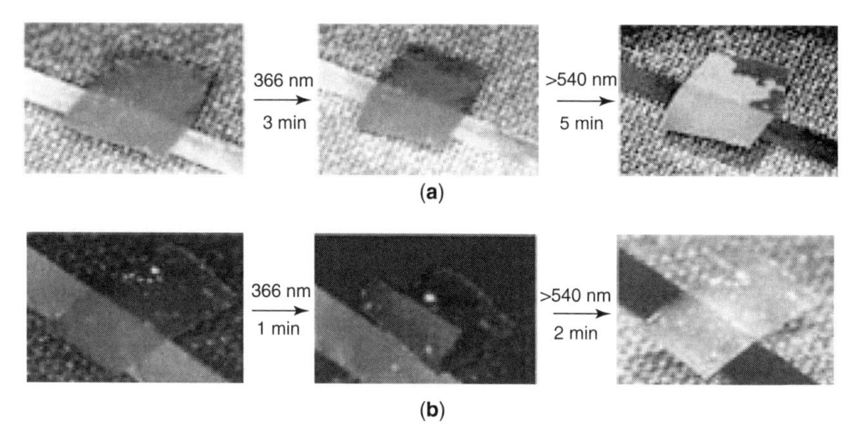

Figure 3.28. Photoresponsive behavior of the hydrogen-bonded LCE films of (**a**) 21a + 21c and (**b**) 21b + 21c. *Source*: Mamiya et al., 2008.

prepared LCE films with a densely cross-linked, twisted configuration of azobenzene moieties (Harris et al., 2005b). The films showed a large amplitude bending and coiling motion upon exposure to UV light, which results from the 90° twisted configuration of the LC alignment (Fig. 3.31).

Most lately, Ikeda and coworkers prepared a continuous ring of the LCE film by connecting both ends of the film (Yamada et al., 2008). The azobenzene mesogens were aligned along the circular direction of the ring. As shown in

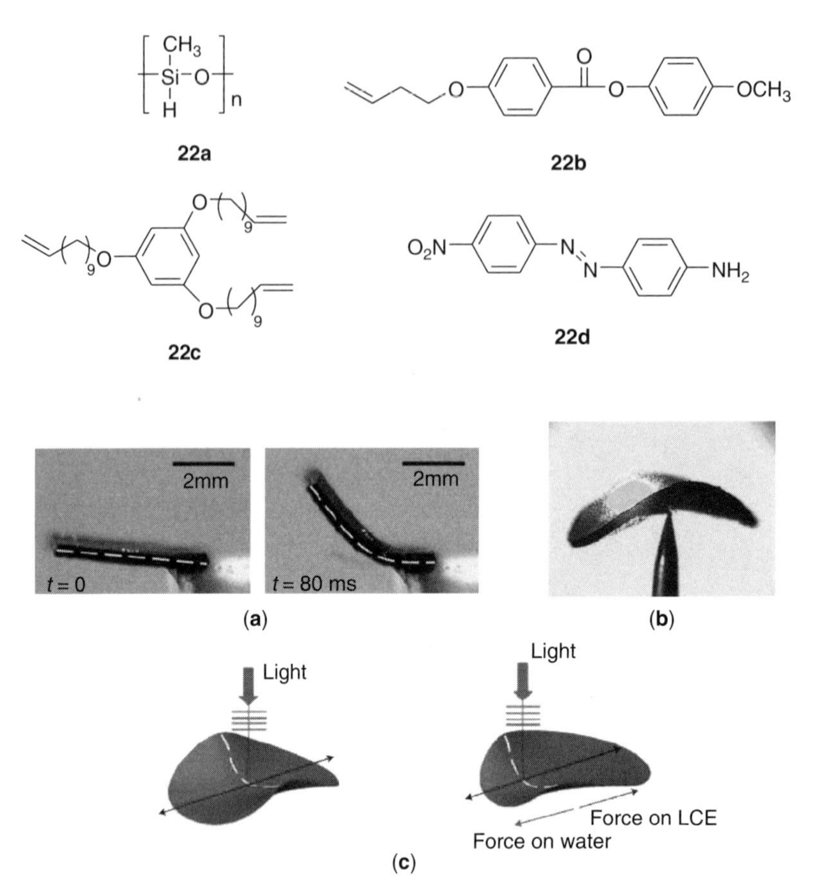

Figure 3.29. (**a**) Photomechanical response of an LCE sample. (**b**) The shape deformation of an LCE sample upon exposure to light at $\lambda = 514$ nm. (**c**) Mechanism of the locomotion of the dye-doped LCE sample. *Source*: Camacho-Lopez et al., 2004.

Fig. 3.32a, upon exposure to irradiation with UV light from the downside right and visible light from the upside right simultaneously, the ring rolled intermittently toward the actinic light source, resulting almost in a 360° roll at room temperature. This is the first example of this kind of photoinduced motion in a single-layer film, although the rolling of the LCE ring shown here was slow and stopped when the ring was broken by irradiation. Furthermore, they prepared a plastic belt of the LCE-laminated film by connecting both ends of the film, and then placed the belt on a homemade pulley system as illustrated in Fig. 3.32b. By irradiating the belt with UV light from top right and visible light from top left simultaneously, a rotation of the belt was induced to drive the two pulleys in a counterclockwise direction at room temperature as shown in Fig. 3.32c. A plausible mechanism of the rotation is as follows: Upon exposure to UV light, a local contraction force is generated at the irradiated part of the belt near the right

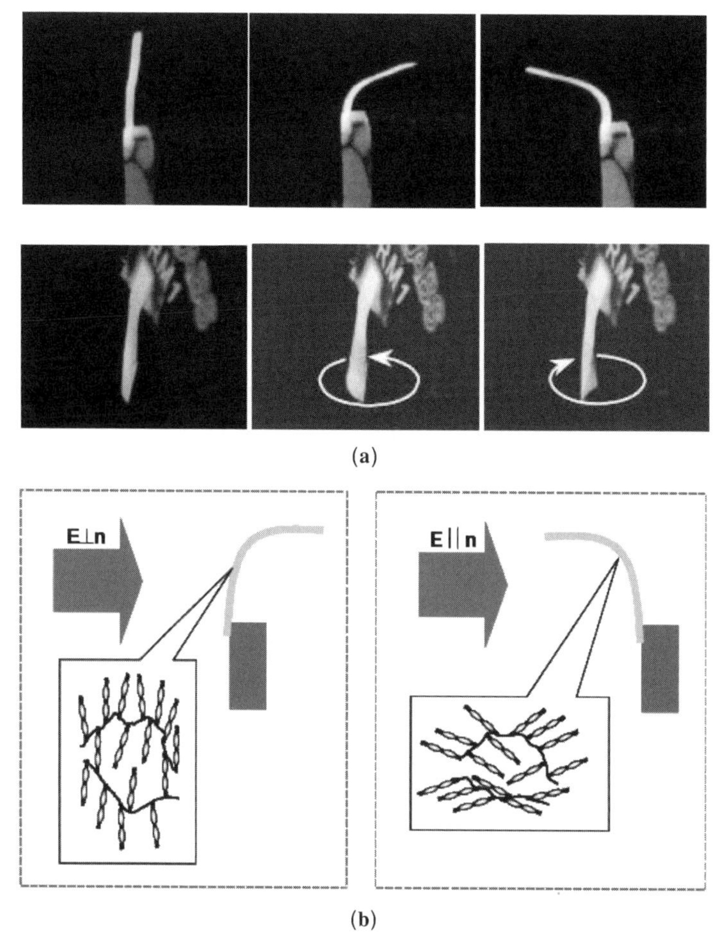

Figure 3.30. (a) Photographs of photoinduced deformation of the polymer film. (b) Schematic illustration of the effect of the photoinduced change in LC alignment (E: polarization direction of laser beam). *Source*: Tabiryan et al., 2005.

pulley along the alignment direction of the azobenzene mesogens, which is parallel to the long axis of the belt. This contraction force acts on the right pulley, leading it to rotate in the counterclockwise direction. At the same time, the irradiation with visible light produces a local expansion force at the irradiated part of the belt near the left pulley, causing a counterclockwise rotation of the left pulley. These contraction and expansion forces produced simultaneously at the different parts along the long axis of the belt give rise to the rotation of the pulleys and the belt with the same direction. The rotation then brings new parts of the belt to be exposed to UV and visible light, which enables the motor system to rotate continuously.

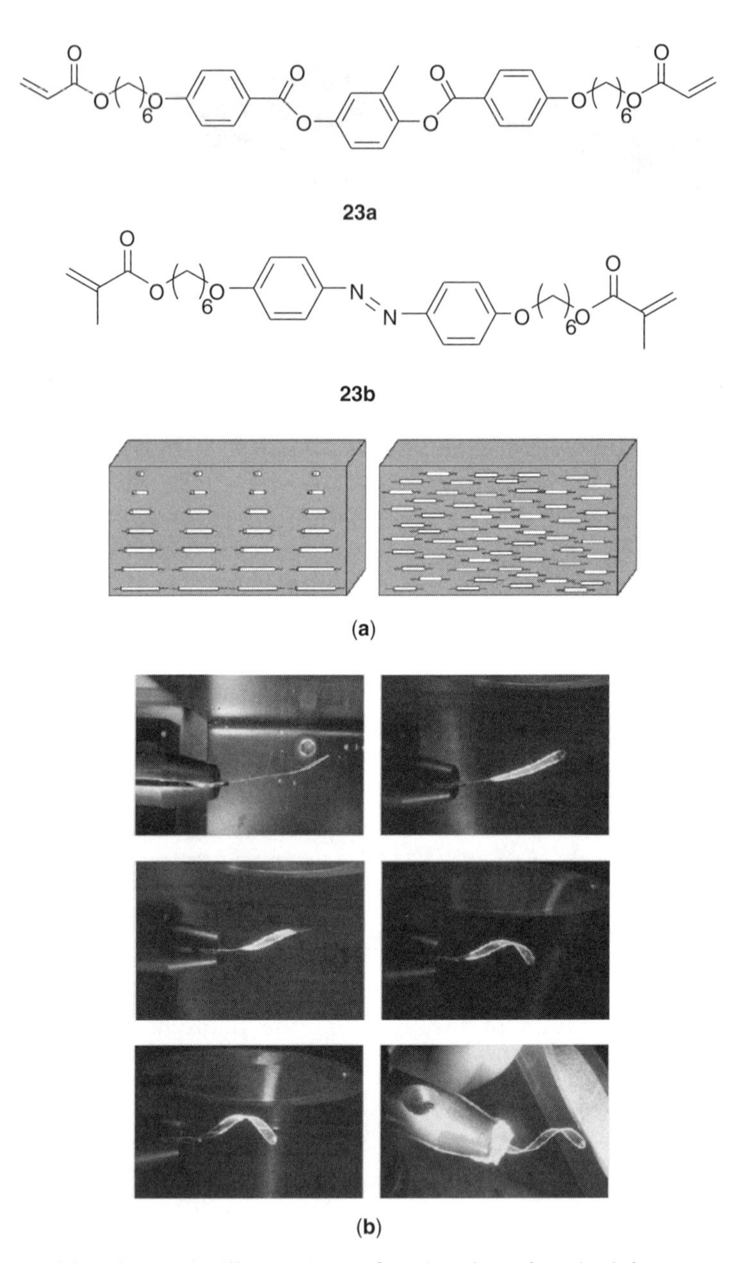

Figure 3.31. (**a**) Schematic illustration of twisted and uniaxial arrangements. (**b**) UV-induced coiling of a film in the twisted configuration. *Source*: Harris et al., 2005b.

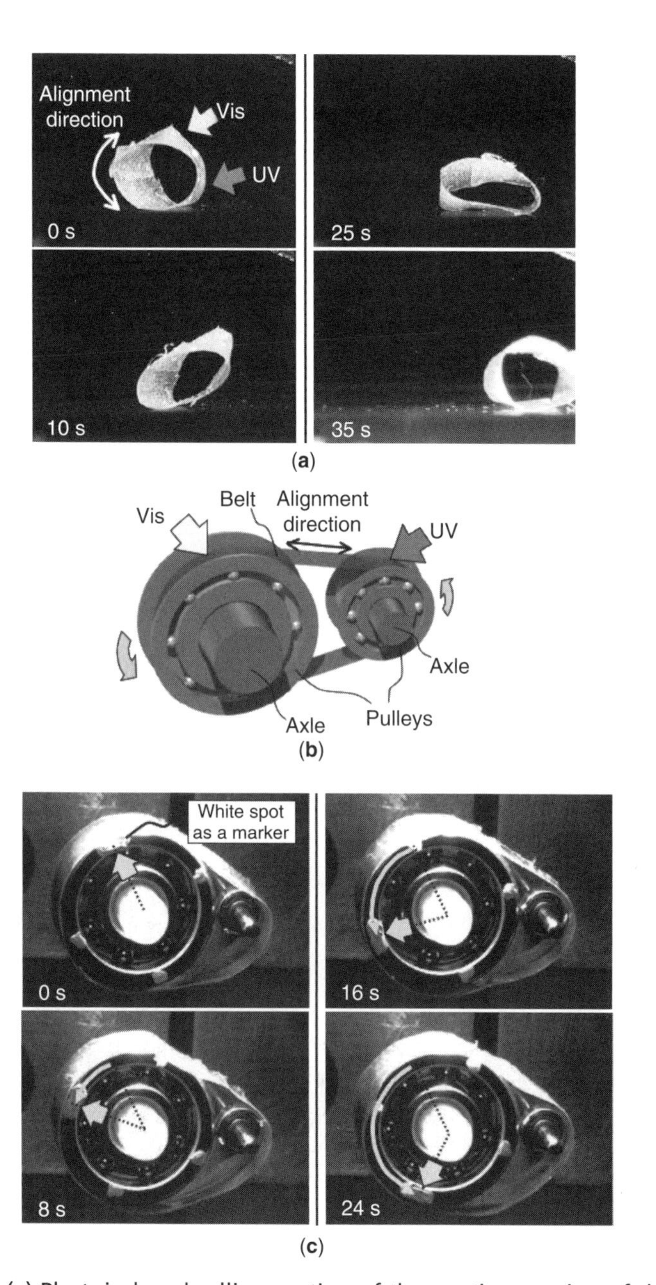

Figure 3.32. (**a**) Photoinduced rolling motion of the continuous ring of the LCE film by simultaneous irradiation with UV and visible light at room temperature. (**b**) Schematic illustration of a light-driven plastic motor system used, showing the relationship between light irradiation positions and a rotation direction. (**c**) Photographs showing time profiles of the rotation of the light-driven plastic motor with the LCE-laminated film induced by simultaneous irradiation with UV and visible light at room temperature. *Source*: Yamada et al., 2008. See color insert.

3.5. SUMMARY AND OUTLOOK

Development of artificial musclelike actuators attracts a growing interest, because they are ideal for the realization of biomimetic movements by changing their shapes and dimensions when a potential is applied. Polymer actuators with such advantages as flexibility, lightweight, low cost, and quiet operation compliance play a leading role in this field. However, light is an ideal stimulus, for it can be localized (in time and space), selective, nondamaging, and allows for remote activation and remote delivery of energy to a system. Therefore, chromophore-containing polymers that can generate light-driven actuation are of great interest.

This chapter describes the all-optical deformation observed in gels, amorphous solid films, and LCEs that contain azobenzenes. Among them, LCEs are promising materials for the construction of artificial muscles driven by light; however, their photomechanical characteristics must be improved, especially the alignment of mesogens, the coupling of LC order with polymer networks, and the change in order by light. Further effort is required in tailoring the different material systems to specific demands of selected applications so as to fully develop the potential of this technology platform. This includes the precise adjustment of mechanical and thermal properties as well as the development of appropriate assembling technology to incorporate photosensitive components into final products. The market potential for light-sensitive polymers is quite remarkable. Examples for potential applications include light-responsive sensors, actuators, and medical devices.

REFERENCES

Barnes NR, Davis FJ, Mitchell GR. 1989. Molecular switching in liquid crystal elastomers. Mol Cryst Liq Cryst 168:13–25.

Baughman RH. 1996. Conducting polymer artificial muscles. Synth Met 78:339–353.

Baughman RH, Cui C, Zakhidov AA, Iqbal Z, Barisci JN, Spinks GM, Wallace GG, Mazzoldi A, De Rossi D, Rinzler AG, Jaschinski O, Roth S, Kertesz M. 1999. Carbon nanotube actuators. Science 284:1340–1343.

Bay L, West K, Sommer-Larsen P, Skaarup S, Benslimane M. 2003. A conducting polymer artificial muscle with 12% linear strain. Adv Mater 15:310–313.

Beyer P, Zentel R. 2005. Photoswitchable smectic liquid-crystalline elastomers. Macromol Rapid Commun 26:874–879.

Brand HR, Finkelmann H. 1998. Physical properties of liquid crystalline elastomers. In: Handbook of liquid crystals. Demus D, Goodby J. Gray GW, Spiess HW, Vill V, editors. Weinheim: Wiley-VCH.

Brehmer M, Zentel R. 1994. Ferroelectric liquid-crystalline elastomers. Macromol Chem Phys 195:1891–1904.

Brodowsky HM, Boehnke UC, Kremer F, Gebhard E, Zentel R. 1999. Mechanical deformation behavior in highly anisotropic elastomers made from ferroelectric liquid crystalline polymers. Langmuir 15:274–278.

Broer DJ, Finkelmann H, Kondo K. 1988. In-situ photopolymerization of an oriented liquid-crystalline acrylate. Makromol Chem 189:185–194.

Broer DJ, Boven J, Mol GN, Challa G. 1989. In-situ photopolymerization of oriented liquid-crystalline acrylates. Makromol Chem 190:2255–2268.

Buguin A, Li MH, Silberzan P, Ladoux B, Keller P. 2006. Micro-actuators: when artificial muscles made of nematic liquid crystal elastomers meet soft lithography. J Am Chem Soc 128:1088–1089.

Camacho-Lopez M, Finkelmann H, Palffy-Muhoray P, Shelley M. 2004. Fast liquid-crystal elastomer swims into the dark. Nat Mater 3:307–310.

Clarke SM, Hotta A, Tajbakhsh AR, Terentjev EM, 2001. Effect of cross-linker geometry on equilibrium thermal and mechanical properties of nematic elastomers. Phys Rev E 64:061702.

Cotton JP, Hardouin F. 1997. Chain conforation of liquid-crystalline polymers studied by small-angle neutron scattering. Prog Polym Sci 22:795–828.

Courty S, Mine J, Tajbakhsh AR, Terentjev EM. 2003. Nematic elastomers with aligned carbon nanotubes: new electromechanical actuators. Europhys Lett 64:654–660.

Cui L, Zhao Y. 2004. Azopyridine side chain polymers: an efficient way to prepare photoactive liquid crystalline materials through self-assembly. Chem Mater 16:2076–2082.

Cui L, Tong X, Yan X, Liu G, Zhao Y. 2004. Photoactive thermoplastic elastomers of azobenzene-containing triblock copolymers prepared through atom transfer radical polymerization. Macromolecules 37:7097–7104.

Cviklinski J, Tajbakhsh AR, Terentjev EM. 2002. UV isomerisation in nematic elastomers as a route to photo-mechanical transducer. Eur Phys J E 9:427–434.

De Gennes PG. 1975. Physique moleculaire. C R Acad Sci B 281:101–103.

Disch S, Finkelmann H, Ringsdorf H, Schuhmacher P. 1995. Macroscopically ordered discotic columnar networks. Macromolecules 28:2424–2428.

Eich M, Wendorff JH. 1987. Erasable holograms in polymeric liquid crystals. Makromol Chem Rapid Commun 8:467–471.

Eich M, Wendorff JH, Reck B, Ringsdorf H. 1987. Reversible digital and holographic optical storage in polymeric liquid crystals. Makromol Chem Rapid Commun 8:59–63.

Eisenbach CD. 1980. Isomerization of aromatic azo chromophores in poly(ethyl acrylate) networks and photomechanical effect. Polymer 21:1175–1179.

Finkelmann H, Nishikawa E. 2001. A new opto-mechanical effect in solids. Phys Rev Lett 87:015501.

Finkelmann H, Rehage G. 1984. The degree of order in liquid crystalline side chain polymers. Adv Polym Sci 60/61:99–172.

Finkelmann H, Kock HJ, Rehage G. 1981. Investigations on liquid crystalline polysiloxanes, 3. Liquid crystalline elastomers—a new type of liquid crystalline material. Makromol Chem Rapid Commun 2:317–322.

Finkelmann H, Kim ST, Muñoz A, Palffy-Muhoray P, Taheri B. 2001. Tunable mirrorless lasing in cholesteric liquid crystalline elastomers. Adv Mater 13:1069–1072.

Fukuda A, Takezoe H. 1990. Structures and Properties of Ferroelectric Liquid Crystals. Tokyo: Corona.

Fukushima T, Asaka K, Kosaka A, Aida T. 2005. Fully plastic actuator through layer-by-layer casting with ionic-liquid-based bucky gel.Angew Chem Int Ed Engl 44:2410–2413.

Gao J, Sansiñena JM, Wang HL. 2003. Tunable polyaniline chemical actuators. Chem Mater 15:2411–2418.

Gebhard E, Zentel R. 1998. Freestanding ferroelectric elastomer films. Macromol Rapid Commun 19:341–344.

Harris KD, Bastlaansen CWM, Lub J, Broer DJ. 2005a. Self-assembled polymer films for controlled agent-driven motion. Nano Lett 5:1857–1860.

Harris KD, Cuypers R, Sscheibe P, Van Oosten CL, Bastiaansen CWM, Lub J, Broer DJ. 2005b. Large amplitude light-induced motion in high elastic modulus polymer actuators. J Mater Chem 15:5043–5048.

He XZ, Zhang BY, Meng FB, Lin JR. 2005. Effect of the length of the carbochain on the phase behavior of side-chain cholesteric liquid-crystalline elastomers. J Appl Polym Sci 96:1204–1210.

Hiraoka K, Sagano W, Nose T, Finkelmann H. 2005. Biaxial shape memory effect exhibited by monodomain chiral smectic C elastomers. Macromolecules 38:7352–7357.

Hogan PM, Tajbakhsh AR, Terentjev EM. 2002. UV manipulation of order and macroscopic shape in nematic elastomers. Phys Rev E 65:041720.

Holland NB, Hugel T, Neuert G, Cattani-Scholz A, Renner C, Oesterhelt D, Moroder L, Seitz M, Gaub HE. 2003. Single molecule force spectroscopy of azobenzene polymers: switching elasticity of single photochromic macromolecules. Macromolecules 36:2015–2023.

Hosono N, Furukawa H, Masubuchi Y, Watanabe T, Horie K. 2007. Photochemical control of network structure in gels and photo-induced changes in their viscoelastic properties. Colloids Surf B Biointerfaces 56:285–289.

Hu Z, Zhang X, Li Y. 1995. Monitoring release of neurotrophic activity in the brains of awake rats. Science 269:525–553.

Hugel T, Holland NB, Cattani A, Moroder L, Seitz M, Gaub HE. 2002. Single-molecule optomechanical cycle. Science 296:1103–1106.

Ikeda T. 2001. Photochemical modulation of refractive index by means of photosensitive liquid crystals. Mol Cryst Liq Cryst 364:187–197.

Ikeda T. 2003. Photomodulation of liquid crystal orientations for photonic applications. J Mater Chem 13:2037–2057.

Ikeda T, Tsutsumi O. 1995. Optical switching and image storage by means of azobenzene liquid-crystal films. Science 268:1873–1875.

Ikeda T, Horiuchi S, Karanjit DB, Kurihara S, Tazuke S. 1988. Photochemical image storage in polymer liquid crystals. Chem Lett 17:1679–1682.

Ikeda T, Horiuchi S, Karanjit DB, Kurihara S, Tazuke S. 1990a. Photochemically induced isothermal phase transition in polymer liquid crystals with mesogenic phenyl benzoate side chains. 1. Calorimetric studies and order parameters. Macromolecules 23:36–42.

Ikeda T, Horiuchi S, Karanjit DB, Kurihara S, Tazuke S. 1990b. Photochemically induced isothermal phase transition in polymer liquid crystals with mesogenic phenyl benzoate side chains. 2. Photochemically induced isothermal phase transition behaviors. Macromolecules 23:42–48.

Ikeda T, Horiuchi S, Karanjit DB, Kurihara S, Tazuke S. 1990c. Photochemically induced isothermal phase transition in polymer liquid crystals with mesogenic cyanobiphenyl side chains. Macromolecules 23:3938–3943.

Ikeda T, Sasaki T, Kim HB. 1991. "Intrinsic" response of polymer liquid crystals in photochemical phase transition. J Phys Chem 95:509–511.

Ikeda T, Sasaki T, Ichimura K. 1993. Photochemical switching of polarizationin ferroelectric liquid-crystal films. Nature 361:428–430.

Ikeda T, Nakano M, Yu Y, Tsutsumi O, Kanazawa A. 2003. Anisotropic bending and unbending behavior of azobenzene liquid-crystalline gels by light exposure. Adv Mater 15:201–205.

Irie M. 1986. Photoresponsive polymer. Reversible bending of rod-shaped acrylamide gels in an electric field. Macromolecules 19:2890–2892.

Irie M. 1990. Photoresponsive polymers. Adv Polym Sci 94:27–67.

Irie M, Kungwatchakun D. 1995. Photoresponsive polymer. Mechanochemistry of polyacrylamide gels having triphenylmethane leuco derivatives. Makromol Chem Rapid Commun 5:829–832.

Ishikawa M, Kitamura N, Masuhara H, Irie M. 1991. Size effect on photoinduced volume change of polyacrylamide microgels containing triphenylmethane leuco cyanide. Makromol Chem Rapid Commun 12:687–690.

Jager EWH, Smela E, Inganas O. 2000. Microfabricating conjugated polymer actuators. Science 290:1540–1545.

Kelly SM. 1995. Anisotropic networks. J Mater Chem 5:2047–2061.

Kim HK, Wang XS, Fujita Y, Sudo A, Nishida H, Fujii M, Endo T. 2005a. Photomechanical switching behavior of semi-interpenetrating polymer network consisting of azobenzene-carrying crosslinked poly(vinyl ether) and polycarbonate. Macromol Rapid Commun 26:1032–1036.

Kim HK, Wang XS, Fujita Y, Sudo A, Nishida H, Fujii M, Endo T. 2005b. Reversible photo-mechanical switching behavior of azobenzene-containing semi-interpenetrating network under UV and visible light irradiation. Macromol Chem Phys 206:2106–2111.

Kim HK, Wang XS, Fujita Y, Sudo A, Nishida H, Fujii M, Endo T. 2005c. A rapid photomechanical switching polymer blend system composed of azobenzene-carrying poly(vinylether) and poly(carbonate). Polymer 46:5879–5883.

Kim P, Lieber CM. 1999. Nanotube nanotweezers. Science 286:2148–2150.

Kishi R, Suzuki Y, Ichijo H, Hirasa O. 1994. Electrical deformation of thermotropic liquid-crystalline polymer gels. Chem Lett 2257–2260.

Kondo M, Yu Y, Ikeda T. 2006. How does the initial alignment of mesogens affect the photoinduced bending behavior of liquid-crystalline elastomers. 2006. Angew Chem Int Ed Engl 45:1378–1382.

Kondo M, Yu Y, Mamiya J, Kinoshita M, Ikeda T. 2007. Photoinduced deformation behavior of crosslinked azobenzene liquid-crystalline polymer films with unimorph and bimorph structure. Mol Cryst Liq Cryst 478:245–257.

Kumar A, Srivastava A, Galaev IY, Mattiasson B. 2007. Smart polymers: physical forms & bioengineering applications. Prog Polym Sci 32:1205–1237.

Kundler I, Finkelmann H. 1998. Director reorientation via stripe-domains in nematic elastomers: influence of cross-link density, anisotropy of the network and smectic clusters. Macromol Chem Phys 199:677–686.

Küpfer J, Finkelmann H. 1991. Nematic liquid single crystal elastomers. Makromol Chem Rapid Commun 12:717–726.

Küpfer J, Finkelmann H. 1994. Liquid crystal elastomers: influence of the orientational distribution of the crosslink's on the phase behaviour and reorientation processes. Macromol Chem Phys 195:1353–1367.

Küpfer J, Nishikawa E, Finkelmann H. 1994. Densely crosslinked liquid single-crystal elastomers. Polym Adv Technol 5:110–115.

Kurihara S, Ikeda T, Sasaki T, Kim HB, Tazuke S. 1990. Time-resolved observation of isothermal phase transition of liquid crystals triggered by photochemical reaction of dopant. J Chem Soc Chem Commun 1751–1752.

Legge CH, Mitchell GR. 1992. Photo-induced phase transitions in azobenzene-doped liquid crystals. J Phys D: Appl Phys 25:492–494.

Lehmann W, Skupin H, Tolksdorf C, Gebhard E, Zentel R, Krüger P, Lösche M, Kremer F. 2001. Giant lateral electrostriction in ferroelectric liquid-crystalline elastomers. Nature 410:447–450.

Lendlein A, Jiang H, Jünger O, Langer R. 2005. Light-induced shape-memory polymers. Nature 434:879–882.

Li MH, Keller P, Li B, Wang X, Brunet M. 2003. Light-driven side—on nematic elastomer actuators. Adv Mater 15:569–572.

Li MH, Keller P, Yang J, Albouy PA. 2004. An artificial muscle with lamellar structure based on a nematic triblock copolymer. Adv Mater 16:1922–1925.

Liu C, Chun BS, Mather PT, Zheng L, Haley EH, Coughlin EB. 2002. Chemically cross-linked polycyclooctene: synthesis, characterization, and shape memory behavior. Macromolecules 35:9868–9874.

Liu Z, Calver P. 2000. Multilayer hydrogels as muscle-like actuators. Adv Mater 12: 288–291.

Mamada A, Tanaka T, Kungwatchakun D, Irie M. 1990. Photoinduced phase transition of gels. Macromolecules 23:1517–1519.

Mamiya J, Yoshitake A, Kondo M, Yu Y, Ikeda T. 2008. Is chemical crosslinking necessary for the photoinduced bending of polymer films? J Mater Chem 18:63–65.

Mao Y, Terentjev EM, Warner M. 2001. Cholesteric elastomers: deformable photonic solids. Phys Rev E 64:041803.

Matějka L, Dušek K. 1981. Photochromic polymers: photoinduced conformational changes and effect of polymeric matrix on the isomerization of photochromes. Makromol Chem 182:3223–3236.

Matějka L, Dušek K, Iiavský M. 1979. The thermal effect in the photomechanical conversion of a photochromic polymer. Polym Bull 1:659–664.

Matějka L, Ilavský M, Dusek K, Wichterle O. 1981. Photomechanical effects in crosslinked photochromic polymers. Polymer 22:1511–1515.

Meng FB, Zhang BY, Xiao WQ, Hu TX. 2005. Effect of nonmesogenic crosslinking units on the mesogenic properties of side-chain cholesteric liquid crystalline elastomers. J Appl Polym Sci 96:625–631.

Merian E. 1966. Steric factors influencing the dyeing of hydrophobic fibers. Textile Res J 36:612–618.

Mol GN, Harris KD, Bastiaansen CM, Broer DJ. 2005. Thermo-mechanical responses of liquid-crystal networks with a splayed molecular organization. Adv Funct Mater 15:1155–1159.

Moriyama T, Kajita J, Takanashi Y, Ishikawa K, Takezoe H, Fukuda A. 1993. Optically addressed spatial light modulator using an antiferroelectric liquid crystal doped with azobenzene. Jpn J Appl Phys 32:L589–L592.

Mun-Sik K, Vinay KG. 2002. Photochromic cross-links in thermoresponsive hydrogels of poly(N-isopropylacrylamide): enthalpic and entropic consequences on swelling behavior. J Phys Chem B 106:4127–4132.

Naciri J, Srinivasan A, Jeon H, Nikolov N, Keller P, Ratna BR. 2003. Nematic elastomer fiber actuator. Macromolecules 36:8499–8505.

Natansohn A, Rochon P. 2002. Photoinduced motions in azo-containing polymers. Chem Rev 102:4139–4175.

Negishi M, Tsutsumi O, Ikeda T, Hiyama T, Kawamura J, Aizawa M, Takehara S. 1996a. Photochemical switching of ferroelectric liquid crystals using a photoswitchable chiral dopant. Chem Lett 25:319–320.

Negishi M, Kanie K, Ikeda T, Hiyama T. 1996b. Synthesis and photochemical switching of the antiferroelectric liquid crystals containing a diazenediyl group. Chem Lett 25:583–584.

Nishikawa E, Finkelmann H. 1997. Orientation behavior of smectic polymer networks by uniaxial mechanical fields. Macromol Chem Phys 198:2531–2549.

Nishikawa E, Finkelmann H. 1999. Smectic-A liquid single crystal elastomers—strain induced break-down of smectic layers. Macromol Chem Phys 200:312–322.

Osada Y, Gong JP. 1998. Soft and wet materials: polymer gels. Adv Mater 10:827–837.

Osada Y, Okuzaki H, Hori H. 1992. A polymer gel with electrically driven motility. Nature 355:242–243.

Otero TF, Cortés MT. 2003. Artifical muscles with tactile sensitivity. Adv Mater 15: 279–282.

Pelrine R, Kornbluh R, Pei Q, Joseph J. 2000. High-speed electrically actuated elastomers with strain greater than 100%. Science 287:836–838.

Saikrasun S, Bualek-Limcharoen S, Kohjiya S, Urayama K. 2005. Anisotropic mechanical properties of thermoplastic elastomers in situ reinforced with thermotropic liquid-crystalline polymer fibers revealed by biaxial deformations. J Polym Sci Part B: Polym Phys 43:135–144.

Sakai T, Murayama H, Nagano S, Takeoka Y, Kidowaki M, Ito K, Seki T. 2007. Photoresponsive slide-ring gel. Adv Mater 19:2023–2025.

Sánchez C, Villacampa B, Alcala R, Martínez C, Oriol L, Piñol M, Serrano JM. 1999. Mesomorphic and orientational study of materials processed by in situ photopolymerization of reactive liquid crystals. Chem Mater 11:2804–2812.

Sasaki T, Ikeda T. 1993. Photochemical switching of polarization in ferroelectric liquid crystals: effect of structure of host FLCs. Ferroelectrics 149:343–351.

Sasaki T, Ikeda T. 1995a. Photochemical control of properties of ferroelectric liquid crystals. 1. Effect of structure of host ferroelectric liquid crystals on the photochemical switching of polarization. J Phys Chem 99:13002–13007.

Sasaki T, Ikeda T. 1995b. Photochemical control of properties of ferroelectric liquid crystals. 2. Effect of the structure of guest photoresponsive molecules on the photochemical switching of polarization. J Phys Chem 99:13008–13012.

Sasaki T, Ikeda T. 1995c. Photochemical control of properties of ferroelectric liquid crystals. 3. Photochemically induced reversible change in spontaneous polarization and electrooptic property. J Phys Chem 99:13013–13018.

Sasaki T, Ikeda T, Ichimura K. 1992. Time-resolved observation of photochemical phase transition in polymer liquid crystals. Macromolecules 25:3807–3811.

Sasaki T, Ikeda T, Ichimura K. 1994. Photochemical control of properties of ferroelectric liquid crystals: photochemical flip of polarization. J Am Chem Soc 116:625–628.

Shandryuk GA, Kuptsov SA, Shatalova AM, Plate NA, Talroze RV. 2003. Liquid crystal H-bonded polymer networks under mechanical stress. Macromolecules 36:3417–3423.

Shirota K, Yamaguchi I. 1997. Optical switching of antiferroelectric liquid crystal with azo-dye using photochemically induced SmC^*_A-SmC^* phase transition. Jpn J Appl Phys 36:L1035–L1037.

Smela E. 2003. Conjugated polymer actuators for biomedical applications. Adv Mater 15:481–494.

Smets G, De Blauwe F. 1974. Chemical reaction in solid polymeric systems. Photomechanical phenomena. Pure Appl Chem 39:225–238.

Spinks GM, Wallace GG, Fifield LS, Dalton LR, Mazzoldi A, De Rossi D, Khayrullin II, Baughman RH. 2002. Pneumatic carbon nanotube actuators. Adv Mater 14:1728–1732.

Stenull O, Lubensky TC. 2005. Phase transitions and soft elasticity of smectic elastomers. Phys Rev Lett 94:018304.

Sung JH, Hirano S, Tsutsumi O, Kanazawa A, Shiono T, Ikeda T. 2002. Dynamics of photochemical phase transition of guest/host liquid crystals with an azobenzene derivative as a photosensitive chromophore. Chem Mater 14:385–391.

Suzuki A, Tanaka T. 1990. Phase transition in polymer gels induced by visible light. Nature 346:345–347.

Suzuki A, Ishii T, Maruyama Y. 1996. Optical switching in polymer gels. J Appl Phys 80:131–136.

Tabiryan N, Serak S, Dai XM, Bunning T. 2005. Polymer film with optically controlled form and actuation. Opt Express 13:7442–7448.

Tammer M, Li J, Komp A, Finkelmann H, Kremer F. 2005. FTIR-spectroscopy on segmental reorientation of a nematic elastomer under external mechanical fields. Macromol Chem Phys 206:709–714.

Tanchak OM, Barrett CJ. 2005. Light-induced reversible volume changes in thin films of azo polymers: the photomechanical effect. Macromolecules 38:10566–10570.

Tazuke S, Kurihara S, Ikeda T. 1987. Amplified image recording in liquid crystal media by means of photochemically triggered phase transition. Chem Lett 16:911–914.

Terentjev EM. 1999. Liquid-crystalline elastomers. J Phys Condens Matter 11:R239–R257.

Thomsen DL, Keller P, Naciri J, Pink R, Jeon H, Shenoy D, Ratna BR. 2001. Liquid crystal elastomers with mechanical properties of a muscle. Macromolecules 34: 5868–5875.

Tsutsumi O, Shiono T, Ikeda T, Galli G. 1997. Photochemical phase transition behavior of nematic liquid crystals with azobenzene moieties as both mesogens and photosensitive chromophores. J Phys Chem B 101:1332–1337.

Tsutsumi O, Demachi Y, Kanazawa A, Shiono T, Ikeda T, Nagase Y. 1998a. Photochemical phase-transition behavior of polymer liquid crystals induced by photochemical reaction of azobenzenes with strong donor-acceptor pairs. J Phys Chem B 102:2869–2874.

Tsutsumi O, Kitsunai T, Kanazawa A, Shiono T, Ikeda T. 1998b. Photochemical phase transition behavior of polymer azobenzene liquid crystals with electron-donating and accepting substituents at 4,4′-positions. Macromolecules 31:355–359.

Van der Veen G, Prins W. 1971. Photomechanical energy conversion in a polymer membrane. Nature Phys Sci 230:70–72.

Van der Veen G, Prins W. 1974. Photoregulation of polymer conformation by photochromic moieties-I. Anionic ligand to an anionic polymer. Photochem Photobiol 19:191–196.

Van der Veen G, Noguet R, Prins W. 1974. Photoregulation of polymer conformation by photochromic moieties-II. Cationic and neutral moieties of an anionic polymer. Photochem Photobiol 19:197–204.

Warner M, Terentjev EM. 1996. Nematic elastomers—a new state of mater. Prog Polym Sci 21:853–891.

Warner M, Terentjev EM. 2003. Liquid Crystal Elastomers. UK: Oxford University.

Warner M, Gelling KP, Vilgis TA. 1988. Theory of nematic networks. J Chem Phys 88:4008–4013.

Wermter H, Finkelmann H. 2001. Liquid crystalline elastomers as artificial muscles. e-Polymer 013:1–13.

Xie P, Zhang R. 2005. Liquid crystal elastomers, networks and gels: advanced smart materials. J Mater Chem 15:2529–2550.

Yamada M, Kondo M, Mamiya J, Yu Y, Kinoshita M, Barrett CJ, Ikeda T. 2008. Photomobile polymer materials – towards light-driven plastic motors. Angew Chem Int Ed Engl 47:4986–4988.

Yoshida R, Uchida K, Kaneko Y, Sakai K, Kikuchi A, Sakuraui Y, Okano T. 1995. Comb-type grafted hydrogels with rapid de-swelling response to temperature changes. Nature 374:240–242.

Yu Y, Nakano M, Ikeda T. 2003. Directed bending of a polymer film by light. Nature 425:145.

Yu Y, Nakano M, Ikeda T. 2004a. Photoinduced bending and unbending behavior of liquid-crystalline gels and elastomers. Pure Appl Chem 76:1435–1445.

Yu Y, Nakano M, Shishido A, Shiono T, Ikeda T. 2004b. Effect of cross-linking density on photoinduced bending behavior of oriented liquid-crystalline network films containing azobenzene. Chem Mater 16:1637–1643.

Yu Y, Maeda T, Mamiya J, Ikeda T. 2007. Photomechanical effects of ferroelectric liquid-crystalline elastomers containing azobenzene chromophores. Angew Chem Int Ed Engl 46:881–883.

Yusuf Y, Sumisaki Y, Kai S. 2003. Birefringence measurement of liquid single crystal elastomer swollen with low molecular weight liquid crystal. Chem Phys Lett 382:198–202.

Yusuf Y, Cladis PE, Brand HR, Finkelmann H, Kai S. 2004a. Hystereses of volume changes in liquid single crystal elastomers swollen with low molecular weight liquid crystal. Chem Phys Lett 389:443–448.

Yusuf Y, Ono Y, Sumisaki Y, Cladis PE, Brand HR, Finkelmann H, Kai S. 2004b. Swelling dynamics of liquid crystal elastomers swollen with low molecular weight liquid crystals. Phys Rev E 69:021710.

Yusuf Y, Huh JH, Cladis PE, Brand HR, Finkelmann H, Kai S. 2005. Low-voltage-driven electromechanical effects of swollen liquid-crystal elastomers. Phy Rev E 71:061702.

Zanna JJ, Stein P, Marty JD, Mauzac M, Martinoty P. 2002. Influence of molecular parameters on the elastic and viscoelastic properties of side-chain liquid crystalline elastomers. Macromolecules 35:5459–5465.

Zentel R. 1986. Shape variation of cross-linked liquid-crystalline polymers by electric fields. Liq Cryst 1:589–592.

Zentel R. 1989. Liquid crystalline elastomers. Adv Mater 10:321–329.

Zentel R, Reckert G. 1986. Liquid crystalline elastomers based on liquid crystalline side group, main chain and combined polymers. Makromol Chem 187:1915–1926.

Zentel R, Benalia M. 1987. Stress-induced orientation in lightly crosslinked liquid-crystalline side-group polymers. Makromol Chem 188:665–674.

Zhang X, Li Y, Hu Z, Litter CL. 1995. Bending of N-isopropylacrylamide gel under the influence of infrared light. J Chem Phys 102:551–555.

Zhang Y, Iijima S. 1999. Elastic response of carbon nanotube bundles to visible light. Phys Rev Lett 82:3472–3475.

Zriny M, Feher J, Filipcsei G. 2000. Novel gel actuator containing TiO_2 particles operated under static electric field. Macromolecules 33:5751–5753.

4

AMORPHOUS AZOBENZENE POLYMERS FOR LIGHT-INDUCED SURFACE PATTERNING

Kevin G. Yager and Christopher J. Barrett

4.1. SURFACE MASS TRANSPORT

In 1995, a surprising and unprecedented optical effect was discovered in polymer thin films containing the azo chromophore Disperse Red 1 (DR1), shown in Fig. 4.1. The Natansohn/Rochon (Rochon et al., 1995) research team and the Tripathy/Kumar collaboration (Kim et al., 1995b) simultaneously and independently discovered a large-scale surface mass transport when the films were irradiated with a light interference pattern. In a typical experiment, two coherent laser beams, with a wavelength in the azo absorption band, are intersected at the sample surface (Fig. 4.2a, b). The sample usually consists of a thin spin-cast film (10–1000 nm) of an amorphous azopolymer on a transparent substrate. The sinusoidal light interference pattern at the sample surface leads to a sinusoidal surface patterning, that is, a surface relief grating (SRG). These gratings were found to be extremely large, up to hundreds of nanometers, as confirmed by atomic force microscopy (AFM) (Fig. 4.3). The SRGs diffract very efficiently, and in retrospect it is clear that many reports of large diffraction efficiency before 1995, attributed to birefringence, were in fact due to surface gratings. The process occurs readily at room temperature [well below the glass-to-rubber transition temperature (T_g) of the amorphous polymers used] with moderate irradiation (1–100 mW/cm^2) over seconds to minutes. The phenomenon is a reversible mass transport, not irreversible material ablation, since a flat film with the original thickness is recovered upon heating above T_g. Critically, it requires the presence and isomerization of azobenzene chromophores. Other absorbing but nonisomerizing chromophores do not produce SRGs. Many other systems can exhibit optical surface patterning (Yamaki et al., 2000), but the

Smart Light-Responsive Materials. Edited by Yue Zhao and Tomiki Ikeda

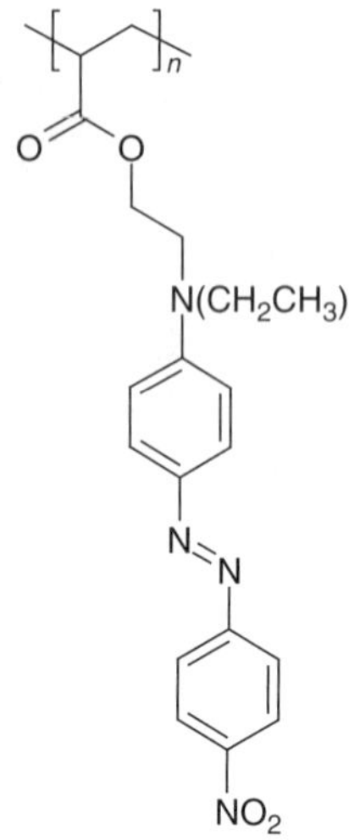

Figure 4.1. Chemical structure of poly(Disperse Red 1) acrylate, pdr1a, a pseudostilbene side-chain azopolymer that generates high quality surface relief structures.

amplitude of the modification is much smaller, does not involve mass transport, and usually requires additional processing steps. The all-optical patterning unique to azobenzenes has been studied intensively since its discovery, yet the mechanism remains controversial. The competing interpretations will be discussed and evaluated here. Many reviews of the remarkable body of experimental results are available (Natansohn and Rochon, 2002; Yager and Barrett, 2001; Delaire and Nakatani, 2000; Viswanathan et al., 1999b).

4.1.1. Experimental Observations

The surface mass patterning unique to azobenzenes is a fundamentally optical process, whereby the incident light pattern is encoded in the material. In an SRG experiment, two beams are intersected at an angle 2θ at the sample surface, giving rise to an SRG with period:

$$\Lambda = \frac{\lambda}{2 \sin \theta} \tag{4.1}$$

where λ is the wavelength of the inscription light. The amplitude (height) of the SRG depends on the inscription angle, displaying a maximum at $\theta \sim 15^\circ$ (Barrett et al., 1996; Kim et al., 1995a). Grating height increases nonlinearly with irradiation time and power, up to a saturation point (Fukuda et al., 2000a; Bian et al., 1999). At moderate fluence, the grating efficiency depends only on the net exposure, not on the temporal distribution of irradiation. Gratings can be formed with intensities as low as 1 mW/cm^2, as long as the inscription wavelength

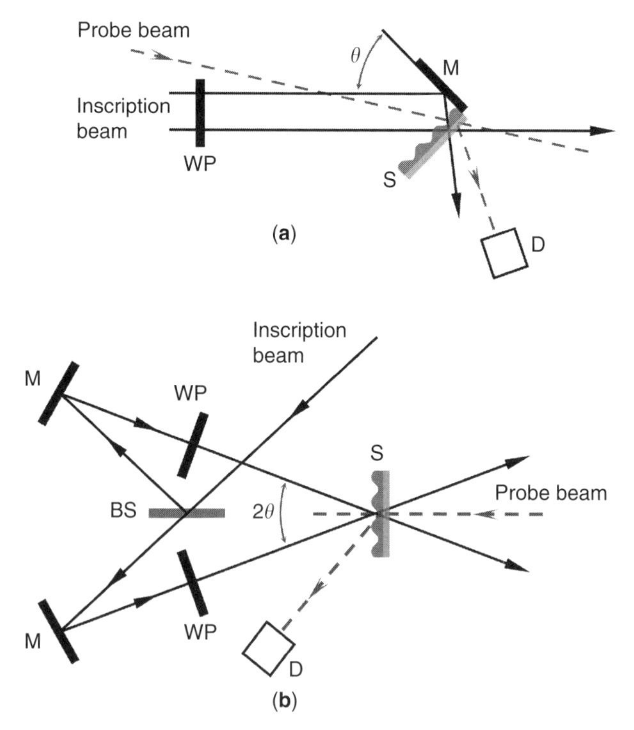

Figure 4.2. Experimental setup for the inscription of a surface relief grating: S refers to the sample, M are mirrors, D is a detector for the diffraction of the probe beam, WP is a waveplate (or generally a combination of polarizing elements), and BS is a 50% beam splitter. The probe beam is usually a HeNe (633 nm), and the inscription beam is chosen on the basis of the chromophore absorption band (often Ar^+ 488 nm). **(a)** A simple one-beam inscription involves reflecting half of the incident beam off of a mirror adjacent to the sample. **(b)** A two-beam interference setup enables independent manipulation of the polarization state of the two incident beams.

is within the azo absorption band. Most chromophores used for SRG formation have a strong overlap of their trans and cis absorption spectra, allowing both isomers to be excited with a single wavelength. However, some experiments have been performed using azobenzenes with trans absorption in the blue, and cis absorption in the red (Jager et al., 2001; Sanchez et al., 2000). Using an interference pattern of red HeNe beams, inscription only occurs if a blue pump beam concurrently irradiates. This biphotonic phenomenon proves that cycling of chromophores, and not simply isomerization, is required for grating formation.

The phase relationship between the incident light field and the resulting surface deformation is crucial in understanding the mechanism of grating formation (Fig. 4.4). Early investigations using the diffraction of an edge (Kumar et al.,

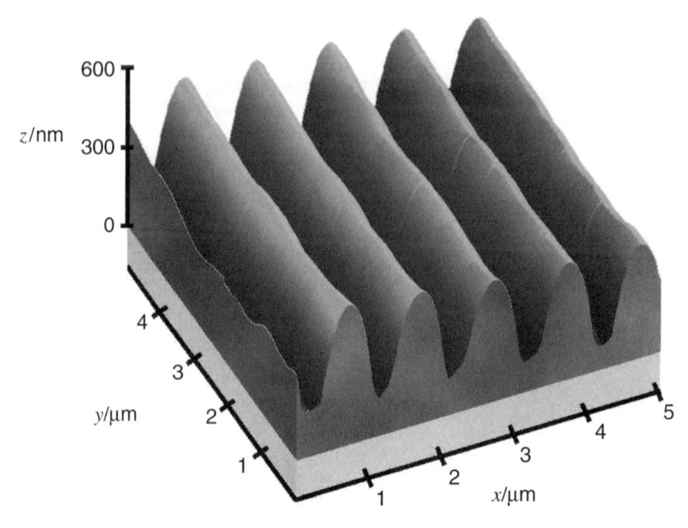

Figure 4.3. AFM image of a typical SRG optically inscribed into an azopolymer film. Grating amplitudes of hundreds of nanometers, on the order of the original film thickness, are easily obtained. In this image, the approximate location of the film-substrate interface has been set to $z = 0$ on the basis of knowledge of the film thickness.

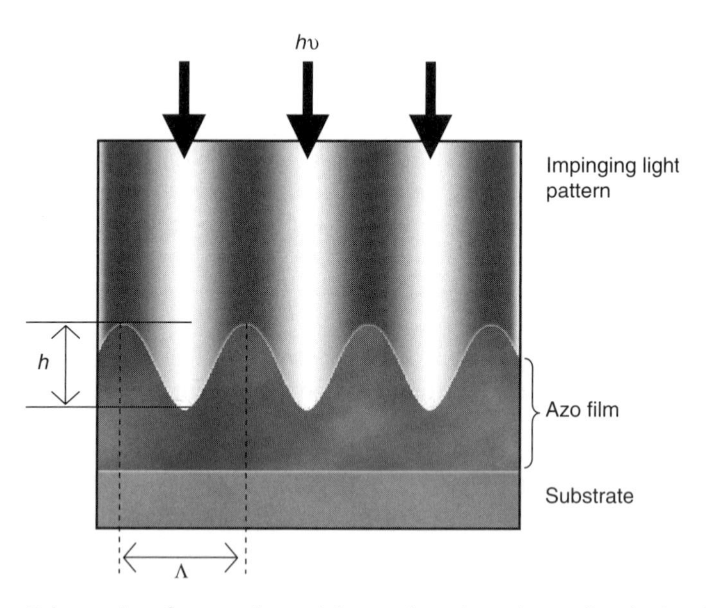

Figure 4.4. Schematic of a grating with spacing Λ and amplitude h. The usual phase relationship is indicated: light intensity maxima correspond to valleys in the surface relief.

1998; Kim et al., 1997; Rochon et al., 1995) and single-beam surface deformations (Bian et al., 1998) convincingly showed that the light and surface relief are 180° out of phase. In other words, light intensity maxima correspond to valleys in the surface relief. In effect, material is moved out of the light and into the dark regions. This rule appears to hold in the majority of cases, yet in a number of systems the phase relationship seems to be exactly inverted, with mass transport into illuminated regions. Specifically, in certain liquid crystalline (LC) systems, the phase behavior is inverted (Helgert et al., 2001; Bublitz et al., 2000; Holme et al., 1999). In one study, an amorphous material exhibited the "usual" phase behavior, and a thin film of this polymer floating on water expanded in the direction of the light polarization. A similar LC azopolymer exhibited inverted phase behavior in SRG experiments, and contracted in the direction of polarized illumination as a floating film. The tempting conclusion is that amorphous and polymeric systems exhibit opposite photomechanical response, which translates into opposite phase behavior in grating inscription. A possibly relevant experiment showed that for LC systems, the cis chromophores may become preferentially oriented along the light polarization direction, instead of being perpendicular to it (Hore et al., 2003). It should be pointed out, however, that some LC systems exhibit the usual phase relationship (Zettsu et al., 2003), and in one LC system, modification of a single ring substituent led to opposite phase behavior (Helgert et al., 2001).

Adding to the complexity of the phase relationship, it was observed that at high irradiation power ($>300\,W/cm^2$), the behavior was inverted in amorphous systems (Bian et al., 1999). Single-beam experiments at high power showed a central peak instead of a depression. By exposing a sample to a gradient two-beam intersection, a film was inscribed with a continuum of laser intensitites. An in-phase grating was found in the high power region, and a conventional out-of-phase SRG was found in the low power region. The intermediate region clearly showed interdigitation of the peaks from the two regimes, resembling a doubling of grating period observed by others. These double-period gratings can be formed in a number of amorphous systems, using the polarization combinations of (p, p) or (+45°, −45°), with indications that even the (s, s) and (p, s) combinations function to a certain extent (Lagugne-Labarthet et al., 2004; Labarthet et al., 2001, 2000; Naydenova et al., 1998a). Observations of a double-frequency orientational grating underneath a normal SRG have also been reported (Schaller et al., 2003). The double-period SRGs were attributed to interference between the diffracted beams from the primary grating, which gives rise to a light modulation, with double the initial frequency, in the material. Whether these double-period gratings are related to the inverted structures observed in high intensity experiments and some LC systems is an open question. It is interesting to note that some of the amorphous polymers that exhibit double-frequency behavior have accessible LC phases at a higher temperature. These phase behavior results strongly suggest that there are (at least) two mechanisms at play during surface patterning. One dominates at low intensity in amorphous systems, whereas another appears to dominate at higher intensity and in LC systems. From an applied standpoint, these double-period gratings are of great interest, as they represent a means of generating patterned structures below the usual diffraction limit of far-field optical lithography.

The fact that the optical inscription process is sensitive to both intensity and polarization is of considerable importance (Jiang et al., 1996). Different polarization combinations lead to different amplitudes, h, of the inscribed SRG, as shown in Table 4.1. (The coordinate system is shown in Fig. 4.5.) An optical field vector component in the direction of light modulation (hence mass transport) appears necessary (Bian et al., 1998). Interestingly, SRGs can even be formed via pure polarization patterns, where the light intensity is uniform over the sample surface (Viswanathan et al., 1999a). In fact, the (s, s) and (RCP, RCP) combinations, which correspond to mainly variations in intensity and little to no polarization contrast, produce very poor SRGs. In contrast, the best gratings are obtained with (+45°, −45°) and (RCP, LCP) combinations, which involves primarily variation in polarization state across the film. It should be noted, however, that the exact polarization pattern present inside the material is not known. The pattern

TABLE 4.1. Polarization Patterns at the Sample Surface During SRG Inscription Using a Variety of Polarized Beam Combinations. The "Quality" of the SRG (as Determined by Grating Height) Is Shown for Comparison

Polarization of beams	Electric field in xy plane					SRG quality
x:	$+\pi$ $+\Lambda/2$	$+\pi/2$ $+\Lambda/4$	0 0	$-\pi/2$ $-\Lambda/4$	$-\pi$ $-\Lambda/2$	
s : s ↕ ↕	•	↕	↕	↕	•	6—Poor
s : p ↕ ↔	⤡	↺	⤢	↻	⤡	4
p : p ↔ ↔ xy Plane	•	↔	↔	↔	•	3
xz Plane	↕	↺	↔	↻	↕	
+45° : +45° ⤢ ⤢	•	⤢	⤢	⤢	•	5
+45° : −45° ⤢ ⤡	↔	↺	↕	↻	↔	2—Good
RCP : RCP ↺ ↺	↺	↺	↺	↺	↺	7—Worst
RCP : LCP ↺ ↻	↔	⤢	↕	⤡	↔	1—Best

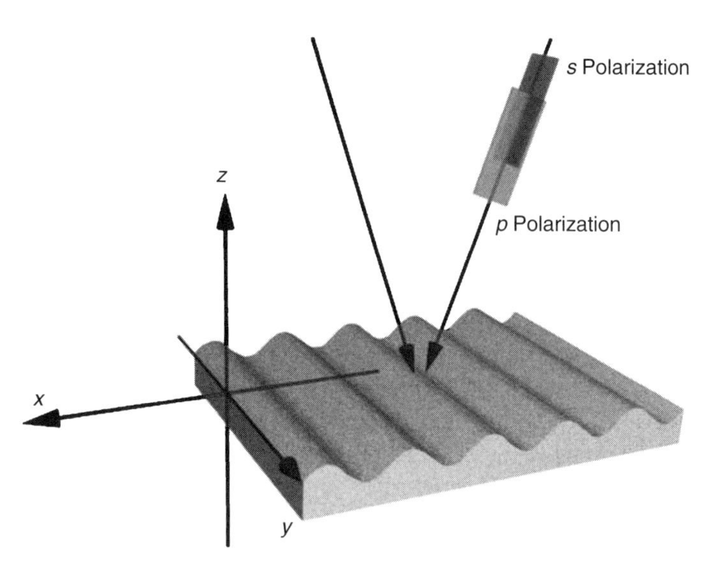

Figure 4.5. Coordinate system used to describe SRG inscription, with x running along the direction of light modulation. The two incident beams have their polarization state controlled. The s polarization is parallel to the surface, whereas the p state is parallel to the plane-of-incidence.

impinging on the sample is readily calculated from knowledge of the input polarizations, yet in the bulk of the material the light pattern will be redirected and repolarized on the basis of the detailed three-dimensional structure of the surface, refractive index, and birefringence. The optical erasure of an SRG, performed by homogenous irradiation, is also polarization-dependent (Lagugne-Labarthet et al., 2002; Jiang et al., 1998). Thus, the gratings possess a memory of the polarization state during inscription, encoded in the orientational distribution of chromophores at various grating positions. Typically, gratings inscribed with highly favorable polarization combinations will be optically erased more quickly. The s-polarization state, which produces poor gratings, is generally optimal for performing the erasure. These results are obviously related to the orientational distribution of chromophores after surface patterning and their subsequent interaction with the erasing polarization.

4.1.2. Patterning

In a typical inscription experiment, a sinusoidally varying light pattern is generated at the sample surface. What results is a sinusoidal surface profile: an SRG. This is the pattern most often reported in the literature, because it is conveniently generated (by intersecting two coherent beams) and easily monitored (by recording the diffraction intensity at a nonabsorbing wavelength, usually using a HeNe laser at 633 nm). However, it must be emphasized that the azo surface mass transport can produce arbitrary patterns. Essentially, the film encodes the impinging light pattern as a topography pattern. Both the intensity and polarization of light are encoded. What appears to be essential is a *gradient* in the intensity or polarization of the incident light field. For instance, a single focused gaussian laser spot will lead to a localized depression and a gaussian line will lead to an elongated trench. (Bian et al., 1998). In principle, any arbitrary pattern could be generated through an appropriate mask, interference/holographic setup, or scanning of a laser spot (Natansohn and Rochon, 2002).

Concomitant with the inscription of a surface relief is a photoorientation of the azo chromophores, which depends on the polarization of the incident beam(s). The orientation of chromophores in SRG experiments has been measured using polarized Raman confocal microspectrometry (Lagugne-Labarthet et al., 2004; Labarthet et al., 2004, 2000). The strong surface orientation is confirmed by photoelectron spectroscopy (Henneberg et al., 2004). What is found is that the chromophores orient perpendicular to the local polarization vector of the impinging interference pattern. Thus, for a ($+45°$, $-45°$) two-beam interference: in the valleys ($x=0$) the electric field is aligned in the y-direction, so the chromophores orient in the x-direction; in the peaks ($x=\Lambda/2$) the chromophores orient in the y-direction; in the slope regions ($x=\Lambda/4$) the electric field is circularly polarized and thus the chromophores are nearly isotropic. For a (p, p) two-beam interference, it is observed that the chromophores are primarily oriented in the y-direction everywhere, since the impinging light pattern is always linearly polarized in the x-direction. Mass transport may lead to perturbations in the orientational distribution, but photoorientation remains the dominant effect.

The anisotropy grating that is submerged below a SRG apparently leads to the formation of a density grating (DG) under appropriate conditions. It was found that upon annealing an SRG, which erases the surface grating and restores a flat film surface, a DG began growing beneath the surface (and into the film bulk) (Geue et al., 2002a; Pietsch et al., 2000). This DG only develops where the SRG was originally inscribed. It appears that the photoorientation and mass transport leads to the nucleation of LC "seeding aggregates" that are thermally grown into larger-scale density variations. The thermal erasure of the SRG, with concomitant growth of the DG, has been measured (Geue et al., 2003) and modeled (Pietsch, 2002). Separating the components because of the surface relief and the DG is described in a later section. Briefly, the diffraction of a visible light laser primarily probes the surface relief, whereas a simultaneous X-ray diffraction experiment probes the DG. The formation of a DG is similar to, and consistent with, the production of surface topography (Watanabe et al., 2000) and surface density patterns (Ikawa et al., 2000), as observed by tapping mode AFM, on an azo film exposed to an optical near field. In these experiments, it was found that volume is not strictly conserved during surface deformation (Keum et al., 2003), consistent with changes in density.

4.1.3. Dependence on Material Properties

For all-optical surface patterning to occur, one necessarily requires azobenzene chromophores in some form. There are, however, a wide variety of azo materials that have exhibited surface mass patterning. This makes the process much more attractive from an applied standpoint: it is not merely a curiosity restricted to a single system, but rather a fundamental phenomenon that can be engineered into a wide variety of materials. It was recognized early on that the gratings do not form in systems of small molecules (for instance, comparing unreacted monomers to their corresponding polymers). The polymer molecular weight (MW), however, must not be too large (Barrett et al., 1996). Presumably a large MW eventually introduces entanglements that act as cross-links, hindering polymer motion. Thus intermediate MW polymers (MW $\sim 10^3$, arguably oligomers) are optimal. That having been said, there are many noteworthy counterexamples. Weak SRGs can be formed in polyelectrolyte multilayers, which are essentially cross-linked-polymer systems (He et al., 2000a,b; Lee et al., 2000; Wang et al., 1998). Efficient grating formation has also been demonstrated using an azo-cellulose with ultrahigh MW (MW $\sim 10^7$) (Yang et al., 2002, 2001). In a high MW polypeptide (MW $\sim 10^5$), gratings could be formed where the grating amplitude was dependent on the polymer conformation (Yang et al., 2003). Restricted conformations (α-helices and β-sheets) hindered SRG formation.

The opposite extreme has also been investigated: molecular glasses (amorphous nonpolymeric azos with bulky pendants) exhibited significant SRG formation (Chun et al., 2003; Kim et al., 2003; Nakano et al., 2002). In fact, the molecular version formed gratings more quickly than its corresponding polymer (Ando et al., 2003). Another set of experiments compared the formation of gratings in two related arrangements: (1) a thin film of polymer and small-molecule azo mixed together and (2) a layered system, where a layer of the small-molecule azo was deposited on top of the pure polymer (Ciuchi et al., 2002). The SRG was negligible in the layered case. Although the authors suggest that "layering" inhibits SRG formation, it may be interpreted that coupling to a host polymer matrix enhances mass transport, perhaps by providing rigidity necessary for fixation of the pattern. A copolymer study did in fact indicate that strong coupling of the mesogen to the polymer enhanced SRG formation (Naydenova et al., 1998b), and molecular glasses with hindered structures also enhanced grating formation (Ishow et al., 2006).

Gratings have also been formed in LC systems (Helgert et al., 2001; Holme et al., 1999). In some systems, it was found that adding stoichiometric quantities of a nonazo LC guest greatly improved the grating inscription (Ubukata et al., 2002, 2000). This suggests that SRG formation may be an inherently cooperative process, related to the mesogenic nature of the azo chromophore. The inscription sometimes requires higher power ($>1\,\text{W/cm}^2$) than in amorphous systems (Ramanujam et al., 1996). In dendrimer systems, the quality of the SRG depends on the generation number (Archut et al., 1998).

Maximizing the content of azo chromophore usually enhances SRG formation (Fukuda et al., 2000a), although some studies have found that intermediate functionalization (50%–80%) created the largest SRG (Borger et al., 2005; Andruzzi et al., 1999). Some attempts have been made to probe the effect of free volume. By attaching substituents to the azo-ring, its steric bulk is increased, which presumably increases the free volume requirement for isomerization. However, substitution also invariably affects the isomerization rate constants, quantum yield, refractive index, etc. This makes any analysis ambiguous. At least in the case of photoorientation, the rate of inscription appears slower for bulkier chromophores, although the net orientation is similar (Ho et al., 1995; Natansohn et al., 1992). For grating formation, it would appear that chromophore bulk is of secondary importance to many other inscription parameters. The mass transport occurs readily at room temperature, which is well below the T_g of the amorphous polymers typically used. Gratings can even be formed in polymers with exceptionally high T_g (Lee et al., 1998), sometimes higher than 370°C (Chen et al., 1999). These gratings can sometimes be difficult to erase via annealing (Wu et al., 2001).

4.1.4. Photosoftening

The formation of an SRG involves massive material motion. It has been suggested that the process is in some way a surface phenomenon since nonazo capping layers tend to inhibit the phenomenon (Viswanathan et al., 1998). Many other experiments, however, confirm that azos deeper in the film (which still absorb light) contribute to the mass transport (Yager and Barrett, 2007; Geue et al., 2002c). It is clear that the azobenzene isomerization is necessary to permit bulk material flow well below the polymer T_g. It is often postulated that repeated trans→cis→trans cycles "photosoften" or "photoplasticize" the polymer matrix, enhancing polymer mobility by orders of magnitude. While compelling, this explanation has been difficult to directly observe. Clearly motion is enabled during isomerization, as demonstrated by mass transport, increases in gas permeability (Kameda et al., 2003), the segregation of some material components to the free surface (Sharma et al., 2002), and the ability to optically erase SRGs (Lagugne-Labarthet et al., 2002; Sanchez et al., 2000; Jiang et al., 1998). The fact that incoherent illumination during SRG inscription enhances grating formation may also be interpreted as evidence that photosoftening is a dominant requirement for mass transport (Yang et al., 2004). Numerous reports have confirmed a reduction in the viscosity of polymer solutions on trans to cis conversion (Moniruzzaman et al., 2004; Bhatnagar et al., 1995; Kumar et al., 1984). This photothinning can be attributed to both chain conformation (reduced hydrodynamic size) and interchain interactions. The extent to which such results can be extended into bulk films is debatable. It is also worth noting that a depression of T_g near the surface of a polymer film is now well established. (Forrest and Dalnoki-Veress, 2001). One might be tempted to explain the mass transport by suggesting that an ultrathin layer of polymer material at the surface is sufficiently mobile (below T_g) to move, thereby exposing "fresh surface," which then becomes mobile. However, the T_g depressions typically measured are not sufficient to account for the process (especially in high T_g samples). Moreover, this does not explain why the mass patterning is only observed in samples that contain azobenzene.

Despite compelling indirect experiments, there are few results that directly suggest photosoftening in bulk samples. Initial hints from AFM response (Kumar et al., 1998), experiments with a quartz crystal microbalance (Srikhirin et al., 2000), and electromechanical spectroscopy (Mechau et al., 2005, 2002; Srikhirin et al., 2000) all indicated that photosoftening occurred. However, the magnitude of the effect was found to be quite small ($<10\%$ change in modulus). Recent AFM modulus (Yager and Barrett, 2006b; Stiller et al., 2004) and tracer diffusion (Mechau et al., 2006) experiments have established that the softening is very small and certainly much smaller than the change in mechanical response that occurs upon heating a polymer to T_g. Thus, although mobility appears to be comparable to a polymer above T_g, other mechanical properties are only slightly modified. It may indeed be that isomerization merely enables localized molecular motion, but that the continual creation of molecular free volume pockets, which are then reoccupied by neighboring isomerizing chromophores, enables a net cooperative movement of material, akin to the displacement of a vacancy defect in a crystal or hole in a semiconductor (Mechau et al., 2002).

4.1.5. Photomechanical Effects

In light of the confounding photosoftening results, an area of increasing research is the direct study of photomechanical effects and macroscopic motion in azobenzene systems. A direct demonstration of macroscopic motion is the bending and unbending of freestanding LC azopolymer films (Ikeda et al., 2003; Yu et al., 2003). This effect is caused by a photocontraction of the free surface, with a decreasing contraction gradient into the depth of the material (as light intensity decreases because of absorption). Polarization can be used to control the direction of bending (Yu et al., 2005), suggesting possible optimization for photoactuation (Tabiryan et al., 2005). Recent measurements using both ellipsometry (Tanchak and Barrett, 2005) and neutron reflectivity (Yager et al., 2006a,b) have identified photoexpansion, on the order of 4%, in amorphous azopolymer films during irradiation. The expansion has both a reversible and irreversible component, suggesting both elastic deformation and viscoelastic flow play a role in azo mass transport. Contraction is observed in LC materials, whereas expansion observed in amorphous polymers is consistent with other investigations. For instance, for thin films floating on a water surface, a contraction in the direction of polarized light was seen in liquid crystal materials, whereas an expansion was observed for amorphous samples (Bublitz et al., 2000). Experiment combining neutron reflectometry and AFM identified both photoexpansion and photocontraction in a single material (Yager and Barrett, 2006b; Yager et al., 2006b). In particular, photoexpansion was observed at low temperatures, whereas photocontraction was observed at higher temperature, with a distinct crossover temperature (at $\sim 50°C$ for pdr1a). This suggests that material motion involves two competing mechanisms: one that causes material expansion (e.g., the internal pressure from the free volume required for azo isomerization) and one that causes material contraction (e.g., dipole pairing and crystallization of azo groups). At low temperatures (in amorphous materials), the rigidity of the matrix prevents the relaxation required for crystallization, and thus photoexpansion is frozen in place. However above a crossover temperature, mobility is sufficient for azo groups to reorganize and crystallize into higher density states. It may thus be that liquid crystal materials exhibit photocontraction simply because they are above their characteristic crossover temperature under normal measurement conditions. These results imply that mass transport can be explained within the framework of photomechanical effects.

4.1.6. Measuring Gratings

The formation of an SRG is typically monitored via the diffraction of a probe laser beam. One must be careful in analyzing this diffraction efficiency, however, because a number of simultaneous gratings will be generated in an azo sample during illumination with a light pattern (Fig. 4.6a–d). A light intensity pattern will lead to a chemical grating, since illuminated regions will be more cis rich than dark regions. The resultant spatial variation of absorbance and refractive index will lead to various optical effects, including diffraction. This grating is temporary,

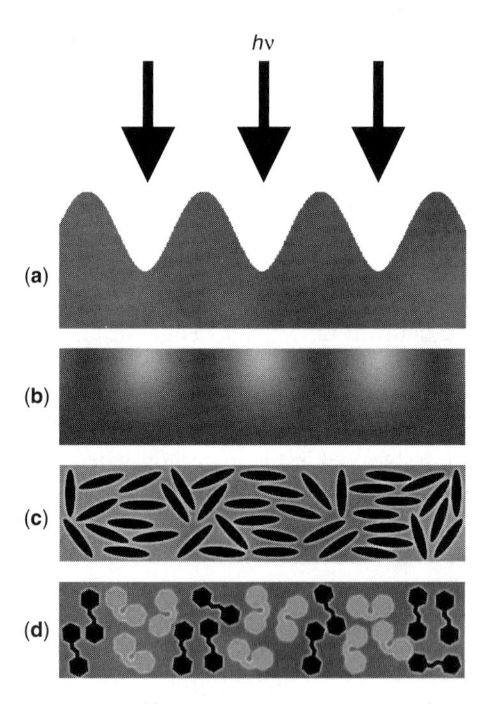

Figure 4.6. Schematic of the different gratings that are formed during illumination of an azo sample. (**a**) SRG. (**b**) DG. (**c**) Birefringence (orientational) grating. (**d**) Chemical (trans/cis) grating. All the gratings contribute to the observed diffraction efficiency, to varying extents. Gratings (**b**), (**c**) and (**d**) are all refractive-index gratings.

persisting only during irradiation. Simultaneously, a birefringence grating will be inscribed. These birefringence patterns are stable and contribute to the observed final diffraction efficiency. An SRG is of course induced if there is any spatial variation in intensity or polarization, and diffraction from this grating is very large, sometimes overwhelming other effects. Lastly, a DG appears to be seeded beneath the material surface, which will also lead to periodic variation of refractive index.

The contributions to diffraction due to the surface relief and birefringence gratings can be deconvoluted using a Jones matrix formalism applied to polarized measurements of the first-order diffracted beams (Helgert et al., 2000; Labarthet et al., 1999, 1998; Naydenova et al., 1998b; Holme et al., 1997). Similarly, scattering theory (Geue et al., 2002b; Pietsch, 2002) was used to fit visible light and X-ray diffraction data (Henneberg et al., 2003; Geue et al., 2002c, 2000; Henneberg et al., 2001b) to deconvolute contributions owing to surface relief and density (refractive index) gratings. For scattering from an SRG, the x and z components of the momentum transfer are

$$\begin{aligned} q_x &= k(\sin\theta_f - \sin\theta_i) \\ q_z &= nk(\cos\theta_i - \cos\theta_f) \end{aligned} \tag{4.2}$$

where θ_i and θ_f are the incident and reflected angles, and $k = 2\pi/\lambda$. For the visible diffracted peak of order m (i.e., when $q_x = m2\pi/\Lambda$), the scattering intensity was derived to be

$$I_{\text{vis}} \approx \left| J_m(q_z h) - e^{iq_z d} J_m(q_z \Delta n_m d) \right|^2 \tag{4.3}$$

where d is the film thickness, h is the grating height, Δn_m is the mth Fourier component of the refractive-index grating, and J_m the mth Bessel function. These Bessel functions can give rise to oscillations in the diffraction signal (Fig. 4.7). Thus, one must be careful not to implicitly assume a linear dependence between grating amplitude and diffraction efficiency. Clearly, both the DG and SRG contribute to the signal. For the X-ray signal, the scattering amplitude is given by the following:

$$A_{\text{x-ray}} \approx J_m(q_z h) - e^{iq_z d} \frac{B_m}{2} \tag{4.4}$$

where one must now use the q_z appropriate for X-rays, and B_m represents the Fourier component of the density modulation (which is presumably identical to the index-modulation). The intensity is not simply the square of A, but can be determined using Fresnel transmission functions. The visible scattering is mostly sensitive to the surface-relief, and the X-ray scattering is due primarily to the density modulation. Using both measurements, the two contributions can of course be determined.

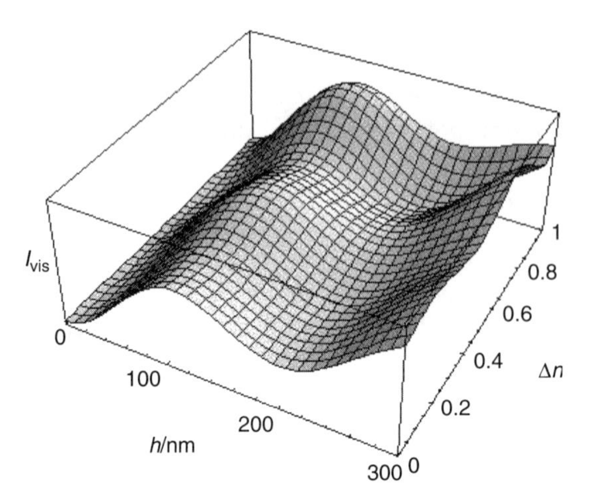

Figure 4.7. Diffraction intensity of a 700 nm thick film as a function of the inscribed grating height, h, and induced refractive-index grating (Δn) based on Equation 4.3. The diffraction intensity is not a linear function of either variable. Oscillations in the signal can complicate analysis of experiments.

It is noteworthy that chromophores may actually be disturbed by probe beams that are well outside of the azo absorption window. For instance, it was found that illumination using red light (outside of the azo absorption band) made DGs (formed underneath SRGs) stable against thermal erasure (Geue et al., 2002a). Others have found that chromophores become slowly aligned even with red laser light, where absorption should be nominal. Luckily the diffraction from azo gratings is intense, enabling the use of heavily attenuated probe beams.

4.1.7. Dynamics

The dynamics of SRG inscription have been the subject of many studies. To a first approximation, the inscription process does not depend on the temporal distribution of laser power, only on the net exposure. Yet this general rule breaks down at high and low power, and for short or interrupted exposures, revealing more complications inherent to the process. The inscription appears to be nonlinear, saturating with inscription time and power, and being dependent on history (Saphiannikova et al., 2004b). Interestingly, the growth of the gratings appears to continue for a short time after illumination has ceased, hinting at a photoinduced stress that persists in the dark. This may be due to the relatively slow decay of the cis population. For short (<2 s) moderate-power laser pulses, no permanent grating can be formed, even after repeated pulse exposure (Henneberg et al., 2001a). For longer (>5 s) pulses, material deformation can be seen after every exposure, and repeated exposure eventually leads to an SRG. These short pulses give rise to localized hills that eventually become a smooth sine wave. This implies that with sufficient power, individual azo chromophores or nanodomains deform, which when summed lead to a grating. Although short exposures do not form permanent gratings, some response is seen, possibly indicating elastic deformation (Geue et al., 2002c; Henneberg et al., 2001b) in addition to plastic flow.

The formation and erasure of surface relief and DGs in azo films has been measured and modeled. Using the formalism described in Section 4.1.6, one can fit the measured scattering by allowing h and Δn (or B_m) to vary in time. For instance, the erasure of the SRG, and concomitant enhancement and then disappearance of the X-ray signal, can be fit using (Geue et al., 2003) the following:

$$
\begin{aligned}
h &= h_{\max} \frac{1}{1 + \exp\left((T - C_{\mathrm{SRG}} T_g)/E_{a,\mathrm{SRG}}\right)} \\
B &= B_{m,\max} \frac{1}{1 + \exp\left((T - C_{\mathrm{DG}} T_g)/E_{a,\mathrm{DG}}\right)}
\end{aligned}
\tag{4.5}
$$

where $E_{a,i}$ and C_i are the activation energy and a fitting parameter, respectively, for the SRG and DG. It is found that the SRG begins disappearing $\sim 15\,\mathrm{K}$ before T_g. In contrast to this thermal erasure of SRGs, it appears that in some LC systems, thermal treatment after SRG formation leads to an enhancement of the grating height (Kawatsuki et al., 2003; Stracke et al., 2000). In these cases, heating may enable motion and aggregation of chromophores. Thin film confinement effects have also been observed. In particular, thin films cause an increase in the erasing temperature, equivalent to an increase of the apparent T_g by as much as $50\,\mathrm{K}$ (Yager and Barrett, 2007). The molecular-scale dynamics (e.g., isomerization) are negligibly affected by confinement, whereas the large-scale mass transport becomes arrested within $\sim 150\,\mathrm{nm}$ of the substrate interface.

4.2. MECHANISM

Several mechanisms have been described to account for the microscopic origin of the driving force in azobenzene optical patterning. Arguments have appealed to thermal gradients, diffusion considerations, isomerization-induced pressure gradients, and interactions between azo dipoles and the electric field of the incident light. Considering the large body of experimental observations, it is perhaps surprising that the issue of mechanism has not yet been settled. At present, no mechanism appears to provide an entirely complete and satisfactory explanation consistent with all known observations. However, viscoelastic modeling of the process has been quite successful, correctly reproducing nearly all experimentally observed surface patterns, without directly describing the microscopic nature of the driving force. Fluid mechanics models provided suitable agreement with observations (Barrett et al., 1998) and were later extended to take into account a depth dependence and a velocity distribution in the film (Fukuda et al., 2000b; Sumaru et al., 1999), which reproduces the thickness dependence of SRG inscription. A further elaboration took into account induced anisotropy in the film and associated anisotropic polymer film deformation (expansion or contraction in the electric field direction) (Bublitz et al., 2001). The assumption of an anisotropic deformation is very much consistent with experimental observations (Bublitz et al., 2000). Such an analysis, remarkably, was able to reproduce most of the polarization dependence, predict phase-inverted behavior at high power, and even demonstrated double-period (interdigitated) gratings. A nonlinear stress-relaxation analysis could account for the nonlinear response during intermittent (pulselike) exposure (Saphiannikova et al., 2004b). Finite-element linear viscoelastic modeling enabled the inclusion of finite compressibility (Saphiannikova et al., 2004a). This allowed the nonlinear intermittent-exposure results, and, critically, the formation of DGs, to be correctly predicted. This analysis also demonstrated, as expected, that surface tension acts as a restoring force that limits grating amplitude (which explains the eventual saturation). Finally, the kinetics of grating formation (and erasure) have been captured in a lattice Monte Carlo simulation that takes into account isomerization kinetics and angular redistribution of chromophores (Mitus et al., 2004; Pawlik et al., 2004, 2003). Thus, the nonlinear viscoelastic flow and deformation (compression and expansion) of polymer material appear to be well understood. What remains to be fully elucidated is the origin of the force inside the material. More specifically, the connection between the azobenzene isomerization and the apparent force must be explained.

4.2.1. Thermal Considerations

Models involving thermal effects were proposed when SRG formation was first observed. Although simple and appealing, a purely thermal mechanism would not account for the polarization dependence that is observed experimentally. The grating formation proceeds at remarkably low laser power, thus thermal mechanisms appear untenable. A more detailed modeling analysis (Yager and Barrett, 2004) showed that the temperature gradient induced in a sample under typical SRG formation conditions was on the order of 10^{-4} K. This thermal gradient is much too small for any appreciable spatial variation of material properties. The net temperature rise in the sample was found to be on the order of 5 K, which again suggests that thermal effects (such as temperature-induced material softening) are negligible. However, high intensity experiments have shown the formation of gratings that could not be subsequently thermally erased (Bian et al., 1999). It is likely that in these cases a destructive thermal mechanism plays a role. In nanosecond-pulsed experiments, gratings can be formed (Si et al., 2002; Leopold et al., 2000; Ramanujam et al., 1999; Schmitt et al., 1997). However, these gratings are because of irreversible ablation of the sample surface, a phenomenon well established in high power laser physics. Moreover, the formation of gratings at these energies does not require azobenzene: any absorbing chromophore will do (Baldus et al., 2001). Computer modeling confirms temperature rises on the order of $\sim$8000 K for nanosecond pulses (Yager and Barrett, 2004), clearly an entirely different regime from the facile room temperature patterning unique to azo chromophores. Although thermal effects should be considered for a complete understanding of SRG formation (especially the phase-inverted structures observed at higher power), they appear to be negligible for typical irradiation conditions at modest laser power.

4.2.2. Asymmetric Diffusion

An elegant anisotropic translation mechanism was developed by Lefin, Fiorini, and Nunzi (Lefin et al., 1998a,b). In this model, material transport occurs essentially because of an (orientational) concentration gradient. It is suggested that the rapid cycling of chromophores between trans and cis states enables transient, random motion of molecules preferentially along their long axis, because of the inherent anisotropy of azo molecules. The probability of undergoing a random-walk step is proportional to the probability of isomerization, which of course depends on the light intensity and the angle between the chromophore dipole and the incident electric vector. This predicts a net flux of molecules out of the illuminated areas and into the dark regions, consistent with experiment. This process would be enhanced by pointing dipoles in the direction of the light gradient (toward the dark regions). This would appear to explain the polarization dependence to a certain extent. In contrast to experiment, however, this model implies the best results when using small molecules, not polymers. For polymer chains laden with many chromophores, random motion of these moeities would presumably lead to a "tug-of-war" that would defeat net transport of the chain. It is at present not clear that the driving force in this model is sufficient to account for the substantial mass transport (well below T_g) observed in experiments.

4.2.3. Mean-Field Theory

Mechanisms based on electromagnetic forces are promising since they naturally include the intensity and polarization state of the incident light field. In the mean-field model developed by Pedersen and coworkers (Pedersen et al., 1998; Pedersen and Johansen, 1997), each chromophore is subject to a potential resulting from all the other chromophore dipoles in the material. Irradiation orients chromophores, and this net orientation leads to a potential that naturally aligns other chromophores. Furthermore, there is an attractive force between side-by-side chromophores that are aligned similarly. This leads to a net force on chromophores in illuminated areas, causing them to order and aggregate. Obviously this model predicts an accumulation of chromophores in the illuminated areas. Thus surface relief peaks will be aligned with light intensity maxima. Although this result does not agree with experiments in amorphous samples, it is consistent with many experiments on LC systems. The mean-field model inherently includes intermolecular cooperativity and orientational order, and it appears natural that it would be manifest in mobile LC systems. The polarization state of incident light is explicitly included in the model, as it serves to align dipoles and thus enhance the mean-field force. Because of this, even pure polarization patterns lead to gratings in this model. This mechanism appeals to the unique properties of azobenzenes only to explain the photoorientation of dipoles. If this mechanism were general, one would expect it to operate on nonisomerizing dipoles that had been aligned by other means, which has not yet been observed.

4.2.4. Permittivity Gradient Theory

A mechanism involving spatial variation of the permittivity, ε, has been suggested by Baldus and Zilker (2001). This model assumes that a spatial modulation of the refractive index, hence permittivity, is induced in the film. This is certainly reasonable, given the well-known photoorientation and birefringence gratings in azo systems. A force is then exerted between the optical electric field and the gradient in permittivity. Specifically, the force is proportional to the intensity of the electric field in the mass transport direction and to the gradient of the permittivity:

$$\vec{f} = -\frac{\varepsilon_0}{2}\vec{E}^2\nabla\varepsilon \tag{4.6}$$

Mass is thus driven out of areas with a strong gradient in ε, which generally moves material into the dark (consistent with the phase relationship in amorphous systems). Here again the mechanism appears general: any system with spatial variation of refractive index should be photopatternable, yet this is not observed. This model would appear to require that adequate photoorientation precede mass transport. Most experiments indicate, however, that both orientational and surface relief phenomena begin immediately and continue concurrently throughout inscription. This model was used to explain SRG formation in pulsed experiments (Baldus et al., 2001; Leopold et al., 2000), where thermal effects were suggested as giving rise to the spatial variation of permittivity, but the resulting force was essentially identical. However, conventional laser ablation appears to be a simpler explanation for those results (Yager and Barrett, 2004).

4.2.5. Gradient Electric Force

Kumar and coworkers proposed a mechanism on the basis of the observation that an electric field component in the direction of mass flow was required (Yang et al., 2006; Bian et al., 2000; Viswanathan et al., 1999a; Kumar et al., 1998). This force is essentially an optical gradient force (Chaumet and Nieto-Vesperinas, 2000; Ashkin, 1997, 1970). Spatial variation of light (electric field intensity and orientation) leads to a variation of the material susceptibility, χ, at the sample surface. The electric field then polarizes the material. The induced polarization is related to the light intensity and local susceptibility:

$$\vec{P}_i = \varepsilon_0\chi_{ij}\vec{E}_j \tag{4.7}$$

Forces then occur between the polarized material and the light field, analogous to the net force on an electric dipole in an electric field gradient. The time-averaged force was derived to be (Viswanathan et al., 1999a) as follows:

$$\vec{f} = \left\langle (\vec{P} \cdot \nabla)\vec{E} \right\rangle \tag{4.8}$$

The grating inscription is related to the spatially varying material susceptibility, the magnitude of the electric field, and the gradient of the electric field. This theory was extended to include near-field optical gradient forces, which have been used for patterning in some experiments (Ikawa et al., 2001). The gradient force model naturally includes the polarization dependence of the incident light, and reproduces essentially all of the polarization features of single-beam and SRG experiments. It has been pointed out (Natansohn and Rochon, 2002), however, that another analysis (Gordon, 1973) of forces exerted on polarizable media suggested a dependence on the gradient of the electric field, but not its polarization direction. The gradient force theory requires azobenzene photochemistry to modulate susceptibility via photoorientation and also implicitly assumes that photoplasticization is enabling mass transport. It would appear, however, that the force density predicted by this model is much too small to account for mass transport in real systems. A straightforward analysis presented by Saphiannikova et al. (2004a) is described here. In the case of two circularly polarized beams (for instance), the force acting in the x-direction, according to the gradient electric force model, would be

$$f = -k\varepsilon_0 \chi E_0^2 \sin\theta(1 + \cos^2\theta)\sin(kx \sin\theta) \tag{4.9}$$

where 2θ is the angle between the two beams and $k = 2\pi/\lambda$. Since $E_0{}^2 = 2Iz_0$, where z_0 is the vacuum impedance and I is the light intensity, the maximum expected force is

$$f_{\max} = \frac{4\pi}{\lambda}\varepsilon_0 z_0 \chi I \tag{4.10}$$

Given that $\varepsilon_0 = 8.854 \times 10^{-12}\,\mathrm{C^2/(Nm^2)}$, $|\chi| \approx 1$ and $z_0 = 377\,\Omega$, and using the typical values $I = 100\,\mathrm{mW/cm^2}$ and $\lambda = 488\,\mathrm{nm}$, a force density of $\sim 100\,\mathrm{N/m^3}$ is obtained. Not only are these two orders of magnitude smaller than the force of gravity (which itself is presumably negligible), but they also fall short of the estimated 10^{11}–$10^{14}\,\mathrm{N/m^3}$ necessary for mass transport in polymer films to occur (Saphiannikova et al., 2004a).

4.2.6. Isomerization Pressure

One of the first mechanisms to be presented was the suggestion by Barrett et al. (1998, 1996) about pressure gradients inside the polymer film. The assumption is that azobenzene isomerization generates pressure both because of the greater free volume requirement of the cis and the volume requirement of the isomerization process itself. Isomerization of the bulky chromophores leads to pressure that is proportional to light intensity. The light intensity gradient thus generates a pressure gradient, which of course leads to material flow in a fluid mechanics treatment. Order-of-magnitude estimates were used to suggest that the mechanical force of isomerization would be greater than the yield point of the polymer, enabling flow. Plastic flow is predicted to drive material out of the light, consistent with observations in amorphous systems. At first it would seem that this mechanism cannot be reconciled with the polarization dependence, since the pressure is presumably proportional to light intensity, irrespective of its polarization state. However, one must more fully take orientational effects into account. Linearly polarized light addresses fewer chromophores than circularly polarized light and would thus lead to lower pressure. Thus, pure polarization patterns can still lead to pressure gradients. Combined with the fact that the polarized light is orienting (and in a certain sense photobleaching), this can explain some aspects of the polarization data. The agreement, however, is not perfect. For instance, the (s, s) and (p, p) combinations lead to very different gratings in experiments. It is possible that some missing detail related to polarization will help explain this discrepancy.

Combining a variety of results from the literature, it now appears that the mechanical argument of a pressure mechanism may be correct. In one experiment, irradiation of a transferred Langmuir–Blodgett film reversibly generated $\sim$5 nm "hills," attributed to nanoscale buckling that relieves the stress induced by lateral expansion (Matsumoto et al., 1998). This result is conspicuously similar to the spontaneous polarization-dependent formation of hexagonally arranged $\sim$500 nm hills seen on an amorphous azopolymer sample irradiated homogeneously (Hubert et al., 2002a,b). In fact, homogeneous illumination of azo surfaces has caused roughening (Mechau et al., 2002), and homogeneous optical erasure of SRGs leads to similar pattern formation (Lagugne-Labarthet et al., 2002). The early stages of SRG formation, imaged by AFM, again show the formation of nanosized hills (Henneberg et al., 2001a). Taken together, these seem to suggest that irradiation of an azo film leads to spontaneous lateral expansion, which induces a stress that can be relieved by buckling of the surface, thereby generating surface structures. In the case of a light gradient, the buckling is relieved by mass transport coincident with the light field that generated the pressure inside the film. In an experiment on main-chain versus side-chain azopolymers, the polarization behavior of photodeformation was opposite (Keum et al., 2003). This may be explained by postulating that the main-chain polymer contracts upon isomerization, whereas the side-chain polymer architecture leads to net expansion. Similarly, the opposite phase behavior in amorphous and LC systems may be due to the fact that the former photoexpand and the latter photocontract (Bublitz et al., 2000). Lastly, many large surface structures were observed in an azo-dye-doped elastomer film irradiated at high power (4 W/cm^2) (Ciuchi et al., 2003). The formation of structures both parallel and perpendicular to the grating direction could be attributed to photoaggregation of the azo dye molecules or buckling of the elastomeric surface. Indeed all of these otherwise contradictory results can be reconciled by considering a more general photomechanical explanation, where both photoexpansion (for materials below a crossover mobility temperature; i.e., amorphous) and photocontraction (for materials above a mobility threshold; i.e., liquid crystal samples) both occur (Yager and Barrett, 2006b). Further investigations into reconciling this model with the polarization dependence of inscription are in order, where presumably photoorientation will play a key role.

4.2.7. Applications of Surface Mass Transport

The single-step, rapid, and reversible all-optical surface patterning effect discovered in a wide variety of azobenzene systems has, of course, been suggested as the basis for numerous applications. Azobenzene is versatile, amenable to incorporation in a wide variety of materials. The mass patterning is reversible, which is often advantageous. However, one may use a system where cross-linking enables permanent fixation of the surface patterns (Zettsu et al., 2001). Many proposed applications are optical and fit well with azobenzene's already extensive list of optical capabilities. The gratings have been demonstrated as optical polarizers (Tripathy et al., 2000), angular or wavelength filters (Stockermans and Rochon, 1999; Rochon et al., 1997), and couplers for optical devices (Paterson et al., 1996). They have also been suggested as photonic band gap materials (Nagata et al., 2001), and have been used to create lasers where emission wavelength is tunable via grating pitch (Rocha et al., 2001; Dumarcher et al., 2000). The process has, of course, been suggested as an optical data storage mechanism (Egami et al., 2000). The high speed and single-step holographic recording has been suggested to enable "instant holography" (Ramanujam et al., 1999), with obvious applications for industry or end consumers. Since the hologram is topographical, it can easily be used as a master to create replicas via molding. The surface patterning also allows multiple holograms to be superimposed, if desired. A novel suggestion is to use the patterning for rapid prototyping of optical elements (Neumann et al., 1999). Optical elements could be generated or modified quickly and during device operation. They could thereafter be replaced with permanent components, if required.

The physical structure of the surface relief can be exploited to organize other systems. For instance, it can act as a command layer, aligning liquid crystals (Kaneko et al., 2002; Parfenov et al., 2001, 2000; Kim et al., 2000; Li et al., 1999). The grating can be formed after the LC cell has been assembled and can be erased and rewritten. Colloids can also be arranged into the grooves of an SRG, thereby templating higher order structures (Yi et al., 2002a, 2001). These lines of colloids can then be sintered to form wires (Yi et al., 2002b). The surface topography inscription process is clearly amenable to a variety of optical-lithography patterning schemes. These possibilities will hopefully be more thoroughly investigated. An advantage of holographic patterning is that there is guaranteed registry between features over macroscopic distances. This is especially attractive as technologies move toward wiring nanometer-sized components. One example in this direction involved evaporating metal onto an SRG, and then annealing. This formed a large number of very long (several millimeter) but extremely thin (200 nm) parallel metal wires (Noel et al., 1996). Of interest for next-generation patterning techniques is the fact that the azo surface modification is amenable to near-field patterning, which enables high resolution nanopatterning by circumventing the usual diffraction limit of far-field optical systems. Proof of principle was demonstrated by irradiating through polystyrene spheres assembled on the surface of an azo film. This results in a polarization-dependent surface topography pattern (Watanabe et al., 2000) and a corresponding surface density pattern (Ikawa et al., 2000). Using this technique, resolution on the order of 20 nm was achieved (Hasegawa et al., 2001). This process appears to be enhanced by the presence of gold nanoislands (Hasegawa et al., 2002). It was also shown that volume is not strictly conserved in these surface deformations (Keum et al., 2003). In addition to being useful as a subdiffraction limit patterning technique, it should be noted that this is also a useful technique for imaging the near field of various optical interactions (Fukuda et al., 2001). The (as of yet not fully explained) fact that subdiffraction limit double-frequency SRGs can be inscribed via far-field illumination (Lagugne-Labarthet et al., 2004; Labarthet et al., 2001, 2000; Naydenova et al., 1998a) further suggests the azopolymers as versatile high resolution patterning materials.

4.3. CONCLUSIONS

The azobenzene chromophore is a unique molecular switch, exhibiting a clean and reversible photoisomerization that induces a reversible change in geometry (Yager and Barrett, 2006a). This motion can be exploited as a photoswitch, and amplified so that larger-scale material properties are switched or altered in response to light. Thus, azo materials offer a promising potential as photomechanical materials. In addition to being useful in a variety of photoswitching roles, the azobenzene isomerization, being fundamentally a geometrical motion, can give rise to many types of motion. These motions can lead to formation of structures at a variety of length scales, from molecular to macroscopic. Furthermore, the formation of these structures is optically controlled, which is very attractive to modern industry. At a molecular level, there is much promise in exploiting azobenzenes to control the organization of nanomaterials and the functioning of nanomechanical devices. At a nanometer level, these materials can be used to form arbitrary surface patterns in a single-step process. At a macroscopic level, the motion could be exploited in a variety of photoactuators and artificial muscles.

With regard to the mechanism of surface mass transport, there is a need for further theory and experiments. An emerging possibility is that two competing mechanisms apply at different power levels. In some systems, notably LC ones, motion is sufficiently free that the "high power" mechanism is readily accessible.

Such an interpretation seems to resolve the apparent conflict between many different results. The nature of the two mechanisms, of course, remains an open question. The high power mechanism may be because of mean-field forces. Again, in mobile LC systems, or at sufficient power, one might expect molecules to align, attract one another, and move cooperatively. In the low power regime, a photomechanical mechanism appears tenable, although polarization questions need to be addressed, presumably by appealing to photoorientation phenomena. From an applied perspective, the azopolymers are well suited for some types of high performance optical lithography. Specifically, these materials have been patterned at the subdiffraction limit level, thus showing that any material resolution limits are below the usual optical limits. The double period gratings that have been produced show that these systems are amenable to subdiffraction limit patterning even with far-field illumination. Ongoing research is evaluating these effects in terms of facile optical formation of nanostructures.

ACKNOWLEDGMENT

This work is dedicated to professors Almeria Natansohn and Sukant Tripathy, teachers and pioneers in the field of azopolymers, who were unable to see the completion of this book.

REFERENCES

Ando H, Takahashi T, Nakano H, Shirota Y. 2003. Comparative studies of the formation of surface relief grating amorphous molecular material vs vinyl polymer. Chem Lett 32(8):710–711.

Andruzzi L, et al. 1999. Holographic gratings in azobenzene side-chain polymethacrylates. Macromolecules 32(2):448–454.

Archut A, et al. 1998. Azobenzene-functionalized cascade molecules: photoswitchable supramolecular systems. Chem Eur J 4(4):699–706.

Ashkin A. 1970. Acceleration and trapping of particles by radiation pressure. Phys Rev Lett 24(4):156–159.

Ashkin A. 1997. Optical trapping and manipulation of neutral particles using lasers. Proc Natl Acad Sci USA 94(10):4853–4860.

Baldus O, Zilker SJ. 2001. Surface relief gratings in photoaddressable polymers generated by cw holography. Appl Phys B-Lasers O 72(4):425–427.

Baldus O, Leopold A, Hagen R, Bieringer T, Zilker SJ. 2001. Surface relief gratings generated by pulsed holography: a simple way to polymer nanostructures without isomerizing side-chains. J Chem Phys 114(3):1344–1349.

Barrett CJ, Natansohn AL, Rochon PL. 1996. Mechanism of optically inscribed high-efficiency diffraction gratings in Azo polymer films. J Phys Chem 100(21):8836–8842.

Barrett CJ, Rochon PL, Natansohn AL. 1998. Model of laser-driven mass transport in thin films of dye-functionalized polymers. J Chem Phys 109(4):1505–1516.

Bhatnagar A, et al. 1995. Azoaromatic polyethers. Polymer 36(15):3019–3025.

Bian S, et al. 1998. Single laser beam-induced surface deformation on azobenzene polymer films. Appl Phys Lett 73(13):1817–1819.

Bian SP, et al. 1999. Photoinduced surface deformations on azobenzene polymer films. J Appl Phys 86(8):4498–4508.

Bian SP, et al. 2000. Photoinduced surface relief grating on amorphous poly(4-phenylazophenol) films. Chem Mater 12(6):1585–1590.

Borger V, et al. 2005. Novel polymers to study the influence of the azobenzene content on the photo-induced surface relief grating formation. Macromol Chem Phys 206(15):1488–1496.

Bublitz D, et al. 2000. Photoinduced deformation of azobenzene polyester films. Appl Phys B: Lasers Opt 70(6):863–865.

Bublitz D, Fleck B, Wenke L. 2001. A model for surface-relief formation in azobenzene polymers. Appl Phys B-Lasers O 72(8):931–936.

Chaumet PC, Nieto-Vesperinas M. 2000. Time-averaged total force on a dipolar sphere in an electromagnetic field. Opt Lett 25(15):1065–1067.

Chen JP, Labarthet FL, Natansohn A, Rochon P. 1999. Highly stable optically induced birefringence and holographic surface gratings on a new azocarbazole-based polyimide. Macromolecules 32(25):8572–8579.

Chun CM, Kim MJ, Vak D, Kim DY. 2003. A novel azobenzene-based amorphous molecular material with a spiro linked bifluorene. J Materials Chem 13(12):2904–2909.

Ciuchi F, Mazzulla A, Cipparrone G. 2002. Permanent polarization gratings in elastomer azo-dye systems: comparison of layered and mixed samples. J Opt Soc Am B 19(11):2531–2537.

Ciuchi F, Mazzulla A, Carbone G, Cipparrone G. 2003. Complex structures of surface relief induced by holographic recording in azo-dye-doped elastomer thin films. Macromolecules 36(15):5689–5693.

Delaire JA, Nakatani K. 2000. Linear and nonlinear optical properties of photochromic molecules and materials. Chem Rev 100(5):1817–1846.

Dumarcher V, et al. 2000. Polymer thin-film distributed feedback tunable lasers. J Opt A: Pure Appl Opt 2(4):279–283.

Egami C, et al. 2000. Two-stage optical data storage in azo polymers. Jpn J Appl Phys 1 39(3B):1558–1561.

Forrest JA, Dalnoki-Veress K. 2001. The glass transition in thin polymer films. Adv Colloid Interface Sci 94(1–3):167–196.

Fukuda T, et al. 2000a. Photofabrication of surface relief grating on films of azobenzene polymer with different dye functionalization. Macromolecules 33(11):4220–4225.

Fukuda T, Sumaru K, Yamanaka T, Matsuda H. 2000b. Photo-induced formation of the surface relief grating on azobenzene polymers: analysis based on the fluid mechanics. Mol Cryst Liq Cryst 345:587–592.

Fukuda T, et al. 2001. Observation of optical near-field as photo-induced surface relief formation. Jpn J Appl Phys 2 40(8B):L900–L902.

Geue T, et al. 2000. X-ray investigations of the molecular mobility within polymer surface gratings. J Appl Phys 87(11):7712–7719.

Geue T, et al. 2002a. Formation of a buried density grating on thermal erasure of azobenzene polymer surface gratings. Coll Surf A 198–200:31–36.

Geue T, Henneberg O, Pietsch U. 2002b. X-ray reflectivity from sinusoidal surface relief gratings. Cryst Res Technol 37(7):770–776.

Geue TM, et al. 2002c. Formation mechanism and dynamics in polymer surface gratings. Phys Rev E 65(5):052801.

Geue TM, et al. 2003. X-ray investigations of formation efficiency of buried azobenzene polymer density gratings. J Appl Phys 93(6):3161–3166.

Gordon JP. 1973. Radiation Forces and Momenta in Dielectric Media. Phys Rev A 8(1):14–21.

Hasegawa M, Ikawa T, Tsuchimori M, Watanabe O, Kawata Y. 2001. Topographical nanostructure patterning on the surface of a thin film of polyurethane containing azobenzene moiety using the optical near field around polystyrene spheres. Macromolecules 34(21):7471–7476.

Hasegawa M, Keum C-D, Watanabe O. 2002. Enhanced photofabrication of a surface nanostructure on azobenzene-functionalized polymer films with evaporated gold Nanoislands. Adv Mater 14(23):1738–1741.

He J-A, et al. 2000a. Photochemical behavior and formation of surface relief grating on self-assembled polyion/dye composite film. J Phys Chem B 104(45):10513–10521.

He J-A, et al. 2000b. Surface relief gratings from electrostatically layered azo dye films. Appl Phys Lett 76(22):3233–3235.

Helgert M, Fleck B, Wenke L, Hvilsted S, Ramanujam PS. 2000. An improved method for separating the kinetics of anisotropic and topographic gratings in side-chain azobenzene polyesters. Appl Phys B-Lasers O 70(6):803–807.

Helgert M, Wenke L, Hvilsted S, Ramanujam PS. 2001. Surface relief measurements in side-chain azobenzene polyesters with different substituents. Appl Phys B: Lasers Opt 72(4):429–433.

Henneberg O, et al. 2001a. Atomic force microscopy inspection of the early state of formation of polymer surface relief gratings. Appl Phys Lett 79(15):2357–2359.

Henneberg O, et al. 2001b. Formation and dynamics of polymer surface relief gratings. Appl Surf Sci 182(3–4):272–279.

Henneberg O, Geue T, Rochon P, Pietsch U. 2003. X-ray and VIS light scattering from light-induced polymer gratings. J Phys D Appl Phys 36(10A):A241–A244.

Henneberg O, Geue T, Pietsch U, Saphiannikova M, Winter B. 2004. Investigation of azobenzene side group orientation in polymer surface relief gratings by means of photoelectron spectroscopy. Appl Phys Lett 84(9):1561–1563.

Ho MS, Natansohn A, Rochon P. 1995. Azo polymers for reversible optical storage 7. The effect of the size of the photochromic groups. Macromolecules 28(18):6124–6127.

Holme NCR, Nikolova L, Ramanujam PS, Hvilsted S. 1997. An analysis of the anisotropic and topographic gratings in a side-chain liquid crystalline azobenzene polyester. Appl Phys Lett 70(12):1518–1520.

Holme NCR, et al. 1999. Optically induced surface relief phenomena in azobenzene polymers. Appl Phys Lett 74(4):519–521.

Hore DK, Natansohn AL, Rochon PL. 2003. Anomalous cis isomer orientation in a liquid crystalline azo polymer on irradiation with linearly-polarized light. J Phys Chem B 107(10):2197–2204.

Hubert C, Fiorini-Debuisschert C, Maurin I, Nunzi JM, Raimond P. 2002a. Spontaneous patterning of hexagonal structures in an azo-polymer using light-controlled mass transport. Adv Mater 14(10):729.

Hubert C, et al. 2002b. Micro structuring of polymers using a light-controlled molecular migration processes. Appl Surf Sci 186(1–4):29–33.

Ikawa T, et al. 2000. Optical near field induced change in viscoelasticity on an azobenzene-containing polymer surface. J Phys Chem B 104(39):9055–9058.

Ikawa T, et al. 2001. Azobenzene polymer surface deformation due to the gradient force of the optical near field of monodispersed polystyrene spheres. Phys Rev B 64(19).

Ikeda T, Nakano M, Yu Y, Tsutsumi O, Kanazawa A. 2003. Anisotropic bending and unbending behavior of azobenzene liquid-crystalline gels by light exposure. Adv Mater 15(3):201–205.

Ishow E, et al. 2006. Structural and photoisomerization cross studies of polar photochromic monomeric glasses forming surface relief gratings. Chem Mater 18(5):1261–1267.

Jager C, Bieringer T, Zilker SJ. 2001. Bicolor surface reliefs in azobenzene side-chain polymers. Appl Opt 40(11):1776–1778.

Jiang XL, Kumar J, Kim DY, Shivshankar V, Tripathy SK. 1996. Polarization dependent recordings of surface relief gratings on azobenzene containing polymer films. Appl Phys Lett 68(19):2618–2620.

Jiang XL, Li L, Kumar J, Kim DY, Tripathy SK. 1998. Unusual polarization dependent optical erasure of surface relief gratings on azobenzene polymer films. Appl Phys Lett 72(20):2502–2504.

Kameda M, Sumaru K, Kanamori T, Shinbo T. 2003. Photoresponse gas permeability of azobenzene-functionalized glassy polymer films. J Appl Polym Sci 88(8):2068–2072.

Kaneko F, et al. 2002. Photo-induced fabrication of surface relief gratings in alternate self-assembled films containing azo dye and alignments of LC molecules. Coll Surf A 198:805–810.

Kawatsuki N, Uchida E, Ono H. 2003. Formation of pure polarization gratings in 4-methoxyazobenzene containing polymer films using off-resonant laser light. Appl Phys Lett 83(22):4544–4546.

Keum CD, Ikawa T, Tsuchimori M, Watanabe O. 2003. Photodeformation behavior of photodynamic polymers bearing azobenzene moieties in their main and/or side chain. Macromolecules 36(13):4916–4923.

Kim DY, et al. 1995a. Polarized laser induced holographic surface relief gratings on polymer films. Macromolecules 28(26):8835–8839.

Kim DY, Tripathy SK, Li L, Kumar J. 1995b. Laser-induced holographic surface relief gratings on nonlinear optical polymer films. Appl Phys Lett 66(10):1166–1168.

Kim DY, et al. 1997. Photo-fabrication of surface relief gratings on polymer films. Macromol Symp 116:127–134.

Kim M-H, Kim J-D, Fukuda T, Matsuda H. 2000. Alignment control of liquid crystals on surface relief gratings. Liq Cryst 27(12):1633–1640.

Kim M-J, Seo E-M, Vak D, Kim D-Y. 2003. Photodynamic properties of azobenzene molecular films with triphenylamines. Chem Mater 15(21):4021–4027.

Kumar GS, DePra P, Zhang K, Neckers DC. 1984. Chelating copolymers containing photosensitive functionalities. 2. Macromolecules 17(12):2463–2467.

Kumar J, et al. 1998. Gradient force: the mechanism for surface relief grating formation in azobenzene functionalized polymers. Appl Phys Lett 72(17):2096–2098.

Labarthet FL, Buffeteau T, Sourisseau C. 1998. Analyses of the diffraction efficiencies, birefringence, and surface relief gratings on azobenzene-containing polymer films. J Phys Chem B 102(15):2654–2662.

Labarthet FL, Rochon P, Natansohn A. 1999. Polarization analysis of diffracted orders from a birefringence grating recorded on azobenzene containing polymer. Appl Phys Lett 75(10):1377–1379.

Labarthet FL, et al. 2000. Photoinduced orientations of azobenzene chromophores in two distinct holographic diffraction gratings as studied by polarized Raman confocal microspectrometry. Phys Chem Chem Phys 2(22):5154–5167.

Labarthet FL, Buffeteau T, Sourisseau C. 2001. Time dependent analysis of the formation of a half-period surface relief grating on amorphous azopolymer films. J Appl Phys 90(7):3149–3158.

Labarthet FL, Bruneel JL, Buffeteau T, Sourisseau C. 2004. Chromophore orientations upon irradiation in gratings inscribed on azo-dye polymer films: a combined AFM and confocal Raman microscopic study. J Phys Chem B 108(22):6949–6960.

Lagugne-Labarthet F, Buffeteau T, Sourisseau C. 2002. Optical erasures and unusual surface reliefs of holographic gratings inscribed on thin films of an azobenzene functionalized polymer. Phys Chem Chem Phys 4(16):4020–4029.

Lagugne-Labarthet F, Bruneel JL, Rodriguez V, Sourisseau C. 2004. Chromophore orientations in surface relief gratings with second-order nonlinearity as studied by confocal polarized Raman microspectrometry. J Phys Chem B 108(4):1267–1278.

Lee S-H, et al. 2000. Azo polymer multilayer films by electrostatic self-assembly and layer-by-layer post azo functionalization. Macromolecules 33(17):6534–6540.

Lee TS, et al. 1998. Photoinduced surface relief gratings in high-Tg main-chain azoaromatic polymer films. J Polym Sci, Part A: Polym Chem 36(2):283–289.

Lefin P, Fiorini C, Nunzi JM. 1998a. Anisotropy of the photo-induced translation diffusion of azobenzene dyes in polymer matrices. Pure Appl Opt 7(1):71–82.

Lefin P, Fiorini C, Nunzi JM. 1998b. Anisotropy of the photoinduced translation diffusion of azo-dyes. Opt Mater 9(1–4):323–328.

Leopold A, et al. 2000. Thermally induced surface relief gratings in azobenzene polymers. J Chem Phys 113(2):833–837.

Li XT, Natansohn A, Rochon P. 1999. Photoinduced liquid crystal alignment based on a surface relief grating in an assembled cell. Appl Phys Lett 74(25):3791–3793.

Matsumoto M, et al. 1998. Reversible light-induced morphological change in Langmuir-Blodgett films. J Am Chem Soc 120(7):1479–1484.

Mechau N, Neher D, Borger V, Menzel H, Urayama K. 2002. Optically driven diffusion and mechanical softening in azobenzene polymer layers. Appl Phys Lett 81(25): 4715–4717.

Mechau N, Saphiannikova M, Neher D. 2005. Dielectric and mechanical properties of azobenzene polymer layers under visible and ultraviolet irradiation. Macromolecules 38(9):3894–3902.

Mechau N, Saphiannikova M, Neher D. 2006. Molecular tracer diffusion in thin azobenzene polymer layers. Appl Phys Lett 89(25):251902–3.

Mitus AC, Pawlik G, Miniewicz A, Kajzar F. 2004. Kinetics of diffraction gratings in a polymer matrix containing azobenzene chromophores: experiment and Monte Carlo simulations. Mol Cryst Liq Cryst 416:113–126.

Moniruzzaman M, Sabey CJ, Fernando GF. 2004. Synthesis of azobenzene-based polymers and the in-situ characterization of their photoviscosity effects. Macromolecules 37(7):2572–2577.

Nagata T, Matsui T, Ozaki M, Yoshino K, Kajzar F. 2001. Novel optical properties of conducting polymer-photochromic polymer systems. Synth Met 119(1–3):607–608.

Nakano H, Takahashi T, Kadota T, Shirota Y. 2002. Formation of a surface relief grating using a novel azobenzene-based photochromic amorphous molecular material. Adv Mater 14(16):1157–1160.

Natansohn A, Rochon P. 2002. Photoinduced motions in azo-containing polymers. Chem Rev 102(11):4139–4176.

Natansohn A, Xie S, Rochon P. 1992. Azo polymers for reversible optical storage 2. Poly[4′-[[2-(acryloyloxy)ethyl]ethylamino]-2-chloro-4-nitroazobenzene]. Macromolecules 25(20):5531–5532.

Naydenova I, et al. 1998a. Diffraction from polarization holographic gratings with surface relief in side-chain azobenzene polyesters. J Opt Soc Am B 15(4):1257–1265.

Naydenova I, et al. 1998b. Polarization holographic gratings with surface relief in amorphous azobenzene containing methacrylic copolymers. Pure Appl Opt 7(4):723–731.

Neumann J, Wieking KS, Kip D. 1999. Direct laser writing of surface reliefs in dry, self-developing photopolymer films. Appl Opt 38(25):5418–5421.

Noel S, Batalla E, Rochon P. 1996. A simple method for the manufacture of mesoscopic metal wires. J Mater Res 11(4):865–867.

Parfenov A, Tamaoki N, Ohnishi S. 2000. Photoinduced alignment of nematic liquid crystal on the polymer surface microrelief. J Appl Phys 87(4):2043–2045.

Parfenov A, Tamaoki N, Ohni-Shi S. 2001. Photoinduced alignment of nematic liquid crystal on the polymer surface microrelief. Mol Cryst Liq Cryst 359:487–495.

Paterson J, Natansohn A, Rochon P, Callendar CL, Robitaille L. 1996. Optically inscribed surface relief diffraction gratings on azobenzene-containing polymers for coupling light into slab waveguides. Appl Phys Lett 69(22):3318–3320.

Pawlik G, Mitus AC, Miniewicz A, Kajzar F. 2003. Kinetics of diffraction gratings formation in a polymer matrix containing azobenzene chromophores: experiments and Monte Carlo simulations. J Chem Phys 119(13):6789–6801.

Pawlik G, Mitus AC, Miniewicz A, Kajzar F. 2004. Monte Carlo simulations of temperature dependence of the kinetics of diffraction gratings formation in a polymer matrix containing azobenzene chromophores. J Nonlinear Opt Phy Mater 13(3–4):481–489.

Pedersen TG, Johansen PM. 1997. Mean-field theory of photoinduced molecular reorientation in azobenzene liquid crystalline side-chain polymers. Phys Rev Lett 79(13): 2470–2473.

Pedersen TG, Johansen PM, Holme NCR, Ramanujam PS, Hvilsted S. 1998. Mean-field theory of photoinduced formation of surface reliefs in side-chain azobenzene polymers. Phys Rev Lett 80(1):89–92.

Pietsch U. 2002. X-ray and visible light scattering from light-induced polymer gratings. Phys Rev B 66(15):155430.

Pietsch U, Rochon P, Natansohn A. 2000. Formation of a buried lateral density grating in azobenzene polymer films. Adv Mater 12(15):1129–1132.

Ramanujam PS, Holme NCR, Hvilsted S. 1996. Atomic force and optical near-field microscopic investigations of polarization holographic gratings in a liquid crystalline azobenzene side-chain polyester. Appl Phys Lett 68(10):1329–1331.

Ramanujam PS, Pedersen M, Hvilsted S. 1999. Instant holography. Appl Phys Lett 74(21):3227–3229.

Rocha L, et al. 2001. Laser emission in periodically modulated polymer films. J Appl Phys 89(5):3067–3069.

Rochon P, Batalla E, Natansohn A. 1995. Optically induced surface gratings on azoaromatic polymer films. Appl Phys Lett 66(2):136–138.

Rochon P, Natansohn A, Callendar CL, Robitaille L. 1997. Guided mode resonance filters using polymer films. Appl Phys Lett 71(8):1008–1010.

Sanchez C, Alcala R, Hvilsted S, Ramanujam PS. 2000. Biphotonic holographic gratings in azobenzene polyesters: surface relief phenomena and polarization effects. Appl Phys Lett 77(10):1440–1442.

Saphiannikova M, Geue TM, Henneberg O, Morawetz K, Pietsch U. 2004a. Linear viscoelastic analysis of formation and relaxation of azobenzene polymer gratings. J Chem Phys 120(8):4039–4045.

Saphiannikova M, Henneberg O, Gene TM, Pietsch U, Rochon P. 2004b. Nonlinear effects during inscription of azobenzene surface relief gratings. J Phys Chem B 108(39):15084–15089.

Schaller RD, Saykally RJ, Shen YR, Lagugne-Labarthet F. 2003. Poled polymer thin-film gratings studied with far-field optical and second-harmonic near-field microscopy. Opt Lett 28(15):1296–1298.

Schmitt K, Benecke C, Schadt M. 1997. Pulsed, laser-induced holographic coupling gratings for waveguides made of cross-linkable polymers. Appl Opt 36(21):5078–5082.

Sharma L, Matsuoka T, Kimura T, Matsuda H. 2002. Investigation into the surface relief grating mechanism via XPS in new azobenzene based optical material. Polym Adv Technol 13(6):481–486.

Si JH, Qiu JR, Zhai JF, Shen YQ, Hirao K. 2002. Photoinduced permanent gratings inside bulk azodye-doped polymers by the coherent field of a femtosecond laser. Appl Phys Lett 80(3):359–361.

Srikhirin T, Laschitsch A, Neher D, Johannsmann D. 2000. Light-induced softening of azobenzene dye-doped polymer films probed with quartz crystal resonators. Appl Phys Lett 77(7):963–965.

Stiller B, et al. 2004. Optically induced mass transport studied by scanning near-field optical- and atomic force microscopy. Phys Low-Dimens Str 1–2:129–137.

Stockermans RJ, Rochon PL. 1999. Narrow-band resonant grating waveguide filters constructed with azobenzene polymers. Appl Opt 38(17):3714–3719.

Stracke A, Wendorff JH, Goldmann D, Janietz D, Stiller B. 2000. Gain effects in optical storage: thermal induction of a surface relief grating in a smectic liquid crystal. Adv Mater 12(4):282–285.

Sumaru K, Yamanaka T, Fukuda T, Matsuda H. 1999. Photoinduced surface relief gratings on azopolymer films: analysis by a fluid mechanics model. Appl Phys Lett 75(13):1878–1880.

Tabiryan N, Serak S, Dai XM, Bunning T. 2005. Polymer film with optically controlled form and actuation. Opt Express 13(19):7442–7448.

Tanchak OM, Barrett CJ. 2005. Light-induced reversible volume changes in thin films of azo polymers: the photomechanical effect. Macromolecules 38(25):10566–10570.

Tripathy SK, Viswanathan NK, Balasubramanian S, Kumar J. 2000. Holographic fabrication of polarization selective diffractive optical elements on azopolymer film. Polym Adv Technol 11(8–12):570–574.

Ubukata T, Seki T, Ichimura K. 2000. Surface relief gratings in host-guest supramolecular materials. Adv Mater 12(22):1675–1678.

Ubukata T, Seki T, Ichimura K. 2002. Surface relief grating in hybrid films composed of azobenzene polymer and liquid crystal molecule. Coll Surf A 198:113–117.

Viswanathan NK, Balasubramanian S, Li L, Kumar J, Tripathy SK. 1998. Surface-initiated mechanism for the formation of relief gratings on azo-polymer films. J Phys Chem B 102(31):6064–6070.

Viswanathan NK, Balasubramanian S, Li L, Tripathy SK, Kumar J. 1999a. A detailed investigation of the polarization-dependent surface-relief-grating formation process on azo polymer films. Jpn J Appl Phys 1 38(10):5928–5937.

Viswanathan NK, et al. 1999b. Surface relief structures on azo polymer films. J Mater Chem 9(9):1941–1955.

Wang X, Balasubramanian S, Kumar J, Tripathy SK, Li L. 1998. Azo chromophore-functionalized polyelectrolytes. 1. Synthesis, characterization, and photoprocessing. Chem Mater 10(6):1546–1553.

Watanabe O, et al. 2000. Transcription of near-field induced by photo-irradiation on a film of azo-containing urethane-urea copolymer. Mol Cryst Liq Cryst 345:629–634.

Wu YL, Natansohn A, Rochon P. 2001. Photoinduced birefringence and surface relief gratings in novel polyurethanes with azobenzene groups in the main chain. Macromolecules 34(22):7822–7828.

Yager KG, Barrett CJ. 2001. All-optical patterning of azo polymer films. Curr Opin Solid State Mater Sci 5(6):487–494.

Yager KG, Barrett CJ. 2004. Temperature modeling of laser-irradiated azo-polymer thin films. J Chem Phys 120(2):1089–1096.

Yager KG, Barrett CJ. 2006a. Novel photo-switching using azobenzene functional materials. J Photochemistry Photobiology A: Chemistry 182(3):250–261.

Yager KG, Barrett CJ. 2006b. Photomechanical surface patterning in azo-polymer materials. Macromolecules 39(26):9320–9326.

Yager KG, Barrett CJ. 2007. Confinement of surface patterning in azo-polymer thin films. J Chem Phys 126(9):094908–094908.

Yager KG, Tanchak OM, Barrett CJ, Watson MJ, Fritzsche H. 2006a. Temperature-controlled neutron reflectometry sample cell suitable for study of photoactive thin films. Rev Sci Instrum 77(4):045106-1–045106-6.

Yager KG, Tanchak OM, Godbout C, Fritzsche H, Barrett CJ. 2006b. Photomechanical effects in azo-polymers studied by neutron reflectometry. Macromol 39(26):9311–9319.

Yamaki S, Nakagawa M, Morino S, Ichimura K. 2000. Surface relief gratings generated by a photocrosslinkable polymer with styrylpyridine side chains. Appl Phys Lett 76(18):2520–2522.

Yang K, Yang SZ, Wang XG, Kumar J. 2004. Enhancing the inscription rate of surface relief gratings with an incoherent assisting light beam. Appl Phys Lett 84(22):4517–4519.

Yang K, Yang SZ, Kumar J. 2006. Formation mechanism of surface relief structures on amorphous azopolymer films. Phys Rev B 73(16).

Yang S, et al. 2002. Azobenzene-modified cellulose. Polymer News 27:368–372.

Yang SZ, Li L, Cholli AL, Kumar J, Tripathy SK. 2001. Photoinduced surface relief gratings on azocellulose films. J Macromol Sci Pure 38(12):1345–1354.

Yang SZ, Li L, Cholli AL, Kumar J, Tripathy SK. 2003. Ambenzene-modified poly (L-glutamic acid) (AZOPLGA): its conformational and photodynamic properties. Biomacromolecules 4(2):366–371.

Ye YH, Badilescu S, Truong VV, Rochon P, Natansohn A. 2001. Self-assembly of colloidal spheres on patterned substrates. Appl Phys Lett 79(6):872–874.

Yi DK, Kim MJ, Kim DY. 2002a. Surface relief grating induced colloidal crystal structures. Langmuir 18(6):2019–2023.

Yi DK, Seo E-M, Kim D-Y. 2002b. Fabrication of a mesoscale wire: sintering of a polymer colloid arrayed inside a one-dimensional groove pattern. Langmuir 18(13):5321–5323.

Yu Y, Nakano M, Ikeda T. 2003. Photomechanics: directed bending of a polymer film by light. Nature 425:145.

Yu YL, Nakano M, Maeda T, Kondo M, Ikeda T. 2005. Precisely direction-controllable bending of cross-linked liquid-crystalline polymer films by light. Mol Cryst Liq Cryst 436:1235–1244.

Zettsu N, Ubukata T, Seki T, Ichimura K. 2001. Soft crosslinkable azo polymer for rapid surface relief formation and persistent fixation. Adv Mater 13(22):1693–1697.

Zettsu N, Fukuda T, Matsuda H, Seki T. 2003. Unconventional polarization characteristic of rapid photoinduced material motion in liquid crystalline azobenzene polymer films. Appl Phys Lett 83(24):4960–4962.

5

AZO POLYMER COLLOIDAL SPHERES: FORMATION, TWO-DIMENSIONAL ARRAY, AND PHOTORESPONSIVE PROPERTIES

Xiaogong Wang

5.1. INTRODUCTION

Azobenzene and its derivatives have been extensively used as dyestuffs, pigments, and pH indicators for a long time (Zollinger, 1991). It is well known that azobenzene can exist as trans or cis isomers in the electronic ground state (Rau, 1990). The trans isomer can be converted into cis isomer upon UV light (365 nm) irradiation. The cis-to-trans isomerization readily occurs at room temperature and can be further accelerated by heating or stimulated by visible light irradiation. The trans–cis isomerization mechanism and related problems have been intensively investigated by both theoretical calculations and experimental analyses for several decades (see, e.g., Cembran et al., 2004; Schultz et al., 2003; Ishikawa et al., 2001; Cattaneo and Persico, 1999). For the photoisomerization, the transition is related to the motion out of the Franck–Condon range on the potential energy surfaces (PESs) of the excited states. Two possible mechanisms of the isomerization have been proposed and tested by different methods. In typical cases, the rotation mechanism attributes the isomerization to the twist around the CNNC dihedral angle, whereas the inversion mechanism ascribes the isomerization to the inversion of one or both of the CNN angles through a linear transition state (Chang et al., 2004; Rau, 1990). Substitution on azobenzene can significantly affect the energy gap between the ground and exited states and the isomerization pathways (Crecca and Roitberg, 2006; Schmidt et al., 2004; Biswas et al., 2002; Åstrand et al., 2000). The former causes the azobenzene derivatives to possess absorption bands at different wavelengths and exhibit many different hues as dyestuffs and pigments.

Smart Light-Responsive Materials. Edited by Yue Zhao and Tomiki Ikeda

The latter can result in a variety of expected and unexpected isomerization behaviors. The problems such as the stable conformation, PESs of the excited states, and isomerization mechanism of the azo compounds are still being explored as active research frontiers. Some puzzling problems as debatable issues still remain in subjects even after persistent study for years. Interested readers can find more comprehensive reports on these issues in some recent publications (see, e.g., Diau, 2004; Satzger et al., 2004).

For real applications, isomerization that can cause color change for dyestuffs had more or less been considered as an unfavorable factor in history for a long time (Zollinger, 1961). Significant effort was devoted to prevent the isomerization through some improved molecular designs. This situation has been dramatically changed after early efforts to covalently introduce azobenzene-type chromophores into polymeric chains (Kumar and Neckers, 1989). The azobenzene-containing polymers (azo polymers for short) can show a variety of unique photoresponsive properties triggered by the photoisomerization (Natansohn and Rochon, 2002; Delaire and Nakatani, 2000; Xie et al., 1993). The polymeric materials are expected for the applications in areas such as data storage, optical switching, sensors, and actuators (Yager and Barrett, 2001; Kawata and Kawata, 2000; Xie et al., 1993). Many types of the azo polymers have been designed and synthesized, which include side-chain, main-chain, dendritic azo polymers, and block copolymers (Lambeth and Moore, 2007; Natansohn and Rochon, 2002; Tian et al., 2002; Junge and McGrath, 1999; Archut et al., 1998; Kumar and Neckers, 1989). The azo polymers have been prepared by radical polymerization, step-growth polycondensation, atomic transfer radical polymerization (ATRP), and ring-opening metathesis polymerization (ROMP) among others. The polymeric backbones not only perform as a good medium with transparent and easy-processable properties but also play an important role in transferring or coordinating the photoisomerization effect of azo chromophores. For azo polymers, light irradiation can cause photoresponsive variations such as phase transition (Ikeda et al., 1990), photoinduced chromophore orientation (Todorov et al., 1984), light-driven thin-film contraction and bending (Li et al., 2003; Yu et al., 2003; Finkelmann et al., 2001), and surface relief grating (SRG) formation (Kim et al., 1995; Rochon et al., 1995). The last but not the least of importance, polymeric chains are able to undergo different self-assembling processes. The structures and photoresponsive properties of the azobenzene-containing materials can be dramatically enriched through the self-assembling approaches.

Self-assembly generally refers to a process in which components, either separate or linked, spontaneously form order aggregates (Whitesides and Boncheva, 2002; Whitesides and Grzybowski, 2002). Self-assembly has been extensively explored to develop a variety of well-defined structures through bottom-up approaches (Förster and Plantenberg, 2002; Philp and Stoddart, 1996). Azo polymers with specially designed functional groups or chain structures can undergo different types of self-assembly. Multilayer thin films composed of azo polyelectrolytes have been obtained through sequential electrostatic layer-by-layer self-assembly (Laschewsky et al., 1997; Lvov et al., 1997; Wang et al., 1997a). The azo chromophores in the films can automatically possess the noncentral symmetric alignment, and the thin films can show the significant second-order nonlinear optical (NLO) effect without electric field poling. The electrostatic layer-by-layer films containing azo chromophores can also show photoresponsive behavior related with the photoisomerization (Wu et al., 2001; Dante et al., 1999). The block copolymer containing mesogenic azobenzene units in one of the blocks has been developed (Morikawa et al., 2006). The polymer can form thin films with well-organized cylinder structures and photoinduced alignment properties. Amphiphilic azo block copolymers have been prepared by ATRP, and light-responsive micellar aggregates have been obtained (Wang et al., 2004). Other recent progresses in the self-assembly of azo polymers can be seen in other chapters of this book.

This chapter reviews some recent research progresses concerning the colloidal spheres constructed from the polydispersed amphiphilic azo polymers. In contrast to the amphiphilic block copolymers being reviewed in Chapters 6 and 7, the polymers discussed in this chapter are polydispersed amphiphilic azo homopolymers and random copolymers. The polymers can be prepared through some relatively simple methods and are easily available from many resources. The polymers can form uniform colloid spheres in selective solvent through self-assembling process. The colloidal spheres can form ordered two-dimensional (2-D) arrays through vertical deposition, and the 2-D arrays can transform to ordered porous films through *in situ* structure inversion. In certain sense, these processes can well disclose the essence of self-assembly in which not only components spontaneously form order aggregates but also order can be produced from disorder. Such self-assembled structures can be expected to show photoresponsive properties related to both the azo chromophores and the architectures at and beyond molecular level. Among them, photoinduced deformations observed on the colloidal sphere is one representative of such properties. In Chapters 3 and 4, light-induced bending of azo polymer films has been reviewed by Yu and Ikeda, and azo polymer surface patterning through photoinduced mass migration has been reviewed by Yager and Barrett. The photoinduced elongation observed on the colloids can be considered as another type of photoinduced motion effects.

5.2. AZO POLYMER SYNTHESIS

Generally, polydispersed amphiphilic azo homopolymers and random copolymers can be prepared through some relatively simple methods. One of the most feasible methods to prepare these polymers is through post-polymerization functionalization. In these synthetic routes, the azo chromophores are incorporated into the polymeric chains through some highly efficient reactions of precursor polymers. Such reaction schemes can show some obvious advantages. As azo groups are inhibitors of radical polymerization, the polymerization of monomers containing azo chromophores could hardly produce a polymer with high molecular weight. The precursor polymers with high molecular weight can be obtained through radical polymerization. Through postpolymerization reaction scheme, azo polymers with high molecular weight can be obtained. Moreover, because azo chromophores are sensitive to some chemical reactions, side reactions could occur during the polymerization of azo monomers. Through postpolymerization reaction scheme, the functional groups are introduced at the last step of the polymer preparation, where azo chromophores will not be exposed to the detrimental environments such as free radical and high temperature. However, postpolymerization reactions should have high reaction yield, otherwise azo polymers cannot possess the high degree of functionalization (DF). This could be one of the main limitations of the postpolymerization reaction scheme.

As typical examples, the polydispersed amphiphilic azo homopolymers and random copolymers can be obtained through postpolymerization azo coupling reaction (Wang et al., 1998, 1997b) and Schotten–Baumann reaction (Wu et al., 2001). In the former, precusor polymers containing alilino moieties are prepared, which can be epoxy-based polymers, polyacrylates, and polyimides among others. After that, the azo polymers can be obtained by the reactions between precursor polymers and the diazonium salts of various anilline derivatives such as 4-carboxylaniline. Figure 5.1 shows the synthetic route to prepare epoxy-based azo polymers through postpolymerization reaction. In this case, the epoxy-based precursor polymer (BP-AN) is prepared through the polycondensation between diglycidyl ether of bisphenol-A and aniline at 110°C for 36 h. Such reaction can produce a linear polymer with the number-averaged molecular weight above 40,000 and polydispersity index approximately 2. The precursor polymer can be easily dissolved in polar organic solvents such as dimethylformamide (DMF). The azo-coupling reactions are carried out in the solutions of BP-AN in DMF. For many types of the diazonium salt, the reaction yield can approach ~100%, that is, the anilino moieties undergo the complete substitution reaction at the para-positions. Figure 5.2 shows the reaction scheme to prepare amphiphilic azo copolymers through postpolymerization Schotten–Baumann reaction. In this case, a very reactive precursor polymer, poly(acryloyl chloride) (PAC), is prepared through the radical polymerization of acryloyl chloride. Various azo polymers can be obtained through the Schotten–Baumann reactions between the precursor polymer and azo compounds containing reactive groups such as HO groups. After the reaction, unreacted acyl groups were transformed to carboxyl groups through hydrolysis.

According to the spectral features and isomerization behavior, the aromatic azo compounds have been classified as azobenzene type, aminoazobenzene type, and pseudostilbene type (Rau, 1990). For the azobenzene-type molecules, the π–π^* transition band (320 nm) appears at shorter wavelength than n–π^* transition band (430 nm) (Kumar and Neckers, 1989). The cis-to-trans isomerization is relatively slow at room temperature, and the existence of cis isomers can be easily identified by the spectroscopic method. For aminoazobenzene-type and pseudostilbene-type molecules, the π–π^* and n–π^* transition bands are overlapped. The cis state of the molecules is unstable, which relaxes back to the trans state quickly. According to the definition, the azo polymers given in Figs. 5.1 and 5.2 can be assigned to contain pseudostilbene-type and azobenzene-type chromophores, respectively.

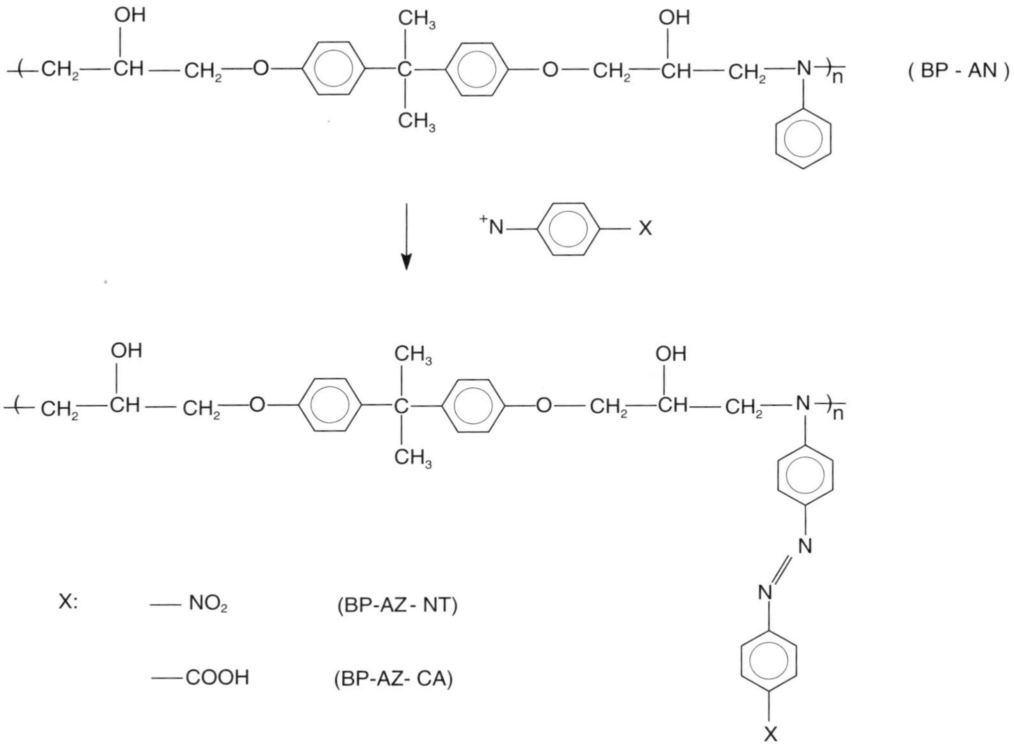

(BP - AN)

X: — NO_2 (BP-AZ- NT)

—COOH (BP-AZ- CA)

Figure 5.1. Chemical structure and synthetic route of some epoxy-based azo polymers prepared by the postpolymerization azo coupling reaction.

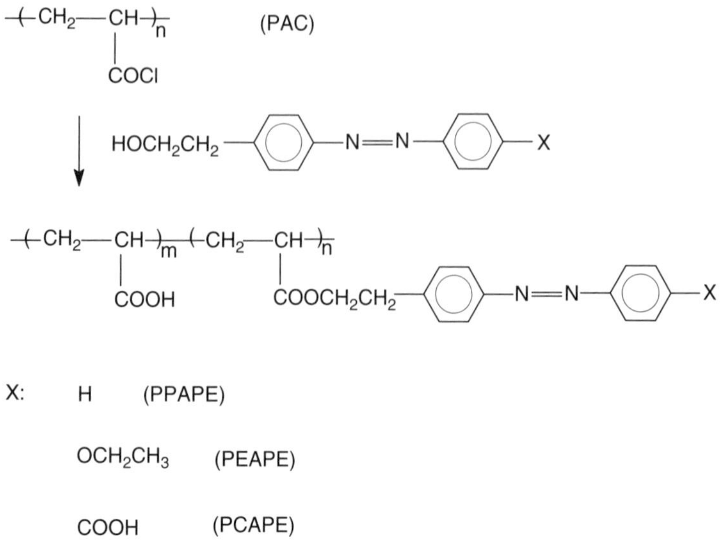

$-(CH_2-CH)_n-$ (PAC)
COCl

$HOCH_2CH_2-C_6H_4-N{=}N-C_6H_4-X$

$-(CH_2-CH)_m-(CH_2-CH)_n-$
COOH $COOCH_2CH_2-C_6H_4-N{=}N-C_6H_4-X$

X: H (PPAPE)

OCH_2CH_3 (PEAPE)

COOH (PCAPE)

Figure 5.2. Chemical structure and synthetic route of acrylate-based azo polymers prepared from the PAC precursor.

5.3. SELF-ASSEMBLY OF POLYDISPERSED AMPHIPHILIC AZO POLYMERS IN SOLUTIONS

According to the properties and size (within the nanometer to micrometer range), self-assembled structures of amphiphilic polymers formed in solutions can be classified as colloids. Colloidal particles have been widely applied in many industrial products such as inks, paints, coatings, cosmetics, and photographic films (Shaw, 1992; Hiemenz, 1986). Recently, a study on polymer-based colloidal particles has aroused broad interests in new areas such as drug delivery, biodiagnostics, and combinatorial synthesis (Xia et al., 2000). Monodispersed colloidal spheres have been widely used to construct 2-D and three-dimensional (3-D) ordered colloidal arrays (see, e.g., Cheng et al., 1999; Jiang et al., 1999a; Park and Xia, 1999; Dimitrov and Nagayama, 1996). The colloid-based materials can potentially be used in sensors, filters, optical switches, photovoltaic devices, soft-lithographic processes, and photonic band gap (PBG) materials among others (see, e.g., Xia et al., 2000; Joannopoulos et al., 1997; Weissman et al., 1996). Uniform colloidal spheres composed of the major types of functional polymers can further expand the functionalities of these colloid-based materials. Polymeric colloids with narrow size distribution can be obtained from emulsion polymerization and self-assembly of polymeric chains in solutions. Micellar aggregates of block or graft copolymers can be considered as one type of the colloidal particles with well-organized inner structure. The approach presented in this section supplies a new method to prepare uniform colloids from polydispersed azo polymers.

5.3.1. Characteristics of Polydispersed Azo Polymer Self-Assembly

Self-assembly of amphiphilic polymers generally refers to the micellar aggregation of the amphiphilic block or graft copolymers in the selective solvents. It is well known that surfactant molecules can well organize to form micelles in solutions when the concentration is above the critical micelle concentration (CMC) (Shaw, 1992; Hiemenz, 1986). The micellization can be attributed to factors such as hydrophobic effect and the intermolecular attraction between the hydrocarbon segments. Amphiphilic block and graft copolymers can be considered as polymeric counterparts of the low-molecular-weight surfactants (Halperin, 1987; Cantor, 1981). Various micellar aggregates have been prepared from block or graft copolymers through self-assembly of the polymeric chains in suitable selective solvents (see, e.g., Kikuchi and Nose, 1996; Kabanov et al., 1995; Zhang and Eisenberg, 1995; Balsara et al., 1991; Wilhelm et al., 1991). Amphiphilic block copolymers can form uniform micellar spheres with a hydrophobic core and a hydrophilic corona in aqueous media or form reverse micelles in organic solvents. Depending on the relative size of the core and corona, the block copolymers can form "starlike" micelles and "crew-cut" micelles (Shen et al., 1997).

A general method has been developed by Eisenberg and coworkers to prepare micellar aggregates from amphiphilic block copolymers (Zhang et al., 1997; Zhang and Eisenberg, 1995; Gao and Eisenberg, 1993). In the process, deionized water is slowly added into a diluted solution of the polymers in an organic solvent such as tetrahydrofuran (THF). When the water content reaches a critical value, the polymeric chains start to aggregate. The water content at this point is defined as the critical water content (CWC). When more water is added into the suspension, the structures of the formed aggregates will be "frozen" and stable suspensions of the micellar aggregates can be obtained. Depending on the compositions of the block copolymers, the micellelike aggregates can be spheres, rods, lamellae, and vesicles (Zhang and Eisenberg, 1995).

Generally, uniform micelles can only be obtained from block copolymers with narrow distribution in both the molecular weight and block length. These well-defined copolymers are prepared by some rather sophisticated polymerization methods, such as the anionic living polymerization and ATRP. The effect of dispersity of the block copolymer, such as distribution of segment length and molecular weight, has been studied by several groups (Terreau et al., 2004; Linse, 1994; Hadziioannou and Skoulios, 1982). Results indicate that the polydispersity can significantly influence the parameters such as CMC and lead to the formation of different morphologies in the same system. It can be imagined that for polymers with wide dispersion in molecular weight and solubility, adding precipitant slowly is more likely to cause the polymer to precipitate according to the solubility. This fractionation process has been widely used to prepare polymers with narrow molecular weight distribution. However, recent studies in this laboratory show that under proper conditions, uniform colloidal spheres can be obtained from the polydispersed amphiphilic azo polymers. The colloid formation, preparation conditions, and possible mechanism will be discussed in the following sections.

5.3.2. Colloidal Sphere Formation and Characterization

The polydispersed amphiphilic azo homopolymers and random copolymers, such as those given in Figs. 5.1 and 5.2, can form uniform colloidal spheres through gradual hydrophobic aggregation process (Li et al., 2006a, 2005a). In a typical process, the colloids are formed by slowly dropping deionized water into diluted solutions of the polymers in a proper solvent such as THF. The method is similar to those developed by Eisenberg and coworkers. to prepare micellelike aggregates from solutions as mentioned earlier. It can be observed that when the water content reaches a critical content, the light-scattering intensity abruptly increases, which indicates that the polymeric chains in the solutions start to aggregate. This water content is also named as the CWC. The aggregation can be completed by gradually increasing the water content in the system. After this, the structure formed in the suspensions can be frozen by adding excess water into the suspensions. Stable water suspensions of the colloidal spheres can be obtained by removing THF from the systems through dialysis. Figure 5.3 shows some typical scanning electron microscopy (SEM) images of the colloidal spheres of PEAPE separated from the suspension. The chemical structure of the polymer can be seen in Fig. 5.2, and the DF of the polymer is 49.9%, defined as the percentage

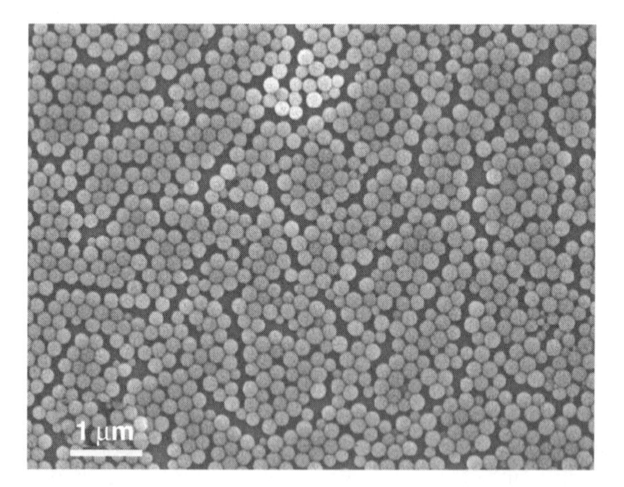

Figure 5.3. A typical SEM image of the PPAPE colloidal spheres obtained from the water suspension, which was prepared by dropping water into the THF solution until 50% (vol%) and then quenching the structures with excess of water and dialyzing against water for 3 days. The initial concentration of PPAPE in THF is 1.0 mg/mL. *Source*: From Li et al., 2006a.

of the structure units bearing azo chromophores among the total units. For other azo polymers, uniform colloidal spheres can also be obtained by this method.

The formation and size of the colloidal spheres depend on the molecular structures and preparation conditions such as the initial concentration of the polymer in THF, the water content in the medium, and the water-dropping rate. As the polymers studied are of different types and have polydispersity in the molecular weight and DF, it is difficult to give a quantitative relationship to correlate the polymer structures with the colloid size and formation details. As a general tendency, CWC decreases as the molecular weight and hydrophobicity increase, and the colloid size increases as the hydrophobicity increases. The influence of the polymer structure can be better understood after discussing the colloid formation mechanism in Section 5.3.3. The effect of the preparation conditions on the colloid formation and size is discussed herein by using PEAPE (DF = 49.9%) as a typical example.

The initial concentration of the polymer in THF has significant influence on the CWC and size of the colloidal spheres. The CWC can be obtained by measuring the variation curves of the scattered light intensity of the samples versus the water contents. The inset of Fig. 5.4 shows a typical plot of the scattered light intensity against the water content. The scattered light intensity is almost zero and remains unchanged before CWC. When the water content reaches CWC, ca. 25% (vol%) in this case, the scattered light intensity increases sharply. For the same polymer, CWC depends on the initial concentration of the polymer in THF. For different azo polymers studied by us, CWC is observed to linearly decrease as

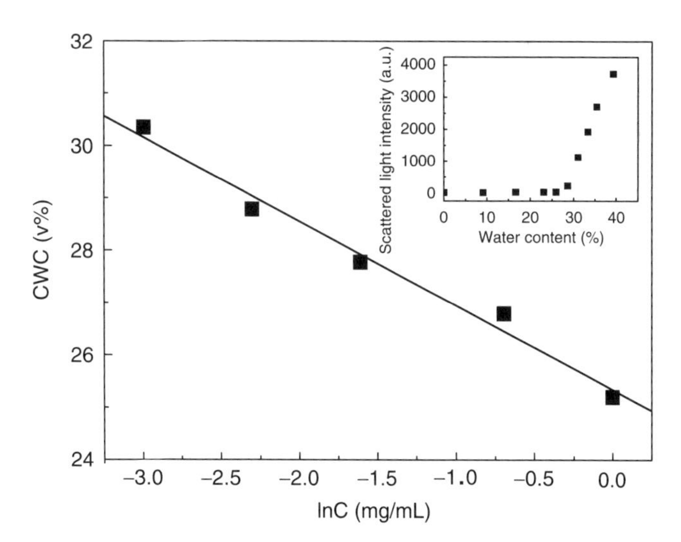

Figure 5.4. Plot of the CWC versus lnC, where C is the initial PPAPE concentration in THF solution. Inset: scattering light intensity as a function of the water content (vol%) in the THF–H_2O dispersion media. The initial concentration of PPAPE in THF is 1.0 mg/mL. *Source*: From Li et al., 2006a.

the logarithm of polymer concentration increases (Fig. 5.4). Such relationship is similar to those observed for amphiphilic block copolymers (Zhang et al., 1997). When the initial concentration increases from 0.05 to 1.0 mg/mL, CWC decreases from 30.4% (vol%) to 25.2% (vol%). The size of the colloidal spheres is also closely related to the initial concentration of the polymers, which increases as the initial concentration increases. By changing the initial concentration in the range aforementioned, the average size of the colloids can be adjusted from 80 to 180 nm.

The water content in the medium plays a key role to control the colloid formation. The influence of the water content can be studied by using dynamic light scattering (DLS) and UV-vis spectroscopy. DLS indicates that when the water content increases above CWC, the hydrodynamic radius (R_h) gradually increases as the water content increases. When the water content is above a certain value, R_h starts to decrease as the water content further increases and then stabilizes at the final value. The structure evolution in the process can be better understood from the UV-vis spectroscopic investigation. It is well known that the photoisomerization rate and isomerization degree at the photostationary state are related to the free volume surrounding the azo chromophores (Kumar and Neckers, 1989). For azobenzene-type molecules, the isomerization behavior can be monitored by UV-vis spectroscopy and used as a molecular "probe" to detect the environmental variation in the systems.

Figure 5.5 shows a typical result obtained for the solutions or suspensions of PEAPE in THF-H_2O media with different water contents. The results were obtained by irradiating the solutions of the polymer or the suspensions of the colloidal spheres with 365-nm UV light for different time intervals and recording the UV-vis spectra of the samples at each moment. Upon the UV light irradiation,

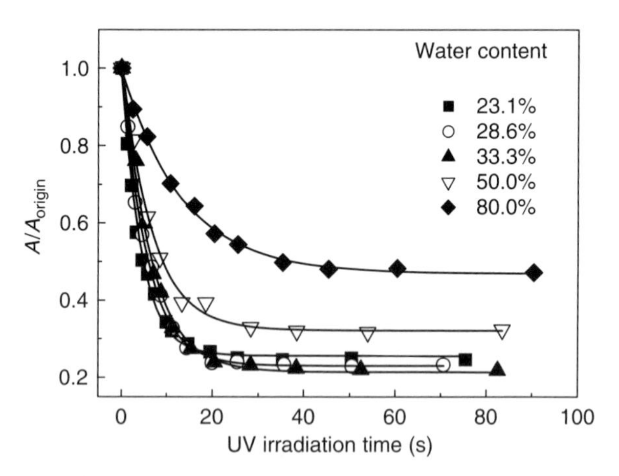

Figure 5.5. A plot of the relative absorbance (A/A_{origin}) versus the irradiation time for solutions or suspensions with different water contents. *Source*: From Li et al., 2006a.

the absorbance of the π–π* transition band at 360 nm decreases and the absorbance of the *n*–π* transition band at 450 nm increases gradually. The relative absorbance $A(t)$ is defined as A/A_{origin}, where A_{origin} and A are the absorbance at 360 nm before the light irradiation and after the irradiation for different time periods. The figure shows the relative absorbance of the solutions and suspensions as a function of the UV light irradiation time and the fitting curves of the first-order exponential decay function. The suspensions have an initial polymer concentration of 0.5 mg/mL and different water contents. When the water content is in the range 23.1%–33.3% (vol%), the trans-to-cis photoisomerization kinetics is similar for the systems. As CWC is 26.8% (vol%) obtained from light-scattering measurement, the observation indicates that the isomerization rate does not change significantly just above CWC. When the water content further increases to 40%–50% (vol%), the photoisomerization rate starts to decrease obviously. When the water content is even higher (80%, vol%), the isomerization rate decreases drastically, but it is still faster than that of the spheres obtained from the "quenching" and dialysis procedure.

The trans-to-cis isomerization degree at the photostationary state indicates a similar tendency (Fig. 5.6). The isomerization degree at the photostationary state is estimated from (Victor and Torkelson, 1987)

$$\text{isomerization degree} = 1.05 \times \left(1 - A_s/A_{origin}\right) \tag{5.1}$$

where A_s is the 360 nm absorbance measured at the photostationary state. When the water content increases in range from 25% to 37.5% (vol%), the isomerization degree is 80% (vol%) and almost keeps constant with the water content increase.

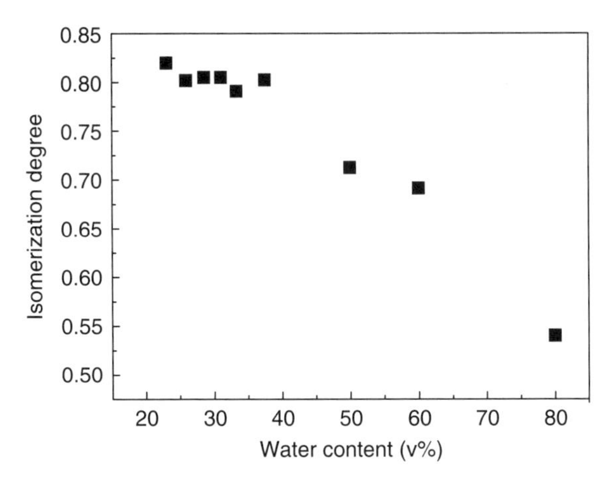

Figure 5.6. The relationship between the isomerization degree at the photostationary state and the water content. The suspension was prepared from a THF solution with the initial polymer concentration of 0.5 mg/mL. *Source*: From Li et al., 2006a.

When the water content further increases, the isomerization degree starts to decrease. When the water content increases from 37.5% to 80% (vol%), the isomerization degree decreases from 80% to ca. 55%. For the spheres in the water suspension, obtained by quenching with an excessive amount of water and dialysis, the isomerization degree decreases to even 33%. The preceding results obtained from the UV-vis spectroscopic measurements indicate that in the water addition process, the aggregates undergo a collapse process after the water content is above 40% (vol%).

In the water addition process, the water-adding rate shows a critical influence on the colloid formation. The slow addition rate is necessary to obtain uniform colloids. In a broad water-dropping rate, the sizes of the colloids obtained at the final stage are almost the same. However, when the water-dropping rate is extremely slow, the colloid size will abruptly increase as the dropping rate further decreases and even the aggregates will precipitate from the suspension.

5.3.3. Colloidal Sphere Formation Mechanism

Similar to the aggregation of block and graft copolymers in the selective solvents, the colloid formation can be attributed to an enthalpy-driven process, which is related to the hydrophobic effect and intermolecular attraction between the hydrocarbon segments. It has been pointed out that the micellization of the amphiphilic diblock copolymers occurs through phase separation caused by the hydrophobic interaction (Zhang et al., 1997). When the precipitant (H_2O) is added into THF solution of the block copolymer, the solubility of the solvent to the hydrophobic blocks decreases and the Flory–Huggins interaction parameter (χ) increases as the water content increases. At a critical χ value, which decreases with the increase of the hydrophobic block length, the polymeric chains start to aggregate and form micelles. The aggregation will be counterbalanced in the Gibbs energy by factors such as electrostatic repulsion and surface tension.

Some important differences can be seen between the block copolymers and polydispersed amphiphilic azo polymers in the aggregate formation mechanism. For the polymers discussed in this chapter, polymeric chains possess significant polydispersity in the molecular weight or hydrophobicity (for copolymers with different DFs). Because of these features, the aggregates undergo a gradual formation process as the deionized water is slowly added into the solutions. When the water content is below CWC, the polymeric chains are all soluble in the THF–H_2O mixture. When the water content reaches CWC or is marginally above CWC, only a fraction of the polymer chains meets the phase separation condition because of the polydispersity. The polymeric chains or segments start to aggregate and form the cores of the colloidal spheres. As the water content further increases, more and more polymer chains meet the phase-separation condition and transfer from the solution into the aggregates. These polymer chains loosely associate and gradually assemble on the cores, which is the aggregate growth process. When the water content further increases and most polymeric chains have been involved in the aggregates, the aggregates gradually collapse and the colloidal structures are

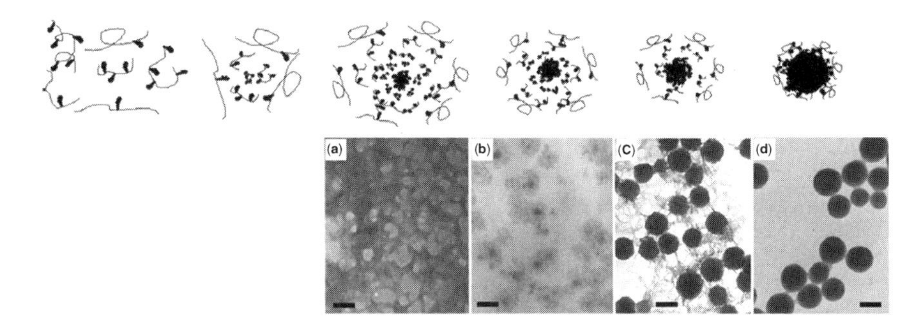

Figure 5.7. Schematic representation of the sphere formation process and some TEM images of the samples obtained from the suspensions with different water contents. Water content (vol%): **(a)** 20%, **(b)** 30%, **(c)** 40%, and **(d)** 50%. The scale bars in the figures are 100 nm. *Source*: From Li et al., 2006a.

frozen in the process. In this nucleation and gradual growth process, polymeric chains organize themselves from cores to shells in the hydrophobicity decrease order. The self-assembling process is schematically illustrated in Fig. 5.7. Because of this formation process, the colloidal spheres have cores formed from the most hydrophobic chains and coronas composed of the most hydrophilic chains.

The self-assembly mechanism is supported by the studies on the polymers given in Figs. 5.1 and 5.2 as well as some others with similar structures. The information about the growth process has been obtained from DLS measurements and in situ UV-vis spectroscopic study as mentioned earlier. The structures formed in the suspensions at different aggregation stages can also be obtained by quenching the aggregates with excess water and characterized by transmission electron microscopy (TEM) after separation from the suspension. The results also indicate that colloidal size increases with the increase in water content (Li et al., 2005a). The water content at which polymeric chains are completely involved in the aggregates can be obtained by using the parameters obtained for plotting, such as Fig. 5.4 and the initial polymer concentration in THF solution (Li et al., 2006a; Zhang et al., 1997). The result shows that this water content is much higher than CWC, which is consistent with the gradual assembly scheme proposed earlier. The aggregate collapse stage has been evidenced by the photoisomerization study aforementioned, which indicates the free-volume decrease during the process.

The average sizes of the colloidal spheres in the systems depend on the relative rates of the nucleation and gradual growth processes. The influence of the water-dropping rate can be explained by considering this factor. When the water addition rate is in a proper range, sufficient amount of nuclei can be formed in the initial stage. The colloids with proper sizes can be obtained after the growth process and will be stable in the suspension because of the electrostatic repulsive force between the colloids, which can result from the overlap of the similar-charged electric double layers. However, when the water addition rate is too slow, the colloids will become too large to be stably dispersed in the medium and will precipitate from the system.

5.3.4. Hybrid Colloids Composed of Two Types of Amphiphilic Azo Polymers

The method and understanding discussed in the foregoing sections can be used to construct colloidal spheres composed of more than one type of azo polymers (Deng et al., 2007). As a typical example, the azo polymers BP-AZ-CA and PEAPE (the chemical structure of the polymers can be seen in Figs. 5.1 and 5.2) have been selected to construct the hybrid colloids. According to the spectral characteristics, the azo chromophores in BP-AZ-CA and PEAPE, 4-amino-4′-carboxylazobenzene (ACAZ) moieties and 4-hydroxyl-4′-ethoxyazobenzene (HEAZ) moieties, can be classified as pseudostilbene- and azobenzene-type azo chromophores, respectively. The hybrid colloids can be prepared by dissolving the two polymers in THF and then gradually adding deionized water into the mixture. After the water content reaches some required value (60%, vol%), the structures formed in the suspensions are "quenched" by adding excess deionized water. The composition of the colloids can be adjusted by selecting different ratios of the two polymers.

Figure 5.8 shows some typical TEM images of the colloids, which were obtained from the stable dispersions of PEAPE (Fig. 5.8a), BP-AZ-CA (Fig. 5.8b), and PEAPE/BP-AZ-CA (1:1, wt:wt) colloids (Fig. 5.8c). It can be seen from the images that all three systems can produce uniform colloidal spheres. The hybrid colloidal spheres seem to possess a more uniform shape compared with the monocomponent colloids. The DLS measurements were used to characterize the size of colloids in the stable suspensions of the colloids. For the colloids of PEAPE, BP-AZ-CA, and PEAPE/BP-AZ-CA (1:1, wt:wt), the average hydrodynamic radii (R_h) are 154, 102, and 111 nm. The hybrid colloids show an obviously lower polydispersity (0.061) compared with both the PEAPE colloids (0.138) and the BP-AZ-CA colloids (0.156). UV-vis spectroscopic study on the PEAPE/BP-AZ-CA colloid suspension shows characteristics of both types of azo

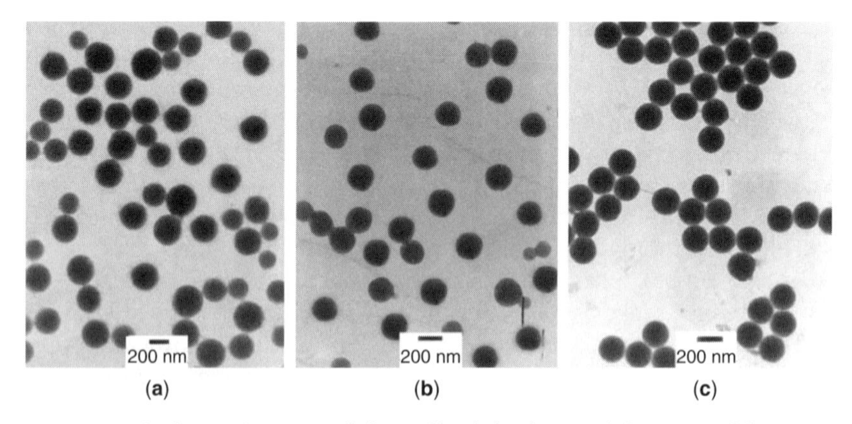

Figure 5.8. Typical TEM images of the colloidal spheres: (a) PEAPE, (b) BP-AZ-CA, and (c) PEAPE/BP-AZ-CA (1:1, wt:wt). *Source*: From Deng et al., 2007.

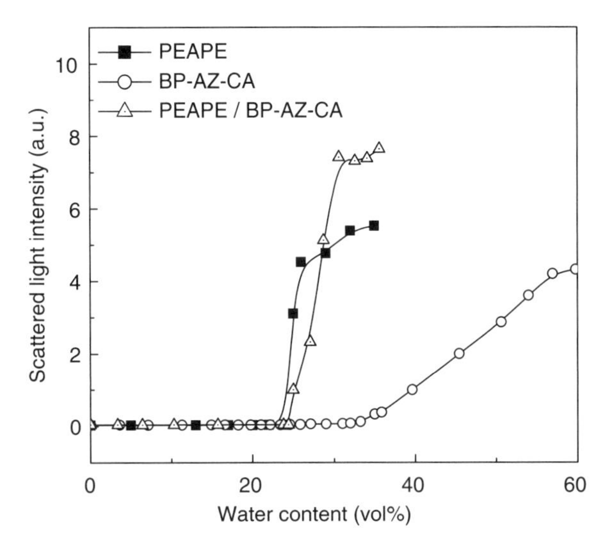

Figure 5.9. Scattering light intensity as a function of the water content (vol%) for PEAPE, BP-AZ-CA, and PEAPE/BP-AZ-CA (1:1, wt:wt) dissolved in THF-H_2O media. The scattered light intensity shown here is a ratio of scattered light intensity to incident light intensity. *Source*: From Deng et al., 2007.

chromophores. The photoinduced deformation study that is discussed in Section 5.4 confirms that the colloids are composed of the two components.

Because of the hydrophilicity difference, the two polymeric components are not uniformly distributed in the colloidal spheres. Figure 5.9 shows light-scattering intensity as a function of the water content (vol%) for PEAPE, BP-AZ-CA, and PEAPE/BP-AZ-CA (1:1, wt:wt) as the water is slowly added into THF-H_2O media. It can be seen that the CWC of PEAPE is much lower than that of BP-AZ-CA (24 vol% for PEAPE and 36 vol% for BP-AZ-CA). CWC of the solution containing both PEAPE and BP-AZ-CA is close to that of PEAPE. By using the method developed by Zhang et al. (1997), the percentage of the associated chains as a function of the water content can be estimated for the single-component system. When the water content reaches 40 vol%, the associated chains of PEAPE are estimated to be 99.5%. For BP-AZ-CA, polymeric chains start to associate at 36 vol% (CWC). Only when the water content reaches 77 vol%, the percentage of the associated chains can reach 99.5%. For the dispersions containing both polymers, the variation of aggregation fraction versus the H_2O increment cannot be directly obtained by this method. Although some deviation could be caused by the interaction between the PEAPE and BP-AZ-CA molecules, the tendency of the components to aggregate in order of their hydrophilicity should be maintained. It means that during the water content increasing process, more hydrophobic PEAPE chains will associate first. A significant amount of PEAPE chains in the dispersion has aggregated when the BP-AZ-CA chains start to aggregate. As the water content further increases, the less hydrophobic BP-AZ-CA chains assemble on the PEAPE-dominated cores, which finally results in the formation of the hybrid colloids. The nonuniform distribution of the two components and its influence on the photoresponsive properties are further discussed in the following section.

5.4. PHOTORESPONSIVE PROPERTIES OF AZO POLYMER COLLOIDAL SPHERES

Depending on the composition and preparation conditions, the azo polymer colloidal spheres can exhibit different photoresponsive properties triggered by the photoisomerization of the azo chromophores. The photoinduced variations are related to both the isomerization features of the azo chromophores and the structures of the colloidal matrices. For colloids containing azobenzene-type chromophores, the photochromic variation caused by isomerization can be measured by UV-vis spectroscopy. As mentioned earlier, the isomerization rate and degree have been used as a tool to monitor the structure variation. Compared with the "isolated" chains and loose aggregates, the photoinduced isomerization rate and degree are much lower for the colloids in their final collapsed state, which is due to the compact inner structure after the collapse. For the colloids containing relatively small amount of pseudostilbene-type azo chromophores, the shape of the colloids remains unaltered after the linearly polarized Ar^+ laser light irradiation, but the azo chromophores in colloids can be driven by the linearly polarized light to align perpendicularly to the polarization direction. This property is discussed in Section 5.5 together with those observed for the 2-D colloidal arrays. Most significant photoresponsive variation is the shape deformation induced by Ar^+ laser irradiation, which can be observed for the colloids containing pseudostilbene-type chromophores. This photoresponsive property is discussed in this section in detail.

5.4.1. Deformation Induced by Interfering Ar^+ Laser Beams

The colloidal spheres formed from the polymers containing pseudostilbene-type chromophores, such as BP-AZ-CA, can be significantly elongated upon irradiation of interfering Ar^+ laser beams (Li et al., 2005b). In a typical case, an Ar^+ laser beam at 488 nm with intensity of $\sim 150\,mW/cm^2$ was used as the light source. The colloidal spheres on substrates were prepared by methods such as casting the water suspension of the colloidal spheres on a silicon wafer or copper grid and then drying in a vacuum oven at 30°C for 12 h. The light irradiation experiments were carried out at room temperature under an ambient condition. When the colloidal spheres are exposed to the interference pattern of the *p*-polarized Ar^+ laser beams, significant shape deformation can be induced. The shape deformation can be observed by SEM, atomic force microscopy (AFM), and TEM. Figure 5.10 shows typical TEM and SEM images of the colloidal spheres after the laser

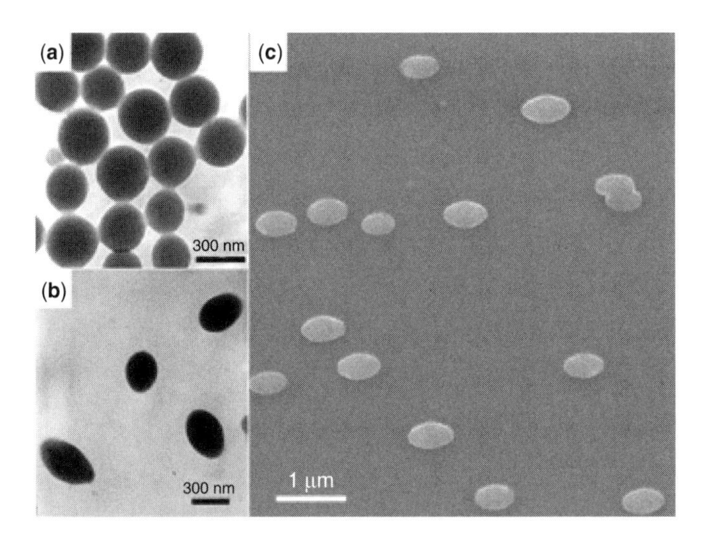

Figure 5.10. (a) TEM image of the spherical colloids before the irradiation. (b) TEM and (c) SEM images of the deformed colloids obtained from the azo polymer colloidal spheres after they are exposed to interfering polarized Ar^+ laser beams for 10 min.The averaged size of the spheres is 295 nm, and the scale bar in the TEM picture is 300 nm. *Source*: From Li et al., 2005b.

irradiation for 10 min, and TEM image of the colloids before light irradiation is also given for comparison. It can be seen that after the irradiation, the spheres are deformed to ellipsoids, which can also be confirmed by AFM observations. The elongation occurs in the polarization direction of the laser beam, which is also the grating vector direction of the interference patterns.

In the testing period (15 min), the colloids can be continuously elongated by the light irradiation. The average major-to-minor axis ratio (l/d) of the colloids can be used to indicate the deformation degree of the colloids. Figure 5.11 shows the SEM images of colloidal particles observed before the irradiation and after the irradiation for different time periods. After 5-min irradiation, obvious change in the shape can be observed. As the irradiation time increases, the elongation can be observed to increase continuously. The relationship between the average major-to-minor ratio (estimated statistically from SEM images of 100 colloidal particles) and the irradiation time is given in Fig. 5.12. The major-to-minor ratio increases almost linearly with the irradiation time increase in the testing range.

The colloidal sphere deformation could be attributed to the optically induced electric field gradient, which has been used by Kumar et al. (1998) to explain the SRG formation. The polymers used to construct the photodeformable colloids can also show sensitive properties to form SRG. Photoinduced mass migration has been used to account for the SRG formation, where the polymer chains are driven by light to move from bright to dark regions. The colloid deformation could also be caused by the mass migration related to the electric field gradient force.

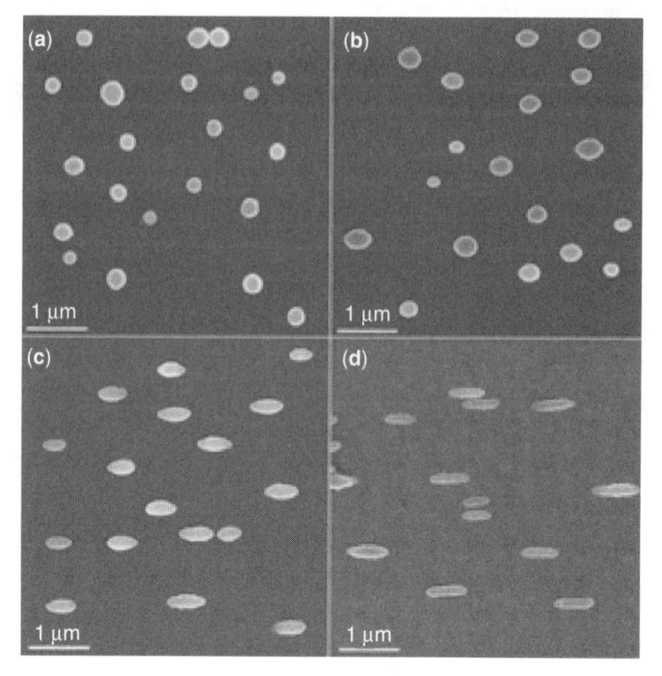

Figure 5.11. SEM images of colloidal spheres before irradiation (**a**) and after irradiation for different time periods: (**b**) 5 min, l/d=1.31; (**c**) 12 min, l/d=2.03; (**d**) 15 min, l/d=2.35. *Source*: From Li et al., 2005b.

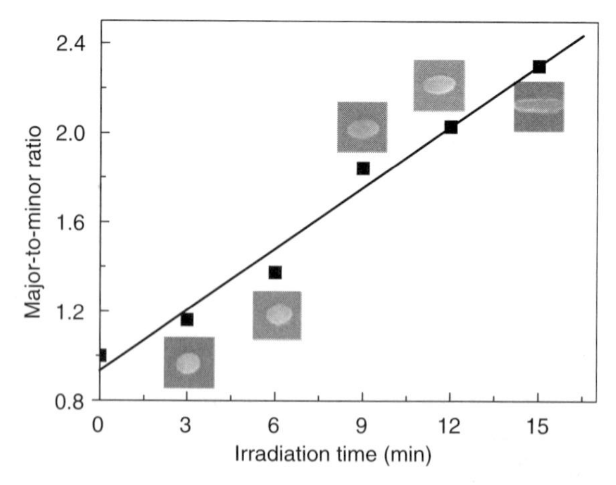

Figure 5.12. Relationship between the average major-to-minor ratio of the colloidal particles and the irradiation time. *Source*: From Li et al., 2005b.

5.4.2. Deformation Induced by a Single Ar^+ Laser Beam

An important difference can be seen by comparing the colloid deformation and SRG formation. SRGs can only be induced by light irradiation of interfering laser beams, which produce periodic fringes with variation in the light intensity or light polarization direction (Natansohn and Rochon, 2002; Delaire and Nakatani, 2000). As mentioned earlier, the colloid deformation can also be induced by the irradiation of the interfering laser beams. However, it was somewhat surprising to observe that the deformation of the colloidal sphere can also be induced by a linearly polarized Ar^+ laser single beam (Li et al., 2006b).

The samples used for the single laser beam irradiation experiments were the same as those discussed in Section 5.4.1. An Ar^+ laser at 488 nm was used as the light source, and the linearly polarized laser beam was spatially filtered, expanded, and collimated. The intensity of the laser beam was $\sim 150\,mW/cm^2$, and the incident laser beam was perpendicular to the wafer surfaces containing the colloids. The experiments were carried out at room temperature under an ambient condition. In a typical case, colloidal spheres of BP-AZ-CA were used for the test, and the result is as follows.

Figure 5.13 gives some typical TEM and SEM images before and after the light irradiation. Figure 5.13a shows a typical TEM image of the colloidal spheres before light irradiation. The average size of the colloidal spheres is 212 nm, with a

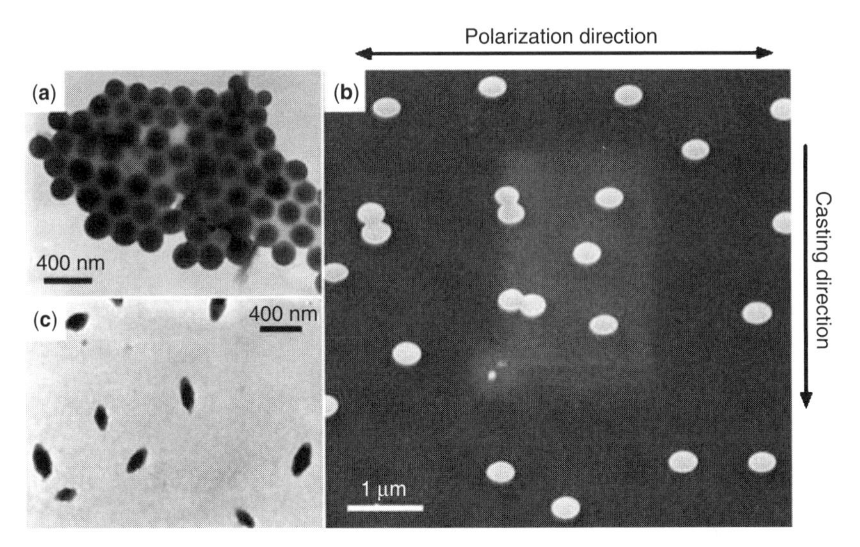

Figure 5.13. (**a**) TEM image of the azo polymer colloidal spheres with an average diameter of 212 nm. (**b**) SEM image of the colloidal spheres after exposure to a linearly polarized Ar^+ laser beam for 10 min.(**c**) TEM image of the deformed colloids after irradiation with the laser beam for 15 min and released from the surface of the silicon wafer by sonication. *Source*: From Li et al., 2006b.

polydispersity index of 0.03 obtained from the DLS measurement. Figure 5.13b shows an SEM image of the colloidal spheres after being irradiated by the Ar^+ laser beam for 10 min. It can be observed that the colloidal spheres are significantly elongated in the polarization direction of the laser beam. To avoid possible influence of the sample preparation, the casting direction is selected to be perpendicular to the polarization direction of the laser beam. Figure 5.13c shows the TEM image of the deformed colloids irradiated by the laser beam for 15 min. The TEM observation confirms that the spheres are deformed to ellipsoids after the irradiation. Figure 5.14 shows SEM images of the colloidal particles observed before irradiation and after irradiation for different time periods. The colloids are continuously stretched along the polarization direction as the irradiation time increases. The average axial ratio (l/d) of the colloids (estimated statistically from SEM images of 100 colloidal particles) almost linearly increases as the irradiation time increases.

Because the stretching relies on the polarization direction, the colloidal spheres can be elongated in multiple directions by adjusting the polarization direction. For example, the colloids can be first stretched in one direction and then in the orthogonal directions by turning the polarization direction around 90°. Figure 5.15 gives a typical SEM image of the colloids irradiated by two laser beams with orthogonally polarized directions, each for 10 min. It can be seen that the colloids are stretched in two orthogonal directions.

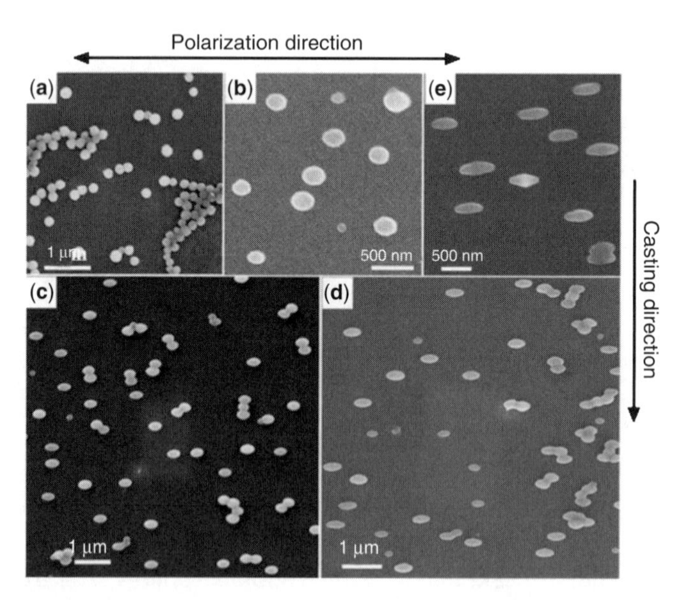

Figure 5.14. SEM images of colloidal spheres before irradiation and (a) after irradiation for different time periods: (b) 5 min, $l/d = 1.28$; (c) 10 min, $l/d = 1.65$; (d) 15 min, $l/d = 1.98$; (e) 20 min, $l/d = 2.35$. *Source*: From Li et al., 2006b.

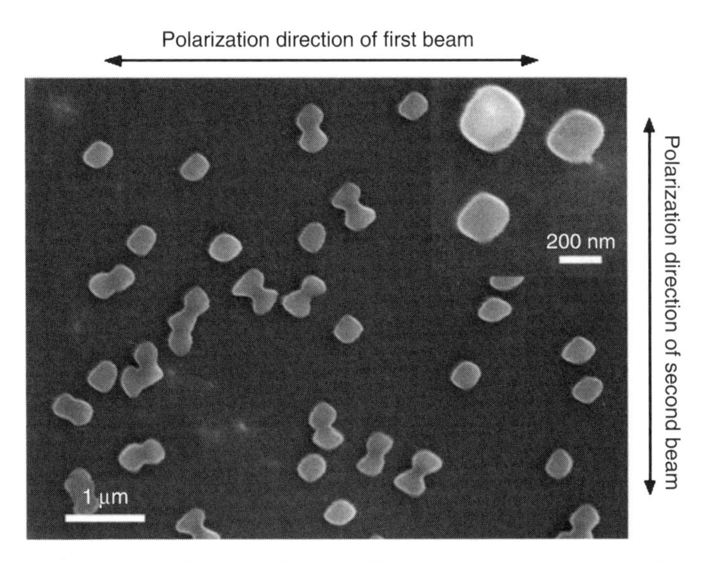

Figure 5.15. Typical SEM image of the colloids after they are irradiated by two laser beams with orthogonally polarized directions, each for 10 min. *Source*: From Li et al., 2006b.

The differentiation between the irradiation effects of the polarized laser single beam and the interfering *p*-polarized laser beams can be observed by using 2-D arrays of the colloidal spheres as the testing samples. The samples were prepared by the vertical deposition method, which is discussed in Section 5.5. By this method, the hexagonally close-packed structure of the colloid spheres can be obtained. Different deformation behavior can be observed after irradiating the colloidal array with the polarized laser single beam for 15 min or the *p*-polarized interfering laser beams for 20 min. For the former, the colloidal spheres are uniformly stretched along the polarization direction of the laser beam. As the restraint of the neighboring colloids, the deformation degree of the colloids in the array is less than that of the isolated colloids. Changing the relative direction of the polarization to the array assembly direction shows no influence on the stretching effect. On the contrary, after being irradiated with the interfering laser beams, only the colloids in bright regions of the interference fringes are changed, whereas those in the dark regions are almost unaffected.

The exact mechanism of the deformation induced by the laser single beam is still not clear. In the experiment, the light intensity of the spatially filtered and collimated laser beam (1.5 cm in diameter) should be uniform in submicrometer size. This phenomenon has been further confirmed in another type of colloid composed of well-defined polymeric chains (Lambeth and Moore, 2007). Such observations cannot be directly explained by using the gradient force of the laser irradiation. More investigations, both through theoretical and experimental approaches, seem to be required to understand this unusual photoresponsive effect.

5.4.3. Photoresponsive Porperties of Hybrid Colloids

Photoresponsive behavior of the hybrid colloids depends on the composition and distribution of the components in the colloids (Deng et al., 2007). The hybrid colloid composed of BP-AZ-CA and PEAPE, whose preparation has been discussed in Section 5.3.4, is further discussed herein concerning its photoresponsive properties. As PEAPE and BP-AZ-CA contain HEAZ moieties and ACAZ moieties, the colloids can show photoinduced variation related to the azobenzene-type and pseudostilbene-type chromophores.

Figure 5.16 shows the UV-vis spectra of the stable suspensions of the PEAPE/BP-AZ-CA (1:1, wt:wt) hybrid colloids before and after UV light (365 nm) irradiation. For single-component colloids of PEAPE and BP-AZ-CA, the π–π* transition bands (λ_{max}) appear at 365 and 470 nm. For the hybrid colloids, the absorption bands at 365 and 455 nm can be assigned to the HEAZ and ACAZ moieties of the polymers. The new band appearing at 387 nm could be the absorption band because of the HEAZ/ACAZ charge-transfer complex. Upon the light irradiation, the absorption at 365 and 387 nm decreases gradually as the result of the trans-to-cis isomerization of the HEAZ chromophores. No variation is observed for absorption band at 455 nm because trans–cis isomerization of the ACAZ moieties (a pseudostilbene-type chromophore) is too fast to be detected at the timescale given here (Rau, 1990). The relative absorbance can be given as A_t/A_{origin}, where A_{origin} and A_t are the absorbance at 365 nm before the light irradiation and after the irradiation for different time periods. A_t/A_{origin} can be used to indicate the relative amount of the trans isomers remaining at t time, and its variation represents the kinetics of the photoisomerization. The variations can

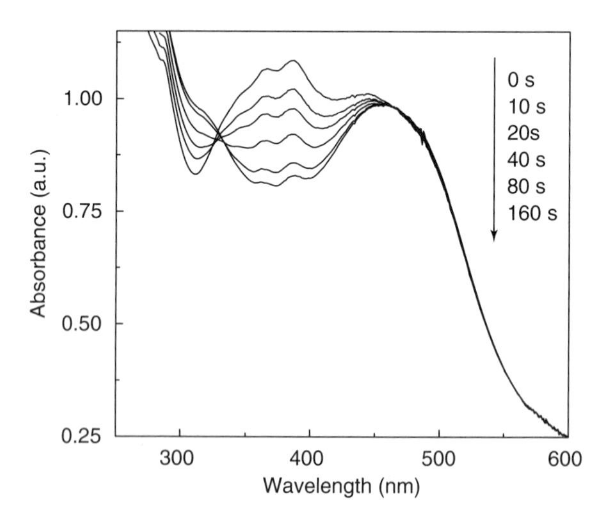

Figure 5.16. Variation of the UV-vis spectra of the colloidal dispersions induced by the UV light irradiation. The dispersions were obtained by adding water into a THF solution of PEAPE/BP-AZ-CA (1:1, wt:wt). *Source*: From Deng et al., 2007.

be best fitted by the first-order exponential decay function

$$A_t/A_{\text{origin}} = A_0 + A_1 \exp(-t/T_1) \tag{5.2}$$

where T_1 is the characteristic time of the decay process. For comparison, the photoisomerization behavior of a stable dispersion of the PEAPE colloids can be obtained in the same way. Both experimental data and fitting curves are shown in Fig. 5.17. T_1 obtained from the curve fitting are 43.4 and 37.3 s for the hybrid colloids and PEAPE colloids, respectively. Comparing with the monocomponent PEAPE colloids, the trans-to-cis isomerization rate is far slower for the hybrid colloids.

The hybrid colloids can show photoinduced deformation, and deformation ability is dependent on the percentage of the BP-AZ-CA component in the colloids. The experimental conditions for the sample preparation and light irradiation are the same as those mentioned earlier. The colloidal spheres were exposed to the spatially filtered and collimated laser beam incident perpendicularly on the substrate surfaces. The morphologies of colloidal spheres were observed by TEM before and after the laser irradiation. For comparison, the hybrid colloids and the monocomponent colloids were studied by the same light irradiation condition.

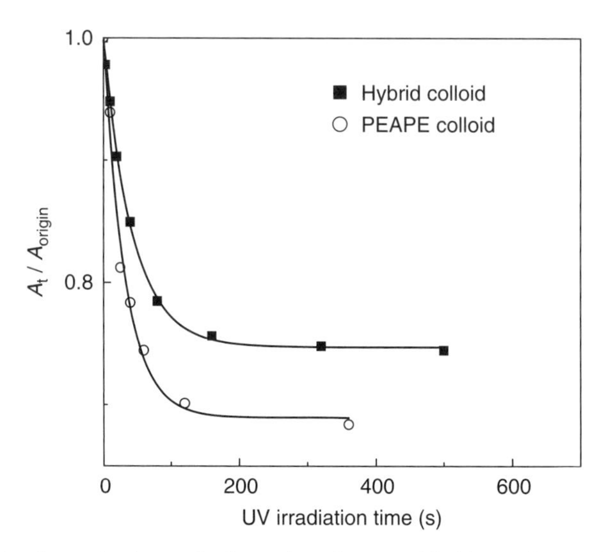

Figure 5.17. A plot of the relative absorbance (A_t/A_{origin}) varying with the irradiation time and the fitting curve for the dispersion containing PEAPE/BP-AZ-CA (1:1, wt:wt) colloids, prepared by adding water into the THF-H_2O solution. The data and fitting curve for PEAPE colloidal dispersion are also given for comparison. *Source*: From Deng et al., 2007.

Figure 5.18 gives the TEM images of the colloidal spheres after the light irradiation. The TEM images of the colloidal spheres before light irradiation have been given in Fig. 5.8. PEAPE colloids do not show observable deformation after being irradiated for 2 h (Fig. 5.18a). After being irradiated for 20 min, the BP-AZ-CA colloidal spheres can be significantly elongated along the light polarization direction (Fig. 5.18b), which has been discussed in Section 5.4.2. The hybrid colloids show unique morphologies after the light irradiation, which are characterized by the "spindlelike" or "tadpolelike" shapes with dark spherical cores (Fig. 5.18c).

Figure 5.19 shows the TEM images of the hybrid colloids containing different amounts of BP-AZ-CA before (Fig. 5.19a–d) and after the light irradiation of the linearly polarized Ar^+ laser beam for 2 h (Fig. 5.19e–h). After the light irradiation, the colloids show different deformation degrees related to the BP-AZ-CA content. The hybrid colloids containing 80 wt% of BP-AZ-CA show a similar shape deformation as BP-AZ-CA colloids (Fig. 5.19e), although the degree of deformation is less for the same irradiation time period. The hybrid colloids containing 20 wt% of BP-AZ-CA can only show an insignificant deformation under the same light irradiation condition (Fig. 5.19h). For the hybrid colloids containing 40% and 60 wt% of BP-AZ-CA (Fig. 5.19f, g), the shape deformation degrees are between the two cases aforementioned.

As discussed earlier, the colloidal spheres formed from BP-AZ-CA can exhibit significant deformation upon Ar^+ laser irradiation, whereas PEAPE colloids do not show such deformation under the same condition. Since every hybrid colloidal sphere shows a very similar deformation upon the light irradiation, it indicates that the colloid spheres are all composed of both BP-AZ-CA and PEAPE components. The fact that the deformation degree is lower than that of the

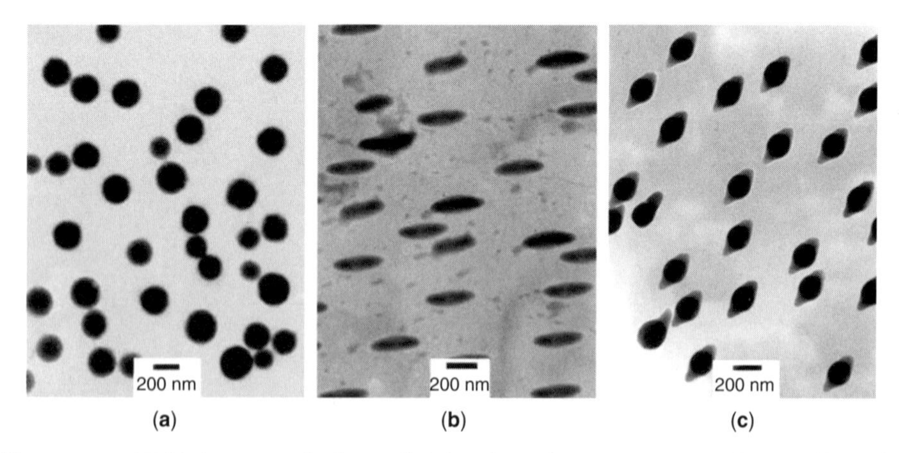

Figure 5.18. TEM images of the colloids after they are exposed to a linearly polarized Ar^+ laser beam for different time periods: (**a**) PEAPE, 2 h; (**b**) BP-AZ-CA, 20 min; (**c**) PEAPE/BP-AZ-CA (1:1, wt:wt), 2 h. *Source*: From Deng et al., 2007.

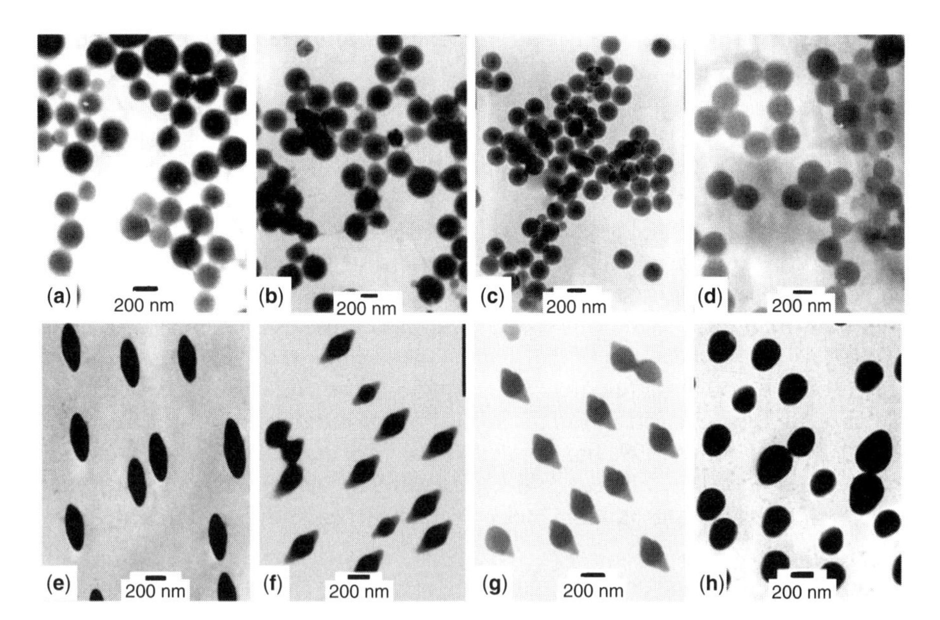

Figure 5.19. Typical TEM images of the PEAPE/BP-AZ-CA colloids containing a different percentage of PEAPE before and after the Ar^+ laser irradiation for 2 h: (a) 20 wt%, before irradiation; (b) 40 wt%, before irradiation; (c) 60 wt%, before irradiation; (d) 90 wt%, before irradiation; (e) 20 wt%, after irradiation; (f) 40 wt%, after irradiation; (g) 60 wt%, after irradiation; (h) 90 wt%, after irradiation. *Source*: From Deng et al., 2007.

BP-AZ-CA colloids can also lead to the same conclusion. It has been discussed in Section 5.3.4 that the hybrid colloids are formed by gradual assembly from the cores to shell, where the more hydrophobic PEAPE chains compose the inner parts and the BP-AZ-CA chains form the shells. The photoisomerization study is consistent with this core–shell structure model. As the HEAZ moieties are confined in the cores, the isomerization cannot have the same free volume as the HEAZ moieties in the shells. Compared with the PEAPE colloids, where a fraction of HEAZ chromophores is distributed in the shells, the free volume for the photoisomerization in hybrid colloids is reduced. This factor can cause the photoisomerization rate of the hybrid colloids to become obviously lower than that of the PEAPE colloids.

Although the hybrid colloids can be deformed by the light irradiation, TEM observation shows that the deformed colloids reflect a morphology obviously different from that of the elongated BP-AZ-CA colloids. Typical appearance possesses a circle core with higher contrast and tails stretched from the core. Considering the photoresponsive behavior of both polymers, it can be concluded that the core and tails are composed of PEAPE and BP-AZ-CA components, respectively. The result implies an interesting effect that the light force can cause the separation of PEAPE and BP-AZ-CA components in case they are not mixed with each at molecular level.

5.5. PHOTORESPONSIVE 2-D COLLOIDAL ARRAY AND ITS *IN SITU* STRUCTURE INVERSION

Recently, construction of 2-D and 3-D periodic structures from monodispersed colloidal spheres has aroused tremendous research enthusiasm for their self-assembled architecture and fascinating applications (Xia et al., 2000). The colloidal crystals can be obtained by methods such as transferring the colloidal array formed at the air–liquid interface, exploiting the capillary forces to drive the assembly, and using electrophoretic deposition or sedimentation in a force field (see, e.g., Cheng et al., 1999; Jiang et al., 1999a; Park and Xia, 1999; Dimitrov and Nagayama, 1996). The colloidal spheres discussed in the preceding sections can be used as a new type of the building block to construct colloidal crystal through self-assembling methods.

5.5.1. Colloidal Array and Photoinduced Dichroism

One of the most interesting photoresponsive properties of azo polymers is the photoinduced birefringence and dichroism (Xie et al., 1993). The photoinduced anisotropy is caused by the disparity of the repeated trans–cis isomerization of azo chromophores under linear polarized light irradiation. The most efficient excitation occurs in the polarization direction, which can force the chromophores to continually change their orientation and to be eventually stabilized at the direction perpendicular to the polarization (Natansohn and Rochon, 2002). The effect shows potential applications in areas such as reversible optical data storage, optical switching and sensors.

The colloids composed of azo polymers can be a new type of the colloid-based material that exhibits the photoinduced birefringence and dichroism (Li et al., 2005a). Figure 5.20 shows the chemical structure of an azo polymer (PNANT) that has been prepared for the purpose. PNANT contains both pseudostilbene-type azo chromophores and ionizable carboxyl groups in the side chain in a random sequence. PNANT was synthesized from a PAC sample with the average degree of polymerization (DP) of 239 and a polydispersity index of 1.53. The precursor polymer containing anilino moieties was synthesized by the reaction between a proper amount of *N*-ethyl-*N*-(2-hydroxyethyl)aniline and PAC through the Schotten–Baumann reaction. Unreacted acyl chloride groups were then hydrolyzed to obtain the COOH groups after the reaction. Finally, PNANT was obtained by azo coupling reaction between the precursor polymer and diazonium salt of 4-nitroaniline. In this reaction, the molar ratio of the diazonium salt to the anilino moieties of the precursor polymer was 1.1 in order that the anilino moieties were completely reacted. The details of these reactions can be seen in Li et al. (2005a), Wu et al. (2001), and Wang et al. (1997b). The average DF,

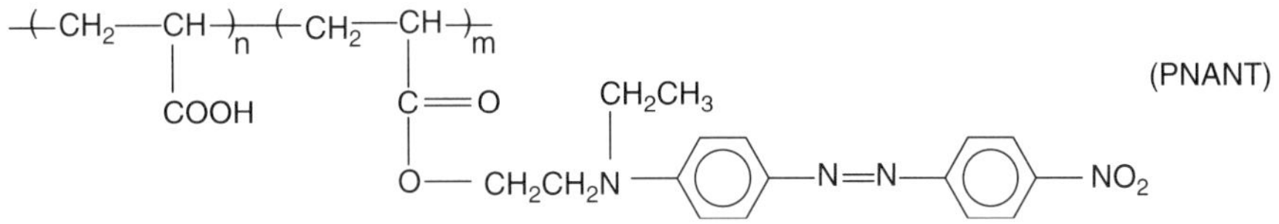

$$-\!\left(CH_2-\underset{\displaystyle COOH}{\underset{|}{CH}}\right)_n\!\!-\!\left(CH_2-\underset{\displaystyle O-CH_2CH_2\overset{\displaystyle CH_2CH_3}{\overset{|}{N}}-C_6H_4-N{=}N-C_6H_4-NO_2}{\underset{|}{\underset{\displaystyle C{=}O}{\underset{|}{CH}}}}\right)_m \quad \text{(PNANT)}$$

Figure 5.20. Chemical structure of azo polymer PNANT.

defined as the percentage of the structure units bearing azo chromophores among the total structure units, was estimated by the elemental analysis to be 20%. PNANT can form uniform colloidal spheres by the gradual hydrophobic aggregation method discussed earlier.

2-D arrays of the colloidal spheres can be obtained by the vertical deposition method (Dimitrov and Nagayama, 1996). In the process, a pretreated hydrophilic silicon wafer was immersed vertically in a suspension containing colloidal spheres (about 0.2 mg/mL). The dispersion medium was gradually evaporated ($\sim$0.05 mL/h) to let the liquid surface slowly move down. During the decline of the surface, the colloids deposited onto the substrate and the attractive capillary force at the meniscus forced the colloidal spheres to organize into close-packed 2-D arrays. Figure 5.21 shows a typical SEM image of the colloidal array.

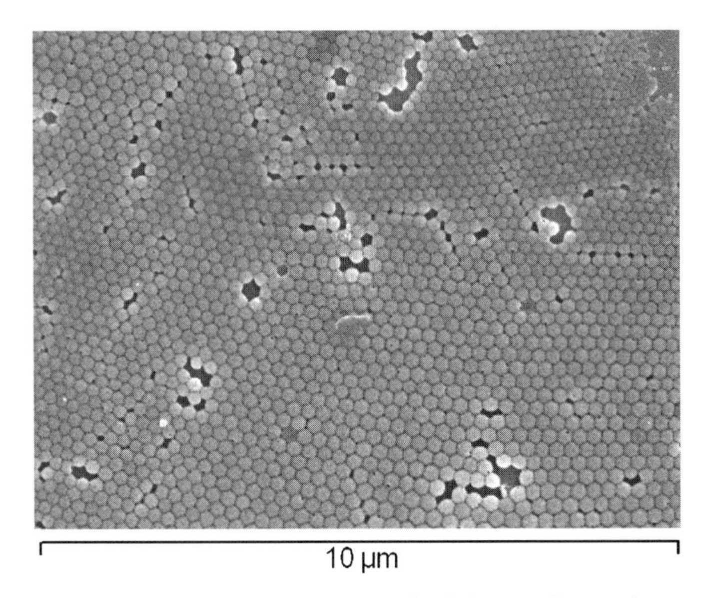

Figure 5.21. A typical SEM image of the colloidal monolayer obtained by the vertical deposition method. The D_h and polydispersity index of the spheres are 195 nm and 0.04, respectively. *Source*: From Li et al., 2005a.

After irradiation with a linearly polarized Ar^+ laser beam (488 nm, 150 mW/cm^2) for 10 min, significant dichroism can be observed for the colloidal array (Fig. 5.22). The figure gives the maximum absorbance measured at different rotation angles with respect to the polarization of the writing beam. The absorbance in the direction parallel to the polarization of the laser beam is obviously weaker than that in the direction perpendicular to the polarization of the laser beam. The result shows that the light irradiation has induced the azo chromophores to take a preferential orientation in the direction perpendicular to the electric vector of the laser beam. The orientation order parameter S can be estimated from the dichroic ratio (Wu et al., 1998). As the irradiation time increases, the orientation order parameter increases sharply at the beginning and levels off gradually. After irradiating for about 10 min, the photoinduced chromophore orientation is almost saturated. The orientation order parameter can be estimated to be ~0.09 at the photostationary state, which is comparable to those reported for spin-coating films (Natansohn and Rochon, 2002).

SEM observation shows that after 15-min irradiation, the colloids do not show observable deformation, which could be due to the low DF and the strong interaction between acrylic acid chains. As the colloidal spheres are uniform, the degree of the photoinduced chromophore orientation should approximately be the same for each sphere. The colloidal spheres and the structures constructed from them can potentially be used as a new-type material for photo-storage. The storage and readout process could be performed on each single sphere in the array by using light source with high resolution.

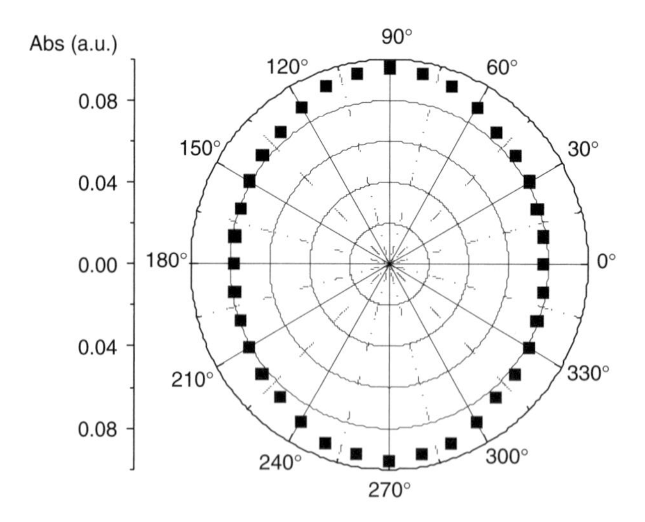

Figure 5.22. The UV-vis absorbance measured at λ_{max} as a function of the rotation angles to the polarization direction of the laser beam, after irradiation with a linearly polarized Ar^+ laser beam for 10 min. *Source*: From Li et al., 2005a.

5.5.2. Porous Structure from *in situ* Colloidal Array Structure Inversion

Polymeric porous structures with pore sizes in the range from nanometer to micrometer scale have attracted considerable attention in the past decades because of their possible applications in selective transportation, as catalytic substrates, in biosensors, and as photonic band gap materials (Wang and Caruso, 2001; Jiang et al., 1999b; Subramanian et al., 1999). As a widely used methodology, the porous structures are prepared by using 2-D and 3-D arrays of colloidal spheres as templates (Xia et al., 2000). In the process, the void spaces among the colloidal spheres are fully infiltrated with a liquid precursor such as a UV or thermally curable prepolymer or an initiator-containing monomer. Highly ordered porous structures can be obtained by solidification of the precursor and subsequent removal of the colloidal spheres. Recently, by using block copolymer films or a dual-component (soluble/insoluble) colloid system, the nanoporous or macroporous films have been prepared by selective interaction of solvents with one of the components (Xu et al., 2003; Yi and Kim, 2003). Solvent-induced structure inversion has also been reported for colloidal crystals of polymeric latex particles. For 3-D colloidal crystals of poly(styrene-*co*-hydroxyethyl methacrylate) spheres, the solvent treatment can yield an ordered crystal with colloidal cores embedded in a polymer matrix [an interconnected colloidal array (ICA)] (Rugge et al., 2003; Chen et al., 2000). In these processes, the porous structures are formed through a self-assembling process without the infiltration treatment.

As discussed in Section 5.3, the colloids of the amphiphilic polydispersed azo polymers possess a hydrophobic core and hydrophilic corona. It is interesting to observe that polar organic solvents such as THF can also induce *in situ* structure inversion of the colloidal array of the azo homopolymer (Li et al., 2006c). Porous structures with pore sizes in submicrometer scale can be directly obtained from the colloidal arrays of BP-AZ-CA through the structure inversion. Moreover, by exploiting the photoresponsive properties of BP-AZ-CA, films with ordered elliptical pores can be feasibly prepared from the colloidal arrays of the ellipsoidal colloids obtained after the laser light irradiation.

The chemical structure of BP-AZ-CA can be seen in Fig. 5.1. The sample used in the following discussion has a number-averaged molecular weight of 41,000, with a polydispersity index of 2.2. The colloidal spheres of the azo polymer were prepared by gradual hydrophobic aggregation scheme as discussed earlier. The sizes of the colloidal spheres were estimated by both TEM and DLS measurement. The average hydrodynamic diameter (D_h) was measured to be 223 nm, with a polydispersity index of 0.04. The 2-D arrays of the close-packed colloidal spheres were fabricated by the vertical deposition method.

The porous films can be obtained by inducing the structure inversion of the 2-D colloidal sphere arrays in an enclosed chamber through solvent annealing for 6–9 h. The chamber is filled with THF vapor from the solvent reservoir, and the temperature of the system is controlled to be 30°C. For microscopic observations, the structures formed are dried in a 30°C vacuum oven for 12 h. A typical SEM

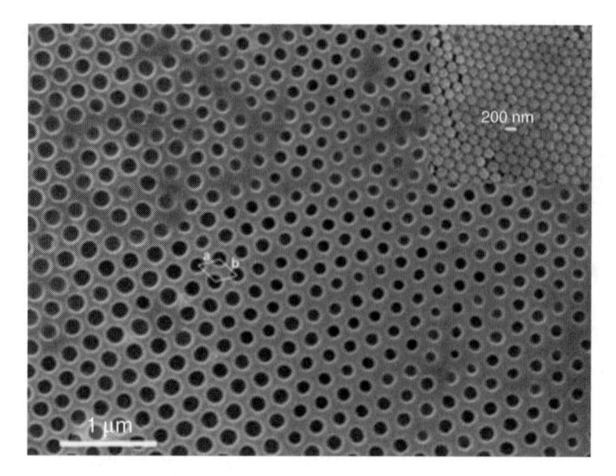

Figure 5.23. Typical SEM image of the mesoporous structures formed by the in situ sphere–pore inversion of the azo polymer colloidal sphere array, induced by solvent annealing. Inset: SEM image of the colloidal sphere array before the solvent treatment. *Source*: From Li et al., 2006c.

image of the porous films obtained from the *in situ* structure inversion is shown in Fig. 5.23. The inset in the same figure shows the SEM image of the sphere arrays before the solvent treatment. After the solvent treatment, the sphere array directly transforms to the porous film, which also exhibits the characteristic pattern of the hexagonal close-packed array. The average pore diameter (150 ± 24 nm), which is estimated from 540 randomly selected pores in SEM images, is obviously smaller than the average sphere size. The distance between the pore centers is 250 ± 20 nm. The AFM image of the porous films is shown in Fig. 5.24, which further confirms the morphology observed by SEM. The pore sizes and the distance between the pore centers are consistent with those observed by SEM. The spectroscopic methods show that the porous structures and the colloidal arrays are composed of the same polymer (Li et al., 2006c).

As discussed in Section 5.4, the colloidal spheres can be stretched to form nonspherical colloids such as ellipsoidal colloids by Ar^+ laser irradiation. The array of the ellipsoidal colloids is obtained by exposing the colloidal sphere array to a polarized Ar^+ laser beam for 10 min. The films with elliptical pores can be obtained from the array of the ellipsoidal colloids after the same annealing treatment. Figure 5.25 shows the SEM images of the array of the ellipsoidal colloids and the corresponding porous film formed from the structure inversion. The average axial ratio of the colloids and pores are 1.46 and 1.25. The smaller axial ratio for the pores could be attributed to the stress relaxation occurring in the structure inversion process.

As discussed earlier, the porous structures and the colloidal arrays are composed of the same polymer indicated by spectroscopic analysis. This confirms that the porous films are formed through a solvent-induced *in situ* structure

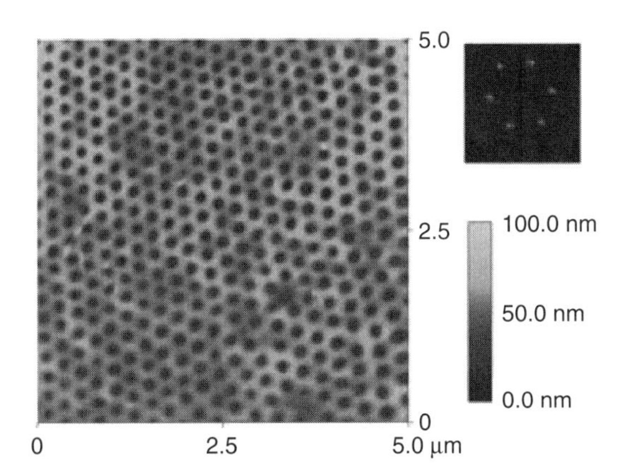

Figure 5.24. Typical AFM image and fast Fourier transform (FFT) plot of the mesoporous films formed by the *in situ* sphere–pore inversion. *Source*: From Li et al., 2006c.

inversion. The sphere–pore inversion can also be indicated by SEM observation on the transition of some isolated spheres. Moreover, the porous structure can be obtained by simply dropping THF on the colloidal array surface, although the edges of the pores are less regular. The formation of the porous structures is closely related with the amount of THF dropped. SEM observation indicates that when the THF amount is less than the proper amount, some pores just begin to form and some spheres can still be seen. When the THF amount is large, parts of the colloidal sphere array are transformed to a flat film without the porous character (Li et al., 2006c).

Although the exact mechanism for the sphere–pore inversion is still unclear at the current stage, some basic characteristics of the process can be discussed as

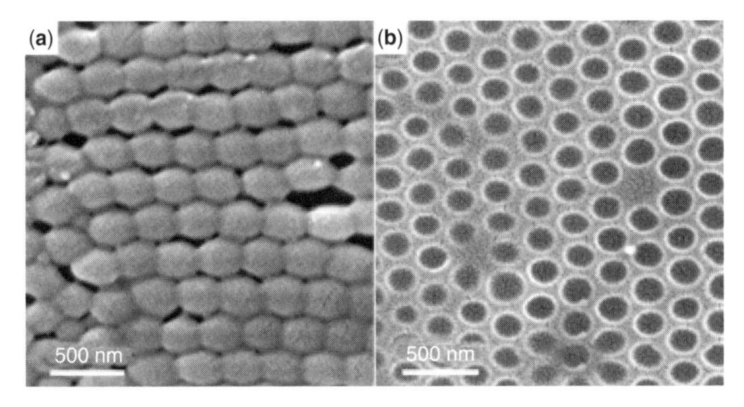

Figure 5.25. SEM images: (a) array of the ellipsoidal colloids, (b) the porous film with elliptical pores. *Source*: From Li et al., 2006c.

follows. The colloidal spheres possess the "core-shell" structure with a hydrophobic core and a hydrophilic corona. When the colloids are exposed to the THF atmosphere for some time period, the hydrophobic core can be swollen and the colloidal arrays gradually transform to an inhomogeneous thin film. When the film is further interacted with THF, the inhomogeneous film can convert into the porous structures. In the process, the parts with poorer solubility in THF will shrink and the polymeric chains are rearranged to decrease the contact area between these parts and THF. The high regularity of the final structure seems to indicate that this process is a highly coordinated process. When the hydrophobic fraction further moves out of the original core regions of the colloidal array, it will not only release the tension caused by chain collapse during the colloid formation but will also minimize the surface tension by forming an interfacial layer between the solvent and the hydrophilic segments. The perfect circular shape of the pores could be an evidence of the process to minimize the surface tension. With the shrinkage of the hydrophilic parts at the same time, the intermediate structure can finally transform to the ordered porous film.

5.6. CLOSING REMARKS

The preceding investigation results demonstrate that the polydispersed amphiphilic azo polymers can be used to construct various well-organized structures through self-assembling processes at different levels. The azo polymers can form uniform colloidal spheres, 2-D arrays of colloidal spheres, and porous films based on different self-assembling processes. Novel photoresponsive properties have been observed for the colloids and colloidal arrays, which include photoinduced elongation of colloids, photoinduced separation of two components from hybrid colloids, and photoinduced dichroism of the colloid array. These observations can give some new insights into the self-assembly nature and photoresponsive properties concerning azo polymers. Some new potential applications can be expected on the basis of the understandings.

REFERENCES

Archut A, Azzellini GC, Balzani V, Cola LD, Vögtle F. 1998. Toward photoswitchable dendritic hosts: interation between azobenzene-functionalized dendrimers and eosin. J Am Chem Soc 120:12187–12191.

Åstrand PO, Ramanujam PS, Hvilsted S, Bak KL, Sauer SPA. 2000. Ab initio calculation of the electronic spectrum of azobenzene dyes and its impact on the design of optical data storage materials. J Am Chem Soc 122:3482–3487.

Balsara NP, Tirrell M, Lodge TP. 1991. Micelle formation of BAB triblock copolymers in solvents that preferentially dissolve the A block. Macromolecules 24:1975–1986.

Biswas N, Abraham B, Umapathy S. 2002. Investigation of short-time isomerization dynamics in p-nitroazobenzene from resonance Raman intensity analysis. J Phys Chem A 106:9397–9406.

Cantor R. 1981. Nonionic diblock copolymer as surfactants between immiscible solvents. Macromolecules 14:1186–1193.

Cattaneo P, Persico M. 1999. An ab initio stydy of the photochemistry of azobenzene. Phys Chem Chem Phys 1:4739–4743.

Cembran A, Bernadi F, Garavelli M, Gagliardi L, Orlandi G. 2004. On the mechanism of the cis-trans isomerization in the lowest electronic states of azobenzene: S_0, S_1 and T_1. J Am Chem Soc 126:3234–3243.

Chang CW, Lu YC, Wang TT, Diau EWG. 2004. Photoisomerization dynamics of azobenzene in solution with S1 excitaion: a femtosecond fluorescence anisotropy study. J Am Chem Soc 126:10109–10118.

Chen YY, Ford WT, Materer NF, Teeters D. 2000. Facile conversion of colloidal crystals to ordered porous polymer nets. J Am Chem Soc 122:10472–10473.

Cheng ZD, Russel WB, Chalkin PM. 1999. Controlled growth of hard-sphere colloidal crystals. Nature 401:893–895.

Crecca CR, Roitberg AE. 2006. Theoretical study of the isomerization mechanism of azobenzene and disubstituted azobenzene derivatives. J Phys Chem A 110:8188–8203.

Dante S, Advincula R, Frank CW, Stroeve P. 1999. Photoisomerization of polyionic layer-by-layer films containing azobenzene. Langmuir 15:193–201.

Delaire JA, Nakatani K. 2000. Linear and nonlinear optical properties of photochromic molecules and materials. Chem Rev 100:1817–1845.

Deng YH, Li N, He YN, Wang XG. 2007. Hybrid colloids composed of two amphiphilic azo polymers: fabrication, characterization, and photoresponsive properties. Macromolecules 40:6669–6678.

Diau EWG. 2004. A new trans-to-cis photoisomerization mechanism of azobenzene on the $S_1(n, \pi^*)$ surface. J Phys Chem A 108:950–956.

Dimitrov AS, Nagayama K. 1996. Continuous convective assembling of fine particles into two-dimensional arrays on solid surfaces. Langmuir 12:1303–1311.

Finkelmann H, Nishikawa E, Pereira GG, Warner M. 2001. A new opto-mechanical effect in solids. Phys Rev Lett 87:015501.

Förster S, Plantenberg T. 2002. From self-organizing polymers to nanohybrid and biomaterials. Angew Chem Int Ed Engl 41:688–714.

Gao ZS, Eisenberg A. 1993. A model of micellization for block copolymers in solutions. Macromolecules 26:7353–7360.

Hadziioannou G, Skoulios A. 1982. Structural study of mixtures of styrene/isoprene two- and three-block copolymers. Macromolecules 15:267–271.

Halperin A. 1987. Polymeric micelles: a star model. Macromolecules 20:2943–2946.

Hiemenz PC. 1986. Principles of Colloid and Surface Chemistry. 2nd ed. New York: Marcel Dekker Inc.

Ikeda T, Horiuchi S, Karanjit DB, Kurihara S, Tazuke S. 1990. Photochemically induced isothermal phase transition in polymer liquid crystals with mesogenic phenyl benzoate side chains. 2 photochemically induced isothermal phase transition behaviors. Macromolecules 23:43–48.

Ishikawa T, Noro T, Shoda T. 2001. Theoretical study on the photoisomerization of azobenzene. J Chem Phys 115:7503–7512.

Jiang P, Bertone JF, Hwang KS, Colvin VL. 1999a. Single-crystal colloidal multilayers of controlled thickness. Chem Mater 11:2132–2140.

Jiang P, Hwang KS, Mittleman DM, Bertone JF, Colvin VL. 1999b. Template-directed preparation of macroporous polymers with oriented and crystalline arrays of voids. J Am Chem Soc 121:11630–11637.

Joannopoulos JD, Villeneuve PR, Fan SH. 1997. Photonic crystals: putting a new twist on light. Nature 386:143–149.

Junge DM, McGrath DV. 1999. Photoresponsive azobenzene-containing dendrimers with multiple discrete states. J Am Chem Soc 121:4912–4913.

Kabanov AV, Nazarova IR, Astafieva IV, Batrakova EV, Alakhov VY, Yaroslavov AA, Kabanov VA. 1995. Micelle formation and solubilization of fluorescent probes in poly(oxyethylene-b-oxypropylene-b-oxyethylene) solutions. Macromolecules 28:2303–2314.

Kawata S, Kawata Y. 2000. Three-dimensional optical data storage using photochromic materials. Chem Rev 100:1777–1788.

Kikuchi A, Nose T. 1996. Unimolecular micelle formation of poly(methyl methacrylate)-graft-polystyrene in mixed selective solvents of acetonitrile/acetoacetic acid ethyl ether. Macromolecules 29:6770–6777.

Kim DY, Tripathy SK, Li L, Kumar J. 1995. Laser-induced holographic surface-relief gratings on non-linear-optical polymer-films. Appl Phys Lett 66:1166–1169.

Kumar GS, Neckers DC. 1989. Photochemistry of azobenzene-containing polymer. Chem Rev 89:1915–1925.

Kumar J, Li L, Jiang XL, Kim DY, Lee TS, Tripathy S. 1998. Gradient force: the mechanism for surface relief grating formation in azobenzene functionalized polymers. Appl Phys Lett 72:2096–2098.

Lambeth RH, Moore JS. 2007. Light-induced shape changes in azobenzene functionalized polymers prepared by ring-opening metathesis polymerization. Macromolecules 40:1838–1842.

Laschewsky A, Wischerhoff E, Kauranen M, Persoons A. 1997. Polyelectrolyte multilayer assemblies containing nonlinear optical dyes. Macromolecules 30:8304–8309.

Li MH, Keller P, Li B, Wang XG, Brunet M. 2003. Light-driven side-on nematic elastomer actuators. Adv Mater 15:569–572.

Li YB, Deng YH, He YN, Tong XL, Wang XG. 2005a. Amphiphilic azo polymer spheres, colloidal monolayers, and photoinduced chromophore orientation. Langmuir 21:6567–6571.

Li YB, Deng YH, Tong XL, Wang XG. 2006a. Formation of photoresponsive uniform colloidal spheres from an amphiphilic azobenzene-containing random copolymer. Macromolecules 39:1108–1115.

Li YB, He YN, Tong XL, Wang XG. 2005b. Photoinduced deformation of amphiphilic azo polymer colloidal spheres. J Am Chem Soc 127:2402–2403.

Li YB, He YN, Tong XL, Wang XG. 2006b. Stretching effect of linearly polarized Ar^+ laser single-beam on azo polymer colloidal spheres. Langmuir 22:2288–2291.

Li YB, Tong XL, He YN, Wang XG. 2006c. Formation of ordered mesoporous films from in situ structure inversion of azo polymer colloidal arrays. J Am Chem Soc 128:2220–2221.

Linse P. 1994. Micellization of poly(ethelene oxide)-poly(propylene oxide) block copolymers in aqueous solution: effect of polymer polydispersity. Macromolecules 27:6404–6417.

Lvov Y, Yamada S, Kunitake T. 1997. Non-linear optical effects in layer-by-layer alternate films of polycations and an azobenzene-containing polyanion. Thin Solid Films 300:107–112.

Morikawa Y, Nagano S, Watanabe K, Kamata K, Iyoda T, Seki T. 2006. Optical alignment and patterning of nanoscale microdomains in block copolymer thin film. Adv Mater 18:883–886.

Natansohn A, Rochon P. 2002. Photoinduced motions in azo-containing polymers. Chem Rev 102:4139–4175.

Park SH, Xia YN. 1999. Assembly of mesoscale particles over large areas and its application in fabricating tunable optical filters. Langmuir 15:266–273.

Philp D, Stoddart JF. 1996. Self-assembly in nature and unnature systems. Angew Chem Int Ed Engl 35:1154–1196.

Rau H. 1990. Photoisomerization of azobenzene. In: Rabek JF, editor. Photochemistry and Photophysics. Vol. 2. Florida: CRC Press. p. 119–142.

Rochon P, Batalla E, Natansohn A. 1995. Optically induced surface gratings on azoaromatic polymer-films. Appl Phys Lett 66:136–139.

Rugge A, Ford WT, Tolbert SH. 2003. From a colloidal crystal to an interconnected colloidal arrays: a mechanism for a spontaneous rearrangement. Langmuir 19:7852–7861.

Satzger H, Root C, Braun M. 2004. Excited-state dynamics of trans- and cis-azobenzene after UV excitation in the $\pi\pi^*$ band. J Phys Chem A 108:6265–6271.

Schmidt B, Sobotta C, Malkmus S, Laimgruber S, Braun M, Zinth W, Gilch P. 2004. Femtosecond fluorescence and absorption dynamics of an azobenzene with a strong push-pull substitution. J Phys Chem A 108:4399–4404.

Schultz T, Quenneville J, Levine B, Toniolo A, Martinez TJ, Lochbrunner S, Schmitt M, Shaffer JP, Zgierski MZ, Stolow A. 2003. Mechanism and dynamics of azobenzene photoisomerization. J Am Chem Soc 125:8098–8099.

Shaw DJ. 1992. Introduction to Colloid and Surface Chemistry. 4th ed. Oxford: Butterworth-Heinemann.

Shen HW, Zhang LF, Eisenberg A. 1997. Thermodynamics of crew-cut micelle formation of polystyrene-b-poly(acrylic acid) diblock copolymers in DMF/H_2O mixtures. J Phys Chem B 101:4697–4708.

Subramanian G, Manoharan VN, Thorne JD, Pine DJ. 1999. Ordered macroporous materials by colloidal assembly: a possible route to photonic bandgap materials. Adv Mater 11:1261–1265.

Terreau O, Bartels C, Eisenberg A. 2004. Effect of poly(acrylic acid) block length distribution on polystyrene-b-poly(acrylic acid) block copolymer aggregates in solution. 2. A partial phase diagram. Langmuir 20:637–645.

Tian YQ, Watanabe K, Kong XX, Abe J, Iyoda T. 2002. Synthesis, nanostructure, and functionality of amphiphilic liquid crystalline block copolymers with azobenzene moieties. Macromolecules 35:3739–3747.

Todorov T, Nikolova L, Tomova N. 1984. Polarization holography. 1: A new high-efficiency organic material with reversible photoinduced birefringence. Appl Opt 23:4309–4312.

Victor JG, Torkelson JM. 1987. On measuring the distribution of local free volume in glassy polymers by photochromic and fluorescence techniques. Macromolecules 20:2241–2250.

Wang DY, Caruso F. 2001. Fabrication of polyaniline inverse opals via templating ordered colloidal assemblies. Adv Mater 13:350–354.

Wang G, Tong X, Zhao Y. 2004. Preparation of azobenzene-containing amphiphilic diblock copolymers for light-responsive micellar aggregates. Macromolecules 37:8911–8917.

Wang XG, Balasubramanian S, Li L, Jiang XL, Sandman DJ, Rubner MF, Kumar J, Tripathy SK. 1997a. Self-assembled second order nonlinear optical multilayer azo polymer. Macromol Rapid Commun 18:451–459.

Wang XG, Balasubramanian S, Tripathy SK, Li L. 1998. Azo chromophore-functionalized polyelectrolytes. 1. Synthesis, characterization, and photoprocessing. Chem Mater 10:1546–1553.

Wang XG, Kumar J, Tripathy SK, Li L, Chen JI, Marturunkakul S. 1997b. Epoxy-based nonlinear optical polymers from post azo coupling reaction. Macromolecules 30:219–225.

Weissman JM, Sunkara HB, Tse AS, Asher SA. 1996. Thermally switchable periodicities and diffraction from mesoscopically ordered materials. Science 274:959–960.

Whitesides GM, Boncheva M. 2002. Beyond molecules: self-assembly of mesoscopic and macroscopic components. Proc Natl Acad Sci USA 99:4769–4774.

Whitesides GM, Grzybowski B. 2002. Self-assembly at all scales. Science 295:2418–2421.

Wilhelm M, Zhao CL, Wang YC, Xu RL, Winnik MA, Mura JL, Riess G, Croucher MD. 1991. Poly(styrene-ethylene oxide) block copolymer micelle formation in water: a fluorescence probe study. Macromolecules 24:1033–1040.

Wu LF, Tuo XL, Cheng H, Chen Z, Wang XG. 2001. Synthesis, photoresponsive behavior, and self-assembly of poly(acrylic acid) based azo polyelectrolytes. Macromolecules 34:8005–8013.

Wu YL, Demachi Y, Tsutsumi O, Kanazawa AT, Shiono A, Ikeda T. 1998. Photoinduced alignment of polymer liquid crystals containing azobenzene moieties in the side chain. 1 Effect of light intensity on alignment behavior. Macromolecules 31:349–354.

Xia YN, Gates B, Yin YD, Lu Y. 2000. Monodispersed colloidal spheres: old materials with new applications. Adv Mater 12:693–713.

Xie S, Natansohn A, Rochon P. 1993. Recent developments in aromatic azo polymers research. Chem Mater 5:403–411.

Xu T, Stevens J, Villa JA, Goldbach JT, Guarini KW, Black CT, Hawker CJ, Russell TP. 2003. Block copolymer surface reconstruction: a reversible route to nanoporous films. Adv Funct Mater 13:698–702.

Yager KG, Barrett CJ. 2001. All-optical patterning of azo polymer films. Curr Opin Solid State Mater Sci 5:487–494.

Yi DK, Kim DY. 2003. Novel approach to the fabrication of macroporous polymers and their use as a template for crystalline titania nanorings. Nano Lett 3:207–211.

Yu YL, Nakano M, Ikeda T. 2003. Direct bending of a polymer film by light. Nature 425:145–145.

Zhang LF, Eisenberg A. 1995. Multiple morphologies of "crew-cut" aggregates of polystyrene-b-poly(acrylic acid) block copolymers. Science 268:1728–1731.

Zhang LF, Shen HW, Eisenberg A. 1997. Phase separation behavior and crew-cut micelle formation of polystyrene-b-poly(acrylic acid) copolymers in solutions. Macromolecules 30:1001–1011.

Zollinger H. 1961. Azo and Diazo Chemistry: Aliphatic and Aromatic Compounds. New York: Interscience Publishers.

Zollinger H. 1991. Color Chemistry: Synthesis, Properties, and Applications of Organic Dyes and Pigments. 2nd ed. Weinheim: VCH.

6

AZOBENZENE-CONTAINING BLOCK COPOLYMER MICELLES: TOWARD LIGHT-CONTROLLABLE NANOCARRIERS

Yue Zhao

6.1. WHAT IS THE USE OF LIGHT-CONTROLLABLE POLYMER MICELLES?

The incorporation of azobenzene into small amphiphilic molecules to make photosensitive surfactants has been much studied (Eastoe and Vespertinas, 2005; Lee et al., 2004; Shang et al., 2003; Orihara et al., 2001; Kang et al., 2000; Einaga et al., 1999; Sakai et al., 1999; Shin and Abbott, 1999; Yang et al., 1995; Shinkai et al., 1982). The main interest of such surfactants is to cause property changes of aqueous solutions in response to illumination as a result of the trans–cis photoisomerization of the chromophore. Not surprisingly, many light-induced property changes are related to a disruption of micelles self-associated by azobenzene surfactant molecules (Lee et al., 2004; Shang et al., 2003; Kang et al., 2000; Shin and Abbott, 1999; Yang et al., 1995). For example, using 4,4′-bis (trimethylammoniumhexyloxy) azobenzene bromide, Shin and Abbot observed a significant decrease in dynamic surface tension upon illumination of UV light. They attributed this to a decrease in the number of micelles in solution that increases the rate of mass transport of surfactant molecules to the interface. In another study, Lee et al., demonstrated that the solution viscosity and gelation behavior of mixtures of azobenzene trimethylammonium bromide and a hydrophobically modified polymer could be reversibly controlled with UV and visible light. A light-induced dissolution of micelles formed by trans-azobenzene surfactants with the hydrophobic side groups of the polymer, acting as cross-links of polymer chains, was believed to be at the origin of the phenomenon. Generally, azobenzene-containing surfactants display a higher critical micelle concentration (cmc) with

Smart Light-Responsive Materials. Edited by Yue Zhao and Tomiki Ikeda

azobenzene moieties in the cis form (upon UV illumination) (Lee et al., 2004; Orihara et al., 2001; Shin and Abbott, 1999; Yang et al., 1995). Among the rare reports focused on the photoinduced disruption of micellar aggregates of azobenzene surfactants, most noticeable works came from Sakai's group (Orihara et al., 2001; Sakai et al., 1999). They showed direct evidence that micelles and vesicles formed in aqueous solution of 4-butylazobenzene-4′-(oxyethyl)trimethylammonium bromide (or its mixtures with another surfactant) could be reversibly disrupted by alternating UV and visible light irradiation. These systems allowed them to demonstrate the possibility of photocontrolling the release of either hydrophobic or hydrophilic compounds solubilized by micelles and vesicles, respectively. Another interesting study was reported by Hashimoto's group (Einaga et al., 1999). They designed an azobenzene-containing multibilayer vesicle with an intercalated inorganic magnetic compound (Prussian blue) and showed that the magnetic property of such a nanomaterial could be reversibly altered as a result of a geometrically confined structural change induced by the photoisomerization of azobenzene. In small-molecule photosensitive surfactants, azobenzene moiety is generally part of the hydrophobic tail or positioned close to the polar (ionic or nonionic) head. In either case, the different polarity of azobenzene in the elongated trans and bent cis forms can cause changes in the hydrophilic–hydrophobic balance that affects the self-association of surfactant molecules. The important molecular shape change of azobenzene accompanying the trans–cis photoisomerization can also contribute to the destabilization of micelles self-assembled by small-molecule surfactants.

The incorporation of azobenzene into amphiphilic block copolymers (BCPs) to make light-controllable *polymeric* micelles has not been reported until 2004 (Wang et al., 2004). The interest of such azobenzene-containing BCP micelles resides in the context of developing polymer nanocarriers for controlled delivery applications, which has been the subject of intensive research effort worldwide because of the high stakes (Gaucher et al., 2005; Haag, 2004; Discher and Eisenberg, 2002; Kwon and Kataoka, 1995). BCP micelles have some important advantages over their small-molecule counterparts with regard of nanocarriers. These include the following: (1) BCP micelles can be more stable in solution and thus afford more stable encapsulation of guest molecules, owing to a much lower cmc and the macromolecular nature (e.g., a high T_g or the crystallization of the hydrophobic block can render the core of micelle more compact and their dissolution slower) and (2) their sizes can easily be adjusted, by changing the lengths of the blocks, to be in the 20–100 nm range to allow their preferential accumulation in porous cancer tissues through the enhanced permeation and retention mechanism (Haag, 2004). To understand the interest of azobenzene BCP micelles, the question about the utility of light-controllable polymer micelles needs to be answered.

Using BCP micelles as a nanocarrier of drugs, an effective target-specific delivery of drugs should consist in three steps. The first step is a stable encapsulation of the drug, meaning that after administration in the body, the BCP micelle protects the drug and prevents it from leaking out quickly.

The second step is a site-specific transportation of the drug, meaning that the drug-loaded BCP micelle should be selectively captured by the target (pathological sites). The third step is that once arrived on the target the BCP micelle "opens the door" to allow the drug to be released. While the stable encapsulation of drug is crucial for giving the nanocarrier the time required to ferry the drug to the target, it could also render the drug release on the target more difficult at a later time. For each of the three steps, there has been much research effort aimed at developing effective strategies. For the stable encapsulation, the crosslinking of BCP micelles is useful to provide the structural integrity and to slow down the leakage of the drug through crosslinked polymer chains (Thurmond et al., 1996). For the site-specific transportation, the functionalization of the micelle shell with a targeting ligand that can be recognized by a receptor highly exposed by the target remains the most appealing approach (Nasongkla et al., 2006). Finally, to release the drug quickly or in a controlled fashion after the micelle arrives on the target, the micelle should be structurally disrupted. To this regard, a large number of BCP micelles have been designed to be able to react to an external stimulus, most of which are sensitive to pH change (Bellomo et al., 2004; Bae et al., 2003; Gillies and Fréchet, 2003). This exploits the fact that tumor tissues are known to be slightly acidic (pH ~6.8) and the endosomal and lysosomal compartments of cells have even lower pH (pH ~5–6) with respect to the physiological pH 7.4 (Haag, 2004; Gillies and Fréchet, 2003). If BCP micelles contain acid-labile bonds in their structures, inside the cells, the acidic pH can break those chemical bonds and the structural disruption of the micelle can lead to the release of the loaded drug (Gillies and Fréchet, 2003). Thermosensitive BCP micelles have also been much studied (Qin et al., 2006; Schilli et al., 2004; Chung et al., 2000). Like pH, change in temperature is also an event that polymer micelles can encounter in the body. If the BCP contains, for example, a poly(*N*-isopropylacrylamide) (PNIPAAm)-based hydrophilic block whose lower critical solution temperature (LCST) can be adjusted to be above 37°C through copolymerization with a hydrophilic comonomer (Taillefer et al., 2000), its micelle can be stable at $T<\mathrm{LCST}$, but disrupted by the phase separation at $T>\mathrm{LCST}$. Other external stimuli exploited for the disruption of BCP micelles include changes in ionic strength (Zhang and Eisenberg, 1996), exposure to oxidation reaction (Napoli et al., 2004) and ultrasound (Rapoport et al., 2003). Following the same line of thought, light is obviously another external stimulus that can be used to cause the disruption of BCP micelles.

The concept of using light to trigger the disruption of BCP micelles and thus to control the release of encapsulated drugs or other bioagents is attractive. In addition to be complementary to the aforementioned stimuli, a light source can be manipulated from the outside of the body. Should the release essentially occur when the BCP micellar nanocarrier is disrupted by light, the most appealing feature would be the great selectivity in terms of the time and the site of release, since the release must take place when illumination is applied and only in the region exposed to light. A parallel can also be drawn between the use of light to control the release of drugs from such polymer nanocarriers and the photodynamic therapy where light is used to activate the drugs (photosensitizers). In what follows, we first discuss how to design light-controllable polymer micelles using azobenzene-containing BCPs and how to apply the principles revealed by studies on azobenzene BCPs to design light-controllable polymer micelles using other chromophores. We then review the research works reported so far regarding light-controllable azobenzene-containing BCP micelles. Finally, we discuss future directions and perspectives in this field.

6.2. HOW TO DESIGN AZOBENZENE BLOCK COPOLYMERS FOR LIGHT-CONTROLLABLE MICELLES?

Generally, a diblock (AB) or triblock (ABA) copolymer in solution can self-assemble into micelles if the solvent is good for one block but bad for the other, i.e., being block selective. The micellar aggregates can have various morphologies (star, crew cut, rod, and vesicles), depending on a number of parameters such as the relative lengths of the blocks, polymer concentration, and the used solvent (Discher and Eisenberg, 2002). If the BCP is amphiphilic, i.e., being composed of a hydrophilic and a hydrophobic (lipophilic) polymer, its micelles in aqueous solution have a compact core region formed by the hydrophobic block and a shell (corona) formed by the hydrophilic block, which ensures their solubility (or dispersion) in water. Here we emphasize that while core–shell micelle formed by a diblock copolymer will be used in the discussion hereafter, the principles are applicable to other morphologies of micellar aggregates and to triblock copolymers as well. As with micelles of small-molecule surfactants, the decisive parameter that controls the noncovalent micellar association of BCPs is the hydrophilic–hydrophobic balance, i.e., the relative importance of the two blocks, which is mainly determined by their relative lengths (BCP composition) and their different solubilities in water (intermolecular interactions with water determined by the chemical structures of the two blocks). In the context of polymer nanocarriers for controlled delivery applications as explained earlier, light-controllable BCP micelles are those that can exhibit a structural or morphological change, such as dissociation, disintegration, and swelling, upon interaction with light to allow the release of loaded guest molecules into the aqueous solution. To this end, the BCP micelles under illumination should undergo a photoreaction that shifts the hydrophilic–hydrophobic balance to make the micelles structurally or morphologically unstable in aqueous solution.

With azobenzene-containing amphiphilic BCPs, the photoreaction concerned is the reversible trans–cis photoisomerization of the chromophore upon UV and visible light illumination, respectively. From both the design and synthetic considerations, it is better to link azobenzene moieties onto the hydrophobic polymer. In this case, as schematically illustrated in Fig. 6.1, the hydrophobic block, bearing pendant azobenzene moieties in the more stable trans isomer can form the hydrophobic core of the core–shell micelle or the membrane of the vesicle for the encapsulation of hydrophobic and hydrophilic agents, respectively (polymer vesicles can also load hydrophobic agents in the membrane, which is not depicted in Fig. 6.1). Upon illumination with UV light, azobenzene moieties can be converted to the cis isomer. To shift the hydrophilic–hydrophobic balance towards the destabilization (dissociation) of the micelles, obviously the cis isomer of azobenzene must be significantly more polar than the trans isomer. This is possible by appropriate choice of substitutions on azobenzene. A hint comes from azobenzene without substitution, for which it is known that the elongated trans form has a zero dipole moment, while the bent cis form has a dipole moment of ~3 D (Kumar and Neckers, 1989). To obtain a small (near zero) dipole moment for trans-azobenzene linked to the polymer chain backbone, the use of the same substitution groups on the *para* positions of azobenzene would be most useful since this substitution pattern gives a nearly symmetrical structure for the trans isomer, with no separation of positive and negative charge centers. On the other hand, the cis isomer should have an increased dipole moment owing to the conformational (shape) change.

On the contrary, azobenzene structures with strong electron-donor and electron-acceptor groups on the *para* positions, which would have a higher dipole

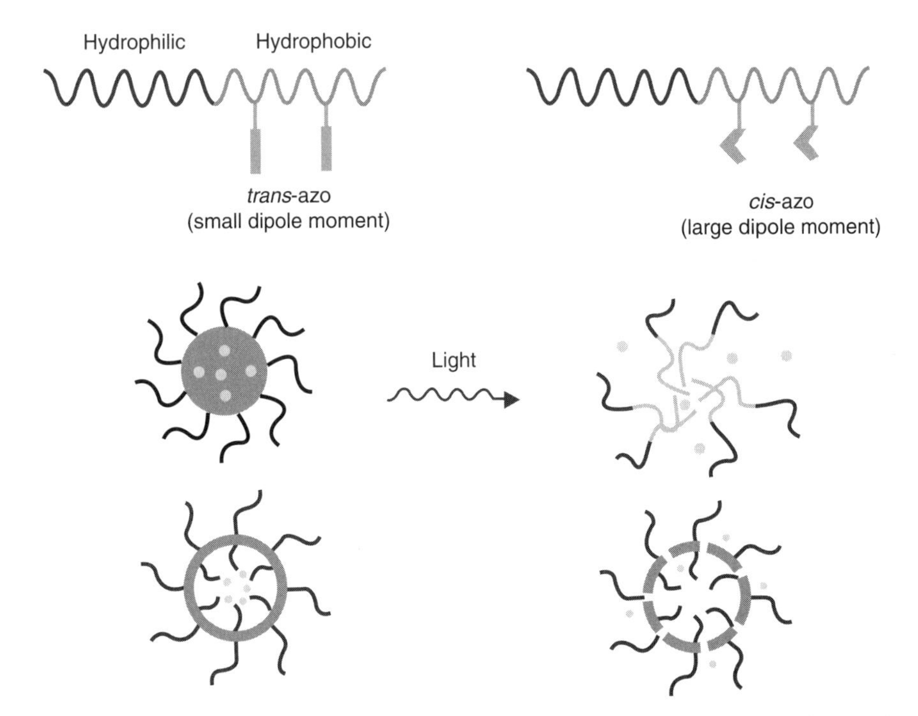

Figure 6.1. Schematic illustration of light-controllable micelles and vesicles formed by azobenzene-containing block copolymers displaying the trans–cis photoisomerization. See color insert.

moment for the trans isomer than the cis isomer (Simmons et al., 2007), should not be used because of the opposite change in dipole moment upon UV light illumination. Another consideration also does not favor the use of azobenzene derivatives with strong electron-donor and electron-acceptor groups. This pattern of substitution on azobenzene usually results in overlap of the absorption of trans-azobenzene (π-π*) with that of cis-azobenzene (n-π*), in the 440–460 nm region. Not only does the cis isomer have a fast thermally activated relaxation back to the trans isomer, photons absorbed by trans-azobenzene, inducing the trans–cis isomerization, can also be significantly absorbed by cis-azobenzene that induces the reverse cis–trans isomerization. The consequence of the fast thermal relaxation of cis-azobenzene and the simultaneous photoisomerization in the two directions is that few azobenzene moieties in the cis form can stay. Obviously, for light-controllable polymer micelles, a bistable photoswitch is preferred. As far as azobenzene is concerned, its derivatives that can undergo a thorough trans–cis isomerization upon UV light illumination and have thermally stable cis isomer are mostly interesting.

From the above analysis, using azobenzene BCPs, a significant increase in dipole moment accompanying the trans–cis isomerization of the chromophore under UV light can increase the entire polarity of the hydrophobic polymer block and, consequently, shift the hydrophilic–hydrophobic balance in the direction thermodynamically unfavorable to the micellar association. Although the micellar association is very sensitive to the balance, such a shift may not be necessarily enough to cause the dissociation of BCP micelles; other factors need to be taken into consideration. This can be illustrated using a classical small-molecule surfactant as example. The anionic surfactants of $CH_3(CH_2)_nNHCOO^-$ has a cmc sensitive to the hydrophilic–hydrophobic balance; the cmc is 1.8×10^{-4} mol dm^{-3} for $n = 13$, while it is reduced by an order to 1.8×10^{-5} mol dm^{-3} for $n = 15$ (longer hydrocarbon tail) (Everett DH, 1988). Now, imagine a micellar solution of the surfactant with $n = 15$ exposed to light, if somehow two methylene groups were removed from the hydrocarbon tail, the micelles would be dissociated if the surfactant concentration in solution was below 1.8×10^{-4} mol dm^{-3} (cmc of the surfactant with $n = 13$), while they would remain if the surfactant concentration was above this cmc. This simple example points to the fact that an important shift in the hydrophilic–hydrophobic balance alone may not be enough to result in the dissociation of the micelles and that other conditions need to be fulfilled. As will be discussed later, this indeed is the case for the micelles formed by azobenzene-containing BCPs. In order to achieve an effective dissociation under UV light, a significant increase in the dipole moment for azobenzene while converting from the trans to the cis isomer should be conjugated with a hydrophilic block that is weakly hydrophilic, yet enough to induce the formation of BCP micelles in a selective solvent (Tong et al., 2005). The first BCP micelles exhibiting reversible light-induced control in dissociation and formation were prepared using azobenzene BCPs (Tong et al., 2005; Wang et al., 2004). More important is the fact that inspired by the principles revealed from these studies on azobenzene BCPs using other chromophores, photosensitive BCP micelles with more effective light-induced dissociation were obtained, and light-controlled release of model compounds loaded in the micelles was also demonstrated (Lee et al., 2007; Jiang et al., 2006, 2005). To achieve more straightforward destabilization of BCP micelles upon illumination, the hydrophobic block can be designed to bring a chromophore as pendant groups, whose photoreaction such as photocleavage (Jiang et al., 2006, 2005) and isomerization (Lee et al., 2007) simply transforms the hydrophobic block to a hydrophilic one. This obviously leads to the dissociation of BCP micelles since the amphiphilicity required for the micellar association is gone. The same strategy has also been extended by Zhang's group to amphiphiles composed of a polymer and a photosensitive chromophore to make photodissociable vesicles (Jiang et al., 2007).

In principle, azobenzene moieties can also be incorporated into the structure of the hydrophilic block of amphiphilic BCPs. Synthetically speaking, it may be more difficult, but it is feasible by using, for example, azobenzene-containing polyelectrolytes as the hydrophilic block. However, if the trans–cis photoisomerization under UV light irradiation results in an increase in dipole moment of azobenzene, the increased polarity of the hydrophilic block obviously cannot destabilize the micelles. Using appropriate substitution patterns for azobenzene, the trans–cis isomerization can lead to a decrease in dipole moment of azobenzene. The resulting decrease in polarity of the hydrophilic block can eventually shift the hydrophilic–hydrophobic balance in the direction unfavorable for BCP micelles. This may result in either morphological changes of the micellar aggregates or simply precipitation of the polymer if the hydrophilicity of the hydrophilic block can no longer ensure the solubility or the stability of the aggregates. Clearly, in the context of polymer nanocarriers for controlled delivery applications, it would be less interesting to use hydrophilic blocks carrying azobenzene moieties.

6.3. SYNTHESIS OF AZOBENZENE-CONTAINING AMPHIPHILIC BLOCK COPOLYMERS

For the synthesis of azobenzene-containing amphiphilic BCPs, the most used method, which is also the most convenient one, is the atom transfer radical polymerization (ATRP). The first work was reported in 2002 by Tian et al. who used a hydrophilic poly(ethylene oxide) (PEO) macroinitiator to grow a hydrophobic block of poly{11-[4-(4-butylphenylazo)phenoxy]undecyl methacrylate} that is a side-chain liquid crystalline (LC) polymer (SCLCP) (structure a in Fig. 6.2) (Tian et al., 2002). Even though this amphiphilic BCP was not synthesized and investigated for its self-assembly behavior in solution, PEO is actually a good choice as the hydrophilic block for BCP micelles having interest for possible biomedical applications. PEO is highly soluble in water, enables steric stabilization of the micelle and can inhibit the surface adsorption of biological substances, thus protecting the loaded guest (Kang et al., 2005). Following this work, a number of amphiphilic azobenzene BCPs have been synthesized using PEO macroinitiators with different azobenzene polymers (Ding et al., 2008; Yu et al., 2007a,b, 2006a,b;

Sin et al., 2005; Yu et al., 2005). Using a dibromo initiator, a dibromo PEO macroinitiator (Br-PEO-Br) was prepared and subsequently employed to grow an azobenzene polymer, yielding an amphiphilic ABA-type triblock copolymer (Tang et al., 2007).

When other water-soluble polymers are used as the hydrophilic block, the direct synthesis of the hydrophilic macroinitiator using ATRP may be problematic, in contrast to PEO. For instance, ATRP cannot be used to prepare a poly(acrylic acid) macroinitiator because of the reaction between the monomer and the metal complexes (Patten and Matyjaszewski, 1998). The general approach employed to bypass the problem is to synthesize first a diblock copolymer that is not amphiphilic and, subsequently, convert one block into the desired hydrophilic polymer through a postreaction. As a matter of fact, the first light-controllable azobenzene BCP micelle was demonstrated using a BCP whose hydrophilic block was obtained in this fashion (structure b in Fig. 6.2) (Wang et al., 2004). A macroinitiator of poly(*tert*-butyl acrylate) (P*t*BA) was first synthesized and then used to grow poly{6-[4-(4-methoxyphenylazo)phenoxy]hexyl methacroylate} (PAzoMA). The resulting diblock copolymer P*t*BA-*b*-PAzoMA was rendered amphiphilic by subsequent hydrolysis of P*t*BA giving rise to the hydrophilic poly(acrylic acid) (PAA). The selective cleavage of *t*BA groups is possible because

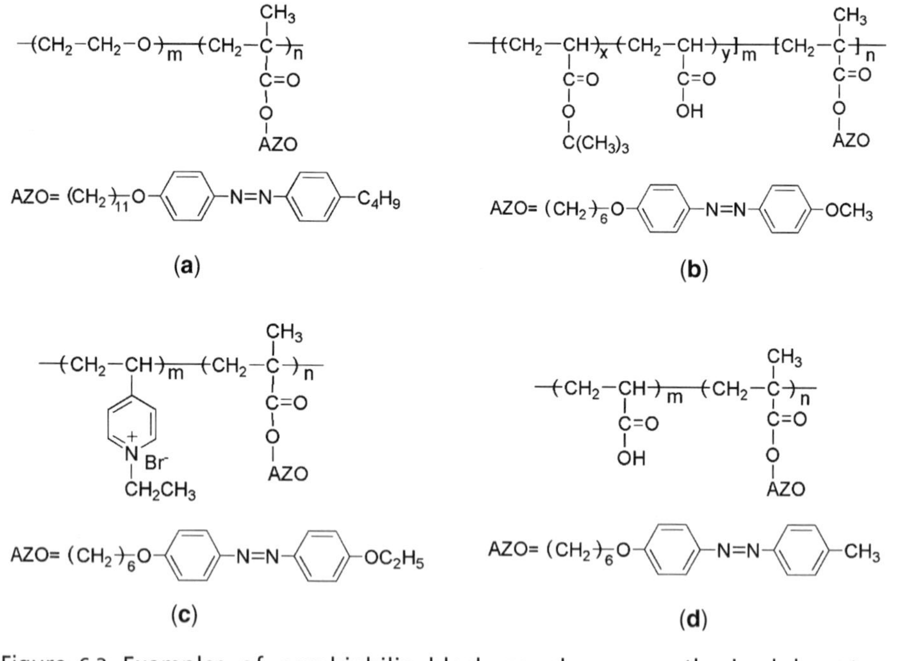

(a) (b) (c) (d)

Figure 6.2. Examples of amphiphilic block copolymers synthesized by atom transfer radical polymerization (ATRP) and reversible addition–fragmentation chain transfer (RAFT) polymerization (end groups and block-linking groups are omitted).

its hydrolysis is much easier than the ester groups linked to azobenzene moieties. However, it was found that the reaction is not trivial. Under the used conditions, with which PAzoMA was not affected by the hydrolysis reaction, a complete (100%) removal of *t*BA groups is not possible. The hydrolysis of P*t*BA-*b*-PAzoMA actually results in a hydrophilic block that is a random copolymer of *tert*-butyl acrylate and acrylic acid, P(*t*BA-*co*-AA). The obtained diblock copolymer P(*t*BA-*co*-AA)-*b*-PAzoMA can form core–shell micelles and vesicles depending on the preparation conditions. And the incomplete hydrolysis of P*t*BA, which weakens the hydrophilicity of the hydrophilic block, was believed to play an important role in the effective dissociation and reformation of the micellar aggregates upon UV and visible light exposures. By means of ATRP, the hydrolysis of P*t*BA remains the general route to the use of PAA as the hydrophilic block. Another example of azobenzene amphiphilic BCP whose hydrophilic block was prepared through two steps is also depicted in Fig. 6.2 (structure c) (Qi et al., 2008), namely, quaternized poly(4-vinyl pyridine)-*b*-poly {6-[4-(4-ethoxyphenyl-azo)phenoxy]hexyl methacrylate} (QP4VP-*b*-PAzoMA). To obtain this BCP, chlorine-terminated poly(4-vinyl pyridine) (P4VP) was first synthesized and then used to initiate the polymerization for PAzoMA. The resulting diblock of P4VP-*b*-PAzoMA was subjected to a quaternization reaction using bromoethane to obtain the amphiphilic BCP. Other azobenzene amphiphilic BCPs prepared using ATRP whose hydrophilic block is not PEO were also reported (Ravi et al., 2005). We note that this chapter concerns only amphiphilic BCPs and does not cover the numerous works of azobenzene BCPs, including the early syntheses using other polymerization techniques, which was reviewed elsewhere (Zhao, 2007a).

It is of obvious interest to explore the use of other polymerization techniques that, being more tolerant to the experimental conditions and monomers, can produce amphiphilic azobenzene BCPs with no need for post reactions. Notably, Su et al. have recently reported the synthesis of such an amphiphilic diblock copolymer with PAA as the hydrophilic block using reversible addition–fragmentation transfer (RAFT) polymerization (structure d in Fig. 6.2) (Su et al., 2007). Using RAFT, they prepared PAA capped with dithiobenzoate and used it as the macro-RAFT transfer agent to polymerize the hydrophobic azobenzene polymer successfully. It can be expected that more amphiphilic azobenzene BCPs will be synthesized using the controlled radical polymerization techniques (ATRP and RAFT) because of their simplicity, versatility, and efficiency.

6.4. REVERSIBLE DISSOCIATION AND FORMATION OF AZOBENZENE BLOCK COPOLYMER MICELLES

The block copolymer of P(*t*BA-*co*-AA)-*b*-PAzoMA can form core–shell micelles and vesicles depending on the composition and the preparation conditions (Tong et al., 2005; Wang et al., 2004). With the same length of the hydrophobic block of PAzoMA, the actual hydrolysis degree of *t*BA on the hydrophilic P(*t*BA-*co*-AA) random copolymer turns out to be important since it determines the hydrophilic–

hydrophobic balance. At some block copolymer compositions, the micellar aggregates respond to light exposure in a sensitive way and display clearly reversible dissociation and reformation upon alternating UV and visible light irradiation that switches the dipole moment of azobenzene pendant groups and thus the entire polarity of the PAzoMA block. Unless otherwise stated, the results obtained with the sample of P(tBA$_{46}$-co-AA$_{22}$)-b-PAzoMA$_{74}$ are used for discussion, with micellar aggregates prepared using the method developed for amphiphilic BCPs that are not soluble in water. The method consists in first dissolving the BCP in a good organic solvent for both blocks and then adding slowly a certain amount of water with which the organic solvent is miscible, to trigger the aggregation of the hydrophobic polymer chains in the mixed solvent to form the core of micelle or the membrane of vesicle (Fig. 6.1). Usually, a large volume of water (e.g., 10-fold) is subsequently added to quench the micellar aggregates thus formed. In case an aqueous micellar solution is needed, the organic solvent can be removed either by evaporation or through dialysis against water.

Figure 6.3 shows the scanning electron microscope (SEM) images of a core–shell micellar solution before UV, after UV and after subsequent visible light irradiation (all solutions were cast on a silicon wafer and dried prior to the SEM observation). Core–shell micelles with diameters around 15 nm were formed in the initial solution; after exposure of the solution to UV light (360 nm, 35 mW cm^{-2}), most micelles were disappeared. However, when this UV irradiated solution was subsequently exposed to visible light (440 nm, 39 mW cm^{-2}), micelles reappeared in the cast sample. This observation clearly shows the reversible dissociation and formation of core–shell micelles of P(tBA$_{46}$-co-AA$_{22}$)-b-PAzoMA$_{74}$ as a result of the trans–cis photoisomerization of azobenzene groups on PAzoMA.

The same conclusion was also reached for larger vesicles of this BCP (diameter ~200 nm) that can be formed by adding more water into the dioxane BCP solution. In addition to SEM observations, the larger micellar aggregates in solution scatter more light, which makes possible the *in situ* monitoring of the

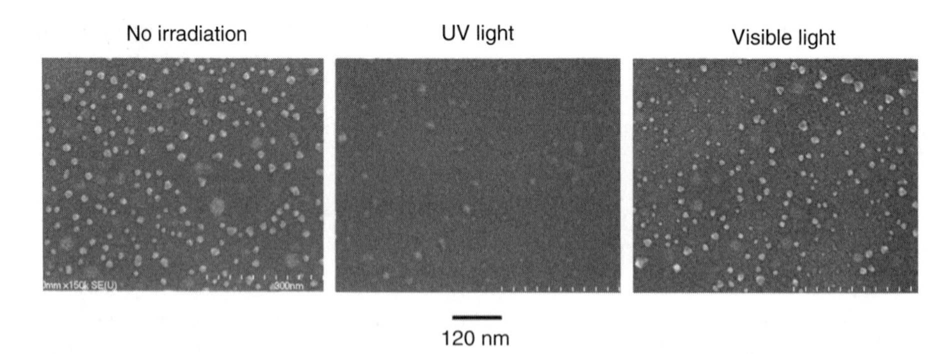

Figure 6.3. SEM images of core-shell micelles of P(tBA$_{46}$-co-AA$_{22}$)-b-PAzoMA$_{74}$ formed by adding 9%, in volume, of water in dioxane solution with initial polymer concentration of 1 mg mL^{-1}: before irradiation, after UV and visible light irradiation. *Source*: Wang et al., 2004. Reprinted with permission.

kinetic processes of light-induced dissociation and formation of the vesicles through measurements of the transmittance of a probe light (a He–Ne laser at 633 nm) whose wavelength is far from the absorption of the chromophore. An example of results is given in Fig. 6.4. In this specific case, the vesicular solution was prepared by adding 16% of water (v/v) in a dioxane solution of P(tBA$_{46}$-co-AA$_{22}$)-b-PAzoMA$_{74}$ (1 mg mL^{-1}). When UV light (360 nm, 18 mW cm^{-2}) was applied vertically from the top of the cuvette, in which was placed the vesicular solution (~1.5 mL) under stirring, the optical transmittance increased quickly

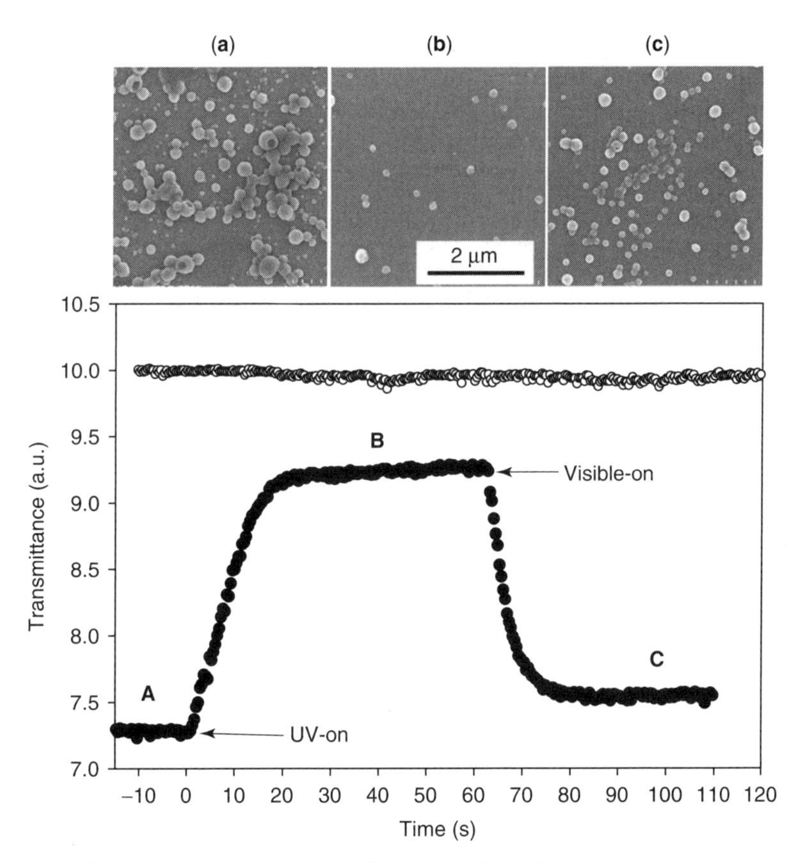

Figure 6.4. Changes in transmittance for a vesicle solution of P(tBA$_{46}$-co-AA$_{22}$)-b-PAzoMA$_{74}$ exposed to UV (360 nm, 18 mW cm^{-2}) and visible (440 nm, 24 mW cm^{-2}) light irradiation, vesicles being formed by adding 16%, in volume, of water in a dioxane solution with initial polymer concentration of 1 mg mL^{-1}. Typical SEM images for samples cast from the solution at different times indicated in the figure show the vesicles before irradiation, their dissociation by UV irradiation and reformation by visible irradiation. For comparison, also shown is the transmittance of the dioxane solution (no water added to induce the aggregation of the polymer) subjected to the same conditions of UV and visible light irradiation. *Source*: Tong et al., 2005. Reprinted with permission.

because of the dissociation of vesicles. On the basis of the transmittance change, this photoinduced dissociation process was completed after about 20 s. Simply by switching the irradiation to visible light (440 nm, 24 mW cm^{-2}), the opposite process occurred immediately, the transmittance dropped as a result of the reformation of vesicles in the solution. These reversible changes were confirmed by SEM observations made on samples cast from the solution before UV irradiation (marked by A), after 40 s UV irradiation (B) and after 40 s visible irradiation (C). The solution under visible light has a transmittance slightly higher than that of the initial vesicular solution, reflecting the difference between the photoinduced vesicle formation kinetics and the initial preparation procedure. Several subsequent alternating UV and visible light irradiations gave rise to reversible changes in transmittance without significant differences. Also presented in the figure is the result of a control test aimed at providing more unambiguous evidence. The higher transmittance was obtained with the same BCP solution without addition of water (i.e., polymer chains remained in dissolved state with no aggregation), subjected to exactly the same UV and visible light irradiation; the occurrence of the reversible trans–cis photoisomerization of azobenzene moieties on PAzoMA did not change the transmittance of the solution.

The first light-controllable BCP micelle exhibiting reversible dissociation and formation in solution was thus demonstrated with the help of an azobenzene-containing BCP. As discussed above, the reversible switch of the dipole moment of the azobenzene chromophore appears to be the decisive factor that explains the morphological changes of the micelles and, in particular, the reversibility of these changes. A quick look at the azobenzene moiety in P(*t*BA-*co*-AA)-*b*-PAzoMA (structure b in Fig. 6.2) can already find an explanation. The azobenzene has substitutions on the *para* positions, a methoxy end group (OCH_3) and oxygen linked to the methylene group of the flexible spacer; its trans conformation should have a very small dipole moment because of this near-symmetrical structure, similar to azobenzene without substitution; whereas the bent cis isomer should have a notable dipole moment because of the charge separation. This indeed was confirmed by a Density Functional Theory (DFT) calculation on the model compound of 4,4′-dimethoxyazobenzene that is close to the azobenzene moiety on PAzoMA. As shown in Fig. 6.5, the computed dipole moment of the cis isomer averaged over four energetically stable conformations is 4.4 D, which is larger than unsubstituted azobenzene ($\sim$3 D). It is now understandable why this azobenzene BCP displays light-switchable polymer micellar aggregation.

6.5. FACTORS INFLUENCING THE REVERSIBLE DISSOCIATION AND FORMATION PROCESSES

Through the real-time monitoring of the transmittance change of a vesicular solution in response to UV and visible light exposure, the effects of a number of variables on the processes of photoinduced dissociation and formation were investigated. In this section, these factors are discussed on the basis of obtained results.

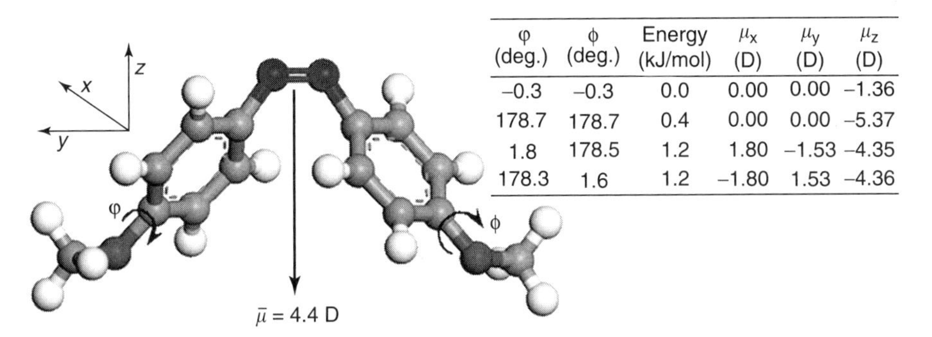

φ (deg.)	ϕ (deg.)	Energy (kJ/mol)	μ_x (D)	μ_y (D)	μ_z (D)
−0.3	−0.3	0.0	0.00	0.00	−1.36
178.7	178.7	0.4	0.00	0.00	−5.37
1.8	178.5	1.2	1.80	−1.53	−4.35
178.3	1.6	1.2	−1.80	1.53	−4.36

Figure 6.5. Lowest energy conformation of the cis isomer of 4,4′ dimethoxyazobenzene. The calculated relative energies and dipole moment components of the four stable conformations are shown, with the total dipole moment being computed as the average over the four conformations. *Source*: Tong et al., 2005. Reprinted with permission.

6.5.1. Effect of Solution Stirring

When UV light is applied to a vesicular (or micellar) solution for the photoinduced dissociation of the polymer aggregates, it is easy to imagine that the mechanical stirring of the solution may affect the rate of dissociation. Besides the change in thermodynamic stability as discussed above, other factors may intervene. First, with the copolymer of P(*t*BA-*co*-AA)-*b*-PAzoMA, the membrane of vesicle is formed by a SCLCP bearing azobenzene mesogens, for which the trans–cis isomerization could result in a photochemical LC-to-isotropic phase transition that may have an optical plasticization effect on the compact hydrophobic membrane. Should this happen, the softened vesicles would become mechanically weaker and thus be more easily broken by the shearing force of the stirring. Secondly, under the used low intensity UV light exposed to a large volume of solution (~1.5 mL), the strong absorption of the chromophore means that it is likely that only azobenzene moieties in the vesicles located nearby the surface of the solution would be excited and undergo the trans–cis photoisomerization; the stirring should help the diffusion of polymer vesicles toward the surface region. A greater shearing force or/and a faster mixing of the solution at a higher stirring rate should therefore speed the photoinduced dissociation process throughout the solution. This indeed is the case.

Using the same UV irradiation intensity (18 mW cm^{-2}), changes in transmittance of the vesicular solution with no stirring and with stirring at different rates by adjusting the rotation speed of the magnetic bar between about 50 and 300 rmp were recorded and compared in Fig. 6.6a, Note that before the UV irradiation was turned on, the vesicular solution had the same transmittance regardless of the stirring rate, indicating that the shear effect could not break the polymer vesicles. It is clear that the process of the vesicle photodissociation became faster as the stirring rate increased. Actually, most interesting is the result obtained for the still solution. After the UV irradiation was applied on, it showed no change in transmittance

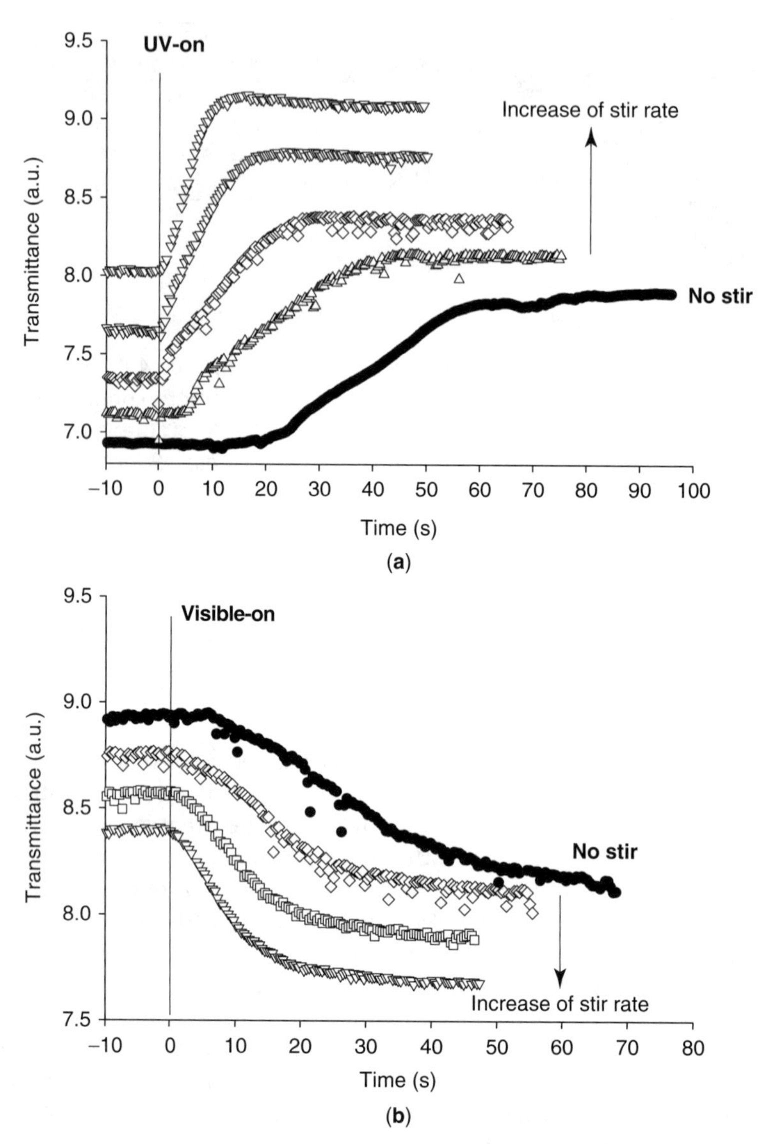

Figure 6.6. (a) Changes in transmittance of the vesicle solution of P(tBA$_{46}$-co-AA$_{22}$)-b-PAzoMA$_{74}$ subjected to UV irradiation (18 mW cm^{-2}), the solution being still (with no stirring) and stirred by a magnetic bar at different speeds of rotation (about 50, 100, 200, and 300 rpm). (b) Changes in transmittance of the clear solution of P(tBA$_{46}$-co-AA$_{22}$)-b-PAzoMA$_{74}$ (after UV-induced dissociation of vesicles) subjected to visible irradiation (24 mW cm^{-2}), the solution being still and stirred by a magnetic bar at different speeds of rotation (about 100, 200, and 300 rpm). Data of transmittance for different stirring rates are shifted along the ordinate for the sake of clarity. *Source*: Tong et al., 2005. Reprinted with permission.

during the first 20 s; then the transmittance increased gradually over the next 40 s or so to reach the plateau. The occurrence of UV light–induced dissociation of polymer vesicles in the solution with no stirring, i.e., in the absence of any shear effect, clearly indicates that the main driving force for the dissociation of vesicles is the thermodynamic instability due to the shifted hydrophilic–hydrophobic balance.

A similar effect of the stirring rate on the opposite process, i.e., the photoinduced reformation of polymer vesicles upon visible light irradiation inducing the reverse cis–trans isomerization of azobenzene mesogens, was observed for the same reason as helping the diffusion of polymer chains toward the surface region. The results in Fig. 6.6b were obtained by applying the visible light irradiation of the same intensity (24 mW cm^{-2}) on the solution containing photodissociated BCP, with no stirring and under different stirring rates. Similar to the dissociation process, before visible light was turned on, the transmittance of the clear solution showed no change under the various stirring rates, while it decreased because of the reformation of the vesicles after visible light was applied on. The rate of the reformation process increases with increasing the stirring rate of the solution. Again, even in the still solution, polymer vesicles can reform, but it takes longer than in solutions under stirring.

6.5.2. Effect of Irradiation Light Intensity

It is no surprise that the intensity of UV and visible light is another parameter that can be used to control the rates of the dissociation and formation processes of the polymer vesicles. Figure 6.7 shows the results obtained with three different intensities of UV irradiation for the dissociation (Fig. 6.7a) and three different intensities of visible irradiation for the formation process (Fig. 6.7b), respectively, using a constant stirring rate ($\sim$200 rpm of the magnetic bar). For both processes, the kinetics becomes faster with increased irradiation intensity. This can easily be understood by the fact the number of azobenzene mesogens being excited and undergoing the trans–cis or the reverse cis–trans isomerization should increase when there are more photons available for absorption by increasing the irradiation intensity, which speeds up any changes associated with the photoisomerization process. In Fig. 6.7b, it is also interesting to note that in the absence of visible light irradiation while keeping the probe He–Ne laser on the solution, no noticeable change in transmittance occurred over 160 s after the UV irradiation was turned off. This indicates that the probe light (633 nm) did not induce the reverse cis–trans isomerization of azobenzene mesogens so that polymer chains, fallen apart from the vesicles upon UV irradiation, remained in the dissolved state. Even after 1 h of illumination with the He–Ne laser, only $\sim$3% decrease in transmittance of the solution was observed, revealing the appreciable thermal stability of azobenzene moieties in the cis form. Significant decrease in transmittance was recorded only after much longer times as a result of thermally activated cis–trans backisomerization ($\sim$50% decrease after 4 h). Of course, the stability of dissolved polymer chains in the solution depends on the substitution pattern on the azobenzene moiety and the solvent that determine the barrier of the thermal cis–trans relaxation.

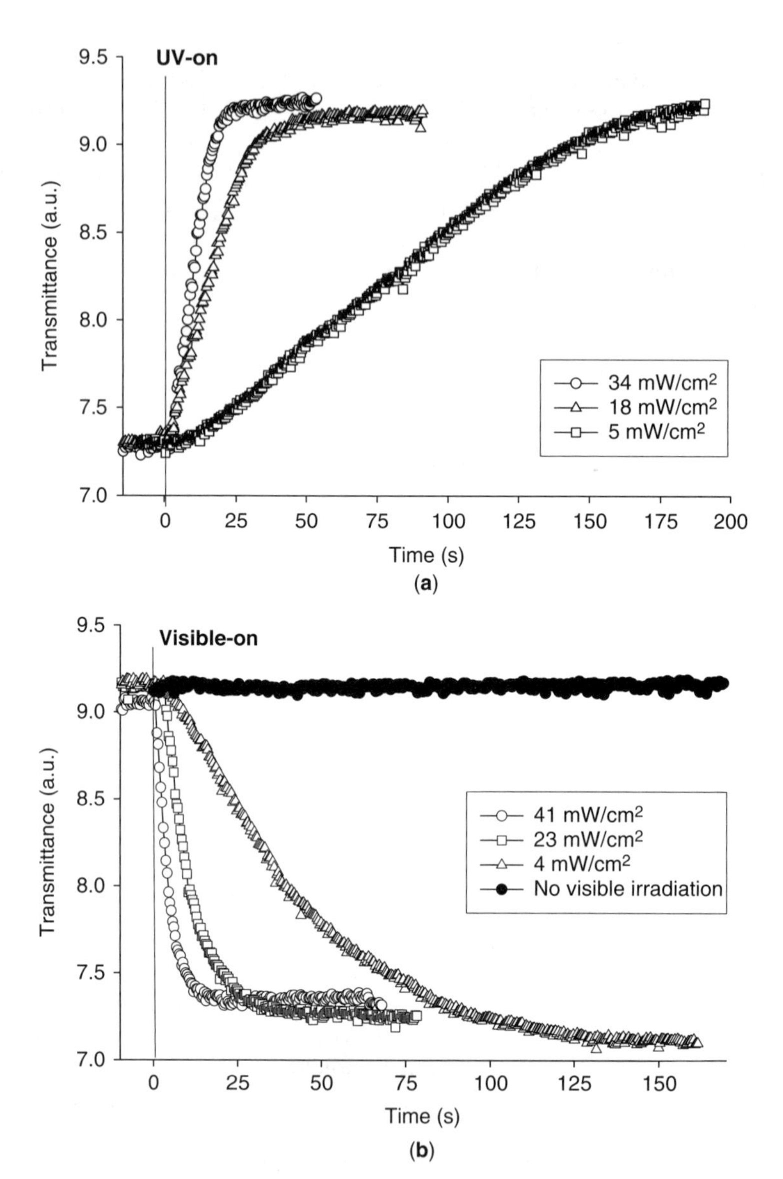

Figure 6.7. Transmittance vs. time for the vesicle solution of P(tBA$_{46}$-*co*-AA$_{22}$)-*b*-PAzoMA$_{74}$ at the same stirring rate of about 200 rpm: (**a**) under UV irradiation of different intensities (for the dissociation of vesicles), and (**b**) under visible irradiation of different intensities (for the reformation of vesicles). For comparison, also shown in (**b**) are changes in transmittance for the solution under only the probe light (He–Ne laser, 633 nm) illumination with no visible light applied. *Source*: Tong et al., 2005. Reprinted with permission.

6.5.3. Effects of Solvent and Block Copolymer Composition

Like P(*t*BA-*co*-AA)-*b*-PAzoMA, most amphiphilic BCPs are not soluble in water, in contrast to small-molecule surfactants. As aforementioned, to prepare their micelles or vesicles, the BCP needs to be dissolved first in an organic solvent that is good for both blocks and then by adding water into the solution. The aggregation of the hydrophobic blocks can occur at a critical water content, leading to the formation of the micellar aggregates. A convenient way to investigate the effect of solvent on the photoinduced processes is to measure and compare the change in transmittance of the BCP solution as a function of added water content in the absence of and under UV irradiation. Using a same solvent, such experiment can also be carried out to reveal the effect of the BCP composition by comparing the behaviors of two BCP samples subjected to the same experimental conditions. Fig. 6.8 shows an example of the results.

Figures 6.8a and b compare P(tBA$_{46}$-*co*-AA$_{22}$)-*b*-PAzoMA$_{74}$ dissolved in dioxane and tetrahydrofuran (2 mL, 1 mg mL^{-1}), respectively, prior to the addition of water (% in volume with respect to the volume of the organic solvent). With and without UV irradiation (18 mW cm^{-2}), the polymer solution following each addition of water was stirred for 3 min before the transmittance was measured. In the case of the dioxane solution, without UV irradiation, the transmittance starts to decrease significantly after addition of about 13% of water, which corresponds to the formation of polymer vesicles. Under UV light, the polymer aggregation process begins at a much higher water content of about 21%, and the transmittance drops more quickly than the solution without UV irradiation. This result shows that in the mixed solvent of dioxane–water, the increase in polarity of the PAzoMA block accompanying the conversion of trans-azobenzene onto cis-azobenzene has shifted the hydrophilic–hydrophobic balance in such a way that the critical water content for the aggregation of PAzoMA block forming the vesicle is increased from 13% to 21%. And the magnitude of the change determines the window in which the photoinduced dissociation and reformation of the vesicles can be observed. In other words, only the vesicles formed at water content between about 15% and 20% could display a clear light-controlled dissociation and reformation. Vesicles formed at higher water content cannot be disrupted by UV light because of the aggregation of this BCP under UV irradiation (with azobenzene moieties in the cis form). Now, using tetrahydrofuran (THF) as the solvent instead of dioxane (all other experimental conditions were kept the same), upon UV irradiation, the increase in the critical water content for the formation of vesicles is even greater, from 15% to 30%. This means a larger window for the photoinduced dissociation and reformation of polymer vesicles. The effect of solvent can be understood by the fact that different solvents have different interactions with the two blocks forming the BCP. In the present case, mixed THF–water seems to be a better solvent (less block selective) for P(tBA$_{46}$-*co*-AA$_{22}$)-*b*-PAzoMA$_{74}$, especially with azobenzene moieties in the cis form (under UV light) so that a larger amount of water is required to make the mixed solvent block selective for the BCP aggregation.

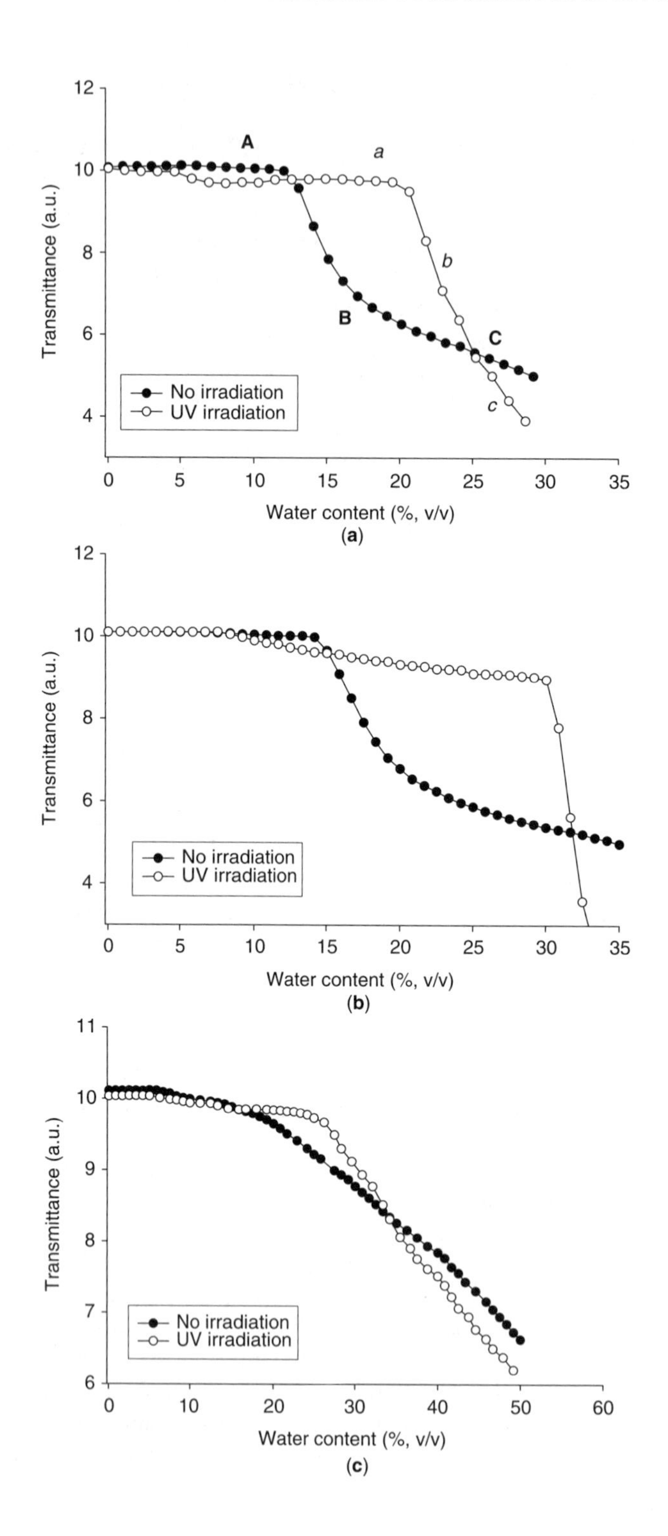

Transmittance (a.u.)
Water content (%, v/v)
A
a
b
B
C
c
No irradiation
UV irradiation
(a)
(b)
(c)

The comparison of the results in Figs. 6.8a and c reveals the effect of BCP composition on the photoinduced processes. Presented in Fig. 6.8c are the results obtained with a different BCP sample, namely, P(tBA$_{19}$-co-AA$_{33}$)-b-PAzoMA$_{31}$. As compared with P(tBA$_{46}$-co-AA$_{22}$)-b-PAzoMA$_{74}$, this sample has a much shorter hydrophobic PAzoMA block and a more hydrophilic block because of the relative length and the higher ratio of the number of AA units to that of tBA units. Under the same conditions, the increase in the critical water content for the aggregation in the dioxane solution upon UV irradiation is much smaller. Indeed, no clear photoinduced dissociation and reformation of vesicles were observed for this sample. SEM observation found that only large micelle-like aggregates ($\sim$200 nm) coexisting with core–shell micelles ($\sim$15 nm) were formed by adding water into the dioxane solution; UV and visible light irradiation only resulted in changes in the number of the aggregates (Wang et al., 2004). This result suggests that a large increase in dipole moment of azobenzene moieties on the hydrophobic block upon UV irradiation is necessary but may not be enough to lead to the dissociation of micellar aggregates. The same increase in dipole moment for azobenzene mesogens takes place in the two samples upon UV light irradiation, but for P(tBA$_{19}$-co-AA$_{33}$)-b-PAzoMA$_{31}$, the resulting increase in polarity of the PAzoMA block is not important enough to alter significantly the hydrophilic–hydrophobic balance because of the stronger hydrophilicity of the P(tBA-co-AA) block. Therefore, it appears that the prominent light-induced dissociation and reformation observed for the vesicles of P(tBA$_{46}$-co-AA$_{22}$)-b-PAzoMA$_{74}$,were the result of an increase in polarity of the hydrophobic PAzoMA block in conjunction with the weakness of hydrophilicity of the hydrophilic block (the relative length and chemical nature). Under such a conjugation, the reversible trans–cis isomerization of azobenzene moieties can alter reversibly the hydrophilic–hydrophobic balance in the two directions, which determines the state of aggregation of the diblock copolymer.

Strictly speaking, the origin of the observed effect is more complicated than it may appear to be. The difference between the two BCP samples may come not only from the difference in composition of the two blocks, but can also be contributed by the different compositions of the P(tBA-co-AA) blocks, which actually changes the chemical nature of the hydrophilic block. In order to disclose the effect of BCP composition, it would be more straightforward to utilize BCPs that differ only in the relative lengths of the two blocks, such as PAA-b-PAzoMA

Figure 6.8. (**a**) Transmittance vs. water content added to the dioxane solution of P(tBA$_{46}$-co-AA$_{22}$)-b-PAzoMA$_{74}$ (initial concentration, 1 mg mL^{-1}) without and with UV irradiation. (**b**) Transmittance vs. water content added to the tetrahydrofuran solution of P(tBA$_{46}$-co-AA$_{22}$)-b-PAzoMA$_{74}$ (initial concentration, 1 mg mL^{-1}) without and with UV irradiation. (**c**) Transmittance vs. water content added to the dioxane solution of P(tBA$_{19}$-co-AA$_{33}$)-b-PAzoMA$_{31}$ (initial concentration, 1 mg mL^{-1}) without and with UV irradiation. *Source*: Tong et al., 2005.

or PEO-*b*-PAzoMA. Moreover, the effect of BCP composition is related to the used solvent. The more the increase in polarity of the hydrophobic block due to the trans–cis isomerization of azobenzene reduces the difference in solubility of the two blocks in a given solvent, the more efficient the photoinduced dissociation of the micellar aggregates should be.

6.6. OTHER LIGHT-RESPONSIVE AZOBENZENE-BASED POLYMER MICELLES

The interest of using photoswitchable polarity of azobenzene to design light-responsive polymer micelles goes beyond BCPs. A nice example can be found in a study of Liu and Jiang who designed light-controllable, noncovalently connected polymer micelles involving no block copolymers (Liu and Jiang, 2006). They showed that the hydrogen (H)-bonded complex between an azobenzene-containing random copolymer, poly(4-phenylazo-maleinanil-co-4-vinylpyridine) (AzoMI-VPy), and carboxy group-terminated polybutadiene (CPB) was soluble in toluene, with azobenzene moieties in the trans form. Under UV light irradiation, the increase in polarity with cis azobenzene renders AzoMI-VPy insoluble, leading to the formation of micelles with the AzoMI-VPy core. The CPB shell, however, remains soluble in the solvent. Upon visible light irradiation, the micelles were dissociated. The change between the micellar and soluble polymer complex under UV and visible light was reversible. Moreover, when the AzoMI-VPy micelle core was chemically cross-linked to afford the structural integrity, visible light irradiation could no longer dissociate the micelles, instead the core–shell micelles ($\sim$250 nm) were transformed into larger hollow spheres ($\sim$900 nm) due to the solubility of AzoMI-VPy with azobenzene in the trans form. As shown in Fig. 6.9, dynamic light scattering (DLS) measurements found that the large change in size between core–shell micelles and hollow spheres is also reversible upon alternating UV and visible light irradiation. This striking optical switching of self-assembly displayed by H-bonded azobenzene-containing polymers is schematized in Fig. 6.10.

Other interesting photoisomerization-induced phenomena were also reported with polymer micellar aggregates formed by azobenzene-containing amphiphilic BCPs. Using a diblock copolymer composed of PEO and poly{2-[ethyl(4-(4-cyanophenylazo)-phenyl)amino]ethyl methacrylate} (PEO-*b*-PEPAEMA), Wang et al. showed photoinduced deformation of spherical micelles under linearly polarized Ar^+ laser irradiation (488 nm) (Wang et al., 2007). Su et al. synthesized a diblock copolymer of poly(*N*-isopropylacrylamide) and poly6-[4-(4-methyl phenylazo)-phenoxyl] hexylacrylate (PNIPM-*b*-PMPHA) that can form giant vesicles (most having diameters of several micrometers) in a water–THF mixture (Su et al., 2007). Deformation and fusion of the giant vesicles upon UV light irradiation (365 nm) was observed and investigated using an optical microscope. A change in polarity and LC order of azobenzene mesogenic groups related to the trans–cis photoisomerization was suggested as the origin of the observed vesicle perturbation.

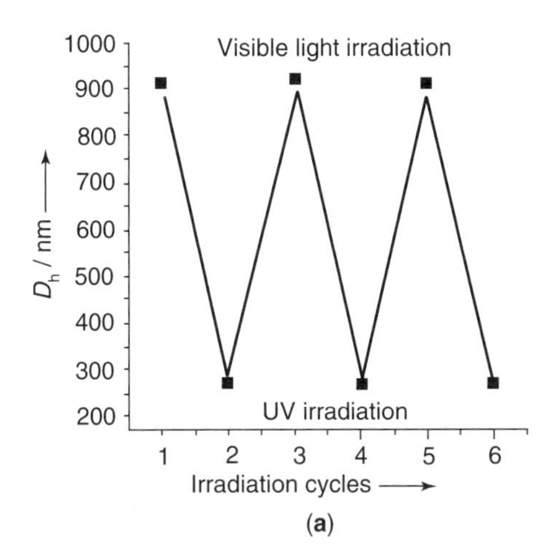

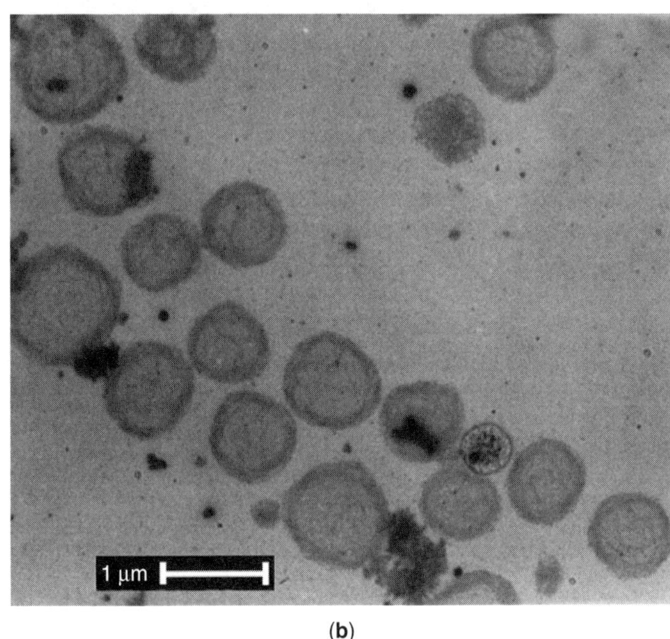

Figure 6.9. (**a**) Reversible large change in size (hydrodynamic diameter) of cross-linked micellar aggregates of AzoMI-VPy/CPB (1.0:0.5) upon alternating UV and visible light irradiation. (**b**) TEM image of visible light-induced hollow spheres. *Source*: Liu et al., 2006. Reprinted with permission.

In another study using a diblock copolymer of the hydrophilic poly-(2-dimethylaminoethyl methacrylate) (PDMAEMA) and a hydrophobic block of azobenzene-containing poly(meth)acrylate (structure in Fig. 6.11), Ravi et al. investigated the effect of photoisomerization on the micellization behavior (Ravi

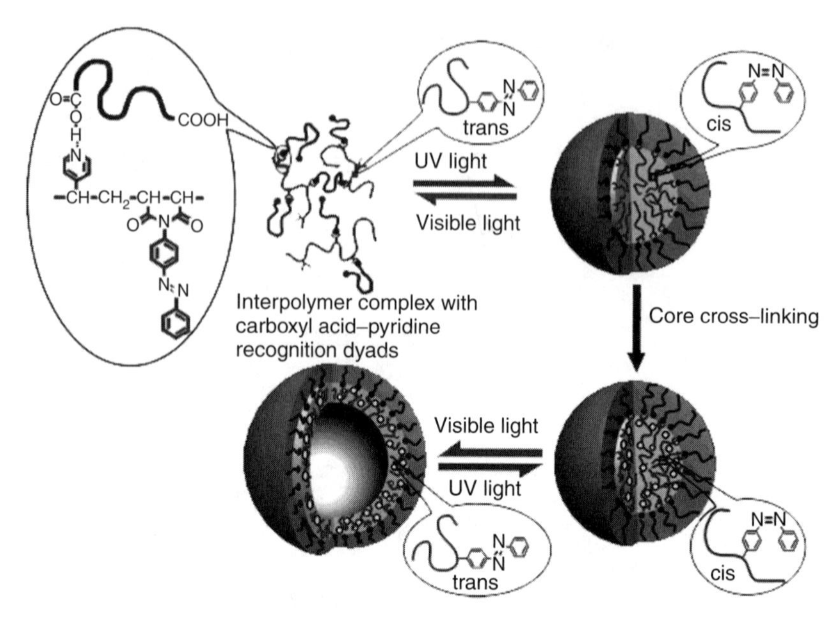

Figure 6.10. Schematic illustration of photoinduced micellization of uncross-linked AzoMI-VPy/CPB and photoinduced micelle-to-hollow sphere transition of cross-linked H-bonded polymers. *Source*: Liu et al., 2006. Reprinted with permission.

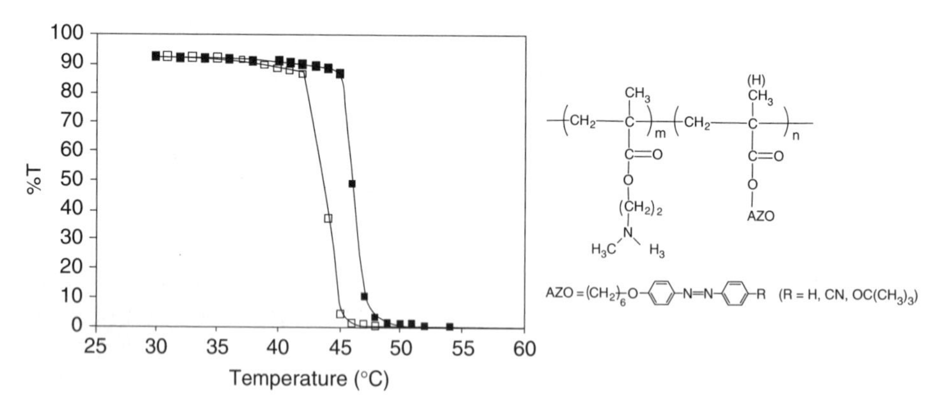

Figure 6.11. Changes in transmission of 0.7 wt% aqueous solution of an azobenzene diblock copolymer composed of 172 units of PDMAEMA and 9 units of azobenzene before (*open square*) and after (*closed square*) UV light irradiation at different temperatures. *Source*: Ravi et al., 2005. Reprinted with permission from Elsevier.

et al., 2005). A significant effect of the trans–cis photoisomerization on the LCST of the thermosensitive PDMAEMA was observed. The phase separation temperature, corresponding to the abrupt decrease in transmittance of the micellar solution, was increased by 3°C after UV irradiation (from 41 to 44°C). Again, this increase of LCST could be attributed to an increase in polarity of cis-azobenzene groups.

6.7. PERSPECTIVES AND FUTURE WORK

The feasibility of using the trans–cis photoisomerization of azobenzene to trigger the reversible dissociation and formation of BCP micelles is now demonstrated. Research on this topic certainly will continue, owing to the great interest from both fundamental and applied points of view. Fundamentally, many exciting and intriguing issues remain to be studied. For instance, what is the substitution pattern on azobenzene that will give rise to the largest increase in polarity upon trans–cis isomerization? For a given azobenzene polymer hydrophobic block, how to adjust the relative length of the hydrophilic block to make the hydrophilic–hydrophobic balance most sensitive to the isomerization-induced change in polarity? Among the azobenzene amphiphilic BCPs known to date, most, if not all, of the used hydrophobic polyacrylates or polymethacrylates are LC polymers bearing azobenzene mesogens. The question remains unanswered as to how the effect of the trans–cis photoisomerization on the LC phases, which can be important due to the photochemical phase transition, could affect the thermodynamic and kinetic stability of polymer micelles. In terms of new designs, BCPs offer a great platform for exciting opportunities. Schematically presented in Fig. 6.12 is such an example.

The idea in Fig. 6.12 is to prepare a light-responsive schizophrenic diblock (AB) copolymer micelle using azobenzene; that is, the core- and shell-forming blocks can be reversed upon illumination at two different wavelengths. Ames et al. first

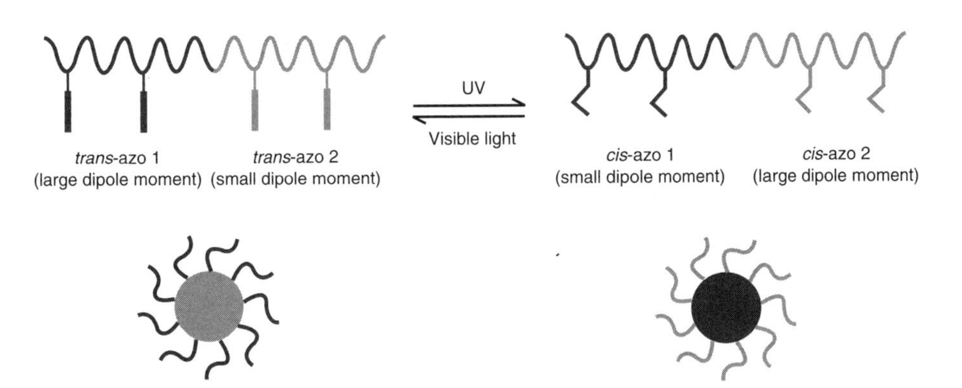

Figure 6.12. A possible design of light-responsive schizophrenic micelle based on azobenzene-containing diblock copolymers.

described pH-responsive schizophrenic BCP micelle composed of blocks of weak polyacid and polybase, which switches between two structures upon pH change (Butun et al., 2001). Later, BCP micelles with temperature change–induced switching were demonstrated with the use of two polymers having respectively an LCST and a upper critical solution temperature (UCST) (Arotcarena et al., 2002). Conceptually, it would be possible to design a diblock copolymer, while both blocks bearing an azobenzene moiety with a very different substitution pattern (azobenzene-1 and azobenzene-2). Azobenzene-1 should have a large dipole moment for the more stable trans isomer and a low dipole moment for the cis isomer, while azobenzene-2 must have exactly the opposite change in dipole moment. In a block-selective solvent, BCP micelles can be formed. In a highly polar solvent, micelles would be formed with the azobenzene-2-containing block forming the core of micelle and the azobenzene-1-containing block forming the shell. In a low polar solvent, the situation should be reversed. In all cases, upon alternating UV and visible light irradiation for the trans–cis and cis–trans isomerization, micelles should reverse their structure as depicted in Fig. 6.12, as a result of the switching polarity of the two azobenzene moieties. By incorporating the two azobenzene moieties in a double hydrophilic BCP, it would be possible to obtain such light-responsive polymer micelles in aqueous solution.

As BCP nanocarriers aiming at controlled delivery applications, azobenzene may not be the best choice of chromophore because of possible concerns of toxicity and biocompatibility and also because of the fact that the trans–cis photoisomerization of azobenzene cannot be induced by near infrared light (NIR, ~700–1000 nm) that is more suitable to biomedical applications (Goodwin et al., 2005). NIR has a deeper penetration through biosubstances as compared to UV and visible light because of reduced absorption and scattering by water and other biomolecules. Nevertheless, as a matter of fact, azobenzene and its derivatives have been widely utilized as photoswitches in biomolecules to accomplish a number of functions (Mayer and Heckel, 2006). One possible strategy that can minimize the risk of toxicity and the biocompatibility concerns using azobenzene-based BCP micelles is to reduce the concentration of azobenzene moieties on the hydrophobic block instead of having an azobenzene group for each monomer unit. In other words, the hydrophobic block should be a random copolymer containing a small number of azobenzene groups. However, the low concentration of the chromophore must still be able to trigger the disruption of BCP micelles upon illumination. To this end, it will be interesting to find a way that can lead to an amplification of the effect of photoisomerization of a few azobenzene moieties to induce a large-scale perturbation of the micelle. For instance, as aforementioned, if the hydrophobic block is a LC random copolymer containing a small amount of azobenzene mesogens, their photoisomerization can be enough to exert a large effect on the micelle. Indeed, the bent shape of cis-azobenzene is known to be incompatible with LC phases formed by elongated mesogenic molecules; consequently, the conversion of trans-azobenzene mesogens into the cis isomer can disorder the surrounding nonazobenzene mesogens to induce an isothermal nematic-to-isotropic phase transition (Ikeda, 2003). This photoplasticization of the core of micelle can be a significant disruption of the polymer micelles. Another related issue that needs to be investigated is that the release of encapsulated guest molecules does not require a complete dissociation of BCP micelles; it may occur upon a disruption to various extents. For example, our recent studies found that swelling of BCP micelles in aqueous solution, as a result of an increased hydrophilicity of the core, is enough to start the release of loaded hydrophobic guest molecules.

But most importantly, the photoisomerization of azobenzene and its derivatives is really an exceptional photoreaction (robust, clean, and reversible) whose marriage with BCPs provides a wonderful model system for developing light-controllable polymer nanocarriers. The design principles established through the studies of azobenzene-containing BCPs can be adapted to other bistable photoswitches displaying reversible photoisomerization and to other chromophores undergoing an appropriate photoreaction as well. Since the first study on azobenzene BCPs that has revealed the key role of a photoswitchable or phototunable hydrophilic–hydrophobic balance, various light-dissociable BCPs micelles using other chromophores have been reported (Lee et al., 2007; Jiang et al., 2007; Zhao, 2007b; Jiang et al., 2006, 2005).

ACKNOWLEDGMENTS

I am grateful to all my students and collaborators who worked on and made contributions to our projects on azobenzene-containing block copolymer micelles, particularly, Dr. Guang Wang, Mrs. Xia Tong, and Dr. Bo Qi. I thank Mr. Yi Zhao for his assistance in the preparation of the manuscript. I also want to express my gratitude for the financial support from the Natural Sciences and Engineering Research Council of Canada (NSERC), the Fonds pour la Formation de Chercheurs et l'Aide à la Recherche of Québec (FQRNT), University of Sherbrooke and St-Jean Photochemicals Inc. (St-Jean-sur-Richelieu, Québec, Canada).

REFERENCES

Arotcarena M, Heise B, Ishaya S, Laschewsky A. 2002. Switching the inside and the outside of aggregates of water-soluble block copolymers with double thermoresponsivity. J Am Chem Soc 124:3787–3793.

Bae Y, Fukushima S, Harada A, Kataoka K. 2003. Design of environment-sensitive supramolecular assemblies for intracellular drug delivery: polymeric micelles that are responsive to intracellular pH change. Angew Chem Int Ed Engl 42:4640–4643.

Bellomo EG, Wyrsta MD, Pakstis L, Pochan DJ, Deming TJ. 2004. Stimuli-responsive polypeptide vesicles by conformation-specific assembly. Nat Mater 3:244–248.

Butun V, Armes SP, Billingham NC, Tuzar Z, Rankin A, Eastoe J, Heenan RK. 2001. The remarkable "flip-flop" self-assembly of a diblock copolymer in aqueous solution. Macromolecules 34:1503–1511.

Chung JE, Yokoyama M, Okano T. 2000. Inner core segment design for drug delivery control of thermo-responsive polymeric micelles. J Control Release 65:93–103.

Ding L, Mao H, Xu J, He J, Ding X, Russell TP, Robello DR, Mis M. 2008. Morphological study on an azobenzene-containing liquid crystalline diblock copolymer. Macromolecules 41:1897–1900.

Discher DE, Eisenberg A. 2002. Polymer vesicles. Science 297:967–973.

Eastoe J, Vespertinas A. 2005. Self-assembly of light-sensitive surfactants. Soft Matter 1:338–347.

Einaga Y, Sato O, Iyoda T, Fujishima A, Hashimoto K. 1999. Photofunctional vesicles containing prussian blue and azobenzene. J Am Chem Soc 121:3745–3750.

Everett DH. 1988. Basic Principles of Colloid Science. London: Royal Society of Chemistry.

Gaucher G, Dufresne MH, Sant VP, Kang N, Maysinger D, Leroux JC. 2005. Block copolymer micelles: preparation, characterization and application in drug delivery. J Control Release 109:169–188.

Gillies ER, Fréchet JMJ. 2003. A new approach towards acid sensitive copolymer micelles for drug delivery. Chem Commun 14:1640–1641.

Goodwin AP, Mynar JL, Ma Y, Fleming GR, Frechet JMJ. 2005. Synthetic micelle sensitive to IR light via a two-photon process. J Am Chem Soc 127:9952–9953.

Haag R. 2004. Supramolecular drug-delivery systems based on polymeric core–shell architectures. Angew Chem Int Ed Engl 43:278–282.

Ikeda T. 2003. Photomodulation of liquid crystal orientations for photonic applications. J Mater Chem 13:2037–2057.

Jiang J, Tong X, Morris D, Zhao Y. 2006. Toward photocontrolled release using light-dissociable block copolymer micelles. Macromolecules 39:4633–4640.

Jiang J, Tong X, Zhao Y. 2005. A new design for light-breakable polymer micelles. J Am Chem Soc 127:8290–8291.

Jiang Y, Wang Y, Ma N, Wang Z, Smet M, Zhang X. 2007. Reversible self-organization of a UV-responsive PEG-terminated malachite green derivative: vesicle formation and photoinduced disassembly. Langmuir 23:4029–4034.

Kang HC, Lee BM, Yoon J, Yoon M. 2000. Synthesis and surface-active properties of new photosensitive surfactants containing the azobenzene group. J Colloid Interface Sci 231:255–264.

Kang K, Perron ME, Prud'homme RE, Zhang Y, Gaucher G, Leroux JC. 2005. Stereo complex block copolymer micelles: core-shell nanostructures with enhanced stability. Nano Lett 5:315–319.

Kumar GS, Neckers DC. 1989. Photochemistry of azobenzene-containing polymers. Chem Rev 89:1915–1925.

Kwon GS, Kataoka K. 1995. Block copolymer micelles as long-circulating drug vehicles. Adv Drug Deliv Rev 16:295–309.

Lee CT Jr., Smith KA, Hatton TA. 2004. Photoreversible viscosity changes and gelation in mixtures of hydrophobically modified polyelectrolytes and photosensitive surfactants. Macromolecules 37:5397–5405.

Lee H, Wu W, Oh JK, Mueller L, Sherwood G, Peteanu L, Kowalewski T, Matyjaszewski K. 2007. Light-induced reversible formation of polymeric micelles. Angew Chem Int Ed Engl 46:2453–2457.

Liu X, Jiang M. 2006. Optical switching of self-assembly: micellization and micelle-hollow-sphere transition of hydrogen-bonded polymers. Angew Chem Int Ed Engl 45:3846–3850.

Mayer G, Heckel A. 2006. Biologically active molecules with a light switch. Angew Chem Int Ed Engl 45:4900–4921.

Napoli A, Valentini M, Tirelli N, Muller M, Hubbell JA. 2004. Oxidation-responsive polymeric vesicles. Nat Mater 3:183–189.

Nasongkla N, Bey E, Ren J, Ai H, Khemtong C, Guthi JS, Chin SF, Sherry AD, Boothman DA, Gao J. 2006. Multifunctional polymeric micelles as cancer-targeted, MRI-ultrasensitive drug delivery systems. Nano Lett 6:2427–2430.

Orihara Y, Matsumura A, Saito Y, Ogawa N, Saji T, Yamaguchi A, Sakai H, Abe M. 2001. Reversible release control of an oily substance using photoresponsive micelles. Langmuir 17:6072–6076.

Patten TE, Matyjaszewski K. 1998. Atom transfer radical polymerization and the synthesis of polymeric materials. Adv Mater 10:901–915.

Qi B, Tong X, Zhao Y, Zhao Y. 2008. A micellar route to layer-by-layer assembly of hydrophobic functional polymers. Macromolecules 41:3562–3570.

Qin S, Geng Y, Discher DE, Yang S. 2006. Temperature-controlled assembly and release from polymer vesicles of poly(ethylene oxide)-block- poly(N-isopropylacrylamide). Adv Mater 18:2905–2909.

Rapoport N, Pitt WG, Sun H, Nelson JL. 2003. Drug delivery in polymeric micelles: from in vitro to in vivo. J Control Release 91:85–95.

Ravi P, Sin SL, Gan LH, Gan YY, Tam KC, Xia XL, Hu X. 2005. New water soluble azobenzene-containing diblock copolymers: synthesis and aggregation behavior. Polymer 46:137–146.

Sakai H, Matsumura A, Yokoyama S, Saji T, Abe M. 1999. Photochemical switching of vesicle formation using an azobenzene-modified surfactant. J Phys Chem B 103:10737–10740.

Schilli CM, Zhang M, Rizzardo E, Thang SH, Chong YK, Edward K, Karlsson G, Muller AHE. 2004. A new double-responsive block copolymer synthesized via RAFT polymerization: poly(N-isopropylacrylamide)-block-poly(acrylic acid). Macromolecules 37:7861–7866.

Shang TG, Smith KA, Hatton TA. 2003. Photoresponsive surfactants exhibiting unusually large, reversible surface tension changes under varying illumination conditions. Langmuir 19:10764–10773.

Shin JY, Abbott NL. 1999. Using light to control dynamic surface tensions of aqueous solutions of water soluble surfactants. Langmuir 15:4404–4410.

Shinkai S, Matsuo K, Harada A, Manabe O. 1982. Photocontrol of micellar catalyses. J Chem Soc Perkin Trans 2:1261–1265.

Simmons JM, In I, Campbell VE, Mark TJ, Léonard F, Gopalan P, Eriksson MA. 2007. Optically modulated conduction in chromophore-functionalized single-wall carbon nanotubes. Phys Rev Lett 98:086802(1–4).

Sin SL, Gan LH, Hu X, Tam KC, Gan YY. 2005. Photochemical and thermal isomerization of azobenzene-containing amphiphilic diblock copolymers in aqueous micellar aggregates and in film. Macromolecules 38:3943–3948.

Su W, Zhao H, Wang Z, Li Y, Zhang Q. 2007. Sphere to disk transformation of microparticle composed of azobenzene-containing amphiphilic diblock copolymers under irradiation at 436 nm. Eur Polym J 43:657–662.

Taillefer J, Jones MC, Brasseur N, Van Lier JE, Leroux JC. 2000. Preparation and characterization of pH-responsive polymeric micelles for the delivery of photosensitizing anticancer drugs. J Pharm Sci 89:52–62.

Tang X, Gao L, Fan X, Zhou Q. 2007. ABA-type amphiphilic triblock copolymers containing p-ethoxy azobenzene via atom transfer radical polymerization: synthesis, characterization, and properties. J Polym Sci Part A: Polym Chem 45:2225–2234.

Thurmond KB II, Kowalewski T, Wooley KL. 1996. Water-soluble knedel-like structures: the preparation of shell-cross-linked small particles. J Am Chem Soc 118:7239–7240.

Tian Y, Watanabe K, Kong X, Abe J, Iyoda T. 2002. Synthesis, nanostructures, and functionality of amphiphilic liquid crystalline block copolymers with azobenzene moieties. Macromolecules 35:3739–3747.

Tong X, Wang G, Soldera A, Zhao Y. 2005. How can azobenzene block copolymer vesicles be dissociated and reformed by light? J Phys Chem B 109:20281–20287.

Wang D, Ye G. Wang X. 2007. Synthesis of aminoazobenzene-containing diblock copolymer and photoinduced deformation behavior of its micelle-like aggregates. Macromol Rapid Commun 28:2237–2243.

Wang G, Tong X, Zhao Y. 2004. Preparation of azobenzene-containing amphiphilic diblock copolymers for light-responsive micellar aggregates. Macromolecules 37:8911–8917.

Yang L, Takisawa N, Hayashita T, Shirahama K. 1995. Colloid chemical characterization of the photosurfactant 4-ethylazobenzene 4′-(oxyethyl)trimethylammonium bromide. J Phys Chem 99:8799–8803.

Yu H, Iyoda T, Ikeda T. 2006a. Photoinduced alignment of nanocylinders by supramolecular cooperative motions. J Am Chem Soc 128:11010–11011.

Yu H, Li J, Ikeda T, Iyoda T. 2006b. Macroscopic parallel nanocylinder array fabrication using a simple rubbing technique. Adv Mater 18:2213–2215.

Yu H, Shishido A, Ikeda T, Iyoda T. 2005. Novel amphiphilic diblock and triblock liquid-crystalline copolymers with well-defined structures prepared by atom transfer radical polymerization. Macromol Rapid Commun 26:1594–1598.

Yu H, Shishido A, Iyoda T, Ikeda T. 2007a. Novel wormlike nanostructure self-assembled in a well-defined liquid-Crystalline Diblock Copolymer with Azobenzene Moieties, Macromol. Rapid Commun 28:927–931.

Yu H, Shishido A, Li J, Kamata K, Iyoda T, Ikeda T. 2007b. Stable macroscopic nanocylinder arrays in an amphiphilic diblock liquid-crystalline copolymer with successive hydrogen bonds. J Mater Chem 17:3485–3488.

Zhang L, Eisenberg A. 1996. Morphogenic effect of added ions on crew-cut aggregates of polystyrene-b-poly(acrylic acid) block copolymers in solutions. Macromolecules 29:8805–8815.

Zhao Y. 2007a. Azobenzene-containing block copolymers. In: Nalwa HS, editor. Polymeric Nanostructures and Their Applications. Vol. 2. California: American Scientific Publisher. pp. 281–309.

Zhao Y. 2007b. Rational design of light-controllable polymer micelles. Chem Rec 7:286–294.

7

ASSOCIATION BETWEEN AZOBENZENE-MODIFIED POLYMERS AND SURFACTANTS OR NANOPARTICLES TO AMPLIFY MACROSCOPIC PHOTOTRANSITIONS IN SOLUTION

Christophe Tribet

7.1. LIGHT RESPONSIVENESS OF SOLUTION PROPERTIES: A QUESTION OF AMPLIFICATION

Polymers having photochrome hydrophobic moieties have been designed to undergo obvious molecular responses when exposed to UV-visible light, including variation of their degree of swelling (Irie et al., 1981b), self-assembly/dissociation (Tong et al., 2005; Wang et al., 2004), and self-organization, for instance, in liquid crystal phases (Bo et al., 2005; Shibaev et al., 2003). The macroscopic responses include phase transition (Irie, 1993), viscosity swings, wetting/dewetting of polymer-coated surfaces (Lim et al., 2006; Ichimura et al., 2000), and ion conductivity (Tokuhisa et al., 1998). As compared with systems based on small molecules, the macromolecular nature of a photoresponsive compound brings new properties, such as gelation and multipoint attachment (connectivity). In addition, polymer-specific properties can also provide a first "molecular" stage of amplification and help to fulfill the practical requirement of high magnitude of the photoresponses. This chapter illustrates the conditions and systems achieving

Smart Light-Responsive Materials. Edited by Yue Zhao and Tomiki Ikeda

polymer-based amplifications of the tenuous molecular photoconversions obtained on photochrome groups, namely, the E–Z isomerization of azobenzene.

Among photochrome groups that have been attached to polymer chains, azobenzene groups have several advantages in the context of solutions and gels (Rau, 2003). First, their light-triggered variation of polarity and conformation (cis–trans isomerization) do not significantly depend on the polarity of the solvent, including water and is readily obtained with excellent reversibility (no dimerization, absence of chemical degradation). Second, the chemistry of azobenzene dyes is highly versatile, making the modification of synthetic chains or biomacromolecules readily accessible, as well as copolymerization of azo monomers in large amount. For instance, azo dyes have been grafted on enzymes to trigger enzyme activity with light (Lee et al., 2005; Willner and Rubin, 1996). Light-triggered solubility has been achieved with a variety of polymer structures: peptides, polysaccharides, and polyacrylic chains. Diverse acrylic copolymers have been synthesized to trigger variation of solution viscosity with light (Moniruzzaman et al., 2004 and references therein; Irie, 1993; Irie et al., 1981a). The latter polymers induce significant and reversible change of the viscosity of solutions, provided that the chain comprises a large fraction (above $\sim$30–50 mol%) of photochrome monomers.

In the applications seeking at the control of solution properties, there is however a major drawback due to the absorption of photons by the samples: The light penetrating in solutions, or gels that contain millimolar concentration of photochrome shows an important decrease of intensity with increasing distance from surface, because of the absorbance of the photochrome themselves (the extinction coefficient of the azo photochrome is ca. $2 \times 10^4\,\text{l.mol}^{-1}.\text{cm}^{-1}$ at the near UV absorption peak). Photochrome groups located deep inside centimeter-thick samples are thus hardly stimulated by light, except in the case of very efficient stirring. Homogeneous and complete response of viscous samples or gels that cannot be stirred requires hours-long irradiations accordingly. Except in the case of enzymes, the modification of solution properties in the presence of polymers is achieved at concentrations of the order of a few %wt polymer, typically 0.5%–10%. The practical challenge to face in solutions is accordingly an upper limit of one millimolar azobenzene incorporated in a solution of typical monomer concentration of 500–25 mM (assuming that the molar mass of a monomer is $\sim$200 g/mol). Therefore the integration level azobenzene in polymers should ideally be as low as possible, and in practice below a few mol% of the monomers. Low chromophore concentration enables one to complete the microscopic isomerization process on irradiation for relatively short times ($\ll$1 h) at common intensities ($\sim$1 mW/cm^2). But in these concentration conditions, the amplification of the molecular effect(s) must be amazingly efficient: *the magnitude of response becomes the key point to be improved for applications in solutions.* Various approaches using polymer properties and critical transitions in solution of polymers are described in the following text. The polymer structures that are used in such applications comprise a few mol% azo monomers and long soluble segments, as illustrated in Fig. 7.1.

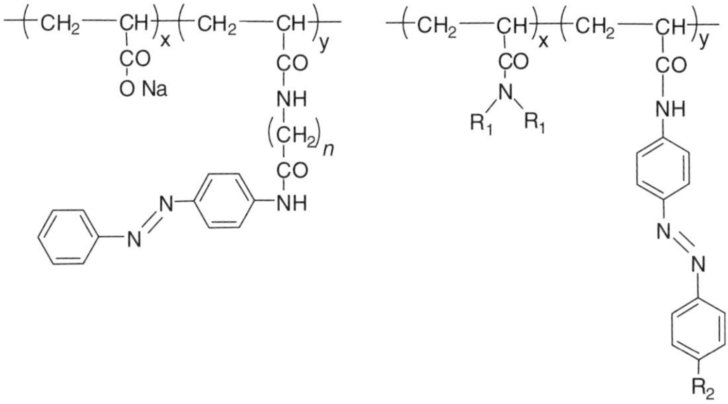

Figure 7.1. Structures of azo-modified polymers (here random copolymers) that have been used in solution to trigger phase separation, physical gelation, or photoswelling of chemical gels. $n = 5$–11 (Pouliquen and Tribet, 2006); $R1 = CH_3$ and $R2 = H$ (Irie, 1993); or $R1 = H$ and $R2 = H$, methyl, ethyl, or butyl (Howley et al., 1997).

7.2. FROM CLOUD POINT TO ASSOCIATIVE PHASE SEPARATION OF PHOTOPOLYMERS

Phase separation and demixing of a polymer from its solution has attracted much attention because of technological importance in coating and encapsulation. Basically, the tendency to demix comes from the weak translational entropy gained upon dissolution of long polymer and macrocomplexes. Phase separation is accordingly achieved when the monomer-solvent interaction energy (namely, the enthalpy of dilution) does not provide strong driving force toward solubilization. The solubility transition takes place abruptly on decreasing the solvent-monomer attraction and can be recognized as a critical transition for infinitely long chains. For this reason, precipitation of stimuli-responsive macromolecules is triggered by relatively weak variations of variable parameters such as temperature, solvent composition (pH, salt), and E–Z isomerization of azobenzene-containing monomers. In practice, the enthalpy of solvatation, specifically of azo monomers, varies with photoswitch of polarity and conformation, because the E and Z states display different Van der Waals and hydrophobic interactions. In turn, light-responsive aggregation or collapse of chains is achieved (Fig. 7.2a). More complicated responses to light have been achieved by mixing the polymers with molecules or nanoparticle that can associate with the azo groups. The phase transition of "simple" polymer/solvent systems is described in Section 7.2.1 and the case of complexes with cyclodextrins (CDs) or surfactants is discussed in Sections 7.2.2 and 7.2.3.

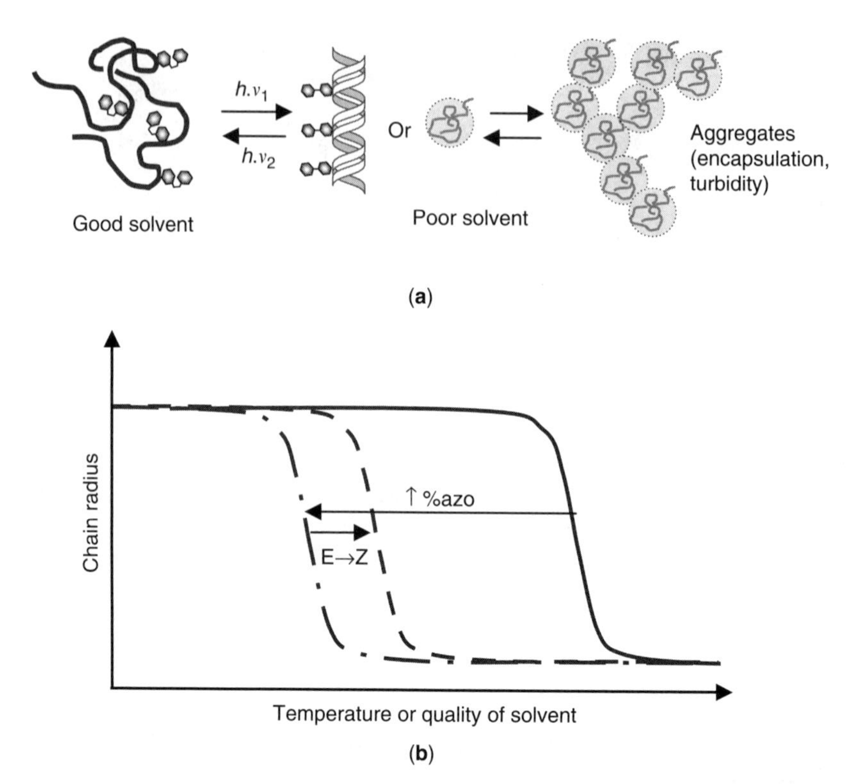

Figure 7.2. Photoswitch of the solubility of chains. (**a**) Schematic drawing of the phototriggered collapse and aggregation of azobenzene-containing polymers in poor solvent conditions or close to low critical solubility temperature (LCST). (**b**) Typical variation of the radius of the chains as a function of solvent parameter, or temperature in the case of chains having a LCST in water. Bold line: parent chain with no azobenzene; dashed and dot-dashed lines: azo-modified chains, respectively, exposed to UV and dark-adapted.

7.2.1. Polymers in Poor Solvents or at Low Critical Solubility Temperature

Despite the variety of polymer/poor solvent pairs that could in principle achieve light-responsive phase separation, the systems investigated to date are not numerous. Polymers in organic solvents were considered as early as in the 1980s by Irie and colleagues, with azobenzene-modified polystyrene in carbon disulfide (Irie and Iga, 1986). Recent reconsideration of polystyrene-based systems have pointed that the desolvatation and precipitation may considerably slow down the chain dynamics ($T \ll T_g$), which in turn traps the photoproduced aggregates in a frozen (insoluble) state (Shulkin and Stover, 2003). Thus the aggregates may not redissolve, but the photocontrol of precipitation kinetics could nevertheless find promising developments for the synthesis of nanoparticles with complex shapes or porosity. Similarly, solvent effects switch azo-containing polypeptide chains between their coil conformation in good solvent and a phototriggered folded structure (helix formation; Fig. 7.2a), although no precipitation occurs. In this case, the desolvatation is usually achieved by mixing water with alcohol or other additives (Pieroni et al., 2001; Willner and Rubin, 1996).

But the vast majority of systems developed to date correspond to phase separation in aqueous solutions, using the low critical solubility temperature (LCST) displayed by most water-soluble chains. In short, macromolecules with known LCST behavior can be modified by inclusions of azobenzene groups to obtain a water-soluble chain containing a few mol% of azo monomers. The hydrophobicity of photochromes causes the LCST to shift toward significantly lower temperature (Fig. 7.2b), but does not change the qualitative properties of high solubility (respectively swelling of a gel or of isolated chains) at low temperature and low solubility (respectively gel shrinkage or collapse of the chain) at temperature beyond the modified LCST. The generality of this effect is reflected by the diverse chemical structures of macromolecules found in literature, including derivatives of poly(*N*-isopropylacrylamide) (Liu et al., 2001; Irie, 1993; Mamada et al., 1990), copolymers of isopropylacrylamide and hydroxyalkylacrylamide (Menzel et al., 1995), methyl and hydroxypropyl cellulose derivatives (Zheng et al., 2004), elastinlike polypeptides (Alonso et al., 2001), collagenlike systems (Kusebauch et al., 2006; Strzegowski et al., 1994).

The LCST of a modified chain decreases typically in proportion to the fraction of azobenzene hydrophobes (Alonso et al., 2001; Menzel et al., 1995) (Eq. 7.1). Fig. 7.2b schematizes the azo-induced shifts under UV exposure and in the dark, respectively, with published magnitudes illustrated in Table 7.1. Because of the polarity of the cis isomer of azobenzene, the dark-adapted form of polymers has most typically a lower LCST than the UV-exposed ones, predominantly in its cis forms.

$$T_{\mathrm{LCST}} = T_0 - \alpha\,\%\mathrm{azo} \quad \text{with } 0 < \alpha_{\mathrm{cis}} < \alpha_{\mathrm{trans}} \tag{7.1}$$

It is clear from Fig. 7.2 and Table 7.1, that the samples are made light-responsive in the temperature range comprised between the LCSTs of the cis and trans forms, respectively. This range might be optimized by proper choice of chemical structures (nature of substituents on azobenzene, nature and length of spacer between the azo and polymer backbone), but except for hydroxypropyl, methyl-cellulose (HPMC), the present data reflect a contribution of azo side groups essentially of the same order of magnitude. A typical value of $\alpha_{\mathrm{UV}}-\alpha_{\mathrm{trans}}$ is, up to now, of 2°C ± 1°C/mol% azo (except for the HMPC data that reach 6–7°C/mol%). Recalling the constraint of using chains with less than a few mol% azo (cf. introduction), this simple estimate shows that a degree of integration of 1 mol% in chains would impart the system with a typical phototriggered LCST-shift of 2°C, only compatible with applications in highly controlled temperature conditions.

TABLE 7.1. Indices of LCST Shifts with Integration Level of Azobenzene Side Groups and Upon Photo-isomerisation

Main chain	NIPA			HPMC	Elastin	
comomoners or remarks	[Irie, 1993]	HEAM [Menzel, 1995]	acrylic/glycolic acids [Liu, 2001]	[Zheng, 2004]	Neutral [Alonso, 2001]	carboxylate-modified [Strzegowski, 1994]
α_{trans}	4.4	16 (resp. 10^{a})	n.a.	$\sim 16^{b}$	5.5	N.A.
α_{UV}-α_{trans}	2.5	2.7 (resp 2.7)	2.0	6.3 ± 0.7	1.7	2.8

α in °C per mol% azo in the polymer. NIPA: poly(N-isopropylacrylamide), HEAM: hydroxyethylacrylamide, HPMC: hydroxypropyl methylcellulose. N.A.: not available.

[a] a more polar azobenzene with 4′methoxyethyloxy substituent.

[b] not shown in the ref, our estimate uses LCST of c.a. 55°C for the unmodified chain.

The basic origins of photovariation of LCST are the photomodulation of hydrophobic binding and Van der Waals interactions. To enhance the effect (at fixed degree of modification with azo, i.e., to achieving larger α values) it is possible to optimize the effect of isomerization on the chromophore–chromophore (attractive) interaction and the chromophore-water (repulsive) interaction. The energy of hydrophobic association is certainly difficult to estimate, but should depend mainly on the number of carbon atoms that can be transferred in apolar microdomains. If the azos are fully accessible in the modified polymers (e.g., flexible chains, long spacers), the isomerization does not markedly affect the hydrophobic binding. In contrast, the E–Z conversion modifies the dipole interactions to an extent that should vary in proportion to the dipole moments of E and Z forms (irrespective of polymer composition), making thus the α_{UV}–α_{trans} coefficient in Equation 7.1 proportional to the dipole variation. We remark that almost optimal photoswitches of dipolar interaction have been reached in the published investigations, because the azobenzenes used were essentially apolar under their trans form and polar under their cis form. Improvements are therefore not obvious, and shall be devised on new basis, for instance, using additional enthalpic contributions to the intrachain interaction.

7.2.2. Complexation and Solubility of Chains

Typical attempts at enhancing the impact of E–Z conversion on intrachain and monomer-solvent interactions use solution of CDs in water. CDs form host–guest inclusion complexes with azobenzene derivatives (Zheng et al., 2002; Sanchez and deRossi, 1996; Yang et al., 1996; Suzuki et al., 1991). The association with CDs of azobenzene-containing chains adds further degrees of complexity (interaction between CDs and nonazo monomers can be attractive or repulsive) to the intrachain interactions. The association of CDs with the chain may depend on the E–Z isomerisation. It was found that αCD and βCDs are best suited than larger CDs to discriminate the E and Z isomers of azobenzene. Naturally, the association of CD modifies the enthalpy of chain folding and collapsing, by masking the hydrophobes in CD cavities, and adding CD-monomers contacts and steric hampering to conformational freedom. All these effects contribute to chain swelling or collapse. The common interpretation of photoresponses in the presence of CDs invokes that the trans isomer associates tightly into the CD cavities, whereas the cis isomer does not bind (Zheng et al., 2004; Rodriguez-Cabello et al., 2002; Ueno et al., 2002). Measurements by NMR and circular dichroism validate this point in some systems containing αCD (Tomatsu et al., 2006; Zheng et al., 2005). In general, things are less obvious, and CDs (especially βCD) may partially dissociate upon transconversion to the cis form, or bind to both the E and Z forms of azobenzene, with a different geometries and stoichiometries (Fig. 7.3a; Inoue et al., 2007; Pouliquen et al., 2007; Tomatsu et al., 2006).

Both phenomena make nevertheless the azo-modified chains more hydrophilic in presence of CDs (upper shift of LCSTs). The dark-adapted (trans) form is even

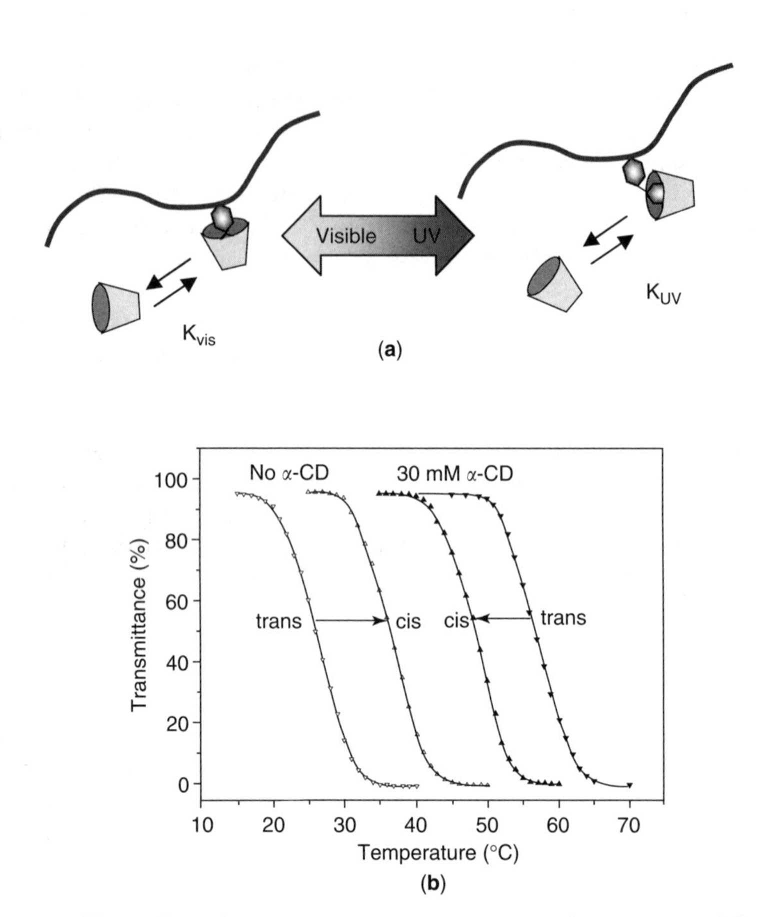

Figure 7.3. Effect of CD host–guest interaction on azobenzene-modified polymers. **(a)** Drawing of the equilibrium between bound and unbound CDs. With the change of geometry under the cis form approaching the CD toward the backbone, Ktrans may be higher than Kcis. **(b)** Experimental variation of turbidity and shift of LCST measured in mixtures of hydroxypropyl methylcellulose in the presence or absence of αCD as quoted in the figure. *Source*: Reprinted with permission from Zheng et al., 2004.

more hydrophilic than the cis one, presumably because of the deeper penetration of the E-form inside the CD cavity. Consequently, the coefficient α_{UV}–α_{trans} in Equation 7.1 can become negative. In such mixed systems, the first stage of amplification is achieved by "recognition" of azobenzene's isomers at the molecular scale. The second amplification stage is brought by chain collapsing in poor solvent. The development of new photocomplexing agents coupled with chain collapsing could certainly help design systems with a large temperature range of photoresponsiveness.

7.2.3. Associative Phase Separation

Finally, other type of phase separation can help achieving marked macroscopic amplification upon crossing a critical condition of demixing. In addition to LCST properties, water-soluble polymer can also be made abruptly insoluble upon association with "nanoparticles," such as a protein and micelles of surfactants (Piculell and Lindman, 1992). Owing to their importance for industrial products (cosmetic, food, personal care), polymer–surfactant demixings have been extensively studied. Making these system responsive to light would open a vast range of applications. Two classes of phase separation have to be distinguished, namely, seggregative (formation of a surfactant-rich phase and a polymer-rich one, when the two partners have no propensity to bind) and associative separation (demixing of a dense phase containing associates of surfactant and polymer). Here we are interested in the associative separation because photoisomerization could significantly affect the association strength. Basically, the association with nanoparticles can evolve enough energy to balance the translation entropies of both the chains and particles. In turn, phase separation (sometimes called coacervation) occurs between a concentrated phase maximizing the particle/polymer contacts and a very dilute phase, possibly containing the excess of unbound nanoparticles or almost particle-free polymer chains. At variance with the chains showing LCST behavior, the demixing can be obtained in good solvent for the bare polymer. These important distinctions lead to phase separation at low temperature and homogeneity at high temperature (upper critical solubility, UCST), as shown in Fig. 7.4a. This type of nonresponsive phase separation has been commonly achieved by hydrophobic association of charged polymers and neutral surfactants (hydrophobe containing polyelectrolytes are more easily water-soluble than neutral amphiphilic chains) (Iliopoulos, 1998; Piculell and Lindman, 1992). It is also possible to use neutral polymer in combination with ionic micelles. When one partner is ionic, however, the contribution of counter-ion to the translational entropy must be reduced to afford the formation of a concentrated, ion-containing phase. The presence of simple salt (e.g., NaCl) at concentrations above 100 mM is usually sufficient to enlarge the two-phase region in the phase map (Howley et al., 1997; Effing et al., 1994).

Because the E to Z photoswitch decreases the propensity of an azo-modified chain to bind to apolar domains, such as the core of micelles, the UCST of mixed surfactant/azobenzene-modified polymers (AMP) is expected and actually shown to decrease upon UV exposure. Temperature drops by 10°C–15°C have been reported (Howley et al., 1997). When the system is investigated in the temperature-window between UCST (UV) and UCST (dark), the phase diagram displays a large biphasic surface for the dark-adapted system, whereas the biphasic domain disappears completely under UV exposure. Fig. 7.4b illustrates the case of an azobenzene-modified polyacrylate mixed with dodecylPEO surfactant (C12E4). At room temperature, solutions well above the CMC of C12E4 were supplemented with increasing amount of polymer. Because of the low density of the surfactant, the C12E4-rich phase formed in the dark with small quantities of

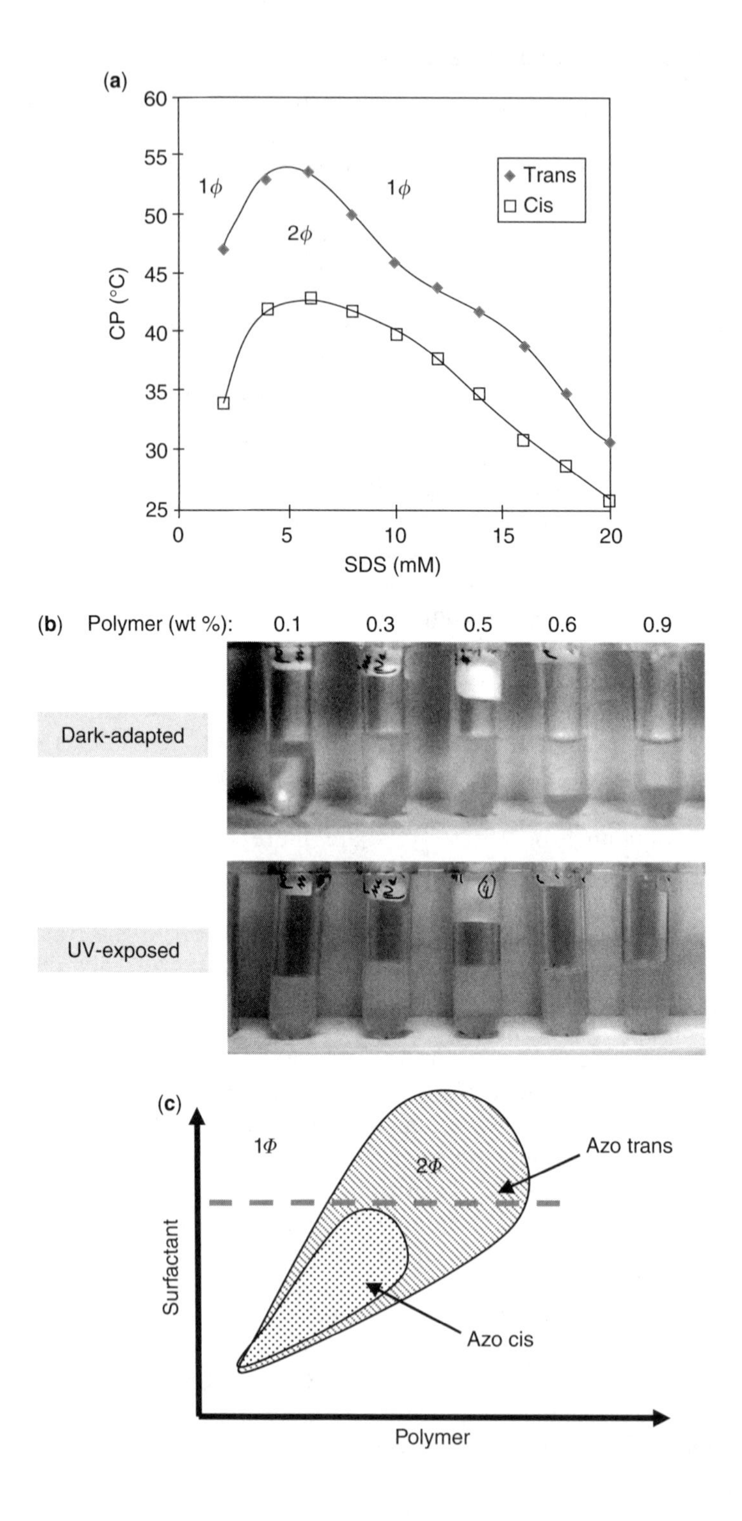
(a)
CP (°C)
SDS (mM)
Trans
Cis
1φ
2φ
1φ
60
55
50
45
40
35
30
25
0
5
10
15
20
(b) Polymer (wt %): 0.1 0.3 0.5 0.6 0.9
Dark-adapted
UV-exposed
(c)
1Φ
2Φ
Azo trans
Azo cis
Surfactant
Polymer

polymer was creaming, whereas the phase formed in the presence of higher polymer concentrations was denser and sedimented. Exposure to UV light made all mixtures homogeneous, and a single phase can be seen after centrifugation, irrespective of the composition. In addition, the range of temperature affording photoresponsiveness is much larger than just the UCST differences between UV and blue exposed (or dark-adapted) samples. Provided that the composition of sample can be adjusted, responsiveness would be found at almost all temperature below UCST (dark). The plot in Fig. 7.4a shows how the downshift of the two-phase region creates a long strip between the dark and UV-adapted boundaries in the (temperature vs. surfactant) diagram. At any temperature lower than UCST (UV), it is possible to shift the composition toward the left-hand side of the diagram (increase of the surfactant concentration slightly above the UV-adapted phase separation point) to bring the system to an homogeneous state under UV exposure that turns into a two-phase system in the dark (Fig. 7.4c) (Howley et al., 1997; Effing and Kwak, 1995).

Basically, the attractive contribution that causes the phase separation comes down to the sharing of one nanoparticle between two or more segments of polymer chains. The particles act as transient "cross-links," which are somehow equivalent to interchain attraction. In semidilute solution, the number of such interchain cross-links could exceed the initial number of interpolymer contacts if the concentration of particles is increased above a critical threshold. This leads to a decrease of the correlation length above this critical concentration of "cross-linker," and in turn to the shrinkage of the network (also called syneresis). Following Leibler and colleagues (Pezron et al., 1988; Keita et al., 1995), the onset of phase separation corresponds to the exact matching between the density of cross-linking particles and the density of interchain contacts in the solution. The equilibrium between bound- and free nanoparticle would consequently determine whether critical solubility conditions are reached or not. Significant effect of light on the degree of binding of particles is thus the key to enlarge the region of light-responsiveness in the phase diagram shown in Fig. 7.4c. The basic principle to designing highly responsive photocomplexes is presented in Section 7.3.

Figure 7.4. Associative phase separation of complexes between surfactants and azobenzene-modified polymers. (**a**) Cloud points (CP) determined at fixed temperature, fixed polymer concentration, and varying concentration of sodium dodecyl sulfate polymer structure cf. Fig. 1). (**b**) Solutions of modified poly-(sodium acrylate) (cf. Fig. 7.1) and the surfactant C12E4 at 2 g/L in 0.3 M NaCl after centrifugation. Polymer concentrations are quoted in the figure. (**c**) Schematic phase diagram at fixed temperature showing the enlargement of the two-phase region (drop of demixing) upon increasing the association strength, that is, while turning cis isomers into trans ones. *Source* of (**a**): Reprinted with permission from Howley et al., 1997. See color insert.

7.3. INTRACHAIN ASSOCIATION WITH COLLOID PARTICLES: PHOTORECOGNITION

Even in monophasic systems, polymer properties can be beneficial to photoresponses and can level up the effects beyond the association–dissociation occurring at the molecular scale. Intrapolymer associations of colloid nanoparticles (e.g., micelles or proteins) most typically exhibit multipoint attachments, that is, several azobenzene groups would simultaneously attach onto the same particle. Chemical equilibrium requires that the energy of transfer of azobenzene from water into their binding sites balances the conformation energy of the chain segments forming loops on the particle (Fig. 7.5a). In addition, if a necklace of bound colloid particles is formed, an additional intrachain particle–particle interaction may also contribute to the association energy. Finally, at concentration above C* for the polymer, the association may transiently cross-link the chains and markedly modify the dynamic and mechanical properties. Altogether, these contributions to association result in complex responses to photovariations of the affinity for azobenzene. This section presents the essential features of such

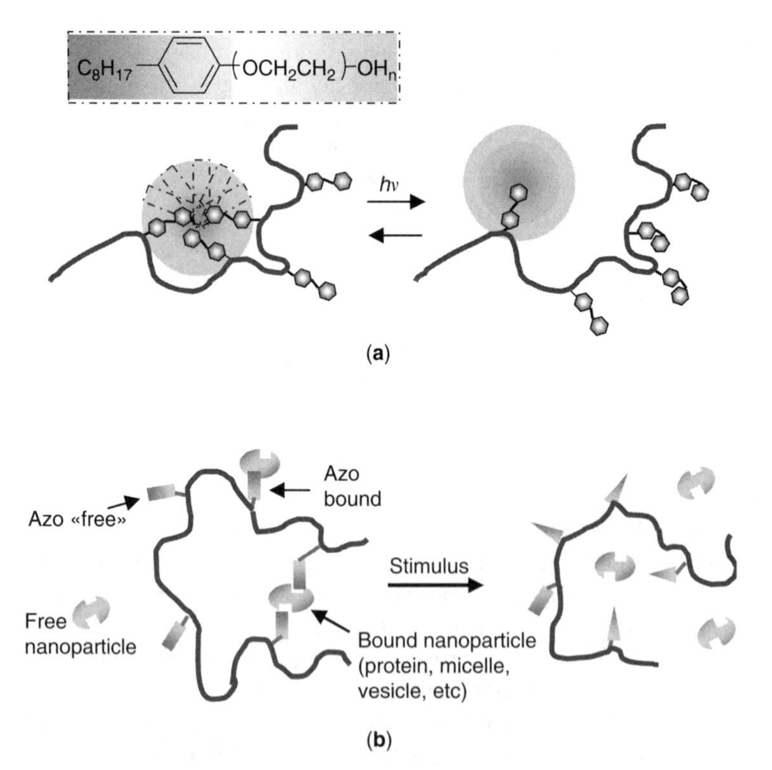

Figure 7.5. Schematic drawing of the light-responsive association between a flexible polymer and nanoparticles. See color insert.

complexes in dilute and semidilute solutions, with emphasis on the origin of amplification of the responses to light.

7.3.1. Complexes with Protein and Micelles in the Dilute Regime

A defined stoichiometry is generally not relevant to describe macromolecular complexes, even in dilute regime (i.e., with only isolated chains and no polymer–polymer aggregates). In practice, the number of bound nanoparticles per polymer chain and the fraction of bound azobenzene are the two primary structural parameters that can be measured (Fig. 7.5b). First, we focus on the binding of the azo chromophores. Proton NMR shifts (May et al., 2005), red-shift on the UV spectra (Labarthet et al., 2000), variation of circular dichroism (Shimomura et al., 2002; Suzuki et al., 1991) are among the experimental signals that can be used to probe the solvatation of the chromophore. Accordingly, it is possible to determine the fraction of free azobenzene groups and whether the association takes place gradually or not. Our measurements show that the fraction of azobenzene side group that are bound into micelles is most typically significantly smaller than 100%. Fig. 7.6a gives representative UV-visible spectra of azo-modified polymers in water, as a function of the concentration of surfactant. The absorbance obviously increases and the main peak is red-shifted with increasing the surfactant concentration, two features that reflect the transfer of azobenzene from water into a less polar environment. The presence of isobestic points betrays that the azo groups are either in contact with water or tightly bound to the surfactant and no intermediate solvatation state is detected (not shown).

The absorbance at a fixed wavelength indicates accordingly the degree of binding of the azobenzene groups. In Fig. 7.6b, the absorbance abruptly increases above the critical micelle concentration (Khoukh et al., 2007). In the case of other nanoparticles, such as proteins or polymers of CDs, the absorbance increases immediately upon addition of the particles (Pouliquen et al., 2007; Pouliquen and Tribet, 2006). At incipient binding, the association was considered very tight and quantitative, and it has not been possible to estimate a binding constant (not shown). The association was presumably strengthened because of multipoint attachments (see later). A saturation plateau can be reached at high concentrations of the nanoparticles. Once normalized by the concentration of azobenzene and a reference shift of extinction coefficient (e.g., as measured with azobenzene in water and dodecane), the plateau value should indicate the fraction of bound chromophore. In practice, the normalized plateau value depends on the integration level of the azobenzene and intrachain repulsion (as modulated by ionic strength in the case of our azo-modified polyelectrolyte). In other words, variability of the molar absorption in the saturated state clearly indicates that a variable fraction of the chromophore remains unbound, even in presence of large excess of surfactant. With neutral micelles of Triton X-100 and C12E4, the saturation could correspond to less than 50 mol% association (Fig. 7.6b, right axis), which means formation of long polymer loops rather than short ones upon binding. Incomplete association of the chromophores is presumably because of

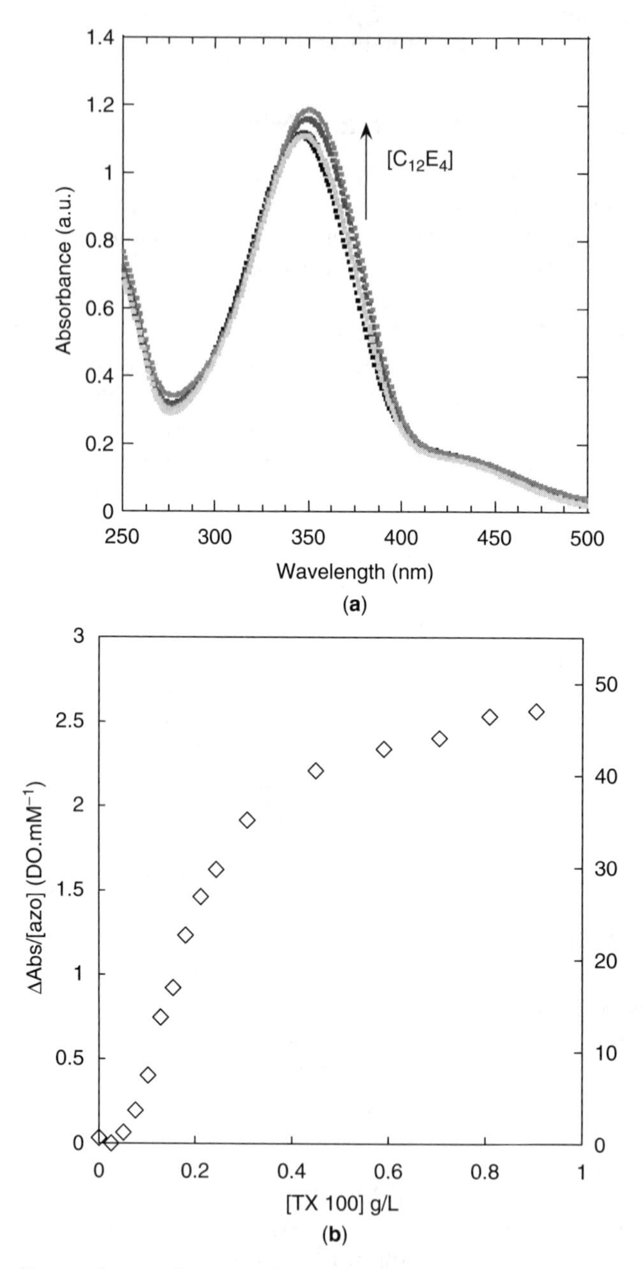

Figure 7.6. Indices of association of azobenzene onto hydrophobic domains. (**a**) Representative spectra of an azo-modified polymer in water (under blue exposure) with increasing concentration of surfactant C12E4 (0–0.4 g/L). Polymer structure: cf. left-hand side in Fig. 7.1, with $n = 11$ and $x = 2\%$. (**b**) Variation of the fraction of bound azobenzene upon addition of Triton X 100 in a solution of polymer at fixed concentration (polymer similar as in (**a**), with $n = 5$ and $x = 4\%$).

electrostatic repulsion that opposes a strong penalty to the formation of loops shorter or closer than the Debye length.

Exposure to UV irradiation triggers the detachment of the bound micelles (or proteins) (Pouliquen and Tribet, 2006; Pouliquen et al., 2002). Likewise, we cannot detect significant sensitivity to the presence of nanoparticles of the absorption spectrum of the predominantly cis polymer (not shown). However, to obtain more accurate determinations of the degree of light-triggered release, the average number of bound micelles (or protein) per chain was measured. Chromatographic methods under continuous injection mode, in this case, capillary electrophoresis, enable one to separate minute amount of free particles with no marked perturbation of the binding equilibrium [the method is also called frontal analysis; see Gao et al. 1997] Representative binding isotherms in Fig. 7.7 obviously point to UV-triggered releases. The degree of photodissociation depends, however, on the composition of the complexes. In the case of the protein bovine serum albumin, we measured an increasing degree of dissociation with the decreasing number of protein per chain (Fig. 7.7b and Pouliquen and Tribet, 2006). Saturated chains with typically one bound protein per one to three azobenzene groups do not respond to light, which points to the lack of specificity of the protein toward the isomerization state of the azobenzene. Nevertheless, chains with less than 2–4 g/g of bound protein (i.e., less than three proteins per 1000 monomers and 10–30 azobenzene groups per bound protein) release up to 90% of their protein content upon exposure to blue light. In contrast, surfactants such as Triton X100 (Fig. 7.7a) and C12E4 (Khoukh et al., 2007) do not photodissociate at low degree of loading of the micelles in the chains (close to the critical micellar concentration, where ~20–40 azobenzene bind to 100 surfactants). The photoresponse increases regularly with increase in the loading of surfactant and accordingly decrease in the amount of bound azo per micelle. A tentative interpretation is proposed that in both cases, an optimal number of ca. 10–20 bound azobenzene per bound nanoparticle must be reached for optimal responses to be reached.

The following arguments present the basic reasons for an optimal azo/particle ratio, and the expected polymer-related amplification of affinity. Binding of an isolated azobenzene on a single nanoparticle should *a priori* be described by using a conventional equilibrium constant as for 1:1 complexation (we will call this incipient binding the "Benessi–Hildebrand" regime). Once the first link onto a micelle (or a protein) is formed, the loss of conformation entropy for loop formation represents the primary penalty for further associations of azobenzenes. Accordingly, the free energy gain for transferring the azobenzene in its binding site (hydrophobic association) must be reduced by a term of the order of $F_{\text{loop}} = \text{kT Ln } N$ [with N being the length between successive chromophores in the chain; Lairez et al. (1997)]. Obviously, short loops should be preferred to long ones. This point retains its validity even when interloop repulsions are taken into account and if a large number of azo can bind to the same particle (Borisov and Halperin, 1995). Consequently, it is expected that the association abruptly changes from a Benessi–Hildebrand regime to multisite and correspondingly tight association if the loop penalty becomes

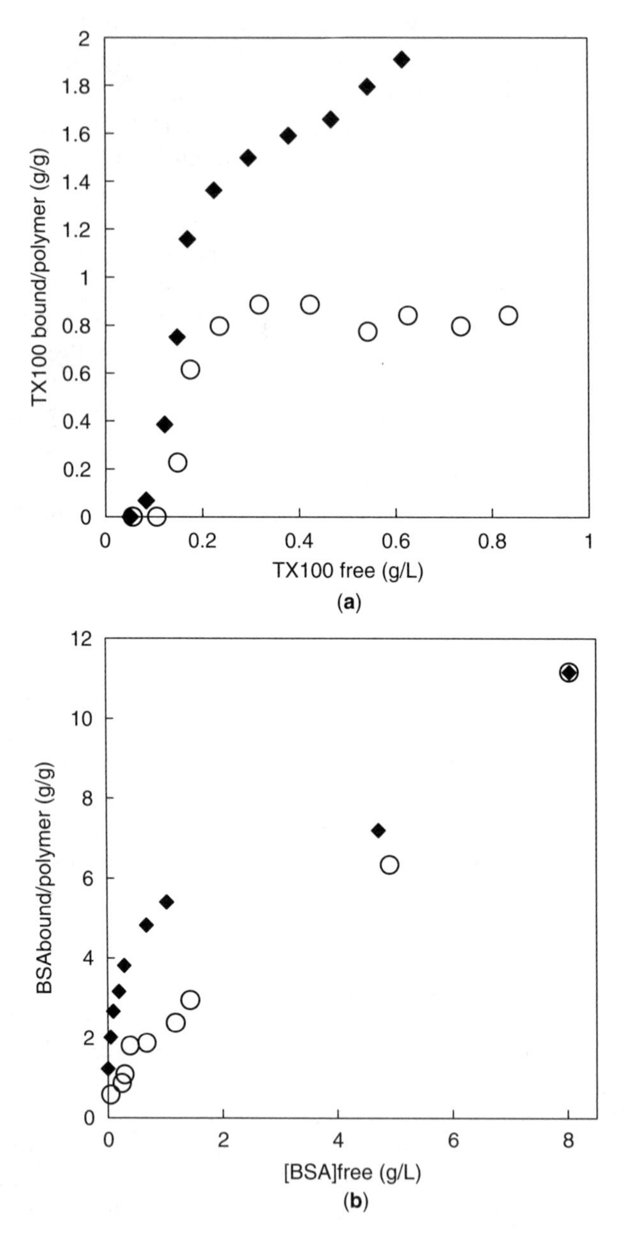

Figure 7.7. Association isotherms of micelles (TX 100) or the protein bovine serum albumin on azobenzene-modified poly (sodium acrylate) in dilute water solution, as measured from capillary electrophoretic analysis. Closed symbols: dark-adapted samples, open symbols: samples exposed to UV light (365 nm). **(a)** Polymer, as shown in Fig. 7.1, with $n = 5$ and $x = 7\%$, **(b)** $n = 12$ and $x = 3\%$.

weaker than the energy of transfer of the azobenzene into the nanoparticles. In other words, a regime of abrupt saturation of the particles is reached when the distance between azobenzenes in the chain is decreased down to the critical value of loop length. The critical transition occurs at $F_{loop} = F_{binding}$. The degree of integration of the azobenzene in the polymer defines the length of the shorter possible loops, and in turn controls the regime of association.

In the regime of tight multipoint association, the saturation should, however, be reached when either all the binding sites on the particle (e.g., number of hydrophobic pockets per protein) are associated or when interloops repulsions damper the energy of association. To estimate the typical number of loops that can be associated on a micelle, we consider that the maximal coverage is almost reached when the loop density brings the bound polymer segments into contact [$N_{loop} \sim 4\ (1 + R\text{mic}/R)^2$, with R being the radius of a loop and Rmic being the radius of micellar core of the order of 1 nm]. At integration levels of the azobenzene significantly lower than ~5 mol% (constrained by the requirement of small absorbance of solutions), the loops typically contain more than 20 monomers and R would be much higher than 1 nm. Accordingly, 4–60 loops could be gathered on a single micelle; a value that matches well with experimental data in Fig. 7.7: 1 g/g bound corresponds here to ~40 azobenzene per 100 surfactants, with ca. 50% unbound azobenzene.

Close to critical transition between the two regimes of association, a small variation of the affinity for the particle triggers the collective detachment of the loops. In other words, when free energy of loop formation is close to balance the energy of hydrophobic binding, the trans to cis isomerization is expected to abruptly make the bound particle almost free. In our experiments, azo-modified polyacrylates containing more than 1 mol% azo were tightly bound to micelles or bovine serum albumin (BSA) in the dark, but the association vanished for chains with lower integration level. The binding transition occurs accordingly at a loop length of the order of 100–200, that is, at F_{loop} #5–6 kT. This energy is comparable to the energy of hydrophobic association, which can be estimated by ~1 kT per hydrocarbon group transferred in a micelle core. Note that because of steric interaction between the backbone and the hydrophilic heads of surfactants, the azobenzene may be partially penetrating in the micelle and the length of the spacer plays an important role. Provided that cis–trans isomerization can modify the energy of hydrophobic association by an order of magnitude of a few kT, light can trigger the transition between the two binding regimes.

Another interesting prediction of this schematic analysis is that an increase of the density of azobenzene must decrease the sensitivity to light (because of the decreasing contribution of the energy of loops). Experimentally, the optimal degree of modification was below 3 mol%, and most typically 1–2 mol%. In the future, the goal of lowering the density of azobenzene in polymers should presumably be approached by designing new azobenzene-modified chains that bind even more tightly than the present ones. Likewise, longer loops would be required to balance the association, but a similar switch by a few kT of the binding energy would still be enough to trigger the detachment. In the case of micelle–polymer complexes, an interesting rational design of azobenzene groups has been proposed by Hatton and coworkers (Shang et al., 2006). Azo-containing amphiphiles have been shown to display optimal photoresponsive self-association when their spontaneous curvature is strongly modified by the transconformation, which is achieved by having the azo moieties in the middle of the hydrophobic tail. Maximizing the photovariation of the energy of association of isolated chromophores would mostly enlarge the window of polymer compositions (range of loop lengths) that can form responsive complexes with surfactants.

There are however practical limitations to the theoretical responses that can be achieved. First, in the common case of random copolymers, the random distribution of the chromophores leads to important fluctuations of the minimum distance between successive azobenzenes (i.e., the minimum loop length). This polydispersity may cause a more gradual association (for instance, only the segments with highest hydrophobicity would contribute to the association at low micelle concentration). This drawback should be limited by using longer chains (decrease of the polydispersity between chains) and presumably by using chromophores that make a significant change of the energy of binding upon irradiation, of the order of several kT. Second, the degree of photoisomerisation is usually not complete under UV exposure (usually ~80% of cis isomer and 20% residual trans isomers) (Morishima et al., 1992). If the association strength is high enough, the 20% residual trans azobenzene still present under UV exposure would cause the formation of loops that are just five times longer than the one formed in the dark, and consequently no detachment of the nanoparticles occurs. Assuming that the regime of strong binding is reached in the dark ($F_{binding}$ (trans) > kT Ln N, with N the length between successive azo in the chain), the additional criterion for detachment in presence of residual trans groups writes $F_{binding}$ (trans) < kT Ln (5 N). Therefore, the effective window for controlling the binding is limited to a range of free energy of 1.6 kT. In the working examples shown earlier, this energy was probably adjusted by proper choice of the chemistry of the azobenzene side groups (for instance, spacer length and hydrophobicity). Among the difficulty added by this constraint, it is indeed expected that the response becomes more sensitive to subtle local interactions, for example, with the micelle hydrophilic head groups.

7.3.2. Sol–Gel Transition in Semidilute Conditions

In the semidilute regime for the particle–polymer complexes, the binding may connect several chains together. Viscosity enhancement and elasticity are the macroscopic signatures of the connectivity. The sol–gel transition occurs abruptly at an average density of about one transient cross-link per chain. In the vicinity of this (critical) transition, both viscosity and the elastic modulus of the solution display a dramatic sensitivity to small variation of the cross-link concentration. The sol–gel transition provides accordingly an additional amplification of the photovariation of binding. Significant magnitude of the photoswitch of viscosity was recently achieved using conventional hydrophobically modified polymers in combination with photochromic micelles (Lee et al., 2004). Micelles with ca. 50–100 azo-surfactants play the role of a photoresponsive cross-link, and their concentration vanishes upon exposure to UV. The system requires, however, that the concentration of nanoparticles (here azo-containing micelles) is made extremely unstable under UV light, which constraints the surfactant concentration to be maintained close to the critical micellar concentration. The principal amplification of the response is because of self-assembly of the surfactant. In addition, the boundary condition of about one cross-link per chain corresponds to the binding of 50–100 azo-surfactants per chain, in equilibrium with possibly a similar amount of unbound ones.

To decrease the amount of azobenzene significantly, we proposed recently, the use of azo-modified polymers connected by nonresponsive nanoparticles. Ideally, one polymer segment with low chromophore density could bind onto the particle and form cross-links that contain just a few azobenzene groups (Fig. 7.8a) instead of several tens in a micelle of azo-surfactant. Extremely low concentrations of cross-links (submillimolar, or even micromol, concentrations in common soft gels) match well the requirement for low concentration of hydrophobic photochromes. Other advantages expected from this approach include (i) the versatility of tuning the viscosity/elasticity by adjusting the concentrations of nanoparticles and azopolymers, with no chemical modification and limited variation of absorbance, (ii) amplification of the response at both the stage of particle binding (see previous paragraph) and by the sol–gel transition. The nanoparticles tested to date include micelle of conventional surfactants (Pouliquen et al., 2002), proteins (Pouliquen and Tribet, 2006), and oligomers of CDs (Pouliquen et al., 2007). In Fig. 7.8b, the elastic and loss moduli of samples containing the same polymer concentration and increasing protein concentration show the change from sol- to gel-like properties: at the critical protein concentration, both moduli display the same dependence on frequency ($G \sim \omega^n$) that betrays the presence of polydisperse clusters of chains. At higher protein concentrations, the gel-like nature of samples is characterized by a large frequency window showing a plateau of G' (naturally for frequency much higher than the reciprocal lifetime of associations). This pseudorubbery plateau is essentially proportional to the density of cross-links. Experimentally, both elastic and loss moduli vary abruptly while crossing the critical gel point composition (Fig. 7.8b). In turn, a photovariation of the fraction of bound nanoparticle would switch the systems between a sol- and gel-like behavior.

The properties of semidilute mixtures of the protein bovine serum albumin and azopolymers match well with the simple idea of a photodestruction of the cross-links. As shown by stress relaxation experiments, the shear modulus (at time zero in Fig. 7.9a) is markedly diminished by exposure to UV, whereas the chain dynamics and relaxation time(s) appear not significantly affected. A decrease by more than a factor of 10 of the modulus indicates clearly the dissociation of most interchain associates (Fig. 7.9b). Nevertheless, the dissociation of the protein–polymer complexes may not occur when the cross-links disappear. As pointed by Leibler and colleagues in the case of interpolymer complexes (Pezron et al., 1988), and by Borrega in the case of protein–polymer physical gels (Borrega et al., 1999),

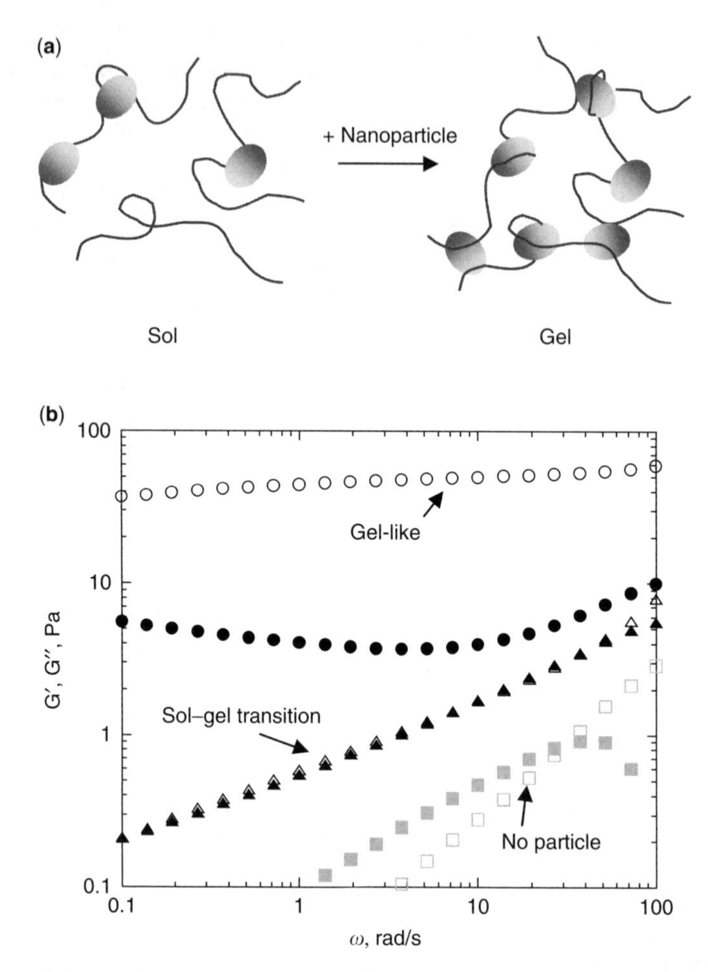

Figure 7.8. Schematic representation of the formation of a physical gel by transient associations with nanoparticles in semidilute solutions of polymer and the corresponding rheological signature. The elastic (G′) and loss (G″) moduli are plotted as a function of the frequency of the oscillating strain for samples containing the same polymer concentration (1 wt%) and bovine serum albumin at 0, 0.03, and 0.2 g/L, respectively.

the concentration of physical cross-links, X, varies in proportion to (i) the density of interchain contacts, (ii) the probability of having on protein-free segment facing a protein-bound segment, and (iii) a "binding constant" that reflects the free energy of association of at least two segments of the chain on one protein. In a simplified version of the theory, it is

$$X \sim k_2 f(1-f) C^{3\upsilon/3\upsilon-1} \tag{7.2}$$

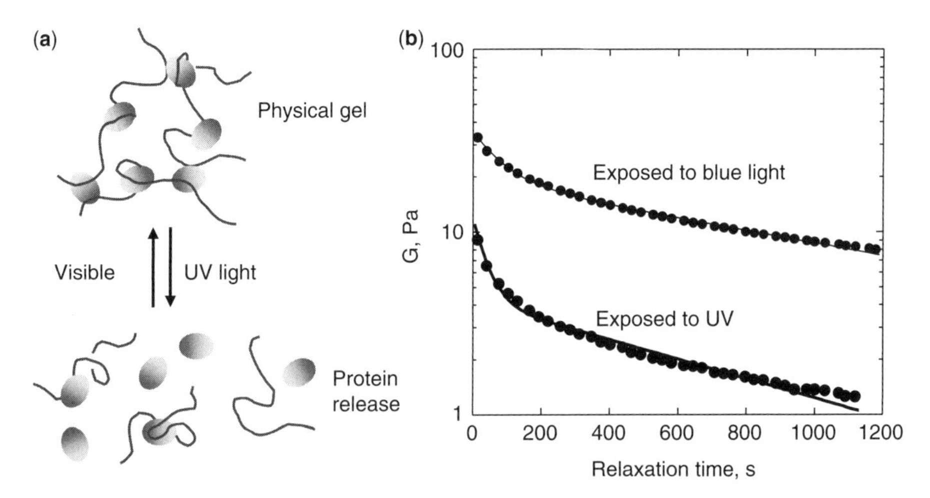

Figure 7.9. Effect of light on the connectivity of physical gels. The relaxation modulus in stress relaxation experiment is plotted after application at time zero of a fixed strain, for the same sample that was either incubated for 24 h in the dark (dark-adapted) or exposed to UV in the rheometer. For details on the samples, see Pouliquen and Tribet (2006).

with k_2 the binding constant, υ the Flory exponent, C the polymer concentration, f the fraction of the polymer's binding sites that contain a bound protein. $f = 1$ would correspond to saturation of all chains; the term $f(1-f)$ corresponds to the probability for contact between "empty" and "protein-bound" segments of chains. E–Z isomerization of the azobenzene hydrophobes could change both k_2 and f (f is essentially the parameter discussed in the preceding paragraph in the case of isolated chains) and it is therefore possible that X is essentially varied because of changes in k_2, irrespective of f (Pouliquen and Tribet, 2006). For instance, at half saturation ($f = 0.5$), the optimal effect of f on connectivity is reached (maximum elastic modulus at fixed C), and small variation of f does not affect the connectivity. In our experiments, it was possible to achieve a constant variation of the modulus irrespective of f. Alternatively, close to the gel point, f [or, respectively, $(1-f)$] is generally small (except when $C \sim C^*$) and the effect of light on the degree of binding of nanoparticle would accordingly markedly affect the cross-link density.

It is in addition possible that exposure to light does not affect the degree of binding, neither the degree of cross-linking, although it changes the dynamic of polymer motion on the bound particles. In other words, the "static" properties including the structure of the physical gel network are not modified even though the rate of motion of the azobenzene attached on the surface of the particle would depend on the isomerization. This case has been encountered with CD–azobenzene association. It has been shown that βCD and αCD can bind azobenzene derivatives with a significant change in the depth of penetration of the dye in the

CD cavity [Fig. 7.3a; Tomatsu et al. (2006); Ravoo and Jacquier (2002)]. A consequence of the different geometries of association is that the hydrodynamic friction of the complexes are modified by UV-visible irradiation. A polymer chain that binds to several CDs is in turn expected to show hydrodynamic properties that depend not only on the integration level of bound CD, but also on CD orientations and distances from the backbone (Pouliquen et al., 2007). In Fig. 7.10a, the schematic drawing of the association/dissociation dynamics point to the fact that UV-visible isomerisation may change the kinetic constants for binding with no or limited effect on their ratio. A clear signature of the photovariation of chain dynamics is produced by viscosity change in the absence of variation of the connectivity (which writes in the Newtonian limit: $\eta = \tau G_0$, with η being the viscosity, τ the slowest relaxation time of the chains being sensitive to irradiation, and G_0 the pseudorubbery plateau being invariant). Stress relaxation measurements carried out under blue or UV lights show in Fig. 7.10b that G_0 can be essentially constant, while the relaxation times are photoresponsive. In such systems, we achieved in practice photoviscosity variation up to a factor of 4 (Pouliquen et al., 2007), but more pronounced magnitude should be obtained with optimized changes of the geometry of the host–guest complexes.

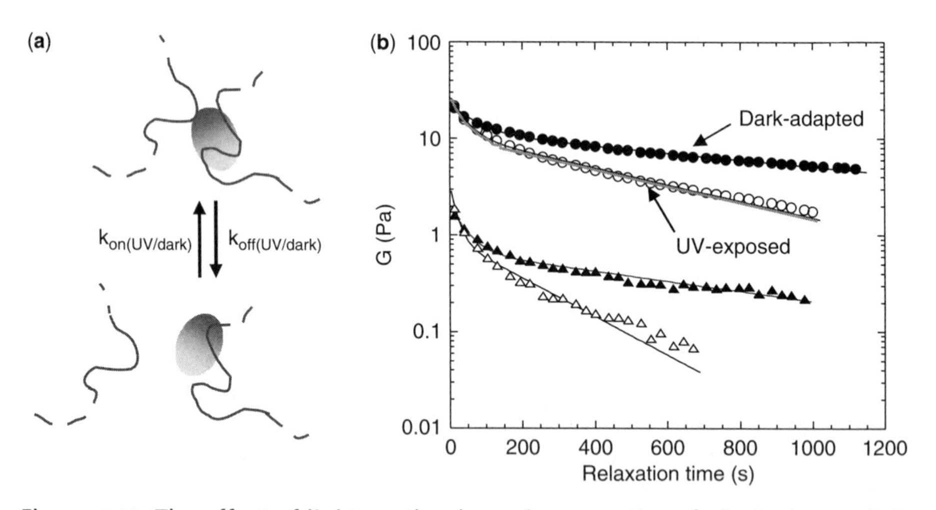

Figure 7.10. The effect of light on the dynamic properties of physical cross-links between azobenzene-modified poly(acrylate) and poly(cyclodextrin). The relaxation modulus in stress relaxation experiments is plotted after application at time zero of a fixed strain to a 0.7% polymer solution in water (polymer structure, cf. Fig. 7.1, with $n = 11$ and $x = 3$) with polycyclodextrin at 0.25% (lower curves) or 0.5% (higher moduli). The sample was either incubated for 24 h in the dark (dark-adapted, closed symbols) or continuously exposed to UV before and after loading in the rheometer (open symbols). Details on the samples composition are given in Pouliquen et al. (2007).

7.4. COMPLEXES ON DISPERSE INTERFACES: PHOTOREVERSIBLE EMULSIONS

The foregoing sections were aimed at clarifying the conditions for achieving ample photoresponses of polymer self-association (including aggregation), or association with a nanoparticle in solution. Once controlled, the formation of such associates open a vast potential for taking advantage of their properties in applications that are not limited to solution turbidity or viscosity. For instance, responsive solubility could readily help the design of light-responsive capsules, for targeted release of active compounds. In this section, we present briefly one extension of photocomplexes to the control of interfacial properties in emulsions by complex emulsifiers, namely, surfactant and photopolymer associates. Being kinetically stable, the preparation of emulsions requires vigorous mixing of oil and water in the presence of emulsifiers (amphiphiles active at oil–water interfaces) such as small-molecule and polymeric surfactants. Once formed, emulsions are composed of droplets of size generally exceeding 0.1 μm dispersed within a continuous liquid. Their lifetime vary in a range of a few seconds to years. Of importance, from the well-established nonequilibrium emulsion stability pattern, we know that one should go through a critical transition region (CTR) of instability to switch the emulsion type (O/W = oil droplets in water-continuous solution, W/O = water droplets in oil continuum). Consequently, the formation of both types of emulsion is attainable from the CTR of emulsion instability by slightly changing one control parameter. This can be practically achieved by altering continuously the interfacial properties of the emulsifier(s) and, hence, their hydrophilic–lipophilic balance. However, numerous attempts to develop light-responsive surfactants have shown the weak impact of the light-triggered molecular modification, such as transconformation or polarity changes on the energetics of interfacial properties (Cicciarelli et al., 2007; Shang et al., 2003). The properties of the surfactant molecules at the interface could, however, be triggered in more complex systems, which would amplify the effect of light. As shown earlier, the association between polymer and surfactant molecules can result in light-triggered composition of complexes, in the formation of gels or dense phases that could contribute to the interface energetics.

Consequently, the following four components were selected to prepare the photoresponsive emulsion system: equal volumes of *n*-dodecane and 0.3 M $NaNO_3$ aqueous solution, the C12E4 surfactant, and an azobenzene-modified poly(acrylate). C12E4 is known to stabilize direct and inverse emulsions below and above the so-called phase inversion temperature (PIT; here #24°C), respectively. Emulsions are unstable in the vicinity of the PIT, a temperature domain corresponding to the CTR of the light-responsive system. Emulsions made of equal oil and water volumes are directly below the PIT (and display a high electric conductivity because of the water continuum), but inverse (and of low conductivity) above the PIT (Khoukh et al., 2005).

In Fig. 7.11, the open triangles show that temperature can be used to generate the conductivity drop at the PIT. The emulsion sequences observed on a

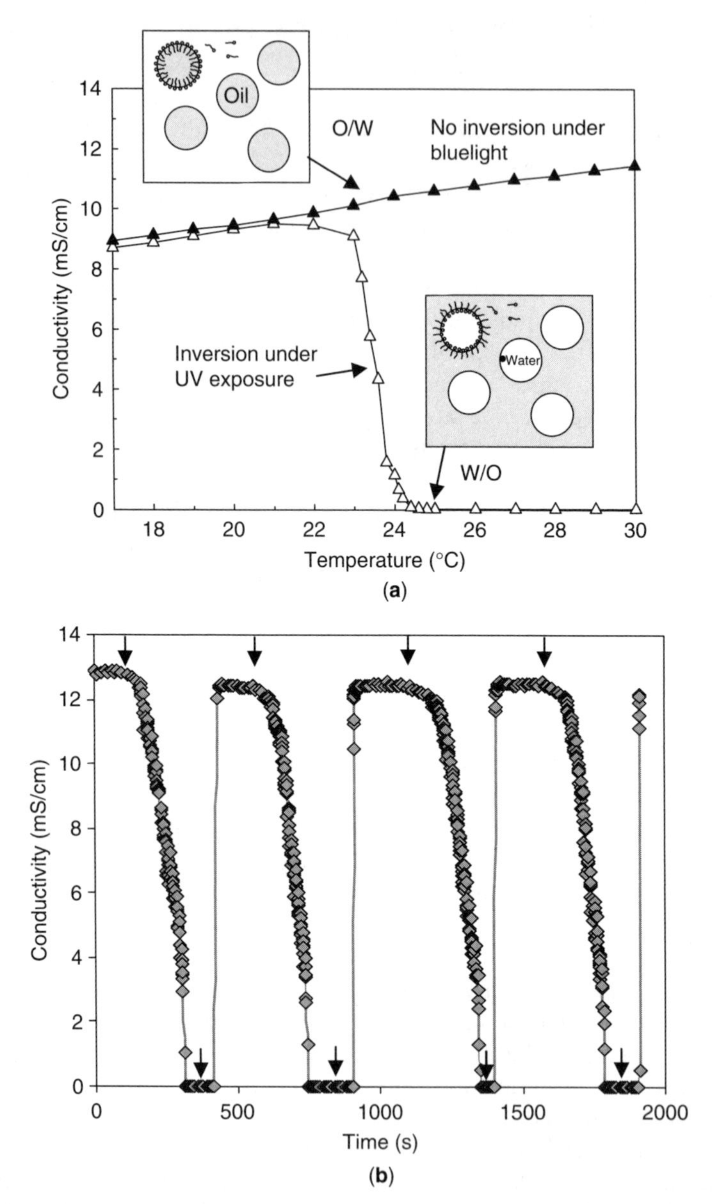

Figure 7.11. Light-responsive and reversible inversion of emulsion (dodecane/water+$NaNO_3$). The conductivity measurements indicate the type of continuous phase (conducting water vs. insulating oil) in samples maintained under gentle agitation (stirring bar). The emulsifier contains an azobenzene-modified polyacrylate ($n=5$, $x=3\%$ in Fig. 7.1) and a temperature-responsive surfactant (C12E4) that in absence of polymer would stabilize inverse emulsion above 24°C. **(a)** Temperature sweep of the same sample exposed to UV or blue light. **(b)** Switches of the wavelength of exposure between UV and blue lights at fixed temperature (25°C) at times pointed by arrows.

temperature sweep at constant surfactant concentration (20 g/L) are essentially the same in the absence of polymer and in the presence of polymer (3 g/L) under permanent UV irradiation (365 nm). In contrast, under continuous visible irradiation (436 nm), only direct emulsions form over the whole range of temperature (Fig. 7.11a). In addition, it was possible to reversibly control the inversion at constant temperature above the PIT, upon switching the wavelength (Fig. 7.11b). A direct emulsion sample prepared by preirradiation at 436 nm is continuously stirred at constant temperature under permanent visible irradiation. The conductivity (values of the order of mS/cm) remains constant with time. Note that in the absence of stirring, the emulsion appears stable for the typical time of experiments (of the order of a few hours) as long as these irradiation conditions are maintained. However, conductivity displays a sharp decrease after the switch of the wavelength from 436 to 365 nm.The conductivity values of the order of μS/cm reveal the formation of inverse emulsions that remain stable with time as long as UV irradiation is maintained. Hence, conductivity can be cycled between two plateau values, which correspond to the formation of water and oil continuous emulsions, respectively. Both emulsion type and emulsion stability are well controlled by light and are switched reversibly (Khoukh et al., 2006, 2005).

7.5. CONCLUSION

In conclusion, we emphasize the originality and versatility of using azobenzene-modified polymers in combination with nanoparticles or micelles. First of all, the various ingredients as well as the methods of preparation and control are now accessible from simple chemistry. Polymer–micelle binding affects a rich variety of properties. The ubiquitous role of surfactants in solubilization of particles, formulation of mixtures, surface treatments should be readily controlled in presence of amphiphilic photopolymers. In addition, the stage of amplification achieved at the scale of the polymer–micelle association could be extended to a more general concept, toward the design of unconventional light-responsive systems of nanoparticles. Minor tailoring efforts are needed to obtain other complexes involving a few azobenzene groups per bound particle, or photochromes other than azobenzenes so as to broaden the field of applications. As an example, the vast potential of using proteins and enzymes instead of micelles of surfactants is underlined here. Light-triggered association with enzymes have been investigated to trigger a biological activity with light (Lee et al., 2005; Willner and Rubin, 1996). The design of photopolymer capable of binding onto hydrophobic pockets onto enzyme deserves investigation because it would enable one to tune enzyme immobilization and activity with no need for a specific synthesis or modification. The triggering of biodigestion or specific consumption of substrates in gels and microchips would become readily accessible. This opens up new opportunities for the development of clean active systems for *in vivo* preparations.

Furthermore, a remarkably low density of photochromes (0.5%–3 mol% of monomers) is sufficient, and actually required to confer appropriate sensitivity to light of polymer–nanoparticle complexes. A low concentration of photochrome (and low absorbances of samples) then becomes compatible with high magnitude of responses. High efficiency at low absorbance opens the way to unexplored short response times in complex fluids (high intensities of light can reach the bulk of samples). For instance, gels and colloid assemblies made photoswitchable on time scale of the order of diffusive times are expected to exhibit new nonequilibrium properties such as photocontrolled morphologies of precipitates (Klajn et al., 2007; Bell and Piech, 2006), or new sorting principles in analytical chemistry (Alcor et al., 2004). The length scale of stimulation can be focused down to micron size. All these features are of great potential benefit in microsystems and microfluidic applications to add a new route toward formation of gradients, modification of flow rates, etc. to the present devices based on local applications of heat or of electric field.

Finally, a second stage of amplification is recruited when the polymer complexes are close to a critical transition, including solubility, sol–gel transition, or adsorption onto interfaces. Light can certainly be used to investigate the dynamics of such complex transitions. Practically, irradiation can be seen as a reversible target-selective tool that requires no extra additive to modify with short pulses of lights (shorter than characteristic diffusion times of the macromolecules) the properties of macromolecules and trigger the variation of interfacial properties (e.g., wettability), release/encapsulation or the continuous phase (e.g., when an emulsion swings between O/W and W/O). Study of the phototriggered dynamics and intermediate stages of phenomena becomes quite readily accessible with azo-containing polymer systems.

ACKNOWLEDGMENTS

Some results and conclusions presented are the outcome of several years of fruitful collaborations with C. Prata (synthesis of azo derivatives), G. Pouliquen, J. Ruchmann and I. Porcar (physical-chemistry investigations on association with proteins or surfactants), and S. Khoukh and P. Perrin (development of photoemulsifiers).

REFERENCES

Alcor D, Croquette V, Jullien L, Lemarchand A. 2004. Stochastic resonance to control diffusive motion in chemistry. Proc Natl Acad Sci USA 101:8276–8280.

Alonso M, Reboto V, Guiscardo L, Mate V, Rodriguez-Cabello JC. 2001. Novel photoresponsive p-phenylazobenzene derivative of an elastin-like polymer with enhanced control of azobenzene content and without pH sensitiveness. Macromolecules 34(23):8072–8077.

Bell NS, Piech M. 2006. Photophysical effects between spirobenzopyran-methyl methacrylate-functionalized colloidal particles Langmuir 22:1420–1427.

Bo Q, Yavrian A, Galstian T, Zhao Y. 2005. Liquid crystalline ionomers containing azobenzene mesogens: phase stability, photoinduced birefringence, and holographic grating. Macromolecules 38(8):3079–3086.

Borisov OV, Halperin A. 1995. Micelles of polysoaps. Langmuir 11:2911–2919.

Borrega R, Tribet C, Audebert R. 1999. Reversible gelation in hydrophobic polyelectrolyte/protein mixtures: an example of cross-links between soft and hard colloids. Macromolecules 32(23):7798–7806.

Cicciarelli BA, Hatton TA, Smith KA. 2007. Dynamic surface tension behavior in a photoresponsive surfactant system. Langmuir 23(9):4753–4764.

Effing JJ, Kwak JCT. 1995. Photoswitchable phase separation in hydrophobically modified poly(acrylamide)/surfactant systems. Angew Chem Int Ed Engl 34(1):88–90.

Effing JJ, McLennan IJ, Kwak JCT. 1994. Associative phase separation observed in a hydrophobically modified poly(acrylamide)/sodium dodecyl sulfate system. J Phys Chem 98(10):2499–2502.

Gao JY, Dubin PL, Muhoberac BB. 1997. Measurement of the binding of proteins to polyelectrolytes by frontal analysis continuous capillary electrophoresis. Anal Chem 69:2945–2951.

Howley C, Marangoni DG, Kwak JCT. 1997. Association and phase behavior of hydrophobically modified photoresponsive poly(acrylamide)s in the presence of ionic surfactant. Coll Polym Sci 275(8):760–768.

Ichimura K, Oh S-K, Nagakawa M. 2000. Light driven motion of liquids on a photoresponsive surface. Science 288:1624–1627.

Iliopoulos I. 1998. Association between hydrophobic polyelectrolytes and surfactants. Curr Opin Coll Int Sci (3):493–498.

Inoue Y, Kuad P, Okumura Y, Takashima Y, Yamaguchi H, Harada A. 2007. Thermal and photochemical switching of conformation of poly(ethylene glycol)-substituted cyclodextrin with an azobenzene group at the chain end. J Am Chem Soc 129(20):6396–6397.

Irie M. 1993. Stimuli-responsive poly(N-isopropylacrylamide)—photoinduced and chemical-induced phase-transitions. Adv Polym Sci 110:49–65.

Irie M, Iga R. 1986. Photoresponsive polymers 9. Photostimulated reversible sol-gel transition of polystyrene with pendant azobenzene groups in carbon-disulfide. Macromolecules 19(10):2480–2484.

Irie M, Hirano Y, Hashimoto S, Hayashi K. 1981a. Photoresponsive polymers. 2. Reversible solution viscosity change of polyamide having azobenzene residues in the main chain. Macromolecules 14:262–267.

Irie M, Hirano Y, Hashimoto S, Hayashi K. 1981b. Photoresponsive polymers. 2. reversible solution viscosity change of polyamide having azobenzene residues in the main chain. Macromolecules 14:262–267.

Keita G, Ricard A, Audebert R, Pezron E, Leibler L. 1995. The poly(vinyl alcohol)-borate system: influence of polyelectrolyte effects on phase diagrams. Polymer 36(1):49–54.

Khoukh S, Oda R, Labrot T, Perrin P, Tribet C. 2007. Light-responsive hydrophobic association of azobenzene-modified poly(acrylic acid) with neutral surfactants. Langmuir 23(1):94–104.

Khoukh S, Perrin P, de Berc FB, Tribet C. 2005. Reversible light-triggered control of emulsion type and stability. Chemphyschem 6(10):2009–2012.

Khoukh S, Tribet C, Perrin P. 2006. Screening physicochemical parameters to tuning the reversible light-triggered control of emulsion type. Colloids Surfaces A Physicochem. Eng Aspects 288(1–3):121–130.

Klajn R, Bishop KJM, Grzybowski BA. 2007. Light-controlled self-assembly of reversible and irreversible nanoparticle suprastructures. Proc Natl Acad Sci USA 104:10305–10309.

Kusebauch U, Cadamuro SA, Musiol HJ, Lenz MO, Wachtveitl J, Moroder L, Renner C. 2006. Photocontrolled folding and unfolding of a collagen triple helix. Angewandte Chemie-International Edition 45(42):7015–7018.

Labarthet FL, Freiberg S, Pellerin C, Pezolet M, Natansohn A, Rochon P. 2000. Spectroscopic and optical characterization of a series of azobenzene-containing side-chain liquid crystalline polymers. Macromolecules 33(18):6815–6823.

Lairez D, Adams M, Carton J-P, Raspaud E. 1997. Aggregation of telechelic triblock copolymers: from animals to flowers. Macromolecules 30:6798–6809.

Lee CT, Smith KA, Hatton TA. 2004. Photoreversible viscosity changes and gelation in mixtures of hydrophobically modified polyelectrolytes and photosensitive surfactants. Macromolecules 37(14):5397–5405.

Lee CT, Smith KA, Hatton TA. 2005. Photocontrol of protein folding: the interaction of photosensitive surfactants with bovine serum albumin. Biochemistry 44(2):524–536.

Lim HS, Han JT, Kwak D, Jin M, Cho K. 2006. Photoreversibly switchable superhydrophobic surface with erasable and rewritable pattern. J Am Chem Soc 128(45):14458–14459.

Liu XH, Wang XG, Liu DS. 2001. Synthesis and characterization of photo-responsive and thermosensitive poly(N-isopropylacrylamide-co-2-[4-(4′-ethoxyphenyl)azophenoxy]ethyl acrylate). Acta Polymerica Sinica (6):773–778.

Mamada A, Tanaka T, Kungwatchakun D, Irie M. 1990. Photoinduced phase-transition of gels. Macromolecules 23(5):1517–1519.

May BL, Gerber J, Clements P, Buntine MA, Brittain DRB, Lincoln SF, Easton CJ. 2005. Cyclodextrin and modified cyclodextrin complexes of E-4-tert-butylphenyl-4′-oxyazobenzene: UV-visible, H-1 NMR and ab initio studies. Org Biomol Chem 3(8):1481–1488.

Menzel H, Kroeger R, Hallensleben M. 1995. Polymers with light controlled water solubility. Macromol Rep A32(5–6):779–787.

Moniruzzaman M, Sabey CJ, Fernando GF. 2004. Synthesis of azobenzene-based polymers and the in-situ characterization of their photoviscosity effects. Macromolecules 37(7):2572–2577.

Morishima Y, Tsuji M, Kamachi M, Hatada K. 1992. Photochromic isomerization of azobenzene moieties compartmentalized in hydrophobic microdomains in a microphase structure of amphiphilic polyelectrolytes. Macromolecules 25(17):4406–4410.

Pezron E, Leibler L, Ricard A, Audebert R. 1988. Reversible gel formation induced by ion complexation. 2. Phase diagrams. Macromolecules 21(4):1126–1131.

Piculell L, Lindman B. 1992. Phase separation in surfactant/polymer mixtures. Adv Colloid Interface Sci 41:149–153.

Pieroni O, Fissi A, Angelini N, Lenci F. 2001. Photoresponsive polypeptides. Acc Chem Res 34(1):9–17.

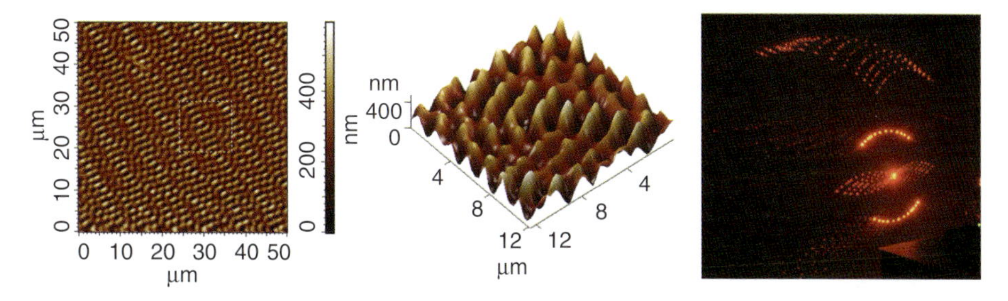

Figure 2.16. Relief structure recorded in film of the (Methyl Red)/PDADMAC material. See page 77 for text discussion of this figure.

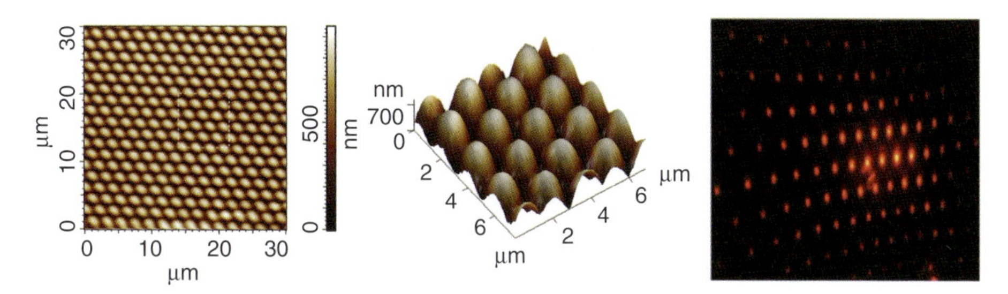

Figure 2.18. Relief structure recorded in film of the Methyl Red sol–gel material. See page 80 for text discussion of this figure.

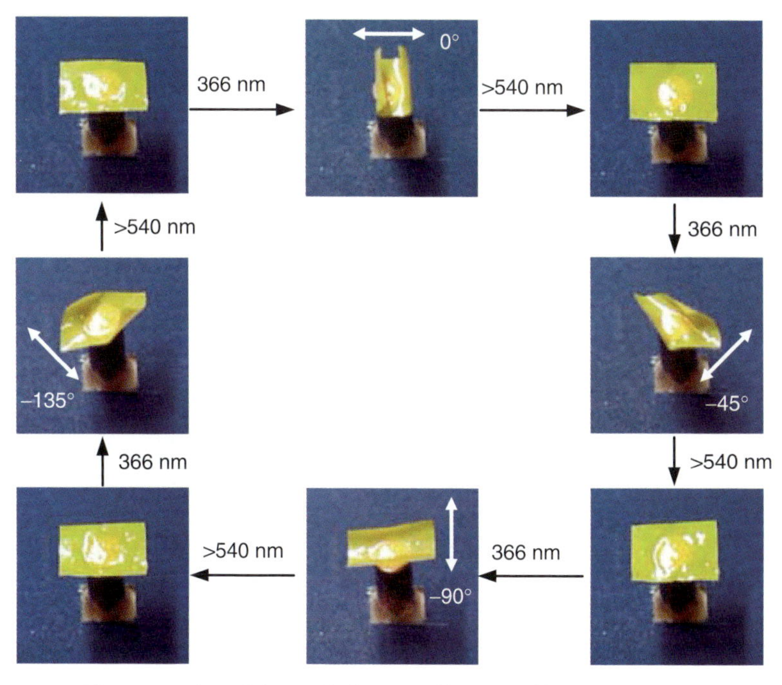

Figure 3.25. Photographs of the polydomain film in different directions in response to irradiation by linearly polarized light at different angles of polarization (*white arrows*) at $\lambda = 366$ nm. See page 128 for text discussion of this figure.

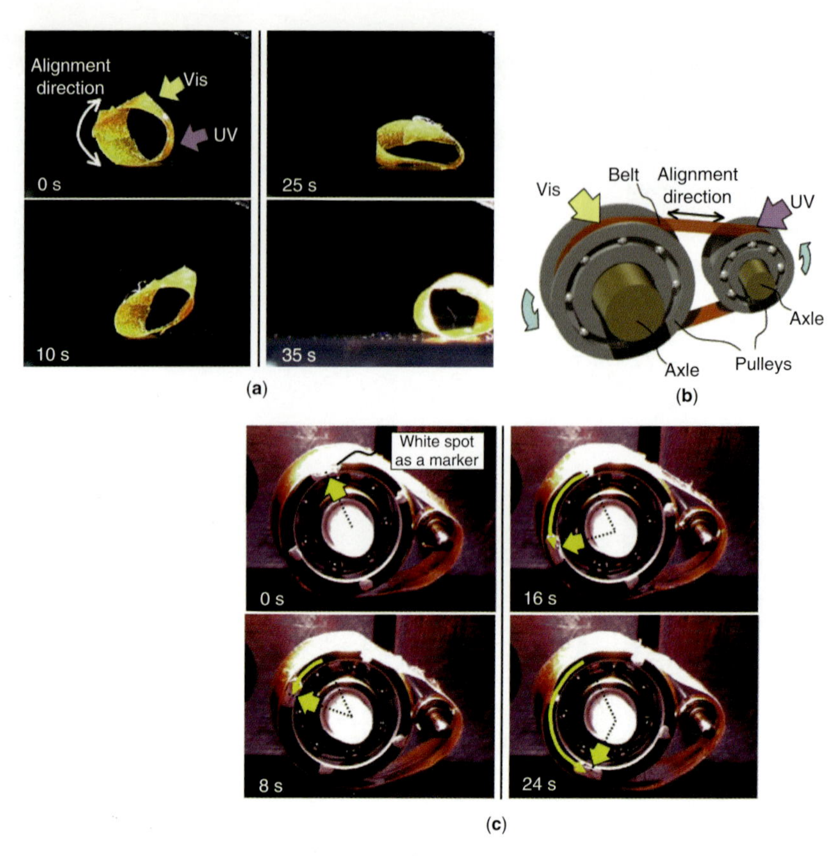

Figure 3.32. See page 132–135 for text discussion of this figure.

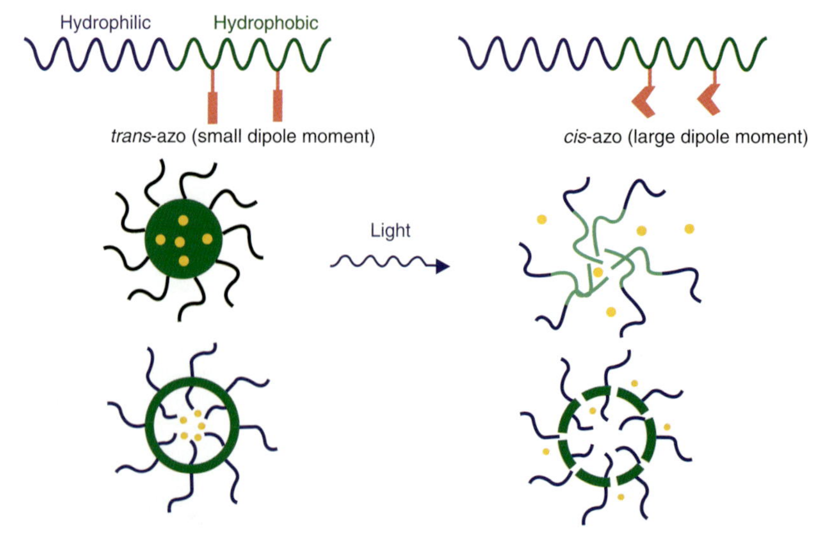

Figure 6.1. Schematic illustration of light-controllable micelles and vesicles formed by azobenzene-containing block copolymers displaying the trans–cis photoisomerization. See page 219 for text discussion of this figure.

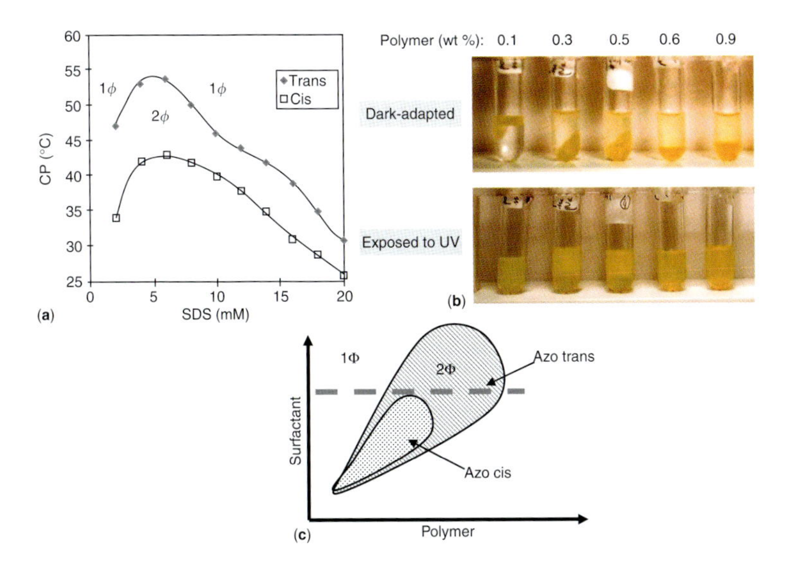

Figure 7.4. Associative phase separation of complexes between surfactants and azobenzene-modified polymers. See page 251–253 for text discussion of this figure.

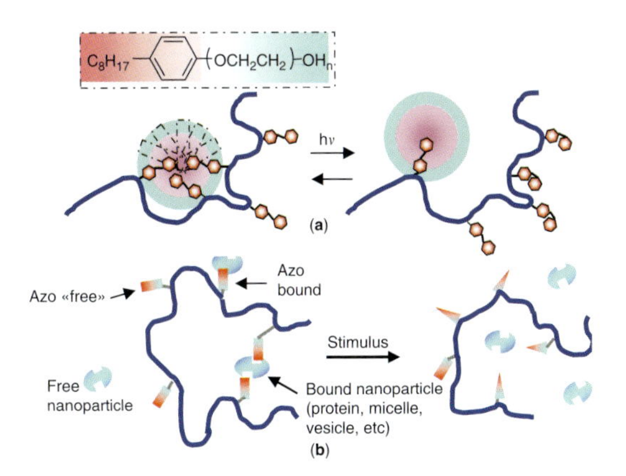

Figure 7.5. Schematic drawing of the light-responsive association between a flexible polymer and nanoparticles. See page 254–255 for text discussion of this figure.

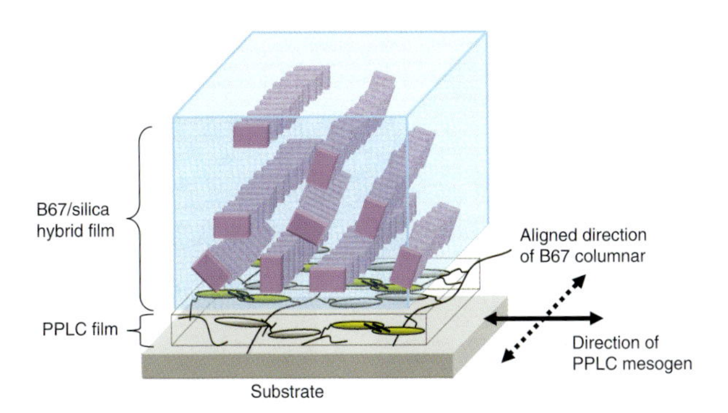

Figure 8.5. Photoalignment of chromonic LC/silica nanohybrid. See page 279 for text discussion of this figure.

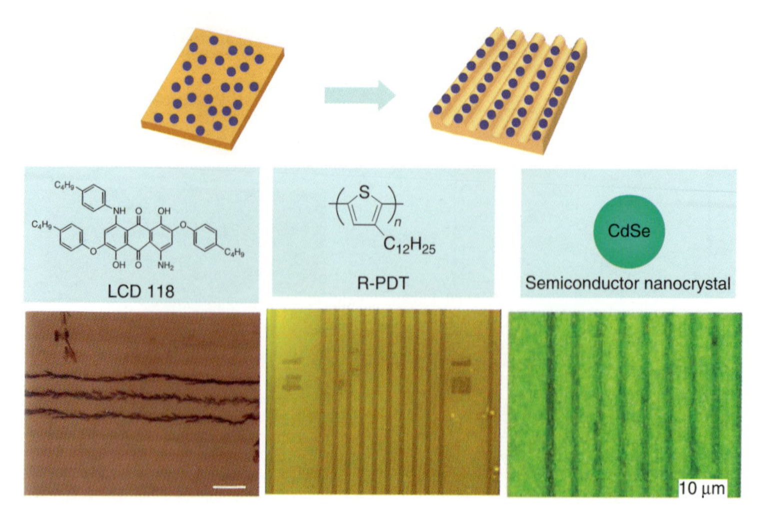

Figure 8.13. Phototactic transport of functional materials (dye, π-conjugated polymer, and nanocrystal) by conveyance actions. See page 288 for text discussion of this figure.

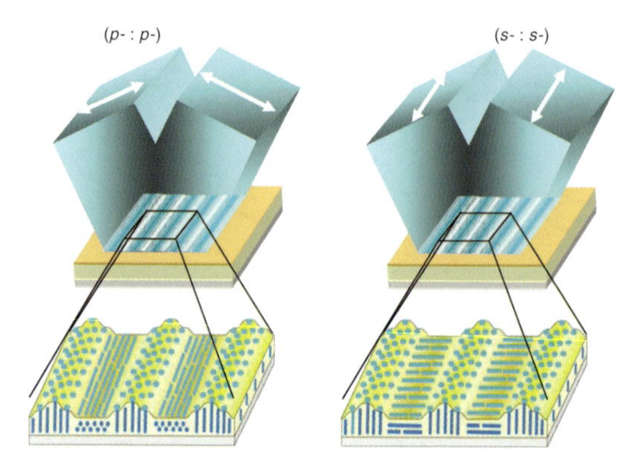

Figure 8.21. Schematic illustration of hierarchical structures involved in photogenerated SRG of a block copolymer after irradiation with (*p*- : *p*-) and (*s*- : *s*-) mode interference laser beam. See page 296–297 for text discussion of this figure.

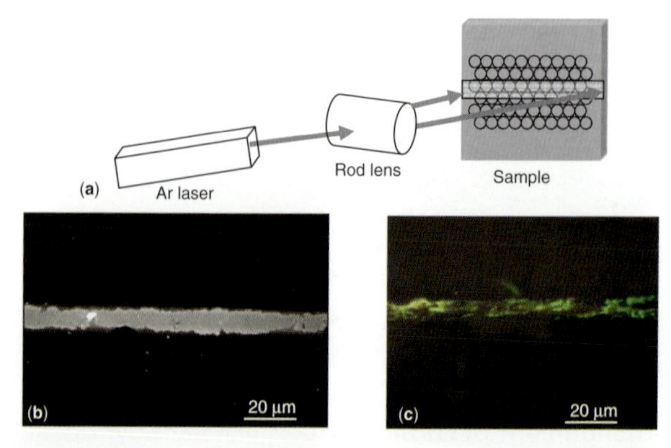

Figure 9.8. See page 310 for text discussion of this figure.

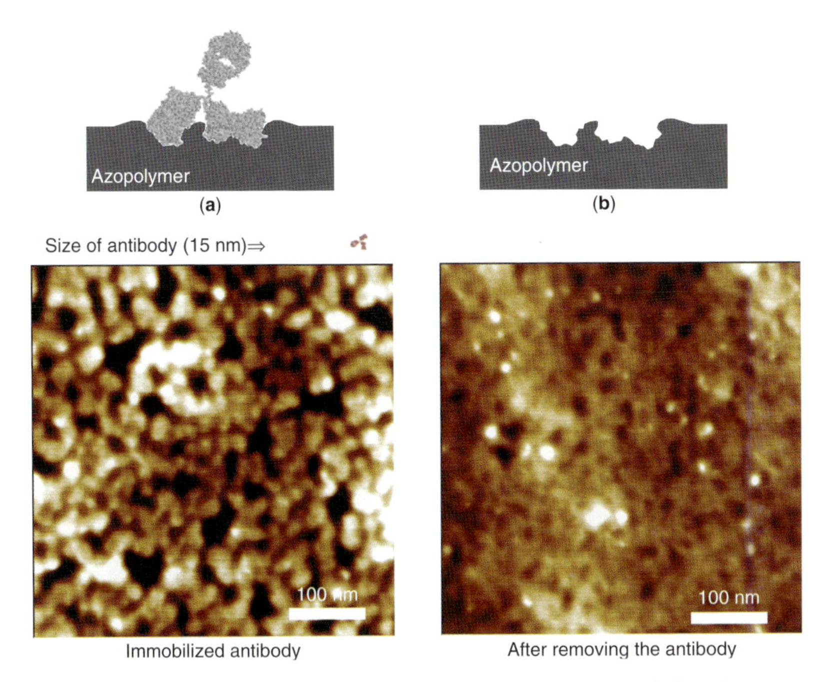

Figure 9.9. AFM images obtained from the photoimmobilization process. See page 312 for text discussion of this figure.

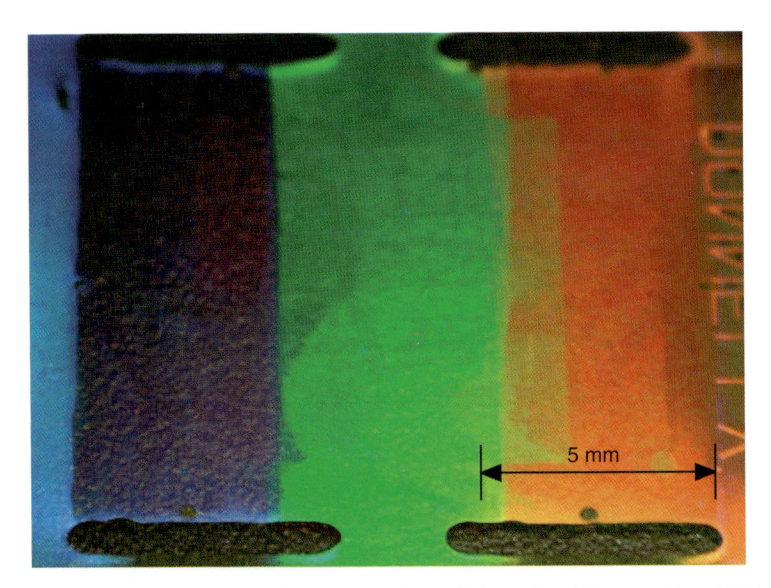

Figure 10.18. Change in the reflection color of the Ch LC by varying UV irradiation time 0 (*left*), 4 (*middle*), and 10 s (*right*). See page 348–350 for text discussion of this figure.

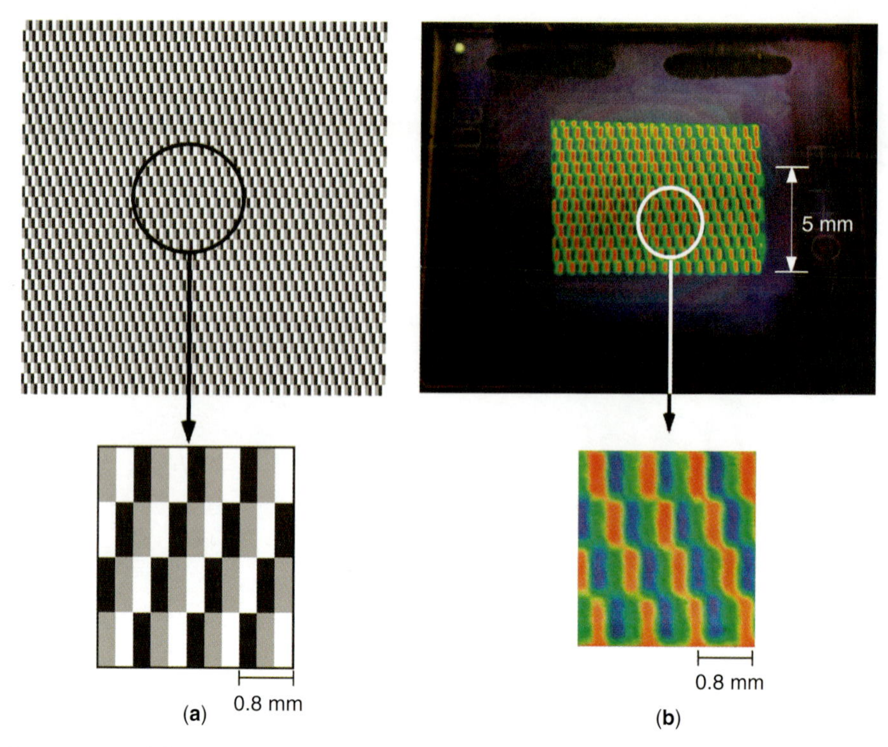

Figure 10.19. (**a**) Gray mask and (**b**) color patterning of the Ch LC containing E44, *m*-azo-8, and chiral obtained by UV irradiation for 10 s through the gray mask at 25°C. See page 348–350 for text discussion of this figure.

Figure 11.16. See page 383 for text discussion of this figure.

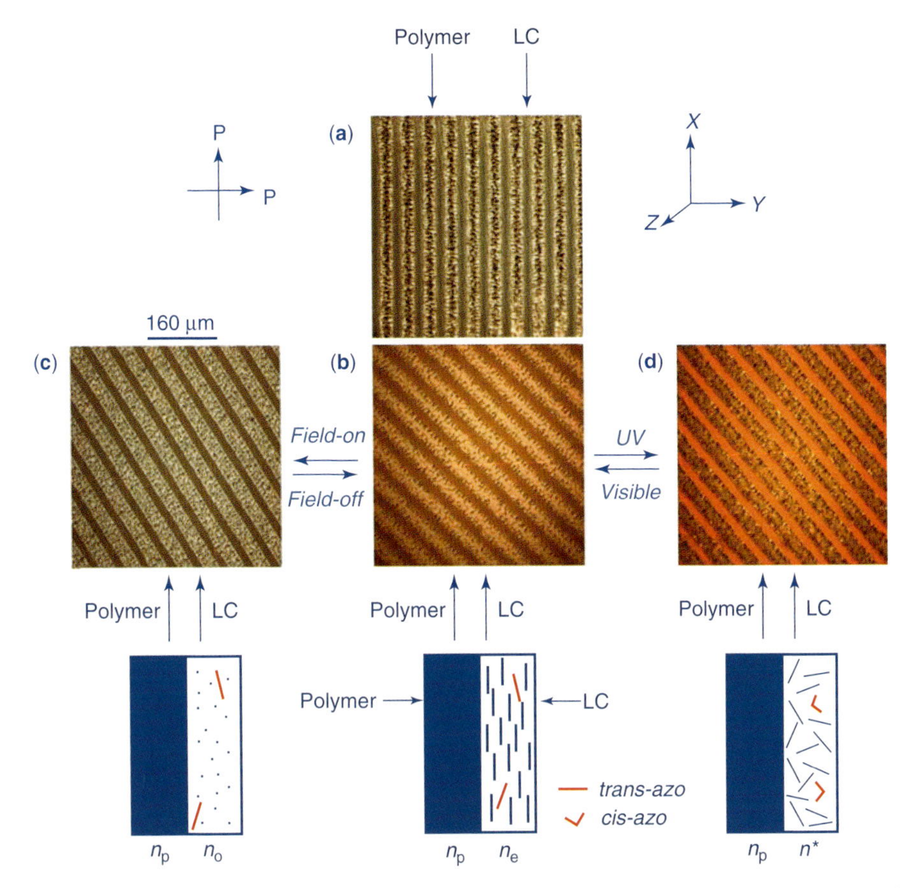

Figure 11.24. Polarizing optical micrographs showing an electrically and optically switchable diffraction grating prepared using an azobenzene polymer-stabilized nematic liquid crystal (15 wt% polymer). See page 393–394 for text discussion of this figure.

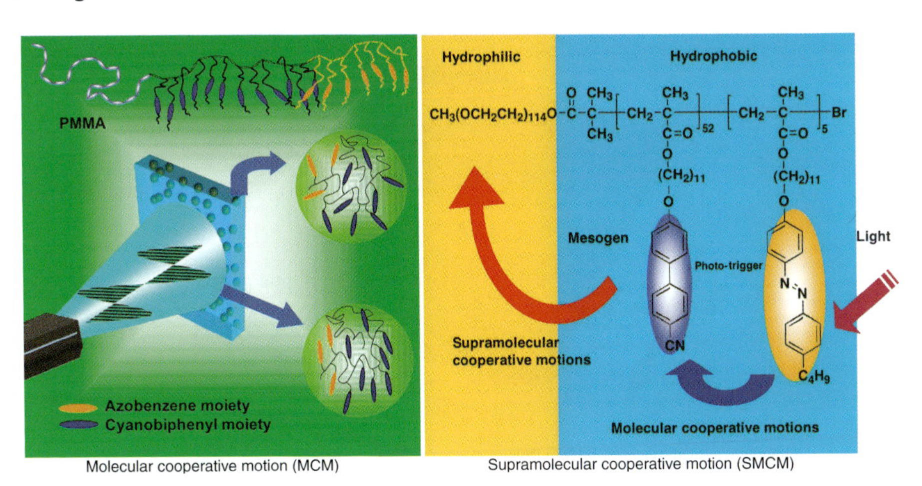

Figure 12.4. Photoinduced cooperative motion in bulk films of azo BCs. Azo in the minority phase (*left*) and azo in the majority phase (*right*). See page 422 for text discussion of this figure.

Figure 12.11. AFM images of PEO nanocylinders in bulk films of azo LCBCs (3.1 μm × 12.4 μm). See page 429–430 for text discussion of this figure.

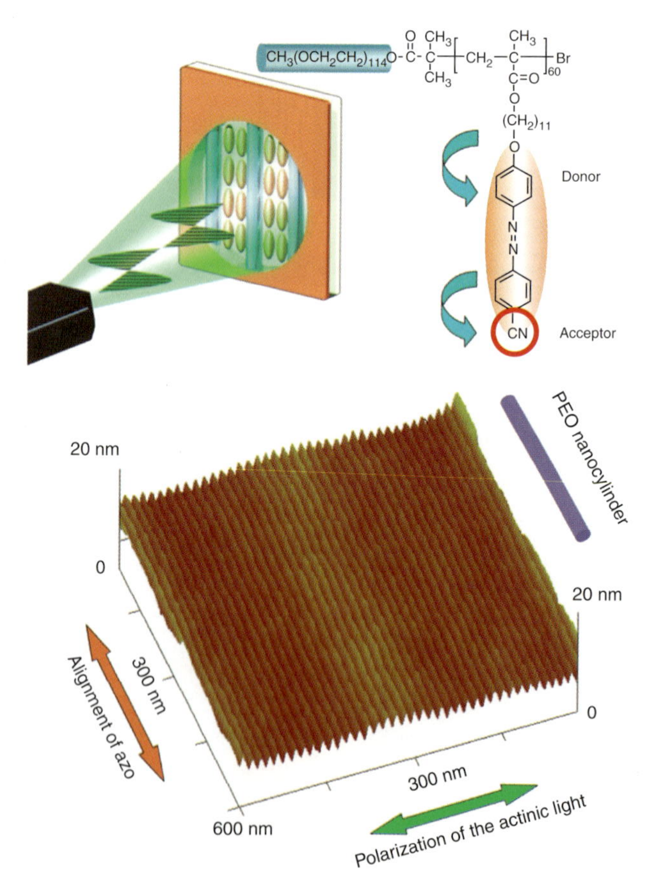

Figure 12.14. Photoinduced alignment of PEO nanocylinders in azo LCBC films as a result of SMCM. See page 433 for text discussion of this figure.

Pouliquen G, Tribet C. 2006. Light-triggered association of bovine serum albumin and azobenzene-modified poly(acrylic acid) in dilute and semidilute solutions. Macromolecules 39(1):373–383.

Pouliquen G, Porcar I, Tribet C, Amiel C. 2002. Photoresponsive thickening in polyamphiphile-based physical gels: the example of micelles, protein and cyclodextrin crosslinkers. In: Bohidar H, Dubin P, Osada Y, editors. ACS Advances in Chemistry. Series 833 (Polymer Gels: Fundamentals and Applications). Chap 18, pp. 262–289.

Pouliquen G, Amiel C, Tribet C. 2007. Photoresponsive viscosity and host-guest association in aqueous mixtures of poly cyclodextrin with azobenzene-modified poly(acrylic)-acid. J Phys Chem B 111(20):5587–5595.

Rau, H. 2003. Azo compounds. In: Dürr H, Bouas-laurent H, editors. Photochromism Molecules and Systems, Amsterdam: Elsevier pp. 165–190.

Ravoo BJ, Jacquier JC. 2002. Host-guest interaction between beta-cyclodextrin and hydrophobically modified poly(isobutene-alt-maleic acid) studied by affinity capillary electrophoresis. Macromolecules 35(16):6412–6416.

Rodriguez-Cabello JC, Alonso M, Guiscardo L, Reboto V, Girotti A. 2002. Amplified photoresponse of a p-phenylazobenzene derivative of an elastin-like polymer by alpha-cyclodextrin: the amplified delta T-t mechanism. Adv Mater 14(16):1151–1154.

Sanchez AM, deRossi RH. 1996. Effect of beta-cyclodextrin on the thermal cis-trans isomerization of azobenzenes. J Org Chem 61(10):3446–3451.

Shang TG, Smith KA, Hatton TA. 2003. Photoresponsive surfactants exhibiting unusually large, reversible surface tension changes under varying illumination conditions. Langmuir 19(26):10764–10773.

Shang TG, Smith KA, Hatton TA. 2006. Self-assembly of a nonionic photoresponsive surfactant under varying irradiation conditions: a small-angle neutron scattering and cryo-TEM study. Langmuir 22(4):1436–1442.

Shibaev V, Bobrovsky A, Boiko N. 2003. Photoactive liquid crystalline polymer systems with light-controllable structure and optical properties. Prog Polym Sci 28(5):729–836.

Shimomura T, Funaki T, Ito K, Choi BK, Hashizume T. 2002. Circular dichroism study of the inclusion-dissociation behavior of complexes between a molecular nanotube and azobenzene substituted linear polymers. J Inclusion Phenom Macrocyclic Chem 44(1–4):275–278.

Shulkin A, Stover HDH. 2003. Photostimulated phase separation encapsulation. Macromolecules 36(26):9836–9839.

Strzegowski LA, Martinez MB, Gowda DC, Urry DW, Tirrell DA. 1994. Photomodulation of the inverse temperature transition of a modified elastin poly(pentapeptide). J Am Chem Soc 116(2):813–814.

Suzuki M, Kajtar M, Szejtli J, Vikmon M, Fenyvesi E, Szente L. 1991. Inclusion-compounds of cyclodextrins and azo dyes 9. Induced circular-dichroism spectra of complexes of cyclomaltooligosaccharides and azobenzene derivatives. Carbohydr Res 214(1):25–33.

Tokuhisa H, Yokoyama M, Kimura K. 1998. Photoinduced switching of ionic conductivity by metal ion complexes of vinyl copolymers carrying crowned azobenzene and biphenyl moieties at the side chain. J Mater Chem 8(4): 889–891.

Tomatsu I, Hashidzume A, Harada A. 2006. Contrast viscosity changes upon photo-irradiation for mixtures of poly(acrylic acid)-based alpha-cyclodextrin and azobenzene polymers. J Am Chem Soc 128(7):2226–2227.

Tong X, Wang G, Soldera A, Zhao Y. 2005. How can azobenzene block copolymer vesicles be dissociated and reformed by light? J Phys Chem B 109(43):20281–20287.

Ueno A, Shimizu T, Mihara H, Hamasaki K, Pitchumani K. 2002. Supramolecular chemistry of cyclodextrin-peptide hybrids: azobenzene-tagged peptides. J Inclusion Phenom Macrocyclic Chem 44(1–4):49–52.

Wang G, Tong X, Zhao Y. 2004. Preparation of azobenzene-containing amphiphilic diblock copolymers for light-responsive micellar aggregates. Macromolecules 37(24):8911–8917.

Willner I, Rubin S. 1996. Control of the structure and functions of biomaterials by light. Angew Chem Int Ed Engl 35(4):367–385.

Yang L, Takisawa N, Kaikawa T, Shirahama K. 1996. Interaction of photosurfactants, [[[4′-[(4-alkylphenyl)azo]phenyl]oxy]ethyl]trimethylammomium bromides, with alpha- and beta-cyclodextrins as measured by induced circular dichroism and a surfactant-selective electrode. Langmuir 12(5):1154–1158.

Zheng PJ, Hu X, Zhao XY, Li L, Tam KC, Gan LH. 2004. Photoregulated sol-gel transition of novel azobenzene-functionalized hydroxypropyl methylcellulose and its alpha-cyclodextrin complexes. Macromol Rapid Commun 25(5):678–682.

Zheng PJ, Wang C, Hu X, Tam KC, Li L. 2005. Supramolecular complexes of azocellulose and alpha-cyclodextrin: isothermal titration calorimetric and spectroscopic studies. Macromolecules 38(7):2859–2864.

Zheng PQ, Li Z, Tong LH, Lu RH. 2002. Study of inclusion complexes of cyclodextrins with Orange II. J Inclusion Phenom Macrocyclic Chem 43(3–4):183–186.

8

LIGHT-RESPONSIVE 2-D MOTIONS AND MANIPULATIONS IN AZOBENZENE-CONTAINING LIQUID CRYSTALLINE POLYMER MATERIALS

Takahiro Seki

8.1. INTRODUCTION

Research on azobenzene (Az)-containing polymeric films has exploded during the past few decades, and it continues to be a very active area of materials chemistry (Ikeda et al., 2007; Seki, 2007, 2004; Ikeda, 2003; Natansohn and Rochon, 2002; Sekkat and Knoll, 2002; Ichimura, 2000; Irie, 2000; Kumar and Neckers, 1989). Azobenzene is perhaps the most frequently used photochromic unit to be incorporated in molecular assemblies and polymer liquid crystals (LCs). It has several advantageous features for fabrication of photoresponsive systems as follows. (i) Tedious synthesis is generally not required, leading to facile preparation of planned derivatives. (ii) The sensitivity of E (trans) form $\rightleftharpoons$ Z (cis) photoisomerization is moderately high for both directions for derivatives having alkyl or alkoxyl substituents. Besides, less expensive light sources, such as a mercury lamp, can be used. (iii) Photofatigue is negligible under ordinary laboratory conditions. (iv) The molecular shape is rodlike and symmetric, which favors incorporation into various molecular assemblies, especially into liquid crystalline systems. The final item is of particular importance to attain facile orientational controls and effectively induce motions. Thus far many efforts have been made to exert various types of performances in the liquid crystalline polymer

Smart Light-Responsive Materials. Edited by Yue Zhao and Tomiki Ikeda

systems (Ikeda et al., 2007; Seki, 2007, 2004; Ikeda, 2003; Natansohn and Rochon, 2002; Sekkat and Knoll, 2002; Ichimura, 2000).

This chapter focuses on recent progress in research of 2-D systems, such as monolayers and ultrathin polymer films; the thickness of which ranges below several 10 nm. Cooperative and amplifying effects are efficiently expressed in such 2-D thin polymer films, where molecules are highly aligned by the confinement of the surface geometry. Topics introduced here involve surface photoalignment of mesostructured materials, preparation of surface-grafted photoresponsive polymer films, phototriggered efficient mass migration and its applications, and photoalignment and patterning of microphase-separated nanostructures formed by block copolymers.

8.2. ALIGNMENT OF FUNCTIONAL MATERIALS BY COMMAND SURFACE

Photochromic molecule changes its polarity (dipole moment) by light, and this property has long been applied to photocontrollable wetability changes of liquids. This is still an active area of investigation (Lim et al., 2007; Ichimura et al., 2000). In cases of anisotropic liquids, namely, thermotropic LCs, the molecular orientation is readily controlled on a photoresponsive monolayer. A pioneering work was reported by Ichimura et al. (1988) (Seki et al., 1993). They showed that the trans–cis photoisomerization reaction of an Az monolayer can reversibly switch the alignment of nematic LCs, and proposed a concept of command surface effect. Recent attempts by Seki and coworkers revealed that photoalignment of organized materials is not limited to typical thermotropic LCs, but a certain polymer (Fukuda et al., 2002a,b; Seki et al., 1999) and even an anisotropic structure of inorganic–organic hybrid materials (Fukumoto et al., 2006, 2005; Kawashima et al., 2004, 2002) are controlled by light. The in-plane control utilizes the angular selective reaction of photoreactive units by irradiation with linearly polarized light (LPL) followed by the ordering owing to the cooperative self-assembling (Seki, 2007, 2004; Ikeda, 2003; Natansohn and Rochon, 2002; Ichimura, 2000).

8.2.1. Photoalignment of Polymer Main Chain of Polysilane

A vast amount of investigations on photoalignment of LCs has been reported during the past one and a half decades. Irradiation with LPL induces photoinduced anisotropy, in which molecules are generally oriented to an unexcitable direction, namely, perpendicular to the electric vector of the polarization direction. This effect is called photoinduced optical anisotropy (Seki, 2004; Natansohn and Rochon, 2002; Ichimura, 2000).

Exposure of the 6Az10-poly(vinyl alcohol) (PVA) monolayer to LPL induces an in-plane orientation of the Az side chain orthogonal to the polarization plane of LPL. A spincast film of poly(di-*n*-hexylsilane) (PDHS) is subsequently prepared onto this photooriented Az monolayer (scheme shown in Fig. 8.1, left). After

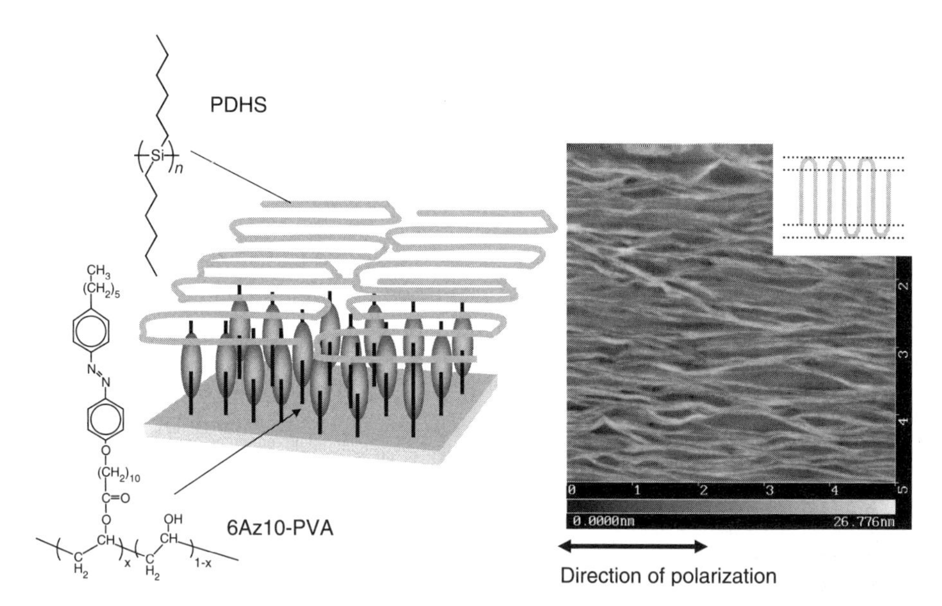

Figure 8.1. Photoalignment of PDHS film by an LPL-irradiated photoresponsive Az monolayer. Schematic illustration of the system (*left*) and topological AFM image (*right*).

crystallization, the PDHS film exhibits a strong in-plane optical anisotropy (Fukuda et al., 2002a,b; Seki et al., 1999). The polarized absorption spectra show that the Si backbone is highly aligned perpendicular to the polarization plane of the actinic light. The aligned direction of the PDHS main chain agrees with that of the Az orientation on the substrate. The crystallization of the PDHS chain occurs on the photooriented Az monolayer in an epitaxial manner. The atomic force microscopic (AFM) observation reveals that fine fibrous structure is evolved along the polarization direction. Most probably, PDHS chains form an edge-on lamella on the photooriented 6Az10-PVA monolayer. The main chains of PDHS are folded orthogonal to the polarization plane of LPL, and thus the resultant fibers are evolved parallel to it (Fig. 8.1, left) (Seki, 2007).

8.2.2. Surfactant–Silica Nanohybrids

Defined-sized mesoporous metal oxide materials are synthesized by templating organic molecular assemblies (Kresge et al., 1992; Yanagisawa et al., 1990). The templating rodlike micelles form lyotropic LC states, and therefore, the hexagonally packed rods are oriented by the structural anisotropy of the surface. The photoorientation of an Az monolayer is transferred to a photooriented spincast film of a polysilane as the first step (Seki et al., 1999), and then deposition of mesoporous silica film is performed (Kawashima et al., 2004, 2002). In this attempt, the photoorientation of the Az monolayer is first transferred to a photooriented spincast film of PDHS as mentioned earlier, and then oriented deposition of surfactant–silica hybrid film is performed (Fig. 8.2, left).

The formation of uniaxially aligned mesochannels is confirmed by transmission electron microscopy (TEM) for the photoaligning layer of 6Az10-PVA–PDHS. The two TEM photographs in Fig. 8.3 depict the cross-sectional structure of the as-synthesized surfactant–silica hybrid film observed in the two directions orthogonal to each other. Figure 8.3a shows the image when the film is sliced parallel to the direction of the actinic polarized light. The hexagonal structure of the mesochannels is clearly visualized through the overall thickness of the film. However, when the film is sliced perpendicular to the direction of the actinic polarized light, the lines with a periodicity of 3–4 nm running parallel to the substrate are observed (Fig. 8.3b). The periodic length coincides well with that of the hexagonal spacing obtained by X-ray diffraction (XRD). The cylindrical structure of the mesochannels is fully stretched straight in the film. Figure 8.3b shows the heterointerface regions composed of the Si wafer (the dark region at the bottom), Az monolayer (2 nm thickness, not clearly shown), PDHS film (40–60 nm thickness, brighter region), and deposited mesochannels with parallel stripes. The orientational information possessed by the Az monolayer (almost invisible in the photograph) is transferred to the alignment of PDHS with a thickness of ca. 50 nm and then to the mesostructured channels with a thickness of ca. 400–500 nm. Thus, large amplifications with respect to the controlled mass amounts are attained.

An alternative simpler way is to use a photocross-linkable LC polymer film containing a cinnamoyl unit and a mesogen in the side chain developed by

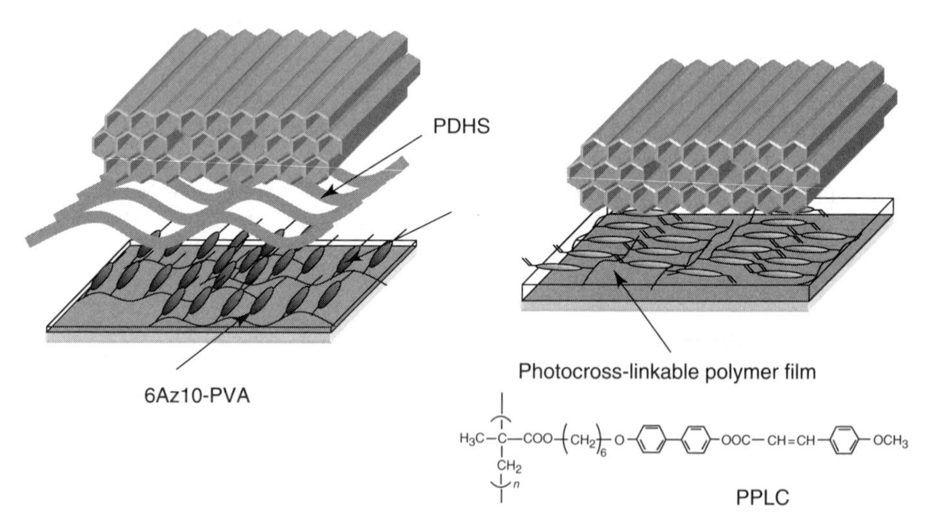

Figure 8.2. Photoalignment of silica mesochannels by photoreactive polymer layers based on the photoisomerization of Az (*left*) and photodimerization of cinnamoyl unit (*right*).

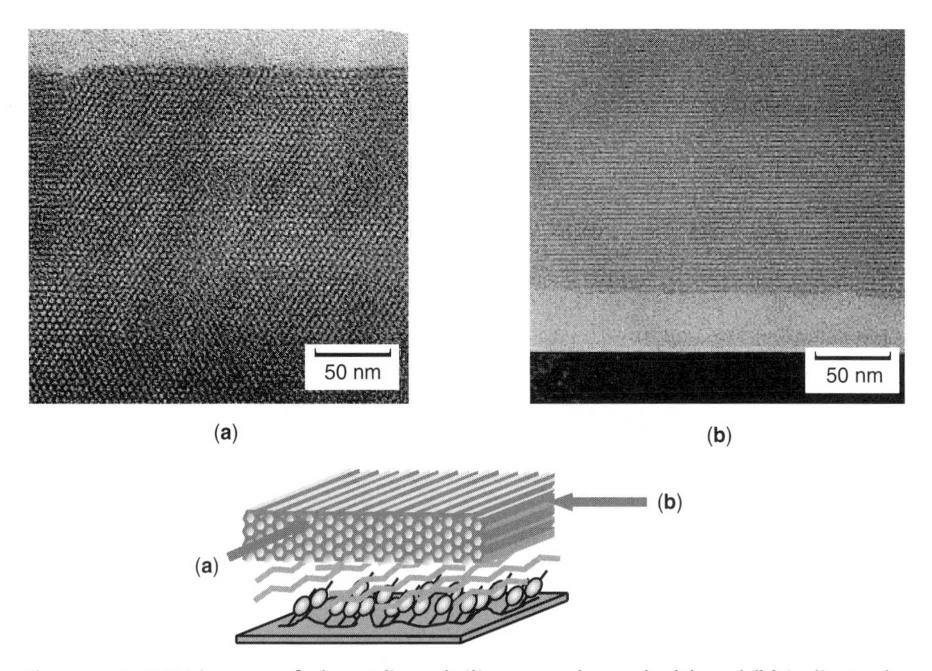

Figure 8.3. TEM images of photoaligned silica mesochannels. (a) and (b) indicate the directions of observation. Reprinted with permission from the Royal Chemical Society.

Kawatsuki (Kawatsuki et al., 2002) (Fig. 8.2, right). Morphological observations, X-ray diffraction measurements, and transmission electron microscopic observations indicate that the surface anisotropy imposed by LPL is transferred to the orientation of the rodlike micelle template; therefore, photoaligned mesochannels silica is obtained after removal of the organic component. Interestingly, even when the dip coating is applied, the channel orientation is predominantly controlled by the direction of photoaligning polymer layer and not by the lifting direction (Fukumoto et al., 2006). Thus the orienting power from the photoaligned polymer film is stronger than that of the flow orientation during the lifting. In a micropatterning experiment, the resolution of ca. 10 μm is obtained (Fig. 8.4) (Fukumoto et al., 2006, 2005).

8.2.3. Photoalignment of Chromonic LC–Silica Nanohybrid

Among vast varieties of lyotropic LCs, chromonic mesophases are the families of new research target starting only about two or three decades ago. The properties are different from those of ordinary lyotropic mesogens of surfactant type. The molecules constitute aromatic(s) in the center, and hydrophilic parts are scattered around the molecules rather than aliphatic structures. Therefore, they are rigid and planar disklike or planklike, rather than micelles of flexible molecular assemblies of ordinary surfactants. The molecules aggregate in solution to form

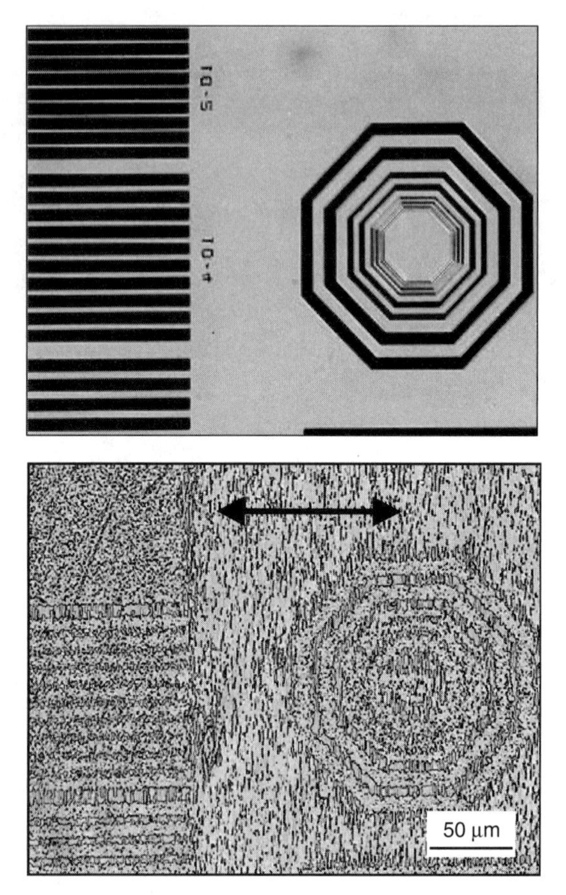

Figure 8.4. Patterned photoalignment of mesochannels using a photomask. Photomask (*top*) and the obtained silica film (*bottom*). Reprinted with permission from the American Chemical Society.

columns via the π–π stacking, rather than rodlike micelles via the hydrophobic effect. The resultant columnar structure of the dye aggregate forms lyotropic mesophases. Recently, such chromonic mesophases have been attracting great attention because of their potential utilities for fabrication of optical elements.

The photoalignment of azo dye aggregate by the command surface was first demonstrated by Ichimura et al. (1995). Further investigations reveal that the photoaligning of the resulting dry film is achieved in the lyotropic state (Fujiwara and Ichimura, 2002; Ichimura et al., 2002). The dried films, via chromonic LC of an anionic bis-azo dye, C.I. Direct Blue 67 (B67), have been found to be aligned and patterned by the alignment layer that is preirradiated with LPL (Ruslim et al., 2004; Matsunaga et al., 2003, 2002). The fundamental phase diagrams and mesomorphic properties of B67 have been reported by Ruslim et al. (2003). These innovative findings are reviewed by Lydon (2004).

Nanohybridization of B67 with silica network was recently attempted by Hara et al. (2007). The chromonic LC of B67 was mixed with silica precursor of tetraethoxysilane (TEOS) together with an anionic surfactant (EmalC20) and 2-(2-aminoethoxyl)ethanol (AEE). Here AEE plays an important role in stabilizing the columnar structure of B67. This molecule works as the mediator of the interface between the anionic B67 aggregate and silica. Without AEE, the columnar structure is transformed to a lamella structure in the process of drying (Hara et al., 2007).

The optical anisotropy of the nanohybrid films obviously obeys the direction of the preirradiated LPL to the photocrosslinkable polymer liquid crystal (PPLC) film, and the effect of flow orientation because the lifting from the sol solution is

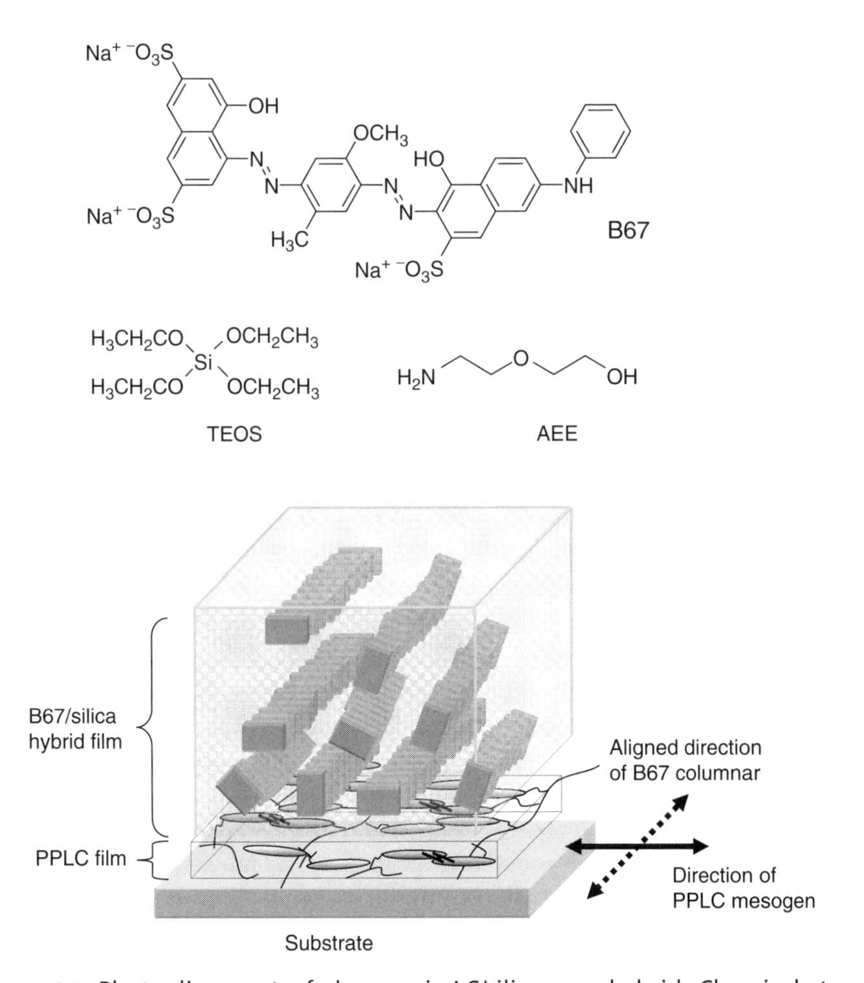

Figure 8.5. Photoalignment of chromonic LC/silica nanohybrid. Chemical structure of the component and a scheme of B67 columns on a photoaligned PPLC film. See color insert.

almost negligible. Polarized optical microscopy indicates the in-plane orientation of the dye aggregates over large areas. The spectroscopic data and in-plane XRD measurements indicate that the B67 molecules are stacked to form a column where the planar B67 molecules (light absorption transition moment) are stacked orthogonal to the column direction. The structure of the photoaligned B67 and columns are schematically indicated in Fig. 8.5. The molecular axis of B67 coincides with the direction of photooriented mesogens (parallel to LPL) of PPLC film, and therefore, the columns are extended orthogonal to the LPL direction. The orientation of the chromonic columns of B67 aligned by the LPL-irradiated PPLC film exactly agrees with that of rodlike micelles formed by ordinary surfactant aggregates (Fukumoto et al., 2006). However, the dip coating procedure *without* LPL irradiation obviously brings about macroscopic alignment of the B67 columns along the lifting direction (Hara et al., 2007). The macroscopic alignment should stem from the flow orientation in this case.

On the basis of these facts, micropatterning of the chromonics–silica hybrid can be attained utilizing a combination of photo- and flow orientation. A patterned irradiation with LPL onto the PPLC film is made through a photomask, and then the dip coating is achieved to prepare the chromonic–silica hybrid film. In this procedure, the directions of LPL and dipping are arranged to be consistent so that the orientation of the columns in the regions of irradiated and non-irradiated areas become orthogonal with each other.

8.3. SURFACE-GRAFTED AZ-CONTAINING LC POLYMER

In recent years, polymer brushes have become fascinating subjects in polymer research. They are defined as dense layers of end-grafted polymer chains confined to a solid surface or interface (Advincula et al., 2004). Recently, well-defined polymer brush structures with a high graft density have been developed using living radical polymerization, such as the atomic transfer radical polymerization (ATRP) method (Tsujii et al., 2006; Edmondoson et al., 2004). Since the ATRP can be performed using various types of monomers, many functional polymer brushes can be designed.

Introduction of LC properties into the surface-graft chains seems of great interest in view of fabrication of "smart" responsive surfaces (Hamelinck and Huck, 2005; Peng et al., 1999). Uekusa et al. (2007) have synthesized LC polymer brushes bearing an Az mesogenic group in the side group by adopting the surface-initiated ATRP method. The polymer brushes have characteristic properties quite different and unpredictable from those of the spincast films. A unique orientation behavior of a smectic LC state in the thin films prepared by the surface-initiated polymerization is indicated. UV-visible absorption spectroscopic measurements reveal that the Az mesogens adopt a parallel orientation in the film, which is in sharp contrast to that in a spincast film of the identical polymer. A normal orientation of the mesogenic groups is commonly observed for the spincast films of smectic LC Az polymer (Han et al., 2000, 1999; Geue et al., 1997).

A 2-D XRD profile of the grafted polymer film recorded on an imaging plate is depicted in the top portion of Fig. 8.6. Diffraction patterns are observed only in the in-plane direction, that is, the periodic structures derived from the smectic LC phase are perpendicular to the substrate plane with LC mesogens being aligned parallel to the substrate. An in-plane orientation of mesogens is often observed in a microphase-separated state of LC block copolymers (Mao et al., 1997); however, to our knowledge, this type of smectic LC orientation in a polymer film was observed for the first time. In the one-dimensional (1-D) pattern, a clear peak is observed at $2\theta = 2.48°$, which corresponds to a layer spacing of 3.56 nm. This spacing corresponds to a long-range ordering of approximately the length of the fully extended Az side chain, suggesting the formation of interdigitated structures of Az side groups. In contrast, a diffraction pattern on the spincast film of the identical homopolymer is observed only in the out-of-plane direction at $2\theta = 2.73°$ ($d = 3.26$ nm) (Fig. 8.6, bottom). On the basis of the spectroscopic and X-ray data, schematic illustrations of the structure of the surface-grafted and spincast polymer film are shown in Fig. 8.7.

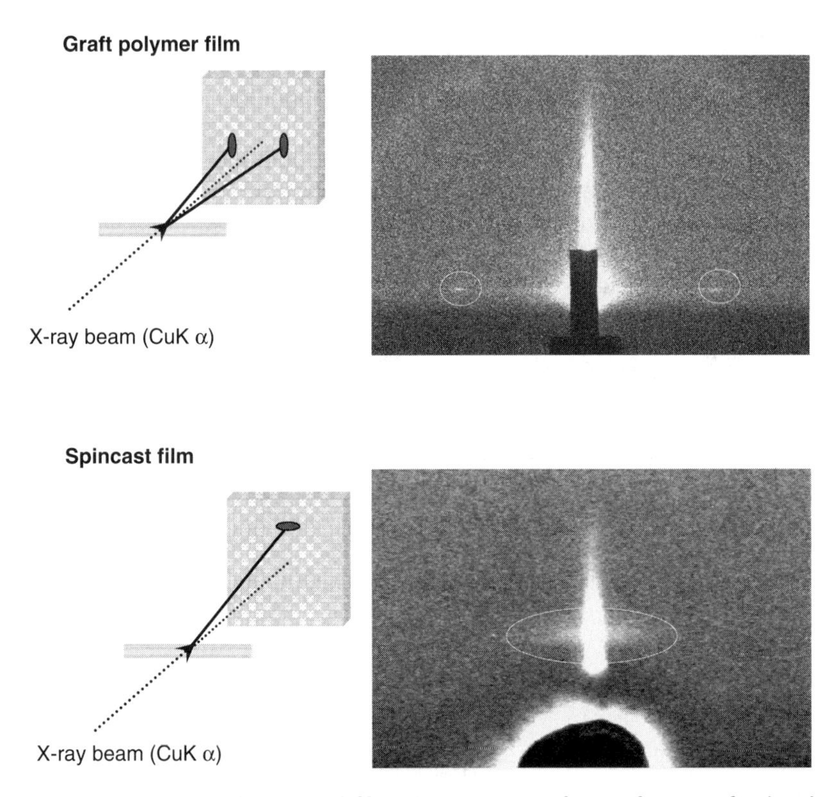

Figure 8.6. Grazing-angle X-ray diffraction patterns for surface-grafted polymer (*top*) and spincast (*bottom*) thin films of LC Az polymer. Reprinted with permission from the American Chemical Society.

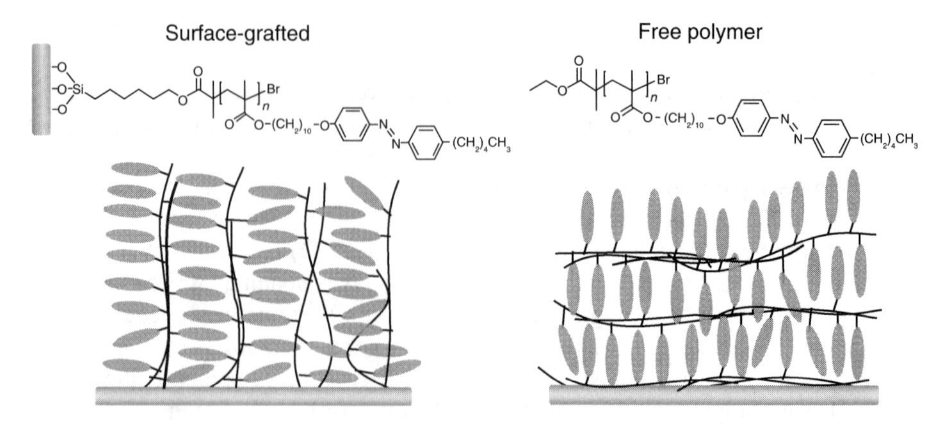

Figure 8.7. Schematic illustration of mesogen and smectic layer orientations of surface-grafted polymer (*left*) and spincast (*right*) thin films of LC Az polymer.

In the surface-grafted LC polymer film, thermophysical properties are similar to those of a spincast film. The in-plane orientation of Az groups in the surface-grafted polymer film may lead to different photoresponsive nature in the film. Since the photoresponsive mesogens are oriented parallel to the substrate plane, the film may be more sensitive to the polarization of LPL. In fact, LPL induces stronger in-plane anisotropy in the surface-grafted film than in the ordinary cast film. It is anticipated that this newly designed light-responsive surface may find many new opportunities and applications.

8.4. PHOTOGENERATED MASS MIGRATIONS

8.4.1. Conventional Type

Surface relief gratings (SRGs) (regular topological surface modification) formed via irradiation with an interference pattern of coherent argon ion laser beam (Fig. 8.8) were found about a decade ago (Kim et al., 1995; Rochon et al., 1995) and is perhaps the most interesting target in the current research of Az polymers. A great deal of data has been accumulated rapidly because of its basic phenomenological interest and its technological applications (Yager and Barrett, 2001;

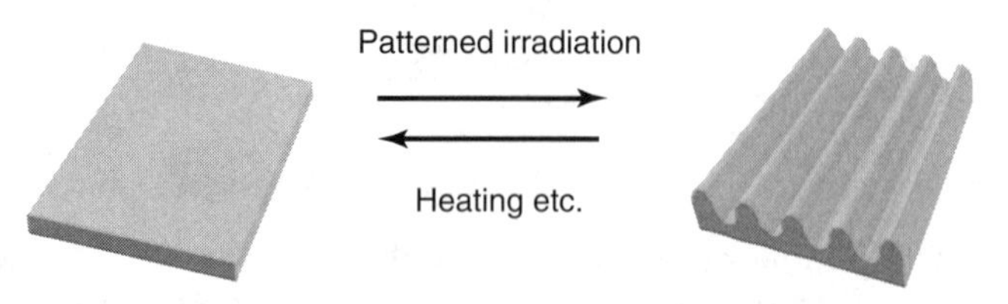

Figure 8.8. Photoinscription of surface relief structure by patterned irradiation.

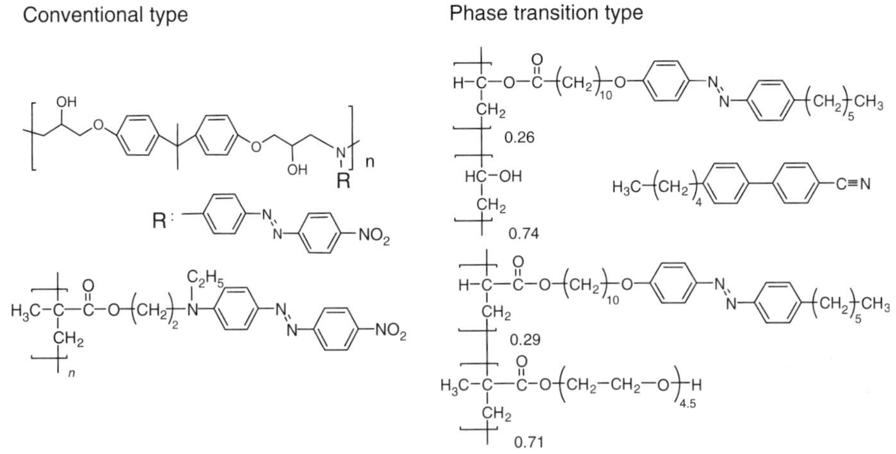

Conventional type

Phase transition type

Figure 8.9. Structure of typical polymers employed for SRG inscription studies.

Viswanathan et al., 1999). Some chapters in this book deal with this fascinating research topic. Features of this phenomenon can be summarized as follows: (i) The film material transfer occurs over micrometer distances, far beyond the dimension of single polymer chain, are achieved, and (ii) the inscribed relief can be erased by heating above a softening temperature or uniform irradiation of circularly polarized light (CPL). Most of the polymers employed for these studies possess a push–pull type Az unit such as Disperse Red 1, and the physical state is amorphous in nature with no particular regular structure in the film (Fig. 8.9). Other characteristics are further listed in Table 8.1. Most investigations have

TABLE 8.1. Features of Mass Transfer Behavior Observed in the Two Types of Az Polymer Systems

Items	Conventional type	PT type
Physical state	Mostly amorphous	Smectic LC (in the trans state)
PT on illumination	No	Yes
Typical exposure energy	ca. 10^4–10^5 mJ cm^{-2}	ca. 10^2 mJ cm^{-2}
Polarization dependence	Yes	No
Substituents of Az	Push–pull type (short-lived cis state)	Electronically small affect (long-lived cis state)
Erasing procedure	Heating above T_g Irradiation with CPL	Heating to isotropic state Irradiation with UV
Motion without irradiation	No	Yes
Motion via sensitization	No data	Yes
Procedure	One-step exposure	Mostly two-step exposure
Essential driving factor	Photon-promoted	Thermal via self-assembly
Suitable terminology	Photoinduced	Phototriggered

employed Az-containing polymer materials, but new systems such as amorphous molecular materials (Nakano et al., 2002) and spiropyran-doped polymer films (Ubukata et al., 2007) have also been reported.

8.4.2. Phase Transition Type

Another type of photoactivated mass migration system has been first proposed by Ubukata (Ubukata et al., 2000) and extended by Zettsu and coworkers (Zettsu and Seki, 2004; Zettsu et al., 2001), using soft LC polymer systems such as binary component materials and random copolymers containing an oligo(ethylene oxide) (EO) segment (Fig. 8.9). The hybrid films are irradiated with nonpolarized UV (365 nm) light in advance to attain a cis-rich photoequilibrated state (UV light treatment). Starting from this state, an argon ion laser beam (488 nm) or a 436-nm line from a mercury lamp, which induces the isomerization to the trans form, is irradiated to the film. The mass migration is completed at surprisingly small dose levels ($<100\,\mathrm{mJ\,cm^{-2}}$), which are three orders of magnitude smaller than those required for the conventional amorphous polymer systems. This fact strongly suggests that the migration mechanism of this type fully differs from that of the conventional type. A very recent study revealed that the photochemical phase change (Ikeda, 2003) of the smectic LC to isotropic phase is essential for the migration (Zettsu et al., 2007a). Therefore, this class of materials can be dubbed as the phase transition (PT) type.

8.4.3. On the Migration Features of the PT Type

With regard to the migration mechanisms for the conventional type, a number of models have been proposed. The details are discussed in the literatures (Natansohn and Rochon, 2002; Sekkat and Knoll, 2002; Yager and Barrett, 2001; Viswanathan et al., 1999). In the conventional type, considerations of electromagnetic force given by photons seem to explain the migration behavior most appropriately. For the PT type, however, no summary has been reported. Therefore, a brief summary will be made here.

The properties of the PT-type motions together with those of the conventional type are summarized in Table 8.1. There are a number of different issues in the migration feature for the two types. The mass migration of PT type has no polarization dependence, which shows a marked contrast with the conventional type. Two beam experiments show that the (*p*-:*p*-) (intensity and polarization holography) and (*s*-:*s*-) (pure intensity holography) mode interference equally cause the efficient migration, but (*s*-:*p*-) mode (pure polarization holography) does not lead to any surface relief formation (Zettsu et al., 2003) (Fig. 8.10). Other notable features in the PT type are the occurrences of continuing motions even after ceasing the illumination (Ubukata et al., 2004), and migrations are also promoted via sensitized excitation from a near infrared absorbing dye (Zettsu et al., 2007a). In these contexts, the essence of the migration in the PT type should be thermal processes via self-assembly of the film material, and the electromagnetic

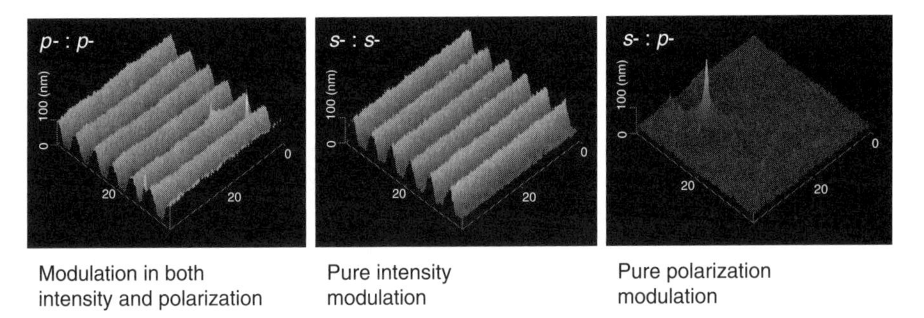

Figure 8.10. SRG formation by double-beam interference of coherent argon ion laser in different polarization modes.

effect (gradient force) considered for the conventional type is, if any, negligible. In this sense, the term, "phototriggered" should be more appropriate than "photoinduced" to express the PT-type system. Since the mass migration is coupled with the PT of the film, the process seems be closely related to the bending behavior observed in the large-sized film (Ikeda et al., 2007; Kondo et al., 2006; Yu et al., 2003). The micropatterned deforming motions on the film surface may explain the relief formation. However, precise understandings of the mechanism for the PT type are still the subject of future investigation.

8.4.4. Extended Studies in the PT-Type Mass Migration

8.4.4.1. Fixation of the Relief Structure. Facile migration and improved shape stability are exclusive features. These contradictory requirements can be fulfilled by subsequent cross-linking. After the surface relief structure is formed for the copolymer with OH-terminated EO chain, the polymer is subjected to chemical cross-linking via formalization (acetal formation with formaldehyde) between the OH groups (scheme shown in Fig. 8.11). The temperature dependency of the first-order diffraction intensity of a monitor beam and AFM observations justify the remarkable stability improvement after the cross-linking (Fig. 8.11). The SRG structure in the untreated film disappeared upon heating at 80°C, whereas the surface modulation pattern in the formalized film remained even at 250°C without any damage. The process employed here can be compared with a simple approach using Az polymers of high glass transition temperature (T_g). Fukuda and Matsuda et al. (2000) have employed maleimide-based high T_g amorphous polymer (T_g = 170–279°C). In their polymer systems, the thermal stability is considerably improved; however, vast amounts of exposure energy are required. The post-cross-linking is a unique strategy, in which mass migration is achieved in a soft state and immobilization can occur at any desired stage, providing marked thermal stability as compared with high T_g polymers. Without the cross-linking, the relief structure can be fully erased by illumination with UV light and regenerated many times essentially with no deterioration (Zettsu and Seki, 2004).

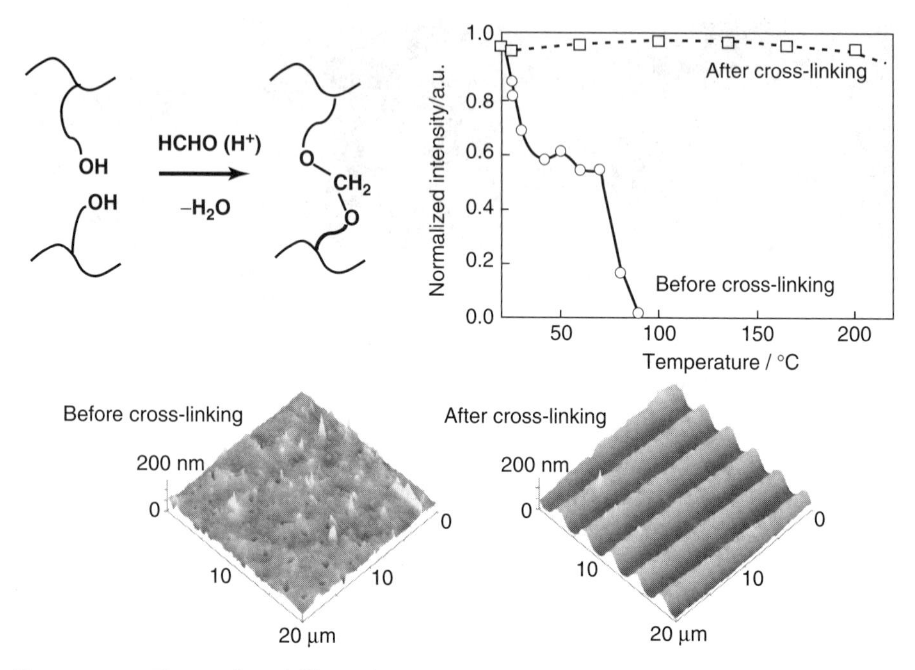

Figure 8.11. Thermal stability of SRG structure before and after cross-linking. Evaluation of temperature dependency (*top*) of light diffraction (*top right*) and topological AFM images after heating at 250°C for 2 h (*bottom*). Reprinted with permission from the American Chemical Society.

Once the cross-link is formed in the SRG film, the photoisomerization to the cis form does not alter the film morphology. This behavior is in marked contrast to the film before cross-linking, in which UV irradiation erases the SRG structure. Reversible photoisomerization occurred with retention of the SRG structure (Fig. 8.12, top). When low-molecular-mass LC was placed on the fixed film, the LC cell reversibly aligned changes between the homeotropic and homogeneous states upon alternate visible–UV light irradiation (Fig. 8.12, bottom) (Zettsu and Seki, 2004). In the homogeneous state, the LC molecules aligned parallel to the undulations of SRG. Here, the in-plane direction was controlled without irradiation with LPL. The surface geometry of the film governs the LC alignment.

8.4.4.2. Conveyer Effect. Thus far applications of photoinduced migrations considered have been almost limited to utilization of the resultant static relief structures. They involve holographic recording, phase mask, LC alignment, waveguide couplers, fabrication of intricate surfaces, etc. In the PT-type mass-migrating system, the motion itself can be a valuable function. An application as a "molecular conveyer" for patterning of light-inert (non-photo-responsive) functional materials can be proposed (Ubukata et al., 2004) (Fig. 8.13).

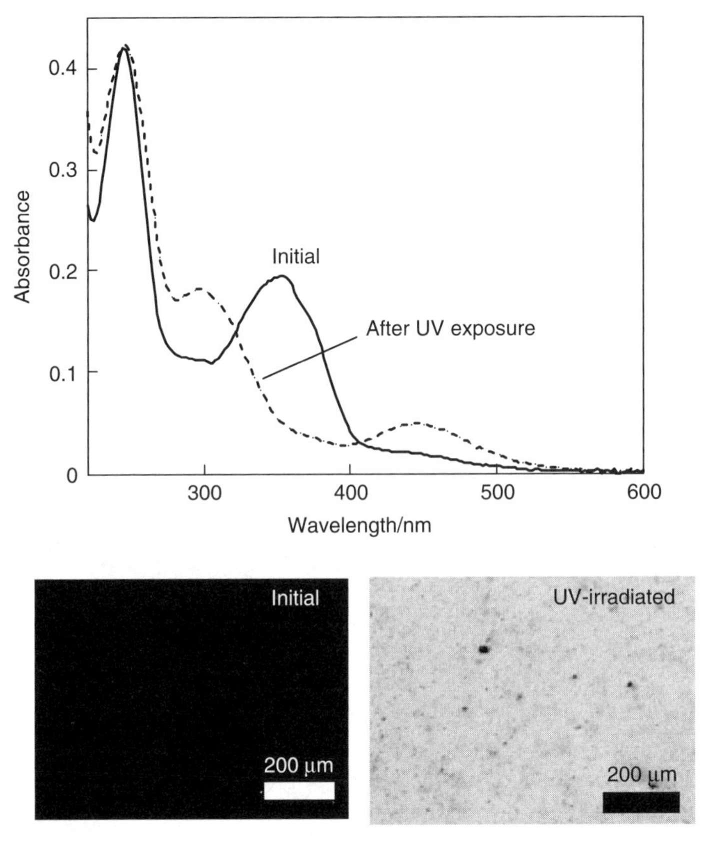

Figure 8.12. UV-visible spectra after cross-linking (*upper*) and LC alignment behavior as the command surface. Upon UV irradiation, LC is aligned in parallel with the SRG structure. Reprinted with permission from the American Chemical Society.

In the dye-conveying system at higher dye content, UV irradiation followed by dark adaptation at room temperature for a day led to a growth of dendritic crystals of the dye. When crystallization was performed after photoinscription of the lined relief (Fig. 8.13, left), the direction of dye organization was restrained and aligned along the relief structure. After the crystallization at room temperature, the protruding line shows a strong optical anisotropy. The polarized optical microscope image indicates that the dichroic dye aligns in the direction of the inscribed line. The absorption moment of the dichroic dye is highly aligned parallel to the direction of the inscribed line. Most probably, the orientational anisotropy is induced by a uniaxial growth in the diffusion-limited crystallization restricted in the 1-D geometry (Ubukata et al., 2004).

Application of the mass conveying principle is not limited to low-molecular-mass molecules; a polymer material, poly(dodecyl thiophene) (Zettsu et al., 2004) (Fig. 8.13, center), and an inorganic material, for example, semiconductor

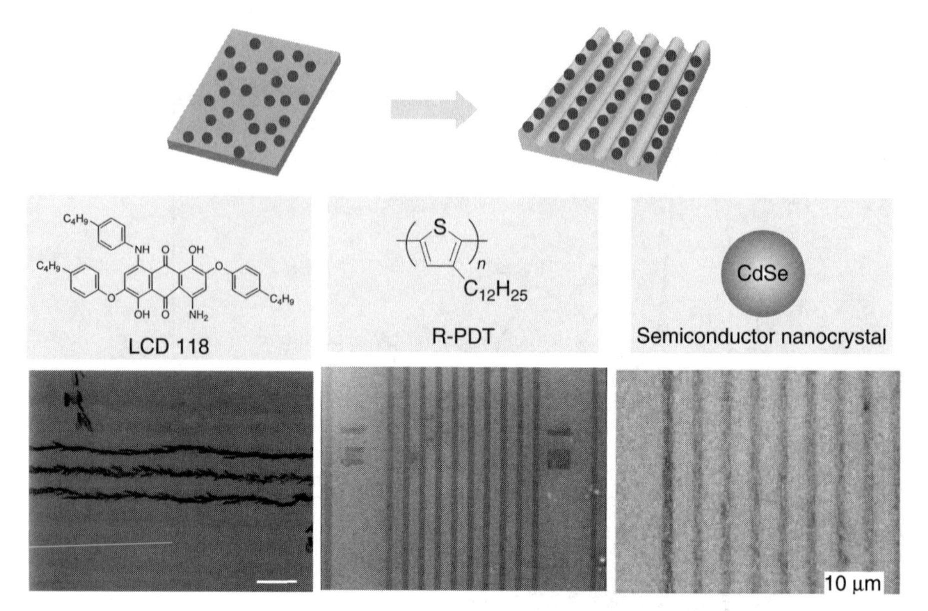

Figure 8.13. Phototactic transport of functional materials (dye, π-conjugated polymer, and nanocrystal) by conveyance actions. See color insert.

nanocrystals (Ubukata et al., 2004) (Fig. 8.13, right) can also be conveyed. The conveying action may be widely applicable to many other kinds of functional materials that are themselves inert to light. A great advantage of using the instant migration system is that only small amounts of light are required, which would not damage the host materials.

8.4.4.3. Supramolecular Mass Migration Systems. Azobenzene unit is essential for the photoinduced mass migration, but after the relief formation, the existence of this strongly light-absorbing chromophore will be a drawback for many optical applications. To overcome these contradictory requirements, a system in which the Az unit can be detached after the relief formation is proposed (Zettsu et al., 2007b). Here the Az unit is not necessarily linked to the polymer backbone, but the supramolecular LC framework via hydrogen bonding (Kato system) (Kato, 2002; Kato and Frechet, 1989) is applied. After the cross-linking mentioned earlier, the Az unit is readily removed from the film with retention of the morphological feature of the relief structure (scheme shown in Fig. 8.14). The structures of the polymer and guest compound used in this study are also shown in Fig. 8.14. An Az compound, 8Az-4Im, with an imidazole base at the terminus and a random copolymer consisting of poly(acrylate) with 4-oxybenzoic acid moiety should work as the H-bonding acceptor and donor, respectively. The H-bonded complex is prepared from a mixed pyridine solution containing 8Az-4Im and pBA6Ac-PE4.5. Infrared spectroscopy and thermal analysis indicate that the hydrogen bonding actually occurs between imidazole and benzoic acid.

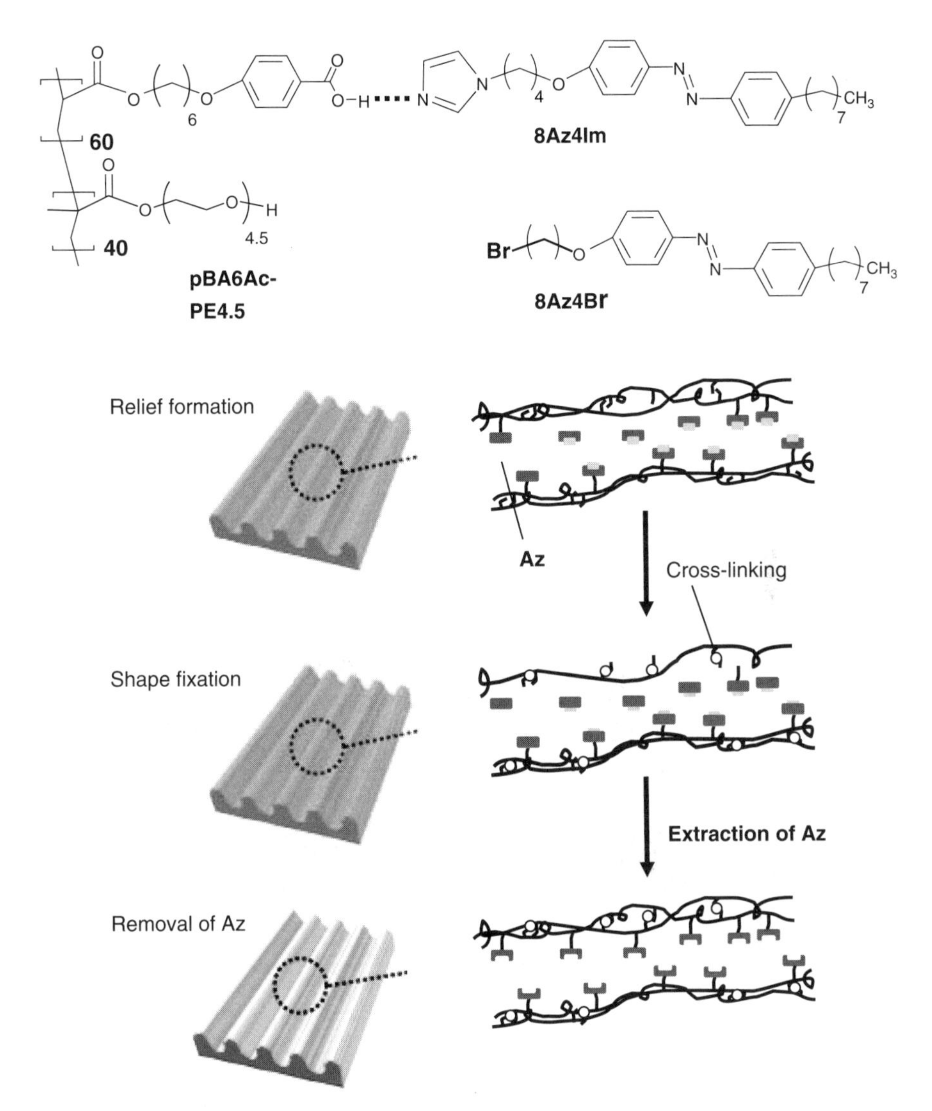

Figure 8.14. Chemical structure of H-bonded supramolecular LC polymer film and scheme of the strategy for selective detachment of Az unit after the relief formation.

The mass migration is readily formed also for this supramolecular system. This fact implies an important issue that the covalent linkage of the Az unit to the polymer chain is not necessary to cause the efficient phototriggered material transfer, but H-bonding can be the exact alternative. As a control experiment, 8Az-4ImBr (Fig. 8.14), a precursor compound of the imidazol compound, is also examined. No SRG formation is observed, obviously showing the importance of H-bond formation.

The selective extraction of 8Az-4Im from the film is performed after the chemical post fixation of the SRG structure. The H-bonded acceptor 8Az-4Im is extracted selectively by ethanol or THF from the film with retention of the periodic structures. Figure 8.15 shows the UV-vis spectrum, and topographical AFM images taken from the sample after the cross-linking procedure (initial) and

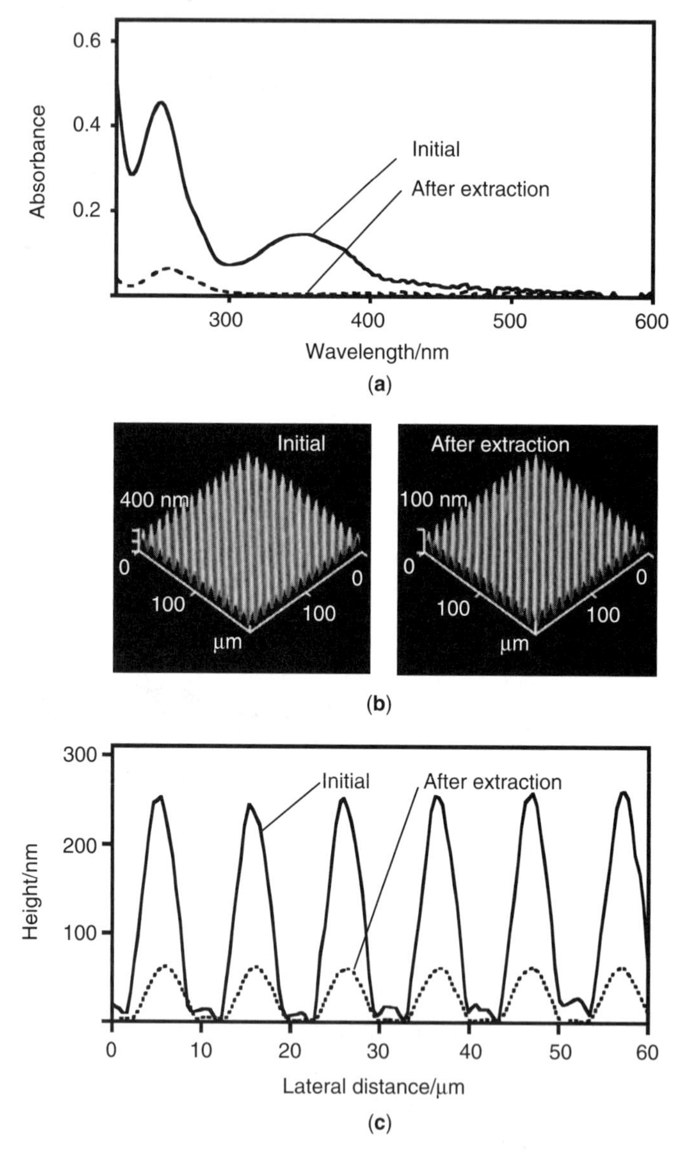

Figure 8.15. Selective extraction of Az unit from the SRG structures. (**a**) UV-vis absorption spectrum and (**b**) topographical AFM images and (**c**) their cross-sectional profiles taken from the sample after the cross-linking procedure (initial) and subsequent rinsing with THF (after extraction).

subsequent rinsing with tetrahydrofuran (THF) (after extraction). As shown, the absorption maximum of the π–π* band in the trans form of Az positioned at 356 nm fully disappears after the extraction. Thus, 8Az-4Im molecules are completely eliminated from the film. AFM scans reveal that the periodic structure is exactly maintained on the film. The selective extraction of the 8Az-4Im molecules gives rise to a drastic reduction in the peak-to-valley height difference of SRG from 250 to 50 nm because of the volume contraction. The degree of volume contraction almost agrees with the loss of the 8Az-4Im mass. The present proposal can be a facile and versatile method for bleaching the SRG film.

8.5. PHOTORESPONSIVE LC BLOCK COPOLYMER SYSTEMS

Microphase separation (MPS) leading to regular patterns at nanometer levels formed by block copolymers in thin films has recently been a subject of intensive study. Such nanostructures allow for fabrication of even smaller feature sizes than those obtained by the conventional photolithography process, and have potentials for future nanofabrications (Lazzari et al., 2006). Attempts to create active and photocontrollable MPS systems are a fascinating challenge.

8.5.1. Monolayer Systems

Earlier investigations revealed that PVAs possessing an Az-containing side chain show large photoinduced area changes on a water surface (Seki et al., 1998, 1996). In a particular case, the monolayer exhibits ca. threefold expansion and contraction by alternating illumination with UV and visible light in a fully reversible manner. This effect can be explained by the polarity change of the Az unit. The molecular packing model exactly explains the degree of area changes, indicating that the macrosize effect correctly reflects the molecular events. In situ X-ray reflectivity (XR) analysis on the water surface by Kago et al. (1999) revealed the actual thickness change on water. Quite recently, Ohe et al. (2007) directly evaluated the photomodulated orientations of the individual chemical bonds of this polymer in detail by sum frequency generation spectroscopy.

It is anticipated that, when one of the blocks is composed of a photoresponsive segment (area-variable component) in the monolayer state, the change in the area fraction of the blocks may alter the MPS nanostructures. By utilizing the photomechanical response of the monolayer mentioned earlier, such light-induced modulations in the 2-D nanostructures have been actually attained (Kadota et al., 2005) (Fig. 8.16). The trans and cis isomeric states of Az result in dot and strip structures, respectively, of the Az-containing domain in a reversible manner. This tunable behavior of the morphology provides an important aspect on the chain conformation on water. The middle block of poly(ethylene oxide) (PEO) chain should adopt predominantly loop conformations rather than bridge ones. This fact coincides with the tendency of segregation rather than interpenetration of polymer chains in the 2-D state (Sato et al., 1998). At highly humidified state, the morphological changes on a hydrophilic solid surface of mica are also observed

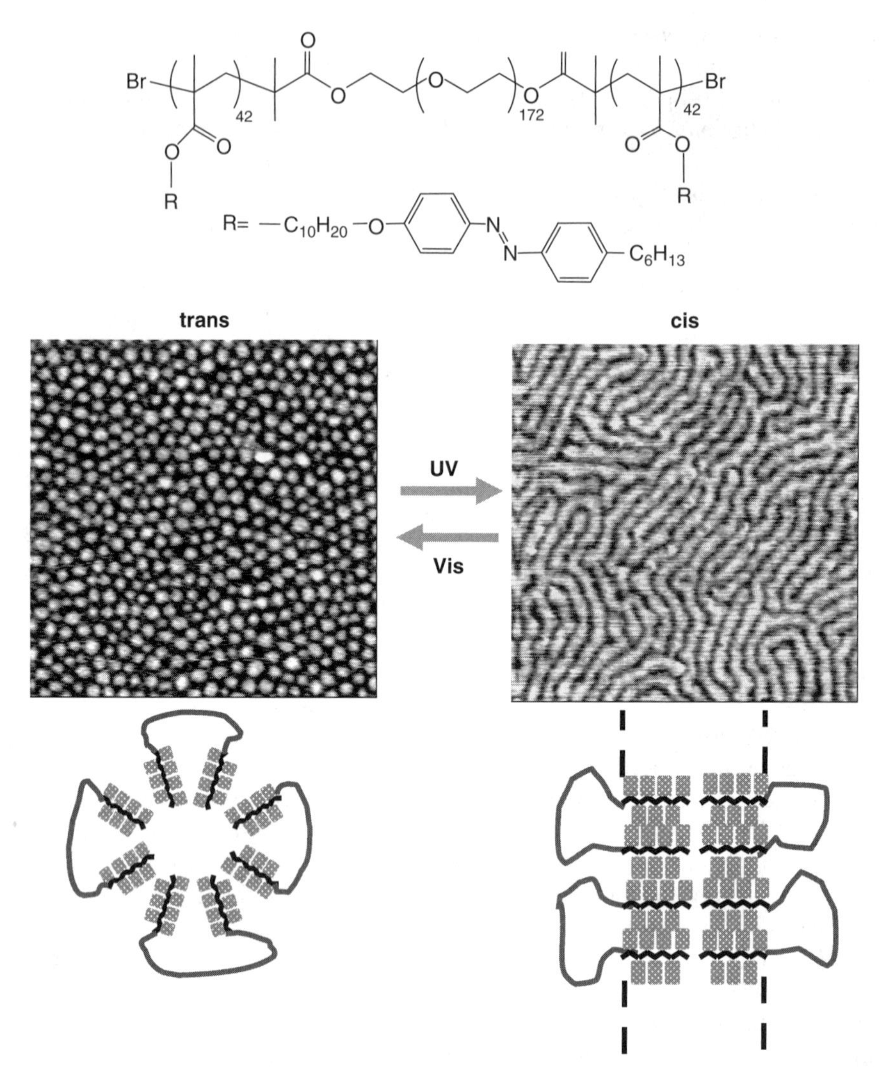

Figure 8.16. Photocontrolled 2-D MPS nanostructure of an ABA triblock copolymer on water and supposed aggregation state. Reprinted with permission from the American Chemical Society.

essentially in a reversible manner (Fig. 8.17) (Kadota et al., 2006), although the morphological alternations are not sufficient because of the suppressed motional freedom on the solid surface.

8.5.2. Photocontrolled Macroscopic Alignment of MPS Structures

Many efforts have been made to induce macroscopic alignment of the MPS patterns (Lazzari and De Rosa, 2006). When block copolymers with LC nature are

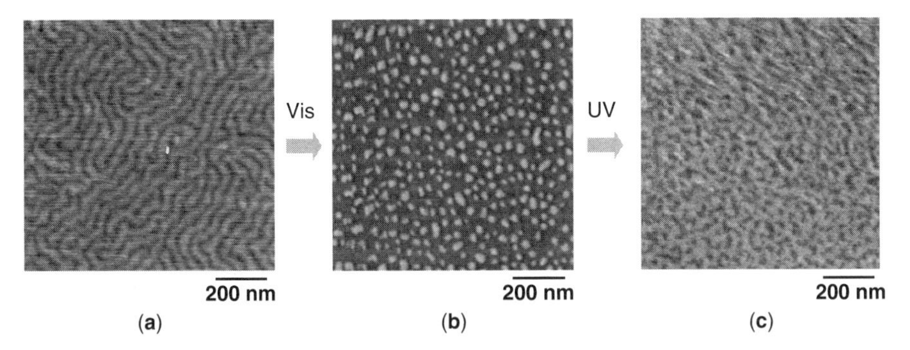

Figure 8.17. Photoinduced modification of 2-D MPS nanostructures of the triblock copolymer monolayer on mica. (a) Initial cis-rich state, (b) trans-rich state after UV irradiation, and (c) successive cis-rich state after visible light irradiation.

employed, the regular MPS structure is formed over large areas by the cooperative effect. The importance of liquid crystallinity for large-scale alignment of MPS structure is demonstrated by Iyoda and coworkers (Tian et al., 2002). This section introduces new attempts to photoalign the MPS structure by utilizing the orientational cooperative motions of LCs.

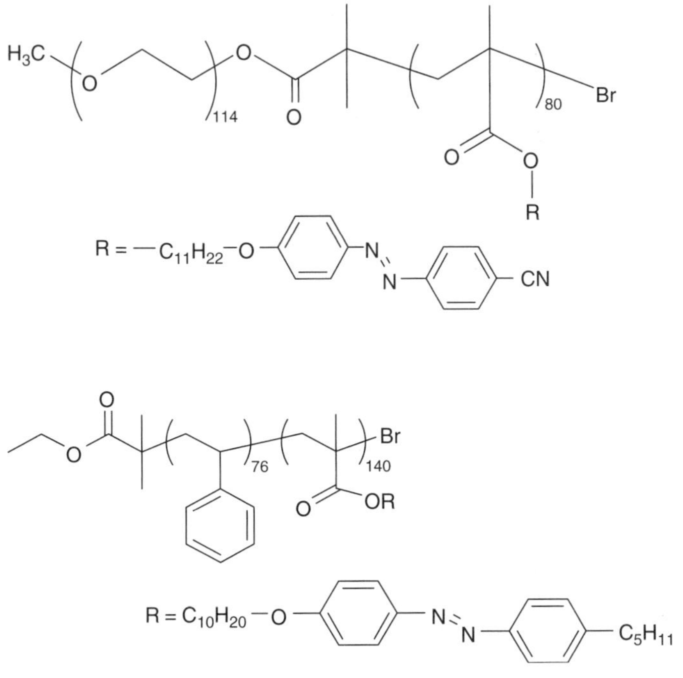

$R = -C_{11}H_{22}-O-C_6H_4-N{=}N-C_6H_4-CN$

$R = C_{10}H_{20}-O-C_6H_4-N{=}N-C_6H_4-C_5H_{11}$

Figure 8.18. Structure of Az-containing LC block copolymers employed for photocontrol of MPS nanocylinders.

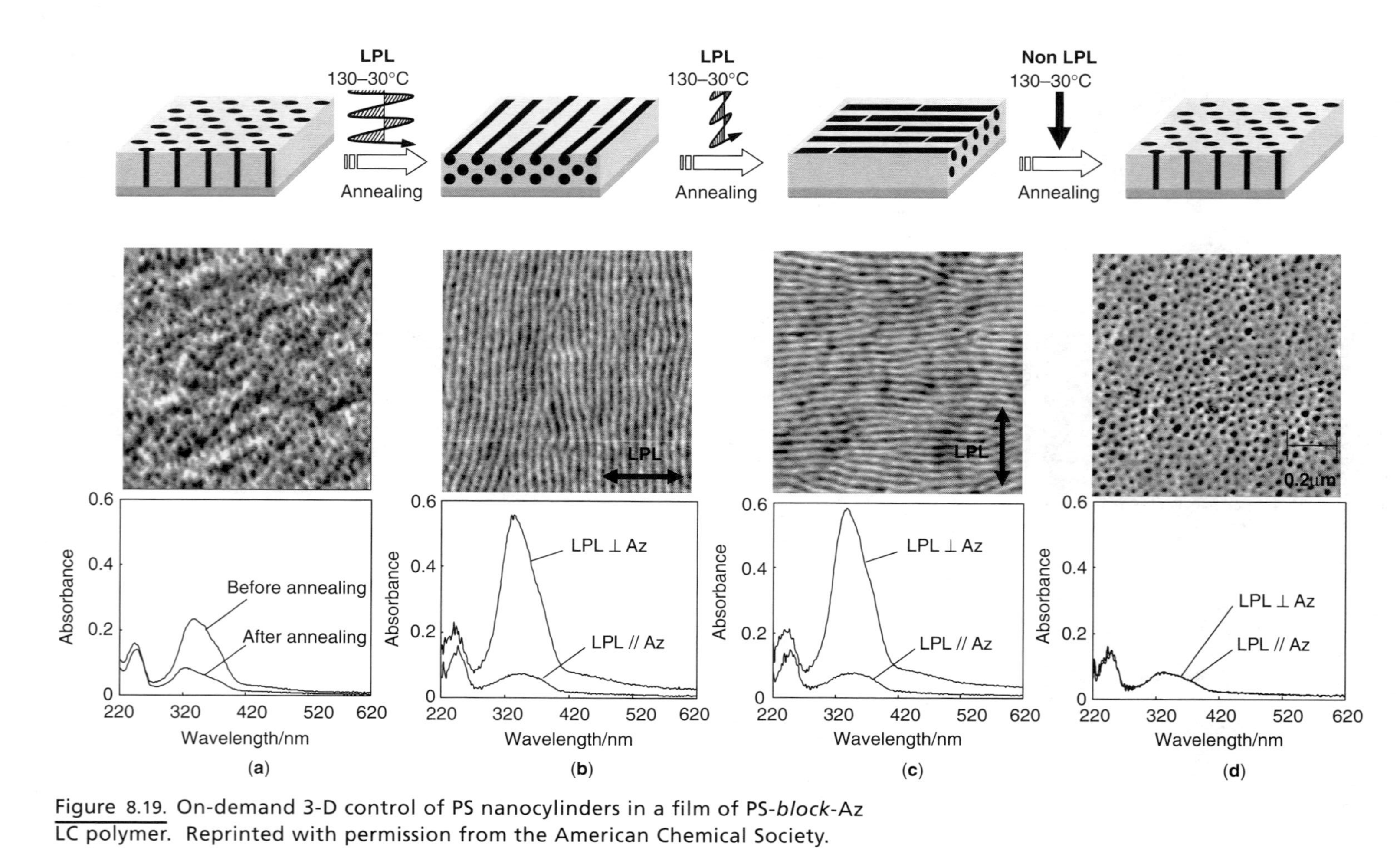

Figure 8.19. On-demand 3-D control of PS nanocylinders in a film of PS-*block*-Az LC polymer. Reprinted with permission from the American Chemical Society.

An obvious effect of LPL irradiation to align MPS in block copolymer films is demonstrated for a soft PEO-based Az-containing LC block copolymer film by Yu et al. (2006) and a polystyrene (PS)-based block copolymer of higher T_g by Morikawa et al. (2007; Fig. 8.18). The MPS nanocylinders of the light-inert blocks aligned orthogonal to the direction of the electric field vector of the irradiated LPL. As a rigid segment polymer, adjustment of annealing temperature is an important factor for the successful alignment. The macroscopically aligned MPS structure evolves at an annealing temperature slightly above T_g of PS block and below the smectic to isotropic transition of the LC Az polymer block. It is of note

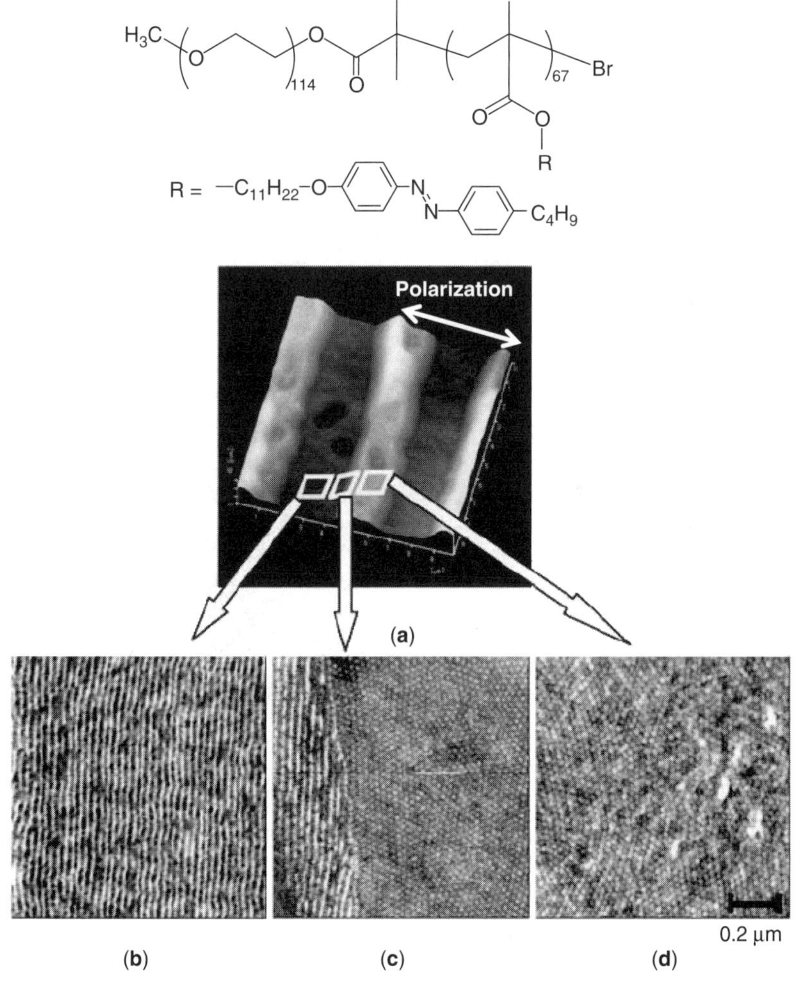

Figure 8.20. Hierarchical structures involved in photogenerated SRG of a block copolymer. (**a**) Topographical image, (**b**) phase images at a trough, (**c**) boundary, and (**d**) crest regions.

that the MPS alignment can be altered successively on demand both in the in-plane and out-of-plane directions (Morikawa et al., 2007). This tunable procedure is a marked feature attained in the photoprocess (Fig. 8.19).

8.5.3. Micropatterning of MPS Structure in the Hierarchical Structure

So far photoinduced or phototriggered mass migrations (see Section 8.4) have been achieved for homopolymers or random copolymers. Quite recently, block copolymers have also been subjected to mass migration (Morikawa et al., 2006; Fig. 8.20). The key for the out-of-plane alignment (whether the cylinders orient normal or parallel to the substrate surface) is the control of film thickness and that for the in-plane control is the adjustment of direction of LPL. This polymer adopts hexagonally packed PEO cylinders orienting normal to the substrate plane over long distance ranges in thin film because of its liquid crystallinity (Tian et al., 2002). However, the nanocylinders orient parallel to the substrate plane when the thickness becomes $<70\,\text{nm}$ (Morikawa et al., 2006, Watanabe et al., 2006). For other systems, thickness change of a polymer film $<100\,\text{nm}$ is found to alter the nanostructures and their orientations for block copolymer systems (Knoll et al., 2002) and also influence the orientation of mesogenic groups with respect to the

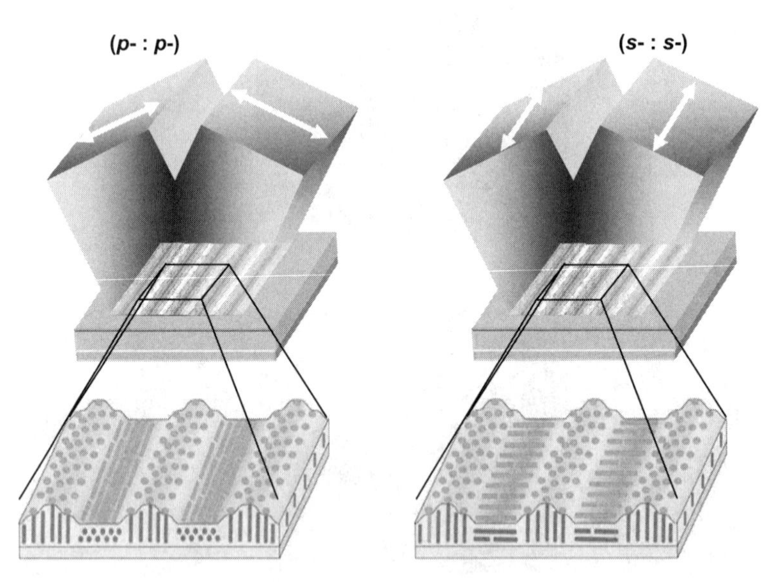

Figure 8.21. Schematic illustration of hierarchical structures involved in photo-generated SRG of a block copolymer after irradiation with (*p*- : *p*-) and (*s*- : *s*-) mode interference laser beam. Note that the in-plane direction of the cylinder is clearly controlled by the polarization mode. The out-of-plane change is because of the thickness variation. See color insert.

substrate (Mensinger et al., 1992) in a smectic polymer film. The thickness-dependent orientational change of the nanocylinders observed here seems to be closely related to these phenomena.

Then the two types of holographic irradiation [(*p*-:*p*-) and (*s*-:*s*-)] are performed. When the thickness after the migration is adjusted as to be above and below 70 nm in the thick and thin regions, respectively, the out-of-plane control is achieved, namely, the cylinders align normal or parallel to the substrate plane depending on the relief thickness. The difference in the holographic irradiation mode of (*p*-:*p*-) and (*s*-:*s*-) leads to contrasting in-plane orientations of the laid cylinders in the thin regions. The (*p*-:*p*-) and (*s*-:*s*-) mode interferences provide cylinders oriented parallel and orthogonal to the relief undulations, respectively (scheme shown in Fig. 8.21). Thus, both out-of-plane and in-plane orientations of MPS structure are controlled by the simple interference irradiation of laser beam. It is stressed here that a single irradiation leads to the formation of three different hierarchy levels, namely, the molecular orientation (subnanometer level), MPS structure (several 10 nm), and SRG (micrometers).

8.6. CONCLUSION AND SCOPE

This chapter summarizes several topics that are currently progressing in the research areas of photoresponsive monolayers and LC ultrathin polymer films.

The surface photoalignment method using "command surface" is now extending to a branch of organic–inorganic nanohybrid systems. Thus, surface-mediated processes provide new opportunities to exert micropatterning of simple components to various composite systems. Such approaches would be of particular use in fabrication of optical devices.

Surface-grafted LC photoresponsive polymer film can be a new target of LC film research. The orientation of the Az mesogenic group strongly influences the photoresponse behavior and sensitivity. Since the planar-arranged Az mesogens should be more sensitive to LPL, fabrications of highly sensitive photorecording media may be realized. Furthermore, smart photoresponse functions such as photoswitchable anisotropic friction properties may be anticipated.

In the LC molecular and polymer assembly systems, the light trigger is effectively converted into the molecular motions to induce orientation and migration in cooperative manners. Motions are basically driven by thermal self-assembly processes of the soft materials. Suggestive examples can be seen in the mass migration systems. In the LC PT type, the relief formation is most likely to be the consequence of the microscopic deformation of the film occurring at the LC–isotropic interface formed within the film. This situation is in marked contrast to the migration mechanism in the conventional amorphous materials. As shown here, the detachment of Az chromophore (bleaching the color) will be a promising strategy to expand optical applications of the photogenerated relief structure.

With respect to the alignment control, the marked research advance in the block copolymer systems is to be emphasized. Here, the photoalignment is not limited at the molecular level but amplified to regulate the MPS structure of the larger feature size. Understanding the interplay between different size hierarchies with retention of dynamic features should be a significant issue to create future soft materials. The nanopatterned surface of block copolymer films can be applied as a template surface to fabricate various functional materials, including biological macromolecules (Kumar and Harn, 2005), conjugated polymers (Goren and Lennox, 2001), inorganic materials (Kim et al., 2004; Spatz et al., 1998), and metals (Fahmi and Stamm, 2005). When such patterns are tuned by light, more complicated and hierarchical structures that will be more suited for device fabrication would be realized.

The topics dealt with in this chapter are mostly limited to 2-D systems, but the ultimate goal would be to create sophisticated dynamic 3-D systems similar to the biological ones. It is anticipated that studies on the motions of photoresponsive films in the assembled states can give proper clues to such future directions.

REFERENCES

Advincula RC, Brittain WJ, Caster KC, Rühe J, editors. 2004. Polymer Brushes. Weinheim: Wiley-VCH.

Edmondoson S, Osborne VL, Huck WTS. 2004. Polymer brushes via surface-initiated polymerizations. Chem Soc Rev 33:14–22.

Fahmi AW, Stamm M. 2005. Spatially correlated metallic nanostructures on self-assembled diblock copolymer templates. Langmuir 21:1062–1066.

Fujiwara T, Ichimura K. 2002. Surface-assisted photoalignment control of lyotropic liquid crystals. Part 2. Photopattterning of aqueous solutions of a water-soluble anti-asthmatic drug as lyotropic liquid crystals. J Mater Chem 12:3387–3391.

Fukuda K, Seki T, Ichimura K. 2002a. Photoorientation of poly(di-n-hexylsilane) by azobenzene monolayer 1. Preparative conditions of poly(di-n-hexylsilane) spincast films. Macromolecules 35:2177–2183.

Fukuda K, Seki T, Ichimura K. 2002b. Photoorientation of poly(di-n-hexylsilane) by azobenzene monolayer 2. Structural optimization of the surface azobenzene monolayer. Macromolecules 35:1951–1957.

Fukuda T, Matsuda H, Shiraga T, Kimura T, Kato M, Viswanathan N, Kumar J, Tripathy SK. 2000. Photofabrication of surface relief grating on films of azobenzene polymer with different dye functionalization. Macromolecules 33:4220–4225.

Fukumoto H, Nagano S, Kawatsuki N, Seki T. 2005. Photo-orientation of mesoporous silica thin films on photo-crosslinkable polymer film. Adv Mater 17:1035–1039.

Fukumoto H, Nagano S, Kawatsuki N, Seki T. 2006. Photoalignment behavior of mesoporous silica thin films synthesized on a photo-crosslinkable polymer film. Chem Mater 18:1226–1234.

Geue T, Zieglar A, Stumpe J. 1997. Light-induced orientation phenomena in Langmuir-Blodgett multilayers. Macromolecules 30:5729–5738.

Goren M, Lennox RB. 2001. Nano scale polypyrol patterns using block copolymer micelles as templates. Nano Lett 1:735–738.

Hamelinck PJ, Huck WTS. 2005. Homeotropic alignment on surface-initiated liquid crystalline polymer brushes. J Mater Chem 15:381–385.

Han M, Morino S, Ichimura K. 1999. Successive occurrence of biaxial reorientation of azobenzene chromophores of a liquid crystalline polymer induced by linearly polarized light. Chem Lett 28:645–646.

Han M, Morino S, Ichimura K. 2000. Factors affecting in-plane and out-of-plane photoorientation of azobenzene side chains attached to liquid crystalline polymers induced by irradiation with linearly polarized light. Macromolecules 33:6360–6371.

Hara M, Mizoshita N, Nagano S, Seki T. 2007. Chromonic/silica nanohybrids. Synthesis and macroscopic alignment. Langmuir 23:12350–12355.

Ichimura K. 2000. Photoalignment of liquid crystal systems. Chem Rev 100:1847–1874.

Ichimura K, Fujiwara T, Momose M, Matsunaga D. 2002. Surface-assisted photoalignment control of lyotropic liquid crystals. Part 1. Characterization and photoalignment of aqueous solutions of a water-soluble dye as lyotropic liquid crystals. J Mater Chem 12:3380–3386.

Ichimura K, Momose M, Kudo K, Akiyama H, Ishizuki N. 1995. Surface-assisted photolithography to form anisotropic dye layers as a new horizon of command surfaces. Langmuir 11:2341–2343.

Ichimura K, Oh SG, Nakagawa M. 2000. Light-driven motion of liquids on a photoresponsive surface. Science 288:1624–1626.

Ichimura K, Suzuki Y, Seki T, Hosoki A, Aoki K. 1988. Reversible change in alignment mode of nematic liquid crystals regulated photochemically by "command surfaces" modified with an azobenzene monolayer. Langmuir 4:1214–1216.

Ikeda T. 2003. Photomodulation of liquid crystal orientations for photonic applications. J Mater Chem 13:2037–2057.

Ikeda T, Mamiya J, Yu Y. 2007. Photomechanics of liquid-crystalline elastomers and other polymers. Angew Chem Int Ed Engl 46:506–528.

Irie M, editor. 2000. Special issue of photochromism: memories and switches. Chem Rev 100(5):1683–1890.

Kadota S, Aoki K, Nagano S, Seki T. 2005. Photocontrolled microphase separation of a block copolymer in two dimensions. J Am Chem Soc 127:8266–8267.

Kadota S, Aoki K, Nagano S, Seki T. 2006. Morphological conversions of nanostructures in monolayers of an ABA triblock copolymer having azobenzene moiety. Colloids Surf A 284/285:535–541.

Kago K, Fuerst M, Matsuoka H, Yamaoka H, Seki T. 1999. Direct observation of photoisomerization of polymer monolayer on water surface by X-ray reflectometry. Langmuir 15:2237–2240.

Kato T. 2002. Self-assembly of phase-segregated liquid crystal structures. Science 295:2414–2418.

Kato T, Frechet JMJ. 1989. New approach to mesophase stabilization through hydrogen-bonding molecular interactions in binary mixtures. J Am Chem Soc 111:8533–8534.

Kawashima Y, Nakagawa M, Ichimura K, Seki T. 2004. Photo-orientation of mesoporous silica materials via transfer from azobenzene-containing polymer monolayer. J Mater Chem 14:328–335.

Kawashima Y, Nakagawa M, Seki T, Ichimura K. 2002. Photoorientation of mesostructured silica via hierarchical multiple transfer. Chem Mater 14:2842–2844.

Kawatsuki N, Goto K, Kawakami T, Yamamoto T. 2002. Reversion of alignment direction in the thermally enhanced photoorientation of photo-cross-linkable polymer liquid crystal films. Macromolecules 35:706–713.

Kim DH, Jia X, Lin Z, Guarini KW, Russell TP. 2004. Growth of silicon oxide in thin film block copolymer scaffolds. Adv Mater 16:702–709.

Kim DY, Tripathy SK, Li L, Kumar J. 1995. Laser-induced holographic surface relief gratings on nonlinear optical polymer films. Appl Phys Lett 66:1166–1168.

Knoll A, Horvat A, Lyakhova KS, Krausch G, Sevink GJA, Zvelindovsky AV, Magerle R. 2002. Phase behavior in thin films of cylinder-forming block copolymers. Phys Rev Lett 89:035501.

Kondo M, Yu Y, Ikeda T. 2006. How does the initial alignment of mesogens affect the photoinduced bending behavior of liquid-crystalline elastomers? Angew Chem Int Ed Engl 45:1378–1382.

Kresge CT, Leonowicz ME, Roth WJ, Varatuli JC, Beck JB. 1992. Ordered mesoporous molecular sieves synthesized by a liquid-crystal template mechanism. Nature 359:710–712.

Kumar GS, Neckers DC. 1989. Photochemistry of azobenzene-containing polymers. Chem Rev 89:1915–1925.

Kumar N, Harn J. 2005. Nanoscale protein patterning using self-assembled diblock copolymers. Langmuir 21:6652–6655.

Lazzari M, De Rosa C. 2006. Method for the alignment and the large-scale ordering of block copolymer morphologies. In: Lazzari M, Liu G, Lecommandoux S, editors. Block Copolymers in Nanoscience. Weinheim: Wiley-VCH. pp. 191–231.

Lazzari M, Liu GJ, Lecommandoux S, editors. 2006. Block Copolymers in Nanoscience. Weinheim: Wiley-VCH.

Lim HS, Han JT, Kwak DH, Cho KW. 2007. Photoreversibly switchable superhydrophobic surface with erasable and rewritable pattern. J Am Chem Soc 128:14458–14459.

Lydon J. 2004. Chromonic mesophases. Curr Opin Colloid Interface Sci 8:480–490.

Mao G, Wang J, Clingman SR, Ober CK, Chen JT, Thomas E. 1997. Molecular design, synthesis, and characterization of liquid crystal-coil diblock copolymers with azobenzene side groups. Macromolecules 30:2556–2567.

Matsunaga D, Tamaki T, Akiyama H, Ichimura K. 2002. Photofabrication of micropatterned polarizing elements for stereoscopic displays. Adv Mater 14:1477–1480.

Matsunaga D, Tamaki T, Ichimura K. 2003. Azo-pendant polyamides which have the potential to photoalign chromonic lyotropic liquid crystals. J Mater Chem 13:1558–1564.

Mensinger H, Stamm M, Boeffel C. 1992. Order in thin films of a liquid crystalline polymer. J Chem Phys 96:3183–3190.

Morikawa Y, Kondo T, Nagano S, Seki T. 2007. 3D Photoalignment and patterning of microphase separated nanostructure in polystyrene-based block copolymer. Chem Mater 19:1540–1542.

Morikawa Y, Nagano S, Watanabe K, Kamata K, Iyoda T, Seki T. 2006. Optical alignment and patterning of nanoscale microdomains in a block copolymer thin film. Adv Mater 18:883–886.

Nakano H, Takahashi T, Kadota T, Shirota Y. 2002. Formation of a surface relief grating using a novel azobenzene-based photochromic amorphous molecular material. Adv Mater 14:1157–1160.

Natansohn A, Rochon P. 2002. Photoinduced motions in azo-containing polymers. Chem Rev 102:4139–4176.

Ohe C, Kamijo H, Arai M, Adachi M, Miyazawa H, Itoh K, Seki T. 2008. Sum frequency generation spectroscopic study on photoinduced isomerization of a poly(vinyl alcohol) containing azobenzene side chain at the air-water interface. J Phys Chem C 112:172–178.

Peng B, Johannsmann D, Rühe J. 1999. Polymer brushes with liquid crystalline side chains. Macromolecules 32:6759–6766.

Rochon P, Batalla E, Natansohn A. 1995. Optically induced surface gratings on azoaromatic polymer films. Appl Phys Lett 66:136–138.

Ruslim C, Hashimoto M, Matsunaga D, Tamaki T, Ichimura K. 2004. Optical and surface morphological properties of polarizing films fabricated from a chromonic dye by the photoalignment technique. Langmuir 20:95–100.

Ruslim C, Matsunaga D, Hashimoto M, Tamaki T, Ichimura K. 2003. Structural characteristics of the chromonic mesophases of C.I. Direct Blue 67. Langmuir 19:3686–3691.

Sato N, Ito S, Yamamoto M. 1998. Molecular weight dependence of shear viscosity of a polymer monolayer: evidence for the lack of chain entanglement in the two-dimensional plane. Macromolecules 31:2673–2675.

Seki T. 2004. Dynamic photoresponsive functions in organized layer systems comprised of azobenzene-containing polymers. Polym J 36:435–454.

Seki T. 2007. Smart photoresponsive polymer systems organized in two dimensions Bull Chem Soc Jpn 80:2084–2109.

Seki T, Fukuda K, Ichimura K. 1999. Photocontrol of polymer chain organization using a photochromic monolayer. Langmuir 15:5098–5101.

Seki T, Fukuda R, Yokoi M, Tamaki T, Ichimura K. 1996. Photomechanical response of azobenzene containing monolayers on water surface. Bull Chem Soc Jpn 69:2375–2381.

Seki T, Sakuragi M, Kawanishi Y, Tamaki T, Fukuda R, Ichimura K. 1993. "Command surfaces" of Langmuir-Blodgett films. Photoregulation of liquid crystal alignment by molecularly tailored surface azobenzene layers. Langmuir 9:211–218.

Seki T, Sekizawa H, Morino S, Ichimura K. 1998. Inherent and cooperative photomechanical motions in monolayers of an azobenzene containing polymer at the air-water interface. J Phys Chem B 102:5313–5321.

Sekkat Z, Knoll K, editors. 2002. Photoreactive Organic Thin Films. San Diego: Academic Press.

Spatz JP, Eibeck P, Mössmer S, Möller M, Herzog T, Ziemann P. 1998. Ultrathin diblock copolymer/titanium laminates-a tool for nanolithography. Adv Mater 10:849–852.

Tian Y, Watanabe K, Kong X, Abe J, Iyoda T. 2002. Synthesis, nanostructures, and functionality of amphiphilic liquid crystalline block copolymers with azobenzene moieties. Macromolecules 35:3739–3747.

Tsujii Y, Ohno K, Yamamoto S, Goto A, Fukuda T. 2006. Structure and properties of high-density polymer brushes prepared by surface-initiated living radical polymerization. Adv Polym Sci 197:1–45.

Ubukata T, Hara M, Ichimura K, Seki T. 2004. Phototactic transport motions of polymer film for micropatterning and alignment of functional materials. Adv Mater 16:220–224.

Ubukata T, Seki T, Ichimura K. 2000. Surface relief gratings in host-guest supramolecular materials. Adv Mater 12:1675–1678.

Ubukata T, Takahashi K, Yokoyama Y. 2007. Photoinduced surface relief structures formed on polymer films doped with photochromic spiropyrans. J Phys Org Chem 20:981–984.

Uekusa T, Nagano S, Seki T. 2007. Unique molecular orientation in a smectic liquid crystalline film attained by surface-initiated graft polymerization. Langmuir 23:4642.

Viswanathan NK, Kim DY, Bian S, Williams J, Liu W, Li L, Samuelson L, Kumar J, Tripathy SK. 1999. Surface relief structures on azo polymer films. J Mater Chem 9:1941–1955.

Watanabe K, Watanabe R, Aoki D, Shoda S, Komura M, Kamata K, Iyoda T. 2006. Alignment of self-organized nanocylinder array structure in amphiphilic liquid crystalline block copolymer film. Trans Mater Res Soc Jpn 31:237–240.

Yager KJ, Barrett CJ. 2001. All-optical patterning of azo polymer films. Curr Opin Solid State Mater Sci 5:487–494.

Yanagisawa T, Shimizu T, Kuroda K, Kato C. 1990. The preparation of alkyltriinethylaininonium–kaneinite complexes and their conversion to microporous materials. Bull Chem Soc Jpn 63:988–992.

Yu HF, Iyoda T, Ikeda T. 2006. Photoinduced alignment of nanocylinders by supramolecular cooperative motions. J Am Chem Soc 128:11010–11011.

Yu Y, Nakano M, Ikeda T. 2003. Photomechanics directed bending of a polymer film by light. Nature 425:145.

Zettsu N, Fukuda T, Matsuda H, Seki T. 2003. Unconventional polarization characteristic of rapid photoinduced material motion in liquid crystalline azobenzene polymer films. Appl Phys Lett 83:4960–4962.

Zettsu N, Ogasawara T, Arakawa R, Nagano S, Ubukata T, Seki T. 2007a. Highly photosensitive surface relief gratings formation in a liquid crystalline azobenzene polymer: new implications for the migration process. Macromolecules 40:4607–1613.

Zettsu N, Ogasawara T, Mizoshita N, Nagano S, Seki T. 2007b. Photo-triggered surface relief gratings formation in supramolecular liquid crystalline polymer system with detachable azobenzene unit. Adv Mater 20:516–531.

Zettsu N, Seki T. 2004. Highly efficient photogeneration of surface relief structure and its immobilization in cross-linkable liquid crystalline azobenzene polymers. Macromolecules 37:8692–8698.

Zettsu N, Ubukata T, Seki T. 2004. Two-dimensional manipulation of poly(3-dodecylthiophene) using light-driven instant mass migration as a molecular conveyer. Jpn J Appl Phys 43:L1169–L1171.

Zettsu N, Ubukata T, Seki T, Ichimura K. 2001. Soft crosslinkable azo polymer for rapid surface relief formation and persistent fixation. Adv Mater 13:1693–1697.

9

PHOTOINDUCED IMMOBILIZATION OF MOLECULES ON THE SURFACE OF AZOBENZENE POLYMERS: PRINCIPLES AND APPLICATION

Osamu Watanabe

9.1. INTRODUCTION

A large number of investigations concerning the photoisomerization of azobenzene derivatives have been reported so far because of their potential applications in optical recording media, holographic technology, and optical components (Yager and Barrett, 2006; Oliveira et al., 2005; Zhao et al., 2003; Hasegawa et al., 2002a; Ichimura, 2000; Andruzzi et al., 1999; Batalla et al., 1995; Natansohn, 1999; Yamamoto et al., 1999; Zilker et al., 1998; Berg et al., 1996; Hvilsted et al., 1995; Ikeda and Tsutsumi, 1995; Kim et al., 1995; Natansohn et al., 1992; Todorov et al., 1984). In particular, the so-called azopolymers, in which azobenzene derivatives are included in a polymer chain, have attracted attention because of the phenomena that they undergo involving molecular reorientation and shape variation by mass transportation, which are induced by combinational movement of the azobenzene derivatives and the polymer chain.

This chapter introduces the photoinduced immobilization of microobjects onto the surface of azopolymers as a newly developed photoresponsive phenomenon of azopolymers. The size of the microobjects can extend over a wide range, from a few nanometers to several micrometers, and can include biological molecules such as DNA, enzymes, immunoglobulin or cells; and microspheres made of polymeric, inorganic, or metallic materials. Figure 9.1 shows the principle of photoinduced immobilization, which represents a very simple technique

Smart Light-Responsive Materials. Edited by Yue Zhao and Tomiki Ikeda

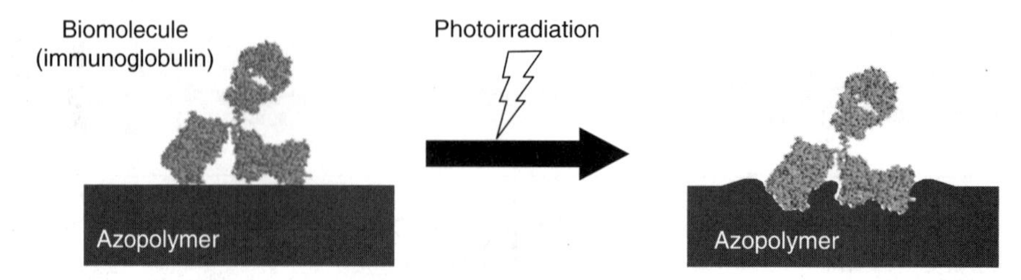

Figure 9.1. Schematic illustration of the photoimmobilization of biomolecules (immunoglobulin) on the surface of an azopolymer. The surface of the azopolymer is deformed to the shape of the immunoglobulin after photoirradiation. *Source*: Narita, 2007.

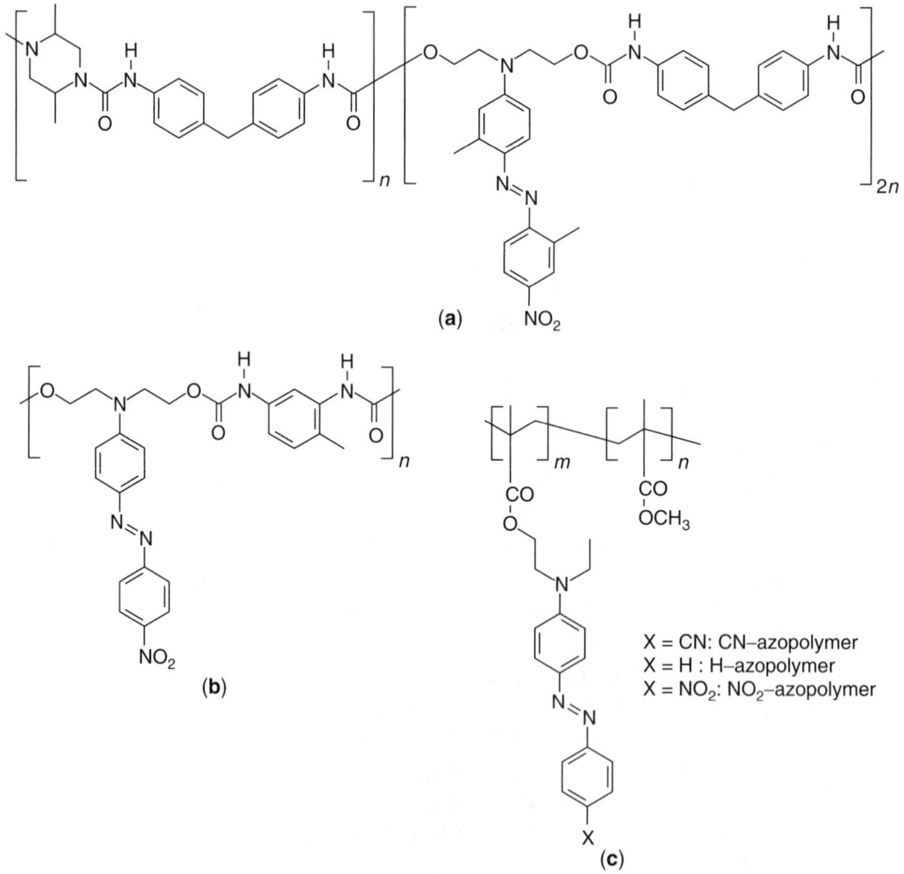

(a)

(b)

(c)

X = CN: CN–azopolymer
X = H : H–azopolymer
X = NO_2: NO_2–azopolymer

Figure 9.2. The chemical structures of the azopolymers described in this chapter.

(Narita et al., 2007; Ikawa et al., 2006; Watanabe, 2004). First, the microobject (immunoglobulin in the case shown in Fig. 9.1) is set on the surface of the azopolymers, which is then photoirradiated from above. The surface of the azopolymer deforms in the presence of the immunoglobulin because the viscoelastic properties of azopolymer surfaces change during photoirradiation. The deformation occurs such that it enfolds the immunoglobulin, and so the contact area between the surfaces of the immunoglobulin and the azopolymer increases. This deformation mainly occurs through the photoplasticization of the azopolymer matrix owing to a trans–cis–trans isomerization cycle of the azobenzene moiety, as shown in Fig. 9.2. The surface of the azopolymer glaciates again and maintains the deformed shape after ceasing the irradiation, as shown in Fig. 9.1 (right-hand side). As a result, the immunoglobulin is effectively immobilized on the surface of the azopolymer without chemical modification.

This novel method is useful for the immobilization of a variety of small particles such as charged proteins, negatively charged DNA, and hydrophobic polystyrene microspheres on azopolymer surfaces, and it has been shown that the immobilized biomolecules can maintain their higher order structure without damage to their functionality. This versatility in terms of immobilization is a significant advantage of this technique.

The photoinduced immobilization technique is closely related to the formation process used for surface relief gratings (SRG) because both phenomena are based on mass transportation of the azopolymer surface. The relief structure on the azopolymer surface is induced by the interference of the two coherent beams that are used for the irradiation, with the same periodic structure as the interference light, as shown in Fig. 9.3. Intensive studies for SRG formation have been reported since it was first developed in 1995 (Batalla et al., 1995; Kim et al., 1995). Various deformed structures that can be induced by photoirradiation

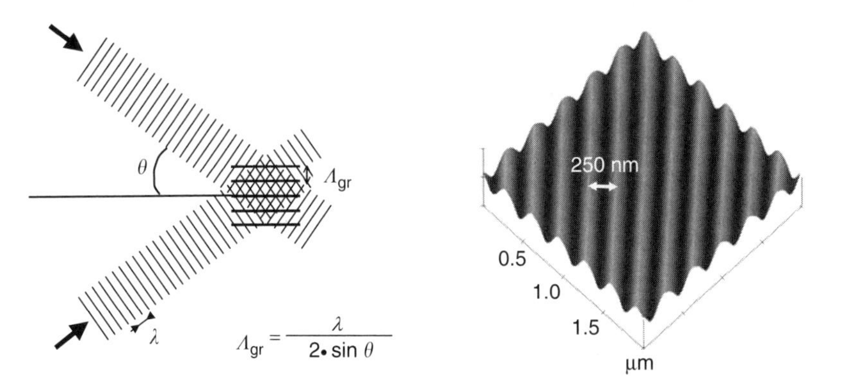

Figure 9.3. Formation of SRG on an azopolymer, as generated by two-beam interference irradiation. The grating pitch, Λ_{gr} is determined by the wavelength and the incident angle of the irradiated light. The figure on the right exhibits a topographical image as measured by AFM.

have been demonstrated in addition to SRG (Fukuda et al., 2001; Ye et al., 2001). The surface deformation mechanism needs to be understood by considering not only photochemical phenomena involving the azobenzene moiety and the mobility of the polymer matrix but also by interactions of the irradiation light with electric fields. Therefore, a large number of researchers are making continued efforts to clarify the complex deformation mechanism (Barada et al., 2006; Barret et al., 1996; Kumar et al., 1998; Pedersen et al., 1998).

9.2. BACKGROUND STUDY: NANOFABRICATION

The authors have recently switched the objectives of their research into SRG formation to now consider interactions between small objects and azopolymer surfaces, such as mass transportation and molecular reorientation, and have investigated photoinduced nanofabrication using a novel approach. They have discovered some interesting phenomena that are applicable to nanometer-scale fabrication by irradiating light onto microobjects set on azopolymer surfaces (Keum et al., 2003; Hasegawa et al., 2002b, 2001; Ikawa et al., 2001a,b, 2000; Watanabe et al., 2006, 2001, 2000). The resolution of recording or fabrication processes that is defined by light is determined by how narrowly the irradiating light can be focused, and, because of to diffraction limits, in practice this equates to about half of the wavelength of the irradiating light. The use of the optical near field can overcome diffraction limits to reach nanometer-scale dimensions, and this has been expected to become a powerful tool for attaining nanometer-scale-manufacturing capability (Knoll and Keilmann, 1999; Ohtsu, 1998; Betzig and Trautman, 1992). This section demonstrates nanoscale deformation phenomena that can be induced by the optical near field when using microobjects set onto an azopolymer; these phenomena were investigated by the authors and triggered their work into photoinduced immobilization.

Various sizes of microspheres (from tens of nanometers to several micrometers) made of various materials such as polystyrene or silica can be easily obtained, and it is possible to place these into an ordered arrangement because of the uniformity of their diameters. If a microsphere is irradiated with light, an optical near field is induced around the microspheres, as shown in Fig. 9.4. The authors selected polystyrene microspheres for use as the near-field light source and demonstrated a topographical nanostructure-patterning technique on the surface of an azopolymer. Nanostructured patterning was carried out as shown schematically in Fig. 9.5 using the azopolymer shown in Fig. 9.2, which has a glass transition temperature of 145 °C and a maximum absorption of 475 nm. A film of azopolymer with a thickness of 0.5 μm was spin coated onto a glass substrate from a pyridine solution, and the surface of the film before irradiation with light showed no regular structural periodicity. An aqueous solution containing polystyrene spheres was dropped onto the surface of the polymer films, and then the spheres were allowed to arrange themselves into a hexagonal-packed monolayer by a self-organization process. After drying the samples, they were irradiated from the side

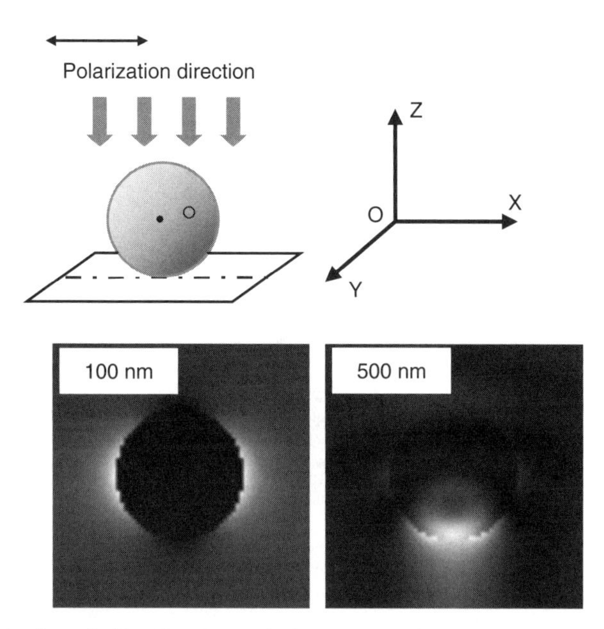

Figure 9.4. Calculated distribution of the optical intensity on the X–Z plane of the polystyrene microspheres, 100 nm (*left*) and 500 nm (*right*). The bright region indicates a relatively strong intensity. *Source*: Ikawa, 2001b.

using a 488-nm Ar-ion laser with an intensity of tens of milliwatt per square centimeters to eliminate the influence of gravity, as shown in Fig. 9.5. After irradiation, the sample was washed with water and benzene to remove the microspheres and then the surface structure of the polymer film was observed using atomic force microscopy (AFM) and scanning electron microscopy (SEM). Figure 9.6 shows AFM images of the resulting polymer surfaces, where hexagonal

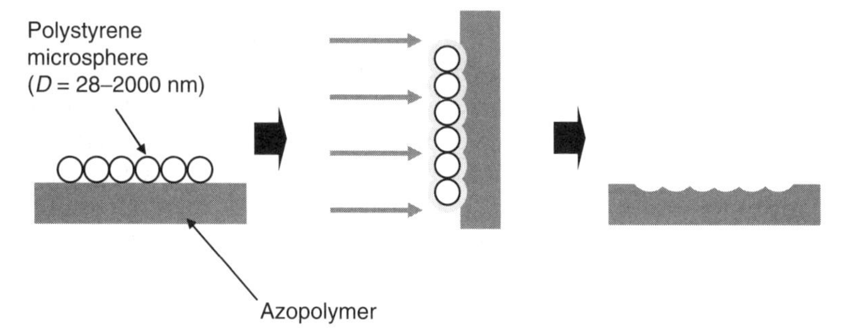

Figure 9.5. Schematic representation of a nanopatterning process formed by using microspheres as the near-field source, showing the alignment of the microspheres (*left*), light irradiation (*center*), and the elimination of the microspheres (*right*). *Source*: Watanabe, 2000.

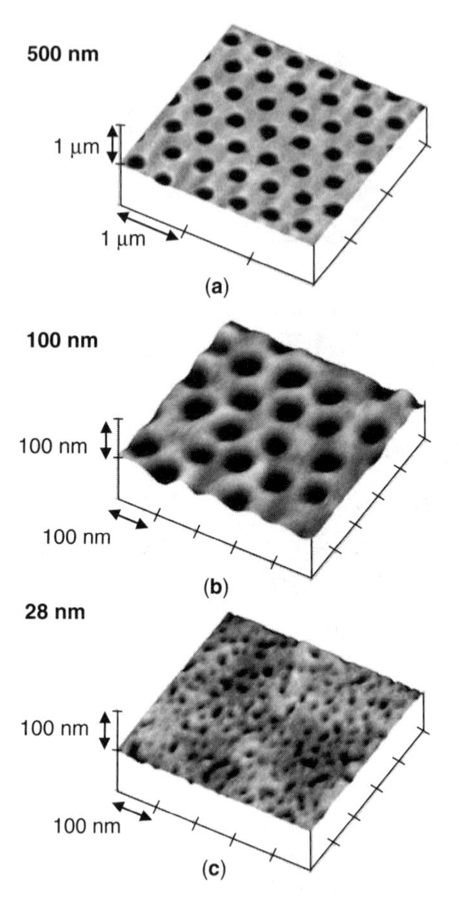

Figure 9.6. AFM images of nanopatterned structures formed on an azopolymer surface using microspheres with diameters of (a) 500 nm, (b) 100 nm, and (c) 28 nm, respectively. *Source*: Watanabe, 2000.

structures (500- and 100-nm microspheres) were directly transcribed onto the polymer surface as a series of indentations. A very fine indented structure was also observed in the case of 28-nm microspheres, although the arrayed structure was distorted. It can be concluded that these structures were induced by the optical near field around the polystyrene microspheres, because the dimensions of the 100- and 28-nm diameter spheres are beyond the diffraction limit.

9.3. PRINCIPLES OF PHOTOINDUCED IMMOBILIZATION

Figure 9.7 also shows SEM images of the resulting polymer surfaces (in addition to those shown in Fig. 9.6), including both the indented structures and the

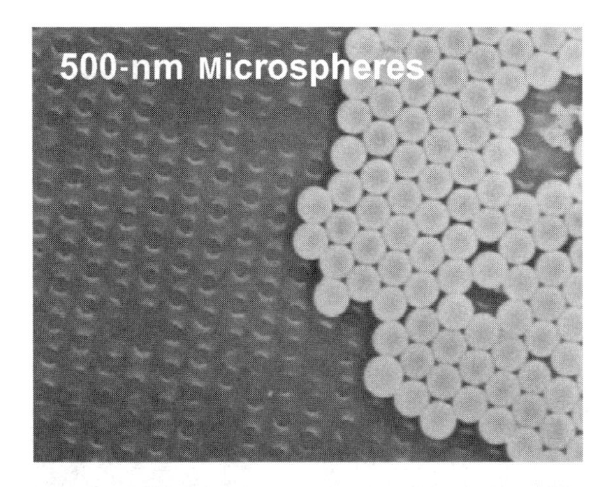

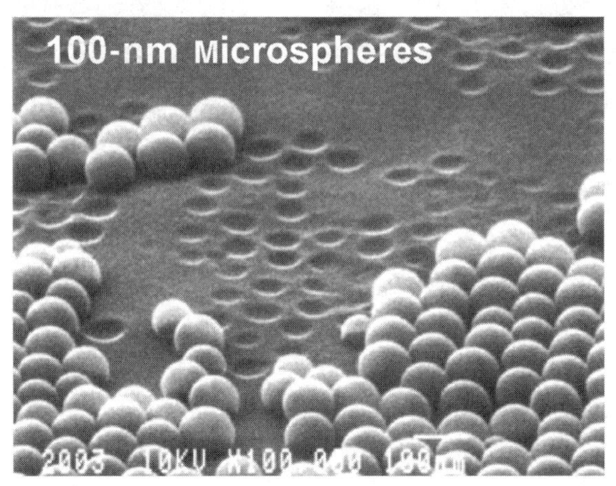

Figure 9.7. SEM images of nanopatterned structures formed on an azopolymer surface formed by using microspheres with diameters of 500 nm (*top*) and 100 nm (*bottom*), respectively. The microspheres remain partially. *Source*: Ikawa, 2001b.

microspheres that remain after the process. These images confirm that the indentations are formed directly below the microspheres. Although the microspheres should obviously be removed for nanofabrication experiments, the removal of the microspheres from the azopolymer has been found to be difficult in the course of these studies. The authors looked at this problem from a different angle, which led them to proactively suggest that this phenomenon could be applied as an "immobilizing" technology. This section demonstrates a photoinduced immobilization technique for microobjects, and introduces the authors' recent experimental results (Narita et al., 2007; Ikawa et al., 2006; Watanabe, 2004).

Polystyrene microspheres were therefore deliberately photoimmobilized onto an azopolymer surface for the first time. A monolayer of 1-μm microspheres that

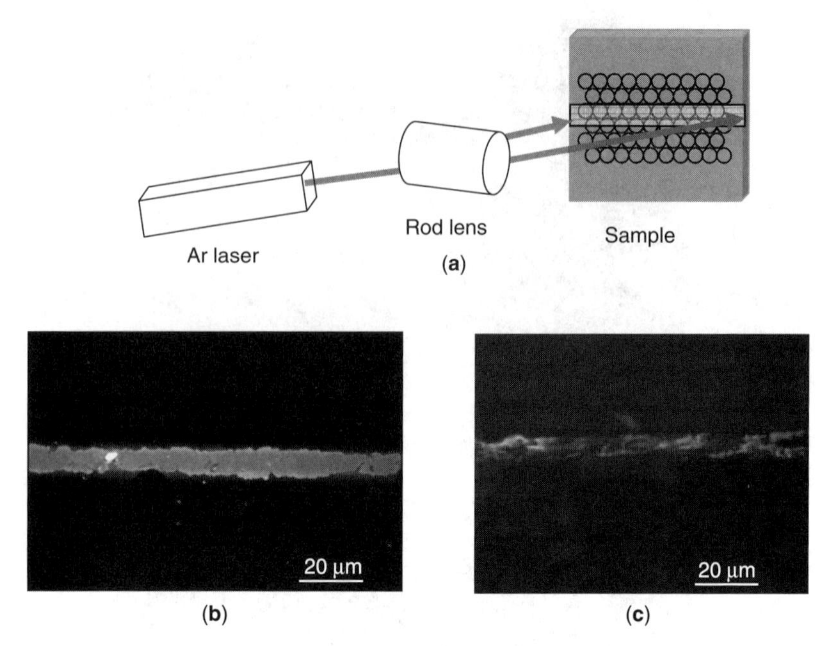

Figure 9.8. **(a)** Experimental setup for patterned immobilization using a linearly shaped laser beam. **(b)** Dark field optical image of immobilized polystyrene microspheres with diameters of 1 μm on the surface of an azopolymer. **(c)** Fluorescence image of the immobilized λ-DNA molecules. *Source*: Ikawa, 2006. See color insert.

had been applied to the azopolymer surface was irradiated with a linear-shaped laser beam of 488-nm wavelength and 10-mW/cm^2 optical power density using a cylindrical lens for 5 min, as shown in Fig. 9.8a.

The surface was washed in an ultrasonic cleaner and was then observed with an optical microscope, as shown in Fig. 9.8b. Only the microspheres in the linearly irradiated region were strongly immobilized, despite the ultrasonic washing and the relatively large size of the microspheres. DNA molecules were then selected as a potential target material for the immobilization of biological macromolecules. An aqueous solution of 1-mg/mL λ-DNA was spotted onto the surface of an azopolymer and covered with a cover glass, where the λ-DNA was stained with a fluorophore (YOYO-1 iodide, Molecular Probe) in advance, and the surface was then irradiated with the same linearly shaped laser beam for 5 min. The surface was washed for 5 min in an aqueous solution and then observed using a conventional fluorescence microscope. Figure 9.8 confirms that the labeled λ-DNA was only immobilized in the irradiated region. The same experiment using green fluorescent protein (GFP) also indicates the immobilization of protein in the irradiated region.

In this way, the authors could demonstrate that an azopolymer can capture micrometer- to nanometer-scaled microobjects, including synthetic polymers and biological molecules, on the photoirradiated area. The provision of an immobilization process is one of the most essential processing steps that is required to obtain practical biomolecule carriers such as biosensors, bioreactors, or biochips. Therefore, a large number of immobilization techniques have already been developed for biological molecules, in which the molecules are immobilized on a carrier using covalent bonds (Feng et al., 2005), ionic bonds (Lee et al., 2003), physical adsorption (Ouyang et al., 2003), cross-linkage of the biomolecules (Levy and Shoseyov, 2004), or microencapsulation (Hartmann, 2005). Chemically induced immobilization methods require optimized processes depending on the structures and properties of the individual biomolecules, which in turn require some complicated procedures (Peluso et al. 2003); yet these techniques are widely used. Azopolymers can immobilize microobjects that possess a variety of surface characteristics, including negatively charged DNA, charged proteins, and hydrophobic polystyrene. The characteristics of the azopolymer make it possible to immobilize a wide variety of biological molecules on the same substrate through a one-step photoirradiation process.

To take advantage of the functionality of these biomolecules, identifying an immobilization process that does not lead to deactivation of the molecules is important. In particular, biomolecules such as proteins show sensitive behavior in terms of changes in environment, as shown by the denaturing of proteins when the surrounding temperature increases even slightly. Since it is possible that damage to biomolecules following photoinduced immobilization could trigger functional degradation, the authors first examined the activity of an immobilized enzyme. An aqueous solution of 1-mg/mL bacterial protease (subtilisin; 27.5 kDa, Sigma) was spotted onto the surface of an azopolymer, and the surface was irradiated with a laser beam of 488-nm wavelength and 80-mW/cm^2 optical power density for 5 min to immobilize the enzyme. As a control experiment, a similar specimen was prepared without photoirradiation. The activity of the subtilisin was verified as the hydrolysis of the artificial substrate (*tert*-butoxycarbonyl-Gly-Gly-Leu-*p*-nitroanilide, Mw = 465.5, Merck). The artificial substrate solution was spotted onto the azopolymer surface in the same area where the subtilisin had been immobilized, and then the specimen was maintained at 37 °C and 85% relative humidity for 1 h. The hydrolysis of the artificial substrate was determined spectroscopically by immediately measuring the absorbance of the reactant at a wavelength of 410 nm. The conversion ratio of the reaction was ~10% for the subtilisin-immobilized sample, whereas it was ~1% for the control sample (without photoirradiation). These results clearly show that biomolecules immobilized on an azopolymer surface can maintain their enzyme functionality during and after the immobilization process.

Next, the authors investigated how deformation of an azopolymer surface can be induced by biomolecules as well as by microspheres. A phosphate-buffered saline (PBS) solution containing Cy-5-linked immunoglobulin, IgG, was spotted onto the surface of an azopolymer. After evaporating the solution, the surface was irradiated for 30 min with light of 470-nm wavelength and 10-mW/cm^2 optical power density from an array of blue light-emitting diodes (LEDs) and then the surface was washed for 30 min with PBS containing 0.01-wt% Tween 20 as a nonionic surfactant. The amount of immobilized IgG was confirmed by the fluorescence intensity of the spot, and the minimum detectable amount was 10 pg. Next, a surface image was obtained by tapping mode AFM (Digital Instruments, Dimension 3100) using a sharp silicon cantilever with a tip radius of <5 nm. In Fig. 9.9a, the azopolymer surface is covered with a layer of small granulated particles of 10–30 nm diameter and ~ 8 nm height, where the height was estimated from the defects and the edge of the layer. The sizes of the particles were nearly equivalent to one subunit of IgG (~ 10 nm), considering that the image includes AFM tip convolution artifacts. The layer is so flat that the IgG monolayer is believed to be located on the azopolymer surface. The sample was subsequently washed with PBS containing 2-wt% sodium dodecyl sulfate to remove the IgG. After confirming that the fluorescence from the spot had disappeared, another AFM image was obtained, which is shown in Fig. 9.9b. Dents of ~ 20 nm diameter

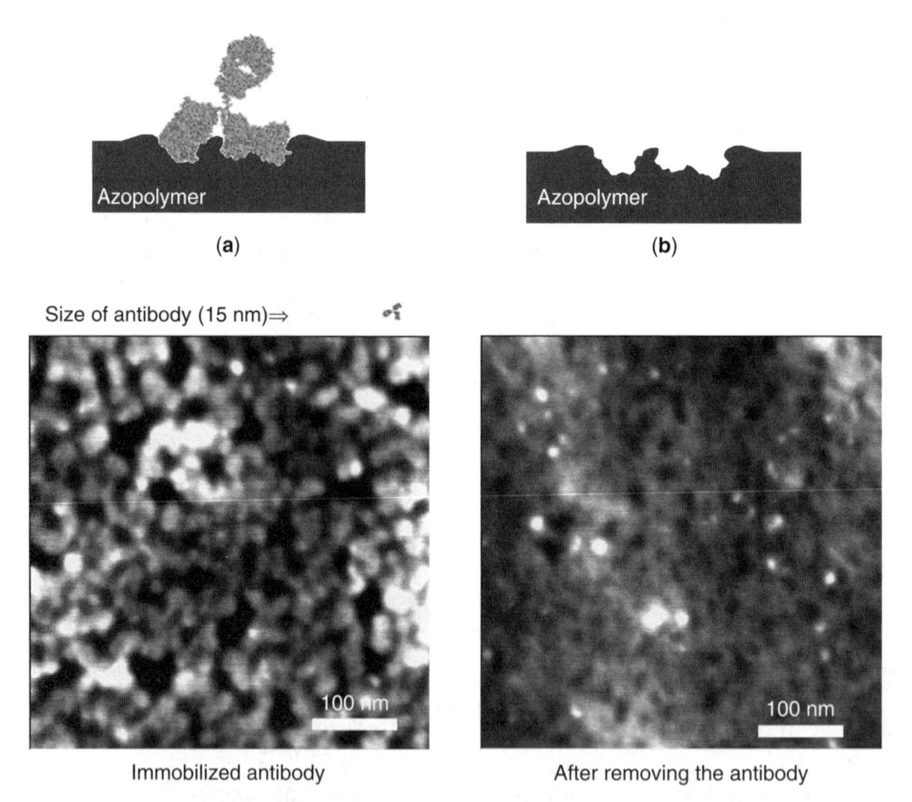

Figure 9.9. AFM images obtained from the photoimmobilization process. (**a**) The surface was observed after the photoimmobilization of immunoglobulin. The real-size immunoglobulin is shown. (**b**) The surface was observed after an elimination process using 2-wt% sodium dodecyl sulfate (SDS) solution. *Source*: Ikawa, 2006. See color insert.

and 2-nm deep can be observed on the surface. In contrast, no dents were formed on the azopolymer surface where no IgG was deposited. Comparing these images, the dents formed on the surface in Fig. 9.9b are considered to mirror the surface shape of the IgG. These findings lead to the conclusion that the azopolymer surface "recognizes" each molecular shape and deforms along the contours of the biomolecules during photoirradiation, as shown in Fig. 9.9b. The results also suggest that the increase in contact area between the azopolymer and biomolecules after photoirradiation restrains desorption from the surface.

As described later, the authors also examined the possibility of antigen–antibody reactions on the surface of the azopolymers. PBS solutions of human serum albumin and bovine serum albumin with different concentrations were spotted onto azopolymer surfaces according to the layout shown in Fig. 9.10a. After evaporating the spotted solution, photoirradiation was performed over the entire surface using an array of blue LEDs and was then washed with PBS containing 0.01 wt% Tween 20 to remove the un-immobilized albumins. After drying the sample again, the obtained sample was reacted with anti-HSA monoclonal mouse antibodies, and then the washed sample was reacted with Cy-5-labeled

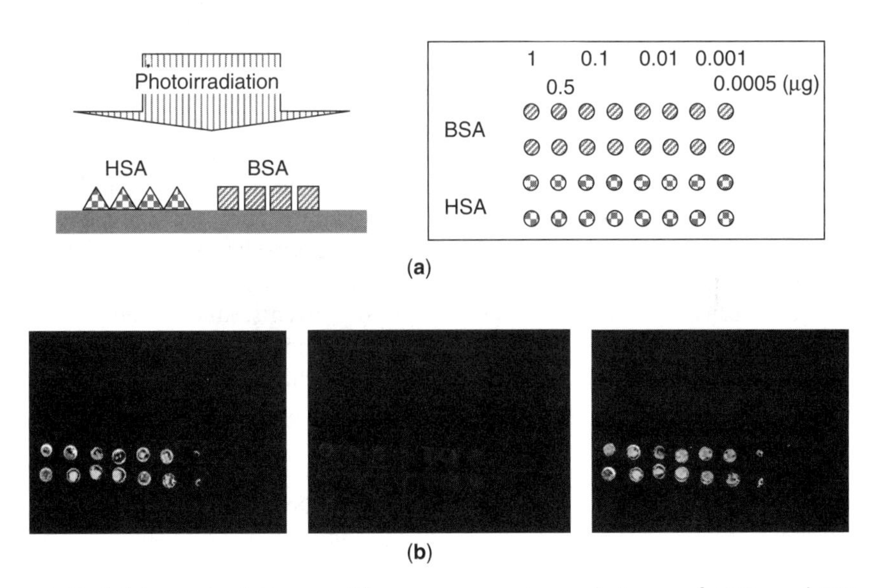

Figure 9.10. (a) Layout of the spotting process using solutions of HSA and BSA with different concentrations on the surface of the azopolymer. (b) Fluorescence images observed after various treatments. The left-hand side image is as observed from the spotted sample after photoimmobilization, washing, and immunological reaction. Center image is as observed from a sample treated with hydrochloric acid following the first observation. The same immunological reaction was repeated after the acid treatment, when the right-hand side image was observed.

antimouse polyclonal goat antibodies as a secondary antibody to detect albumins. Fluorescence emission was only observed from the spot on which the HSA had been immobilized, as shown in Fig. 9.10b, which means that a reaction that was selective to the HSA had occurred on the antigen-immobilized surface of the azopolymer and then was detected. Next, the sample was treated with hydrochloric acid to separate the antibodies from the HSA. The authors confirmed that no fluorescence emission could be observed from the substrate after the HCl treatment. They then repeated the same immunoreaction on the treated sample, such that almost the same fluorescence image as that described earlier was again obtained, as shown in Fig. 9.10b (center and right). This demonstrates that biochips fabricated on an azopolymer surface can be reused.

9.4. APPLICATION FOR IMMUNOCHIPS

In the next phase, the authors applied the photoimmobilization method to try to obtain protein chips, and more specifically, an immunochip. Enzyme-linked immunosorbent assay (ELISA) systems are commonly used (Karlsson et al., 1991; Voller et al., 1978) as a popular method for detecting small amounts of protein in sample solutions such as serums. Although the ELISA system is an excellent method for detection of proteins, it still has some problems; it is comparatively expensive, and it is difficult to detect proteins when small quantities of the sample solution are used. The authors might be able to realize an immunochip that could act as a micro-ELISA system with the capability to deal with small quantities of sample solutions if they could succeed in immobilizing the antibodies on the substrate. Such an immunochip could be used to measure multiple target proteins on the same substrate simultaneously, so it has the potential to become a novel method of replacing conventional ELISA systems in the fields of biochemical indexes, diagnostic agents, and clinical inspection (Kambhampati, 2003). One can say that the photoimmobilization method is one of the most promising prospects for immunochip applications because it can provide immobilization on the substrate surface irrespective of the surface states of the biomolecules.

The authors first examined specific reactions of photoimmobilized antibodies on azopolymer surfaces for the immunochip application. Solutions of anti-goat antibodies (left-hand side) and anti-rabbit antibodies (right-hand side) were spotted onto an azopolymer surface at different concentrations, the layout of which is shown in Fig. 9.11a. After photoimmobilization and washing, the sample was reacted separately with Cy-5-labeled antigens (goat IgG and rabbit IgG). The anti-goat IgG antibody recognized goat IgG when Cy-5-labeled goat IgG was introduced onto the sample, whereas anti-rabbit IgG recognized rabbit IgG when Cy-5-labeled rabbit IgG was introduced, as shown in Fig. 9.11b and c. The specific reactivity of the antibodies was realized by fixing photoimmobilized antibodies on the surface of the azopolymer. They also examined the preservation stability of the photoimmobilized antibodies. Although the reactivity of the antibodies dropped

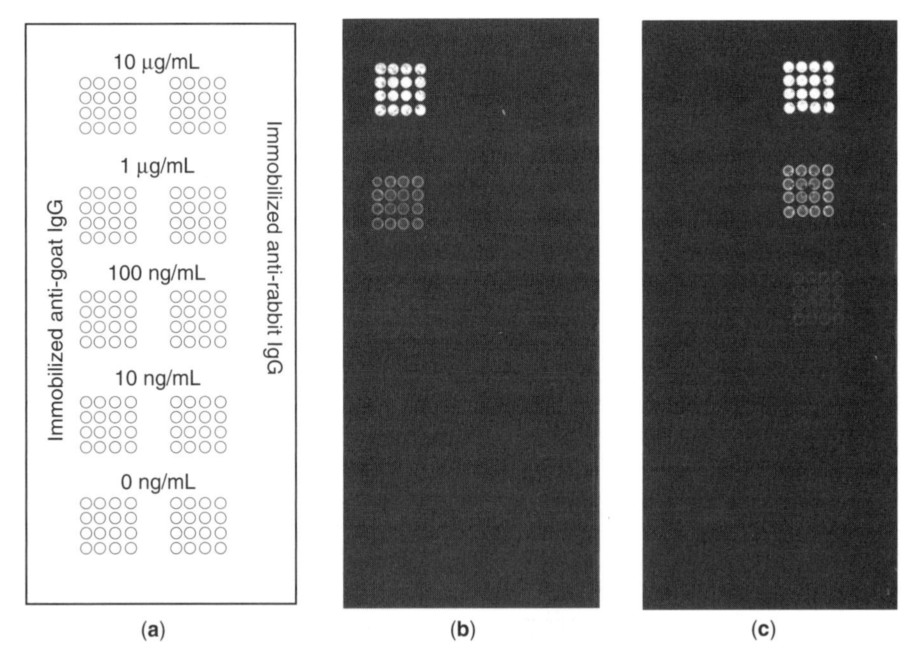

Figure 9.11. (a) Layout of a slide that was spotted and immobilized using the anti-goat IgG rabbit antibody and the anti-rabbit IgG goat antibody. (b) Fluorescent image after incubation of a Cy-5-labeled goat IgG. (c) Fluorescent image after incubation of a Cy-5-labeled rabbit IgG.

away over a period of 10 days when they were stored at room temperature, it was maintained for ~2 months when stored at 4°C. This result is acceptable in terms of commercial viability, though further increases in stability would be preferable.

The authors next examined the sensitivity for the immunochips. An immunochip usually has a two-dimensional (2-D) surface, so the detection limit for antigens can be estimated from the amount of immobilized antibodies that are present. It was difficult to increase the sensitivity of a detection system that uses a photoluminescence probe. However, they succeeded in obtaining higher sensitivity for an immunochip in which they adopted a chemiluminescence detection system using an enzyme reaction. They selected adiponectin, which is a biologically active agent that is excreted from adipose cells and which prevents arteriosclerosis, as the intended biological marker, and they tried to assay it using an enzyme sandwich immunoassay on the azopolymer surface. Anti-adiponectin antibodies were photoimmobilized on the azopolymer surface and then a solution including adiponectin was reacted on the fabricated immunochip. Subsequently, the sample that had captured the adiponectin on its surface using the immobilized antibodies was treated with biotin-labeled anti-adiponectin antibodies (first sandwich process) and then with alkaline phosphatase (ALP)-labeled streptoavidin (second sandwich process) (Kendall et al., 1983). After introducing the chemiluminescent

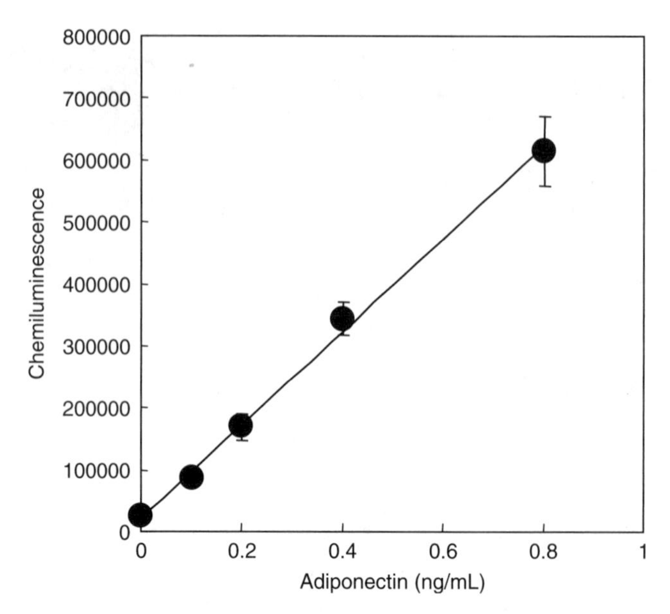

Figure 9.12. Calibration curve for quantifying mouse adiponectin. Each error bar indicates the standard deviation for each data point.

substrate onto the surface, the intensity of the chemiluminescence was measured to determine the concentration of adiponectin. The authors measured the intensity of the chemiluminescence against the concentration of adiponectin using samples with predetermined concentrations. Figure 9.12 exhibits the calibration curve that was obtained in the region of low concentration, and it shows that a linear relationship exists between intensity and concentration. Adiponectin in a sample solution can be detected down to a concentration of at least 0.1 ng/mL, which is almost the same sensitivity as that obtained with ELISA. A conventional ELISA system and the IgG chip system were compared using mouse adiponectin of culture supernatant. Figure 9.13 shows the correlation between the immunochip and conventional ELISA. A high degree of correlation exists ($r^2 = 0.97$), indicating that the use of an immunochip with an azopolymer film is a promising candidate for practical use.

9.5. IMMOBILIZATION DEPENDING ON THE AZOBENZENE MOIETY

This section compares the photoinduced immobilization of IgG on two types of azopolymers (shown in Fig. 9.2) bearing various concentrations of 4-amino-4′-cyanoazobenzene (CN-azopolymer) or amino azobenzene (H-azopolymer). CN-azopolymer and H-azopolymer contain a push–pull-type azobenzene and an amino azobenzene, respectively (Rabek, 1988). These azobenzenes have different

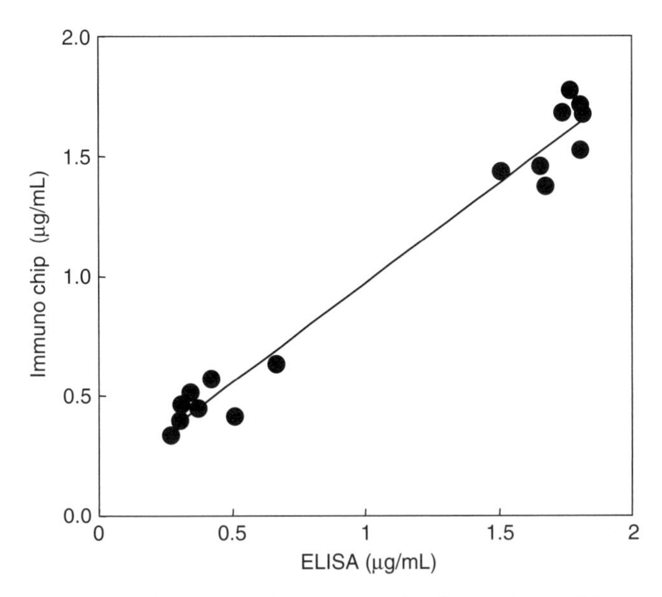

Figure 9.13. Correlation between the two methods, ELISA and immunochip, for quantifying mouse adiponectin at a level of 16 culture supernatant.

adsorption spectra, and they exhibit different deformation and immobilization features under photoirradiation. Therefore, information can be obtained about the photoimmobilization mechanisms by comparing these two types of azopolymers. First, the authors examined the relationship between immobilization efficiency and the indented depth with respect to the photoirradiation time and the specific azobenzene moiety. Second, they compared the relationship between immobilization efficiency and chemical structure, and elucidated how this correlated with the photoisomerization properties and the retention rate of immobilized antibodies.

The photodeformation capabilities of the CN-azopolymer and the H-azopolymer were examined by determining the depth of the indents formed by polystyrene microspheres under LED irradiation. After photoirradiation and removal of the microspheres, regularly arranged indented patterns formed by the microspheres were observed on the surfaces of the azopolymers. The depths of the indents were plotted as a function of irradiation time for several kinds of azopolymers, as shown in Fig. 9.14. The indent depths in the azopolymer increased with increasing irradiation time. The depths of the indents saturated and reached a maximum after 30 min of photoirradiation for each of the azopolymers. The saturated depths were lowest in those azopolymers with the lowest content of azobenzene moieties. These results indicate that the photoresponsive moiety plays an important role in inducing photodeformation and that the indent depth is related to the content of the azobenzene in the azopolymers. There were no differences in photodeformation capabilities between the CN- and

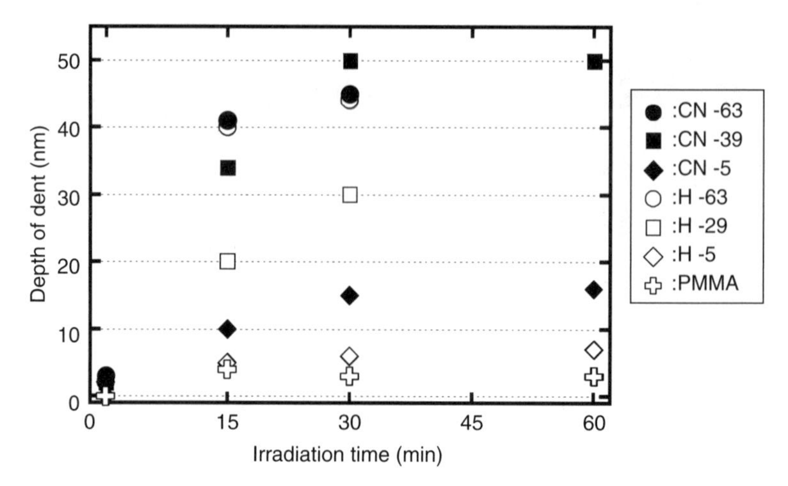

Figure 9.14. Changes in the depth of the dent as a function of photoirradiation time for CN-azopolymers (*solid figures*) and H-azopolymers (*open figures*). The numbers show the weight content of the azo moieties. The open crosses show the control experiment using PMMA. *Source*: Narita, 2007. Reprinted with permission.

H-azopolymers, even though they contained different types of azobenzene. These results show that H-azopolymers could exhibit immobilization capabilities similar to CN-azopolymers, despite the differences in their chemical structures.

Photoimmobilization of IgGs was achieved by performing photoirradiation for 30 min to examine the efficiency of the immobilization process. The relative efficiencies of the photoimmobilization processes on the different azopolymers were plotted as a function of their azobenzene contents, as shown in Fig. 9.15. The immobilization efficiencies of both the CN- and H-azopolymers. increased with azobenzene content up to ~30 wt% and then became saturated, although the saturated values were different between the CN- and H-azopolymers. This result indicates that an azobenzene moiety that exhibits photoisomerization is essential to immobilize IgGs on azopolymers when using a photoimmobilization process.

Next, the authors confirmed the relationship between the immobilization of IgG and the deformation efficiency of the azopolymer. The relative immobilization efficiencies were plotted as a function of indent depth, as shown in Fig. 9.16. Incremental changes in the immobilization efficiency were observed by increasing the depth of the indents. However, the immobilization efficiency of the H-azopolymers was higher than that of the CN-azopolymers across the whole range, and this is also shown in Fig. 9.15. This difference shows that the degree of photoimmobilization is not only affected by the deformation capability but is also a property of the surface of the azopolymer and is related to the chemical structure of the azobenzene that is incorporated in the azopolymer. Whitesides and coworkers have also reported that immobilization is influenced by the properties of the surface (Ostuni et al., 2001).

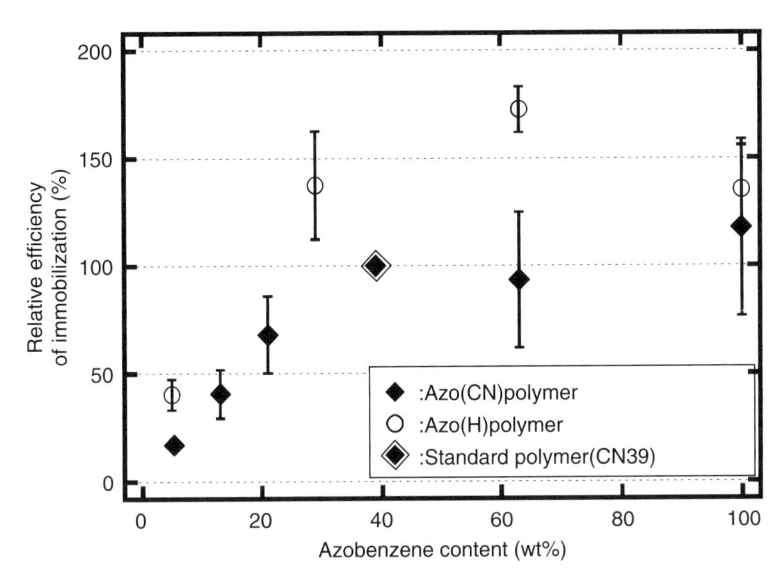

Figure 9.15. Dependence of the relative immobilization efficiency of Cy-5-labeled antibodies on azopolymer content. The solid diamonds and open circles represent CN- and H-azopolymers, respectively. CN-39 was used as the standard polymer, as shown by the solid diamond in an open diamond. *Source*: Narita, 2007. Reprinted with permission.

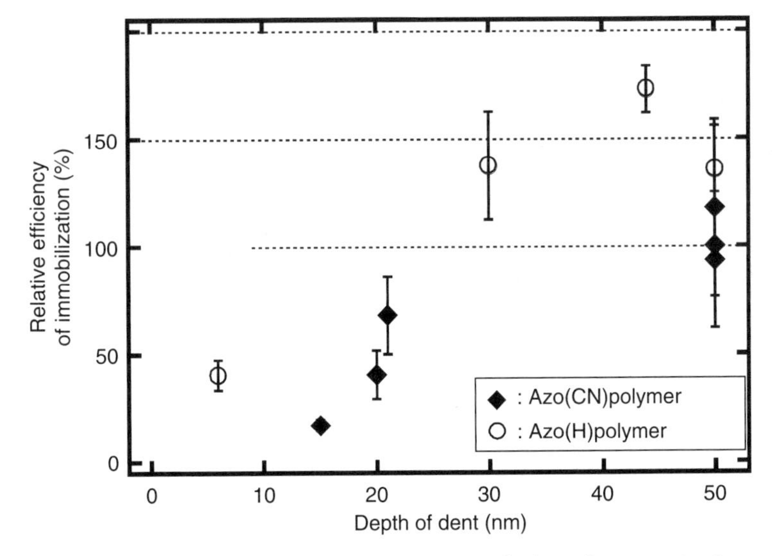

Figure 9.16. Relationship between the depth of the dent and the relative immobilization efficiency of Cy-5-labeled antibodies. The solid diamonds and the open circles represent CN- and H-azopolymers, respectively. *Source*: Narita, 2007. Reprinted with permission.

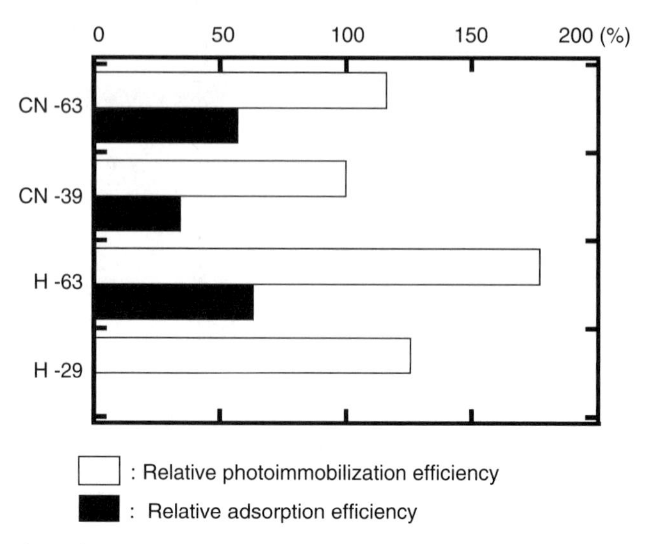

Figure 9.17. The relative photoimmobilization efficiency and the relative adsorption properties of Cy-5-labeled antibodies on the azopolymers. *Source*: Narita, 2007. Reprinted with permission.

Therefore, the authors attempted to examine the efficiency of the adsorption of antibodies onto the surfaces of the azopolymers. The adsorption efficiency for antibodies was determined by the efficiency of the immobilization process without photoirradiation. The value that was obtained for the CN-39 azopolymer was used as a baseline for the relative efficiency of the adsorption of antibodies. The relative adsorption efficiency of Cy-5-IgG on the azopolymers is shown in Fig. 9.17. The relative adsorption efficiency was lower than the relative photoimmobilization efficiency, which also demonstrates that photoirradiation is an important process if one wishes to firmly immobilize most of the antibodies. The relationship between the relative adsorption efficiency and the relative immobilization efficiency of Cy-5-IgG on each of the azopolymers showed that they were almost the same. Although the adsorption efficiency of the H-azopolymer was slightly higher than that of the CN-azopolymer, the difference was not sufficient to explain the differences in the immobilization efficiencies.

The authors conjectured that the adsorption properties of the azopolymers may have originally been equal and that the differences in the photoimmobilization efficiencies could be generated by the photoirradiation process. To characterize these photoprocesses, the adsorption capabilities of the azopolymers after photoirradiation should be considered. Azopolymer films carrying Cy-5-IgG that had been immobilized by photoirradiation were held in stirred PBS to remove the antibodies from the surfaces of the azobenzene films. All the H-azopolymers exhibited much better retention rates than the equivalent CN-azopolymers.

There are remarkable differences between the CN- and H-azopolymers in terms of the photoisomerization phenomena that occur on the azobenzene moieties. The authors also confirmed a stable cis state in films of the H-azopolymer by measuring the changes with time of the absorption capability of the films using a probe light during and after photoirradiation. In the case of the CN-azopolymer films, the cis state was almost totally backisomerized 30 min after the light was turned off. However, the H-azopolymer films showed a stable cis state, and the relaxation time for the cis state was >160 h, which was the time calculated from the recovery curve of the absorbance of the trans state.

It was concluded that the photoimmobilization capability is not only controlled by photodeformation but also by the retention capability, depending on the chemical structure after photoirradiation. Changes in the adsorption properties after the photoimmobilization process are an interesting phenomenon in terms of dynamic changes in the surface properties for adsorption.

9.6. TWO-DIMENSIONAL ARRANGEMENT AND AREA-SELECTIVE IMMOBILIZATION OF MICROSPHERES

Photonic crystals exhibit interesting physical phenomena and enable the manufacture of novel optical devices (Megens et al., 1999). Although a large number of studies directed toward fabricating photonic crystals for photonic applications have been reported, 2-D photonic crystals have attracted a great deal of attention because they provide a more suitable structure for integrated photonic circuit applications such as waveguides (Lončar et al., 2000), channel add/drop filters (Noda et al., 2000), and directional couplers (Koshiba et al., 2000). Several self-assembly approaches for obtaining 2-D colloidal crystals have been reported, such as processes that use capillary force (Kralchevsky and Nagayama, 1994), electrophoretic migration (Giersig and Mulvaney, 1993), and Langmuir–Blodgett films (Lenzmann et al., 1994). Although self-assembled 2-D colloidal crystals are of great interest, several problems remain to be solved before the technique can be applied for practical use; for example, there are the problems of polycrystalline domains, defects, and multilayers in crystals and difficulties associated with designing the arrangement and the intended defect structure. There have been several reports of crystallization on periodically patterned templates for self-assembled 2-D colloidal crystals (Ye et al., 2001). Among these, relief structures fabricated on azobenzene-containing polymer films by photoirradiation with an interference light pattern are one of the most promising approaches for easily forming templates (Ye et al., 2001). Additionally, to apply colloidal crystals to optical devices, it is important to have some form of selective arrangement. An area-selective arrangement of colloidal spheres has been achieved by skillfully managing the surface properties (Masuda et al., 2002). However, no simple method of simultaneously attaining an area-selective and controlled arrangement for immobilized 2-D colloidal crystals has yet been developed.

This section proposes a novel and simple method of solving these problems using two photoinduced phenomena of azobenzene-containing polymers. One is a well-known photodeformation process, which provides an intended template for arranging the microspheres. The other is the newly discovered photoinduced immobilization process described in the preceding sections. First, an indented template with a 2-D lattice structure is formed by repeatedly irradiating with a pattern generated from interfering light beams. Second, colloidal spheres are crystallized on the template structure. Finally, area-selective immobilization provides a 2-D photonic crystal slab that includes waveguides or defects for controlling the light waves.

First, the simple area-selective immobilization of colloidal spheres on an azobenzene-containing urethane polymer was demonstrated. An aqueous solution containing 1-μm diameter polystyrene microspheres (Duke Scientific Corp., 5100A) was dropped onto the flat surface of the polymer film, and the solution was then sucked up with a pipette to form self-organized colloidal crystals. Area-selective photoimmobilization was performed by moving the irradiation site using a confocal laser-scanning microscope with a wavelength of 488 nm (Olympus, OLS1100). The sample was washed with ultrasonic cleaning to remove any un-immobilized and multilayered spheres and was then examined with a microscope after drying. Figure 9.18 shows a checkerboard design of 2-D colloidal spherical crystals immobilized on polymer films. The microspheres were immobilized only in the irradiated region. Photoinduced immobilization provides a simple method by which a patterned monolayer of spheres can be easily and selectively immobilized on the substrate.

Second, the implementation of designed arrays of 2-D colloidal crystals immobilized on a polymer was demonstrated. Two sets of gratings were formed

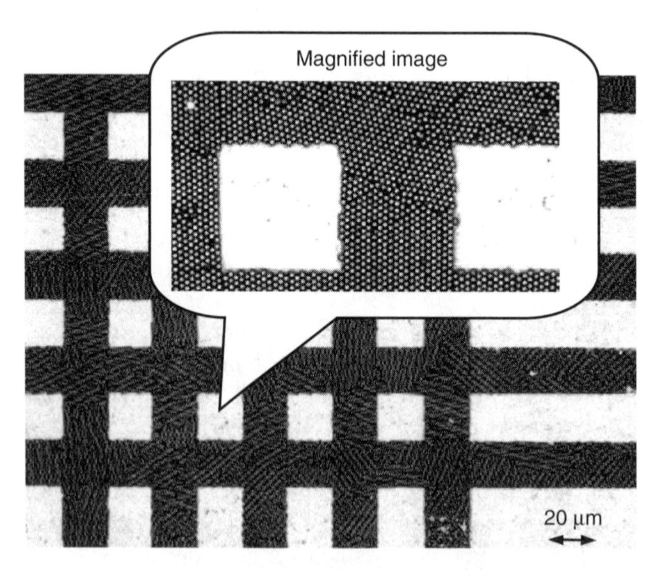

Figure 9.18. Microscope image of area-selective immobilized 1-μm diameter polystyrene microspheres on the flat surface of an azobenzene-containing polymer film. The irradiated area was controlled by moving the sample stage. *Source*: Watanabe, 2006.

on a polymer film by irradiating with an interference pattern generated from an Ar-ion laser beam. Two kinds of cross-grating indented templates with 2-D tetragonal and hexagonal lattices were fabricated on the polymer, as shown in Fig. 9.19 (left-hand side). The self-assembly of colloidal crystals on the polymer was carried out by using the dipping method. This template film was immersed in an aqueous solution of microspheres and was then drawn up at a rate of 1 mm/min. The samples were irradiated with an Ar-ion laser (488 nm) to immobilize the microspheres and were then ultrasonically cleaned. Monolayered tetragonal and

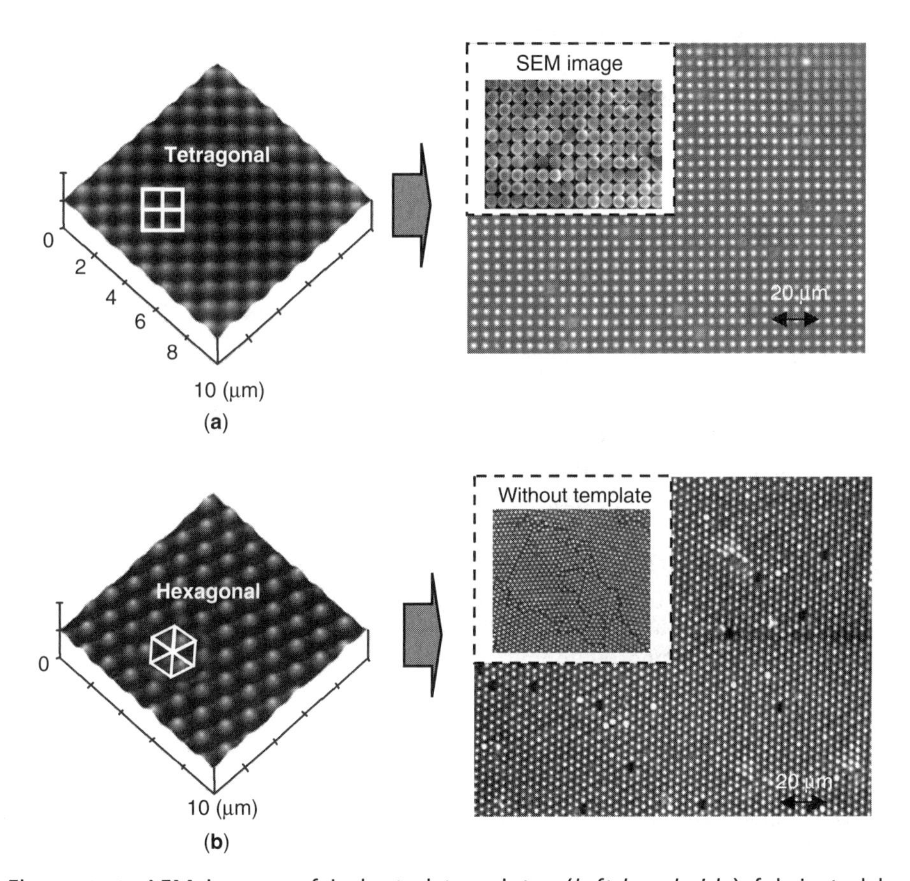

Figure 9.19. AFM images of indented templates (*left-hand side*) fabricated by repeated irradiation of an azopolymer with an interference light pattern. The lattice structures are (**a**) tetragonal and (**b**) hexagonal. The structures were controlled by the incident angle of the interfering light pattern and the rotation angle in the plane before the second irradiation. Microscope images of photo-immobilized 2-D colloidal crystals (*right-hand side*) after arrangement on the templates. The small image in (**a**) shows SEM images of the photoimmobilized tetragonal arrangement, and the small microscope image in (**b**) shows self-assembled colloidal crystals with a multidomain structure on the flat surface. *Source*: Watanabe, 2006.

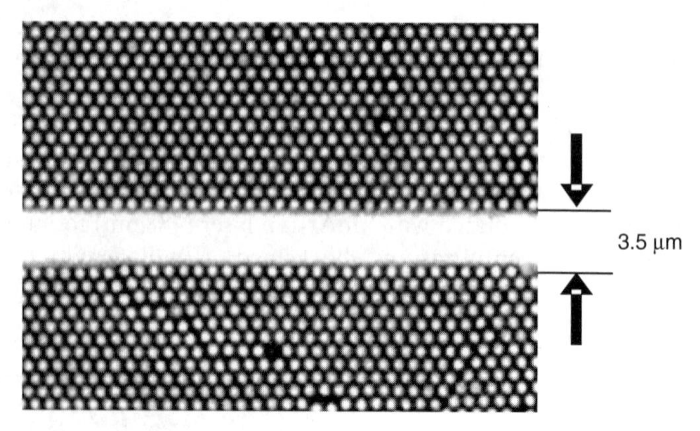

Figure 9.20. Area-selective photoimmobilized 2-D colloidal crystals (waveguide-type pattern). *Source*: Watanabe, 2006.

hexagonal arrangements of the microspheres were obtained from the corresponding templates, as shown in Fig. 9.19 (right-hand side), though there are a few defects present and also a multilayered area. The hexagonal arrangement of colloidal crystals on the template had an approximately unity structure without being multidomain, compared with the self-assembled arrangement on the flat film, as shown in the small image in Fig. 9.19b. The combination of both the arrangement on the template and the photoinduced immobilization provides an excellent method for fabricating large area 2-D colloidal crystals with controlled lattices and low defect densities.

Finally, the area-selective photoimmobilization of 2-D arrays of colloidal spheres on templates formed in azobenzene-containing polymer films was demonstrated. Area-selective photoimmobilization was combined with the process whereby microspheres can be arranged on a template. Similar to the processes described earlier, area-selective photoimmobilization was performed by setting the template containing the array of spheres onto a moveable sample stage. After an ultrasonic wash, the 2-D colloidal crystals were examined. Figure 9.20 shows waveguide-type 2-D crystals with a hexagonal arrangement. Several other types of patterns were also examined, such as bending waveguides and cross-type waveguides.

To conclude, the authors have succeeded in forming 2-D photonic crystal slabs that include deliberately introduced defects or waveguides. This was accomplished by a newly proposed method comprising two processes, the arrangement and the area-selective immobilization of microspheres. These processes were made possible by utilizing two different photoresponsive properties of azobenzene-containing polymers; namely, photodeformation and photoimmobilization.

9.7. SUMMARY

Among a number of photochromic materials, azobenzene derivatives have a distinguished property that induces a spatially extended change in form because of their geometric isomerization. The SRG-related investigations inspired by the Natansohn and Tripathy groups take an advantage of this characteristic property effectively. The authors have introduced the principles of a newly created photoimmobilization technology using photoresponsive azopolymers and reviewed its application to the fabrication of immunochips and the arrangement of microspheres. Although there have been a number of investigations into the photoinduced functionality of azopolymers, further interesting problems remain to be solved from the viewpoint of basic research.

The authors consider two important factors in their approach: the deformation process that is induced by interaction between the microobjects and the azopolymer surface and the immobilization process that resulted from the deformation process. In the first case, a novel aspect should be added, such as surface energy, radiation force, and intermolecular force as the moving force. The established knowledge accumulated by investigations concerning SRG, photoinduced orientation, and photoisomerization is obviously important to understand the mechanism. The photoresponse process would be controlled by the interaction, and this interaction would be affected by photoresponse conversely. In the latter case, the research field could enlarge by considering how to use the immobilized surface such as biochip and bioreactor. In particular, the interaction from the position of adsorption in biological engineering should also be reconsidered. The interaction concerning the adsorption would involve controlling the arrangement and orientation of microobjects. Not only the behavior and application of azopolymer by itself but also the relevant interaction with microobjects on the surface should be considered. The authors believe that there is further potential to develop this technique more widely; for instance, applications involving biological molecules are an intriguing field with high potential for future growth, including aspects of molecular orientation and the formation of organized structures. This novel approach would serve as a steppingstone to further development.

REFERENCES

Andruzzi L, Altomare A, Ciardelli F, Solaro R, Hvilsted S, Ramanujam PS. 1999. Holographic gratings in azobenzene side-chain polymethylmethacrylates. Macromolecules 32:448–454.

Barada D, Fukuda T, Itoh M, Yatagai T. 2006. Numerical analysis of photoinduced surface relief formed on azobenzene polymer film by optical near-field exposure. Jpn J Appl Phys 45:6730–6737.

Barret CJ, Natansohn AL, Rochon PL. 1996. Mechanism of optically inscribed high-efficiency diffraction gratings in azo polymer films. J Phys Chem 100:8836–8842.

Batalla E, Natansohn AL, Rochon PL. 1995. Optically induced surface gratings on azoaromatic polymer films. Appl Phys Lett 66:136–138.

Berg RH, Hvilsted S, Ramanujam PS. 1996. Peptide oligomers for holographic data storage. Nature 383:505–508.

Betzig E, Trautman JK. 1992. Near-field optics: microscopy, spectroscopy, and surface modification beyond the diffraction limit. Science 257:189–219.

Feng CL, Zhang Z, Forch R, Knoll W, Vancso GJ, Schonherr H. 2005. Reactive thin polymer films as platforms for the immobilization of biomolecules. Biomacromolecules 6:3243–3251.

Fukuda T, Sumaru K, Kimura T, Matsuda H, Narita Y, Inoue T, Sato F. 2001. Observation of optical near-field as photo-induced surface relief formation. Jpn J Appl Phys 41:L900–L902.

Giersig M, Mulvaney P. 1993. Preparation of ordered colloid monolayers by electrophoretic deposition. Langumuir 9:3408–3413.

Hartmann M. 2005. Ordered mesoporous materials for bioadsorption and biocatalysis. Chem Mater 17:4577–4593.

Hasegawa M, Ikawa T, Tsuchimori M, Watanabe O, Kawata Y. 2001. Topographical nanostructure patterning on the surface of a thin film of polyuerthane containing azobenzene moiety using the optical near field around polystyrene spheres. Macromolecules 34:7471–7476.

Hasegawa M, Ikawa T, Tsuchimori M, Watanabe O. 2002a. Photochemically induced birefringence in polyurethanes containing donor-acceptor azobenzenes as photoresponsive moieties. J Appl Polym Sci 86:17–22.

Hasegawa M, Keum C-D, Watanabe O. 2002b. Enhanced photofabrication of a surface nanostructure on azobenzene-functionalized polymer films with evaporated gold nanoislands. Adv Mater 14:1738–1741.

Hvilsted S, Andruzzi F, Kulinna C, Siesler HW, Ramanujam PS. 1995. Novel side-chain liquid crystalline polyester architecture for reversible optical storage. Macromolecules 28:2172–2183.

Ichimura K. 2000. Photoalignment of liquid-crystal systems. Chem Rev 100:1847–1873.

Ikawa T, Mitsuoka T, Hasegawa M, Tsuchimori M, Watanabe O, Kawata Y. 2000. Optical near field induced change in viscoelasticity on an azobenzene-containing polymer surface. J Phys Chem B 104:9055–9058.

Ikawa T, Hasegawa M, Tsuchimori M, Watanabe O, Kawata Y, Egami C, Sugihara O, Okamoto N. 2001a. Surface deformation on azobenzene polymer film induced by optical near-field around polystyrene microspheres. Synth Met 124:159–161.

Ikawa T, Mitsuoka T, Hasegawa M, Tsuchimori M, Watanabe O, Kawata Y. 2001b. Azobenzene polymer surface deformation due to the gradient force of the optical near field of monodispersed polystyrene spheres. Phys Rev B 64:195408.

Ikawa T, Hoshino F, Matsuyama T, Takahashi H, Watanabe O. 2006. Molecular-shape imprinting and immobilization of biomolecules on a polymer containing azo dye. Langmuir 22:2747–2753.

Ikeda T, Tsutsumi O. 1995. Optical switching and image storage by means of azobenzene liquid-crystals films. Science 268:1873–1875.

Kambhampati D, editor. 2003. Protein Microarray Technology. Weinheim: Wiley-VCH Verlag GmbH.

Karlsson R, Michaelsson A, Mattsson L. 1991. Kinetic-analysis of monoclonal anitibody-antigen interactions with a new biosensor base analytical system. J Immunol Methods 145:229–240.

Kendall C, Ionescumatiu I, Dresmann GR. 1983. Utilization of the biotin avidin system to amplify the sensitivity of the enzyme-linked immunosorbent. J Immuno Methods 56:329–339.

Keum C-D, Ikawa T, Tsuchimori M, Watanabe O. 2003. Photodeformation behavior of photodynamic polymers bearing azobenzene moities in their main and/or side chain. Macromolecules 36:4916–4923.

Kim DY, Tripathy SK, Li L, Kumar J. 1995. Laser-induced holographic surface relief gratings on nonlinear optical polymer films. Appl Phys Lett 66:1166–1168.

Knoll B, Keilmann F. 1999. Near-field probing of vibrational absorption for chemical microscopy. Nature 399:134–137.

Koshiba M, Tsuji Y, Hikari M. 2000. Time-domain beam propagation method and its application to photonic crystal circuits. J Lightwave Technol 18:102–110.

Kralchevsky PA, Nagayama K. 1994. Capillary forces between colloidal particles. Langumuir 10:23–26.

Kumar J, Li L, Jiang XL, Kim DY, Lee TS, Tripathy S. 1998. Gradient force: the mechanism for surface relief grating formation in azobenzene functionalized polymers. Appl Phys Lett 72:2096–2098.

Lee Y, Lee EK, Cho YW, Matsui T, Kang I-C, Kim T-S, Han MH. 2003. ProteoChip: a highly sensitive protein microarray prepared by a novel method of protein, immobilization for application of protein-protein interaction studies. Proteomics 3:2289–2304.

Lenzmann F, Li K, Kitai AH, Stover HD. 1994. Thin-film micropatterning using polymer microspheres. Chem Mater 6:156–159.

Levy I, Shoseyov O. 2004. Cross bridging proteins in nature and their utilization in bio- and nanotechnology. Curr Protein Pept Sci 5:33–49.

Lončar MD, Doll T, Vučković, J, Sherer A. 2000. Design and fabrication of silicon photonic crystal optical waveguides. J Lightwave Technol 18:1402–1411.

Masuda Y, Itoh M, Yonezawa, T, Koumoto K. 2002. Low-dimensional arrangement SiO2 particles. Langmuir 18:4155–4159.

Megens M, Wijnhoven JEGJ, Lagendijk A, Vos WL. 1999. Fluorescence lifetimes and linewidths of dye in photonic crystals. Phys Rev A 59:4727–4731.

Natansohn A. 1999. Azobenzene-Containing Materials. Weinheim: Wiley-VCH Verlag Gmbh.

Natansohn A, Rochon P, Gosselim J, Xie S. 1992. Azo polymers for reversible optical storage. 1. Poly[4′-[[2-(acryloxy)ethyl]ethylamino]-4-nitrobenzene]. Macromolecules 25:2268–2273.

Narita M, Hoshino F, Mouri M, Tsuchimori M, Ikawa T, Watanabe O. 2007. Photoinduced immobilization of biomolecules on the surface of azopolymer films and its dependence on the concentration and type of the azobenzene moiety. Macromolecules 40:623–629.

Noda S, Chutinan A, Imada M. 2000. Trapping and emission of photons by a single defect in a photonic bandgap structure. Nature 407:608–610.

Ohtsu M. 1998. Near-Field Nano/Atom Optics and Technology. Tokyo: Springer-Verlag.

Oliveira ON, dos Santos DS, Balogh DT, Zucolotto V, Medonca CR. 2005. Optical storage and surface-relief gratings in azobenzene-containing nanostructured films. Adv Colloid Interface Sci 116:179–192.

Ostuni E, Chapman RG, Holmlin RE, Takayama S, Whitesides GM. 2001. A survey of structure-property relationships of surfaces that resist the adsorption of protein Langmuir, 17:5605–5620.

Ouyang Z, Takats Z, Blake TA, Gologan B, Guymon AJ, Wiseman JM, Oliver JC, Davisson VJ, Cooks RG. 2003. Preparing protein microarrays by soft-landing of mass-selected ions. Science 301:1351–1354.

Pedersen TG, Johansen PM, Holme NCR, Ramanujam PS, Hvilsted S. 1998. Mean-field theory of photoinduced formation of surface reliefs in side-chain azobenzene polymers. Phys Rev Lett 80:89–92.

Peluso P, Wilson DS, Do D, Tran H, Venkatasubbaiah M, Quincy D, Heidecker, B, Poindexter K, Tolani N, Phelan M, Witte K, Jung LS, Wagner P, Nock S. 2003. Optimizing antibody immobilization strategies for the construction of protein microarrays. Anal Biochem 312:113–124.

Rabek JF. 1988. Photochemistry and Photophysics Vol. II. Florida: CRC Press.

Todorov T, Nikolova L, Tomova N. 1984. Polarization holography. 1: A new high-efficiency organic material with reversible photoinduced birefringence. Appl Opt 23:4309–4312.

Tripathy SK, Kim DY, Jiang L, Li L, Lee TS, Wang X, Kumar J. 1997. New paper based photonic azopolymers for recording and retrieval. Proc. SPIE Int. Soc. Opt. Eng. 3227:176–183.

Voller A, Bartlett A, Bidwell DE. 1978. Enzyme immunoassays with special reference to ELISA techniques J Clin Pathol 31:507–520.

Watanabe O. 2004. Molecular recognition and immobilization of biomolecules induced by photo-irradiation on the surface of an azopolymer R&D Review of Toyota CRDL 39 46. http://www.tytlabs.co.jp/english/review/index.html

Watanabe O, Ikawa T, Hasegawa M, Tsuchimori M, Kawata Y, Egami C, Sugihara O. 2000. Transcription of near-field induced by photo-irradiation on a film of azo-containing urethane-urea copolymer. Optical near field induced change in viscoelasticity on an azobenzene-containing polymer surface. Mol Cryst Liq Cryst Sci Technol Sect A 345:305–310.

Watanabe O, Ikawa T, Hasegawa M, Tsuchimori M, Kawata Y. 2001. Nanofabrication induced by near-field exposure from a nanosecond laser pulse. Appl Phys Lett 79: 1366–1368.

Watanabe O, Narita M, Ikawa T, Tsuchumori M. 2006. Photo-induced deformation behavior depending on the glass transition temperature on the surface of urethane copolymers containing a push-pull type azobenzene moiety. Polymer 47:4742–4749.

Watanabe O, Ikawa T, Kato T, Tawata M, Shimoyama H. 2006. Area-selective photo-immobilization of a two-dimensional array of colloidal spheres on a photodeformed template formed in photoresponsive azopolymer film. Appl Phys Lett 88:204107.

Yager KG, Barrett CJ. 2006. Novel photo-switching using azobenzene functional materials J Photochem Photobiol A Chem 182:250–261.

Yamamoto T, Hasegawa M, Kanazawa A, Shiono T, Ikeda T. 1999. Phase-type gratings formed by photochemical phase transition of polymer azobenzene liquid crystals: enhancement of diffraction efficiency by spatial modulation of molecular alignment. J Phys Chem 103:9873–9878.

Ye YH, Badilescu S, Truong VV, Rochon P, Natansohn A. 2001. Self-assembly of colloidal spheres on patterned substrates. Appl Phys Lett 79:872–874.

Zhao Y, Bai SY, Asatryan K, Galstian T. 2003. Holographic recording in a photoactive elastomer Adv Funct Mater 13:781–788.

Zilker SJ, Bieringer T, Haarer D, Stein RS, van Egmond JW, Kostromine SG. 1998. Holographic data storage in amorphous polymers. Adv Mater 10:855–857.

10

PHOTOTUNING OF HELICAL STRUCTURE OF CHOLESTERIC LIQUID CRYSTALS

Seiji Kurihara

10.1. INTRODUCTION

Cholesteric liquid crystals (Ch LCs) have attracted the keenest interest of scientists because of their unique helical supramolecular structures, which can reflect light selectively in particular spectral regions (Feringa et al., 2000; De Gennes and Prost, 1993; Finkelmann and Stegemeyer, 1978, 1974). A Ch LC can be induced in a nematic LC system by doping a chiral compound. Ch phases are characterized by the helical packing of the mesogens with a certain sign and a certain pitch. The helical pitch, which is defined as the distance in an LC needed for the director of the individual mesogens to rotate through a full 360°, is a measure of the chirality of the system. The optical properties of the Ch LCs depend, in part, on their pitch, and these can be altered either by changing the structure of LC molecules themselves or by admixing auxiliaries that change the helical arrangements of the molecules of the Ch LCs (Gottarelli and Spada, 1985; Gottarelli et al., 1983, 1981). When illuminated with white light, the Ch LCs reflect light of particular wavelengths dependent on the helical pitch of the LC phase (De Gennes and Prost, 1993). According to the theory, the selective reflection wavelength (λ) of the Ch LCs is governed by the following equation as shown in Fig. 10.1:

$$\lambda = np \tag{10.1}$$

where p is the helical pitch length and n is the refractive index of the LC.

Another important parameter of Ch LC systems is the so-called helical twisting power (HTP), defined as the ability of a chiral group to induce Ch mesomorphism in a nematic host. HTP depends on the dipole–quadruple

Smart Light-Responsive Materials. Edited by Yue Zhao and Tomiki Ikeda

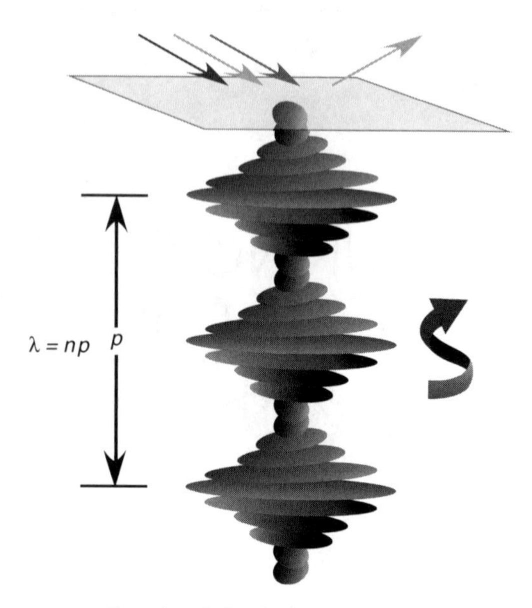

Figure 10.1. Sketch of the helical structure of a Ch LC.

interactions of the chiral molecule with its nematogenic neighbors, the anisotropy of the nematic host phase, and on the order parameter (Goossens, 1971). HTP is related to the helical pitch by the following equation (De Gennes and Prost, 1993):

$$\mathrm{HTP} = \frac{1}{pc} \tag{10.2}$$

where c is the concentration of the dopant. Thus, higher HTP values require less chiral dopant to yield the same λ value. In addition, it is reported that HTP of the chiral dopants depends on their structures, and the chiral dopants with LC-like structure possess relatively higher HTP. Generally, anisotropic molecular shapes, such as rod shape, are required for liquid crystallinity and the appearance of LC phases (Demus, 1998; De Gennes and Prost, 1993). Therefore, if the shapes of molecules in LCs can be changed photochemically, LC properties including helical pitch length can be switched photochemically.

Many studies have been done on photocontrolling of helical strcuture of Ch LCs through photoisomerization of various types of photochromic compounds, with/without chiral groups in molecules, such as azobenzenes (Bobrovsky and Shibaev, 2005; Kumaresan et al., 2005; Pieraccini et al., 2004; van Delden et al., 2004; Mallia and Tamaoki, 2003; Tamaoki et al., 2003; Bobrovsky and Shibaev, 2002; Ruslim and Ichimura, 2002; Moriyama and Tamaoki, 2001; Ruslim and Ichimura, 2001; Lee et al., 2000; Ruslim and Ichimura, 2000, Tamaoki et al., 2000;

Sackmann, 1971) , stilbenes (Haas et al., 1974), menthones (Van de Witte et al., 1999, 1998; Brehmer et al., 1998), diarylethenes (Yamaguchi et al., 2001; Denekamp and Feringa, 1998), overcrowded alkenes (Eelkema et al., 2006; van Delden et al., 2003; Feringa et al., 1995), and flugides (Sagisaka and Yokoyama, 2000). When the photochromic compounds with chiral groups are dissolved in a nematic LC, phototuning of the induced helical structure can be easily achieved by the use of suitable light, leading to a difference in HTP between photoisomers through photochemical change in their molecular shapes. Azobenzene molecules undergo reversible trans–cis photoisomerization by irradiation of light with appropriate wavelength (Ross and Blank, 1971), and when a chiral azobenzene molecule is doped in Ch LCs, its HTP can be changed by reversible photoisomerization. Thus, according to Eq. 10.2, the helical pitch can be controlled through reversible photoisomerization.

Azobenzene systems represent very attractive phototriggers in Ch LCs, thanks to their resistance to photofatigue, the simplicity of the molecules, and the ease of modification of their molecular structures. This chapter first describes helical twisting ability of chiral azobenzene compounds by focusing the structural effects on HTP as well as photochemical change in HTP, and then phototuning of helical structures of Ch LCs through the trans–cis photoisomerization of the chiral azobenzene compounds for applications to optical devices.

10.2. PROPERTIES AND DESIGN OF CHIRAL AZOBENZENES

10.2.1. Effect of Spacer Length

Figure 10.2 shows chiral and nonchiral azobenzene compounds. A Ch phase was induced by mixing each chiral azobenzene compound in a host nemaic LC. In such a binary system consisting of a nonchiral host nematic LC and a chiral compound, the reciprocal of the helical pitch ($1/p$) is known to increase linearly with the concentration (c) of the chiral compound at lower c, and HTP of the chiral compound can be defined as the slope of the $1/p$ versus c (Eq. 10.2). Figure 10.3 shows changes in the $1/p$ of the induced Ch LCs as a function of c in the dark and under ultraviolet (UV) light (365 nm) to cause trans–cis photoisomerization of the chiral azobenzene compounds. The helical pitch was determined by Cano's method (Bahr et al., 1991). The $1/p$ increased almost linearly with the increase of c. HTP of the chiral azobenzenes was on the order of azo-1 > azo-2 > azo-3: 23×10^2 for azo-1, 6.6×10^2 for azo-2, and 2.6×10^2 ($mm^{-1} mol^{-1}$ g of E44) for azo-3, respectively (Kurihara et al., 2000). HTP was decreased by increasing the distance between azobenzene group and chiral center. The distance between the azo group and the chiral center is one of the factors influencing the twisting ability. A decrease in the $1/p$ was brought about photochemically by UV irradiation. This result means that the trans to cis photoisomerization of these azobenzene compounds caused the decrease in the helical pitch length.

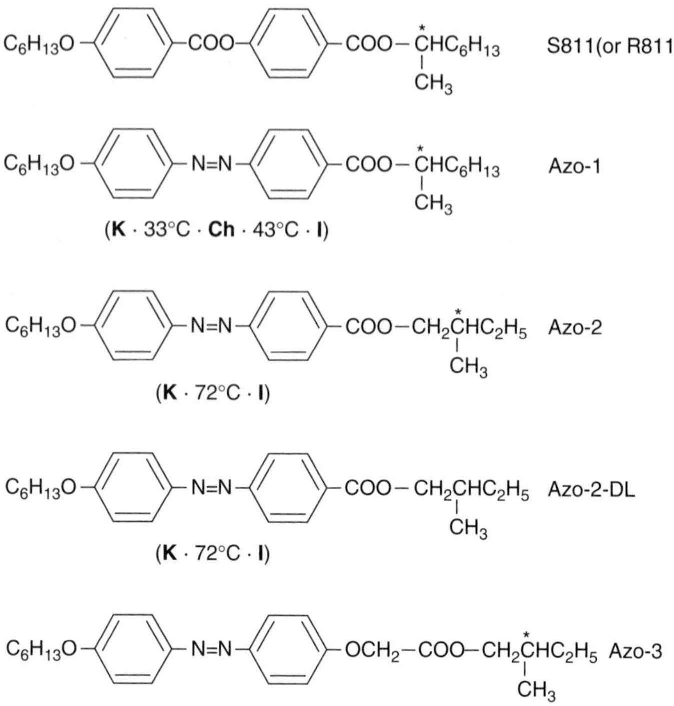

$C_6H_{13}O$–⟨benzene⟩–COO–⟨benzene⟩–COO–$\overset{*}{C}H(CH_3)C_6H_{13}$ S811(or R811)

$C_6H_{13}O$–⟨benzene⟩–N=N–⟨benzene⟩–COO–$\overset{*}{C}H(CH_3)C_6H_{13}$ Azo-1

(**K** · 33°C · **Ch** · 43°C · **I**)

$C_6H_{13}O$–⟨benzene⟩–N=N–⟨benzene⟩–COO–$CH_2\overset{*}{C}H(CH_3)C_2H_5$ Azo-2

(**K** · 72°C · **I**)

$C_6H_{13}O$–⟨benzene⟩–N=N–⟨benzene⟩–COO–$CH_2CH(CH_3)C_2H_5$ Azo-2-DL

(**K** · 72°C · **I**)

$C_6H_{13}O$–⟨benzene⟩–N=N–⟨benzene⟩–OCH_2–COO–$CH_2\overset{*}{C}H(CH_3)C_2H_5$ Azo-3

Figure 10.2. Structures of chiral and achiral compounds.

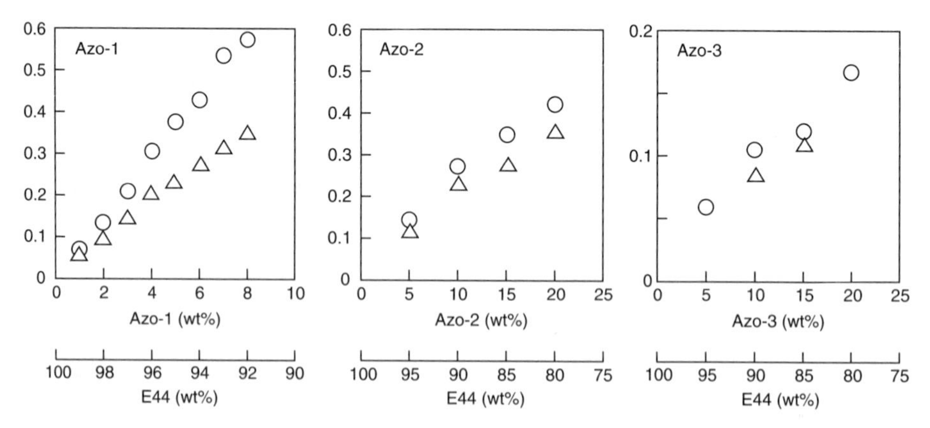

Figure 10.3. Changes in helical pitch as a function of the chiral azobenzene compounds (azo-1, azo-2, and azo-3) in E44 before (○) and after (Δ) UV irradiation at 40°C.

To explore the effects of the change in the molecular shape on HTP, computational study was carried out. Optimized geometry and parameters of 5CB, *trans*-azobenzene and *cis*-azobenzene molecules are shown in Figs. 10.4–10.6. 5CB is a major component of E44, which is used as the host nematic LC. Parameters estimated by ab initio calculation are consistent with parameters determined experimentally. In addition, empirical and semiempirical calculations (MM and MOPAC/PM3) also gave good consistency with the experimental results (Sanpei, 2005; Fukunaga et al., 2004; Stevensson et al., 2003, 2001). These results indicate clearly that the shapes of *trans*-azobenzene and *cis*-azobenzene molecules are rod and bent, respectively. The *trans*-azobenzene molecule usually dissolves in a host LC without extreme perturbation effect on a molecular orientation of the host LC, because the rodlike shape of the *trans*-azobenzene molecule is similar to that of the host LC molecule. However, the bent-shaped *cis*-azobenzene molecule disorganizes the molecular orientation of the host LC (Tazuke, 1987). The photochemical change in the molecular shape may be one of the factors for the photochemical change in HTP of the chiral azobenzene compounds.

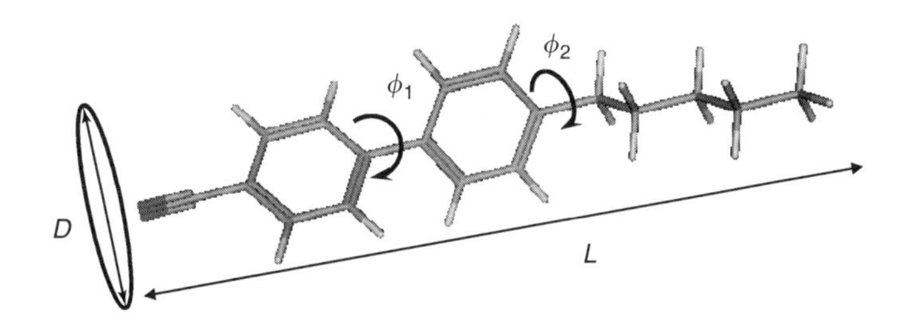

Parameter	Observed	Method			
		Empirical	Semiempirical		ab initio
		MMFF94x	AM1	PM3	B3LYP/6–31G**
ϕ_1, degree	35	52.7	40.2	47.5	36.5
ϕ_2, degree	90	89.2	76.8	84.9	90.3
L, Å	15.6–16.5	16.24	16.14	16.12	16.38
D, Å	4.5–5.0	5.52	5.52	5.52	5.05
L/D	3.1–3.6	2.94	2.92	2.92	3.24
Dipole moment (Debye)	4.34 (In benzene)	—	4.12	4.27	5.96

Figure 10.4. Structure and its parameters of 5CB(4-pentyl-4′-cyanobiphenyl).

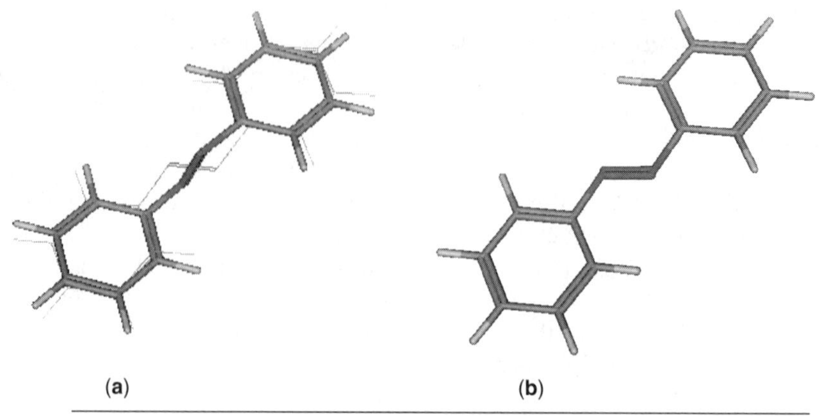

Parameter	Observed	Method		
		Empirical MMFF94x	Semiempirical PM3	ab initio B3LYP/6–31G**
r(N=N), Å	1.247	1.250	1.231	1.261
r(N-C)	1.428	1.405	1.446	1.419
r(C-C)	1.389	1.40	1.40	1.398
a(NNC), degree	113.6	109.8,120.1	119.6	114.8
a(NCC)	115.6,123.7	119.8	115.5	124.8
d(CNNC)	180	180	180	180
μ (Debye)	0.0	—	0.0	0.0
L	—	11.1	11.4	11.3
D	—	7.1	5.0	5.0
L/D	—	1.6	2.3	2.3

Figure 10.5. Comparison of experimental (thin) and optimized geometry of *trans*-azobenzene from (**a**) empirical (stick); thin: experimental, stick: empirical(MMFF94x) and (**b**) semiempirical (stick) method; thin: experimental, stick: semiempirical(PM3).

To clarify the effect of the trans–cis photoisomerization of the chiral azobenzene compounds on the change in HTP, the change in the helical pitches of Ch LCs containing nonchiral azobenzene compound, *p*-azo-7-dl (Fig. 10.7), which was derived with racemic alcohol, was examined. Ch LCs were prepared by adding R811 and chiral or nonchiral azobenzene compounds with the same structure, *p*-azo-7-dl or *p*-azo-7, in E44. In the case of the (R811/*p*-azo-7/E44) sample, the helical pitch was increased by the trans–cis photoisomerization of

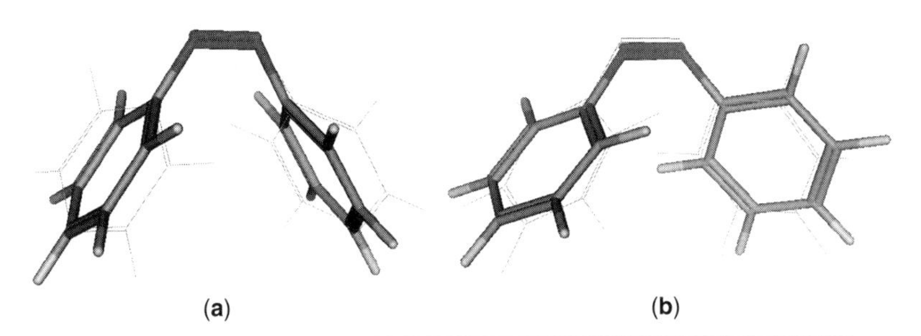

Parameter	Observed	Method		
		Empirical MMFF94x	Semiempirical PM3	ab initio B3LYP/6–31G**
r(N=N), Å	1.253	1.250	1.216	1.249
r(N-C)	1.449	1.405	1.451	1.436
r(C-C)	1.398	1.40	1.394	1.398
a(NNC), degree	121.9	109.8,120.1	128.2	124.1
a(NCC)	122.5	119.8	123.7	123.0
d(CNNC)	8.0	180	0.2	9.4
μ (Debye)	—	—	3.05	1.00
L, Å	—	11.1	9.0	9.0
D, Å	—	7.1	6.2	5.0
L/D	—	1.6	1.5	1.8

Figure 10.6. Comparison of experimental (thin) and optimized geometry of *cis*-azobenzene from **(a)** empirical (stick); thin: experimental, stick: empirical(MMFF94x) and **(b)** semiempirical (stick) method; thin: experimental, stick: semiempirical(PM3).

p-azo-7 under UV light as shown in Fig. 10.8. Contrary to this sample, the (R811/*p*-azo-7-dl/E44) sample showed a decrease in the helical pitch by the trans–cis photoisomerization of *p*-azo-7-dl (Fig. 10.8).

This result indicates that the photochemical change in HTP is related not only to the perturbation effect of the trans–cis photoisomerization on the molecular orientation but also to photochemical change in the intermolecular interaction between the *trans* or the *cis* chiral azobenzene molecule and the host LC molecule.

10.2.2. Effects of Molecular Shape

HTP of some chiral azobenzene compounds in E44 are given in Table 10.1. As described in the preceding section, HTPs of the chiral compounds increased as the distance between azobenzene group and chiral group became short. Therefore, the

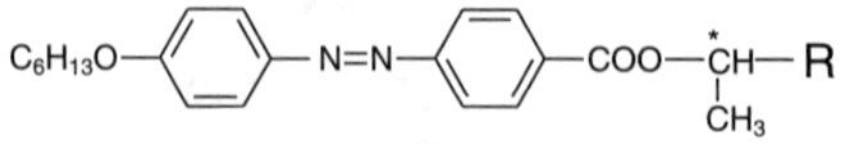

Azo-1 (R=C_6H_{13})
Azo-4 (R=$COOCH_3$)
Azo-5 (R=$COOC_2H_5$)
Azo-6 (R=Ph)

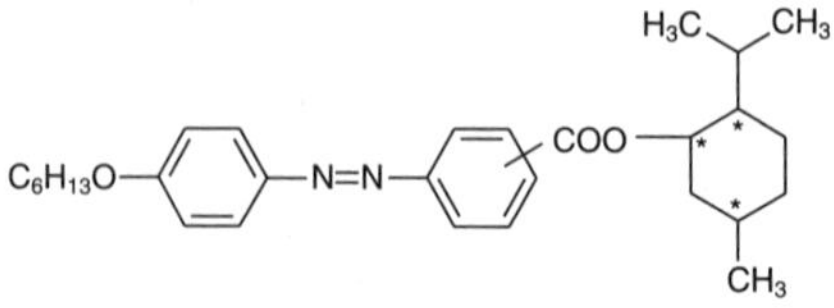

o-, *m*-, *p*-Azo-7

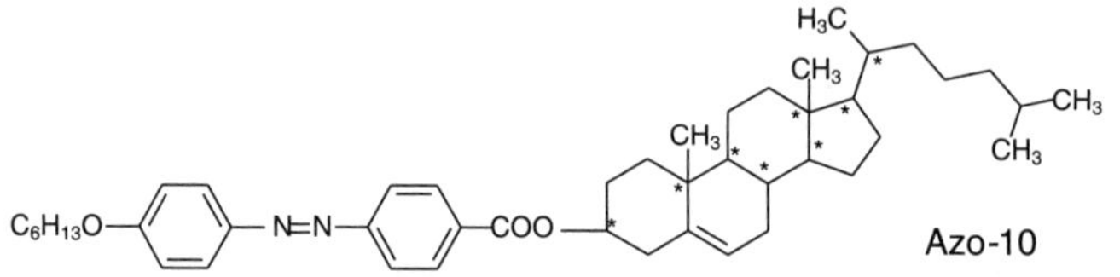

Azo-10

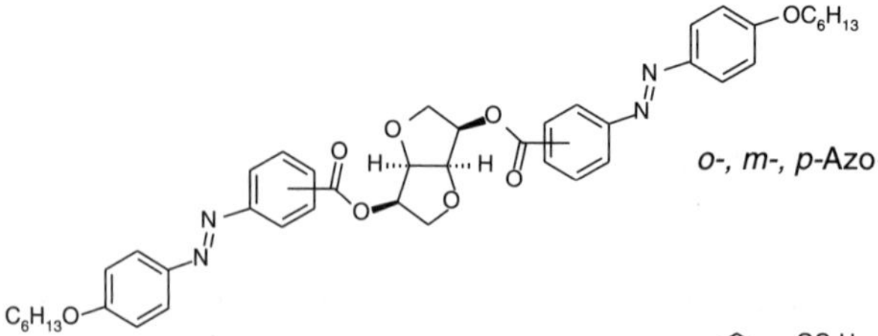

o-, *m*-, *p*-Azo-8

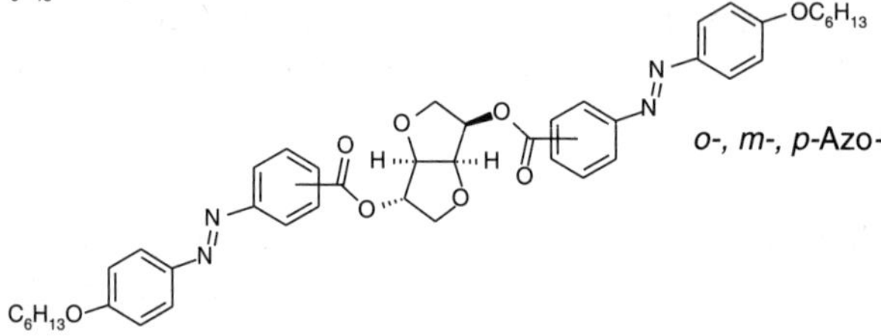

o-, *m*-, *p*-Azo-9

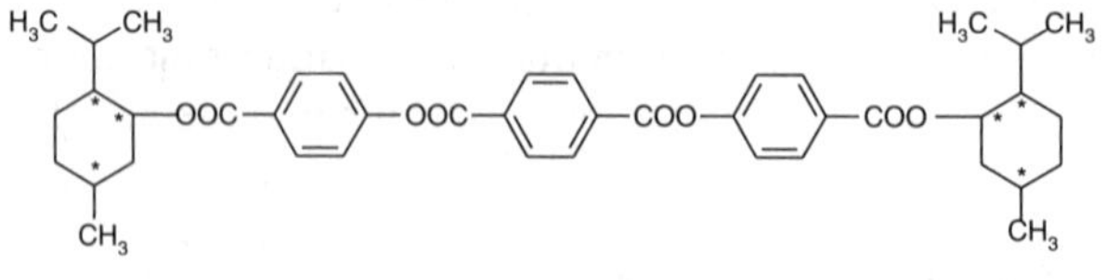

Chiral

Figure 10.7. Structures of chiral azobenzene compounds and a nonphotochromic chiral compound.

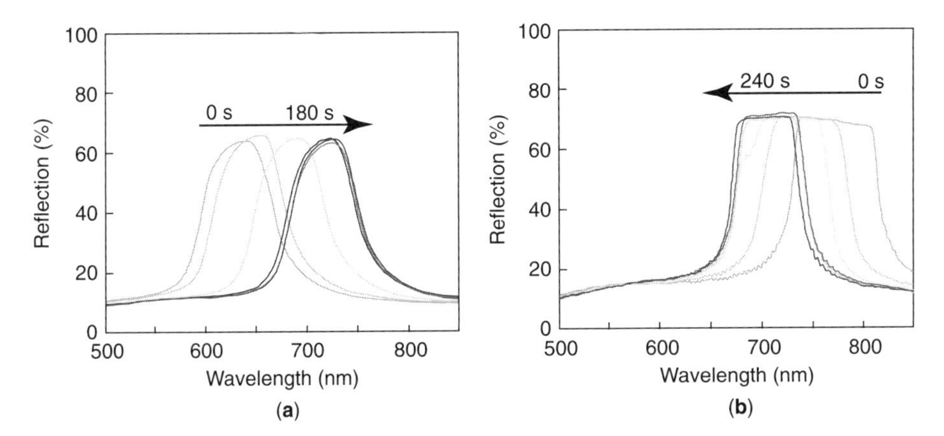

Figure 10.8. Changes in reflection spectra of (a) (E44/R811/*p*-azo-7)(74.9/20.0/5.6 wt%) and (b) (E44/R811/*p*-azo-7-dl)(74.1/20.3/5.6 wt%) by UV light irradiation (45 mW/cm^2).

chiral azobenzene compounds were synthesized by introducing different chiral groups into the same azobenzene group via ester bond without any spacers (Fig. 10.7). HTPs of some chiral compounds without azobenzene group are given in Table 10.2.

It is noteworthy that HTP is increased by the introduction of the azobenzene group into the chiral compounds. This may be due to the enhancement of the intermolecular interaction between chiral molecule and LC molecule by the introduction of the azobenzene group with the rod shape. The molecular long axis and molecular aspect ratio are increased by the introduction of the azobenzene group. The molecular length (L) and diameter (D) estimated from the models obtained by MOPAC/PM3 method are given in Table 10.1. The molecular aspect ratio (L/D), which is defined as the ratio of the molecular length to the molecular diameter of chiral azobenzene molecules, is a measure of anisotropy of the molecular shapes. The anisotropy in the molecular shape is one of the important parameters for appearance of LC phases (Onsager, 1949). In other words, the molecular aspect ratio is related to the anisotropic intermolecular interaction between the chiral and the LC molecules. Therefore, the higher the molecular aspect ratio becomes, the higher HTP of the chiral compounds becomes.

Figure 10.9 shows the snapshot after 100 ps of mixture of 5CB/*p*-azo-9 (98/2 mol%) obtained by MD simulation. It seems likely that $\pi-\pi$ stacking interaction between 5CB molecules is enhanced through the intermolecular interaction between the chiral azobenzene molecule and 5CB molecule. As a result, the order parameter, S, was increased by mixing *trans-p*-azo-9 in 5CB. However, S of 5CB mixed with *cis-p*-azo-9 was almost constant. The molecular orientation of the LC molecules doped with the *trans*-azobenzene is stable relatively to that doped with

TABLE 10.1. HTPs [($\times$ 108 m^{-1} mol^{-1} g-E44), Experimental], the Length of the Long Axis, L, and the Diameter, D [(Å), Measured from Molecular Orbital Calculation with PM3 Method], Data of Chiral Azobenzene Compounds

Compound	Chiral group[a]	Trans form			
		HTP	L	D	L/D
Azo-1	$-\overset{*}{C}H(CH_3)-C_6H_{13}$	32.0	27.2	5.4	5.0
Azo-4	$-\overset{*}{C}H(CH_3)-COOCH_3$	28.9	23.7	4.4	5.1
Azo-5	$-\overset{*}{C}H(CH_3)-COOC_2H_5$	36.6	24.9	4.6	5.4
Azo-6	$-\overset{*}{C}H(CH_3)-C_6H_5$	68.7	22.7	5.0	4.5
Azo-7	H_3C, CH_3, CH_3 (cyclohexyl ring, * * *)	49.0	24.8	6.3	3.9
Azo-8	O, H, H, O (fused bicyclic ring)	141	40.3	9.3	4.4
Azo-9	O, H, H, O (fused bicyclic ring)	202	42.8	7.8	5.4

[a]Azo-chiral: $C_6H_{13}O-C_6H_4-N=N-C_6H_4-C(=O)-O-Chiral$

TABLE 10.2. Structure and HTP of Chiral Compounds

Compound	Structure	HTP	L, Å	D, Å	L/D
Let	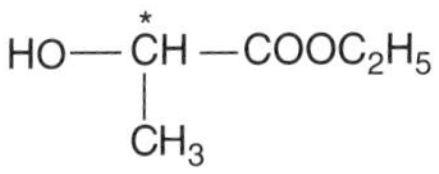 $HO-\overset{*}{C}H(CH_3)-COOC_2H_5$	11.6	5.8	2.9	2.0
Menth	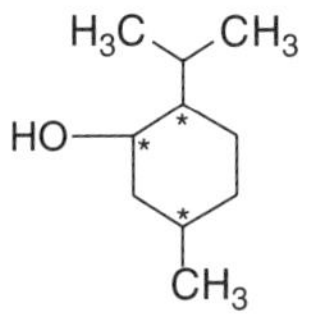 H_3C, CH_3, HO, CH_3	9.3	3.9	7.6	0.5
Isomannide	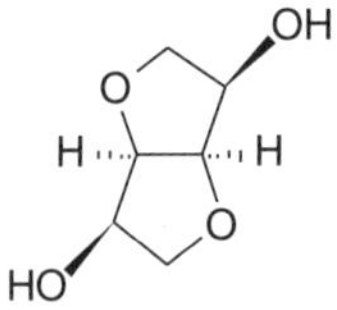 OH, O, H, H, O, HO	29.2	4.4	3.9	1.1
Isosorbide	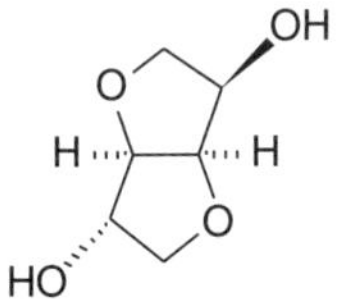 OH, O, H, H, O, HO	30.4	5.1	3.6	1.4

the *cis*-azobenzene. Thus, the shape of the azobenzene molecule influences the molecular orientation of a host LC (Sanpei, 2005).

10.2.3. Effects of Chiral Groups on Photochemical Change in HTP

In binary Ch LC systems consisting of the chiral compound and the host nematic LC, the intermolecular interaction between the chiral azobenzene molecule and LC molecule contributes to HTP of the chiral azobenzene compounds. In this section, photoswitching behavior of some chiral azobenzene compounds (Fig. 10.7) is examined in terms of the influence of the chiral groups on the photochemical change in HTP.

Figure 10.10 shows changes in absorption spectra of *p*-azo-7 in methanol and in E44 by UV irradiation. The absorption spectra are characterized by a strong absorption at 370 nm corresponding to π–π^* transition of trans form and a weak absorption at 450 nm, which originates from the n–π^* transition of cis form. Upon UV irradiation, absorbance at ~370 nm was decreased because of the trans–cis photoisomerization with an increase in low intensity band at around 450 nm.

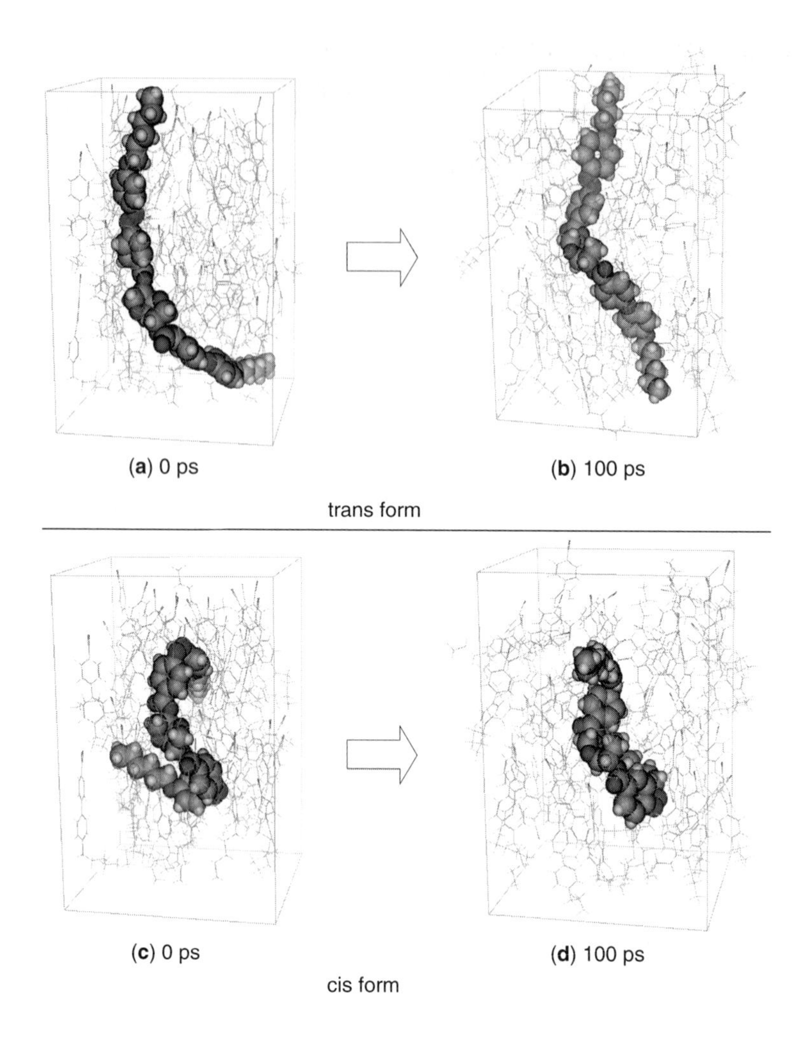

Change in the orientation order **S** by replacing with different chiral dopant.

Host/Dopant	Ratio, mol/mol	**S** Before	**S** After	Δ**S**, %
5CB	50/0	0.53	0.51	–3.8
5CB /*trans-p*-Azo-9	49/1	0.53	0.66	+19.7
5CB /*cis-p*-Azo-9	49/1	0.53	0.54	+1.9

Figure 10.9. Snapshots before [(**a**) trans, (**c**) cis] and after 100 ps [(**b**) trans, (**d**) cis] of mixed LCs (5CB/*p*-azo-9 = 49/1).

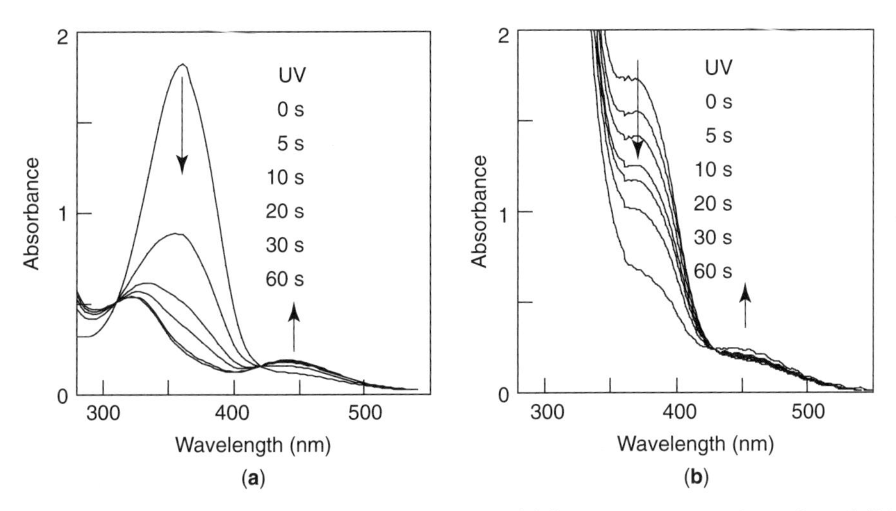

Figure 10.10. Changes in absorption spectra of (**a**) *p*-azo-7 in methanol and (**b**) the mixture of E44 and *p*-azo-7 (95/5 wt%) by UV irradiation (7.6 mW/cm^2).

The photoisomerization ratios determined by comparing ^{1}H NMR spectra of azobenzene compounds before and after UV irradiation are given in Table 10.3. The yields of cis form of all chiral azobenzene compounds at the photostationary state were ~90% (Alam et al., 2007). Therefore, it is assumed that there is no effect of the yield of cis form at photostationary state on the change in HTP under UV light.

The photochromic chiral azobenzene compounds, azo-1, azo-4, azo-5, azo-6, azo-7, and azo-10, only differ in their chiral substituents as can be seen in Fig. 10.7. Their HTPs were decreased upon UV irradiation because of the trans–cis photoisomerization of the chiral azobenzene molecules (Table 10.3).

Azo-8 and azo-9 contain two azobenzene groups in a molecule. Three isomeric (*o*-, *m*-, *p*-) azo-8 and azo-9 were synthesized to investigate the influence of difference in connecting position with an azobenzene group and a chiral group. ΔHTPs of azo-8 and azo-9 are given in Table 10.3. Interestingly, *m*-azo-9 was found to show photochemical increase in HTP upon UV irradiation. HTPs of trans- and cis form of *m*-azo-9 were $15.6 \times 10^{-8}\,m^{-1}\,mol^{-1}$ g-E44 and $79.4 \times 10^{-8}\,m^{-1}\,mol^{-1}$ g-E44, respectively. To explore the effects of molecular structure on the photochemical change in HTP, the structures of chiral azobenzene molecules were predicted by MOPAC/PM3 method (Alam et al., 2007). The models of azo-6, azo-10, azo-8, and azo-9 in gaseous phase obtained by MOPAC method are shown in Figs. 10.11 and 10.12. The cis forms of azo-6 and azo-10 having single azo group in a molecule possess the bent shape. The cis forms of *p*-azo-8 and *p*-azo-9 have a zigzag shape rather than the bent shape as shown in Fig. 10.11. However, *m*-azo-9 was found to show quite different molecular structure compared with others in its isomeric states. The cis form of *m*-azo-9 is in the rod shape, whereas the trans form

TABLE 10.3. Characterization of Chiral Azobenzene Compounds and Chiral

Azo-n	Tmp, °C	HTP (10^8 m^{-1} mol^{-1} g-E44)			Cis form ratio, %	Aspect ratio (L/D)	
		HTP$_{before}$	HTP$_{after}$	ΔHTP[a]		Trans form	Cis-form
Azo-1	33	32.0	20.9	−11.1	96	5.0	1.3
Azo-4	69	28.9	12.6	−16.3	94	5.1	1.2
Azo-5	72	36.6	19.9	−16.7	95	5.4	1.2
Azo-6	83	68.7	36.3	−32.4	97	4.5	1.3
o-Azo-7	50	44.3	12.7	−31.6	87	2.6	1.7
m-Azo-7	65	48.4	11.1	−37.3	90	3.3	1.1
p-Azo-7	79	49.0	10.8	−38.2	94	3.9	1.1
Azo-10	102	10.7	10.5	−0.2	93	5.2	1.4
o-Azo-8	104	59.9	34.5	−25.4	89	3.0	2.2
m-Azo-8	127	129	20.5	−108	90	4.3	1.8
p-Azo-8	187	141	16.5	−124	90	4.4	1.6
o-Azo-9	74	4.9	4.5	−0.1	89	2.1	3.3
m-Azo-9	51	15.6	79.4	+63.8	90	2.5	4.5
p-Azo-9	203	202	48.9	−153	90	5.4	1.4
Chiral	143	128					

[a] ΔHTP = HTP$_{after}$ − HTP$_{before}$.

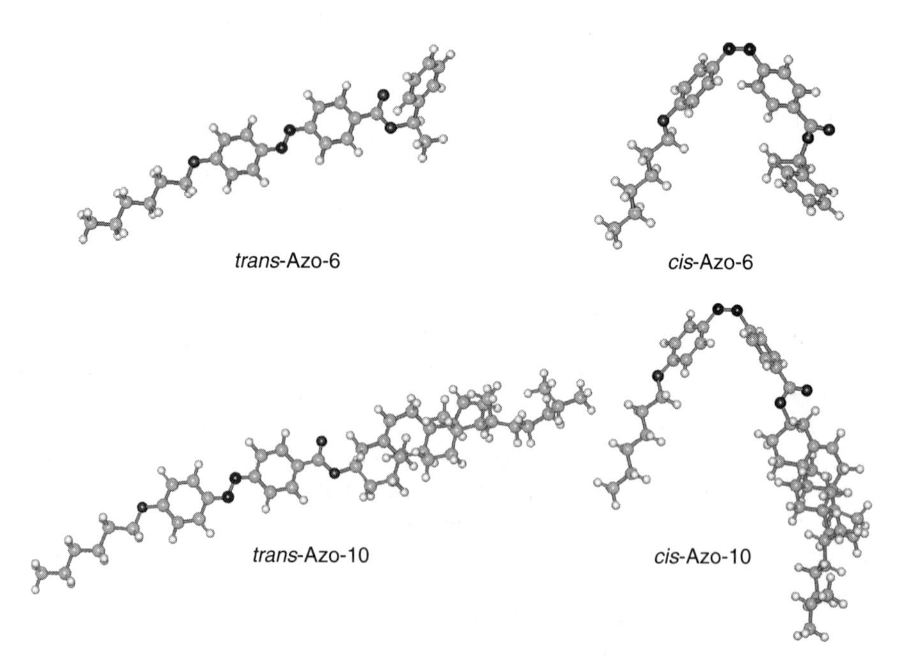

Figure 10.11. Optimized structures for azo-6 and azo-10 by MOPAC at PM3 level.

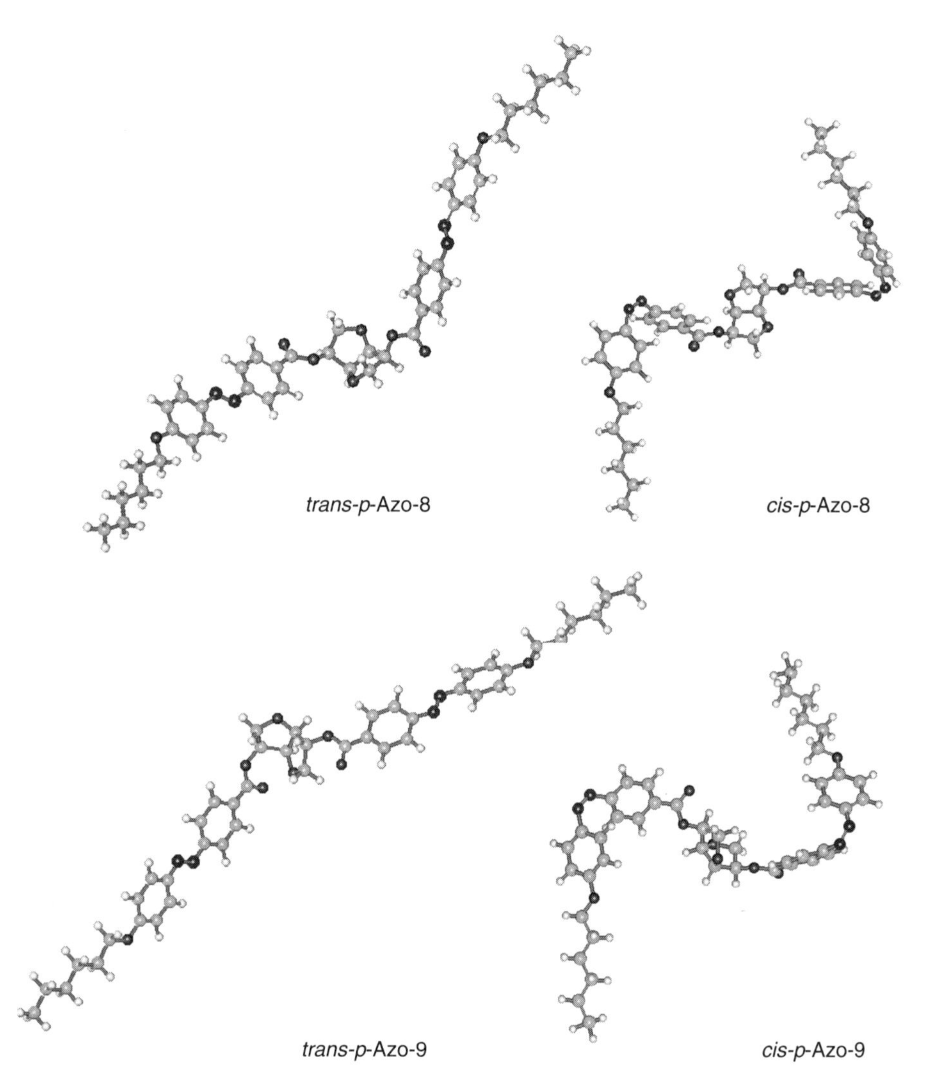

Figure 10.12. Optimized structures for *p*-azo-8 and *p*-azo-9 by MOPAC at PM3 level.

has the bent shape as shown in Fig. 10.13. As a result, HTP of *cis-m*-azo-9 is higher than that of *trans-m*-azo-9.

As mentioned in the preceding section, the anisotropy in the molecular shape influences significantly HTP of chiral compounds. Thus, to discuss the structural effect on the photochemical change in HTP, the molecular aspect ratio, L/D, was estimated with MOPAC/PM3 method. The molecular aspect ratios of the chiral azobenzene compounds in their isomeric states are given in Table 10.3. The data reveal that HTP of the chiral azobenzene compounds in both the isomeric states

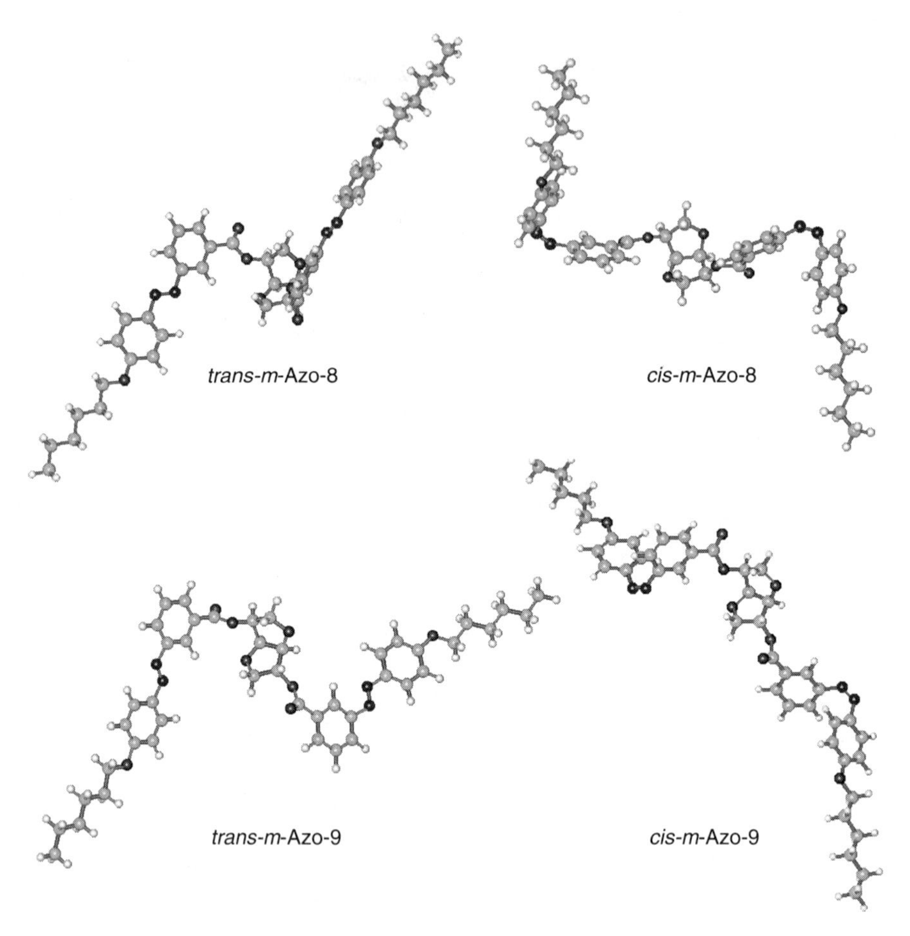

Figure 10.13. Optimized structures for *m*-azo-8 and *m*-azo-9 by MOPAC at PM3 level.

depends on their molecular aspect ratio with an exception of *o*-azo-9 and azo-10. The molecules with larger L/D exhibited larger HTP, regardless of trans form and cis form. However, the molecular aspect ratio as well as HTP of the chiral azobenzene compounds are changed by trans–cis photoisomerization. Therefore, it is worthwhile to explore the effects of change in the shapes of the chiral azobenzene molecules on their HTP, to develop effective phototrigger for photomodulation of helical structure.

In Fig. 10.14, photochemical change in HTP, Δ(HTP), is plotted as a function of photochemical change in molecular aspect ratio, $\Delta(L/D)$; here $\Delta(\mathrm{HTP}) = \mathrm{HTP}_{\mathrm{cis}} - \mathrm{HTP}_{\mathrm{trans}}$; $\Delta(L/D) = (L/D)_{\mathrm{cis}} - (L/D)_{\mathrm{trans}}$. It is observed that Δ(HTP) of the chiral azobenzene compounds is dependent on their $\Delta(L/D)$. The relation of Δ(HTP) and $\Delta(L/D)$ is classified into three groups: Gr-A (azo-1, azo-4, azo-5), Gr-B (azo-6 and

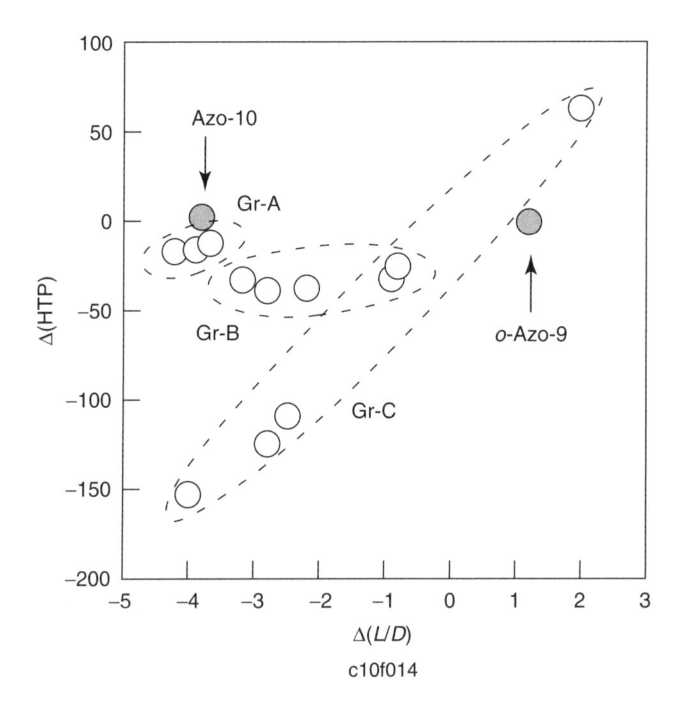

Figure 10.14. HTP as a function of *L/D* of chiral azobenzene compounds.

o-, *m*-, *p*-azo-7), and Gr-C (*o*-, *m*-, *p*-azo-8, *o*-, *m*-, *p*-azo-9). Compounds with a single chiral group in azobenzene core moiety fall in Gr-A and Gr-B. Compounds in Gr-B have higher HTP_{trans} as well as higher Δ(HTP) compared with the compounds in Gr-A. One of the reasons of the difference in Δ(HTP) of the chiral azobenzene compounds may be the molecular size of the chiral groups. Azo-10 with cholesterol group as chiral moiety showed much smaller Δ(HTP), although its $\Delta(L/D)$ was comparable to others. Figure 10.11 shows models of isomers of azo-6 and azo-10. As can be seen in the figure, the trans forms are rodlike, whereas the cis forms possess a bent structure. Comparison of the models revealed that the size of the bent part of azo-10 is much larger than that of azo-6, because of large molecular size of the cholesterol group. Therefore, it can be assumed that the cis form of azo-10 can interact with nematogenic neighbors more efficiently than azo-6 and other azobenzene compounds having smaller chiral groups. Consequently, no significant difference between HTP_{trans} and HTP_{cis} of azo-10 was observed.

However, compounds in Gr-C showed higher HTP and Δ(HTP) as given in Table 10.3 and Fig. 10.14. The introduction and isomerization of plural photochromic groups in a single chiral core is favorable for effective modulation of the helical structure photochemically. The zigzag molecular shape of the compounds in Gr-C contributed to the larger photochemical change in HTP. However, *o*-azo-9 showed a little deviation from the plot of Gr-C. Figure 10.15 shows the models

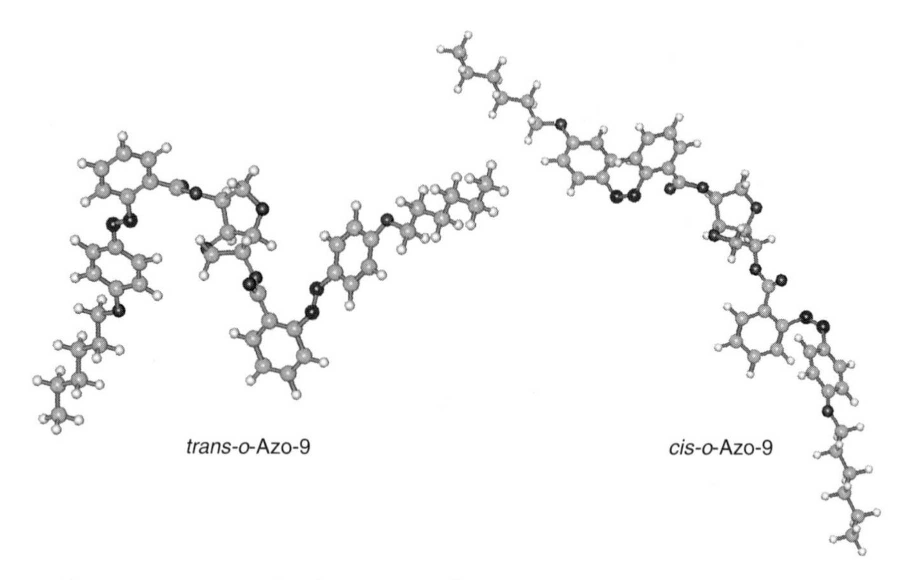

Figure 10.15. Optimized structures for *o*-azo-9 by MOPAC at PM3 level.

of isomers of *o*-azo-9. A zigzag shape of *trans*-*o*-azo-9 contributes to its smaller HTP. In addition, the linearity of core part (–N=N–Ar–COO–) of *cis*-*o*-azo-9 is extremely lower than that of *m*-azo-9 and other rod-shaped molecules, because of bonding at ortho position. Consequently, *o*-azo-9 exhibits smaller HTP as well as smaller Δ(HTP) compared with others. The results suggest that Δ(HTP) of the chiral photochromic compounds is related not only to the aspect ratio but also to the features of the chiral groups such as size, compactness, rigidity, linearity, planarity, and so on.

10.3. APPLICATIONS

It is well known that a Ch LC shows three major textures depending on applied electric field and surface alignment properties, when it is sandwiched between two parallel glass plates; they are planar texture, focal conic texture, and homeotropic texture as shown in Fig. 10.16 (Demus and Richter, 1978). In the planar texture, the Ch LCs reflect light selectively depending on the helical pitch. The Ch LCs in the focal conic texture scatter light in forward directions, because of its polydomain structure with the helical structure. The Ch LCs can exist in both planar texture and focal conic texture at zero field condition (Yang et al., 1997). However, unwinding of helical structure of the Ch LC with an electric field provides homeotropic structure, where the LC director is perpendicular to the cell surface. Consequently, the Ch LCs become transparent. If one can control the helical pitch as well as the textures of the Ch LC photochemically, the photoresponsive Ch LCs will be promising for various types of optical materials.

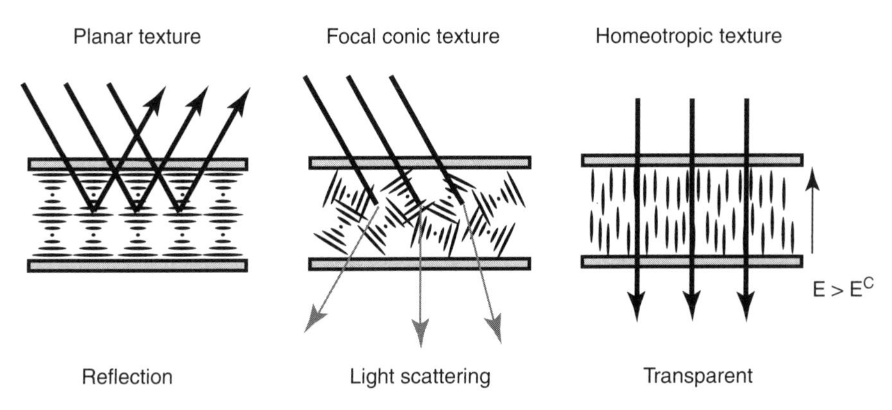

Figure 10.16. Schematic diagrams of optical properties of Ch LCs.

In addition, the Ch LCs in the planar texture can be regarded as a one-dimensional photonic band gap material, because of their periodical helical structure. This unique property will open new possibilities for application of the Ch LCs such as tunable laser and so on (Matsuhisa et al., 2007; Belyakov, 2006; Lin et al., 2005; Funamoto et al., 2003; Furumi et al., 2003; Kopp et al., 2003, 1998; Shibaev et al., 2003; Matsui et al., 2002; Ozaki et al., 2002).

This section discusses the photochemical switching of selective reflection, transparency, helical sense, and lasing through the phototuning of helical structure of the Ch LCs by photoisomerization of chiral azobenzene compounds.

10.3.1. Photochemical Switching of Selective Reflection

To control photochemically the selective reflection over the whole visible region, a trigger molecule with larger HTP as well as larger photochemical change in HTP is required. It seems likely that *p*-azo-8 and *p*-azo-9 are suited for the photochemical switching of the selective reflection. However, a phase separation was brought about by mixing *p*-azo-8 or *p*-azo-9 in E44 even at a concentration of 3 wt% of azo compound. In contrast to the *p*-type dyes, *m*-azo-8 was miscible in E44 even at concentrations higher than 10 wt%.

A mixture of E44/chiral/*m*-azo-8 (80/8/12 wt%) showed a Ch phase above room temperature up to 72°C; Ch•72°C•I. This Ch LC was injected into a glass cell, consisting of two glass plates coated with polyimide and rubbed to provide the planar texture in the glass cell, with a 5-μm cell gap. The Ch LC had a left-handed helical structure. Fig. 10.17 shows change in transmittance spectra of the Ch LC in the glass cell by UV irradiation at 25°C. Before UV irradiation, only an extreme low transmittance was observed, shorter than 400 nm, corresponding to the π–π^* absorption band of azo chromophore. However, purple color could be seen with the naked eye, indicating the reflection of light $\sim$400 nm, as can be seen in Fig. 10.18 (left-hand side). UV irradiation resulted in the shift of the selective

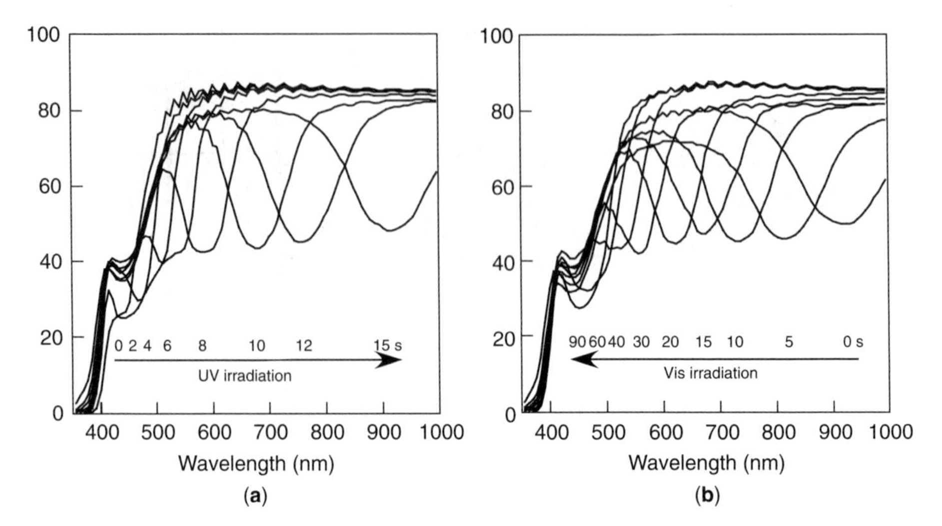

Figure 10.17. Changes in the selective reflection of the Ch LC consisting of E44, chiral, and *m*-azo-8 (80:8:12 wt%) by (**a**) UV irradiation (7.6 mW/cm^2) and (**b**) Vis light irradiation (14.3 mW/cm^2) at 25°C. *Source*: Yoshioka et al., 2005.

reflection to a longer wavelength, and the wavelength became longer than 900 nm after the irradiation for 15 s. The transmission band returned to the initial position by Vis light irradiation as shown in Fig. 10.17. Thus, a reversible shift of the selective reflection over the whole visible region could be achieved by UV and Vis light irradiation (Yoshioka et al., 2005).

Figure 10.18 shows change in reflection color from the Ch LC by using UV irradiation time of 0, 4, and 10 s. Before irradiation, the color was purple, and it turned to green and red, corresponding to the shift of the reflection wavelength (Fig. 10.17). The color could also be adjusted by varying the light intensity with a gray mask, as can be seen in Fig. 10.19. The resolution of the color patterning was estimated to be 70–100 μm by patterning experiments with the use of photomask. See color insert for color versions of these figures.

10.3.2. Control of Transparency

A nonphotochromic chiral compound, chiral (Fig. 10.7), which was derived with (1 s, 2R, 5S)-(+)-menthol, was found to give a right-handed helix by addition in E44. Contrary to chiral, *m*-azo-8 gave a left-handed helical helix within E44. Therefore, when both chiral compounds are added in E44, the opposite helical senses of chiral and *m*-azo-8 each compensate the other's HTP, leading to a compensated nematic phase.

Figure 10.20 shows polarizing microscopic photographs of the mixture of E44/chiral/*m*-azo-8 (80/9.8/10.2 wt%), changing to a Ch phase from a compensated nematic phase on UV irradiation. A schlieren texture in the initial state

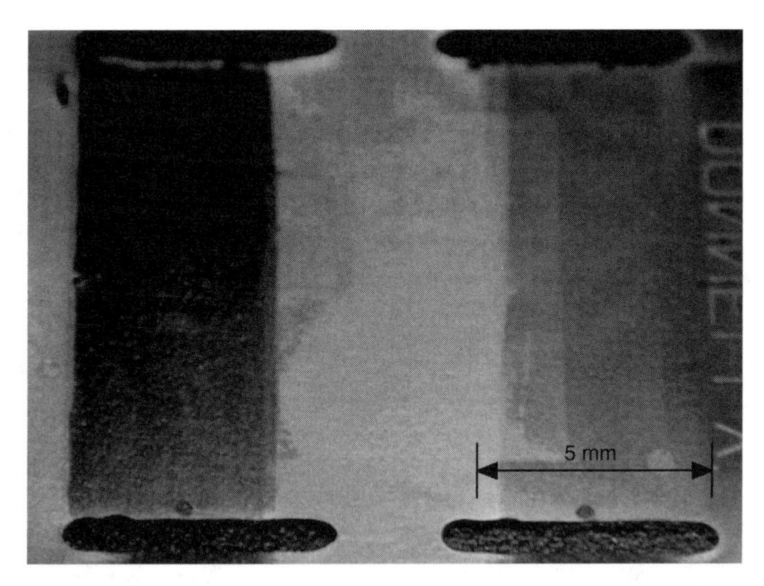

Figure 10.18. Change in the reflection color of the Ch LC by varying UV irradiation time 0 (*left*), 4 (*middle*), and 10 s (*right*). *Source*: Yoshioka et al., 2005. See color insert.

(the left-hand side figure in Fig. 10.20) indicates that the LC exists in the compensated nematic state. The compensated nematic state was destroyed upon UV irradiation, because of the photochemical decrease in HTP of azo-1, with a right-handed Ch phase based on chiral (the right-hand side figure in Fig. 10.20). In addition, Fig. 10.20 indicates that the helical pitch was decreased by UV irradiation time. Reversible photoisomerization from the cis form to the trans form was observed on Vis light irradiation, and HTP was compensated once more, so the compensated nematic phase was recovered again upon Vis light irradiation. The LC mixture was injected into a 5-μm homeotropic glass cell. Figure 10.21 shows an experimental setup for measuring the photochemical switching behavior of the LC mixture in the glass cell under irradiation of UV and Vis light. In the initial state, the LC mixture was in the compensated nematic phase and showed ~90% transmittance as can be seen in Fig. 10.22. The transmittance was decreased and increased by irradiation of UV and Vis light, respectively. The compensated nematic phase was destroyed upon UV irradiation, because of the photochemical decrease in HTP of *m*-azo-8. Consequently, the focal conic texture formed in the homeotropic glass cell scattered light as shown in Fig. 10.23.

10.3.3. Photochemical Inversion of Helix

An excess doping of chiral dopant into a compensated nematic phase breaks the compensated state, and consequently produces a helical structure again.

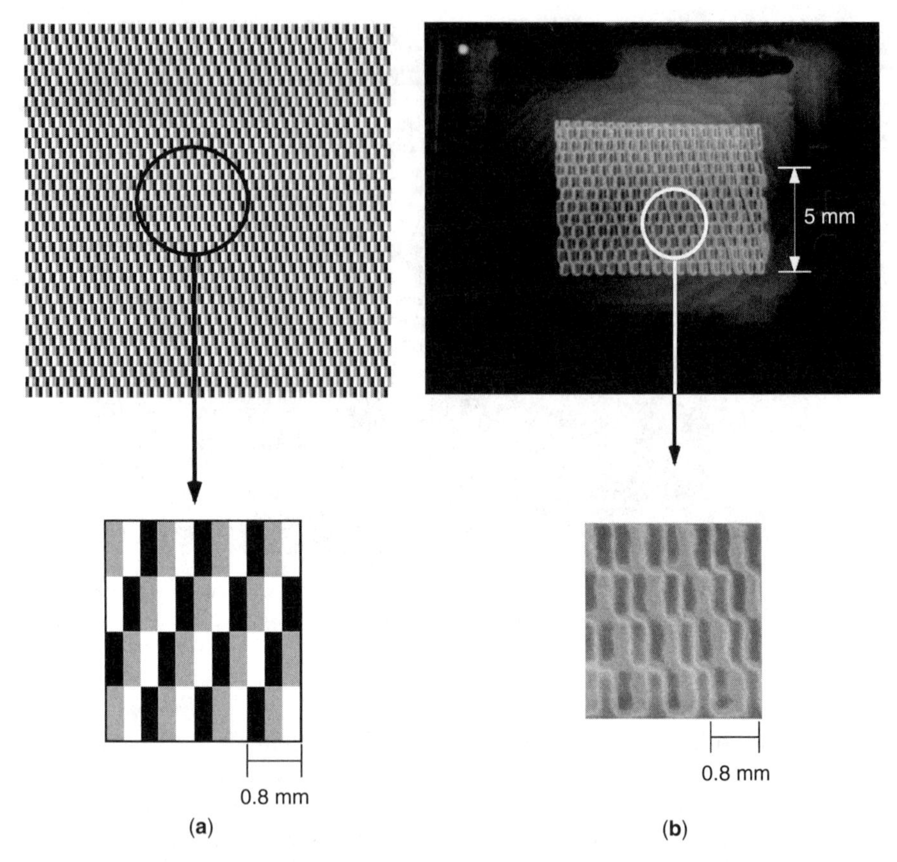

Figure 10.19. (**a**) Gray mask and (**b**) Color patterning of the Ch LC containing E44, *m*-azo-8, and chiral obtained by UV irradiation for 10 s through the gray mask at 25°C. *Source*: Yoshioka et al., 2005. See color insert.

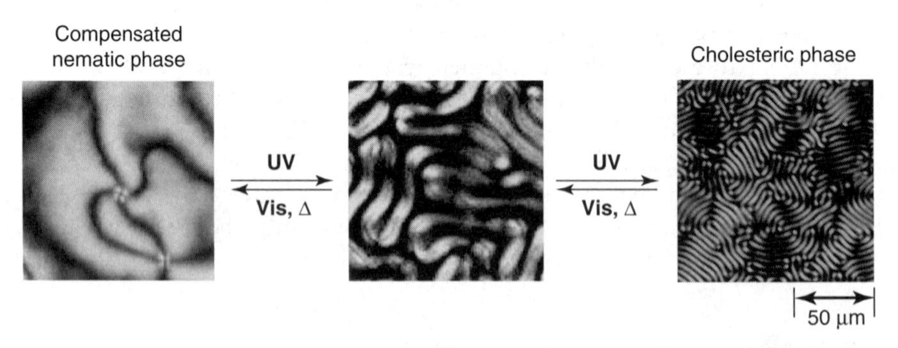

Figure 10.20. Reversible photochemical phase transition between a compensated nematic phase and a Ch phase of E44:chiral:azo-1(80:9.8:10.2 wt%) by UV and visible light irradiation. *Source*: Yoshioka et al., 2005.

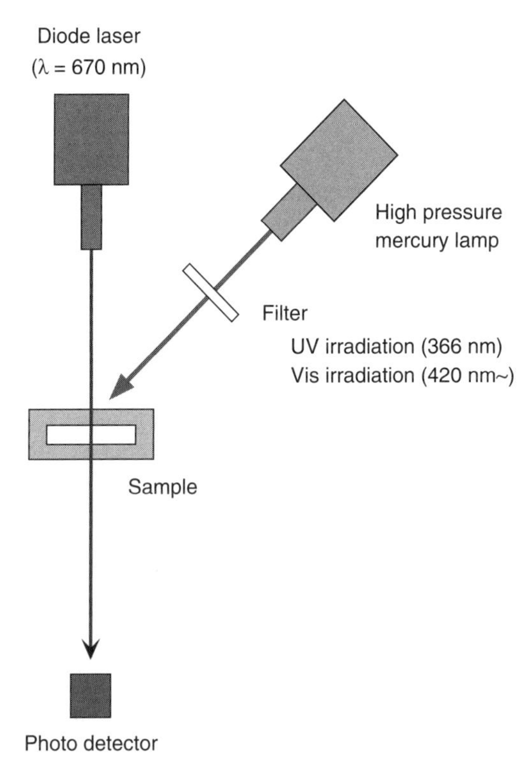

Figure 10.21. Experimental setup for measuring photochemical change in transparency.

Figure 10.24 shows an effect of UV and Vis light irradiation on the texture of E44/R811/azo-1(80/8.4/11.6 wt%) mixture. R811 and azo-1 were found to induce right-handed and left-handed helical structure in E44, respectively. A fingerprint texture observed before UV irradiation indicates that the mixture exists in the left-handed helical structure, because of the excess twisting ability of azo-1. The helical pitch increased and became infinite by UV irradiation, indicating transformation into a nematic phase. Then, the fingerprint texture appeared again by further UV irradiation. The reversal in the texture was caused by successive Vis light irradiation.

Circular dichroic (CD) spectroscopy is one of the effective methods for characterizing helical structure of polymers, polypeptides, and so on (Berova et al., 2000). A cyanine dye, NK-529, shows positive- and negative-induced CD in the right-handed and left-handed helices (E44/R811 and E44/S811), respectively, as shown in Fig. 10.25. Figure 10.26 shows change in the induced CD spectra of NK-529 in the mixture of E44/R811/azo-1 (85/6/9 wt%) in a 5-μm homogeneous glass cell by UV irradiation. Comparison of induced CD spectra shown in Figs. 10.25 and 10.26 reveals that the transformation of the helical structure

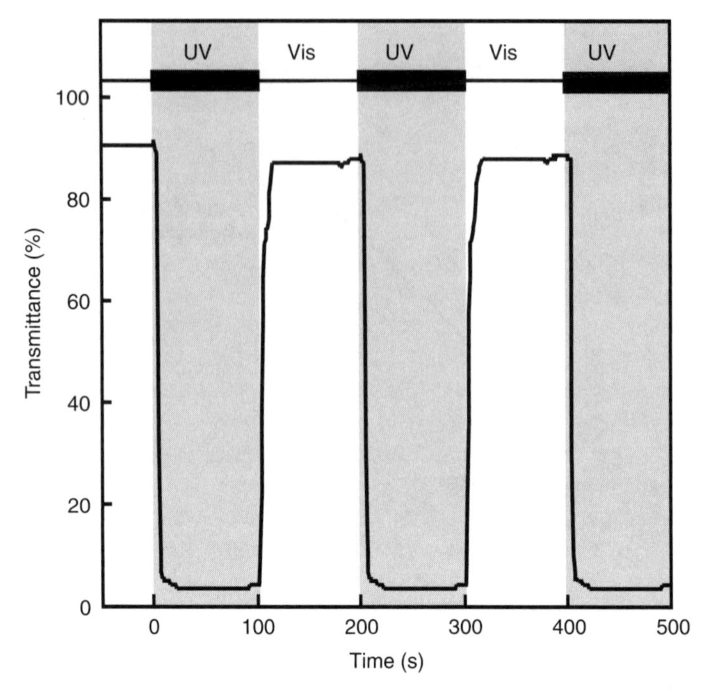

Figure 10.22. Changes in the transmittance of a light from a diode laser (670 nm) through the E44:chiral:*m*-azo-8 = 80:9.8:10.2 wt% mixture in a 25-μm homogeneous glass cell by UV (3.2 mW/cm^2) and Vis (8.3 mW/cm^2) light irradiation at 25°C.

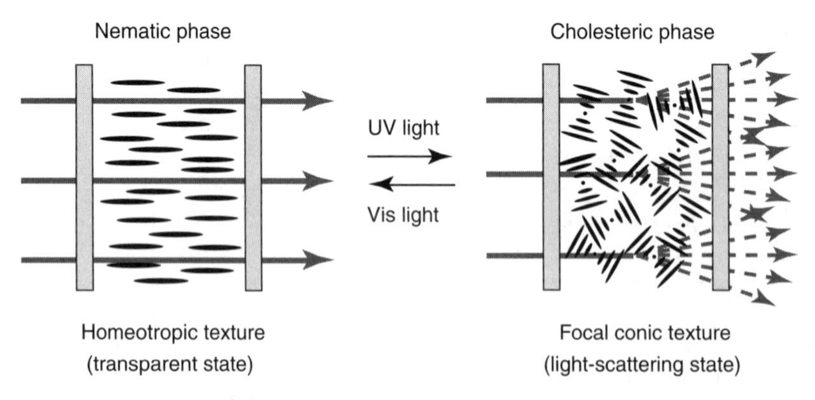

Figure 10.23. Textures of the nematic and the Ch phases in the homeotropic glass cell. *Source*: Alam et al., 2007.

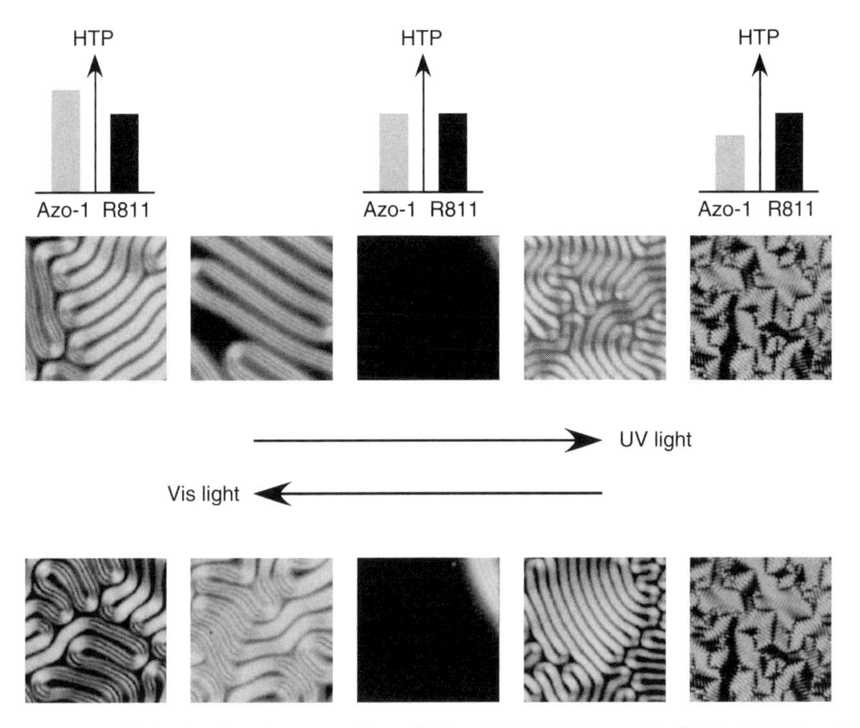

Figure 10.24. Polarized micrographs of the E44/R811/azo-1 (80:8.4:11.6 wt%) mixture in 5-μm glass cell without any alignment treatment upon UV and Vis light irradiation.

from the left-handed helix into the right-handed one is brought about upon UV irradiation (Kurihara et al., 2001). The transformation was reversible by irradiation of UV and Vis light.

10.3.4. Photochemical Control of Lasing

Recently, Ch LCs have attracted interest as tunable photonic band gap materials (Ozaki, 2007), because the Ch LCs possess photonic band gap properties as well as response to the external stimuli (John, 1987; Yablonovitch, 1987).

A Ch mixture of E44/S811/*m*-azo-9/DCM (73.1/22.2/4.3/0.4 in wt%) was prepared (Fig. 10.27). DCM is a laser dye. This mixture showed a Ch phase above room temperature up to 62°C; Ch•62°C•I. *m*-azo-9 and S811 gave left-handed and right-handed helices when they were added in E44.

The mixture was injected into a homogeneous glass cell with 25-μm cell gap to obtain a planar molecular orientation. The longer edge of the reflection band of the Ch LC was adjusted to the emission maximum of DCM as shown in Fig. 10.28

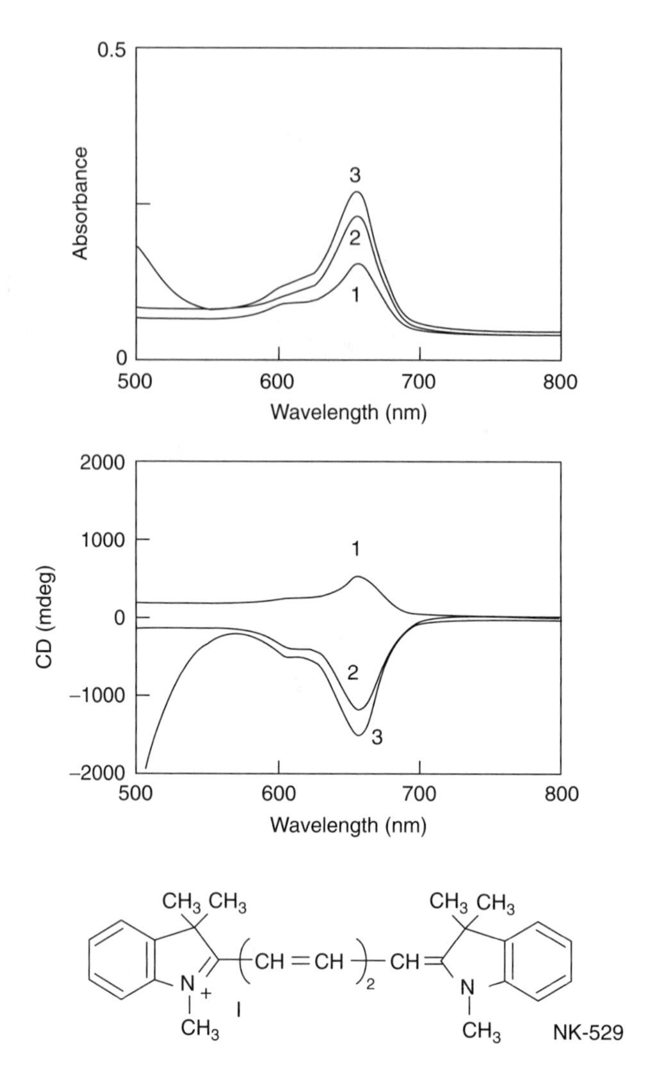

Figure 10.25. Absorption and CD spectra of NK-529 added in E44/R811 (95/5wt%) (1), E44/S811 (95/5wt%) (2), and E44/azo-1 (95/5wt%) (3).

(curves 1 and 2). The laser emission properties were studied by excitation with the second harmonic light at 532 nm from a Q-switched Nd:YAG laser beam (pulse repetition frequency: 10 Hz, the pulse width: 15 ns). The laser beam was focused on the glass cell by using a convex lens. The Ch sample was irradiated with a laser beam at an incident angle of 45° from the surface normal.

By excitation with the second harmonic light of Nd-YAG laser at lower light intensity, a sharp band at 532 nm as well as a broad band at ~565 nm were

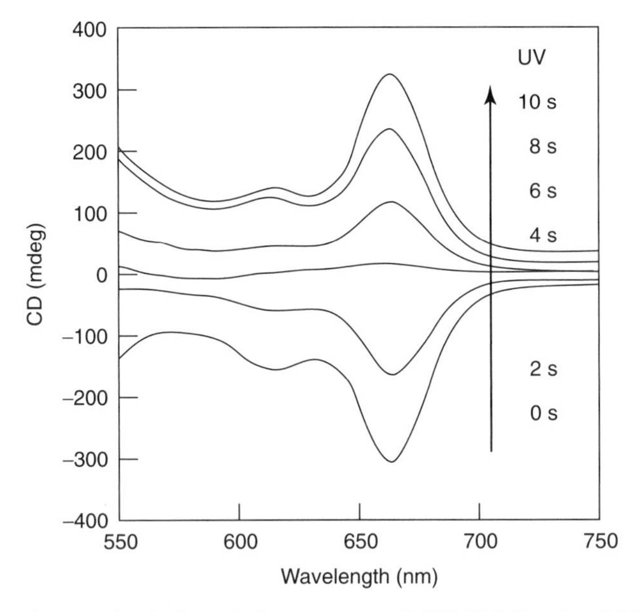

Figure 10.26. Change in induced CD spectra of NK-529 in the E44/R811/azo-1 (85:6:9 wt%) in the homogeneous glass cell by UV light irradiation.

observed. The band at 532 nm is corresponding to the reflection of the excitation light. The band at 565 nm became sharp and intensive by increase in the intensity of the excitation light. Figure 10.29 shows the dependences of the emission intensity and the emission bandwidth on the excitation light intensity. When the light intensity exceeded the pumping energy of 8 μJ/pulse, the emission intensity was significantly enhanced, accompanied by the abrupt narrowing of the band from 45 to 1.5 nm. This indicates that the pumping threshold energy for the lasing is ~8 μJ/pulse.

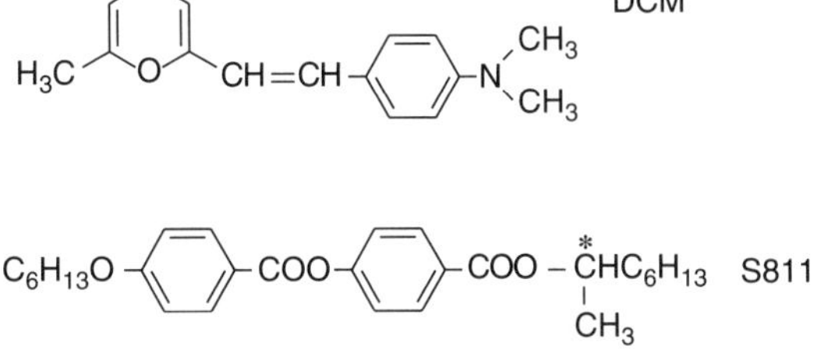

DCM

S811

Figure 10.27. Laser dye (DCM) and chiral compound.

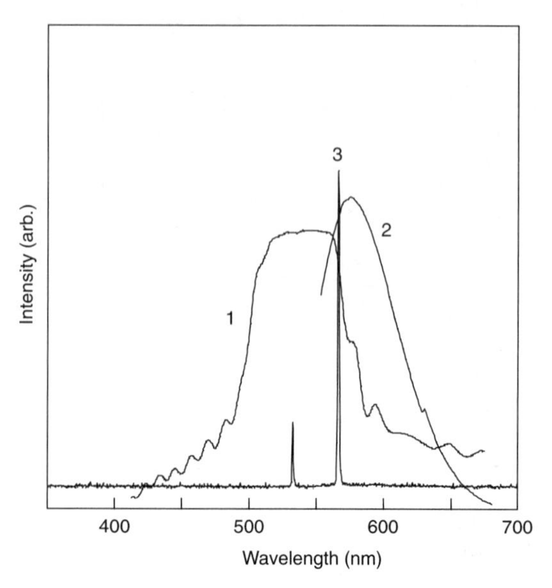

Figure 10.28. Reflection spectrum (1), fluorescent emission spectrum (2) by excitation at 532 nm, and laser emission spectrum (3) by excitation with a second harmonic light of Nd-YAG laser of a Ch mixture of E44, S811, *m*-azo-8, and DCM (73.1/22.2/4.3/0.4 in wt%). *Source*: Kurihara et al., 2006.

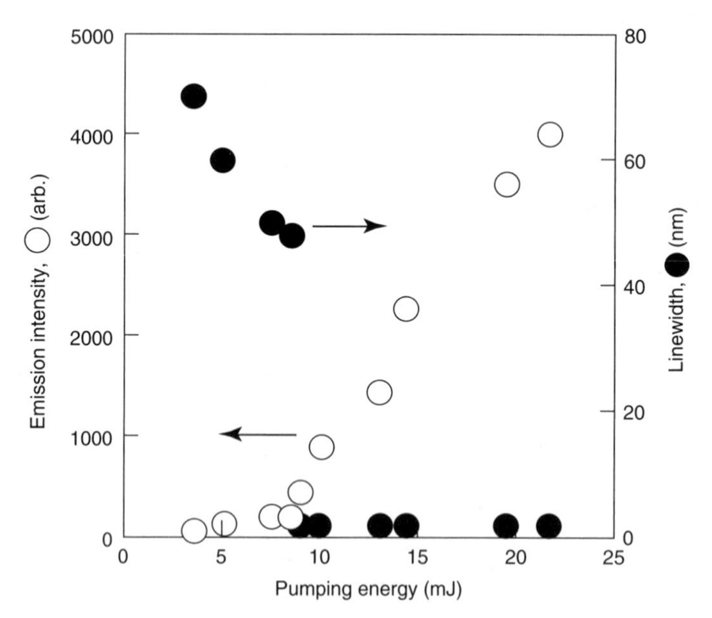

Figure 10.29. Emission intensity and linewidth of laser emission as a function of pumping energy with a second harmonic light of Nd-YAG laser. *Source*: Kurihara et al., 2006.

The effect of the irradiation of UV light at 365 nm on the helical pitch length as well as the laser emission at ambient room temperature is shown in Fig. 10.30. In the transmittance spectra shown in Fig. 10.30, the minimum peaks or plateaus in the range 550–700 nm are corresponding to the selective reflection band of the Ch LC. They were clearly shifted to longer wavelength region by UV irradiation because of the photochemical decrease in HTP of *m*-azo-8. At the same time, the laser emission was also shifted to longer wavelength region by the UV irradiation as can be seen in Fig. 10.30. This is clearly related to the shift of the selective reflection (Kurihara et al., 2006). In the beginning of the UV irradiation, the laser

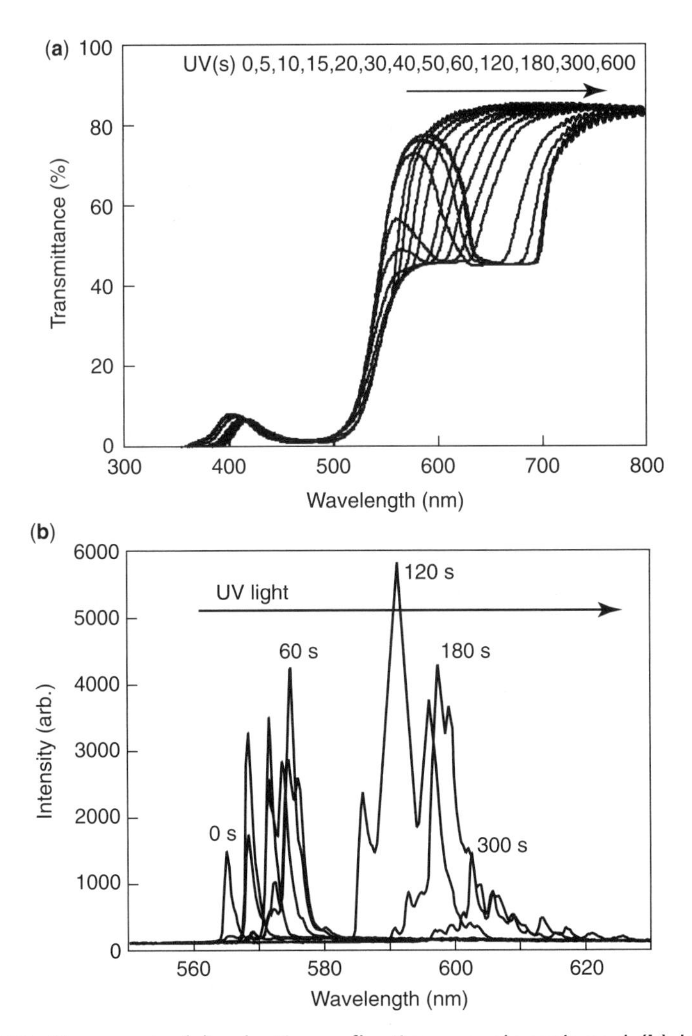

Figure 10.30. Changes in (**a**) selective reflection wavelength and (**b**) laser emission wavelength by UV light irradiation.

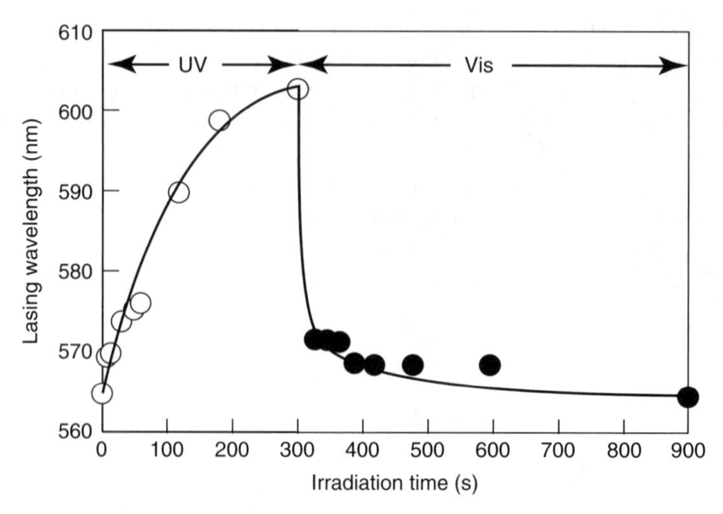

Figure 10.31. Change in the lasing wavelength from the Ch LC containing DCM by irradiation of UV- and Vis light. *Source*: Kurihara et al., 2006.

emission band was single, whereas the multiple bands were observed after longer UV irradiation. This may be perturbation effect of cis form on the molecular orientation of LC molecules. The lasing wavelength could be returned to the initial by visible light irradiation as shown in Fig. 10.31.

10.4. CONCLUSION

Azobenzenes have simple chemical structures and exhibit simple photoisomerization between the trans- and cis form. However, they can really provide various photofunctions by their combination with sophisticated systems such as Ch LCs. The Ch LCs have a good potential for display applications, because they show interesting optical properties depending on the helical structures and the textures. Many types of optical materials can be created by the use of the photoisomerization of chiral azobenzenes in the Ch LCs. New ideas, including design of molecules and systems, may allow us for future materials not only in the field of optics but also in other fields.

REFERENCES

Alam MZ, Yoshioka T, Ogata T, Nonaka T, Kurihara S. 2007. The influence of molecular structure on helical twisting power of chiral azobenzene compounds. Liq Cryst 470:63–70.

Bahr C, Escher C, Fliegner D, Heppke G, Molsen H. 1991. Behavior of helical pitch in cholesteric and chiral smectic C* phases. Ber Bunsen-Ges 95(10):1233–1237.

Belyakov VA. 2006. Low threshold DFB lasing in chiral LC at diffraction of pumping wave. Mol Cryst Liq Cryst 453:43–69.

Berova N, Nakanishi K, Woody RW. 2000. Circular Dichroism. New York: Wiley-VCH.

Bobrovsky AY, Shibaev VP. 2002. Chiral nematic polymer mixture containing crosslinker and photosensitive chiral dopant: new type of materials with tunable photooptical properties. Adv Funct Mater 12(5):367–372.

Bobrovsky A, Shibaev V. 2005. Thermo-, chiro- and photo-optical properties of cholesteric azobenzene-containing copolymer in thin films. J Photochem Photobiol A: Chem 172(2):140–145.

Brehmer M, Lub J, Van de Witte P. 1998. Light-induced color change of cholesteric copolymers. Adv Mater 10(17):1438–1441.

De Gennes PG, Prost J. 1993. The Physical of Liquid Crystals. New York: Oxford University Press.

Demus D, Richter L. 1978. Textures of Liquid Crystals. Weinheim/New York: Deutscher. pp. 115–120.

Demus D, editor. 1998. Handbook of Liquid Crystals. Weinheim: Wiley-VCH Verlag GmbH.

Denekamp C, Feringa BL. 1998. Optically active diarylethenes for multimode photoswitching between liquid-crystalline phases. Adv Mater 10(14):1080–1082.

Eelkema R, Pollard MM, Katsonis N, Vicario J, Broer DJ, Feringa BL. 2006. Rotational reorganization of doped cholesteric liquid crystalline films. J Am Chem Soc 28(44):14397–14407.

Feringa BL, Huck NPM, van Doren HA. 1995. Chiroptical switching between liquid crystalline phases. J Am Chem Soc 117(39):9929–9930.

Feringa BL, van Delden RA, Koumura N, Geertsema EM. 2000. Chiroptical molecular switches. Chem Rev 100(5):1789–1816.

Finkelmann H, Stegemeyer H. 1974. Description of cholesteric mixtures by an extended Goossens theory. Ber Bunsen-Ges 78(9):869–874.

Finkelmann H, Stegemeyer H. 1978. Temperature dependence of the intrinsic pitch in induced cholesteric systems. Ber der Bunsen-Gesellschaft 82(12):1302–1308.

Fukunaga H, Takimoto J, Doi M. 2004. Molecular dynamics simulation study on the phase behavior of the Gay-Berne model with a terminal dipole and a flexible tail. J Chem Phys 120(16):7792–7800.

Funamoto K, Ozaki M, Yoshino K. 2003. Discontinuous shift of lasing wavelength with temperature in cholesteric liquid crystal. Jpn J Appl Phys, Part 2: Letters 42(12B):L1523–L1525.

Furumi S, Yokoyama S, Otomo A, Mashiko S. 2003. Electrical control of the structure and lasing in chiral photonic band-gap liquid crystals. Appl Phys Lett 82(1):16–18.

Goossens WJA. 1971. Molecular theory of the cholesteric phase and of the twisting power of optically active molecules in a nematic liquid crystal. Mol Cryst Liq Cryst 12(3):237–244.

Gottarelli G, Spada GP. 1985. Induced cholesteric mesophases: origin and application. Mol Cryst Liq Cryst 123(1–4):377–388.

Gottarelli G, Samori B, Stremmenos C, Torre G. 1981. Induction of cholesteric mesophases in nematic liquid crystals by some chiral aryl alkyl carbinols. A quantitative investigation. Tetrahedron 37(2):395–399.

Gottarelli G, Hibert M, Samori B, Solladie G, Spada GP, Zimmermann R. 1983. Induction of the cholesteric mesophase in nematic liquid crystals: mechanism and application to the determination of bridged biaryl configurations. J Am Chem Soc 105(25):7318–7321.

Haas WE, Nelson KF, Adams JE, Dir GA. 1974. The uv imaging with nematic chlorostilbenes. J Electrochem Soc 121(12):1667–1669.

John S. 1987. Strong localization of photons in certain disordered dielectric superlattices. Phy Rev Lett 58(23):2486–2489.

Kopp VI, Fan B, Vithana HKM, Genack AZ. 1998. Low-threshold lasing at the edge of a photonic stop band in cholesteric liquid crystals. Opt Lett 23(21):1707–1709.

Kopp VI, Zhang Z-Q, Genack AZ. 2003. Lasing in chiral photonic structures. Prog Quantum Electron 27(6):369–416.

Kumaresan S, Mallia VA, Kida Y, Tamaoki N. 2005. Thermal and photooptical properties of azoxybenzene/alkyloxy-azobenzene-cholesterol dimesogens with alkyl diacetylene linker. J Mater Res 20(12):3431–3438.

Kurihara S, Nomiyama S, Nonaka T. 2000. Photochemical switching between a compensated nematic phase and a twisted nematic phase by photoisomerization of chiral azobenzene molecules. Chem Mater 12(1):9–12.

Kurihara S, Nomiyama S, Nonaka T. 2001. Photochemical control of the macrostructure of cholesteric liquid crystals by means of photoisomerization of chiral azobenzene molecules. Chem Mater 13(6):1992–1997.

Kurihara S, Hatae Y, Yoshioka T, Ogata T, Nonaka T. 2006. Photo-tuning of lasing from a dye-doped cholesteric liquid crystals by photoisomerization of a sugar derivative having plural azobenzene groups. Appl Phys Lett 88(10):103121/1–103121/3.

Lee H-K, Doi K, Harada H, Tsutsumi O, Kanazawa A, Shiono T, Ikeda T. 2000. Photochemical modulation of color and transmittance in chiral nematic liquid crystal containing an azobenzene as a photosensitive chromophore. J Phys Chem B 104(30):7023–7028.

Lin T-H, Chen Y-J, Wu C-H, Fuh AY-G, Liu J-H, Yang P-C. 2005. Cholesteric liquid crystal laser with wide tuning capability. Appl Phys Lett 86(16):161120/1–161120/3.

Mallia VA, Tamaoki N. 2003. Photoresponsive vitrifiable chiral dimesogens: photo-thermal modulation of microscopic disordering in helical superstructure and glass-forming properties. J Mater Chem 13(2):219–224.

Matsuhisa Y, Huang Y, Zhou Y, Wu S-T, Ozaki R, Takao Y, Fujii A, Ozaki M. 2007. Low-threshold and high efficiency lasing upon band-edge excitation in a cholesteric liquid crystal. Appl Phys Lett 90(9):091114/1–091114/3.

Matsui T, Ozaki R, Funamoto K, Ozaki M, Yoshino K. 2002. Flexible mirror-less laser based on a free-standing film of photopolymerized cholesteric liquid crystal. Appl Phys Lett 81(20):3741–3743.

Moriyama M, Tamaoki N. 2001. Photo-controllable and fixative optical properties of non-polymeric liquid crystals with azobenzene chromophore. Chem Lett 2001 (11):1142–1143.

Onsager L. 1949. The effects of shapes on the interaction of colloidal particles. Ann NY Acad Sci 51:627–659.

Ozaki M. 2006. Tunable photonic band gap based on chiral liquid crystals having spiral periodic structure and laser action. Oyo Butsuri 75(12):1445–1452.

Ozaki M, Kasano M, Ganzke D, Haase W, Yoshino K. 2002. Mirrorless lasing in a dye-doped ferroelectric liquid crystal. Adv Mater 14(4):306–309.

Pieraccini S, Gottarelli G, Labruto R, Masiero S, Pandoli O, Spada GP. 2004. The control of the cholesteric pitch by some azo photochemical chiral switches. Chem Eur J 10(22):5632–5639.

Ross DL, Blank J. 1971. Photochromism. New York: John Wiley & Sons. Chapter 5.

Ruslim C, Ichimura K. 2000. Conformational effect on macroscopic chirality modification of cholesteric mesophases by photochromic azobenzene dopants. J Phys Chem B 104(28):6529–6535.

Ruslim C, Ichimura K. 2001. Photocontrolled alignment of chiral nematic liquid crystals. Adv Mater 13(9):641–644.

Ruslim C, Ichimura K. 2002. Photoswitching in chiral nematic liquid crystals: interaction-selective helical twist inversion by a single chiral dopant. J Mater Chem 12(12):3377–3379.

Sackmann E. 1971. Photochemically induced reversible color changes in cholesteric liquid crystals. J Am Chem Soc 93(25):7088–7090.

Sagisaka T, Yokoyama Y. 2000. Reversible control of the pitch of cholesteric liquid crystals by photochromism of chiral fulgide derivatives. Bull Chem Soc Jpn 73(1):191–196.

Sanpei H. 2005. Studies on design and synthesis of chiral azobenzene derivatives, Master Thesis. Kumamoto University, Kumamoto, Japan.

Shibaev PV, Kopp V, Genack A, Hanelt E. 2003. Lasing from chiral photonic band gap materials based on cholesteric glasses. Liq Cryst 30(12):1391–1400.

Stevensson B, Komolkin AV, Sandstrom D, Maliniak A. 2001. Structure and molecular ordering extracted from residual dipolar couplings: a molecular dynamics simulation study. J Chem Phys 114(5):2332–2339.

Stevensson B, Sandstrom D, Maliniak A. 2003. Conformational distribution functions extracted from residual dipolar couplings: a hybrid model based on maximum entropy and molecular field theory. J Chem Phys 119(5):2738–2746.

Tamaoki N, Song S, Moriyama M, Matsuda H. 2000. Rewritable full-color recording in a photon mode. Adv Mater 12(2):94–97.

Tamaoki N, Aoki Y, Moriyama M, Kidowaki M. 2003. Photochemical phase transition and molecular realignment of glass-forming liquid crystals containing cholesterol/azobenzene dimesogenic compounds. Chem Mater 15(3):719–726.

Tazuke S, Kurihara S, Ikeda T. 1987. Amplified image recording in liquid crystal media by means of photochemically triggered phase transition. Chem Lett 1987:911–914.

Van de Witte P, Galan JC, Lub J. 1998. Modification of the pitch of chiral nematic liquid crystals by means of photoisomerization of chiral dopants. Liq Cryst 24(6):819–827.

Van de Witte P, Brehmer M, Lub J. 1999. LCD components obtained by patterning of chiral nematic polymer layers. J Mater Chem 9(9):2087–2094.

van Delden RA, van Gelder MB, Huck NPM, Feringa BL. 2003. Controlling the color of cholesteric liquid-crystalline films by photoirradiation of a chiroptical molecular switch used as dopant. Adv Funct Mater 13(4):319–324.

van Delden RA, Mecca T, Rosini C, Feringa BL. 2004. A chiroptical molecular switch with distinct chiral and photochromic entities and its application in optical switching of a cholesteric liquid crystal. Chem Eur J 10(1):61–70.

Yablonovitch E. 1987. Inhibited spontaneous emission in solid-state physics and electronics. Phys Rev Lett 58(20):2059–2062.

Yamaguchi T, Inagawa T, Nakazumi H, Irie S, Irie M. 2001. Photoinduced pitch changes in chiral nematic liquid crystals formed by doping with chiral diarylethene. J Mater Chem 11(10):2453–2458.

Yang D-K, Huange X-Y, Zhu Y-M. 1997. Bistabel cholesteric reflective displays: materials and drive schemes. Annu Rev Mater Sci 27:117–146.

Yoshioka T, Ogata T, Nonaka T, Moritsugu M, Kim S-N, Kurihara S. 2005. Reversible-photon-mode full-color display by means of photochemical modulation of a helically cholesteric structure. Ad Mater 17:1226–1229.

11

TUNABLE DIFFRACTION GRATINGS BASED ON AZOBENZENE POLYMERS AND LIQUID CRYSTALS

Yue Zhao

11.1. DIFFRACTION GRATINGS CAN EASILY BE RECORDED ON AZOBENZENE-CONTAINING POLYMERS AND LIQUID CRYSTALS

The reversible trans–cis photoisomerization of azobenzene and its derivatives make polymers and liquid crystals (LCs) bearing this chromophore a suitable medium for recording a diffraction grating for light. Using photosensitive materials, a diffraction grating basically is a periodical change of refractive index, with the period being in the order of the wavelength of light to be diffracted. As schematically illustrated in Fig. 11.1, when a film of an azobenzene-containing polymer or LC is exposed to a recording light going through a grating photomask, or without mask, to an interference pattern (high and low intensity) generated by two coherent recording laser beams, a grating structure or pattern can easily be formed because of the change in refractive index in the reactive regions hit by light. Various events in the reactive regions, which are all related to the photoisomerization of azobenzene, can give rise to the index change. First, the simple conversion of *trans*-azobenzene to *cis*-azobenzene can result in an index change, since the two isomers do not have the same index. In this case, the index modulation, which is the difference in refractive index between the reactive and nonreactive regions, is usually small. Second, the photoinduced orientation (alignment) of azobenzene moieties can change the refractive index. In this case, the recording light needs to be linearly polarized to induce orientation of azobenzene in the direction perpendicular to the polarization, through the mechanism of "hole burning" or angular selection (Ikeda, 2003; Natansohn and Rochon, 2002; Ichimura, 2000; Rasmussen

Smart Light-Responsive Materials. Edited by Yue Zhao and Tomiki Ikeda

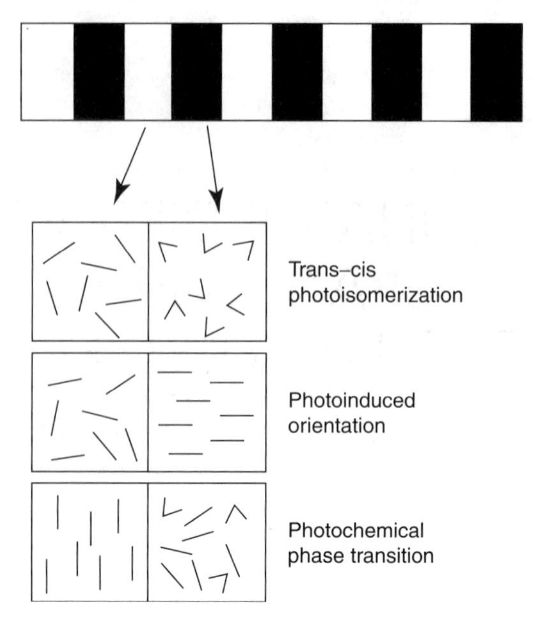

Figure 11.1. Schematic illustration of the possible periodic changes in refractive index related to the photoisomerization of azobenzene. Black strips designate irradiated areas.

et al., 1999; Pham et al., 1997). Third, with azobenzene LCs, the photochemically induced phase transition, generally nematic to isotropic, can also give rise to an index change (Yoneyama et al., 2002; Ikeda and Tsutsumi, 1995). Although azobenzene mesogens in the elongated trans form are compatible with ordered LC phases, the bent cis isomer is usually incompatible with them, and the perturbation effect may be enough to isothermally transform the LC phase into the isotropic state, or induce disorder in the LC phase.

The diffraction efficiency η is an important parameter for a grating. In the case of transmission grating in the Raman–Nath regime, it is related to the refractive index modulation Δn through $\eta \approx (\pi d\Delta n/\lambda)^2$, where d is the thickness and λ the wavelength of the probe light (Yoneyama et al., 2002). The index modulation obviously is governed by the combination of a number of factors, including the used azobenzene materials, the grating formation mechanism involved, and the recording conditions. The first-order diffraction angle Θ is related to the grating period Λ (fringe spacing) through $\Lambda = \lambda/(2\sin\Theta)$.

It should be emphasized here that the grating schematized in Fig. 11.1 is a volume grating, that is, the index change occurs throughout the thickness direction of the film. Upon exposure to an interference pattern with no mask, a remarkable property of azobenzene polymers is the formation of surface relief grating (SRG) as a result of a mass transport triggered by the trans–cis photoisomerization (Kim et al., 1995; Rochon et al., 1995). As reviewed in Chapters 4 and 8 there has been a considerable amount of research work on SRG of azobenzene polymers. Since this chapter focuses on gratings that can be tuned by an external stimulus, the large body of works on SRG (Natansohn and Rochon, 2002) and nontunable volume gratings (Ikeda, 2003) are outside the scope of this review. Moreover, the interference pattern in Fig. 11.1 is the modulation in intensity of light that gives a similar effect to a grating photomask. In the case of photoinduced orientation, an interference pattern with polarization modulation (same light intensity) can well be used to record a grating (Qi et al., 2005). Unless otherwise stated, the tunable diffraction gratings discussed in this chapter are volume gratings, and the recording interference pattern is intensity modulation.

11.2. WHAT ARE TUNABLE DIFFRACTION GRATINGS?

Diffraction gratings are an optical element indispensable for many devices and instruments. Generally, the important characteristics of a grating are the diffraction angle and diffraction efficiency. Tunable diffraction gratings refer to gratings whose characteristics can be reversibly changed (tuned) by an external stimulus. The tunability is an attractive feature that offers possibilities for new applications in such devices as fiber-optic switches and dynamically variable focal-length lenses (Sutherland et al., 1994; Zhang and Sponsler, 1992). This chapter reviews and discusses research works that use azobenzene-containing polymers or LCs to record tunable diffraction grating. Figure 11.2 schematically presents the three types, namely, mechanically, electrically, and optically tunable diffraction gratings. For a mechanically tunable diffraction grating, the diffraction angle can be reversibly changed upon an elastic deformation of the material resulting in a change in the grating period. As the deformation can also lead to a change in refractive index modulation, the diffraction efficiency may also be tuned. For an electrically tunable diffraction grating, the application of an electric field changes the index modulation, usually because of the electric field–induced LC orientation, which allows for switching of the diffraction efficiency between the on- and off state, that is, between high- and low efficiency. In the case of optically tunable diffraction grating, the diffraction efficiency of the probe light can be switched between the on- and off state by exposing the grating to light at two wavelengths. With azobenzene polymers and LCs, it is easy to imagine that light at the two wavelengths should induce the trans–cis and the reverse cis–trans isomerization of azobenzene moieties, respectively, to trigger a reverse change in index modulation of the grating.

11.3. MECHANICALLY TUNABLE DIFFRACTION GRATINGS

The recording of such a grating on an azobenzene polymer film, whose diffraction characteristics can be *reversibly and drastically* changed through deformation of the film, requires the use of an azobenzene-containing elastomer. In principle, any azobenzene polymer in the glassy state (below its T_g) can undergo an elastic

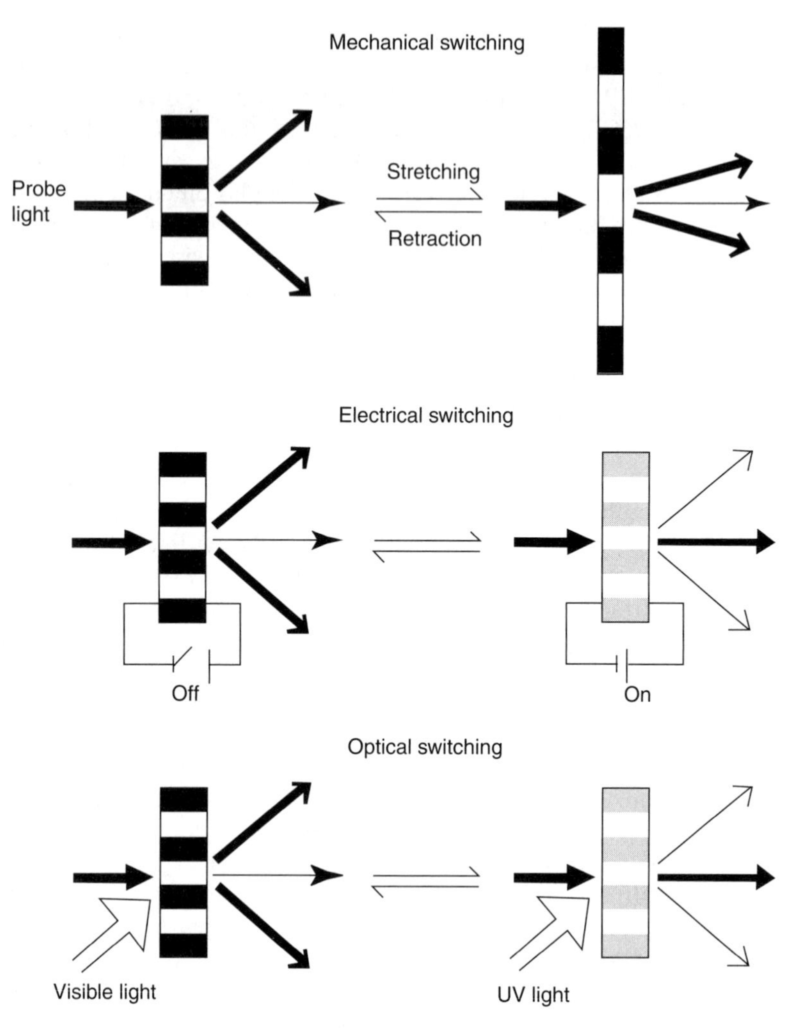

Figure 11.2. Schematic illustration of mechanically, electrically, and optically switchable diffraction gratings.

deformation before the yielding or failure, but the strain generally is very small (a few percent), and the stress is high. In recent years, we have demonstrated a new type of azobenzene polymers, namely, azobenzene thermoplastic elastomers (ATEs) (Cui et al., 2004; Zhao et al., 2003; Bai and Zhao, 2002; Zhao et al., 2002; Bai and Zhao, 2001). As a unique feature, thin films of variable thickness can easily be prepared through their solution casting, and they are capable of undergoing large and reversible deformation at room temperature. Studies by the author's group found that they are interesting photosensitive materials for mechanically tunable diffraction gratings.

11.3.1. Preparation of Azobenzene Thermoplastic Elastomers

ATEs differ from conventional elastomers (Camacho-Lopez et al., 2004) in that their network structure and elastic deformation under stress are supported by physical cross-links rather than chemical (covalent) cross-linking. This is why they can be dissolved in a solvent or be melted at elevated temperatures for processing like a thermoplastic polymer. One important category of thermoplastic elastomers is the ABA-type triblock copolymer, with the central B block being a room temperature elastomer (amorphous, low T_g) and the two end blocks being a high T_g (or high T_m) polymer. Basically, at appropriate copolymer compositions, the microphase separation because of the immiscibility of the two constituent polymers (A and B blocks) can lead to the formation of rigid microdomains of A that, by interconnecting the rubbery chains of B, act as the physical cross-links supporting the elastic deformation under stress. The author's group developed two methods for the synthesis of azobenzene thermoplastic elastomers (ATE), as depicted in Fig. 11.3. The first method (Fig. 11.3a) consists in grafting an azobenzene side-chain liquid crystalline polymer (azo-SCLCP) onto a styrene–butadiene–styrene (SBS) triblock copolymer, which is a commercially available thermoplastic elastomer [with ~30% polystyrene (PS)]. By simply polymerizing an azobenzene acrylate (or methacrylate) monomer in a toluene solution with dissolved SBS, using benzoyl peroxide as the free radical initiator, the grafting can be achieved (Bai and Zhao, 2002, 2001). The reaction mechanisms may involve both a radical transfer to the methylene group of the polybutadiene (PB) block that initiates the polymerization of the azobenzene monomer and a direct reaction of propagating radicals of azo-SCLCP with the double bonds of PB. Depending on the experimental conditions and the monomer used, the amount of azo-SCLCP grafts varies from ~8% to 22%. Figure 11.4 shows the photos of solution-cast films of the starting SBS and an ATE sample containing 8% of azo-SCLCP. The film of ATE is transparent as SBS. After drying, the ATE film can be peeled off from the glass slide and repeatedly stretched to a strain as high as 500%. The elasticity of ATE comes from the SBS, in which the rubbery PB chains are interconnected by the glassy PS cylindrical microdomains acting as the physical cross-links. In ATE, the grafts of azo-SCLCP are embedded in the rubbery PB matrix and, in some samples, may form their own microdomains.

A controlled free radical polymerization technique was also used, namely, atom transfer radical polymerization (ATRP), to synthesize ATE on the basis of ABA triblock copolymers (Fig. 11.3b) (Cui et al., 2004). The triblock copolymer was designed to have a rubbery midblock of poly(*n*-butyl acrylate) (PnBA) and two end blocks of poly{6-[4-(4-methoxyphenylazo)phenoxy]hexyl methacrylate} (PAzoMA) that is azo-SCLCP. For synthesis, a dibromo initiator, namely, 1,1′-biphenyl-4,4′-bis(2-bromoisobutyrate), can first be used to prepare the dibromo PnBA macroinitiator, which is then used to polymerize the azobenzene methacrylate monomer to yield the two end blocks of PAzoMA. This ATE is different from azo-SCLCP-grafted SBS. It is a thermoplastic elastomer, in which

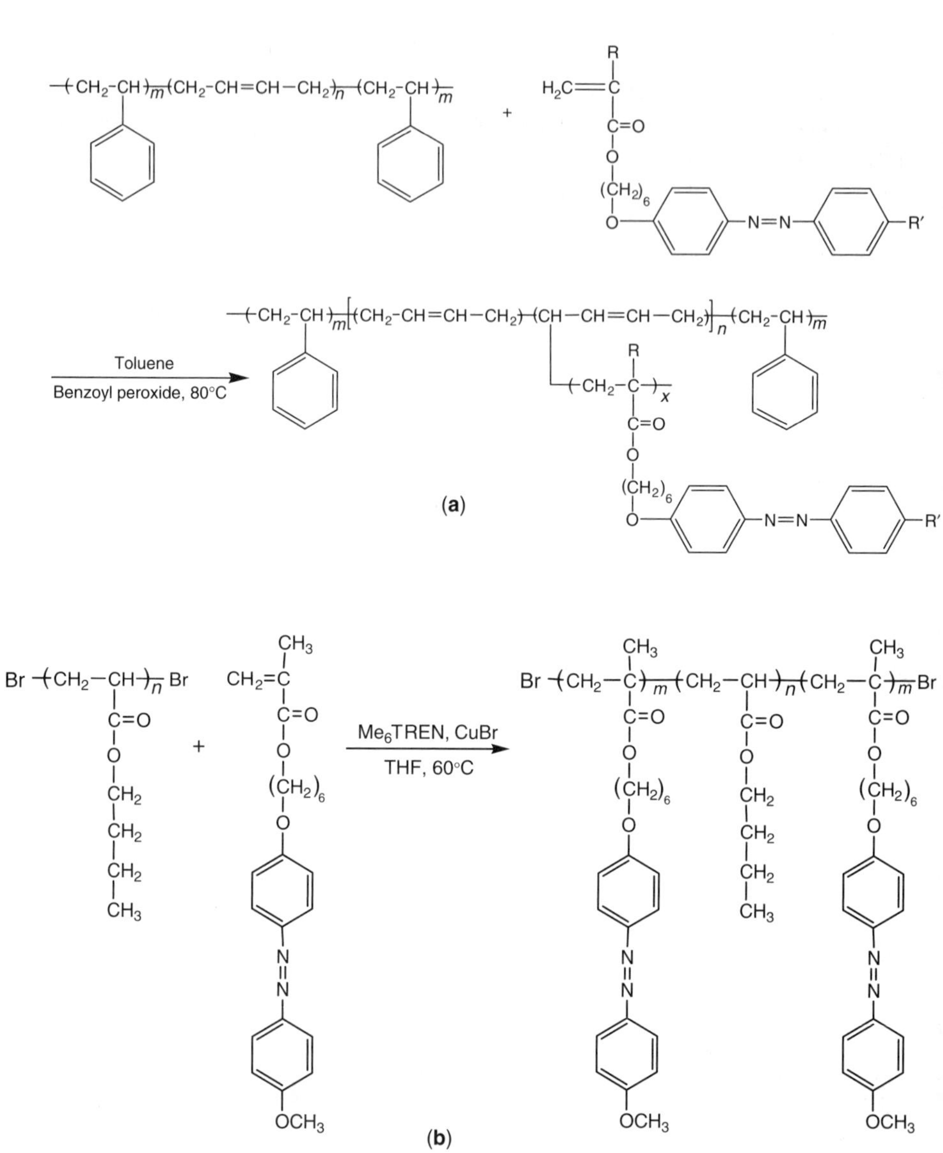

Toluene

Benzoyl peroxide, 80°C

(a)

Me_6TREN, CuBr

THF, 60°C

(b)

Figure 11.3. Two methods for the synthesis of azobenzene thermoplastic elastomers: (**a**) grafting of an azobenzene-containing side-chain liquid crystalline polymer (azo-SCLCP) onto the thermoplastic styrene–butadiene–styrene (SBS) triblock copolymer using radical polymerization, and (**b**) triblock copolymer composed of a rubbery central block and two azobenzene SCLCP end blocks synthesized using atom transfer radical polymerization (ATRP).

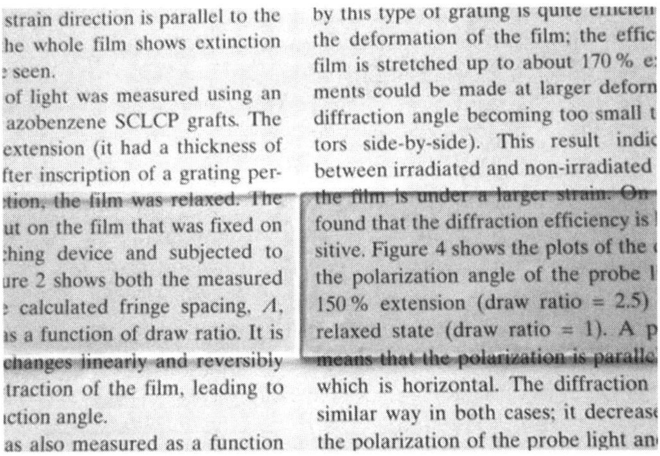

strain direction is parallel to the
he whole film shows extinction
seen.
of light was measured using an
azobenzene SCLCP grafts. The
extension (it had a thickness of
fter inscription of a grating per-
tion, the film was relaxed. The
ut on the film that was fixed on
hing device and subjected to
ure 2 shows both the measured
calculated fringe spacing, Λ,
is a function of draw ratio. It is
changes linearly and reversibly
traction of the film, leading to
iction angle.
as also measured as a function

by this type of grating is quite efficien
the deformation of the film; the effic
film is stretched up to about 170 % e
ments could be made at larger deforn
diffraction angle becoming too small t
tors side-by-side). This result indic
between irradiated and non-irradiated
the film is under a larger strain. On
found that the diffraction efficiency is
sitive. Figure 4 shows the plots of the
the polarization angle of the probe l
150 % extension (draw ratio = 2.5)
relaxed state (draw ratio = 1). A p
means that the polarization is paralle
which is horizontal. The diffraction
similar way in both cases; it decrease
the polarization of the probe light an

Figure 11.4. Photographs of a solution-cast film of SBS (*left band*) and a film of azo-SCLCP-grafted SBS (*right band*). *Source*: Zhao et al., 2003. Reprinted with permission.

the microphase-separated domains of azo-SCLCP (i.e., PAzoMA) act as cross-links supporting the extension of PnBA chains. Since the physical cross-links are not formed by an amorphous polymer such as PS, but by a liquid crystalline and photosensitive polymer, new interesting properties can emerge from the coupled elasticity, liquid crystallinity, and photoactivity. In particular, as compared to conventional thermoplastic elastomers such as SBS, an intermediate state exists for PAzoMA-*b*-PnBA-*b*-PAzoMA between the normal elastic ($T_g^{PB} < T < T_g^{PS}$) and thermoplastic ($T > T_g^{PS}$) regimes. In other words, when stretched at $T_{ni} > T > T_g$ of azo-SCLCP (T_{ni} is the nematic-to-isotropic phase transition temperature), a long-range orientation of azobenzene mesogens can be induced inside the deformed LC microdomains that can still support part of the elastic extension of PnBA chains. This molecular orientation of azobenzene can subsequently be retained in the relaxed film at room temperature, resulting in thermoplastic elastomers containing glassy microdomains with oriented mesogens. However, despite the well-defined structure of this type of ATE, its mechanical strength and deformability are not as good as azo-SCLCP-grafted SBS. This is the main reason for which azo-SCLCP-grafted SBS was utilized for the study of mechanical tunable diffraction gratings as described later.

11.3.2. Coupled Mechanical and Optical Effects

When a solution-cast film of ATE is stretched at room temperature, the deformation gives rise to a long-range molecular orientation of azobenzene groups along the strain direction (SD), which can be switched between two orientational states upon alternating UV and visible light irradiation as a result of the coupled mechanical and optical effects. This orientational switching in a stretched film can easily be detected by polarized infrared or UV–Vis spectroscopy. Figure 11.5 shows an example of the reversible change in orientation of azobenzene mesogens in an ATE film stretched to 500% deformation and, under strain, subjected to UV and visible light exposure, the order parameter being determined from the infrared dichroism (Bai and Zhao, 2001). Under unpolarized UV irradiation, the mechanically induced orientation can be erased because of the trans–cis isomerization, while it is recovered upon unpolarized visible irradiation that drives the reverse cis–trans isomerization. The switching is observed with eight cycles in two separate experiments differing in the irradiation time. Figure 11.6 schematically illustrates the main features of the coupling of mechanical and optical effects, where four different states of azobenzene groups in the film are depicted, and for each state two polarized UV–Vis spectra of the film are shown (solid and dashed lines represent the polarization parallel and perpendicular to the SD, respectively). In the unstretched film, **1**, the elongated *trans*-azobenzene groups, whose maximum absorption is ~360 nm, are in the LC phase but have no long-range orientation, as seen from the spectra displaying no dichroism. If the unstretched film is exposed to UV light, the trans–cis isomerization occurs, and the photochemical phase transition results in disordered *cis*-azobenzene groups (bent form), **2**, absorbing at ~450 nm and showing no dichroism. This process can be reversed by the cis–trans back-isomerization of azobenzene either upon visible

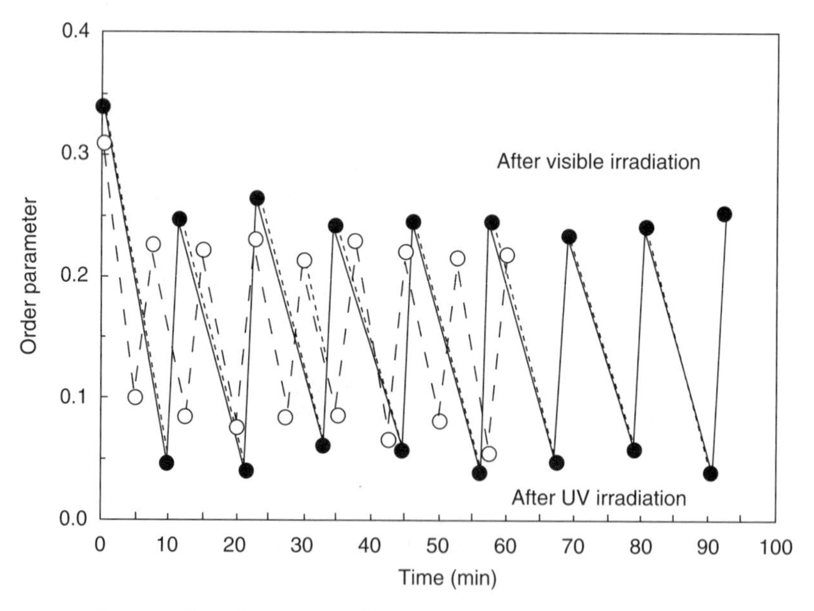

Figure 11.5. Changes in orientation of azobenzene mesogens for two films of azo-SCLCP-grafted SBS, under 500% strain, subjected to repeated cycles of unpolarized UV irradiation for orientation loss and unpolarized visible irradiation for orientation recovery. *Source*: Bai and Zhao, 2001. Reprinted with permission.

irradiation or through thermal relaxation. From **1**, if the film is stretched, *trans*-azobenzene groups are aligned along the SD, **3**, resulting in the dichroism of the absorption band (this mechanically induced orientation is enhanced by the liquid crystallinity of the azobenzene polymer). However, if the stretched film is exposed to UV light, either polarized or unpolarized, the orientation of *trans*-azobenzene groups is lost as they are converted into the cis isomer, **4**, as revealed by the spectra. Upon visible light irradiation, the mechanically induced orientation of *trans*-azobenzene mesogens is recovered. All the processes depicted in Fig. 11.6 are reversible as indicated by the arrows. Stretched ATE films can be used to record mechanically tunable diffraction gratings.

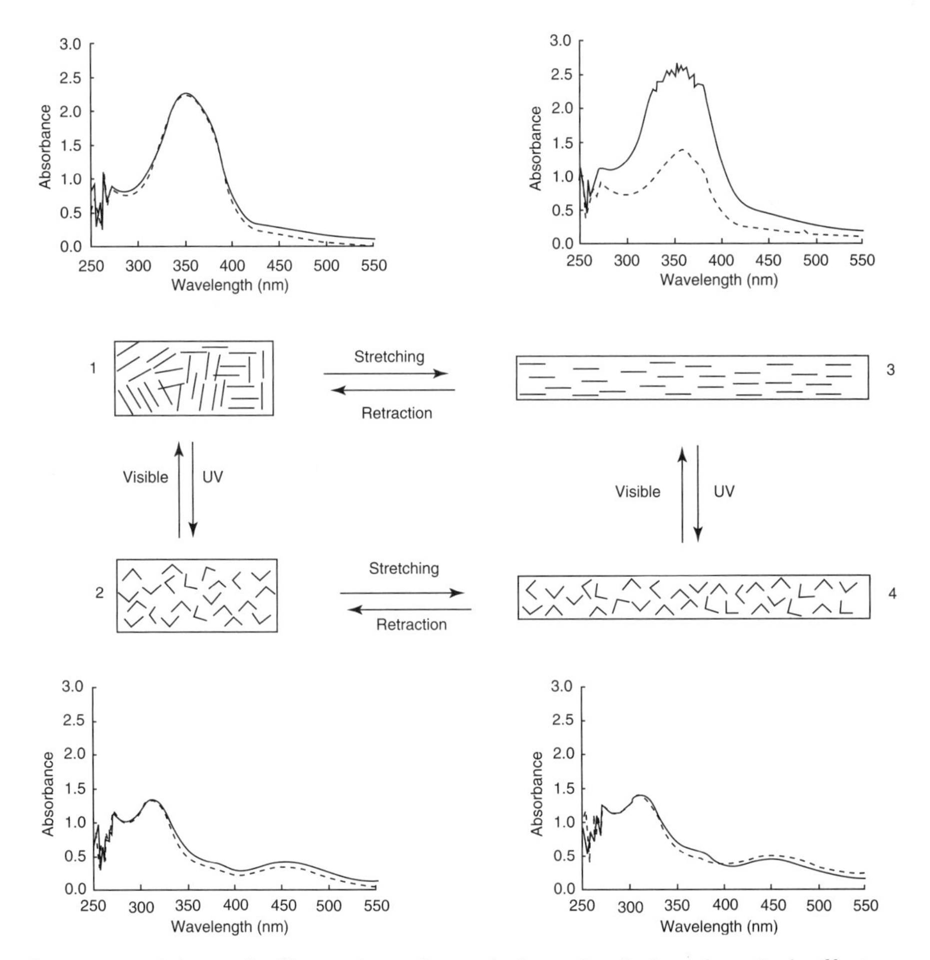

Figure 11.6. Schematic illustration of coupled mechanical and optical effects on azobenzene mesogens in ATE films, the polarized spectra being recorded with the spectrometer's beam polarized parallel (*solid lines*) and perpendicular (*dashed lines*) to the strain direction. *Source*: Zhao et al., 2003. Reprinted with permission.

11.3.3. Elastic Diffraction Gratings Recorded Using a Photomask

From the coupled mechanical and optical effects, a grating can readily be inscribed on a stretched ATE film on the basis of a periodical variation in azobenzene orientation, as schematically illustrated in Fig. 11.7. In a stretched film, azobenzene groups are aligned along the SD. When UV light is applied through a grating photomask placed in front of the film, oriented *trans*-azobenzene groups in irradiated regions are converted into randomly aligned *cis*-azobenzene, while *trans*-azobenzene remains oriented in nonirradiated areas. A grating can have its fringes perpendicular or parallel to SD. The actual situation is more complicated because azo-SCLCP counts only ~10% of the material, being dispersed in an SBS matrix. Under strain, SBS is also highly anisotropic as depicted, with oriented PB chains being supported by PS cylindrical microdomains inclined to SD. Considering

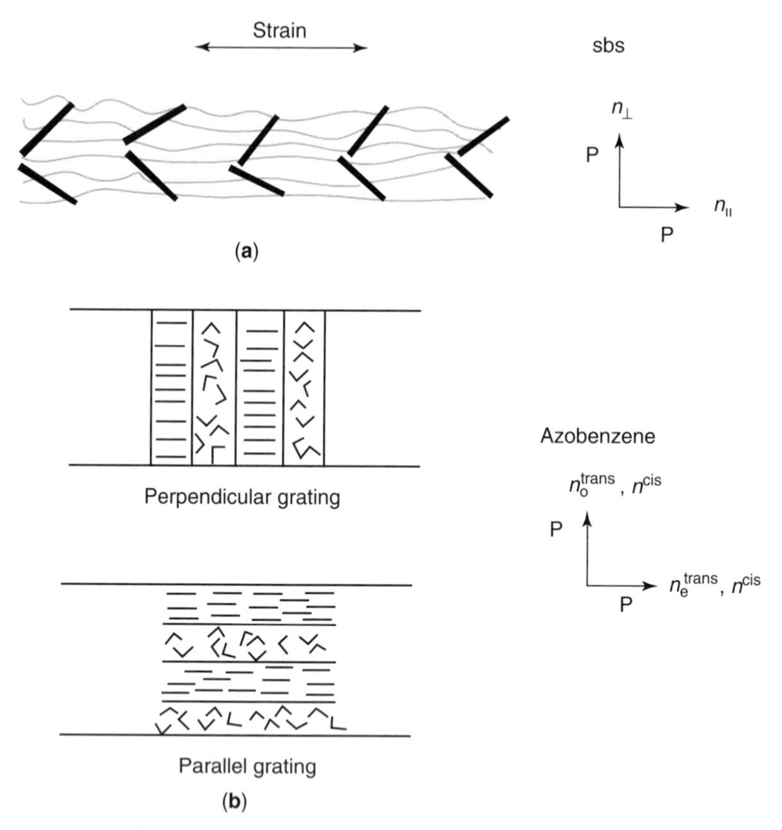

Figure 11.7. Schematic illustration of (**a**) the anisotropic morphology of stretched SBS and (**b**) the periodic presence of disordered *cis*-azobenzene and oriented *trans*-azobenzene in irradiated and nonirradiated fringes, respectively, for both perpendicular and parallel gratings recorded on azo-SCLCP-grafted SBS. *Source*: Bai and Zhao, 2002. Reprinted with permission.

the SBS as a background, the probe light sees different refractive indices, $n_{\parallel}$ and $n_{\perp}$, with its polarization parallel or perpendicular to the strain, respectively. As for the azobenzene polymer, in both parallel and perpendicular gratings, irradiated areas mainly contain disordered *cis*-azobenzene and nonirradiated areas contain oriented *trans*-azobenzene. If the SBS background undergoes no changes initiated by photoisomerization of azobenzene in irradiated areas, the difference in refractive index between irradiated and nonirradiated fringes should mainly arise from differences between disordered *cis*-azobenzene and oriented *trans*-azobenzene. However, the diffraction efficiency should also be polarization-dependent. When the probe light is polarized with its electric vector parallel to the strain, light sees the extraordinary refractive index of *trans*-azobenzene, n_e^{trans}, in nonirradiated areas and the average refractive index of *cis*-azobenzene, n^{cis}, in irradiated areas. In contrast, if the polarization of light is perpendicular to the strain, light senses the ordinary refractive index of *trans*-azobenzene, n_o^{trans}, in nonirradiated areas and, still, n^{cis} in irradiated areas. In the latter case, the diffraction efficiency is smaller because the difference between n_o^{trans} and n^{cis} is smaller than the difference between n_e^{trans} and n^{cis}. As is shown later, this feature indeed was observed.

The tunable feature of such an "elastic" grating is shown by the polarizing optical micrographs (POMs) in Fig. 11.8. On a film stretched to 300% extension (thickness ~10 μm under strain), two gratings with 10-μm fringe spacing were recorded parallel and perpendicular, respectively, to the SD using a photomask (360-nm UV irradiation of intensity ~2 mW/cm^2, 10-min exposure) (Fig. 11.8a). The irradiated areas, including an uncovered region between the two photomasks, appear dark. The stretched film was then allowed to retract to its half-length (Fig. 11.8b); the fringe spacing of the perpendicular grating decreases, while it increases for the parallel one. Gratings can be inscribed at any alignment with respect to SD, which result in different changes in fringe spacing upon elastic extension and retraction of the film. Using the grating with fringes perpendicular (grating vector parallel) to SD, the first-order diffraction of light was measured as a function of draw ratio defined as the film length after stretching over that before stretching. Figure 11.9 shows the measured diffraction angle, Θ, as well as the fringe spacing, Λ, calculated according to $\Lambda = \lambda/(2 \sin \Theta)$, λ being the wavelength of the used probe light (an He–Ne laser operating at 633 nm). It is visible that the fringe spacing changes linearly and reversibly upon elastic extension and retraction of the film, leading to reversible changes in the diffraction angle.

The diffraction efficiency was also measured as a function of draw ratio using a probe light polarized parallel to the SD. The first-order diffraction efficiency (+1) was calculated as the ratio of I_d over I_t^o, where I_d is diffracted intensity (+1) and I_t^o is the transmitted intensity before the writing of grating. Within experimental errors, the result in Fig. 11.10 shows a quite reversible change in the diffraction efficiency upon extension and retraction. The efficiency increases as the film is stretched up to ~170% extension, indicating that the index modulation becomes greater in the film under a larger strain. Moreover, it was found that the diffraction efficiency is highly polarization-sensitive. Figure 11.11 shows the plots of the diffraction efficiency vs. the polarization angle of the probe light for a film under a 150% extension

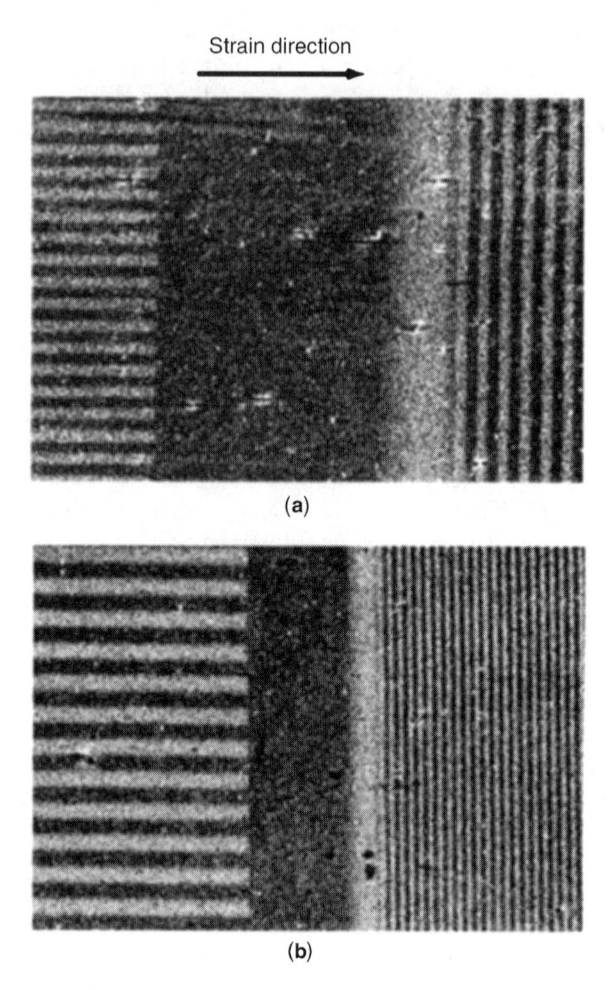

Figure 11.8. Polarized optical micrographs for (**a**) two gratings (10-μm fringe spacing) inscribed on an ATE film stretched to 300% extension, with their fringes parallel and perpendicular, respectively, to the strain direction; and (**b**) the two gratings on the same film retracted to the half-length. *Source*: Zhao et al., 2002. Reprinted with permission.

(draw ratio = 2.5) and at the completely relaxed state (draw ratio = 1). A polarization angle of 0° means that the polarization is parallel to the SD. The diffraction efficiency changes in a similar way in both cases; it decreases as the angle between the polarization of the probe light and SD increases. With respect to the grating, efficiency is maximum when the polarization is perpendicular to the fringes, whereas minimum efficiency is observed when the polarization is parallel to the fringes.

In the completely relaxed state, as nonirradiated areas should be isotropic, the result in Fig. 11.11 suggests that irradiated areas could be anisotropic. The probe

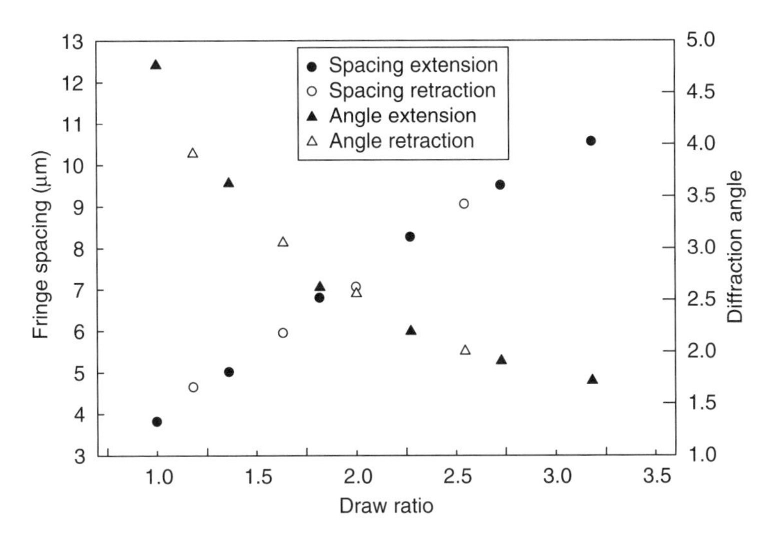

Figure 11.9. First-order diffraction angle (*triangles*) and calculated fringe spacing (*circles*) vs. draw ratio for a grating inscribed on an ATE film subjected to extension (*closed symbols*) and retraction (*open symbols*). *Source*: Zhao et al., 2002. Reprinted with permission.

light senses a larger difference between reactive and nonreactive areas when the polarization is perpendicular to the fringes than when it is parallel. When a film is under strain, nonirradiated areas also become anisotropic because of the stretching-induced orientation of both PB chains and azobenzene mesogens. However, this

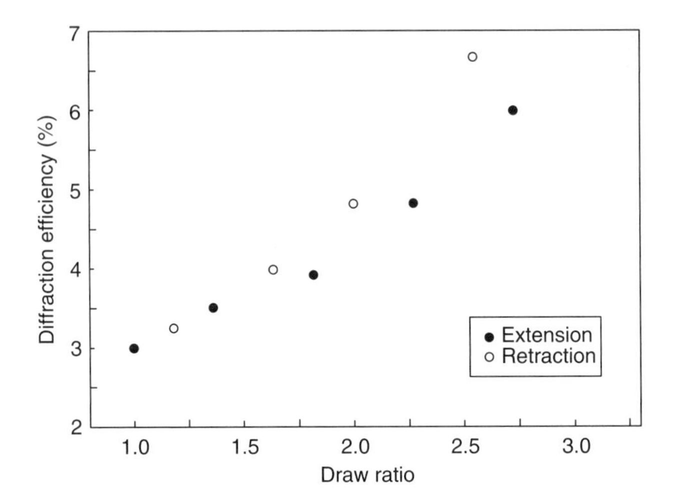

Figure 11.10. First-order (+1) diffraction efficiency vs. draw ratio for a grating inscribed on an ATE film subjected to extension (*closed symbols*) and retraction (*open symbols*). *Source*: Zhao et al., 2002. Reprinted with permission.

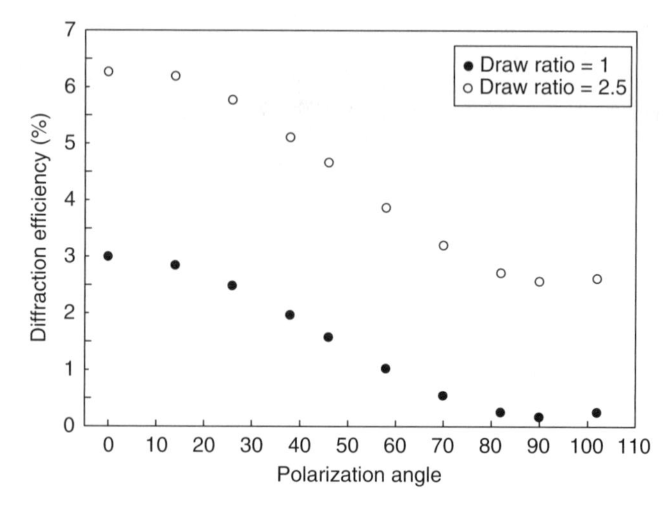

Figure 11.11. First-order (+1) diffraction efficiency vs. polarization angle of the probe light for a grating inscribed on an ATE film stretched to 150% extension and in the completely retracted state. *Source*: Zhao et al., 2002. Reprinted with permission.

orientation also enhances the anisotropy in the irradiated areas, and surprisingly, the polarization dependence remains the same as in the relaxed state, although the diffraction efficiency increases. Upon UV light exposure, irradiated areas lose the orientation of azobenzene mesogens because of the trans–cis photoisomerization, which should initiate the formation of the grating. However, when the film is relaxed, both irradiated and nonirradiated areas lose the orientation of azobenzene, and irradiated areas recover to the trans-rich state after ~10 h at room temperature. In other words, if the refractive index modulation originates only from differences in azobenzene conformation and orientation, it should disappear. Therefore, the stable grating observed in the relaxed film implies that some permanent structural or morphological changes occurred during the photoisomerization of azobenzene in the irradiated regions. More information on the grating formation mechanisms can be obtained by holographic recording as described later in following text.

11.3.4. Grating Formation Dynamics and Mechanisms

Instead of using a UV lamp and a grating photomask, an interference pattern generated by two coherent beams of a UV laser (krypton, $\lambda = 350$ nm) can be used to write a grating on a stretched ATE film. While placing the probe light (633 nm) at the same exposed area, the grating formation dynamics can be monitored by measuring the change in diffraction efficiency. The authors used such an optical setup to get more insight into the mechanism of grating formation. In the experiments, the two recording beams producing the interference pattern were both linearly polarized perpendicular to the SD (*s–s* geometry), with the crossing angle adjusted to yield various periods. The holographic gratings recorded were in the Raman–Nath diffraction regime, the fringes were perpendicular to the SD, and the polarization of the probe light was parallel to it. Again, since azobenzene mesogens are oriented along the SD in the film before the exposure, the grating a priori should be a grating whose spatial modulation of refractive index arises mainly from the difference between oriented *trans*-azobenzene in nonirradiated regions and disordered *cis*-azobenzene in irradiated regions.

The dynamic process of grating formation was monitored for a series of films stretched to various degrees to reveal the role of deformation. Figure 11.12 presents the results (the recording beams were switched on at the zero time). In this example, the period of grating is 2 μm, and the power per recording beam is 160 mW/cm^2. The recording dynamics for an unstretched film is also shown for comparison. A grating is formed even in the unstretched film, which is expected because of the difference in refractive index between disordered *trans*-azobenzene and *cis*-azobenzene. The diffraction efficiency η is, however, low because of the small index modulation and increases very slightly up to an exposure time of 300 s. When the film is stretched to a strain of only 20%, η increases already, showing the contribution from oriented azobenzene groups in the stretched film. Moreover, it can be noted that after a decrease following the peak value within the first 40-s exposure, the increase of η is more significant over time. These features become much more prominent with the film stretched to a strain of 300%. At this large deformation, *trans*-azobenzene groups are well oriented, and the index mismatch, and thus the modulation depth, is more important, resulting in higher diffraction.

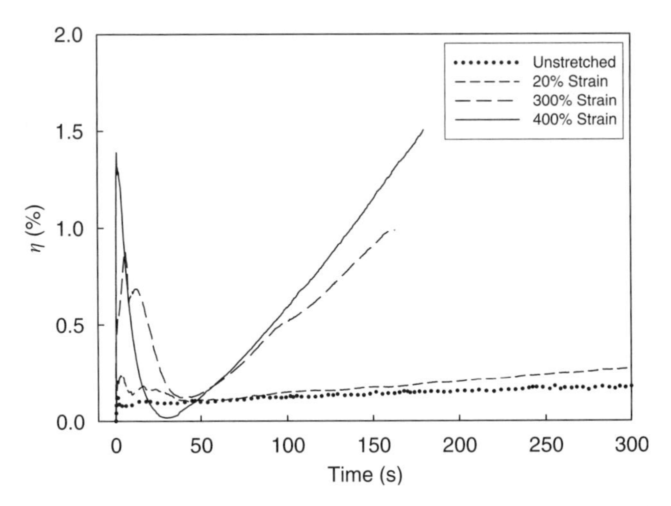

Figure 11.12. First-order (+1) diffraction efficiency vs. exposure time for ATE films stretched to various strains. The period of grating is 2 μm, and the power per recording beam is 160 mW/cm^2. *Source*: Zhao et al., 2003. Reprinted with permission.

However, this azobenzene grating appears unstable as η drops after ~20 s. But interestingly, another grating starts to develop at longer exposure times and η rises again. The effect of film deformation on the grating formation dynamics continues to amplify with the film stretched to a strain of 400%. Two things become clear on the basis of these results: first, the deformation enhances the diffraction efficiency and second, two grating formation processes appear to take place upon exposure of the stretched film to the interference pattern.

Using the intensity of 160 mW/cm^2 per recording beam, a long exposure time can result in a grating with very high diffraction efficiency. Figure 11.13 shows an example, using an interference pattern of 5-μm period and a film stretched to 400% deformation. It is seen that η rises continuously and reaches >30% after 25-min exposure; the inset shows the grating formation dynamics for the first 40 s, where the first and unstable grating formation is visible. Moreover, there is a stable grating that remained in the relaxed film 7 months after the recording, as shown by the POMs in Fig. 11.14. Because the initial grating, recorded on the film stretched to 400% strain, has a period of 5 μm, the grating in the relaxed state has a period of 1 μm. The film shows no signs of degradation and remains highly elastic. Subsequent stretching, in the direction perpendicular to the fringes, increases the period by an amount that is linearly proportional to the deformation, as can be noticed from the example of 150% strain that gives a period of 2.5 μm. The diffraction efficiency η of this remained holographic grating was measured upon extension and retraction of the film. Figure 11.15 shows the change in η as a function of draw ratio, Γ, as well as the corresponding change in index modulation

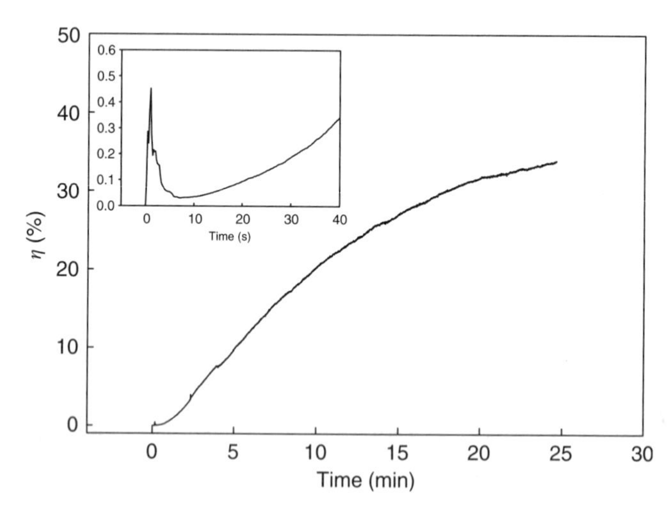

Figure 11.13. Increase in diffraction efficiency with exposure time for the second grating recorded on an ATE film stretched to 400% strain. The inset shows the dynamic process of the first 40 s. The period of grating is 5 μm, and the power per recording beam is 160 mW/cm^2. *Source*: Zhao et al., 2003. Reprinted with permission.

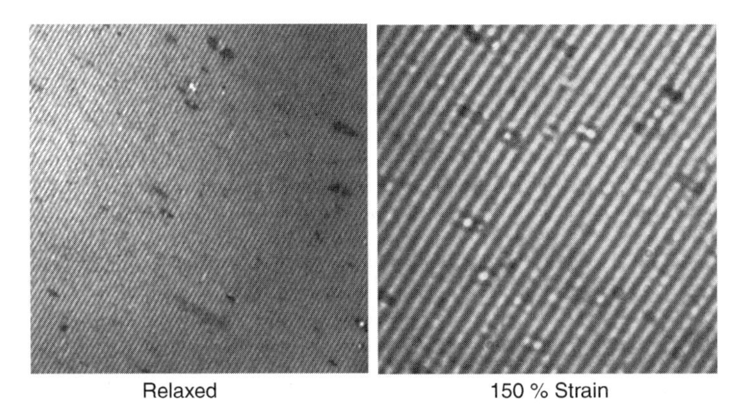

Figure 11.14. Polarizing optical micrographs showing the grating remained in the film 7 months after the recording. The period in the relaxed state is 1 μm, and it increases to 2.5 μm under a strain of 150%. *Source*: Zhao et al., 2003. Reprinted with permission.

Δn. Within the Raman–Nath regime, the grating index modulation can be estimated from $\eta \approx (\pi d \Delta n/\lambda)^2$ using the measured η and the film thickness d calculated from $d = d_o \Gamma^{-0.5}$ at each draw ratio (initial film thickness $d_o = 26$ μm) (Zhao et al., 2003). The results show that η changes in a reversible fashion within experimental errors, increasing slightly with the deformation up to a strain of ~50% while decreasing at larger strains. At large film deformations, the sharp decrease in η is caused by a diminishing index modulation and the decrease in film thickness.

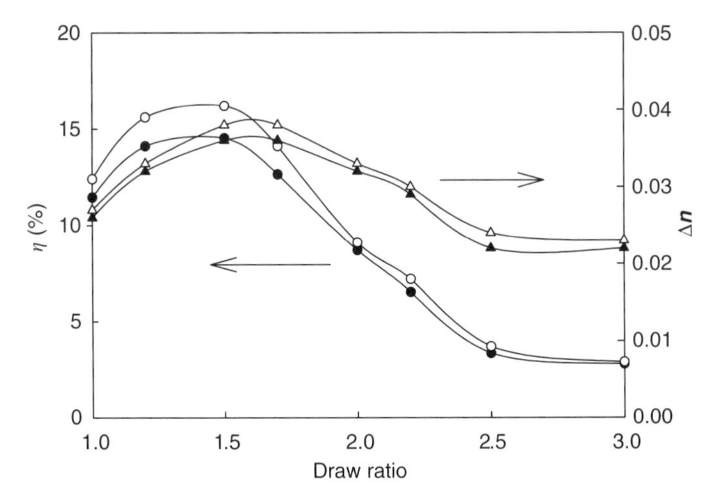

Figure 11.15. Changes in both diffraction efficiency and modulation of refractive index as a function of draw ratio, for the grating remained in the film 7 months after the recording (Fig. 11.14). *Source*: Zhao et al., 2003. Reprinted with permission.

The two gratings formed upon holographic recording have different natures. The first grating appearing quickly upon exposure comes mainly from, as expected, oriented *trans*-azobenzene and disordered *cis*-azobenzene groups. This is supported by experimental evidence, including the strong dependence of η on the polarization of the probe light. Indeed, if the writing beams are turned off before the diffraction reaches the transient peak and starts to fall, which happens a few seconds after the exposure, η may be stable over time, but drops by $\sim$90% when the polarization of the probe light is changed by 90° to be perpendicular to the SD. This is consistent with the index modulation based on azobenzene orientation and conformation as depicted in Fig. 11.7. With parallel polarization, the probe light sees the extraordinary refractive index of *trans*-azobenzene and the mean index of *cis*-azobenzene in the nonirradiated and radiated areas, respectively, and the difference is large; in contrast, with perpendicular polarization, the difference is between the ordinary refractive index of *trans*-azobenzene and the mean index of *cis*-azobenzene, which is much smaller.

The second and stable grating formed at longer exposure times is the result of a structural reorganization and rearrangement in the ATE film under strain. Though the slow dynamics is reminiscent of the formation of SRG in azobenzene polymers (Kim et al., 1995; Rochon et al., 1995), no SRG was found by AFM, which is no surprise because the used *s*–*s* polarization geometry for the two recording beams is known to be ineffective for inducing SRG (Viswanathan et al., 1999), and the confinement of azo-SCLCP in its microdomains should hinder the mass transport (Breiner et al., 2007). The different nature of the second grating is also hinted by the less important polarization dependence of the diffraction efficiency. For samples with a well-developed second grating, changing the polarization of the probe light to be perpendicular to the strain decreases η by only $\sim$30%, in contrast with the 90% drop for the first grating. The strong index modulation remained 7 months after the recording also implies that the grating was not caused by any difference in orientation or conformation of the $\sim$8% azobenzene polymer in the ATE film; rather it was formed through structural changes of the whole elastomer. If azobenzene groups were the sole factor responsible for the second grating, the grating structure would disappear days after the holographic recording as a result of the thermally activated cis–trans backisomerization, restoring the equilibrium concentration of *trans*-azobenzene throughout the whole film.

What could be the structural rearrangement in a stretched ATE film subjected to holographic recording that is responsible for the formation of the second grating? On the basis of what is known about SBS, the following analysis can be proposed. The ATE sample contains some 8% azobenzene polymer grafts chemically connected to the PB chains, with the azobenzene polymer forming microdomains embedded in the PB matrix. In a stretched film, in addition to the glassy PS cylindrical microdomains (Fig. 11.7), the azo-SCLCP microdomains may also act as cross-links and support part of the extensional stress. When the writing UV laser is turned on, the photoisomerization in the excited (reactive) regions converts oriented *trans*-azobenzene into disordered *cis*-azobenzene, corresponding to an isothermal photochemical phase transition from the LC to the isotropic phase. As the azobenzene polymer in the isotropic phase is more fluid than that in the LC state, its microdomains become less effective cross-linking points, which results in relaxation of oriented PB chains and release of the stress. As this happens, the stress is no longer uniformly distributed in space as before irradiation, and local structural reorganization may follow to rebalance the stress along the SD. This perturbation might be enough to disorder *trans*-azobenzene groups in nonexcited areas, which explains the erasure of the first grating. As the irradiation continues, the excited areas may undergo further structural changes because of the photoinduced isotropic state of the azobenzene polymer. In other words, similar to the effect of thermal annealing below T_g of PS, in the optically plasticized areas (excited areas), greater plastic flow of PS chains takes place and leads to greater PB chain orientation relaxation and rearrangement of the PS cylindrical domains (Pakula et al., 1985a,b). The formation dynamics of the second grating, arising from such structural changes, should be slow and dependent on the power of the writing beam, which indeed is the case.

Such a structural rearrangement induced by the photoisomerization of azobenzene mesogens in a stretched ATE film paints a complicated picture for the possible changes in diffraction efficiency upon elastic deformation. In principle, the index modulation depth between irradiated and nonirradiated regions could come from three contributions: different PB chain orientations, different arrangements of PS cylindrical microdomains, and different azobenzene conformations and orientations. Residual PB chain orientation is known to exist in relaxed SBS films (Zhao, 1992; Pakula et al., 1985a,b). Therefore, in a relaxed ATE film after writing of the second grating, different residual PB orientations and different arrangements for PS cylindrical domains may exist in the irradiated and nonirradiated regions. Since the form birefringence of SBS with perfectly oriented cylindrical domains ($\sim 6 \times 10^{-4}$) is much smaller than the refractive index modulation remained in the relaxed ATE film ($\Delta n = 0.026$), different residual PB chain orientations could be the leading factor for the stable grating in the relaxed film. When the film is stretched, the index modulation depth may actually increase at modest strains, since different arrangements of PS cylindrical domains (anisotropic cross-link points) may lead to different PB chain orientations, whereas under large strains the difference in PB chain orientation would become less important, resulting in a decrease in Δn. This would explain the results in Fig. 11.15.

11.4. ELECTRICALLY TUNABLE DIFFRACTION GRATINGS

11.4.1. Use of Liquid Crystals

Diffraction gratings that can be switched on and off by an electric field is of interest for applications, including fiber-optic switches and lenses with a dynamically variable focal length (Sutherland et al., 1994; Zhang and Sponsler, 1992).

Since LCs can readily be aligned by an electric field resulting in large birefringence change, they are materials of choice for making electrically tunable diffraction gratings (Zhang and Sponsler, 1992). Among the known LC materials, polymer-dispersed LCs (PDLCs) can be used for recording holographic gratings through photopolymerization-induced segregation of the polymer from the LC droplets (Sutherland et al., 1994). Cholesteric LCs can form patterns under the effect of a voltage. This was also explored for recording electrically tunable diffraction gratings (Kang et al., 2001; Subacius et al., 1997). However, in case no polymer is used to stabilize the pattern, a spatially controlled electric field, using patterned electrodes on the surface of LC cell, is required to orient LC molecules selectively to afford a periodic variation in refractive index (Scharf et al., 1999). Therefore, it is of fundamental and applied interest to discover how to use the photoisomerization of azobenzene to make LC-based electrically tunable diffraction gratings.

11.4.2. Grating Formation in Photosensitive Self-Assembled Liquid Crystal Gels

By doping an LC with an azobenzene compound, the trans–cis photoisomerization of the azo dopant can cause a change in LC orientation, in addition to the conformational change of azobenzene molecules. It is conceivable that a diffraction grating can be recorded in an azobenzene-doped LC by exposing it to UV light through a grating photomask or to an interference pattern. However, not only is the grating unstable because of the thermal cis–trans backisomerization that should erase the periodic variation in refractive index but it is also not tunable by an electric field since the LC doped with azobenzene either in the trans or cis form can align with the field in a similar manner. To obtain a tunable diffraction grating using azobenzene-containing LCs, the index modulation caused by the photoisomerization should be stable and change in response to a voltage as a result of different electric field–induced LC orientation. Self-assembled LC gels with an azobenzene-containing gelator can fulfill these conditions (Zhao and Tong, 2003).

A self-assembled LC gel is a small-molecule LC gelled by a physical network of fibrous aggregates formed by a gelator compound (Guan and Zhao, 2001, 2000; Yabuuchi et al., 2000; Kato et al., 1998a,b). The usual way to prepare an LC gel consists in dissolving a small amount of a gelator (1–3 wt%) in an LC host at elevated temperatures (generally in the isotropic phase of the LC) and then cooling slowly the homogeneous mixture, upon which gelator molecules tend to aggregate (crystallize) against the dissolution. In case gelator molecules can develop strong specific and highly unidirectional intermolecular interactions such as hydrogen bonding, which favors one-dimensional crystal growth, the gelation of the LC host can occur at certain temperatures (either in the isotropic or LC phase) because of the formation of a physical network of fibrous aggregates with diameters in the range of tens or hundreds of nanometers. The gelation temperature depends on a number of parameters, including gelator concentration, cooling rate, and gelator miscibility in the LC host. The presence of nanofibers can affect the electro-optical behaviors of the LC, changing the threshold voltage for electric field–induced LC orientation and the switching times (Guan and Zhao, 2001, 2000; Yabuuchi et al., 2000; Kato et al., 1998a,b).

If the gelator molecule contains azobenzene group in its structure, the formed fibrous aggregates in a LC gel become photosensitive. Figure 11.16 shows the chemical structure of such a compound denoted as azo gelator and a scanning electron microscope (SEM) image (Fig. 11.16a) of the nanofibers formed by 1% of the gelator in a cholesteric LC host composed of a nematic LC, BL006 (T_{ni} = 115°C), and 5% of a chiral dopant R811 (Merck). The image, recorded after removal of the LC host in hexane, reveals that the physical network is built up by very thin H-bonded fibrous aggregates with diameters of ~100 nm. This means a large specific surface of the fibers and thus a significant amount of azobenzene molecules on surface. Such LC gels are indeed photosensitive because

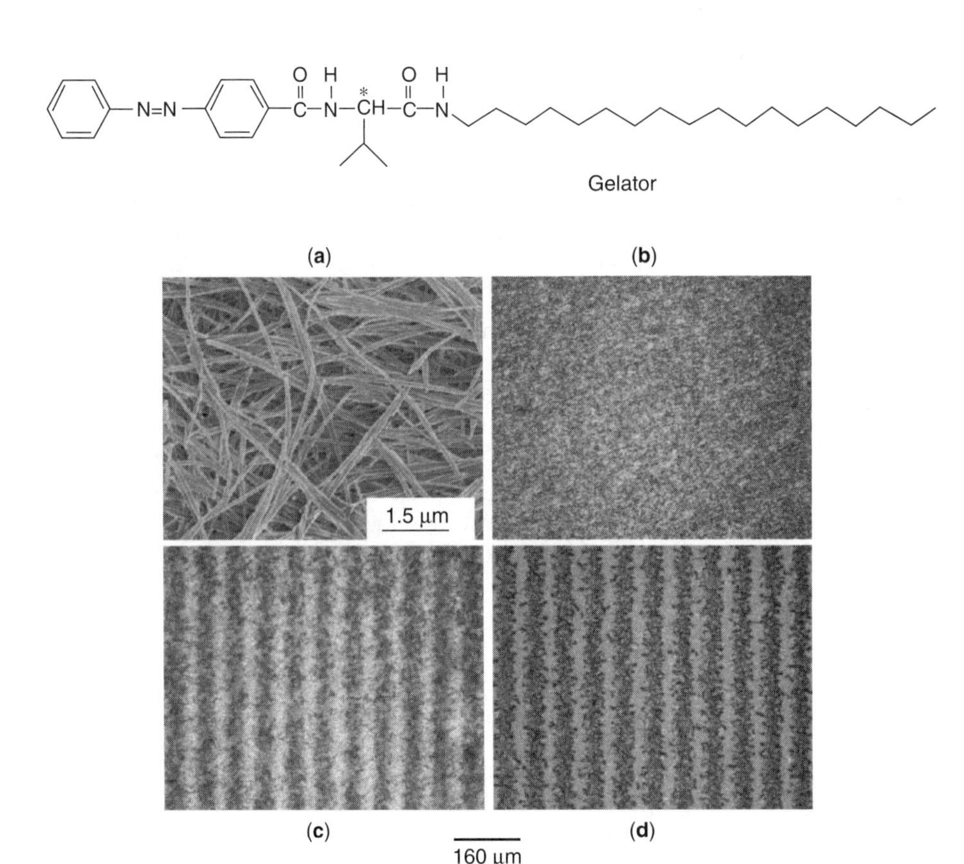

Figure 11.16. (a) SEM image of the fibrous aggregates formed in the self-assembled LC gel with 1% of the shown azobenzene gelator and 5% chiral dopant; (b) photomicrograph of the LC gel before irradiation through a photomask; (c) after 5-min irradiation; and (d) after 10-min irradiation. *Source*: Zhao and Tong, 2003. Reprinted with permission. See color insert.

of the photoisomerization of the chromophore, and they can be used to inscribe stable diffraction gratings whose diffraction efficiency can be reversibly switched using an electric field.

Also presented in Fig. 11.16 are three POMs showing the grating formation process in the LC gel filled in a nonrubbed, indium–tin oxide (ITO)-coated LC cell with a 10-μm gap. To record a grating, the LC cell with a grating photomask fixed on one side was placed inside a thermostat hot stage, heated to ~95°C and exposed to UV light (360 nm, ~100 mW/cm^2) through the photomask (40-μm fringe spacing). Figure 11.16b shows the LC gel before irradiation that has a homogeneous morphology throughout the sample, whereas Fig. 11.16c and d show the formation of a grating structure in the LC gel after 5- and 10-min irradiation, respectively. It is clear that a light-induced reorganization takes place. As the irradiation goes on, an apparent transport of the aggregates from irradiated to nonirradiated regions becomes increasingly important. With a sufficiently long exposure time, 10 min, the aggregates can be completely removed from irradiated areas, leaving essentially LC in those regions. Whereas, at the same time, the LC gel in nonirradiated areas appear to have an increased network density. Once the grating is formed, subsequent cooling to room temperature under or without irradiation results in no reappearance of aggregates in irradiated areas, the grating is retained. To erase the grating, the LC gel needs to be heated to the isotropic phase of the LC (>110°) to redissolve gelator molecules, followed by cooling to room temperature.

The photoisomerization of azobenzene groups is believed to be at the origin of this light-induced grating formation in self-assembled LC gels. However, the process appears to be effective only at recording temperatures close to the melting of the fibrous aggregates of the gelator, implying that a gel state with the gelator molecules starting to dissolve in the LC may be crucial for the observed light-induced reorganization. On the basis of this observation, the proposed grating formation mechanism is schematically illustrated in Fig. 11.17. Before irradiation, there is a homogeneous physical network of thin aggregates dispersed in the LC, with gelator molecules being on the verge of dissolving into the LC; it is likely that a small amount of gelator molecules are already dissolved in the LC. Upon UV irradiation through a photomask, light may alter the equilibrium in irradiated areas by bringing more gelator molecules from the aggregate surface to the LC. In other words, the action of light may increase slightly the solubility of the gelator in the LC host through some mechanisms. Once this happens, the dissolved gelator molecules would diffuse from irradiated areas into the neighboring nonirradiated areas to make the concentration uniform in the whole sample. In nonirradiated areas, an increased concentration of dissolved gelator molecules would shift the equilibrium toward the formation of more aggregates as the gelator molecules condense into the existing aggregates, leading to a denser physical network. This process of light-triggered dissolving, diffusion, and aggregation would continue until the aggregates in irradiated areas are completely dissolved and reformed in the nonirradiated areas. Therefore, after irradiation, a grating is formed with regions containing essentially the pure LC and regions of the LC gel with more

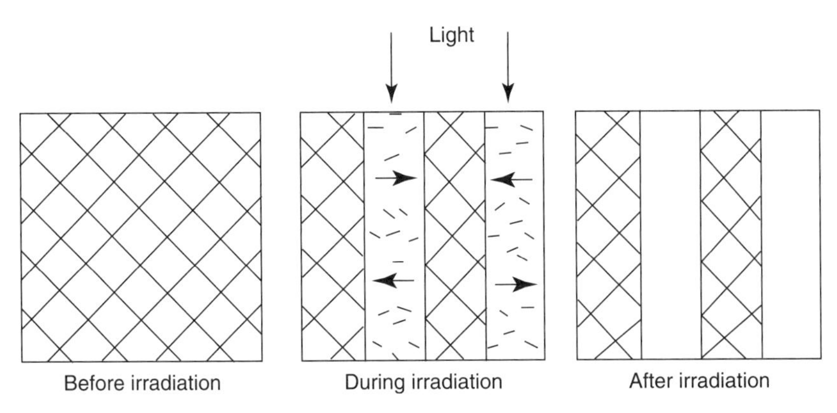

Figure 11.17. Schematic illustration of the mechanism for the light-induced reorganization in self-assembled LC gels and the grating formation (LC host is not shown). *Source*: Zhao and Tong, 2003. Reprinted with permission.

fibrous aggregates. How can UV irradiation increase the solubility of the aggregates of azo gelator in the LC host? Even though the trans–cis photoisomerization cannot be detected with the LC gel, presumably because of the extremely low concentration of azobenzene groups involved (only those on the surface of the 1% aggregates), it is believed that azobenzene group in the gelator is responsible. When the LC gel is heated to temperatures close to the melting of the aggregates, the gelator molecules on the surface would be mobile enough to undergo the trans–cis photoisomerization. In other words, it is likely that the gelator molecules on the surface are converted from the trans to the cis isomer by light, which may increase the surface tension of the aggregates (Tong et al., 2005a,b; Ichimura et al., 2000) and, as a result, increase the dissolution of gelator molecules in the LC.

It should be mentioned that with the use of a photomask on the side of the LC cell exposed to UV light, only gratings with a large period could be obtained. The commercially available cell used (from E.H.C, Japan) is made with two glass plates having a total thickness of ~2250 μm. The probe light is diffracted when passing through the mask, and travels some 1125 μm before hitting the LC gel sample. This means that with masks having a period below 30 μm, the recording light actually irradiates the whole film, and no grating can be formed. Gratings with smaller periods might be obtained using holographic recording.

Kato and coworkers also achieved patterning with self-assembled LC gels (Moriyama et al., 2003). Using an azobenzene-containing gelator (structure in Fig. 11.18) and a low clearing temperature nematic LC, 4-cyano-4′-pentyl biphenyl (5CB), they demonstrated changes in LC gel states induced by the photoisomerization of azobenzene. As depicted in Fig. 11.18, a nematic gel is first obtained; then it is exposed to UV light through a photomask for the trans–cis photoisomerization. They found that the aggregates of gelator in the irradiated regions could be dissolved because of the solubility of azo gelator molecules in the

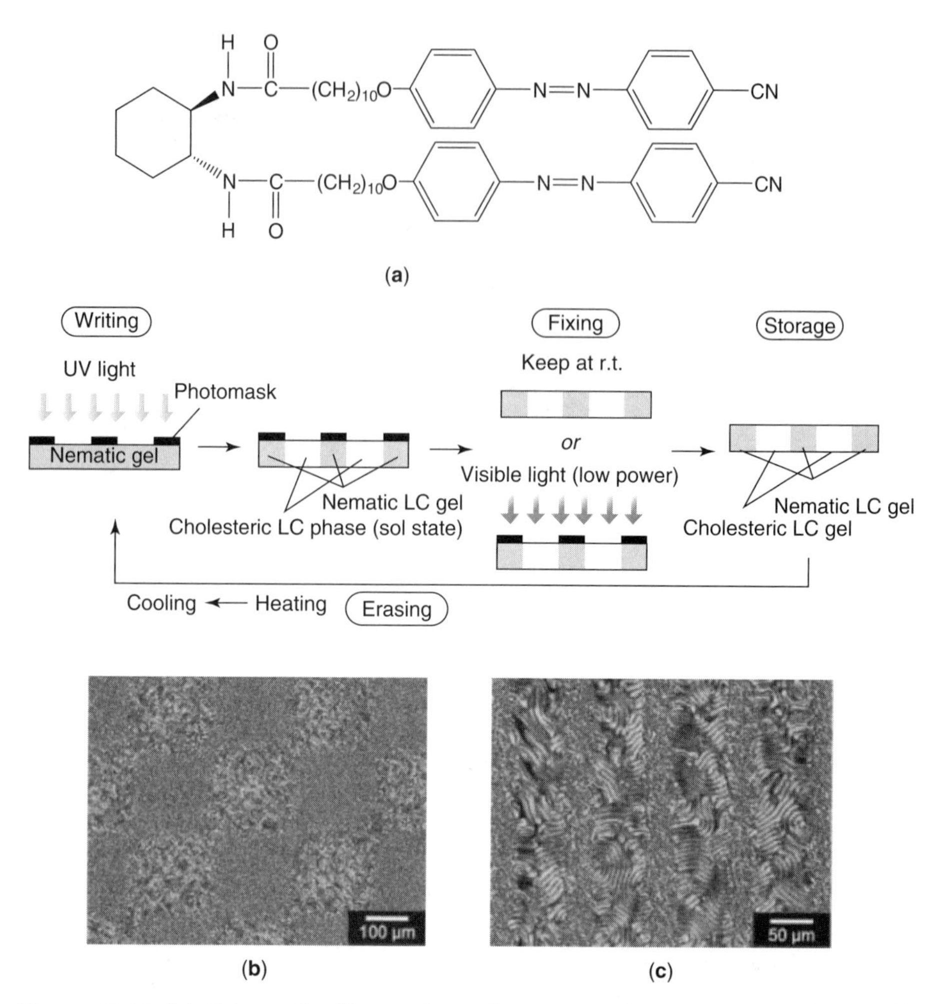

Figure 11.18. **(a)** Schematic illustration of a patterning process with a self-assembled LC gel prepared using the shown azobenzene gelator. **(b,c)** Polarized optical micrographs of the self-assembled LC gel patterned by exposure to UV light through a photomask of 200-μm lattice and a photomask of 50-μm fringe spacing, respectively. *Source*: Moriyama et al., 2003. Adapted with permission.

cis form. Being chiral, dissolved gelator molecules induce the formation of a cholesteric phase displaying the fingerprint texture in irradiated regions, while the nematic gel remains in nonirradiated areas. Upon removal of UV light, a slow cis–trans backisomerization, either thermally activated or by low intensity visible light irradiation, results in the reaggregation of azo gelator molecules in the trans form. The aggregates grow along and stabilize the fingerprint structure, giving rise to a stable cholesteric gel in irradiated areas. Though no diffraction data were presented, different responses of nematic and cholesteric gels to electric field may be explored for electrically tunable diffraction gratings.

11.4.3. Electrical Switching

In Fig. 11.16, the grating structure, for which electrical switching behavior was investigated, constitutes areas of almost pure cholesteric LC and areas of LC contained by fibrous aggregates of the azo gelator. LC molecules in the two areas respond to an electric field differently, which makes the grating display electrically switchable diffraction efficiency. For the measurements discussed later, an He–Ne laser (633 nm) was used as the probe light, and the first-order (+1) diffraction signal, I_d, was measured using a high speed photodetector connected to a digital oscilloscope, whereas a high voltage waveform generator was used to apply the AC (1000 Hz), square wave, or pulse electric fields through the cell. Figure 11.19 shows the changes in the first-order (+1) diffraction efficiency as function of applied voltage for a grating obtained from the LC gel containing 1% gelator and 5% chiral dopant (period ~40 μm). When the electric field is applied for the first time, that is, during the first voltage increase, the diffraction efficiency changes in an apparently complex fashion. It drops first at ~7 V and then rises at ~16 V. To understand this behavior, it should be noted that after the grating is formed at 90°C and cooled to room temperature, the cholesteric LC in the irradiated areas, where the aggregates are moved away by light, has a planar texture (Fig. 11.16d),

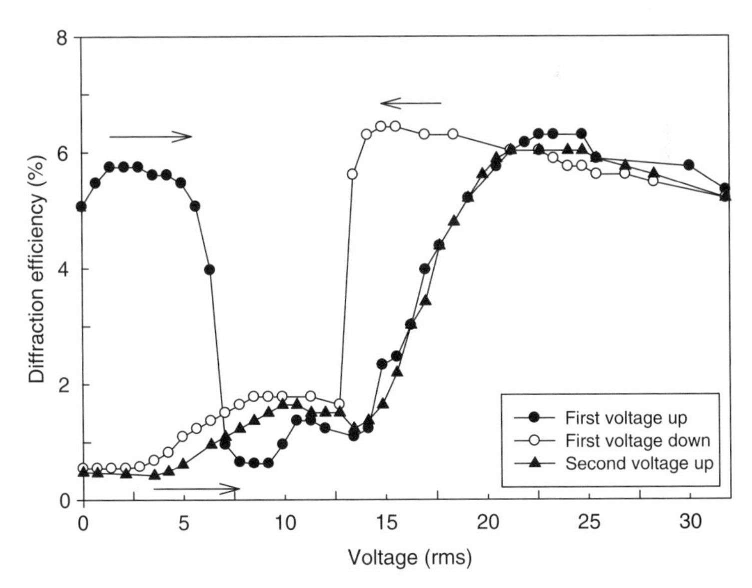

Figure 11.19. First-order (+1) diffraction efficiency vs. applied voltage for a grating formed in the self-assembled LC gel with 1% gelator and 5% chiral dopant. *Source*: Zhao and Tong, 2003. Reprinted with permission.

for which the helices formed by the director (orientation direction) of the LC molecules are normal to the plates of the cell. In nonirradiated areas, where fibrous aggregates are present, the LC should have randomly aligned helices, that is, the focal conic texture. This difference explains the initial refractive index modulation leading to the diffraction at the field-off state. As the electric field is applied, it destabilizes the planar texture into a disordered state (focal conic texture) as in nonirradiated areas, resulting in the drop of diffraction at ~7 V. When the voltage is increased to 16 V, the electric field is strong enough to unwind the helices and align the LC molecules along the field direction (homogeneous texture) in irradiated areas only. LC molecules in nonirradiated areas are not aligned because of the higher threshold voltage of the cholesteric LC caused by the physical network of nanofibers (Tong and Zhao, 2003). This difference results in the change from low diffraction to high diffraction. The high diffraction shows little changes until near 30 V when LC molecules in the regions with the aggregates start to align along the field direction, which reduces the index modulation. On decrease in voltage, the diffraction switches from high to low efficiency at ~14 V, which is caused by the usual hysteresis effect of the LC. At low voltages, the initial high diffraction efficiency is not recovered because the initial planar texture cannot be formed in irradiated areas. On subsequent voltage scans, there is only one stable switch of diffraction efficiency at ~16 V, as shown by the curve obtained during the second voltage increase in Fig. 11.19, which is associated with the transition between the focal conic and homogeneous texture of LC molecules in irradiated areas. Before switching, the low diffraction is due to the small index modulation mainly contributed by the physical network of aggregates, since the LC molecules are not aligned throughout the sample and incident light sees essentially an admixture of the ordinary (n_o) and the extraordinary (n_e) refractive index of the LC. As the electric field aligns the LC molecules in the irradiated area, the index modulation rises because the probe light now sees the ordinary refractive index in these regions. This interpretation is also supported by the observation that the high diffraction efficiency shows no dependence on the polarization direction of the probe light. In essence, this electrical switchability of diffraction efficiency is made possible by the delayed orientation of LC molecules in the gel areas because of the interaction with the fibrous aggregates. The reason for which cholesteric LC is used is that the effect of the physical network on the threshold voltage is more important for cholesteric LCs than for nematic LCs. For instance, grating can also be recorded with the LC gel prepared from pure BL006 (with no chiral dopant), but the sharp increase in diffraction efficiency occurs only over a narrow range of voltages (~2 V) before dropping to the low level, because of the similar threshold voltages of BL006—with and without the network of the gelator.

Figure 11.20 shows the dynamic behavior of the electrically switchable diffraction grating formed in the LC gel with 1% gelator and 2% chiral dopant (period ~40 μm). The repeated switching of the first-order (+1) diffraction efficiency between 0 and 20 V of an applied square-wave electric field (10 s duration for field-off and field-on states) is relatively stable (Fig. 11.20a). Moreover, the change in diffraction efficiency in response to a pulse wave of

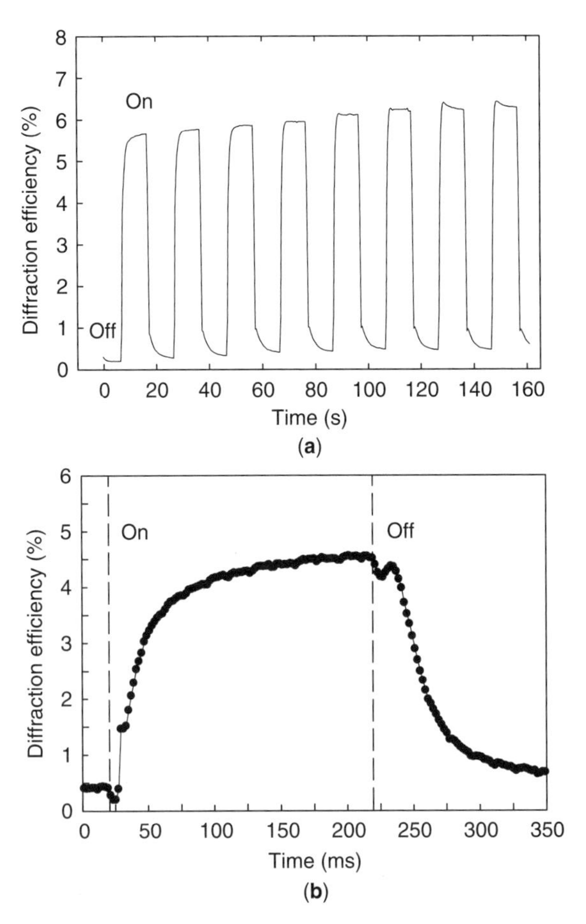

Figure 11.20. Switching of diffraction efficiency for a grating formed in the self-assembled LC gel with 1% gelator and 2% chiral dopant: (**a**) between 0 V (off-state) and 20 V (on-state) of a square-wave electric field and (**b**) dynamic response to a pulse of 200-ms duration. *Source*: Zhao and Tong, 2003. Reprinted with permission.

200 ms can be used to estimate the high diffraction turn-on time ($\sim$65 ms) and the turn-off time ($\sim$50 ms) (Fig. 11.20b).

11.5. OPTICALLY TUNABLE DIFFRACTION GRATINGS

11.5.1. Dynamic Holographic Gratings

How to use azobenzene-containing LCs or polymers to record a grating that can be switched on and off using light? Ikeda and coworkers showed that this could be realized using holographic recording on SCLCPs containing azobenzene mesogens (Yamamoto et al., 2001). When the writing beams (488 nm) are applied to a

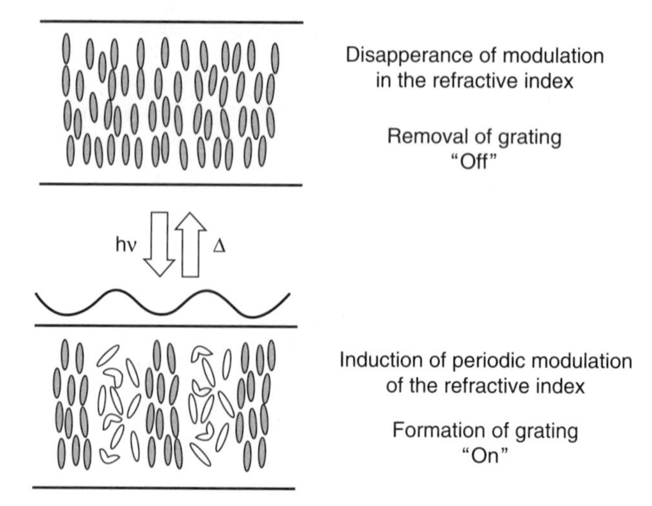

Figure 11.21. Schematic illustration of the use of azobenzene-containing side-chain liquid crystalline polymers to record light-controlled dynamic grating on the basis of photochemical phase transition. *Source*: Yamamoto et al., 2001. Reprinted with permission.

polymer film at a temperature close to the T_{ni}, diffraction of a probe beam (633 nm) occurs because of the appearance of a refractive index modulation in the film, although the diffraction is gone when the writing beams are turned off. The mechanism responsible for this switching proposed by Ikeda's group is shown in Fig. 11.21. In irradiated areas, corresponding to bright fringes of the interference pattern, the trans–cis photoisomerization of azobenzene mesogens induces the nematic-to-isotropic phase transition, resulting in the index modulation. Once the writing laser beams are turned off, the fast thermally activated cis–trans back-isomerization brings azobenzene mesogens back to the nematic phase, which erases the index difference and thus turns off the diffraction of the probe beam. Figure 11.22 shows an example of the repeated writing beam–controlled diffraction efficiency. This result was obtained using a film of poly{6-[4-(4-ethoxyphenylazo)phenoxy]hexyl methacrylate} (following structure): at 143°C, with a power of 127 mW/cm^2 for the writing beams and an exposure time of 62.5 ms, the diffraction switching-on and switching-off times were estimated to be 50 and 600 ms, respectively, and the index modulation was ~0.08. The switching times were found to depend on a number of parameters including temperature, writing beams' intensity, and chemical structure of the used azobenzene SCLCP.

CH_3

$-(CH_2-C)_n-$

$C{=}O$

$O-(CH_2)_6-O-C_6H_4-N{=}N-C_6H_4-OC_2H_5$

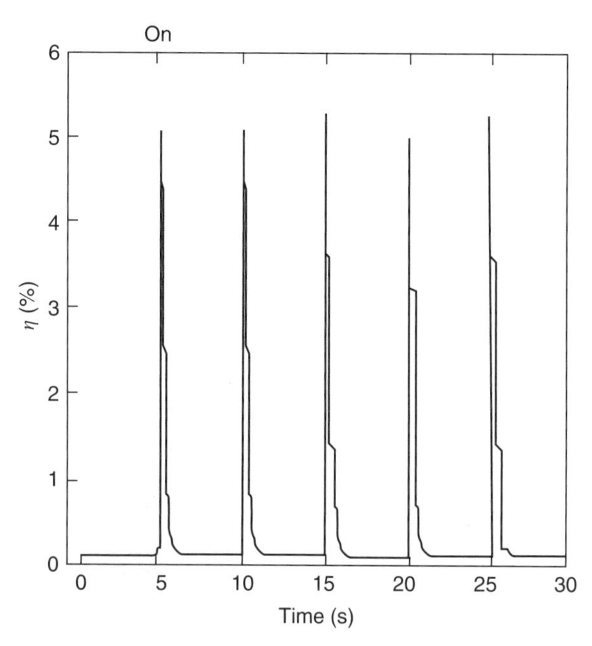

Figure 11.22. Switching behavior of diffracted probe light with a film of the shown azobenzene-containing side-chain liquid crystalline polymer at 143 °C exposed to writing beams (127 mW/cm^2) for 62.5 ms. *Source*: Yamamoto et al., 2001. Reprinted with permission.

The principle of such writing beam–controlled dynamic holographic gratings should be applicable to many azobenzene polymers or LCs. The grating formation time, related to the trans–cis photoisomerization, should always be fast, but the grating erasure time, determined by the thermal relaxation of *cis*-azobenzene, may be much longer depending on the material and experimental conditions. The interest of Ikeda's work is the use of the photoisomerization-induced phase transition (termed "photochemical phase transition") in azobenzene SCLCPs to amplify the difference in refractive index that is required for highly efficient diffraction. Nevertheless, the polymer needs to be held at temperatures close to T_{ni} to ensure fast photochemical phase transition and thermal relaxation of *cis*-azobenzene, which determine the diffraction switching-on and switching-off times.

11.5.2. Optically Tunable Diffraction Gratings in Polymer-Stabilized Liquid Crystals

11.5.2.1. Materials Design and Switching Mechanism. The dynamic holographic grating recorded with azobenzene SCLCPs is erased once the writing beams are turned off. It is of interest to develop materials with which one can record a stable diffraction grating that, after formation, can be repeatedly switched using light. Using photopolymerization-induced phase segregation, we demonstrated that this could be done with an azobenzene polymer–stabilized nematic LC (Tong et al., 2005b). The diffraction efficiency of probe light through such a grating can be switched between high and low values with UV and visible light as a result of the photoisomerization of the azobenzene chromophore. In addition, since the grating is made with LC-based materials, the switching of diffraction efficiency can also be achieved by an electric field that controls the LC orientation. Therefore, the grating displays a dual-mode (optical and electrical) tunability.

The design principle for the material leading to the grating formation is schematized in Fig. 11.23. An azobenzene monomer is dissolved in a nematic LC, together with a photoinitiator (not shown). The homogeneous mixture is then exposed to light through a grating photomask for photopolymerization. Similar to the grating formation using PDLCs (Sutherland et al., 1994), as polymerization takes place only in irradiated areas, formation of polymer results in diffusion of the monomer from nonirradiated areas into the irradiated areas. This process expels the LC from irradiated areas as more polymers are formed and solidified, the consequence of which should be the segregation of the polymer from the LC host leading to the formation of a permanent volume grating. However, it is likely that a small amount of the azobenzene monomer would remain dissolved in the

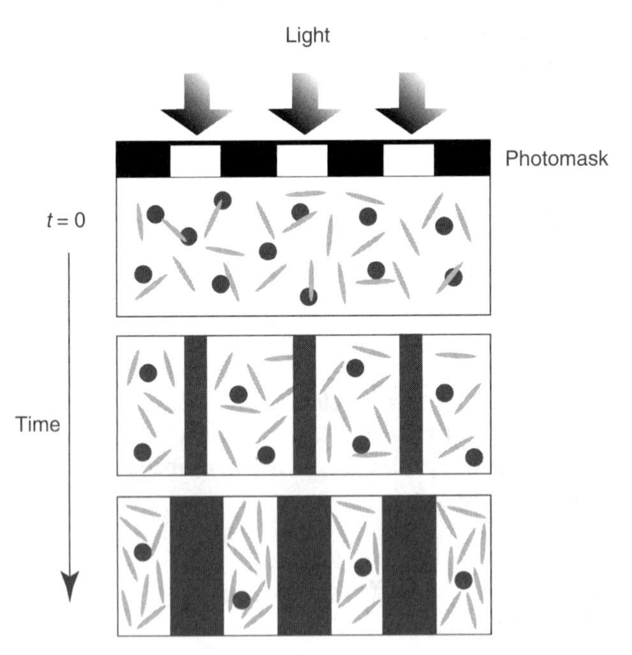

Figure 11.23. Schematic illustration of the preparation of an optically switchable diffraction grating through photopolymerization-induced phase separation, using a liquid crystalline azobenzene monomer (*dots*).

LC in nonirradiated areas and can be further polymerized upon flood irradiation (photomask removed). Basically, the grating is formed by alternating polymer-rich and LC-rich regions. As shown in Fig. 11.24, the grating can not only be switched by an electric field because of different responses of the two regions but is also sensitive to UV and visible light by virtue of the photoisomerization of the small amount of azobenzene polymer in nonirradiated areas.

With this design, it is easy to understand that a concentration of azobenzene monomer before polymerization (>10 wt%) is needed. Generally, azobenzene compounds have a limited solubility in LCs ($<3\%$). To overcome this problem, we prepared a dimethacrylate azobenzene monomer that is liquid crystalline on its own (following structure), with the following phase transition temperatures (°C) on heating: Cr 70 S 99 N 145 Iso.

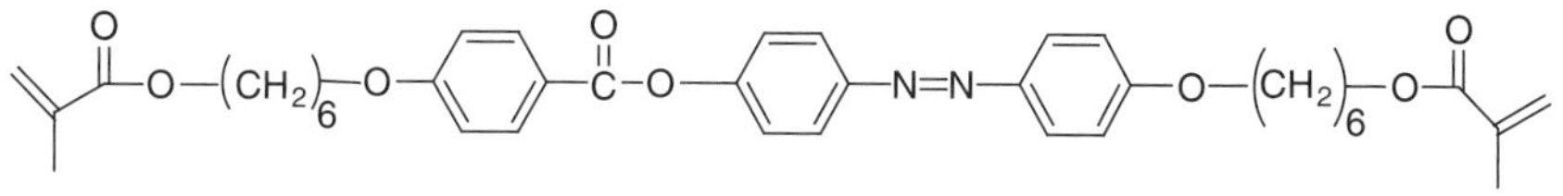

The trans–cis photoisomerization of this azobenzene monomer in THF is almost 100%, and because of the LC-enhanced solubility, as much as 30% of the monomer can readily be dissolved in nematic LCs. The nematic LC used to test the design was BL006 ($T_{ni} = 115$°C), characteristics of which are suitable to the tunable grating: an ordinary refractive index $n_o = 1.53$, a large birefringence $\Delta n = 0.286$ (589 nm, 20°C), and a positive dielectric anisotropy $\Delta\varepsilon = 17.3$. The photoinitiator used was Irgacure 784 that can initiate polymerization upon visible light absorption at $\lambda > 450$ nm, which is important because the long-wavelength initiation reduces the interference by the absorption of the LC host and the azobenzene monomer ($\lambda_{max} = 354$ nm in THF). This materials combination indeed allows for the preparation of optically and electrically switchable gratings. Using the example of a grating formed with a mixture containing 15% azobenzene monomer and 2% photoinitiator, the origin of the dual-mode switching of diffraction efficiency is explained with Fig. 11.24.

The grating was obtained by applying a beam of visible light (540 nm, ~ 2.8 mW/cm^2) through the photomask (40-μm stripes) placed on one side of an ITO-coated parallely rubbed cell (10-μm gap) filled with the LC/monomer/initiator mixture, the mask stripes being aligned parallely to the rubbing direction. Using a microscope hot stage, the photopolymerization proceeded at 125°C (in the isotropic phase of BL006) for 1 h before being cooled to room temperature under visible irradiation; after removal of the mask, a flood irradiation was applied to the whole sample to complete polymerization of the remaining monomer. Figure 11.24a shows the POM of the cell placed with the rubbing direction parallel to one of crossed polarizers (along the x-axis). The grating arising from the polymerization-induced segregation is visible. Polymer strips are found in irradiated areas and well separated from the LC concentrated in nonirradiated areas. The orientation of LC molecules along the rubbing direction in the LC regions is revealed by the extinction in these areas. As illustrated in the schematics under the

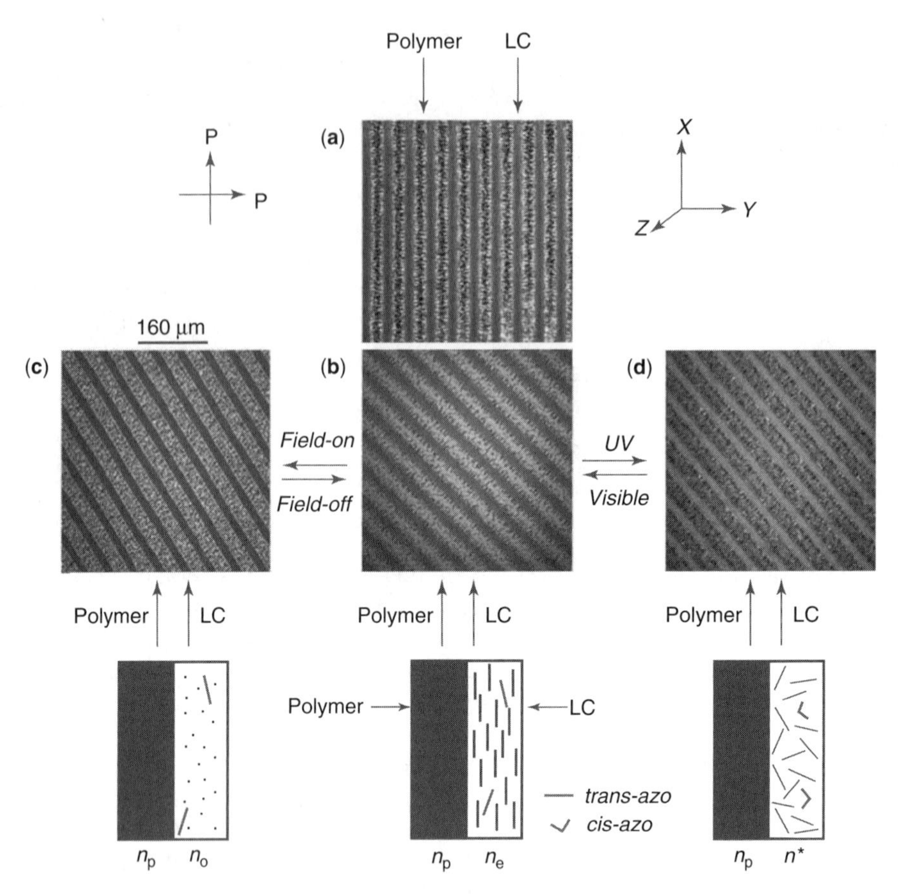

Figure 11.24. Polarizing optical micrographs showing an electrically and optically switchable diffraction grating prepared using an azobenzene polymer-stabilized nematic liquid crystal (15 wt% polymer): (**a**) the grating with fringes parallel to one of crossed polarizers; LC molecules being oriented in the rubbing direction (*x*-axis); (**b**) the same grating rotated by an angle; (**c**) the grating in (**b**) subjected to a voltage (AC field, 1000 Hz, 40-V peak-to-peak) with LC molecules being oriented in the field direction (*z*-axis); and (**d**) the grating in (**b**) exposed to UV light ($\lambda = 365$ nm, 20 mW/cm^2) resulting in the nematic-to-isotropic phase transition of LC. The sketches illustrate the different orientation states of LC molecules and the corresponding refractive index modulation for a probe light propagating along the *z*-axis and polarized along the *x*-axis. *Source*: Tong et al., 2005b. Reprinted with permission. See color insert.

photos, this grating should diffract the probe light (633 nm) propagating along the z-axis (normal incidence) with the electric vector parallel to the fringes (x-axis). This is because the probe sees essentially the extraordinary refractive index of the LC, n_e, in the LC regions and the mean index of the polymer, n_p, in the polymer regions, while n_e is larger than n_p that should be similar to the ordinary refractive index of the LC, n_o, that is, ~1.5. Figure 11.24b shows the POM image of the same grating rotated at ~35°, with respect to the crossed polarizers to better explain why this large index modulation can be controlled both optically and electrically. At this alignment, oriented LC molecules in nonirradiated areas appear bright by allowing light to pass through, and the remaining azobenzene polymer network is visible as denser zones under the optical microscope. When an electric field is applied across the cell, LC molecules distant from polymer walls are aligned along the field direction, that is, perpendicular to the plane of the cell (homeotropic orientation) and thus appear dark under crossed polarizers, which can be seen in Fig. 11.24c. In this case, the probe light sees essentially n_o and n_p in the alternating regions, and the sharp decrease in index modulation should result in a drop of the diffraction efficiency. The electrical switching of index modulation should be reversible, since the initial planer orientation of LC can be recovered when the electric field is turned off.

More interestingly, the index modulation can also be reversibly changed by light using two different wavelengths. Illuminating the grating with a UV light (360 nm), azobenzene groups on the polymer remaining in the LC regions undergo the trans–cis photoisomerization that can induce the photochemical nematic–isotropic phase transition at room temperature. This can be observed in Fig. 11.24d where the LC stripes appear significantly darker because of disordered LC molecules. When this happens, the probe light experiences mainly the average refractive index of the LC, $n^* = (2n_o + n_e)/3$, which should be close to n_p, leading to a drop of the diffraction efficiency. While illuminating the grating with a visible light (440 nm), the back cis–trans photoisomerization of azobenzene occurs that allows the initial surface-induced LC orientation to be recovered.

As already mentioned, the polymer and LC strips forming the grating are actually polymer- and LC-rich regions. Figure 11.25 shows a magnified POM image of the grating, where azobenzene polymers remained in the LC regions are clearly noticeable. On the basis of the morphology of the azobenzene polymer-stabilized LCs revealed on POM (Leclair et al., 2003), the estimated amount of polymer in the LC regions is ~3%. Likewise, the polymer regions also confine LC molecules whose switching at the field-on state is responsible for the darker appearance of the polymer strips in Fig. 11.24c. Nevertheless, as shown in Section 11.5.2.2, experimental data of both electrical and optical switching of the diffraction grating corroborates quite well the analysis of index modulation changes without taking into account the remaining polymer or LC, suggesting that the residual amount of polymer or LC in the LC or polymer regions, respectively, is minor and does not determine the switching behavior of the grating.

11.5.2.2. Optical and Electrical Switching Behaviors. The grating in Fig. 11.24 was used to investigate the switching behavior of diffraction efficiency using light at two wavelengths (UV and visible) as well as an electric field. The setup was the same as described in Section 11.4. The results are shown in Fig. 11.26. In Fig. 11.26a, the multiple-cycle switching of the first-order (+1)

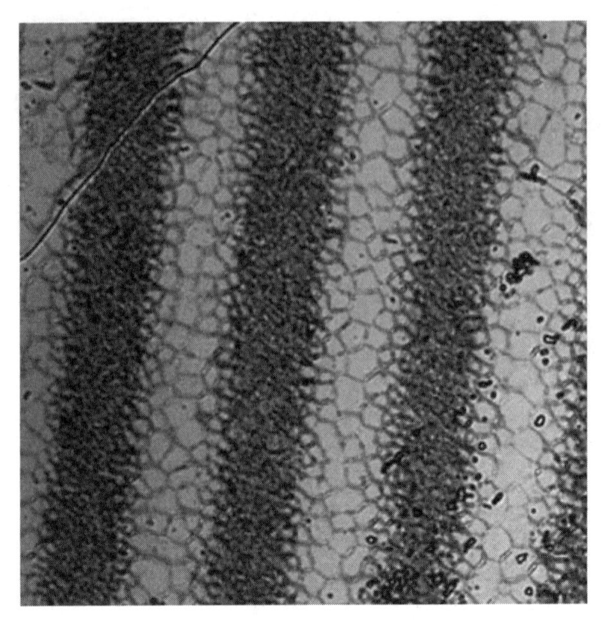

Figure 11.25. A magnified polarizing optical micrograph of the grating in Fig. 11.24, showing the polymer-rich (*dark*) and liquid crystal-rich regions (*clear*). A remaining polymer network in liquid crystal regions is visible. The width of the fringes is $\sim 40\,\mu m$.

diffraction efficiency between the high (6.5%) and low (0.5%) state is observed under a square-wave electric field between 0 V (field-off) and 40 V (field-on). The low diffraction state is somewhat unstable, which is likely related to some dynamic instability of the LC in the DC electric field. Under the used conditions, at the field-on state, the time for the diffraction efficiency to drop from 90% to 10% of the initial value is ~ 110 ms, whereas at the field-off state, the diffraction recovery from 10% to 90% takes ~ 1 s. The electrical switching can be repeated many times. At field-on state, the speed of switching from high to low diffraction efficiency is determined by the speed at witch LC molecules change from the surface-induced in-plane orientation to the homeotropic (or tilted) orientation. On removal of the field, the switching from low to high diffraction is governed by the relaxation rate of the LC returning to the initial in-plane orientation. Similar to azobenzene-containing self-assembled LC gels discussed earlier in the chapter, the electrical switching is based on the change in index modulation resulting from electric field–induced reorientation of LC molecules.

The appealing feature of the grating is the switching of diffraction efficiency controlled by light. Figure 11.26b shows the reversible optical switching of diffraction efficiency using UV and visible lights. In the experiment, the irradiation light was directed to the grating at an angle to the normal of the cell to prevent it from entering the photodetector used to measure the intensity of the diffracted

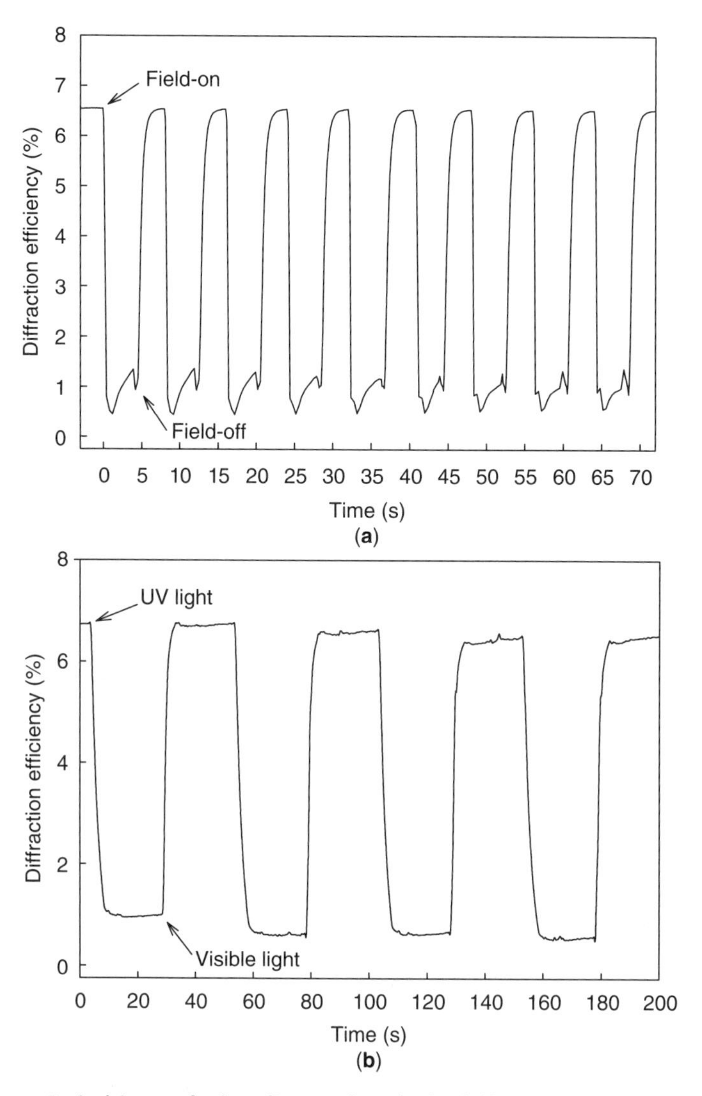

Figure 11.26. Switching of the first-order (+1) diffraction efficiency for the grating in Fig. 11.24 (80 μm for polymer line+liquid crystal line) by means of (a) a square-wave electric field between 0 V (off-state) and 40 V (on-state); and (b) UV light (360 nm, 20 mW/cm^2) and visible light (~440 nm, 26 mW/cm^2), the irradiation time being 5 s.

probe light. With UV light (~20 mW/cm^2) turned on for 5 s, the trans–cis isomerization of azobenzene is enough to induce the nematic–isotropic phase transition, resulting in the drop of diffraction efficiency. Subsequently, after 5 s of visible light exposure (~26 mW/cm^2), the high diffraction state is recovered, because the cis–trans backisomerization of azobenzene allows the oriented nematic phase to be recovered. Under the used 5-s irradiation, the actual diffraction drop and recovery times are ~3.4 and 1.4 s, respectively. Similar to electrical switching, the optical switching can be realized for many cycles without significant degradation of switching performance. From the switching mechanism, it is easy to understand that under UV light, the switching speed is determined by the rate of the photochemical nematic–isotropic phase transition. The reverse switching under the action of visible light is determined by the isotropic–nematic phase transition and the concurrent reorientation of LC molecules induced by rubbed surfaces. Increasing the intensities of UV and visible lights can increase the speed of optical switching. The similar low and high diffraction efficiencies for both electrical and optical switching are in accordance with the analysis of the index modulation changes as discussed in Fig. 11.24. The results thus confirm that both an electrical field and light at two wavelengths can be used to switch reversibly the diffraction efficiency of gratings prepared using the designed azobenzene polymer-stabilized LCs. For both modes of switching, varying the voltage or the intensity of light can switch the diffraction efficiency to different levels. While a voltage needs to be held constantly to keep the low diffraction state, the use of UV light requires no continuous excitation. But in the latter case, the stability is limited by the thermally induced cis–trans backisomerization. For the LC azobenzene monomer used, the thermal relaxation of the cis isomer is slow, taking hours to complete in solution. Consequently, the low diffraction state in the absence of UV excitation is quite stable. Should a long-period stability of the low diffraction state be required, an intermittent short-time UV excitation can be applied to prevent the returning of *cis*-azobenzene to the trans form.

As already explained, using LC-based materials filled in a cell, gratings with periods $<30\,\mu m$ cannot be obtained with a grating photomask because of the diffraction of light by the mask before reaching the photosensitive sample (Tong and Zhao, 2003). Holographic recording is needed to make gratings with smaller periods. A 5-μm period holographic grating was obtained using the same material (BL006/azobenzene monomer/photoinitiator) and an interference pattern generated by an Ar^+ laser (514 nm, 85 mW/cm^2 per pumping beam) for photopolymerization. Figure 11.27 shows the electrical and optical switching of the first-order (+1) diffraction efficiency of the holographic grating. The same switching conditions as in Fig. 11.26 were used except that the UV and visible lights were applied for 2 s. It is seen that the diffraction can be switched both electrically and optically, though the diffraction efficiency and the switching performance are not as good as the large-period grating obtained using a photomask. The grating formation dynamics measurement found that the holographic grating came from the same origin. On cooling the cell from the isotropic phase of BL006 (125°C) after 20-min exposure to the interference pattern, diffraction occurs once the LC enters the nematic phase; the diffraction efficiency increases with lowering the temperature because of increasing difference between n_e and n_p. Moreover, similar to the large-period grating, the diffraction efficiency is highly dependent on the polarization of the probe light, being decreased with increasing the angle between

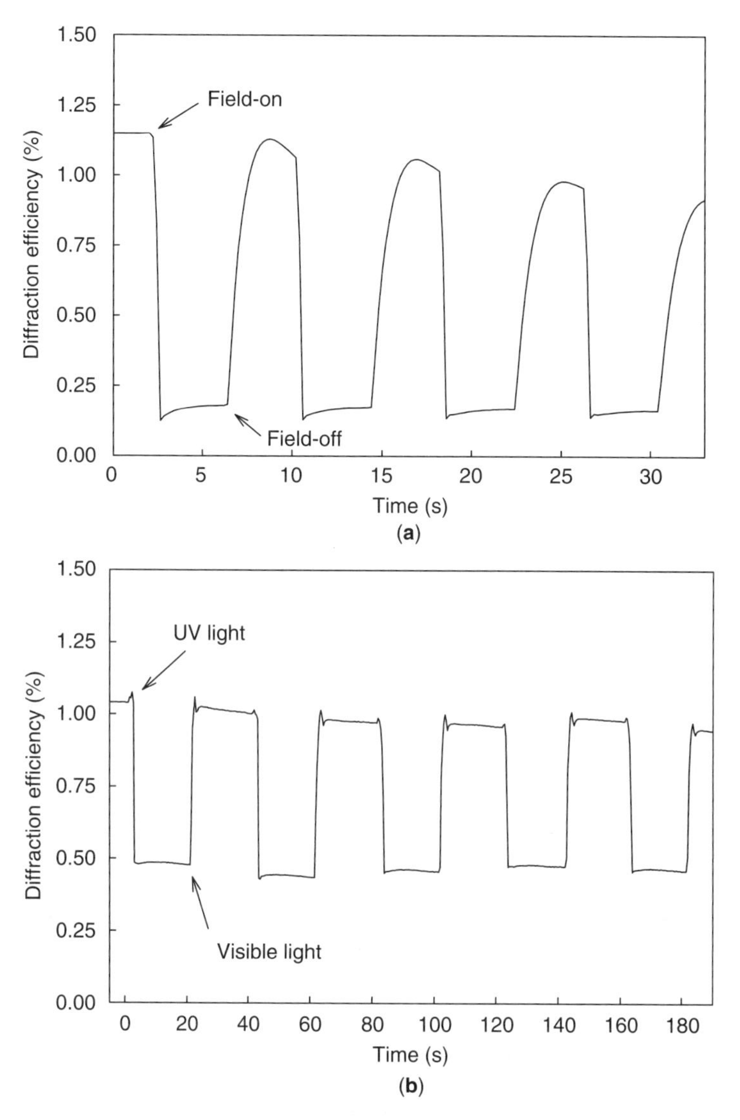

Figure 11.27. Switching of the first-order (+1) diffraction efficiency for a 5-μm period holographic grating, prepared using the same materials, using (**a**) a square-wave electric field between 0 V (off-state) and 40 V (on-state); and (**b**) UV light (360 nm, 20 mW/cm^2) and visible light ($\sim$440 nm, 26 mW/cm^2), the irradiation time being 2 s.

the polarization and the fringes of grating and reaching the lowest level when the polarization is perpendicular to the fringes. Viewed on POM, the 5-μm fringes are visible, but the sample appears quite heterogeneous, with the presence of droplets and defect regions much larger than the period of grating, which led to the diminished performance of grating. The conditions for recording holographic gratings remain to be optimized. It is expected that using more efficient visible light photoinitiators for faster photopolymerization may improve the segregation between the azobenzene polymer and the LC.

11.5.2.3. What Determines the Switching Performance? Using the preceding design of materials, the diffraction efficiency and switching performance in terms of speed and stability could be improved by choosing appropriate materials and adjusting the preparation conditions. First, the nematic LC should have a large birefringence with its n_o equal to n_p, given a good planar alignment. This will give rise to large index modulation and thus high diffraction efficiency for the grating and no or negligible diffraction when an electric field or UV light is applied. Second, upon photopolymerization, the polymer should segregate as thoroughly as possible from the LC to form the periodic grating structure. For the investigated system, there is a small amount of azobenzene polymer that remained in the LC regions, which, as argued, is responsible for the optical switching based on photoinduced nematic–isotropic phase transition. However, it is likely that even with complete separation of the polymer from the LC, mesogenic azobenzene groups in the interface between the two regions could accomplish the optical switching. While ensuring the optical switching of diffraction, the absence of azobenzene polymer in the LC regions is desirable because homogeneous LC orientation can be induced by rubbed surfaces, which determines the high diffraction state, and the electric field–induced reorientation of LC molecules can develop without interaction with the polymer, allowing the electrical switching of diffraction to be governed by the behavior of pure nematic LC. Third, experimental conditions can be optimized by taking into account several factors that were found to affect the grating formation and switching performance: (1) Polymerization in the isotropic phase generally results in a better grating, probably because of the high fluidity that favors the diffusion or transport of the monomer and thus the segregation between polymer and LC. (2) Long polymerization time (1–2 h) is necessary to complete the reaction with the used visible light photoinitiator, whereas photoinitiators absorbing longer wavelengths may afford faster polymerization by reducing the interference of azobenzene monomers and polymer. (3) With the used materials, a concentration of 15%–20% for the azobenzene monomer is necessary. At higher concentrations, the segregation between polymer and LC is not thorough and the polymer remained in LC regions is too much to allow good surface alignment for the LC, whereas at lower concentrations, the amount of polymer in irradiated areas is not sufficient to have large polymer strips.

11.5.3. Optically Switchable Reflection Gratings

In all the studies discussed, the grating is transmission-type, with the grating vector lying in the sample plane. Using azobenzene-containing polymer-stabilized and polymer-dispersed LCs (PSLCs and PDLCs), Bunning and coworkers demonstrated the preparation of reflection gratings that can be switched by light (Urbas et al., 2004a,b). By filling a photosensitive LC/monomer mixture in an ITO-coated cell, they succeeded in using an interference pattern to record a grating structure whose vector is perpendicular to the sample plane. In this case, the periodic modulation of refractive index goes along the plane normal and the grating acts as a Bragg reflector exhibiting wavelength selective reflectivity. With the LC host doped with an LC azobenzene compound, the reversible trans–cis photoisomerization can be used to optically change the orientational state of the LC and thus the modulation of index, making it possible to switch on and off the reflected (diffracted) light.

In one case (Urbas et al., 2004a), an interference pattern of 532-nm visible light is applied to a reactive mixture filled in a 15-μm cell for photopolymerization, the mixture being composed of a diacrylate LC monomer (1,4-phenylene bis{4-[6-(acryloyloxy)hexyloxy]benzoate}), a trifunctional monomer (cross-linker), a nematic LC (ML-1009 from Merck), an LC azobenzene compound (4-butyl-4-methyoxy-azobenzene), and a photoinitiator (Rose Bengal with *n*-phenyl glycine). The concentration of azobenzene dopant is 7%, whereas the LC host is the major component (>75%) in the rest of the mixture. Similar to the system discussed in Section 11.5.2, the grating structure is formed upon exposure to the visible light interference pattern, since photoinduced polymerization of the monomers in high intensity regions, accompanied by monomer diffusion from dark (low intensity) regions, results in the formation of cross-linked polymer separated from the LC-rich phase in dark regions. By changing the angle of intersection of the two laser beams, the holographic recording technique can control the thickness of the repeated bilayer (alternating polymer-rich and LC-rich regions), which determines the reflected wavelength (Bragg wavelength). The authors prepared a reflection grating with ~70 bilayers stacked in the 15-μm cell, each of which has a thickness of 210 nm, corresponding to the selective reflection at 635 nm. The mechanism of optical switching is schematized in Fig. 11.28, supported by experimental date. Before UV light exposure, the index modulation is small because all LC components (nematic host, polymer, and azobenzene dopant) are aligned by rubbed cell surfaces and no selective reflection around 635 nm is observed. When UV light is applied, the trans–cis photoisomerization of azobenzene dopant in the LC-rich regions results in disordering of LC molecules, whereas in polymer-rich regions the perturbation effect is weaker because of the cross-linked polymer. This change increases the modulation depth of refractive index and thus gives rise to the observed reflection peak ~635 nm. Upon visible light exposure, the cis–trans backisomerization brings LC molecules back to the initial order, and the reflection disappears. In the reported case, the peak diffraction efficiency is 58%.

In another study (Urbas et al., 2004b), the formulation of the mixture is different. It contains 35% of a nematic LC (E7) to which 3.5% of the LC azobenzene dopant (4-butyl-4-methyoxy-azobenzene) is added, with the rest being photopolymerizable monomer syrup. With such a formulation, polymerization is known to result in PDLC, in which phase-separated LC droplets are dispersed in the polymer matrix. When the photosensitive reactive mixture filled in an

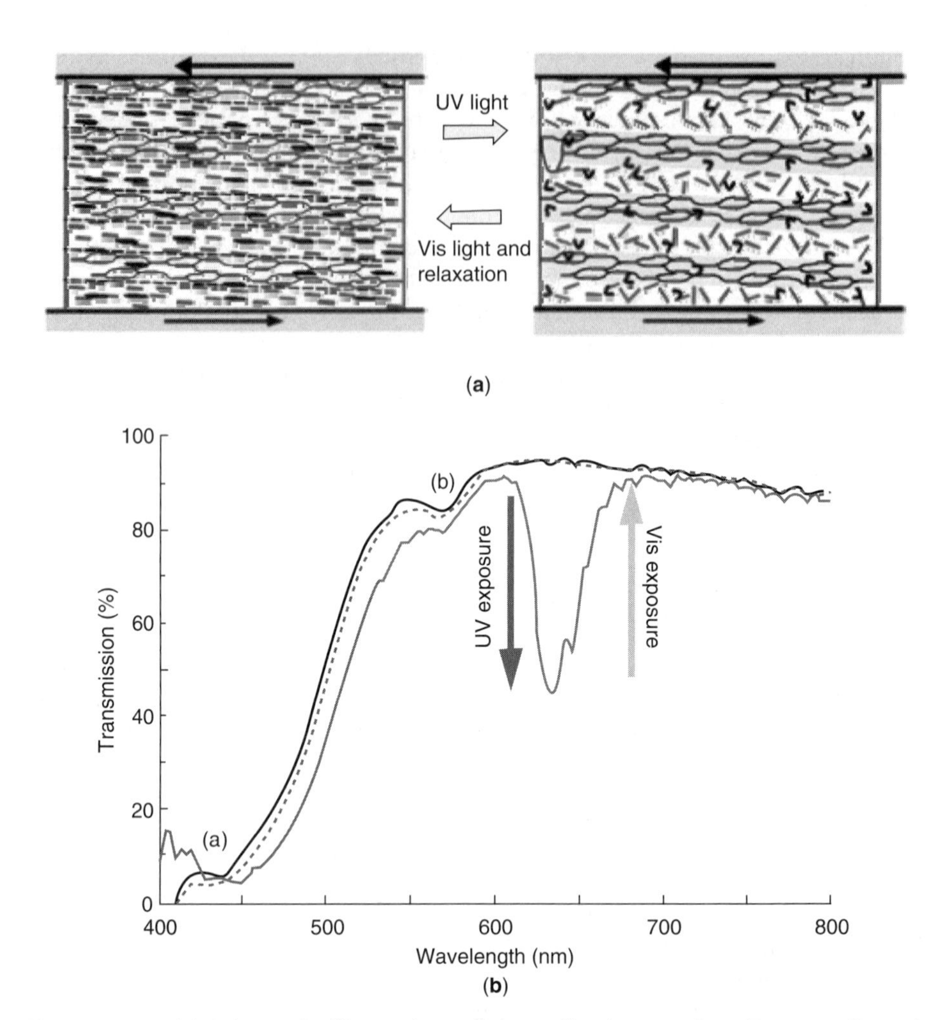

Figure 11.28. (**a**) Schematic illustration of the reflection grating (Bragg reflector) prepared with holographically patterned, polymer-stabilized liquid crystals. Upon its photoisomerization, the azobenzene compound in the LC regions affects more the LC order than that in the regions of cross-linked polymer, which is the origin of optically changeable modulation of refractive index. (**b**) Transmission spectra of the reflection grating showing the optically switchable selective reflection, with the reflection peak appearing upon UV irradiation while disappearing upon visible irradiation. *Source*: Urbas et al., 2004a. Adapted with permission.

ITO-coated cell (8- to 10-μm gap) is subjected to holographically induced photopolymerization, grating can be formed with alternating regions of polymer and LC droplets. Figure 11.29 is a schematic illustration of the optical switching mechanism based on the reversible photoisomerization of azobenzene dopant mixed with the LC host (in droplets) as well as the supporting experimental data.

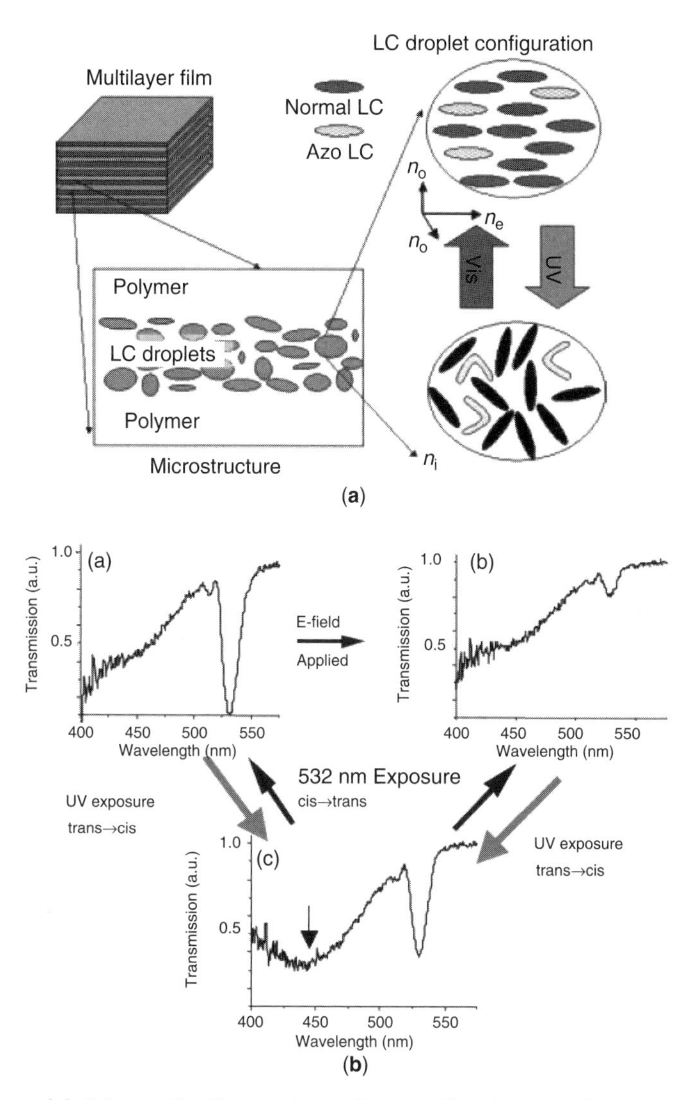

Figure 11.29. **(a)** Schematic illustration of the reflection grating prepared with HPDLCs. The azobenzene compound dissolved in the LC droplets changes the LC order state upon UV and visible light irradiation. **(b)** Transmission spectra of an HPDLC-based reflection grating showing the optically switchable selective reflection in and without the absence of an electric field. *Source*: Urbas et al., 2004b. Adapted with permission.

The as-prepared grating has a high index modulation between polymer regions and the regions containing confined LC droplets in which LC molecules are aligned; a reflection peak at $\sim 530\,\text{nm}$ can be observed. In this state, exposure of the grating to UV light induces the nematic–isotropic phase transition inside the droplets, and the reduced index modulation results in a decrease in diffraction efficiency. Subsequent visible light exposure recovers the initial reflection efficiency. However, the reflection of the as-prepared grating can be turned off by applying an electric field that aligns LC molecules inside the droplets to change the index modulation. With the grating under a voltage, the reflection can be switched on using UV light and switched off using visible light.

11.6. CONCLUDING REMARKS AND PERSPECTIVES

The reversible trans–cis photoisomerization of azobenzene can be explored in designing materials for tunable diffraction gratings. On one hand, the photoisomerization can be used as the trigger for the structural rearrangement in the material leading to the periodic change in refraction index required for the diffraction grating. This is the case of ATEs used for mechanically tunable diffraction gratings and self-assembled LC gels containing an azobenzene gelator used for electrically switchable gratings. On the other hand, the photoisomerization-induced LC-to-isotropic phase transition can be utilized for optically controllable diffraction gratings based on both PDLC and PSLC.

For mechanically controllable diffraction gratings, the advantage of thermoplastic elastomers is the easy film preparation through solution casting or melt-state processing. To obtain a stable diffraction grating, a permanent structural rearrangement or reorganization in the film needs to be induced by the photoisomerization. In the case of azo-SCLCP-grafted SBS, the authors' study found that such structural rearrangement is more likely to occur in a film under large strain, in which the optical plasticization of azobenzene polymer microdomains could facilitate the rearrangement of PS cylindrical microdomains acting as physical cross-links. In light of this, if azo-SCLCP constitutes the rigid cylindrical microdomains (like the triblock copolymer in Fig. 11.3), the photoisomerization, under strain, may give rise to a more important softening effect on the physical cross-links and thus trigger the structural rearrangement. It would be interesting to design ABA triblock copolymers, whose azo-SCLCP microdomains (A block) can undergo important optical plasticization upon illumination, and to investigate the resulting structural changes. One challenge here is to obtain ABA-type ATE that can exhibit large reversible deformation like that of SBS.

Another future study of interest is to investigate in detail the reversible transition between Raman–Nath (transmission) and Bragg (reflection) diffraction regimes. For instance, to prepare a grating with a period of 1 μm in a nonstretched film of, say, 200-μm thickness, a grating with 6-μm period can be recorded on the film stretched to 500% deformation with the thickness decreased to ~82 nm. Therefore, under 500% strain, the grating should be in the thin film regime (Raman–Nath diffraction), whereas in the relaxed state, it should be in the thick film regime (Bragg). The reversible transition between the two regimes can easily be investigated using ATE. Our preliminary observation found that the transition might occur over a range of film deformations, in which both diffraction regimes coexist.

As emphasized earlier, the structural rearrangement occurs throughout the sample of ATE, and the resulting mechanically tunable diffraction grating is a volume grating. However, if SRG can be obtained with an ATE as a result of photoisomerization-induced mass transport on surface, the grating should also be tunable because the surface relief should be affected by surface deformation. It can be imagined that upon extension of the film, the height between trough and top will be flattened, whereas upon retraction the opposite effect emerges. With block copolymer-based ATE, the confinement of azobenzene polymer inside microdomains would hamper the development of SRG. Nevertheless, other approaches can be envisaged. For instance, a thin layer of an azobenzene polymer could be cast on the surface of an elastomer containing no azobenzene; if SRG can be inscribed on surface of the azobenzene polymer, mechanical deformation of the supporting elastomer may control the SRG deformation and thus change its diffraction properties.

The mechanical tunability of gratings on ATE films is also worth exploitation for possible applications. One possibility would be to incorporate this sort of grating into a structure (architecture, instruments) in such a way that a small deformation of the structure can induce a change in grating period and thus be detected by a changing diffraction angle. In principle, elastomers can undergo larger thermal expansion because of polymer chain flexibility. Such a grating might be used as a sensor for the detection of abrupt temperature change, since a large thermal expansion can cause changes in index modulation and grating period.

In the case of electrically switchable gratings, LCs remain the material of choice because of the ease of molecular reorientation under the effect of a voltage. The photoisomerization of azobenzene should play the role of trigger for the grating formation. Self-assembled LC gels combine the property of electric field orientation of LCs and the possibility of a structural rearrangement induced by the photoisomerization of the azobenzene gelator. Even though the grating obtained from the cholesteric LC gel as described in Section 11.4 falls short for practical applications, the interest of exploiting self-assembled LC gels is clear, since this approach may be generalized for many systems made from designed gelators and various types of LCs. The key to the grating formation process in self-assembled LC gels is the dissolution of the physical network in the LC host in irradiated areas. This may be enhanced by using azobenzene gelators whose cis form has a significantly higher polarity than the trans form, which increases the surface tension of the fibrous aggregates. It also helps if the gelator could form very thin nanofibrous aggregates giving rise to a large specific surface and thus a significant number of gelator molecules on the surface undergoing the trans–cis photoisomerization.

Finally, further research effort is worth being made on materials for recording optically switchable diffraction gratings. Optical control of optical properties, such as the diffraction, is not only academic curiosity but required for the development of advanced photonic devices. The dynamic control (switching) of local optical properties by light instead of an electric field offers the advantage of having no need for patterning transparent electrodes, which can be useful for certain applications. The discussed switching mechanism based on the photochemical phase transition (Section 11.5) induced by azobenzene's photoisomerization is particularly appealing. However, a common problem with both PSLCs (Tong et al., 2005a,b; Urbas et al., 2004a) and PDLCs (Urbas et al., 2004b) is that a thorough segregation of azobenzene molecules from the polymer areas is difficult to achieve, so that the photoisomerization not only affects the LC host as desired but also affects the polymer. This may limit the index modulation depth and thus the switching performance. For future studies, the grating fabrication method should not be limited to the photopolymerization-induced phase separation (holographically or not). For instance, if a method allows for efficient alternative deposition of two SLCPs with controlled layer thickness in the order of 200 nm, an effective reflection grating (Bragg reflector) can be designed to have one SCLCP having a low T_g (below or close to room temperature) and bearing azobenzene mesogens, whereas the other SCLCP can be chosen to have a high T_g and contain nonazobenzene mesogens. In this case, after aligning both polymers (e.g., by rubbed surfaces at temperatures above both T_gs), upon UV light irradiation at room temperature, the photochemical LC-to-isotropic phase transition can occur only in layers of the azobenzene SCLCP, which should have no effect on the other SCLCP remaining aligned. This would resolve the phase separation problem as mentioned earlier.

ACKNOWLEDGMENTS

The author is grateful to all the students and collaborators who worked on and made contributions to the projects on tunable diffraction gratings, particularly, Dr. Shuying Bai, Xia Tong, Dr. Li Cui, Dr. Guan Wang, and Professor Tigran Galstian (Laval University, Québec City, Canada). The author thanks Jie He for his assistance in the preparation of the manuscript. The author is also grateful for the financial support from the Natural Sciences and Engineering Research Council of Canada (NSERC), the Fonds pour la Formation de Chercheurs et l'Aide à la Recherche of Québec (FQRNT), and University of Sherbrooke and St-Jean Photochemicals Inc. (St-Jean-sur-Richelieu, Québec, Canada).

REFERENCES

Bai S, Zhao Y. 2001. Azobenzene-containing thermoplastic elastomers: coupling mechanical and optical effects. Macromolecules 34:9032–9038.

Bai S, Zhao Y. 2002. Azobenzene elastomers for mechanically tunable diffraction gratings. Macromolecules 35:9657–9664.

Breiner T, Kreger K, Hagen R, Haeckel M, Kador L, Mueller AHE, Krame EJ, Schmidt HW. 2007. Blends of poly(methacrylate) block copolymers with photoaddressable segments. Macromolecules 40:2100–2108.

Camacho-Lopez M, Finkelmann H, Palffy-Muhoray P, Shelley M. 2004. Fast liquid crystal elastomer swims into the dark. Nat Mater 3:307–310.

Cui L, Tong X, Yan X, Liu G, Zhao Y. 2004. Photoactive thermoplastic elastomers of azobenzene-containing triblock copolymers prepared through atom transfer radical polymerization. Macromolecules 37:7097–7104.

Guan L, Zhao Y. 2000. Self-assembly of a liquid crystalline anisotropic gel. Chem Mater 12:3667–3673.

Guan L, Zhao Y. 2001. Self-assembled gels of liquid crystals: hydrogen-bonded aggregates formed in various liquid crystalline textures. J Mater Chem 11:1339–1344.

Ichimura K. 2000. Photoalignment of liquid crystal systems. Chem Rev 100:1847–1874.

Ichimura K, Oh SK, Nakagawa M. 2000. Light-driven motion of liquid on a photosensitive surface. Science 288:1624–1626.

Ikeda T. 2003. Photomodulation of liquid crystal orientations for photonic applications. J Mater Chem 13:2037–2057.

Ikeda T, Tsutsumi O. 1995. Optical switching and image storage by means of azobenzene liquid-crystal films. Science 268:1873–1875.

Kang SW, Sprunt S, Chien LC. 2001. Switchable diffraction gratings based on inversion of the dielectric anisotropy in nematic liquid crystals. Appl Phys Lett 78:3782–3784.

Kato T, Kondo G, Hanabusa K. 1998a. Thermoreversible self-organized gels of a liquid crystal formed by aggregation of trans-1,2-bis(acylamino)cyclohexane containing a mesogenic moiety. Chem Lett 3:193–194.

Kato T, Kutsuna T, Hanabusa K, Ukon M. 1998b. Gelation of room-temperature liquid crystals by the association of a trans-1,2-bis(amino)cyclohexane derivative. Adv Mater 10:606–608.

Kim DY, Tripathy SK, Li L, Kumar J. 1995. Laser-induced holographic surface relief gratings on nonlinear optical polymer films. Appl Phys Lett 66:1166–1168.

Leclair S, Mathew L, Giguere M, Motallebi S, Zhao Y. 2003. Photoinduced alignment of ferroelectric liquid crystals using azobenzene polymer networks of chiral polyacrylates and polymethacrylates. Macromolecules 36:9024–9032.

Moriyama M, Mizoshita N, Yokota T, Kishimoto K, Kato T. 2003. Photoresponsive anisotropic soft solids: liquid-crystalline physical gels based on a chiral photochromic gelator. Adv Mater 15:1335–1338.

Natansohn A, Rochon P. 2002. Photoinduced motions in azo-containing polymers. Chem Rev 102:4139–4157.

Pakula T, Saijo K, Hashimoto T. 1985a. Structural changes in polystyrene-polybutadiene-polystyrene block polymers caused by annealing in highly oriented state. Macromolecules 18:2037–2044.

Pakula T, Saijo K, Kawai H, Hashimoto T. 1985b. Deformation behavior of styrene-butadiene-styrene triblock copolymer with cylindrical morphology. Macromolecules 18:1294–1302.

Pham VP, Galstian T, Granger A, Lessard R. 1997. Novel azo dye-doped poly(methyl methacrylate) films as optical data storage media. Jpn J Appl Physt 36 (part 1B):429–438.

Qi B, Yavrian A, Galstian T, Zhao Y. 2005. Liquid crystalline ionomers containing azobenzene mesogens: phase stability, photoinduced birefringence and holographic grating. Macromolecules 38:3079–3086.

Rasmussen PH, Ramanujam PS, Hvilsted S, Berg RH. 1999. A remarkably efficient azobenzene peptide for holographic information storage. J Am Chem Soc 121: 4738–4743.

Rochon P, Batalla E, Natansohn A. 1995. Optically induced surface gratings on azoaromatic polymer films. Appl Phys Lett 66:136–138.

Scharf T, Fontannaz J, Bouvier M, Grupp J. 1999. An adaptive microlens formed by homeotropic aligned liquid crystal with positive dielectric anisotropy. Mol Cryst Liq Cryst 331:235–243.

Subacius D, Bos PJ, Lavrentovich OD. 1997. Switchable diffractive cholesteric gratings. Appl Phys Lett 71:1350–1352.

Sutherland RL, Tondiglia VP, Natarajan LV, Bunning TL, Adams WW. 1994. Electrically switchable volume gratings in polymer-dispersed liquid crystals. Appl Phys Lett 64:1074–1076.

Tong X, Zhao Y. 2003. Self-assembled cholesteric liquid crystal gels for electro-optical switching. J Mater Chem 13:1491–1495.

Tong X, Wang G, Soldera A, Zhao Y. 2005a. How can azobenzene block copolymer vesicles be dissociated and reformed by light? J Phys Chem B 109, 20281–20287.

Tong X, Wang G, Yavrian A, Galstian T, Zhao Y. 2005b. Dual-mode switching of diffraction gratings based on azobenzene-polymer-switched liquid crystals. Adv Mater 17:370–374.

Urbas A, Tondiglia V, Natarajan L, Sutherland R, Yu H, Li JH, Bunning T. 2004a. Optically switchable liquid crystal photonic structures. J Am Chem Soc 126:13580–13581.

Urbas A, Klosterman J, Tondiglia V, Natarajan L, Sutherland R, Tsutsumi O, Ikeda T, Bunning T. 2004b. Optically switchable Bragg reflectors. Adv Mater 16:1453–1456.

Viswanathan NK, Kim DY, Bian S, Williams J, Liu W, Li L, Samuelson L, Kumar J, Tripathy SK. 1999. Surface relief structures on azo polymer films. J Mater Chem 9:1941–1955.

Yabuuchi K, Rowan AE, Nolte RJM, Kato T. 2000. Liquid-crystalline physical gels: self-aggregation of a gluconamide derivative in mesogenic molecules for the formation of anisotropic functional composites. Chem Mater 12:440–443.

Yamamoto T, Ohashi A, Yoneyama S, Hasegawa M, Tsutsumi O, Kanazawa A, Shiono T, Ikeda T. 2001. Phase-type gratings formed by photochemical phase transition of polymer azobenzene liquid crystals: 2. Rapid switching of diffraction beams in thin films. J Phys Chem B 105:2308–2313.

Yoneyama S, Yamamoto T, Tsutsumi O, Kanazawa A, Shiono T, Ikeda T. 2002. High-performance material for holographic gratings by means of a photoresponsive polymer liquid crystal containing a tolane moiety with high birefringence. Macromolecules 35:8751–8758.

Zhang J, Sponsler MB. 1992. Switchable liquid crystalline photopolymer media for holography. J Am Chem Soc 114:1506–1507.

Zhao Y. 1992. Structural changes upon annealing in a deformed styrene-butadiene-styrene triblock copolymer as revealed by infrared dichroism. Macromolecules 25:4705–4711.

Zhao Y, Tong X. 2003. Light-induced reorganization in self-assembled liquid crystal gels: electrically switchable diffraction gratings. Adv Mater 15:1431–1435.

Zhao Y, Bai S, Dumont D, Galstian TV. 2002. Mechanically tunable diffraction gratings recorded on an azobenzene elastomer. Adv Mater 14:512–514.

Zhao Y, Bai S, Asatryan K, Galstian T. 2003. Holographic recording in a photoactive elastomer. Adv Funct Mater 13:781–788.

12

AZO BLOCK COPOLYMERS IN THE SOLID STATE

Haifeng Yu and Tomiki Ikeda

12.1. INTRODUCTION

Functional block copolymers (BCs) are among novel types of macromolecules of industrial and academic interest because they provide opportunities to study the nanostructure formation and control under the influence of more than one driving force, for example, the effect of microphase-separated nanostructures on functionality, and vice versa, the influence of a specific functionality on the formation of diverse nanostructures. Introducing an azo block as one of the constituent segments into the functional BCs, azo BCs can be easily obtained, which are expected to bring the well-characterized photoisomerization of the azo moiety into the microphase separation of BCs (e.g., Natansohn and Rochon, 2002; Seki, 2004). Moreover, if the azo block can form mesogenic phases, the regular periodicity of liquid crystalline (LC) ordering influences the microphase-separated nanostructures, making it possible to self-assemble into periodic nanostructures on the macroscopic scale. In bulk films of liquid crystalline block copolymers (LCBCs), the interplay between the microphase separation and the elastic deformation of LC ordering, known as supramolecular cooperative motion (SMCM), enables one to form more hierarchical structures with photoresponsive properties, which offer novel opportunities to control supramolecular self-assembled nanostructures (e.g., Yu et al., 2006a,b). Combining the excellent properties of azo polymers with microphase separation, azo BCs may find diverse applications in advanced technology as well as newly promising nanotechnology.

Because of the incompatibility of the azo block and other non-azo segments in bulk films, the azo block gives rise to a rich variety of nanostructures. For instance, AB-type azo diblock copolymers may form separated phases (e.g., spheres,

Smart Light-Responsive Materials. Edited by Yue Zhao and Tomiki Ikeda

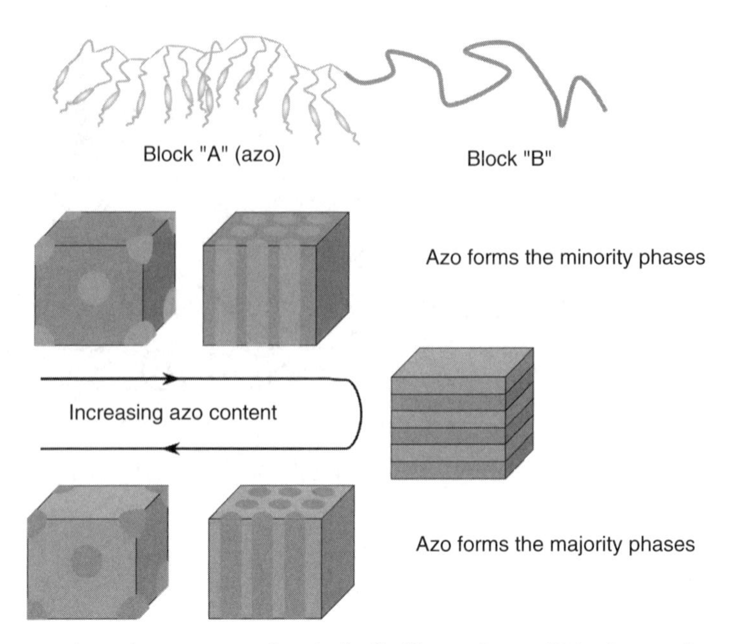

Figure 12.1. Microphase separation in bulk films of azo diblock copolymers with well-defined structures. The light domains represent the azo segment "A," whereas the dark parts are formed from another non-azo block "B."

cylinders), lamellae, or continuous phases by gradually increasing the azo content in the BC composition, which makes their properties different from those of azo polymer blends or azo random copolymers. Figure 12.1 shows a schematic representation of the microphase separation of an AB-type diblock copolymer, in which block "A" is an azo polymer segment and block "B" represents a non-azo polymer. Even if the azo block constituent is the minority phase in BC films on the nanoscale (<100 nm), it retains its photochemical sensitivity and can be photo-manipulated into an ordered state with the transition moments of azo moieties almost perpendicular to the polarization direction of the actinic light according to the Weigert effect (e.g., Ikeda, 2003a), which improves its optical performance by eliminating the scattering of visible light. Azo groups in the majority phase show a behavior similar to that of azo homopolymers. Additional performances such as hydrophilicity (e.g., Yu et al., 2006a,b), crystallization, optical transparency (e.g., Yu et al., 2007a,b), thermoplasticity (Cui et al., 2004), ionization (Qi et al., 2005), and water solubility (Ravi et al., 2005) can be acquired by special molecular design of the non-azo block "B."

Since the fascinating azo BC offers an effective and convenient method to design advanced materials with more than two functionalities, it has become one of the emerging topics in novel azo materials. This chapter tries to introduce this field from the preparation method and basic properties to the control of their supramolecular self-assembly and the microphase separation in bulk films of azo BCs, as well as their applications.

12.2. PREPARATION METHOD

To obtain regularly ordered nanostructures of microphase separation in BC films, BCs should have well-defined structures, narrow molecular weight polydispersities, and each block has to be larger than a certain minimum molecular weight. Several polymerization methods, such as anionic, cationic, free-radical, and metal-catalyzed polymerizations, have been explored to build BCs that meet these requirements. Owing to the existence of azo groups (Ph–N=N–Ph) with a long conjugation of π electrons, not all the living polymerization techniques are appropriate for synthesizing azo BCs with a suitable molecular weight. Therefore, several modified ways have been developed.

12.2.1. Direct Polymerization of Azo Monomers

Living polymerization of azo monomers is one of the most effective ways to prepare well-defined azo BCs. Generally, a monodispersed macroinitiator should be prepared first. It is then used as an initiator for the subsequent polymerization of azo monomers. Finkelmann and Bohnert (1994) first reported the synthesis of LC–side chain AB azo BCs by direct anionic polymerization of an azo monomer. As shown in Scheme 12.1, the polymerization of polystyrene (PS)-based diblock copolymers was carried out from a PS-lithium capped with 1,1-diphenylethylene (DPE), whereas the poly(methyl methacrylate) (PMMA)-based diblock copolymers were prepared by addition of methyl methacrylate (MMA) monomers to the "living" azo polyanion, obtained by reaction of 1,1-diphenyl-3-methylpentylithium (DPPL) with the azo monomer in tetrahydrofuran (THF) at lower temperature. By this method, a series of well-defined azo BCs were obtained with controlled molecular weights and narrow polydispersities (Lehmann et al., 2000).

Although the living anionic polymerization technique is the most acceptable method for synthesis of well-defined azo BCs because of the living nature of the end groups, this technique has some disadvantages too:

1. Monomers and reagents should be highly pure.
2. Very stringent drying is required.
3. Purification of the living macroinitiator is difficult.
4. A low temperature is often needed (e.g., $-70°C$).

Because of the aforementioned difficulties, many commercial polymers and copolymers (almost 50% of synthetic polymers) are prepared by radical processes. The major success of radical polymerization is the large number of monomers that can undergo free-radical polymerization, the convenient temperature range, and the

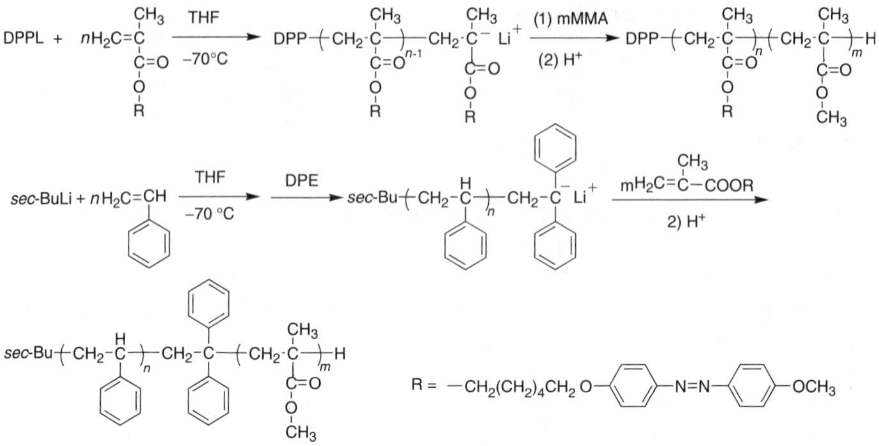

DPPL = 1,1-diphenyl-3-methylpentylithium
DPE = 1,1-diphenylethylene

Scheme 12.1 Direct preparation of azo BCs by anionic living polymerization. *Source*: Reproduced with modifications from Finkelmann and Bohnert, 1994.

minimal requirement for purification of monomers and solvents, which should only be deoxygenated. Atom transfer radical polymerization (ATRP) method is one of the newly developed controlled/living radical polymerization, whose products always have well-defined structures and low molecular-weight polydispersities (Matyjaszewski and Xia, 2001; Coessensm et al., 2001). Since the ATRP method allows for a control over the chain topology, composition, and end functionality for a large range of radically polymerizable monomers (Matyjaszewski and Xia, 2001), many azo BCs with specified structures have been synthesized by this simple approach (e.g., Yu et al., 2005c; Tang et al., 2007; Cui et al., 2005; Forcen et al., 2007; Han et al., 2004; He et al., 2003; Scheme 12.2).

Recently, Iyoda and coworkers developed a modified ATRP method to prepare novel amphiphilic LC diblock copolymers consisting of a flexible poly-(ethylene oxide) (PEO) as a hydrophilic segment and poly(methacrylate) containing an azo moiety in the side chain as a hydrophobic LC segment (Tian et al., 2002). The unique characteristic of the amphiphilic azo LCBC films is the nanoscaled microphase separation of a PEO nanocylinder (10–20 nm) regular array with a periodicity of ~20 nm in an azo matrix. Starting from commercially available PEO, Yu et al. (2005c) applied this approach to prepare several well-specified azo triblock copolymers with excellent photocontrol performances, as shown in Scheme 12.3.

Besides the ATRP method, other controlled radical polymerization techniques such as reversible addition/fragmentation chain transfer polymerization (RAFT) (Zhang et al., 2007) and nitroxide-mediated polymerization (NMP) (Yoshida and Ohta, 2005), have also been explored to synthesize azo BCs.

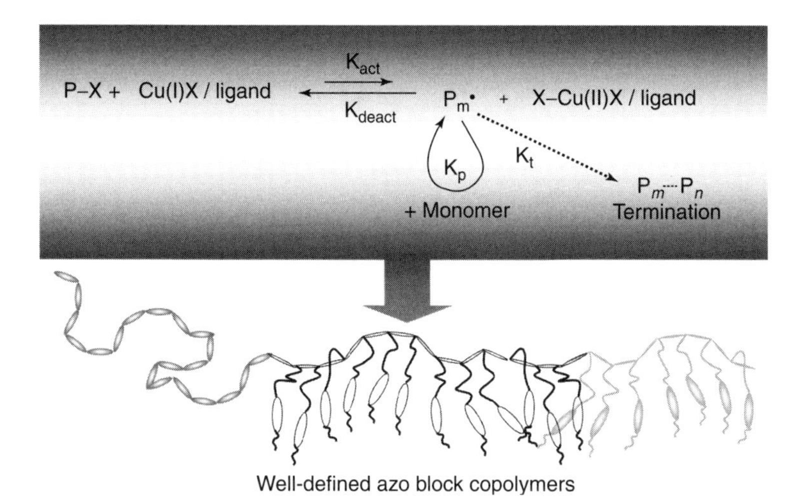

Scheme 12.2 Illustration of well-defined azo BCs prepared by ATRP. K_{act} and K_{deact} are rate constants of activation and deactivation steps, respectively. K_p and K_t are rate constants of propagation and termination of generated free radicals. X is a halogen atom. P represents a polymer chain.

12.2.2. Polymer Analogue Reaction

The azo BCs can be prepared also by postfunctionalization of active groups in BC precursors as shown in Scheme 12.4. In 1989, Adams and Gronski were the first to prepare LC–side chain BCs with cholesteryl groups as mesogens by postpolymerization reaction (Adams and Gronski, 1989). Then Ober and coworkers synthesized a family of well-defined azo BCs by a polymer analog reaction starting from poly(styrene-*b*-1,2-and-3,4-isoprene) with a high content of pendent vinyl (and methyl vinyl) (Mao et al., 1997). Quantitative hydroboration chemistry was used to convert the pendent double bonds of the isoprene block into hydroxyl groups, to which the azo groups were attached via acid chloride coupling, as shown in Scheme 12.5. The intermediate BC of 2-hydroxyethyl methacrylate (HEMA) in this synthetic route is interesting because of the high polarity of the polyHEMA segments and because of the many possibilities to functionalize the hydroxyl group, which has been used to prepare azo BCs by other groups using a similar polymer analog way (e.g., Frenz et al., 2004; Hayakawa et al., 2002).

Similar to the postesterification method, a post azo coupling reaction (PACR) has been used to design amorphous azo BCs, as shown in Scheme 12.6 (Wang et al., 2007). Generally, three kinds of azo chromophores have been discussed by Natansohn and Kumar (Kumar and Neckers, 1989). The first "azo" has relatively poor π–π^* and n–π^* absorbance overlap, and the lifetime of the cis-isomer is relatively long. The second one is "amino-azo"; there is significant overlap of the two bands, and the cis-isomer lifetime is shorter. The third azo is

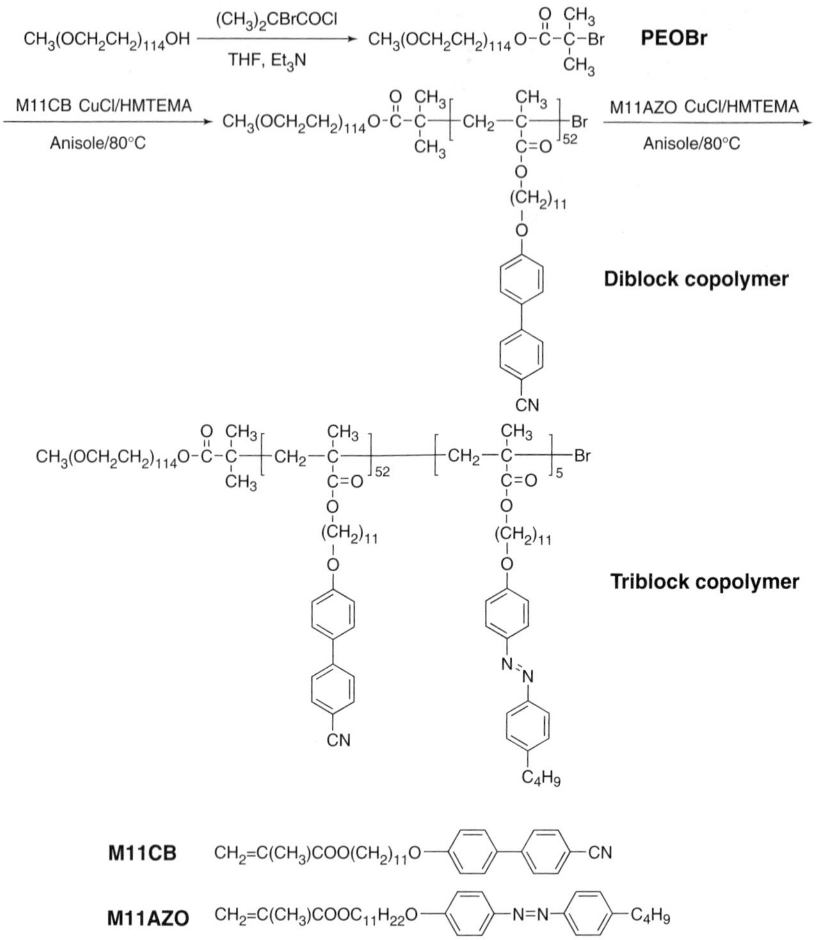

$CH_3(OCH_2CH_2)_{114}OH$ —(CH$_3$)$_2$CBrCOCl / THF, Et$_3$N→ $CH_3(OCH_2CH_2)_{114}O\text{-}C(=O)\text{-}C(CH_3)_2\text{-}Br$ **PEOBr**

—M11CB CuCl/HMTEMA / Anisole/80°C→ Diblock copolymer —M11AZO CuCl/HMTEMA / Anisole/80°C→ Triblock copolymer

Diblock copolymer

Triblock copolymer

M11CB $CH_2{=}C(CH_3)COO(CH_2)_{11}O$–C$_6H_4$–C$_6H_4$–CN

M11AZO $CH_2{=}C(CH_3)COOC_{11}H_{22}O$–C$_6H_4$–N=N–C$_6H_4$–$C_4H_9$

Scheme 12.3 Preparation of a PEO-based azo triblock copolymer by a modified ATRP method. *Source*: Reproduced from Yu et al., 2005c.

"pseudostilbene," where the azo is usually substituted with electron donor and electron acceptor substituents. It is difficult to directly synthesize BCs containing amino-azo or their push–pull derivatives by ATRP because of the inhibition effect of the amino-azo-containing monomers toward free radicals (Wang et al., 2007). The PACR is a convenient way to introduce such azo groups into macromolecular chains, as shown by Wang et al. (1997a,b). Recently, they prepared amorphous azo BCs by the PACR of an amphiphilic PEO-based precursor and then studied the photoinduced shape change of the spherical aggregates formed by the postfunctionalized azo BCs.

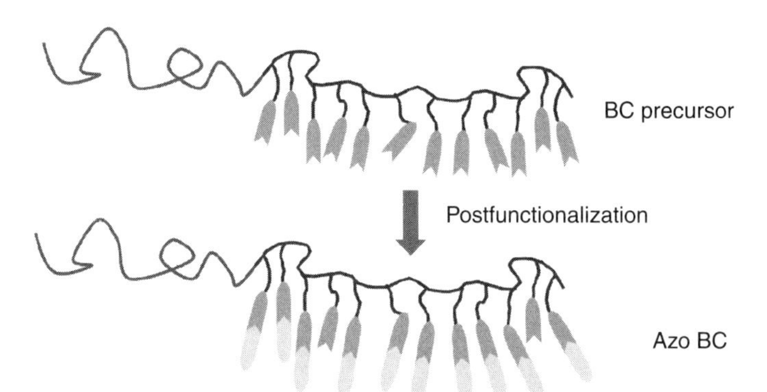

Scheme 12.4 Schematic illustration of the synthesis of well-defined azo BCs by a polymer analog reaction.

12.2.3. Supramolecular Self-Assembly

Upon supramolecular self-assembly, low molecular-weight additives can be used to adjust the properties of BC bulk films by hydrogen bonds. Ikkala and coworkers first introduced this concept in PS-*b*-poly(4-vinylpyridine) (P4VP) stoichiometrically complexed with pentadecylphenol molecules to form the supramolecules by hydrogen bonding (Makinen et al., 2000). The advantages of this method are as follows:

1. Selective hydrogen bonding between one of the blocks (P4VP) of BCs and the additive
2. Obvious microphase separation in bulk films
3. Easy extraction of the additive by selective solvents

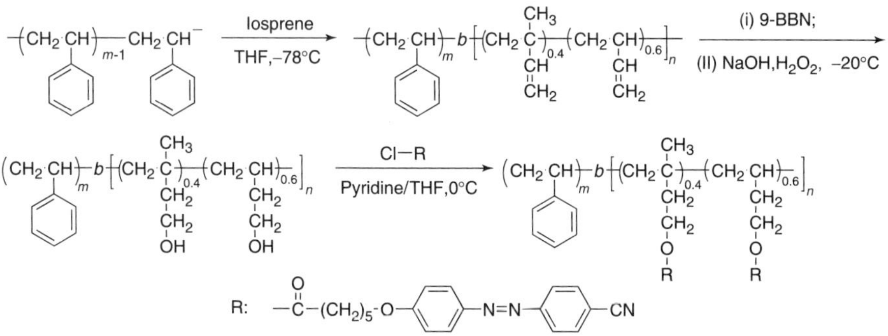

Scheme 12.5 Synthesis of azo BCs by a polymer analog reaction. *Source*: Reproduced with modifications from Mao et al., 1997.

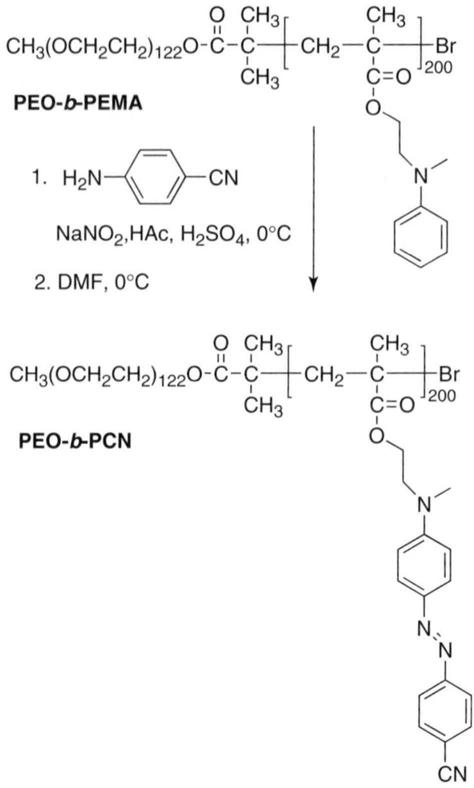

Scheme 12.6 An amorphous azo BC prepared by PACR. *Source*: Reproduced with modifications from Wang et al., 2007.

4. Ability to form microphase-separated films
5. Simple characterization of the nanostructures obtained

Sidorenko et al. succeeded in fabrication of well-ordered nanostructures in thin polymer films by supramolecular assembly of PS-*b*-P4VP and 2-(4-hydroxybenzeneazo) benzoic acid (HABA), consisting of cylindrical nanodomains formed by P4VP-HABA associates in a matrix of PS (Sidorenko et al., 2003). As shown in Scheme 12.7, extraction of HABA with a selective solvent results in nanochannel membranes with a hexagonal lattice of hollow channels in the diameter crossing the membrane from top to bottom.

The supramolecular self-assembly between PS-*b*-poly(acrylic acid) (PAA) and imidazole-terminated hydrogen-bonding mesogenic groups was also used to prepare non-azo LCBCs (Chao et al., 2004). Owing to the attached LC properties, the nanostructures in the LCBC films obtained can be oriented by using an alternating current (AC) electric field, in a direction parallel to the electrodes.

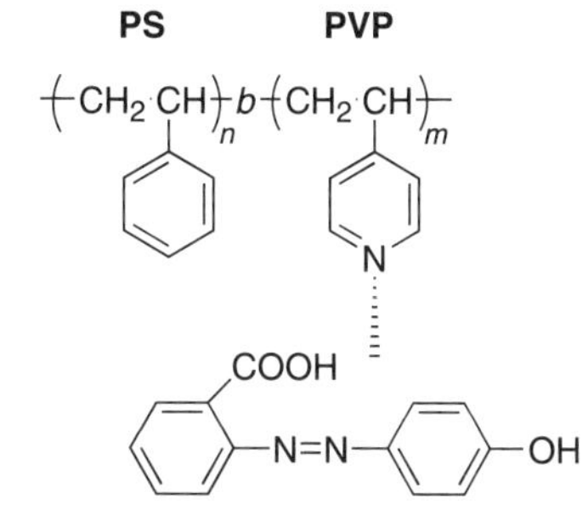

PS PVP

c12s007

Scheme 12.7 Preparation of azo BCs by supramolecular self-assembly between a BC and a low molecular weight azo compound. *Source*: Reproduced from Sidorenko et al., 2003.

12.2.4. Special Reactions

Some special reaction with a particularly designed route can be used to synthesize azo BCs. For instance, a series of poly(vinyl ether)-based azo LCBCs were synthesized by using living cationic polymerization and free-radical polymerization techniques (Serhatli and Serhatli, 1998). As shown in Scheme 12.8, 4.4′-azobis(4-cyano pentanol) (ACP) was used to couple quantitatively two well-defined polymers of LC living poly(vinyl ether), initiated by the methyl trifluoromethane sulfonate/tetrahydrothiophene system. Then the ACP in the main chain was thermally decomposed to produce polymeric radical, which was used to initiate the polymerization of MMA or styrene to obtain PMMA-based or PS-based azo BCs (AB or ABA types).

12.3. PROPERTIES

12.3.1. Basic Properties

Azo BCs are multifunctional polymeric materials, which combine the photochemical properties of azo polymers and the microphase separation of functional BCs. On the one hand, the azo block and the other segments cannot be miscible, unlike statistically random copolymers in which the azo moieties are dispersed homogeneously over the whole bulk films. On the other hand, the azo BCs cannot form macroscopically phase-separated structures like polymer blends (Bates and Fredrichson, 1999).

The azo moiety may play the roles of both mesogen and photosensitive chromophore, when it is attached to the polymer main chain by a longer soft spacer. Both the photoisomerization and the phase transition of the LC to the isotropic phase are involved in the process of microphase separation because of

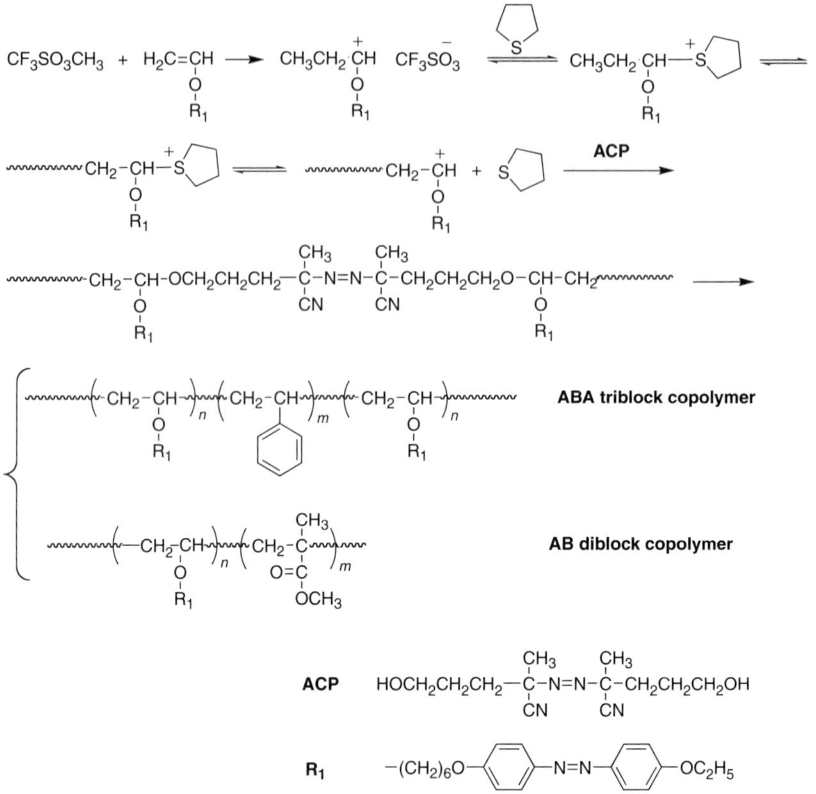

Scheme 12.8 Preparation of azo BCs by using living cationic polymerization and free-radical polymerization techniques. *Source*: Reproduced from Serhatli and Serhatli, 1998.

the immiscibility between the azo blocks and non-azo blocks. As shown in Fig. 12.2, all the processes in the microphase separation at different temperatures might be influenced by the azo photoisomerization, andthe microphase-separated nanostructures in azo BC films could have effect on the photochemical behavior of azo blocks and vice versa (Cui et al., 2003). The photochemical control and supramolecular self-assembly make the azo BCs in solid state superior to that of azo homopolymers or azo random copolymers.

On the one hand, the azo BCs inherit most of the excellent properties of azo homopolymers (Ichimura, 2000), such as trans–cis photoisomerization, photoselective alignment with azo transition moments almost perpendicular to the polarization direction of the actinic light, and photochemical phase transition, since the *trans*-azo can be a mesogen because its molecular shape is rodlike, whereas the *cis*-azo never shows any LC phase because of its bent shape. All the illustrations of the photosensitive performances are shown in Fig. 12.3 (Yu et al., 2005a).

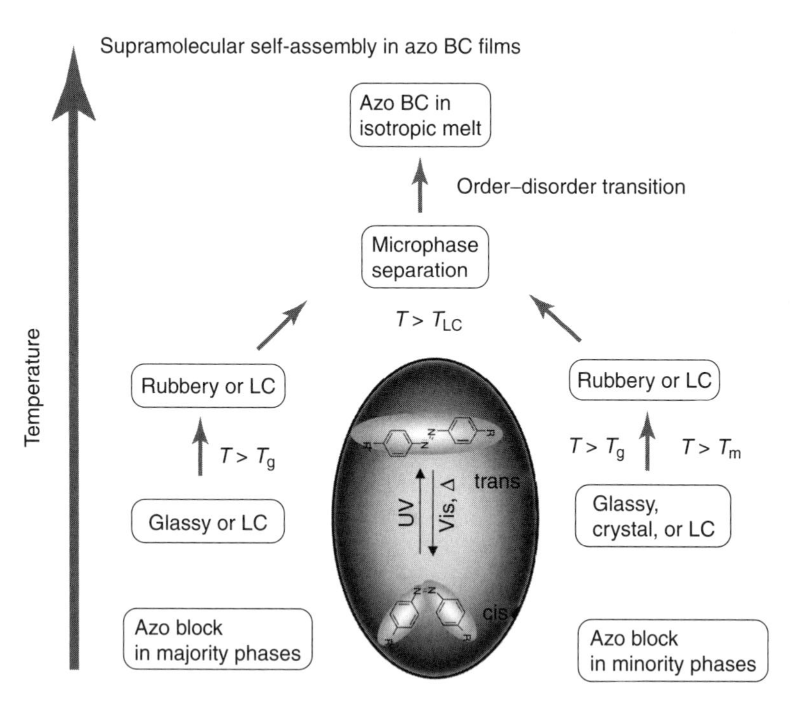

Figure 12.2. Microphase separation and azo photoisomerization in bulk films of well-defined azo BCs. T_{LC} is LC–isotropic phase transition temperature.

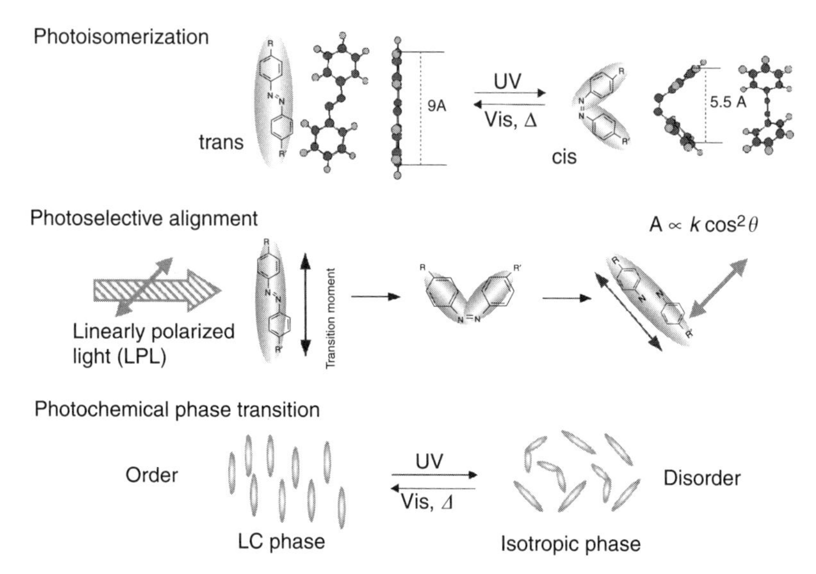

Figure 12.3. Properties of azo BCs inherited from azo homopolymers. A is the absorption of azo chromophores, θ represents the angle between the polarization direction of the linearly polarized light and the transition moment of an azo moiety.

On the other hand, the well-defined azo BCs show excellent features different from azo random copolymers. Recently, the performance of molecular cooperative motion (MCM) between azo moieties and other photoinert groups has been studied in azo triblock copolymers with specifically designed structures (Yu et al., 2007). Triggered by a linearly polarized laser beam at 488 nm, the azo chromophores became aligned by the Weigert effect, then the photoinert groups were oriented together with the azo moieties under the function of MCM, although they do not absorb the actinic light (Wu et al., 1998a,b). When the azo and the photoinert mesogens form the majority phase (continuous matrix), the photocontrol orientation can be transferred to the microphase-separated nanodomains inside the LC matrix owing to SMCM (see Fig. 12.4; Yu et al., 2005c). Interestingly, the photoinduced MCM can be confined in nanoscaled regions by microphase separation when the azo and the other mesogenic blocks are situated in the minority phases (separated phases) (Yu et al., 2007a). No obvious influence on the substrate was detected since the photoinduced alignment occurred in the intermittent phases, and the phototriggered oriented motion was disrupted by the glassy substrates of PMMA or PS. Thanks to the nanoscaled MCM, the scattering of visible light can be avoided by confining the mesogenic domain to the nanoscale, which improves the optical properties of azo BC films. Accordingly, thick films (~200 μm) with high transparency and low absorption based on the microphase separation of well-defined azo BCs have been obtained, enabling them to record Bragg-type gratings for volume storage. Figure 12.5 gives the possible scheme of Bragg diffraction based on transparent thick films (~200 μm) of the well-defined azo triblock copolymer. Furthermore, the stability of the photoinduced orientation in films was greatly improved by decreasing the photosensitive azo content in BC composition.

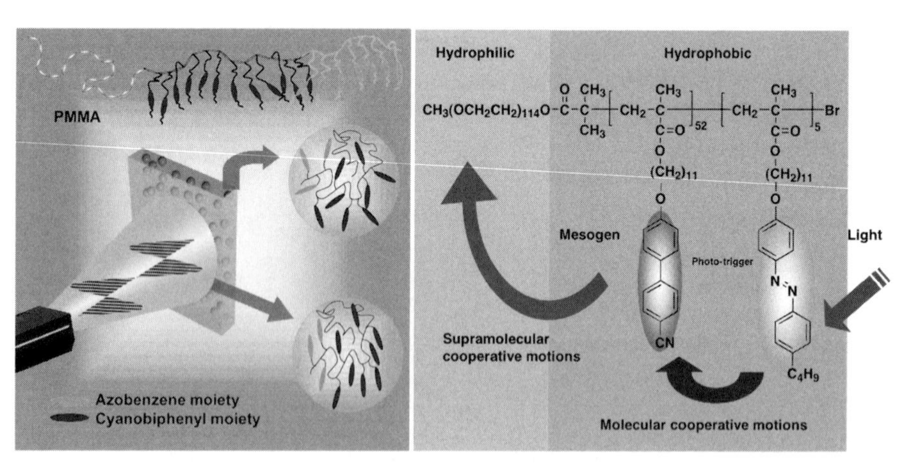

Figure 12.4. Photoinduced cooperative motion in bulk films of azo BCs. Azo in the minority phase (*left*) and azo in the majority phase (*right*). *Source*: Reproduced from Yu et al., 2005c; 2007a. See color insert.

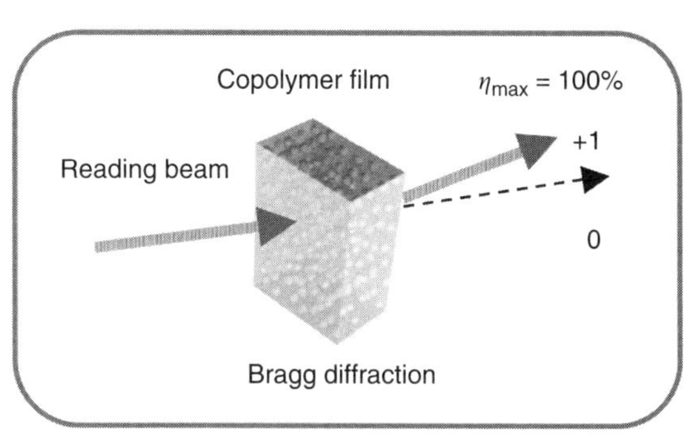

Figure 12.5. Photograph of a triblock copolymer film with 200 μm thickness (*left*) and scheme of the Bragg grating obtained from a thick film (*right*). *Source*: Reproduced from Yu et al., 2007a.

12.3.2. Properties from Non-Azo Blocks

Generally, the non-azo blocks also influence the final properties of azo BCs in the solid state. For instance, introduction of PEO as one block may endow the azo BCs with hydrophilicity, ion conductivity, and crystallization of PEO (e.g., Li et al., 2007a,b; Lin et al., 2007). As shown in Fig. 12.6, both PEO crystals and azo LC phases were observed clearly by polarizing optical microscopy (POM). When the number of repeated units of PEO is far higher than that of the azo block, typical spherulite structures with birefringence were observed under POM because of the high degree of order and crystallinity contained within the PEO domains. In general, the process of spherulization starts on a nucleation site and continues to extend radially outward until a neighboring spherulite is reached, leading to the spherical shape of the spherulite shown in Fig. 12.6. When the azo mesogenic blocks form the majority phases, LC textures of focal conic fan were obtained at a higher temperature, indicating smectic LC phases. As shown on the right-hand side of Fig. 12.6, further experimental results of wide-angle X-ray power diffraction (WXRD) indicated that LC phases of smectic A, C, and X (SmA, SmC, SmX) were obtained at different temperatures, corresponding to the LC textures in POM images, respectively (Tian et al., 2002). A possible schematic illustration of the layer structures of the LC phases is also given.

Poly(2-(dimethylamino)ethyl methacrylate) p(DMAEMA) may bring water solubility to the azo BCs when the azo segment is short enough (Ravi et al., 2005). Using PMMA as one block, the azo BC films may show good optical transparency (e.g., Yu et al., 2007b; Breiner et al., 2007). Introduction of PS provides the designed azo BCs with a high glass transition temperature (Tg) (Morikawa et al., 2007) and offers confinement effects on the photoinduced alignment of azo LC side chains (Tong et al., 2004). Putting rubbery poly(n-butyl acrylate) (PnBA) as middle block in ABA-type azo triblock copolymers, thermoplastic elastomers were obtained (Cui et al., 2004), which contrast with conventional thermoplastic

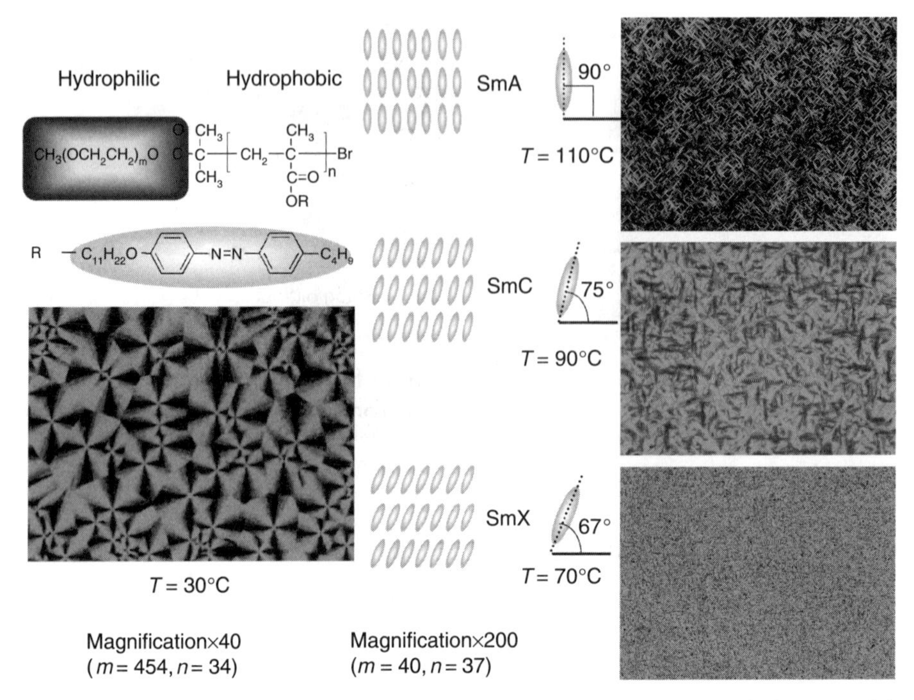

Figure 12.6. POM images of PEO-based azo BCs. Illustration of the possible smectic LC layer structures is given on the left of each POM image. *Source*: Reproduced with modifications from Tian et al., 2002.

elastomers (such as styrene-butadiene-styrene triblock copolymer). Semifluorinated alkyl substituents of the azo moieties may decrease the surface energy of azo BC films. Moreover, it is possible for sufficiently long perfluoroalkyl units to stabilize the surface so that azo photoisomerization does not change the surface properties (Paik et al., 2007). In addition, poly(2-methoxyethyl vinyl ether) [poly(MOVE)]-base azo BC can supply the polymer surface with the thermally controlled water wettability because of the presence of the poly(MOVE) block (Yoshida, 2005).

12.3.3. Properties Originating from Microphase Separation

Azo BCs are one fascinating class of soft materials, showing a rich variety of microphase-separated nanostructures in films because the strongly bonded azo segments and the non-azo blocks are thermodynamically incompatible. As shown in Fig. 12.1, the microphase-separated morphologies can be controlled to be spheres, cylinders, or lamellae, depending on the length, chemical nature, architecture, and number of repeated units in each block (e.g., Thomas and Lescanec, 1994).

When the azo block shows LC properties, the inherent mesogenic ordering might act on the thermodynamic process of microphase separation by SMCM.

The investigation of azo LCBCs can undoubtedly fertilize the detailed illustration of microphase separation, and completely novel nanostructures and phase behaviors are expected by the supramolecular self-assembly. Recently, a novel wormlike nanostructure composed of azo mesogens was observed in azo LCBC films formed by $PMMA_{155}$-*b*-$PMA(11Az)_{25}$, as shown in Fig. 12.7. Although nanoscale wormlike morphologies have been obtained in micelles formed by BC solutions (Discher and Eisenberg, 2002), they have never been reported in bulk films. Because of the difference in elastic modulus between amorphous PMMA and the azo LC phases, wormlike microphase domains were clearly observed in atomic force microscopic (AFM) images of the LCBC films (Yu et al., 2007b).

According to the molecular component of the azo BC, the wormlike domains with a width of 25 nm should be segregated by the azo blocks, leading to the disappearance of an LC texture after microphase separation because of the limited POM resolution. The novel wormlike nanostructures might be caused by the balance between microphase separation and LC self-assembly, which makes it different from the phase-segregated morphologies of BCs with weak molecular interactions (Yu et al., 2007b).

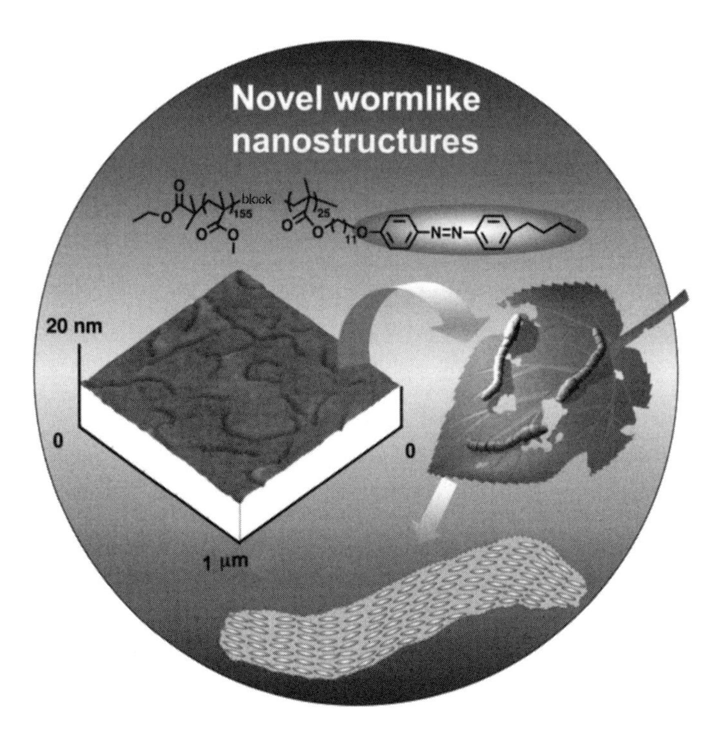

Figure 12.7. Wormlike microphase-separated nanostructures obtained by supramolecular self-assembly in bulk films of a well-defined azo LCBC. *Source*: Reproduced with modifications from Yu et al., 2007b.

Upon microphase separation, the azo moieties of relatively lower content aggregate together, which enables them to form easily an LC order in local areas, whereas a random copolymer of a similar azo content shows a statistical molecular structure, in which the azo moiety is homogeneously dispersed within the matrix of non-azo blocks. No microphase separation can be observed since the azo block and the other blocks are completely miscible, resulting in an amorphous phase (Yu et al., 2008a). That is to say, the azo BC shows an LC phase more easily than an azo statically random copolymer with a similar low content of azo mesogens (Naka et al., 2009). As shown in Fig. 12.8, although the random copolymer has a similar azo composition of $\sim$22 mol% to that of the well-defined azo BC, it shows no LC phase. The acquired LC property might endow the azo LCBC with advanced performances such as physical anisotropy, self-organization, long-range ordering, MCM, and SMCM. Since the segregated azo block is incompatible with the PMMA segments in microphase-separated films, a cooperative effect occurs in the process of photoalignment, leading to a larger photoinduced birefringence than in the case of the random copolymer (Naka et al., 2009). Also, H- or J-aggregates are easily formed since the azo blocks exhibit a high local concentration in the microphase-separated domains (Tian et al., 2002).

The microphase-separated morphologies also exert an influence on the LC properties of the azo blocks (Mao et al., 1997). For instance, a PS-based azo LCBC with mesogenic nanocylinders embedded within the PS matrix exhibited a clearing point, 22°C higher than that of the azo LCBC in an LC lamellar structure, even though the former sample has a slightly lower molecular weight (Mn) than the latter. Ober and coworkers proposed that the cylindrical nanodomain structures in azo LCBCs might stabilize the smectic mesophase within it rather than the lamellar microphase morphology (Mao et al., 1997).

12.4. CONTROL OF MICROPHASE SEPARATION

The microphase separation is one of the most important properties of azo BCs in the solid state. It is well known that classical BC films can microphase separate into nanoscale periodic structures that may be ideal candidates for many applications (Hamley, 2003). The driving force for the ordered nanostructure is to achieve the required balance of minimizing the interfacial energy and maximizing the conformational entropy of the BCs. However, thin films of microphase-separated BCs typically do not have long-range ordering, which limits their further utilization. Generally, LC polymers and BCs are two types of ordered noncrystalline materials that can undergo self-assembly (Finkelmann and Walther, 1996; Mao and Ober, 1996). From the viewpoint of molecular design, azo LCBCs integrate their unique characteristics of both materials into a single system, which also bears photocontrol properties inherited from the azo chromophore. Such a SMCM is regarded as one of the most effective approaches to control microphase-separated nanostructures in azo LCBC films (Yu et al., 2006a,b), as shown in Fig. 12.9. Here we show several newly developed approaches to manipulate nanostructures in azo LCBCs.

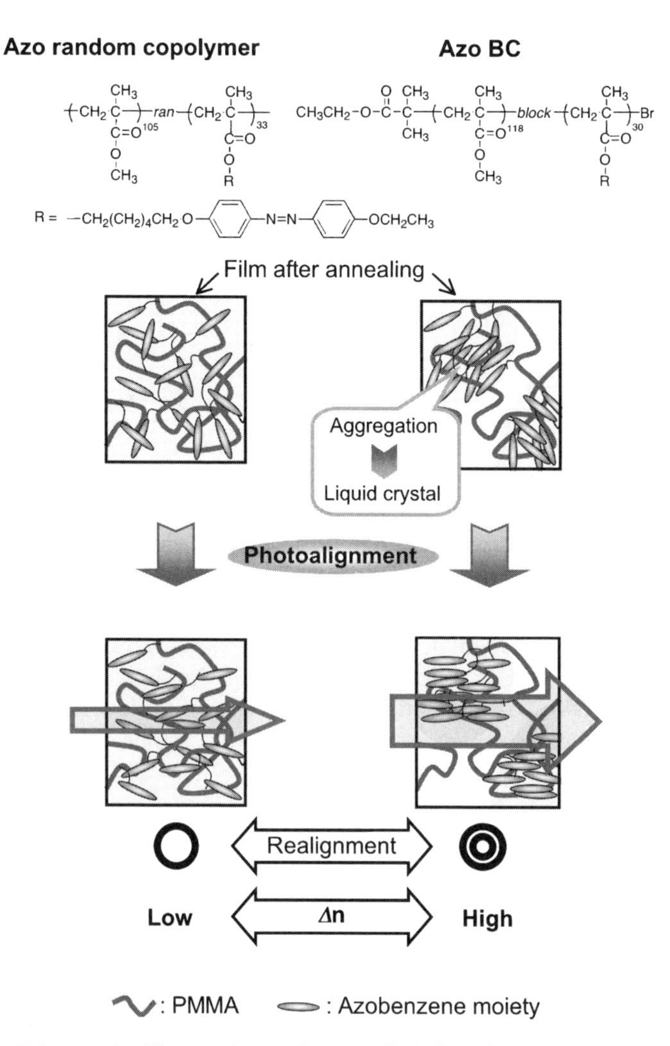

Figure 12.8. Schematic illustration of a well-defined azo BC and a random copolymer of similar azo content (~22 mol%) and their photoinduced alignment upon irradiation with a linearly polarized beam at 488 nm. As a result of the microphase separation, the azo BC shows LC phase, whereas the random copolymer is amorphous.

12.4.1. Thermal Annealing

Fabrication of large-area periodic nanoscale structures using supramolecular self-organizing systems is of great interest because of the simplicity and low cost of the fabrication process (Whitesides and Grzybowski, 2002). Although macroscopically ordered microphase separation has been successfully prepared in amorphous BC films (e.g., Cheng et al., 2006; Darling, 2007), both high reproducibility and

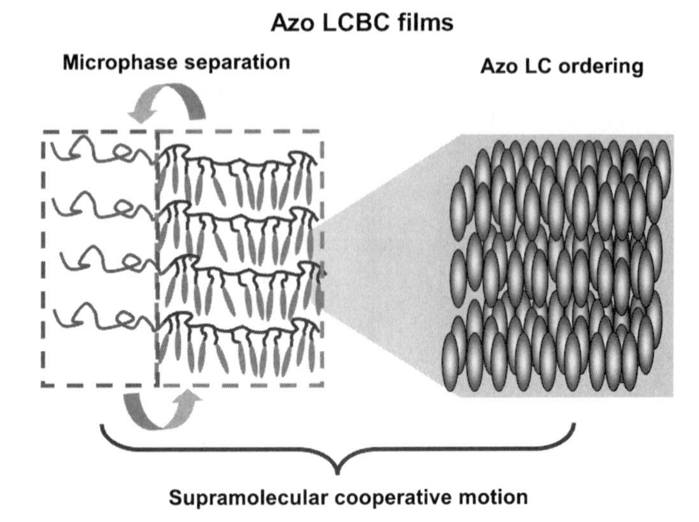

Figure 12.9. Schematic illustration of the SMCM in bulk films of azo LCBCs.

mass production of such regularly ordered nanostructures through self-assembling nanofabrication processes still remain challenging. Specially designed PEO-based azo LCBCs with SMCM are good candidates to produce low cost materials with self-assembled nanostructures, leading to industrial applications in future engineering plastics.

As shown in Fig. 12.10, transmission electron microscopy (TEM), AFM, and field emission scanning electron microscopy (FESEM) imaging revealed beautiful PEO nanocylinder nanostructures with ordered alignment perpendicular to the substrates [silicon wafers, mica, glass, or poly(ethylene terephthalate) (PET) films] and with hexagonal packing (e.g., Komura and Iyoda, 2007). No external driving forces were exerted on the PEO-based LCBC films. Such regular periodic arrangements of nanocylinders were not limited to the film surface; cross-sectional images confirmed the formation of three-dimensional (3-D) arrays. It has been proven that the 3-D arranged nanostructures should be assisted by the out-of-plane orientation of azo mesogens, which constitute the continuous phase in the BC films (e.g., Komura and Iyoda, 2007). In the TEM cross-sectional image, the smectic layer structures of azo blocks were observed also normally to the PEO nanocylinders (parallel to the substrate), which was also confirmed by small-angle X-ray-scattering (SAXS) measurements (Watanabe et al., 2005). Such cooperative effect between PEO nanocylinders and the azo mesogenic orientation is a result of SMCM at a temperature higher than the clearing point of LC-to-isotropic phase), which decreases the viscosity of LCBC films and enables the interaction between the microphase separation and the smectic LC ordering to proceed completely.

The film thickness and the substrate properties exert a great influence on the nanomorphologies of LCBCs. By carefully setting the experimental conditions, well-ordered arrays of nanoscale phase domains may be extended to a

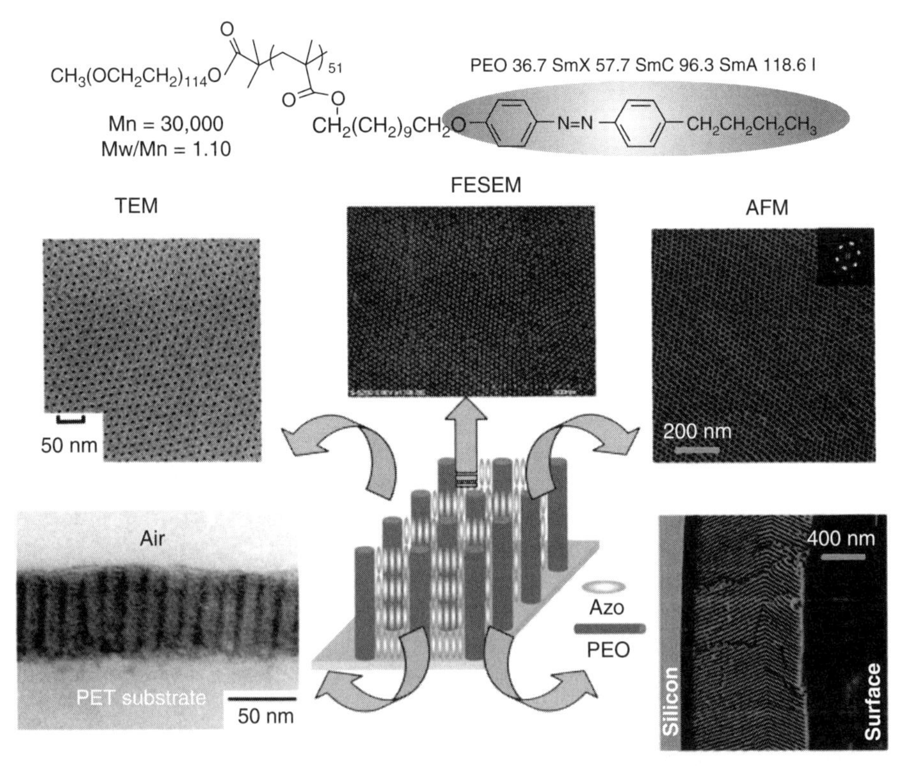

Figure 12.10. Perpendicular array of PEO nanocylinders and azo mesogens in PEO-based azo BC films by thermal annealing. Orientation of the nanocylinders is assisted by the out-of-plane alignment of the smectic azo LC molecules.

macroscopic scale, which can be regarded as analogous to "a polymeric single crystal." An AFM image of perpendicularly arranged PEO nanocylinders in an area of 3.1 μm × 12.4 μm is shown in Fig. 12.11. In fact, such dotted patterning of PEO nanocylinders can be achieved over several centimeters, which has been confirmed by acquiring AFM images spaced by 100 μm over the whole substrate.

12.4.2. Rubbing Method

According to the SMCM in azo LCBCs, a long-range order of PEO nanocylinders coinciding with the azo LC alignment direction can be obtained, which suggests the application of other LC alignment techniques to control the microphase-separated nanocylinders. Although there are several LC alignment methods such as rubbing, optical, electric or magnetic field, Langmuir–Blodgett films, silicon oxide–treated surface, and oblique evaporation, only the rubbing polyimide technology is widely applied in commercial production of liquid crystal displays (LCDs).

Figure 12.11. AFM images of PEO nanocylinders in bulk films of azo LCBCs (3.1 μm × 12.4 μm). In the left panel, the PEO nanocylinders are aligned parallel to the rubbing direction, whereas the PEO nanocylinders are oriented perpendicular to the substrate in the right panel. *Source*: Reproduced with modifications from Yu et al., 2006b. See color insert.

In fact, parallel processes for patterning densely packed nanostructures are required in diverse areas of nanotechnology (Kim et al., 2003). The regular nanostructures self-assembled in bulk films of PEO-based azo LCBCs provide the opportunity to prepare such parallel stripe pattern by SMCM. To achieve this, a commercialized rubbing technique was chosen to homogeneously align azo mesogens in bulk LCBC films. The periodic ordering of oriented azo mesogens might be transferred to microphase-segregated PEO nanocylinders by SMCM. Generally, LC molecules can be controlled to align along the rubbing direction on surfaces of rubbed polyimide films because of the lower energy level along the rubbing-induced grooves (Berreman, 1972). Recently, this mechanical rubbing method has been extensively investigated with low molecular-weight compounds, LC polymers, and cross-linked LC gels (Ikeda, 2003b; Demus et al., 1998). It has been shown to have a strong influence on the bulk films of azo LCBCs in Fig. 12.12.

After rubbing and annealing treatments, perfect alignment of PEO nanocylinders, coinciding with the azo LC orientation, was clearly observed by AFM, as shown in Fig. 12.12. Overall, an area of 500 nm × 500 nm, ~21 PEO nanocylinders with a diameter of 10 nm were regularly oriented along the rubbing direction. This defect-free periodic nanocylinder array in the azo LCBC films can be acquired over arbitrarily large areas on the surface of rubbed polyimide films (Fig. 12.11). Similar to other methods of controlling microphase-segregated domains in amorphous BC films, such as electric field (Morkved et al., 1996),

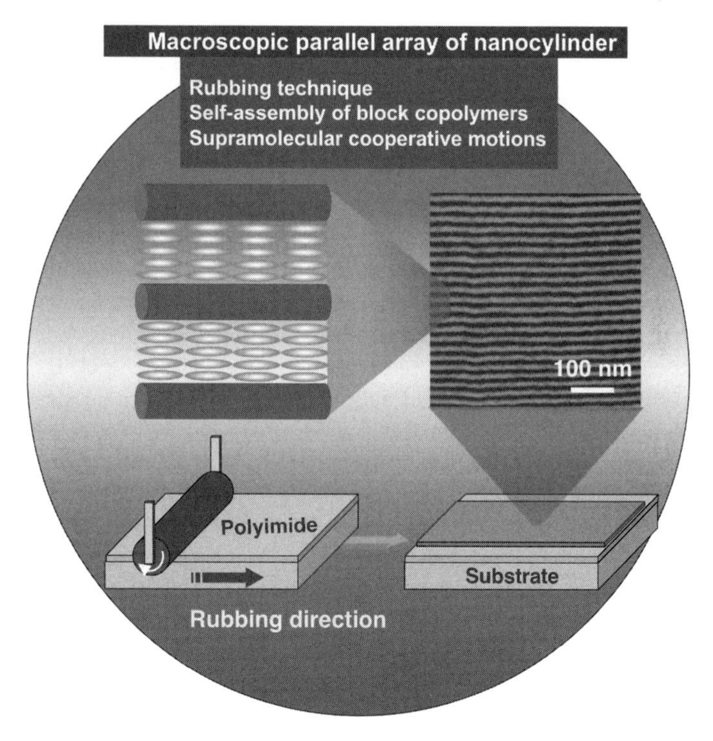

Figure 12.12. Fabrication of a parallel nanocylinder array in bulk films of a PEO-based azo LCBC by the mechanical rubbing technique. All the PEO nanocylinders are oriented along the rubbing direction. *Source*: Reproduced with modifications from Yu et al., 2006b.

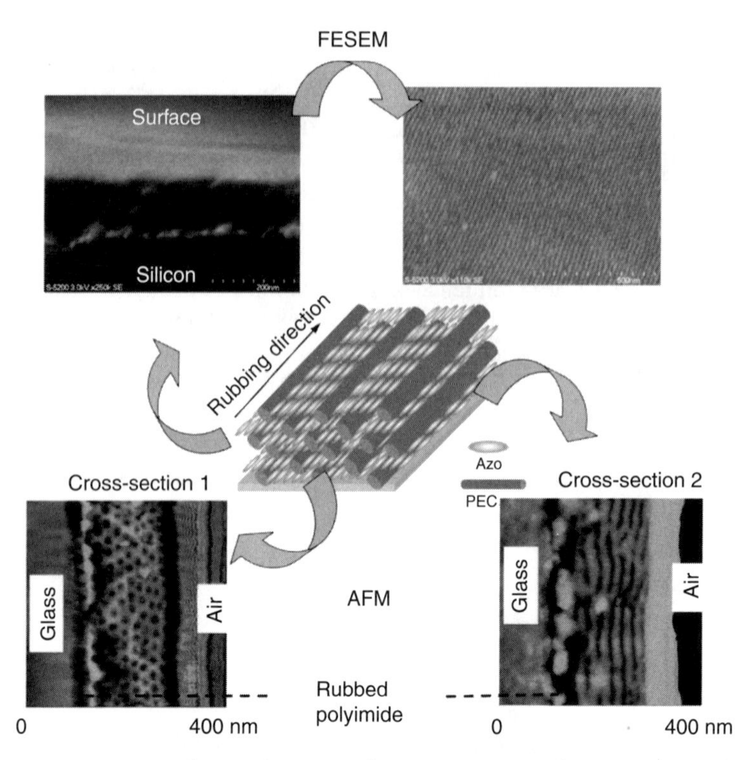

Figure 12.13. FESEM and AFM images of cross-sections of a 3-D array of parallel PEO nanocylinders on the surface of rubbed polyimide films in azo LCBC films. *Source*: Reproduced with modifications from Yu et al., 2006b.

crystallization (Rosa et al., 2003), controlled interfacial interaction (Huang et al., 1998), chemically or topologically patterned substrates (Cheng et al., 2004), this rubbing technique exerted a forceful action on 3-D array of nanostructures, and appeared in the cross-sectional images (AFM and FESEM in Fig. 12.13).

Most interestingly, the sample showed a regular surface relief of ~ 3 nm with a periodicity of 24 nm, possibly due to the confined crystallization of the PEO blocks embedded in the ordered LC phases (Reiter et al., 1999). Obviously, both the nanocylinder dimension and the surface periodicity can be easily adjusted by the volume fraction of PEO or azo blocks in the LCBCs. The macroscopic 3-D nanocylinder array in a long-range order with arbitrarily controlled direction in plane was successfully achieved by the SMCM, opening a novel pathway of controlling the nanoscopic domains over large areas (Yu et al., 2006b).

The industrialized rubbing technique to regulate macroscopic nanoarray in azo LCBC films has the advantages of simplicity, convenience, and low cost over other approaches (e.g., Morkved et al., 1996), which creates opportunities for manufacturing nanoscopic device components (Fasolka and Mayes, 2001). In the processes of nano-cylinder alignment, the rubbing acts on the microphase domains by the bridge of the

LC media, implying that the microphase domains can be controlled by manipulating LC alignment under SMCM. The present rubbing results technologically implicated that other LC manipulating ways can be used to regulate the azo LCBC films.

12.4.3. Photoalignment

In the mechanical rubbing process to control LC alignment, dust or static electricity might be produced, which could induce defects in the macroscopic array of microphase-separated nanostructures in azo LCBC films, and this method can only be applied for a flat surface. Therefore, noncontact methods, such as photocontrol approaches, have been explored (Yu et al., 2006a). Light is one of the most convenient and cheap energy sources, whose intensity, wavelength, polarization direction as well as interference patterns can be manipulated simply. More perfect and simpler fabrication of microphase-separated nanostructures in azo LCBC films is expected by photocontrol.

Upon irradiation with linearly polarized light, azo chromophores are known to undergo photoinduced alignment with transition moments almost perpendicular to the polarization direction by repetition of trans–cis–trans isomerization cycles (e.g., Viswanathan et al., 1999). Such ordering can be transferred directly to other photoinert mesogens by MCM coinciding with the ordered azo moieties. By incorporating the photoinduced alignment of azo into LCBC with SMCM, the molecular ordering of the azo can be transferred to a supramolecular level. Therefore, well-ordered nanostructures of the azo LCBC films might be obtained by photocontrol (Ikeda and Tsutsumi, 1995).

As shown in Fig. 12.14, a linearly polarized laser beam at 488 nm was used to control PEO nanocylinders self-assembled in an amphiphilic azo LCBC of well-defined structure. To enhance the absorption at 488 nm, a pseudostilbene-type azo with a cyano group as an electron acceptor substituent was used in preparation of the azo LCBC. Upon annealing without photoirradiation, hexagonal packing of the PEO cylinders perpendicular to the glass substrates was obtained, because of the out-of-plane orientation of the azo mesogens, which is similar to that in Fig. 12.10. The photoinduced alignment of azo moieties was carried out at room temperature, and then the anisotropic azo LCBC films were thermally annealed at a temperature just lower than the phase transition temperature of smectic LC-to-isotropic phase. Perfect array of PEO nanocylinders aligned perpendicularly to the polarization direction of the laser beam because of SMCM was achieved.

Recently, Seki and coworkers reported a PEO nanocylinder control method by a periodic change in film thickness induced by mass transfer to form holographic gratings (Morikawa et al., 2006). In their preparation, the film thickness was strictly modulated, and the azo BC was mixed with a low molecular-weight LC (5CB) to photoinduce a large mass transfer, and 5CB was eliminated after grating formation. To simplify the process, they adopted a polarized beam to control the nanocylinders in PS-based azo LCBC films (Morikawa et al., 2007). Defects appeared in the microphase-separated nanostructures, probably caused by incomplete microphase separation resulting from the high T_g of the PS block.

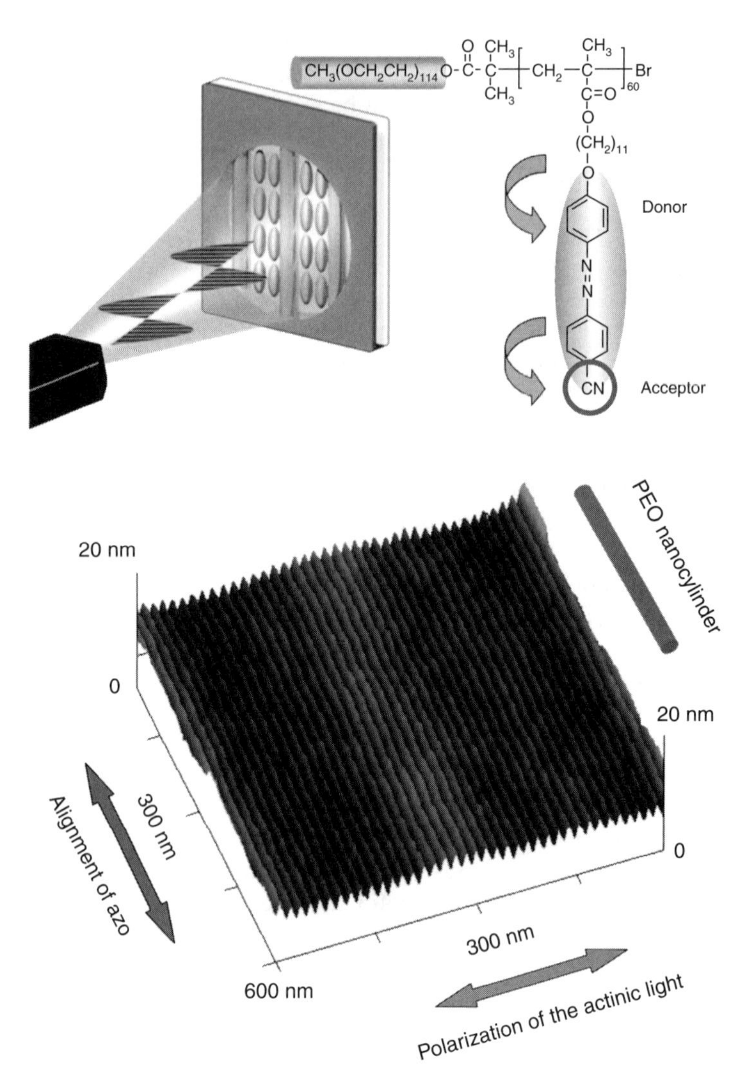

Figure 12.14. Photoinduced alignment of PEO nanocylinders in azo LCBC films as a result of SMCM. *Source*: Reproduced with modifications from Yu et al., 2006a. See color insert.

In nondoped films of PEO-based azo LCBCs with well-defined structures, macroscopic parallel patterning of PEO nanocylinders can be obtained easily in an arbitrary area by simple and convenient photocontrol. Furthermore, the noncontact method might provide the opportunity to control nanostructures even on curved surfaces. On the basis of the principle of SMCM, the orientation of microphase-separated nanocylinders dispersed in azo LC matrix should agree with the alignment direction of azo mesogens. The azo molecules can be 3-D

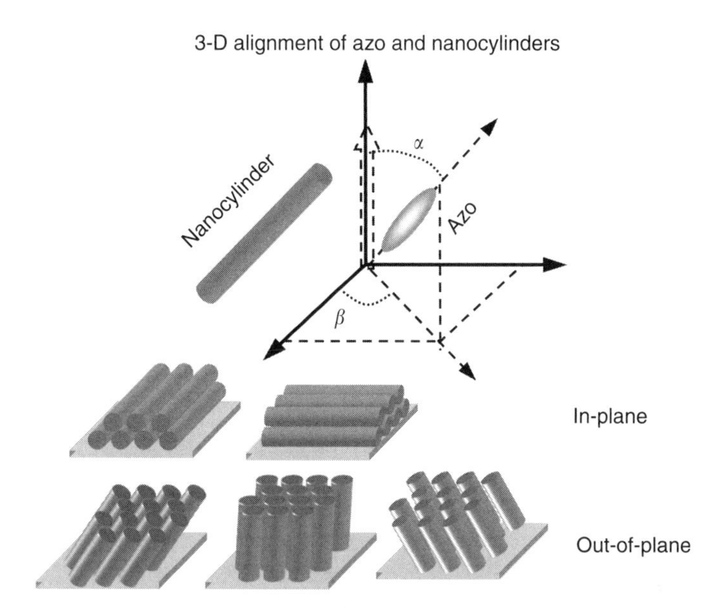

Figure 12.15. Illustration of possible photomanipulated 3-D alignments of azo moieties and nanocylinders in azo LCBC films by SMCM.

manipulated by unpolarized light, as shown in Fig. 12.15 (Wu et al., 1999). Both in-plane and out-of-pane alignment of nanocylinders coinciding with azo orientation might be precisely photocontrolled by SMCM, which is expected to provide complicated nanotemplates for top–down-type nanofabrications such as lithography and beam processing.

12.4.4. Electric Field

Recently, Kamata et al. (2004) developed an electrochemical method to control alignment of PEO nanocylinders normal to the substrate in PEO-based azo LCBC films. As shown in Fig. 12.16, the azo BC films were prepared by spin coating from a toluene solution on an indium tin oxide (ITO) glass substrate kept at 50°C for 2 days. Using the ITO glass as a working electrode, a sandwich-type cell was assembled with a Teflon spacer and an injected KBr aqueous solution as electrolyte. Under the function of an electrolytic potential in the potentiostatic mode using Hokuto HZ-3000 with Pt counter electrode and Ag/AgCl reference electrode, all the PEO nanocylinders were oriented parallel to the electrolytic field as the lowest energy alignment (Morkved et al., 1996). Since the hydrophilic PEO nanocylinders are ion conductive, they can be manipulated normal to the substrate by ion diffusion locally induced in the vicinity of the electrode (Kamata et al., 2004).

The control process was carried out at 50°C. All the PEO nanocylinders were aligned along the electrolytic field, in spite of the non-microphase-separated state,

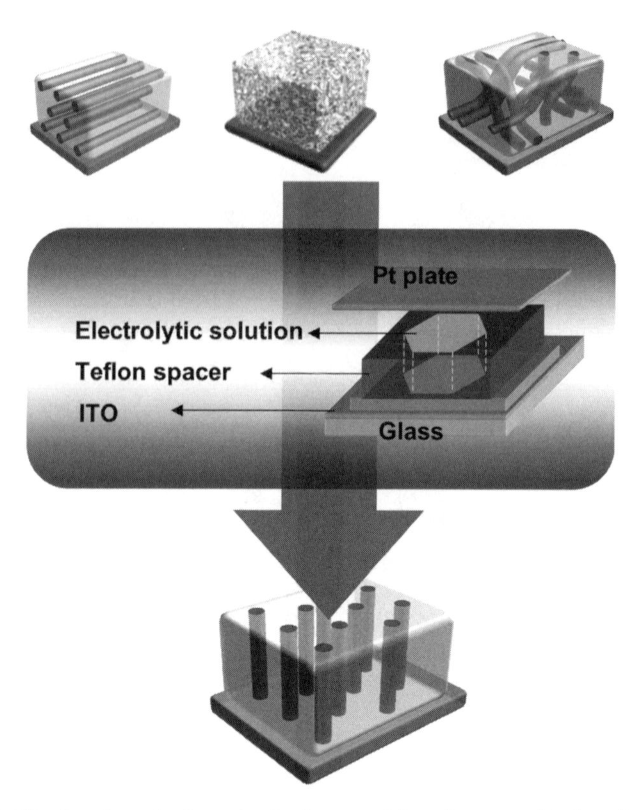

Figure 12.16. Electrochemical control of assembly of nanocylinders in PEO-based azo LCBC films. All the PEO nanocylinders were aligned along the electrolytic field, in spite of a nonmicrophase-separated state, parallel or random alignment of PEO nanocylinders.

parallel or random alignment of PEO nanocylinders. This control using electrochemical potentials is a good candidate for the fabrication of nanocomposites.

For a supramolecular system with hydrogen-bonding non-azo LCBCs, an AC electric field was used to rapidly align the nanostructures at temperatures below the order–disorder transition but above T_g (Chao et al., 2004). The low molecular-weight mesogens play an important role in controlling microphase separation. The fast orientation switching of the nanostructures was attributed to the dissociation of hydrogen bonds, which might be used to control nanostructures in azo supramolecular BC systems.

12.4.5. Magnetic Field

Similar to the electric field, the magnetic field can be used to manipulate the microphase separation in azo LCBCs by SMCM. The noncontact orientation

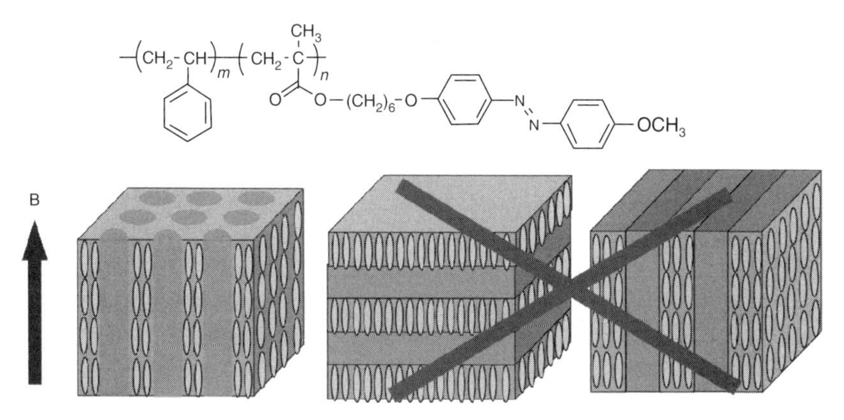

Figure 12.17. Magnetically induced alignment of microphase-separated nanostructures in PS-based azo LCBCs. All the PS nanocylinders were oriented along the magnetic field, which shows no effect on the lamellar morphologies. *Source*: Reproduced with modifications from Tomikawa et al., 2005.

method provides a higher degree of freedom for sample shapes than the mechanical orientation method, and no danger is present, such as the dielectric breakdown that can be encountered in the electrical orientation approach. Moreover, the uniform orientation of LC polymers can be obtained over the whole region of a sample, regardless of the macroscopic shape of the sample and the strength of the magnetic field (Tomikawa et al., 2005). As shown in Fig. 12.17, hexagonally packed PS nanocylinders dispersed in an azo mesogenic matrix were aligned along the magnetic field upon annealing for a longer time (>2 h) in the nematic LC phase. Strangely, the magnetic field shows no function on the lamellar nanostructures in films, possibly because ordered lamellar microdomains with a long correlation length are only rearranged very little (Osuji et al., 2004). Therefore, although the LC was magnetically aligned in nanoscale layers, it showed no influence on the inverse continuous phase (Hamley et al., 2004).

12.4.6. Shearing Flow and Other Methods

The orientation of lamellar and cylindrical microdomains in azo LCBCs has been triggered with a shearing flow yielding highly oriented samples (e.g., Osuji et al., 1999; Osuji et al., 2000). During oscillatory shearing of PS-based azo LCBCs at LC temperature or in an isotropic phase, the LC phase has a distinct effect on the orientation of microphase-separated nanodomains. Upon transformation from the LC phase to the isotropic phase, the LC orientation is lost, whereas the orientation of nanocylinders is sustained. Interestingly, the uniaxial planar orientation of the mesogens is recovered completely on cooling from an isotropic melt. This spontaneous reorientation of the LC phase takes place after repeating the heating and cooling cycles above and below the LC–isotropic phase transition temperature, showing that the oriented nanocylinders act as an anchoring substrate for the LC mesogens (Tokita et al., 2007; Tokita et al., 2006). Undoubtedly, other approaches for controlling microphase separation in films of amorphous BCs, such as solvent evaporation, film thickness, modified substrates, mixture with homopolymers, and roll casting, can also be used with azo LCBC.

12.5. APPLICATIONS

12.5.1. Enhancement of Surface Relief Gratings

Holographic gratings have potential applications in information technology, which can be recorded on both amorphous and LC polymers through photoisomerization or photochemical phase transition of azo chromophores (e.g., Rochon et al., 1995; Kim et al., 1995). Diffraction efficiency (DE) is one of the most important parameters of holographic gratings. In amorphous materials, surface relief grating (SRG) contributes mainly to DE. Azo BCs are good candidates to control DE by SRG enhancement upon microphase separation (Yu et al., 2009, 2005b). As shown in Fig. 12.18, after a preirradiation treatment, both SRGs and refractive index gratings (RIGs) were recorded in PEO-based azo LCBC films upon irradiation of an interference pattern obtained with two coherent laser beams, in which selective

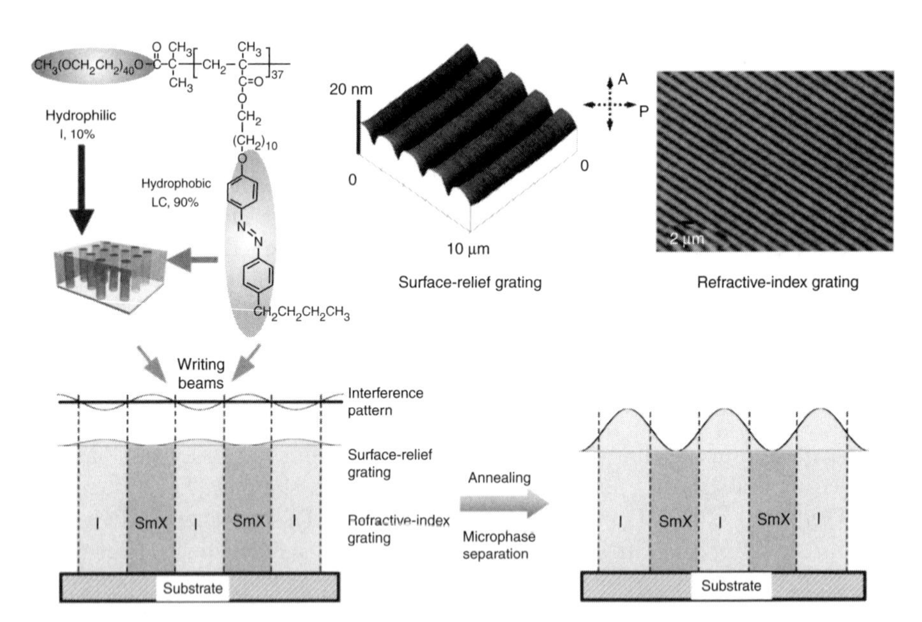

Figure 12.18. Holographic gratings and their enhancement by microphase separation upon annealing. An amphiphilic PEO-based azo BC was used as recording material. Both SRG and RIG were recorded at $\theta = 7^\circ$, leading to a fringe spacing of 2 μm. *Source*: Reproduced with modifications from Yu et al., 2005b.

TABLE 12.1. Grating Structures in Amphiphilic Azo LCBC Films

Interference pattern	Relief of SRG	Observation of POM	Phase in RIG
Low intensity	Peak	Dark stripe	Isotropic phase
High intensity	Valley	Bright stripe	LC

cis–trans photoisomerization of the azo moieties and the isotropic–LC phase transition and homogenous alignment of mesogens were induced in the bright areas of the interference pattern. The DE of the gratings depended strongly on the polarization of the reading beam because of the photoinduced alignment of azo mesogens. Table 12.1 gives schematic illustration of possible grating structures.

The LC alignment induced by the writing beams was monitored quantitatively by measuring the transmittance passing through the grating area of a film, placed between two crossed polarizers, as a function of the rotation angle. The angular-dependent transmittance exhibited a sinusoidal shape with a periodicity of 90°, indicating that the *trans*-azo mesogens obtained from the phase transition are homogenously aligned. The SRG structure can be clearly observed from the AFM images (Fig. 12.18), in which standard sinusoidal curves were obtained. The grating constant of the SRG was 2 μm, identical to that of RIG.

Then nanoscale microphase separation was induced thermally by annealing the grating samples at 100°C for 24 h without writing beams. The surface relief was increased to ~110 nm (18.3% of the film thickness), almost one order of magnitude larger than the value before annealing. The peak-to-valley contrast became more explicit after annealing because of the enhancement of the surface modulation. Furthermore, the sinusoidal shape of the surface profile became a little irregular, indicating that the LC alignment had been disturbed to a certain degree by the microphase separation. Together with the enhancement of SRG, the DE increased to ~9% after annealing, almost two orders of magnitude larger than the DE before annealing. This increase in DE may be ascribed mainly to the enhancement of surface modulation. The plausible grating structures before and after annealing are depicted in Fig. 12.18.

Compared with other methods to control DE, such as gain effects, mechanical stretch, electrical switch, self-assembly, mixture with LC, and cross-linking, the present microphase separation method has the advantage of being simple and convenient (Yu et al., 2009, 2005b). That is to say, holographic gratings can be inscribed at room temperature by two coherent laser beams of low intensity within a short time, and subsequent annealing improves the DE by almost two orders of magnitude. These holographic gratings with enhanced effect might be applied to secure information storage since the information can be easily read out by the thermally induced microphase separation.

To control precisely the DE of gratings recorded in films of amphiphilic azo LCBCs with microphase-separated nanostructures, the effect of the recording time on grating formation and enhancement was studied systematically. Both AFM and POM pictures exhibited a great change after annealing, as shown in Fig. 12.19. The drive of the SRG enhancement might be ascribed to the difference

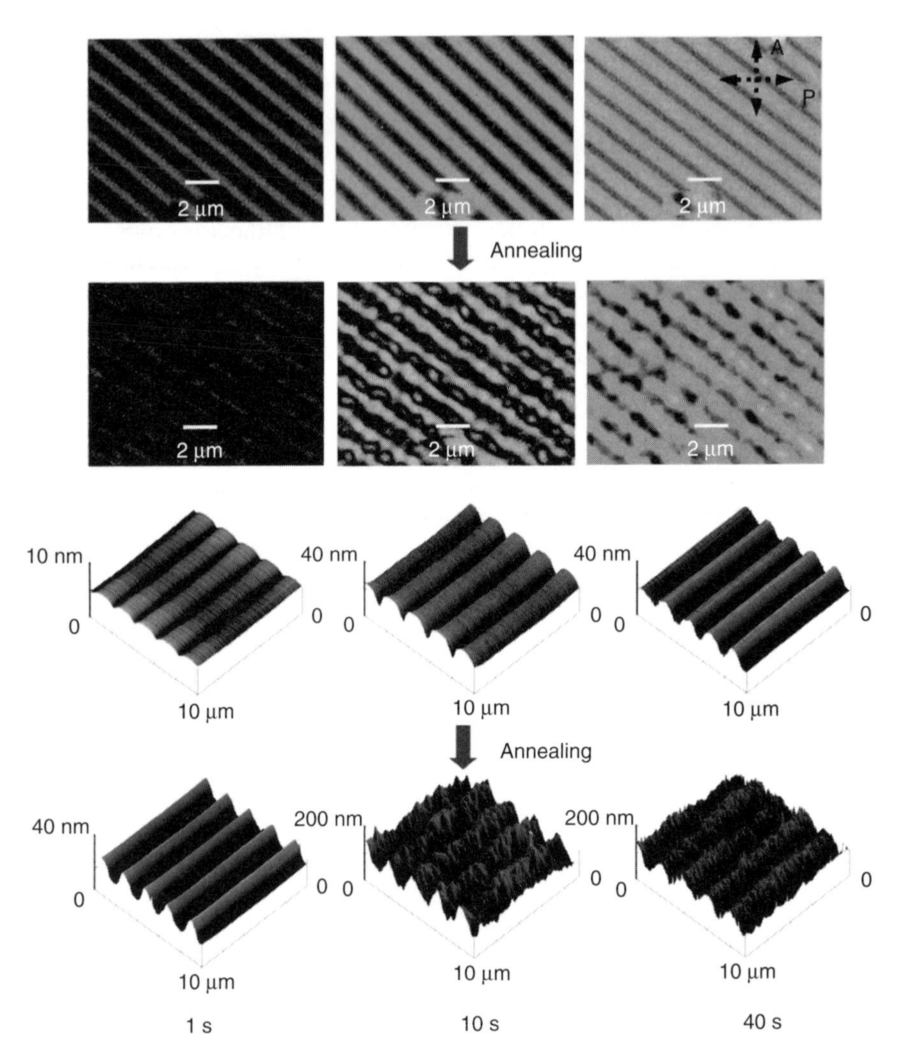

Figure 12.19. Effect of recording time on the enhancement of SRGs. The POM and AFM pictures of the samples were recorded at 1, 10, and 40 s, respectively. An interference pattern from two coherent *s*-polarized laser beams with equal intensity of 50 mW/cm^2 was used. A is the analyzer, and P, the polarizer.

in pressure (ΔP) between bright stripes and dark stripes (POM pictures of gratings in Fig. 12.19) produced by the microphase separation, which is different from other enhancement mechanisms. The PEO blocks have a better compatibility with the isotropic polymer blocks in the dark area than the LC polymer blocks in the bright area after grating recording. Thus, better microphase separation can be obtained in the bright stripes upon annealing, inducing a higher pressure in the bright stripes compared with the dark areas, leading to the uplift of the dark stripes (the peaks of the SRG). Furthermore, the molecular mobility of the azo LCBC can be improved in an isotropic phase, which enables the process of mass transport to form the surface relief enhancement. Obviously, the higher the ΔP induced by the microphase separation, the stronger the enhancement effect caused. As a result of the higher surface relief, a larger DE of the grating was obtained.

The enhancement coefficient with recording time was estimated. The best enhancement effect was obtained at 10-s recording upon microphase separation. By precisely adjusting the recording time, the DE can be finely controlled from 0.13% to 10% (Yu et al., 2005b). All the movements in the grating formation and enhancement are summarized in Table 12.2.

12.5.2. Enhancement of Refractive Index Modulation

By the cooperative effect between photoactive azo moieties and photoinert groups, a small external stimulus can induce a large change in the refractive index of the materials, a phenomenon that has been widely used in holographic recording. This is especially useful in bulk films of azo BCs with azo mesogens in the minority phase within glassy PMMA substrates; for instance, photoinduced mass transfer is greatly prohibited because of the microphase separation in the process of holographic recording, which led to lack of surface relief structures (Frenz et al., 2004). Thus, refractive index modulation plays an important role in the grating formation in this case.

As shown in Fig. 12.20, two holographic gratings were recorded in films of two PMMA-based azo BCs, one of which is a well-defined azo diblock copolymer; the other sample is a diblock random copolymer. Here the diblock random copolymer is formed by two blocks. One segment is well defined, and the other mesogenic block is statistically random. After grating formation, neither film showed a surface relief structure, and only RIGs were obtained. Upon irradiation of two coherent laser beams, RIG in the azo diblock copolymer was recorded by alignment of the azo moieties in microphase-separated domains. In contrast, the photoinduced alignment of the azo was amplified by the photoinert cyanobiphenyl moieties as a result of the cooperative effect in the diblock random copolymer (Yu, et al., 2008a). This resulted in a similar refractive index modulation as in the case of the left one in Fig. 12.20, although the azo content was lower in the azo BC. The cooperative motion was confined within the nanoscale phase domains, unlike the case of random copolymers (Wu et al., 1998).

12.5.3. Nanotemplates

Fabrication of well-arranged metal nanoparticle arrays by using nanocylinder structured template films is one of the important topics in nanotechnology. The size and the periodicity of the nanoparticle arrays can be controlled independently by choosing appropriate templates. This control has been achieved by using well-ordered nanotemplate films of PEO-based azo LCBCs (Li et al., 2007b). As shown in Fig. 12.21, a well-ordered array of Ag nanoparticles was fabricated successfully

TABLE 12.2. All the Movements in Grating Formation and Enhancement

Type of movement	Photoisomerization	H-aggregation	Phase transition	Phase separation	Mass transportation
Process	Cis to trans	Dipole function	I to SmX	LC and I	Whole molecule
Molecular scale	Single molecule	Several molecules	Many molecules	Nanoscale	Macroscale
Annealing effect	Yes	Yes	LC: Yes I: Yes	Yes	Yes
Kinds of gratings	RIG	No	RIG	No	SRG

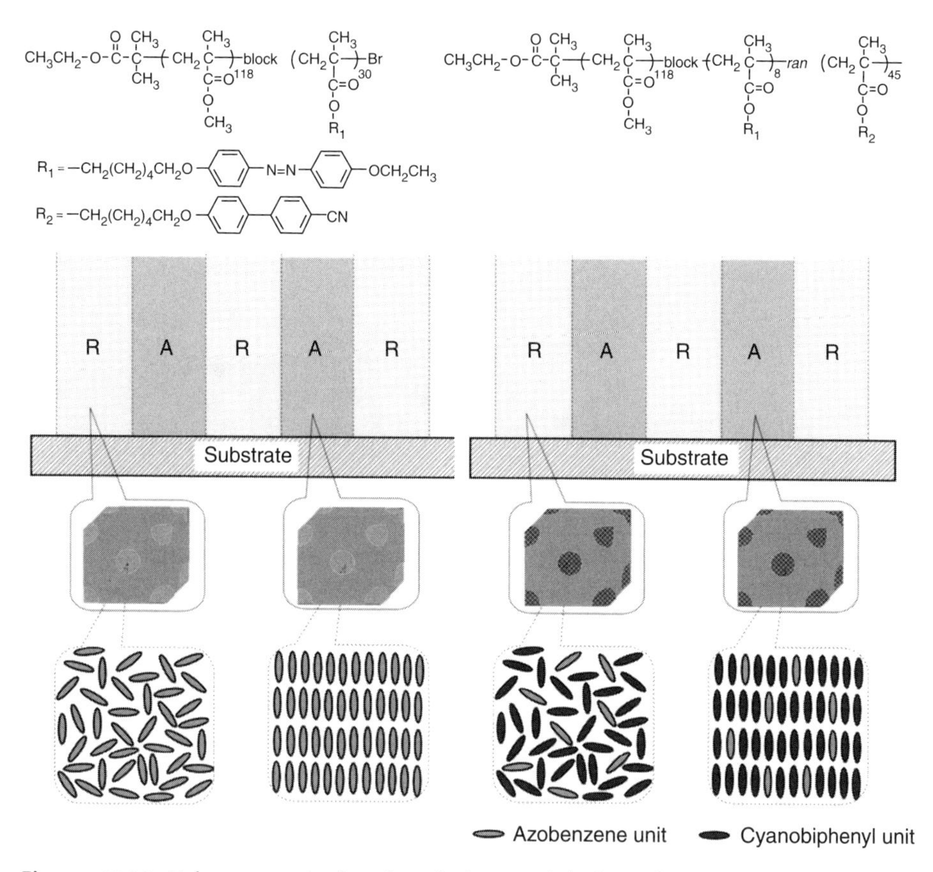

Figure 12.20. Enhancement of surface index modulation of azo BCs by cooperative effect. A is aligned, and R is random. *Source*: Reproduced from Yu et al., 2008a.

over a large area on soft or hard substrates (PET sheets, Si wafers, and quartz plates) via the selective Ag^+ doping of the hydrophilic PEO domains in a microphase-separated azo LCBC film and an associated vacuum ultraviolet UV (VUV) treatment to eliminate the LCBC templates to reduce simultaneously the Ag^+. Obviously, the periodicity of the highly dense Ag nanoparticles was precisely controlled by the nanotemplates of the azo LCBC films.

The simple and facile fabrication of a metal nanoparticle array can be used as a novel type of BC photolithography to overcome the size limitations of conventional top–down lithography, with the goal of macroscopically fabricating hierarchical nanopatterns with controlled ordering for potential applications ranging from photonics and plasmonics to metal wiring in molecular electronics (Suzuki et al., 2007). The self-assembly from azo LCBC film templates also provides a good method to modify the nanoscale shape of various kinds of functional materials, such as electric conducting RuO_2, magnetic Fe, or organic conducting polymers (Suzuki et al., 2007).

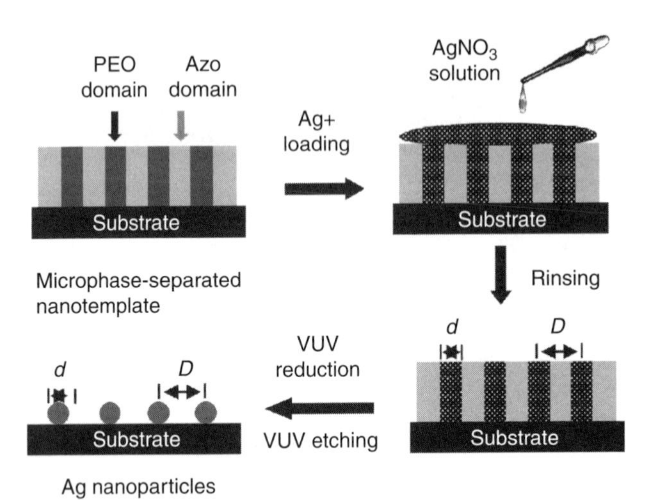

Figure 12.21. Fabrication of periodic array of Ag nanoparticles on either flexible or rigid substrates. The gray closed circles represent the photoreduced metallic Ag nanoparticles with regular particle size and periodicity. *Source*: Reproduced with modifications from Li et al., 2007b.

Figure 12.10 presents a PEO-based azo LCBC film, which exhibits well-ordered hydrophilic PEO nanocylinders with hexagonal packing embedded in an azo LC matrix. The anisotropic PEO nanocylinders can be used as ion-conductive channels since PEO has been widely used as a solid electrolyte (Fig. 12.22).

By incorporating $LiCF_3SO_3$ into the PEO nanocylinders, a supramolecular complexed structure and anisotropic ion transportation were achieved on the basis of the azo LCBC nanotemplate (Li et al., 2007a). Highly ordered ion-conducting PEO nanocylinder arrays with perpendicular orientation were formed by coordination between the lithium cations and the ether oxygens of the PEO blocks. At low- and medium salt concentration, selective complexation of $Li+$ with the PEO phase leads to the formation of an ordered array of ion-conducting PEO nanocylinders, which are perpendicular to the substrate surface, and on the right-hand side of Fig. 12.22, there is a 3-D illustration of the corresponding phase-segregated structure, in which anisotropic ion transport is observed. At high salt concentration, the lithium salt is dissolved in both PEO and azo domains. This decreases LC ordering and disturbs microphase separation. An array of tilted and distorted nanocylinders is formed with poor regularity. Consequently, the anisotropic value of ion conductivity is reduced (Breiner, 2007).

On the surface of PEO-based azo LCBC films, each hydrophilic PEO domain appears as a circular hollow surrounded by the hydrophobic azo matrix. These amphiphilic properties enable the selective absorption of Au nanoparticles with hydrophilic or hydrophobic surface modifications. Of course, the surface properties of the gold nanoparticles are a critical factor in this nanofabrication (Watanabe et al., 2007). To extend site coverage further and favor high selectivity,

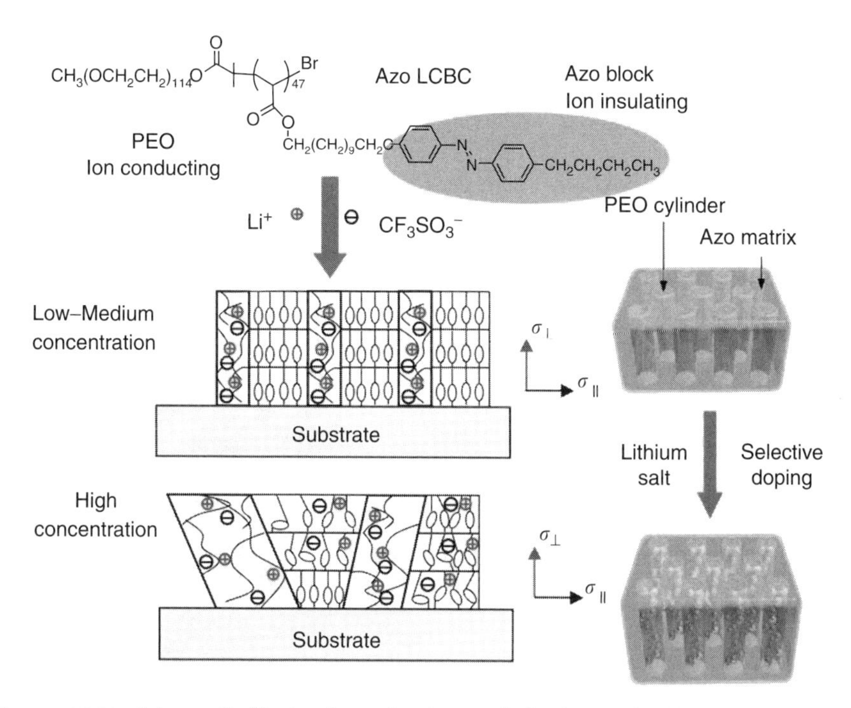

Figure 12.22. Schematic illustration of anisotropic ionic conduction in nanochannels with nanotemplates of azo LCBC films. The complexes $PEO+LiCF_3SO_3$ were prepared at low-, medium-, and high salt concentration. *Source*: Reproduced from Li et al., 2007a.

the gold nanoparticles should be modified by additional functional ligands, which provide electrostatic and hydrogen bond interactions. As shown in Fig. 12.23, site-specific recognition of gold nanoparticles was obtained in the nanocylinder domains of PEO blocks or the azo continuous domains. The on-site coverage and selective effects of the surface properties, the concentration of gold nanoparticles, the dipping time, and the composition of the dispersion medium were investigated by a simple dip coating method. Then the ordering of the gold nanoparticles from the template was transferred to the substrate by the inexpensive VUV approach, which was effective in removing the templates without destroying the regularity of the assembled nanoparticles. Such facile transfer is characteristic of soft (organic) templates compared with hard (inorganic) templates.

The sol–gel process can also be incorporated with azo BC film lithography. A hexagonally ordered SiO_2 nanorod array with mesochannels aligned along the longitudinal axes was obtained for the first time, as shown in Fig. 12.24 (Chen et al., 2008). The mesochannels inside the SiO_2 nanorods were aligned perpendicularly to the substrate and had a diameter of ~ 2 nm. The aspect ratio of the SiO_2 nanorods was controlled by the immersion time and the film thickness. A height of several hundreds of nanometers can be achieved. This hierarchically ordered

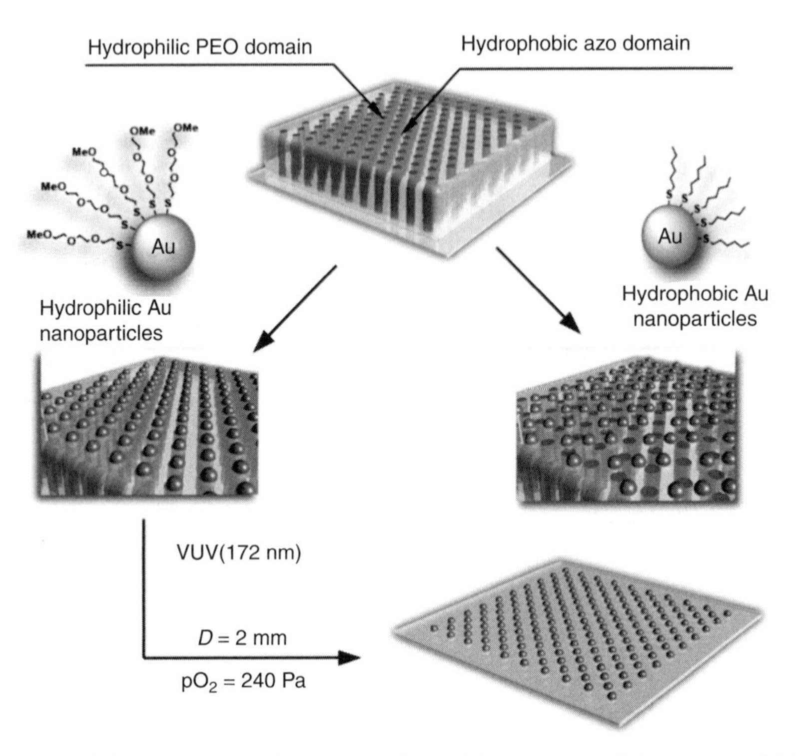

Figure 12.23. Scheme of selective absorption of Au nanoparticles on amphiphilic surfaces of azo LCBC films with well-ordered microphase separation. D is the diameter of irradiated area. *Source*: Reproduced with modifications from Watanabe et al., 2007.

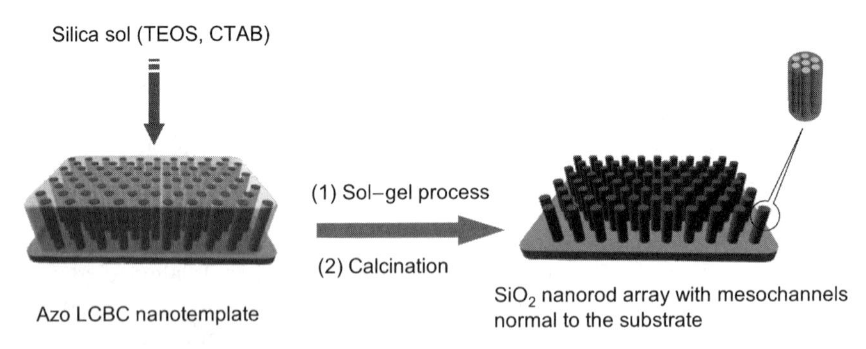

Figure 12.24. Potential representation of the preparation of SiO_2 nanorod arrays with mesochannels normal to the substrate, combining the sol–gel process with azo LCBC film lithography.

mesoporous silica materials have potential applications as innovative materials for waveguides, lasers, and biomacromolecular separation systems.

The PEO-based azo LCBC films with well-ordered microphase separation show excellent reproducibility and mass production through molecular or supramolecular self-assembly. Both parallel and perpendicular patterns of nanostructures with high regularity can be precisely manipulated, which guarantees the nanotemplate-based nanofabrication processes and results in diverse self-assembled nanostructures, leading to industrial use in plastics engineering.

12.5.4. Volume Storage

In recent years, advanced recording media with fast data transfer and high density have been developed. Optical holography provides unique opportunities for the next-generation storage technique, as shown in Fig. 12.25. The basis of the storage process is a photoinduced refractive index change in materials bearing chromophores. Desirable organic materials for holographic storage should exhibit high DE, fast response, high resolution, stable and reversible storage, low energy consumption during the recording and reading processes, as well as easy mass production. Azo-containing LC or amorphous polymers, photorefractive polymers, polymer-dispersed LC, and LC or glassy oligomers have been extensively investigated. But none of them meets all the aforementioned requirements.

To increase storage density, Bragg-type gratings are often required. Generally, thick films ($>100\,\mu m$) with low absorption and no scattering of visible light are ideal for volume holograms. But this cannot be applied directly in Bragg gratings, since it is difficult for visible light to pass through azo-containing thick films (e.g., $100\,\mu m$) because of the large molar extinction coefficient of azo chromophores ($\sim 10^4\,L\,mol^{-1}\,cm^{-1}$). One of the approaches related to solve this problem is to use the cooperative effect (Minabe et al., 2004). Ikeda and coworkers reported the formation of holographic gratings in a thick film by modulation of the refractive index ($\Delta n'$), which was induced by the orientation of azo moieties and mesogens such as cyanobiphenyl or tolane moieties (e.g., Ishiguro et al., 2007). The random copolymers showed a maximum DE (η_{max}) of 97% in the Bragg regime. Although

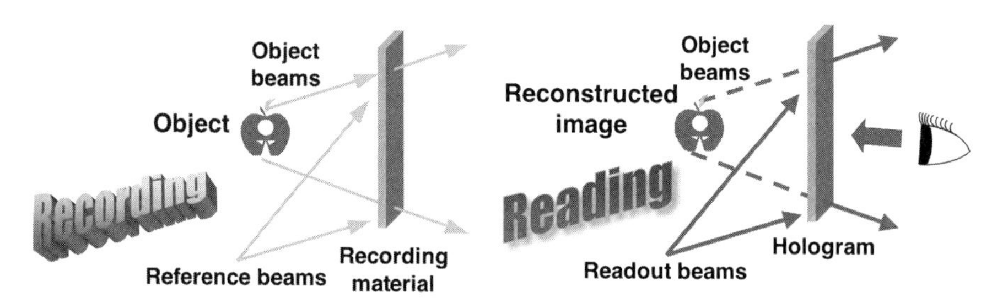

Figure 12.25. Schematic illustrations of recording and reading processes in holographic storage.

55 holograms with angular multiplicity were successfully recorded (Saishoji et al., 2007), the scattering could not be completely avoided.

Recently, particular interest has been paid to the microphase separation of azo BC films, because the azo chromophores can be constrained in one microphase-separated domain on the nanoscale by the microphase separation, thus eliminating the scattering of visible light, while maintaining the photoinduced change in refractive index (Yu et al., 2007). Schmidt and coworkers first used this concept for holographic data storage in PS-based azo diblock random copolymers (Häckel et al., 2005). In the minority phase containing the mesogenic segments, the azo and benzoylbiphenyl mesogenic side groups were in a statistical distribution. Thus, it was possible to decrease the overall optical density, which plays an important role in the grating recording, while increasing the local refractive index difference between illuminated and unirradiated volume elements and improving the stability of the orientation. Furthermore, they prepared blends of the azo BCs and PS homopolymers and were able to obtain thick transparent films (1.1 mm) (Häckel et al., 2007). On the basis of these films, they performed angular multiplexing of 80 holograms at the same spatial position, which showed a long-term stability at room temperature.

12.5.5. Other Applications

Other than the optoelectronics and nanotechnology, the azo BCs have been widely explored in other fields. Zhao and coworkers combined optical and mechanical properties by designing thermoplastic elastomers using azo ABA-type BCs, with which they were able to prepare freestanding films by simple casting (Cui et al., 2004). Recently, azo BCs with a low content of azo were used as surface-modifying agents of polymer substrates to allow dynamic control over surface properties, such as water wettability (Yoshida, 2005). Other applications of ABA-type LC triblock copolymers with lamellar structures are envisaged in the creation of artificial muscle (Li et al., 2004); azo BCs with well-defined structures can be utilized as photomobile materials (e.g., Yu et al., 2003; Ikeda et al., 2007), which is under way in Ikeda laboratory in Tokyo Institute of Technology.

12.6. OUTLOOK

The microphase-separated nanostructures in bulk films of well-defined azo BCs have fascinated one to understand the correlation between their ordered structures and the polymer structures with optical control properties from photosensitive azo blocks. One of the most exquisite advantages of introducing azo moieties into well-defined BCs is precise control of the supramolecular structures in BC films. With the development of information technology, new waves are surging, driving such self-assembled nanostructures leading to industrial use as future engineering plastics for optoelectronics and nanotechnology. Expected as one of the powerful counterparts of top–down-type nanofabrication, the central focus should be placed on their high reproducibility and mass production as well as precise manipulation of these ordered nanostructures.

It is not enough to obtain macroscopically ordered nanostructures only through self-assembling nanofabrication processes, and the introduction of azo blocks gives a good chance to adjust and obtain reliable nanotemplate for advanced applications. In azo BCs, azo shows different behaviors to the nanostructures depending on its volume ratio. When the azo block forms a substrate in films, its alignment or disordered states have great influence on nanostructures, like an object in a hand shown in Fig. 12.26. On the contrary, the azo block in the minority phase cannot change the reverse majority phase, just as a big tree cannot be moved by one hand. But the azo mesogens can be photoaligned even in a nanoscale space, which completely eliminates the scattering of visible light and demonstrates the optical applications of azo BCs.

According to SMCM, more complicated nanostructures in macroscopic ordering might be fabricated by combining self-organization and photocontrol of the azo group in the azo domains. In addition to in-plane and out-of-plane alignment, circular anisotropy of the azo chromophores can be induced by irradiation with circularly polarized light (Choi et al., 2007), which might endow microphase-separated nanostructures with novel properties such as supramolecular chirality for optical-switching, optical-storage, and light-driven devices. Besides, the optical processing of azo has been carried out by interference pattern from subwavelength to micrometer scale (e.g., Yu et al., 2008b), which might introduce hierarchical structures into azo BC films. Although the research on azo

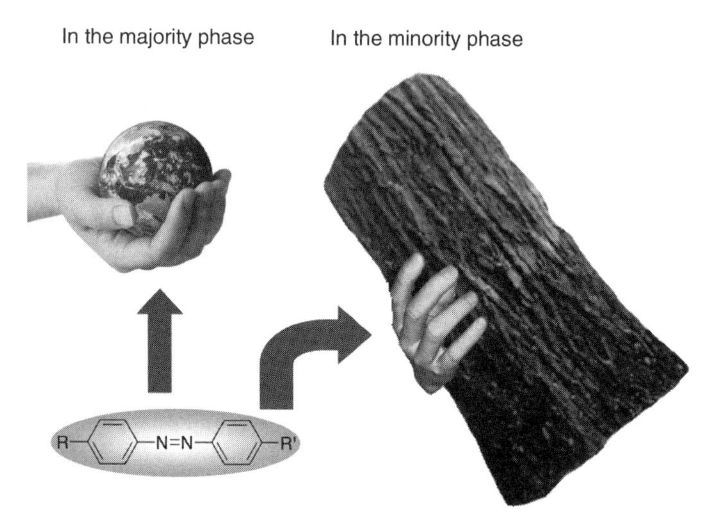

Figure 12.26. Cartoons illustrating azo functions in BC films with the azo block in the majority phase or the minority phase. (The azo chromophore in azo BCs can be used as a controller in the role of a hand.)

BCs is still at a primary stage, many groups are involved in this novel field of azo materials, which would broaden our knowledge of upsurging functional BCs and push ahead to find their diverse applications in optoelectronics, information storage, nanotechnology, as well as biotechnology.

REFERENCES

Adams J, Gronski W. 1989. LC side chain AB-block copolymers with an amorphous A-block and a liquid-crystalline B-block. Macromol Rapid Commun 10:553–557.

Bates FS, Fredrichson GH. 1999. Block copolymer-designer soft materials. Phys Today 52(2):32–38.

Berreman DW. 1972. Solid surface shape and the alignment of an adjacent nematic liquid crystal. Phys Rev Lett 28:1683–1686.

Breiner T, Kreger K, Hagen R, Hackel M, Kador L, Muller AHE, Kramer EJ, Schmidt H-W. 2007. Blend of poly(methacrylate) block copolymers with photo addressable segments. Macromolecules 40:2100–2108.

Chao C, Li X, Ober C, Osuji C, Thomas EL. 2004. Orientational switching of mesogens and microdomains in hydrogen-bonded side-chain liquid-crystalline block copolymers using AC electric fields. Adv Funct Mater 14:364–370.

Chen A, Komura M, Kamata K, Iyoda T. 2008. Highly ordered arrays of mesoporous silica nanorods with tunable aspect ratios from block copolymer thin films. Adv Mater 20:763–767.

Cheng JY, Mayers AM, Ross CA. 2004. Nanostructure engineering by templated self-assembly of block copolymers. Nat Mater 3:823–828.

Cheng JY, Ross CA, Smith HI, Thomas EL. 2006. Templated self-assembly of block copolymers: top-down helps bottom-up. Adv Mater 18:2505–2521.

Choi SW, Kawauchi S, Ha NY, Takezoe H. 2007. Photoinduced chirality in azobenzene-containing polymer systems. Phys Chem Chem Phys 9:3671–3681.

Coessensm V, Pintauer T, Matyjaszewski K. 2001. Functional polymers by atom transfer radical polymerization. Prog Polym Sci 26:337–377.

Cui L, Dahmane S, Tong X, Zhu L, Zhao Y. 2005. Using self-assembly to prepare multifunctional diblock copolymers containing azopyridine moiety. Macromolecules 38:2076–2084.

Cui L, Tong X, Yan X, Liu G, Zhao Y. 2004. Photoactive thermoplastic elastomers of azobenzene-containing triblock copolymers prepared through atom transfer radical polymerization. Macromolecules 37:7097–7104.

Cui L, Zhao Y, Yavrian A, Galstian T. 2003. Synthesis of azobenzene-containing diblock copolymers using atom transfer radical polymerization and the photoalignment behavior. Macromolecules 36:8246–8252.

Darling SB. 2007. Directing the self-assembly of block copolymers. Prog Polym Sci 32:1152–1204.

Demus D, Goodbye J, Gray GW, Spiess HW, Vill V. 1998. Handbook of Liquid Crystals. Weinheim, Germany: Wiley-VCH Verlag.

Discher DE, Eisenberg A. 2002. Polymer vesicles. Science 297:967–974.

Fasolka MJ, Mayes AM. 2001. Block copolymer thin films: physics and applications. Annu Rev Mater Res 31:323–355.

Finkelmann H, Bohnert R. 1994. Liquid-crystalline side-chain AB block copolymers by direct anionic polymerization of a mesogenic methacrylate. Macromol Chem Phys 195:689–700.

Finkelmann H, Walther M. 1996. Structure formation of liquid crystalline block copolymers. Prog Polym Sci 21:951–979.

Forcen P, Oriol L, Sanchez C, Alcala R, Hvilsted S, Jankova K, Loos J. 2007. Synthesis, characterization and photoinduction of optical anisotropy in liquid crystalline diblock azo-copolymers. J Polym Sci Part A: Polym Chem 45:1899–1910.

Frenz C, Fuchs A, Schmidt HW, Theisscn U, Haarer D. 2004. Diblock copolymers with azobenzene side-groups and polystyrene matrix: synthesis, characterization and photoaddressing. Macromol Chem Phys 205:1246–1258.

Hamlcy IW. 2003. Nanostructure fabrication using block copolymers. Nanotechnology 14:R39–R54.

Hamley IW, Castelletto V, Lu Z, Imrie C, Itoh T, Al-Hussein M. 2004. Interplay between smectic ordering and microphase separation in a series of side-group liquid-crystal block copolymers. Macromolecules 37:4798–4807.

Han YK, Dufour B, Wu W, Kowalewski T, Matyjaszewski K. 2004. Synthesis and characterization of new liquid-crystalline block copolymers with p-cyanoazobenzene moieties and poly(n-butyl acrylate) segments using atom-transfer radical polymerization. Macromolecules 37:9355–9365.

Hayakawa T, Horiuchi S, Shimizu H, Kawazoe T, Ohtsu M. 2002. Synthesis and characterization of polystyrene-b-poly(1,2-isoprene-ran-3,4-isoprene) block copolymers with azobenzene side groups. J Polym Sci Part A: Polym Chem 40:2406–2414.

Häckel M, Kador L, Kropp D, Schmidt H. 2007. Polymer blends with azobenzene-containing block copolymers as stable rewritable volume holographic media. Adv Mater 19:227–231.

Häckel M, Kador L, Kropp D, Frenz C, Schmidt H. 2005. Holographic gratings in diblock copolymers with azobenzene and mesogenic side groups in the photoaddressable dispersed phase. Adv Funct Mater 15:1722–1727.

He X, Zhang H, Yan D, Wang X. 2003. Synthesis of side-chain liquid-crystalline homopolymers and triblock copolymers with p-methoxyazobenzene moieties and poly(ethylene glycol) as coil segments by atom transfer radical polymerization and their thermotropic phase behavior. J Polym Sci Part A: Polym Chem 41:2854–2864.

Huang E, Rockford L, Russell TP, Hawker CJ. 1998. Nanodomain control in copolymer thin films. Nature 395:757–758.

Ichimura K. 2000. Photoalignment of liquid-crystal systems. Chem Rev 100:1847–1873.

Ikeda T. 2003a. Photomodulation of liquid crystal orientations for photonic applications. J Mater Chem 13:2037–2057.

Ikeda T, Mamiya J, Yu Y. 2007. Photomechanics of liquid-crystalline elastomers and other polymers. Angew Chem Int Ed Engl 46:506–528.

Ikeda T, Nakano M, Yu Y, Tsutsumi O, Kanazawa A. 2003b. Anisotropic bending and unbending behavior of azobenzene liquid-crystalline gels by light. Adv Mater 15:201–205.

Ikeda T, Tsutsumi O. 1995. Optical switching and image storage by means of azobenzene liquid-crystal films. Science 268:1873–1875.

Ishiguro M, Sato D, Shishido A, Ikeda T. 2007. Bragg-type polarization gratings formed in thick polymer films containing azobenzene and tolane moieties. Langmuir 23:332–338.

Kamata K, Watanabe R, Iyoda T. 2004. Orientation of hydrophilic nanocylinder in amphiphilic block copolymer film. 40th IUPAC World Polymer Congress, July 4, Paris, France.

Kim DY, Tripathy SK, Li L, Kumar J. 1995. Laser-induced holographic surface relief gratings on nonlinear optical polymer films. Appl Phys Lett 66:1166–1168.

Kim S, Solak H, Stoykovich M, Ferrier N, Pablo J, Nealey P. 2003. Epitaxial self-assembly of block copolymers on lithographically defined nanopatterned substrates. Nature 424:411–414.

Komura M, Iyoda T. 2007. AFM cross-sectional imaging of perpendicularly oriented nanocylinder structures of microphase-separated block copolymer films by crystal-like cleavage. Macromolecules 40:4106–4108.

Kramer EJ, Schmidt HW. 2007. Blends of poly(methacrylate) block copolymers with photoaddressable segments. Macromolecules 40:2100–2108.

Kumar GS, Neckers DC. 1989. Photochemistry of azobenzene-containing polymers. Chem Rev 89:1915–1925.

Lehmann O, Forster S, Springer J. 2000. Synthesis of new side-group liquid crystalline block copolymers by living anionic polymerization. Macromol Rapid Commun 21:133–135.

Li J, Kamata K, Komura M, Yamada T, Yoshida H, Iyoda T. 2007a. Anisotropic ion conductivity in liquid crystalline diblock copolymer membranes with perpendicularly oriented PEO cylindrical domains. Macromolecules 40:8125–8128.

Li J, Kamata K, Watanabe S, Iyoda T. 2007b. Template- and vacuum-ultraviolet-assisted fabrication of a Ag-nanoparticle array on flexible and rigid substrates. Adv Mater 19:1267–1271.

Li MH, Keller P, Yang J, Albouy PA. 2004. Artificial muscle with lamellar structure based on a nematic triblock copolymer. Adv Mater 16:1922–1925.

Lin S, Lin J, Nose T, Iyoda T. 2007. Micellar structures of block-copolymers with ordered cores in dilute solution as studied by polarized and depolarized light scattering. J Polym Sci Part B: Polym Phys 45:1333–1343.

Makinen R, Ruokolainen J, Ikkala O, Moel K, Brinke G, Odorico W, Stamm M. 2000. Orientation of supramolecular self-organized polymeric nanostructures by oscillatory shear flow. Macromolecules 33:3441–3446.

Mao G, Ober CK. 1996. Block copolymers containing liquid crystalline segments. Acta Polym 48:405–422.

Mao G, Wang J, Clingman S, Ober C, Chen J, Thomas EL. 1997. Molecular design, synthesis, and characterization of liquid crystal-coil diblock copolymers with azobenzene side groups. Macromolecules 30:2556–2567.

Matyjaszewski K, Xia JH. 2001. Atom transfer radical polymerization. Chem Rev 101:2921–2990.

Minabe J, Maruyama T, Yasuda S, Kawano K, Hayashi K, Ogasawara Y. 2004. Design of dye concentrations in azobenzene-containing polymer films for volume holographic storage. Jpn J Appl Phys 43:4964–4967.

Morikawa Y, Kondo T, Nagano S, Seki T. 2007. Photoinduced 3D ordering and patterning of microphase-separated nanostructure in polystyrene-based block copolymer. Chem Mater 19:1540–1542.

Morikawa Y, Nagano S, Watanabe K, Kamata K, Iyoda T, Seki T. 2006. Optical alignment and patterning of nanoscale microdomains in a block copolymer thin film. Adv Mater 18:883–886.

Morkved T, Lu M, Urbas A, Ehrichs E, Jaeger H, Mansky P, Russell T. 1996. Local control of microdomain orientation in diblock copolymer thin films with electric fields. Science 273:931–933.

Naka Y, Yu HF, Shishido A, Ikeda T. 2009. Photoresponsive and holographic behavior of an azobenzene-containing block copolymer and a random copolymer. Mol Cryst Liq Cryst 498:118–130.

Natansohn A, Rochon P. 2002. Photoinduced motions in azo-containing polymers. Chem Rev 102:4139–4175.

Osuji C, Chen JT, Mao G, Ober CK, Thomas EL. 2000. Understanding and controlling the morphology of styrene–isoprene side-group liquid crystalline diblock copolymers. Polymer 41:8897–8907.

Osuji C, Ferreira P, Mao G, Ober C, Vander J, Thomas EL. 2004. Alignment of self-assembled hierarchical microstructure in liquid crystalline diblock copolymers using high magnetic fields. Macromolecules 37:9903–9908.

Osuji C, Zhang Y, Mao G, Ober CK, Thomas EL. 1999. Transverse cylindrical microdomain orientation in an LC diblock copolymer under oscillatory shear. Macromolecules 32:7703–7706.

Paik MY, Krishnan S, You F, Li X, Hexemer A, Ando Y, Kang SH, Fischer DA, Kramer EJ, Ober CK. 2007. Surface organization, light-driven surface changes, and stability of semifluorinated azobenzene polymers. Langmuir 23:5110–5119.

Qi B, Yavrian A, Galstian T, Zhao Y. 2005. Liquid crystalline ionomers containing azobenzene mesogens: phase stability, photoinduced birefringence and holographic grating. Macromolecules 38:3079–3086.

Ravi P, Sin S, Gan L, Gan Y, Tam K, Xia X, Hu X. 2005. New water soluble azobenzene-containing diblock copolymers: synthesis and aggregation behavior. Polymer 46: 137–146.

Reiter G, Gastelein G, Hoerner P, Riess G, Blumen A, Sommer J. 1999. Nanometer-scale surface patterns with long-range order created by crystallization of diblock copolymers. Phys Rev Lett 83:3844–3847.

Rochon P, Batalla E, Natanhson A. 1995. Optically induced surface gratings on azoaromatic polymer films. Appl Phys Lett 66:136–138.

Rosa CD, Park C, Thomas EL, Lotz B. 2003. Microdomain patterns from directional eutectic solidification and epitaxy. Nature 405:433–437.

Saishoji A, Sato D, Shishido A, Ikeda T. 2007. Formation of Bragg gratings with large angular multiplicity by means of photoinduced reorientation of azobenzene copolymers. Langmuir 23:320–326.

Seki T. 2004. Dynamic photoresponsive functions in organized layer systems comprised of azobenzene-containing polymers. Polym J 36:435–454.

Serhatli IE, Serhatli M. 1998. Synthesis and characterization of amorphous-liquid crystalline poly(vinyl ether) block copolymers. Turk J Chem 22:279–287.

Sidorenko A, Tokarev I, Minko S, Stamm M. 2003. Ordered reactive nanomembranes/nanotemplates from thin films of block copolymer supramolecular assembly. J Am Chem Soc 125:12211–12216.

Suzuki S, Kamata K, Yamauchi H, Iyoda T. 2007. Selective doping of lead ions into normally aligned PEO cylindrical nanodomains in amphiphilic block copolymer thin films. Chem Lett 36:978–979.

Tang X, Gao L, Fan X, Zhou QF. 2007. ABA-type amphiphilic triblock copolymers containing p-ethoxy azobenzene via atom transfer radical polymerization: synthesis, characterization, and properties. J Polym Sci Part A: Polym Chem 45:2225–2234.

Thomas EL, Lescanec RL. 1994. Phase morphology in block copolymer systems. Phil Trans R Soc Lond A 348:149–166.

Tian Y, Watanabe K, Kong X, Abe J, Iyoda T. 2002. Synthesis, nanostructures, and functionality of amphiphilic liquid crystalline block copolymers with azobenzene moieties. Macromolecules 35:3739–3747.

Tokita M, Adachi M, Masuyama S, Takazawa F, Watanabe J. 2007. Characteristic shear-flow orientation in LC block copolymer resulting from compromise between orientations of microcylinder and LC mesogen. Macromolecules 40:7276–7282.

Tokita M, Adachi M, Takazawa F, Watanabe J. 2006. Shear flow orientation of cylindrical microdomain in liquid crystalline diblock copolymer and its potentiality as anchoring substrate for nematic mesogens. Jpn J Appl Phys 45:9152–9156.

Tomikawa N, Lu Z, Itoh T, Imre CT, Adachi M, Tokita M, Watanabe J. 2005. Orientation of microphase-segregated cylinders in liquid crystalline diblock copolymer by magnetic field. Jpn J Appl Phys 44:L711–L714.

Tong X, Cui L, Zhao Y. 2004. Confinement effects on photoalignment, photochemical phase transition and thermochromic behavior of liquid crystalline azobenzene-containing diblock copolymers. Macromolecules 37:3101–3112.

Viswanathan N, Kim D, Bian S, Williams J, Liu W, Li L, Samuelson L, Kumar J, Tripathy S. 1999. Surface relief structures on azo polymer films. J Mater Chem 9:1941–1955.

Wang D, Ye G, Wang X. 2007. Synthesis of aminoazobenzene-containing diblock copolymer and photoinduced deformation behavior of its micelle-like aggregates. Macromol Rapid Commun 28:2237–2243.

Wang X, Chen J, Marturunkakul S, Li L, Kumar J, Tripathy S. 1997a. Epoxy-based nonlinear optical polymers functionalized with tricyanovinyl chromophores. Chem Mater 9:45–50.

Wang X, Kumar J, Tripathy S, Li L, Chen J, Marturunkakul S. 1997b. Epoxy-based nonlinear optical polymers from post azo coupling reaction. Macromolecules 30:219–225.

Watanabe K, Yoshida H, Kamata K, Iyoda T. 2005. Direct TEM observation of perpendicularly oriented nanocylinder structure in amphiphilic liquid crystalline block copolymer thin films. Trans Mater Res Soc Jpn 30:377–381.

Watanabe S, Fujiwara R, Hada M, Okazaki Y, Iyoda T. 2007. Site-specific recognition of nanophase-separated surfaces of amphiphilic block copolymers by hydrophilic and hydrophobic gold nanoparticles. Angew Chem Int Ed Engl 46: 1120–1123.

Whitesides GM, Grzybowski B. 2002. Self-assembly at all scales. Science 295:2418–2418.

Wu Y, Demachi Y, Tsutsumi O, Kanazawa A, Shiono T, Ikeda T. 1998a. Photoinduced alignment of polymer liquid crystals containing azobenzene moieties in the side chain. 1. Effect of light intensity on alignment behavior. Macromolecules 31:349–354.

Wu Y, Demachi Y, Tsutsumi O, Kanazawa A, Shiono T, Ikeda T. 1998b. Photoinduced alignment of polymer liquid crystals containing azobenzene moieties in the side chain. 3. Effect of structure of photochromic moieties on alignment behavior. Macromolecules 31:4457–4463.

Wu Y, Zhang Q, Ikeda T. 1999. Three-dimensional manipulation of azo polymer liquid crystal by unpolarized light. Adv Mater 11:300–302.

Yoshida E, Ohta M. 2005. Preparation of micelles with azobenzene at their coronas or cores from nonamphiphilic diblock copolymers. Colloid Polym Sci 283:521–531.

Yoshida T, Doi M, Kanaoka S, Aoshima S. 2005. Polymer surface modification using diblock copolymers containing azobenzene. J Polym Sci Part A: Polym Chem 43:5704–5709.

Yu HF, Asaoka A, Shishido A, Iyoda T, Ikeda T. 2007a. Photoinduced nanoscale cooperative motion in a novel well-defined triblock copolymer. Small 3:768–771.

Yu HF, Iyoda T, Ikeda T. 2006a. Photoinduced alignment of nanocylinders by supramolecular cooperative motions. J Am Chem Soc 128:11010–11011.

Yu HF, Iyoda T, Okano K, Shishido A, Ikeda T. 2005a. Photoresponsive behavior and photochemical phase transition of amphiphilic diblock liquid-crystalline copolymer. Mol Cryst Liq Cryst 443:191–199.

Yu HF, Li J, Ikeda T, Iyoda T. 2006b. Macroscopic parallel nanocylinder array fabrication using a simple rubbing technique. Adv Mater 18:2213–2215.

Yu HF, Naka Y, Shishido A, Ikeda T. 2008a. Well-defined liquid-crystalline diblock copolymers with an azobenzene moiety: synthesis, photoinduced alignment and their holographic properties Macromolecules 41:7959–7966.

Yu HF, Okano K, Shishido A, Ikeda T, Kamata K, Komura M, Iyoda T. 2005b. Enhancement of surface relief gratings recorded in amphiphilic liquid-crystalline diblock copolymer by nanoscale phase separation. Adv Mater 17:2184–2188.

Yu HF, Shishido A, Ikeda T. 2008b. Subwavelength modulation of surface relief and refractive index in pre-irradiated liquid-crystalline polymer films. Appl Phys Lett 92:103117/1–103117/3.

Yu HF, Shishido A, Ikeda T, Iyoda T. 2005c. Novel amphiphilic diblock and triblock liquid-crystalline copolymers with well-defined structures prepared by atom transfer radical polymerization. Macromol Rapid Commun 26:1594–1598.

Yu HF, Shishido A, Iyoda T, Ikeda T. 2007b. Novel wormlike nanostructure self-assembled in a well-defined liquid-crystalline diblock copolymer with azobenzene moieties. Macromol Rapid Commun 28:927–931.

Yu HF, Shishido A, Iyoda T, Ikeda T. 2009. Effect of recording time on grating formation and enhancement in an amphiphilic diblock liquid-crystalline copolymer. Mol Cryst Liq Cryst 498:29–39.

Yu Y, Nakano M, Ikeda T. 2003. Directed bending of a polymer film by light. Nature 425:145–145.

Zhang Y, Cheng Z, Chen X, Zhang W, Wu J, Zhu J, Zhu X. 2007. Synthesis and photoresponsive behaviors of well-defined azobenzene-containing polymers via RAFT polymerization. Macromolecules 40:4809–4817.

13

PHOTORESPONSIVE HYBRID SILICA MATERIALS CONTAINING AZOBENZENE LIGANDS

Nanguo Liu and C. Jeffrey Brinker

13.1. INTRODUCTION

In general, the definition of stimuli-responsive materials is that material properties (color, volume, adsorption, refractive index, stress, etc.) undergo predictable changes in response to external stimuli, such as light, heat, and pH. Because of their "intelligent" response, such materials are of potential interest for applications in microfluidics, microvalves, controlled drug release, sensors, optical switches, photomemories, light-driven displays, optical storage, and optomechanical actuation. Azobenzene-containing materials are some of the most commonly used stimuli-responsive materials because of their well-known responses to light and heat (Natansohn, 1999; Rau, 1990b; Kumar and Neckers, 1989). Ultraviolet (UV) irradiation of the trans isomer causes its transformation to the cis isomer. Removal of the UV light, heating, or irradiation with a longer wavelength light switches the system back to the trans form. Trans$\leftrightarrow$cis isomerization of azobenzene moieties changes not only the molecular dimension, but also the dipole moment (0.5–3.1 Debye).

Most of the work on photoresponsive azobenzene-containing materials is based on polymer matrices. Numerous chemistries have been utilized to graft azobenzene ligands to various polymer chains. The azobenzene chromophores transfer light energy into conformational changes upon photoirradiation, which can be used to control chemical and physical properties of the materials, such as viscosity, conductivity, pH, solubility, wettability, permeability, transport properties, mechanical properties, and structural properties.

Azobenzene ligands have also been incorporated into inorganic silica matrices. The robust inorganic silica framework does have some advantages in terms

Smart Light-Responsive Materials. Edited by Yue Zhao and Tomiki Ikeda

of thermal and mechanical stabilities, easy processing, and easy integration into devices. Preparation of the photoresponsive hybrid silica materials requires synthesis of azobenzene-containing organosilanes because they are not commercially available. The azobenzene-containing organosilanes were represented by the formula R-Si(OR′)$_3$ in which R contains an azobenzene ligand and R′ is an ethyl or methyl group. After sol–gel hydrolysis and condensation of the azobenzene-containing organodilanes with other silica precursors, the azobenzene-containing R groups are covalently bonded to the silica matrix. Other than the conventional sol–gel chemistries toward polysiloxane gels, cohydrolysis and cocondensation of organosilanes and alkyl silicate in the presence of acid or base catalysts and amphiphilic surfactant or block copolymer templates lead to self-assembled highly ordered hexagonal, cubic, or lamellar nanostructures. The nanostructures are controlled by several factors, such as the concentration of silicon precursors and catalyst, amounts of the template molecules, temperature, and aging time. After selective removal of the template molecules, the organofunctional hybrid nanoporous silica materials are created. This surfactant-directed self-assembly technique is utilized to synthesize photoresponsive nanoporous silica materials with azobenzene ligands positioned on the pore surfaces. Under UV and visible light irradiation, the azobenzene ligands undergo trans↔cis isomerization and create novel properties of the hybrid silica materials, such as photoregulated mass transport and photocontrolled liquid adsorption.

This chapter focuses on azobenzene-containing organosilanes that are precursors for preparing photoresponsive silica materials, mesoporous silica materials with azobenzene moieties on the pore surfaces, and azobenzene-modified polysilsesquioxane gels.

13.2. AZOBENZENE-CONTAINING ORGANOSILANES

Azobenzene-containing organic polymers have been widely investigated in the past several decades for their photoresponsive properties. However, research in azobenzene-containing inorganic polymers, such as sol–gel materials, is very limited. As proposed by Ozin and colleagues, a wide range of organic and organometallic precurors are potentially suited for incorporation of organic groups with novel chemical and structural properties into sol–gel materials (Maclachlan et al., 2000). Among these precursors, azobenzene-containing organosilane is of particular interest for the preparation of photoresponsive nanocomposite materials. Several efforts have been made to synthetically prepare pendant and bridged azobenzene-containing organosilanes, which are essential for the creation of photoresponsive nanocomposite sol–gel materials.

Ichimura and coworkers synthesized two organosilanes, 4-methoxy-4′-(*N*-(3-triethoxysilylpropyl)carbamoylmethoxy)azobenzene (MTAB) and 4,4′-bis-(*N*-(3-triethoxysilylpropyl)carbamoylmethoxy)azobenzene (DTAB), using the condensation reaction between carboxylic acid and amine in the presence of dicyclohexylcarbodiimide (DCC; Ueda et al., 1993). However, this synthetic route is not preferable because of the high toxicity of the DCC compound and the low yields of MTAB and DTAB.

Corriu and colleagues also synthesized an azobenzene-bridged organosilane, 4,4′-bis(3-triisopropoxysilylpropoxy)azobenzene (BSPA), and prepared the corresponding photoresponsive hybrid materials (Besson et al., 2005). BSPA was synthesized by reacting 4,4′-dihydroxyazobenzene with 3-iodopropyltriisoproxysilane in the presence of NaH at an elevated temperature of 60°C with 70% yield after purification. However, two extra steps need to be adopted for the synthesis of 3-iodopropyltriisoproxysilane, which increases the complexity of the synthetic process.

Brinker and coworkers designed and synthesized two organosilane precursors, 4-(3-triethoxysilylpropylureido)azobenzene (TSUA) and 4,4′-bis(3-triethoxysilylpropylureido)azobenzene (BSUA), for the preparation of photoresponsive nanocomposite materials (Liu et al., 2003b, 2002). The synthetic routes for TSUA and BSUA are more successful than those for other organosilanes in terms of purity, yield, and simplicity. TSUA and BSUA are new compounds. Some of their novel properties have also been studied.

13.2.1. Synthesis and Photoisomerization of TSUA and BSUA

TSUA and BSUA were synthesized using isocyanato–amino coupling reactions (Fig. 13.1) and characterized using several methods, including elemental analysis, nuclear magnetic resonance (NMR) spectra, and Fourier transform infrared (FTIR) spectroscopy (Liu, 2004). For the synthesis of TSUA compound, hexane was added after the reaction and the solution was cooled to −20°C. Shiny needlelike orange crystals precipitated from the solution. The yield was >80%. The BSUA compound, a needlelike yellow solid, precipitated from the solution after refluxing the solution for 24 h under room light. The BSUA yield was >90%.

UV/visible spectroscopy was used to investigate the reversible photo and thermal isomerization behaviors of TSUA and BSUA in solution (Fig. 13.2). For TSUA compound, ε_{max} was estimated to be $2.59 \times 10^4\ M^{-1}\,cm^{-1}$ on the basis of absorption band at 362 nm (the π–π^* transition of the trans isomer). UV irradiation with an Hg arc lamp ($\lambda_{max} = 350$ nm) decreased the intensity of the 362-nm band and slightly increased the intensity of the 450-nm band, which is attributed to the n–π^* transition of the cis isomer. A photostationary state was reached within ca. 10 min. Exposure to room light caused the reverse cis→trans isomerization. Extended room light exposure (12 h) increased the trans isomer to ca. 96%. The cis↔trans isomerization process was observed to be reversible over three cycles. In a detailed study, the first-order kinetic constant of the cis→trans isomerization under room light exposure was estimated to be $1.75 \times 10^{-4}\,s^{-1}$ for TSUA in ethanol solution at room temperature. Similar photoisomerization was observed for the BSUA compound in THF solution. UV irradiation ($\lambda_{max} =$ 350 nm) decreased the intensity of the 378-nm band (the π–π^* transition) and slightly increased the intensity of the 470-nm band (the n–π^* transition attributed

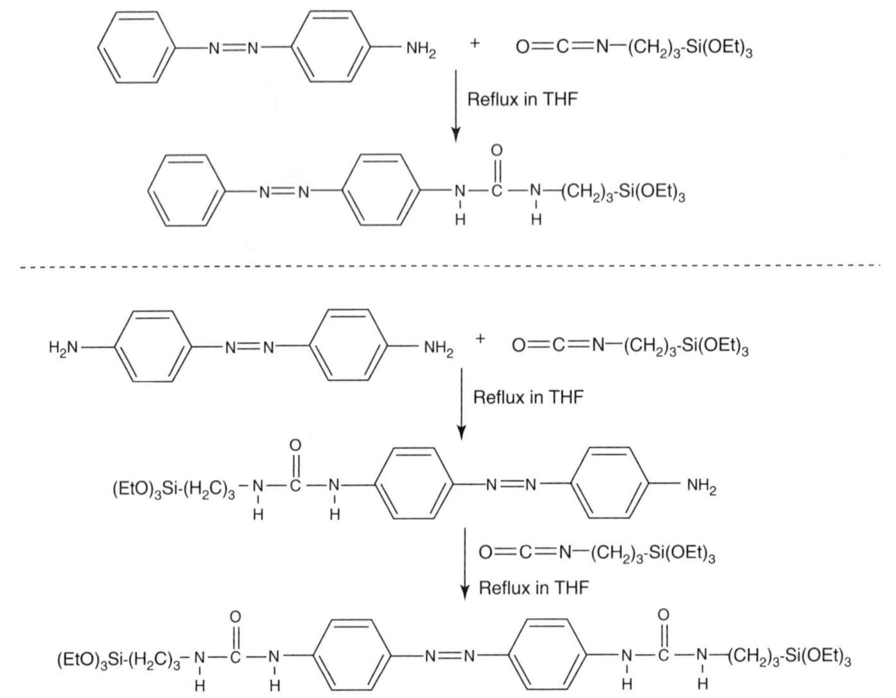

Figure 13.1. Schematic illustration of the synthesis of the TSUA (*top*) and BSUA (*bottom*) compounds. TSUA and the intermediate product in BSUA synthesis are highly soluble in THF, and BSUA is weakly soluble in THF.

to the cis isomer). A photostationary state was reached within ca. 10 min. Exposure to room light or heat caused the reverse isomerization (cis → trans). These results demonstrated the facile photoisomerization characteristics of the TSUA and BSUA compounds in solution.

13.2.2. Crystallography of the TSUA Compound

The as-synthesized TSUA compound is composed of shiny, orange needlelike crystals. The thermogravimetric analysis revealed that TSUA has two kinds of crystal structures (trans and cis crystals) with melting points at 127.5°C and 99.5°C, respectively. The majority of the TSUA crystals are trans crystals. Large, orange, single TSUA trans crystals (ca. 2 mm × 3 mm × 14 mm) were grown using a recrystallization method. The crystallographic structure of TSUA trans crystals was resolved using single crystal X-ray diffraction (XRD). There are four molecules in the unit cell (M1–M4, from left to right in Fig. 13.3). The stable six-membered ring formed by two H-bonds between two neighboring urea groups

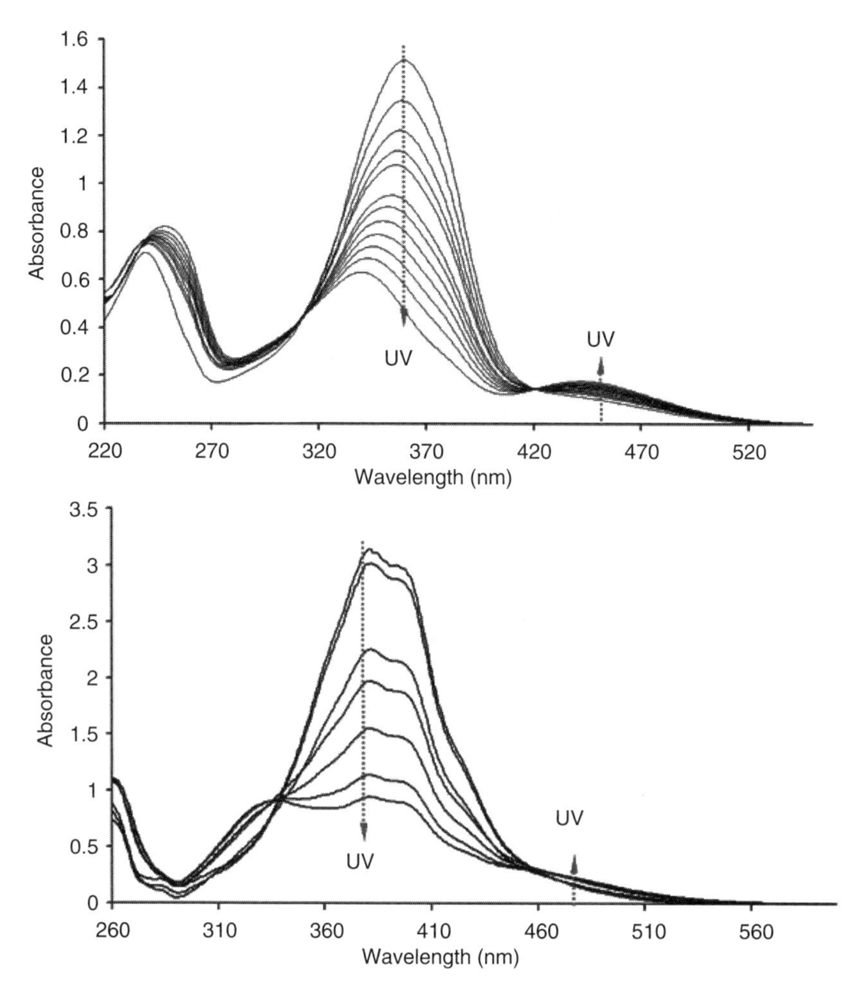

Figure 13.2. UV/visible absorption spectra for TSUA in EtOH (*top*) and BSUA in THF (*bottom*).

is energetically favorable, thus providing the main intermolecular interaction during the crystallization process. This is direct evidence of the hydrogen-bonding capability of urea ligands. The azobenzene ligands are all in the trans conformation. Azobenzene planes in M1 and M2 are parallel to those in M4 and M3, respectively, with the angle of the two sets of planes at 81.2°. The supplementary crystallographic data (CCDC 192903) for TSUA compound can be obtained free of charge through www.ccdc.cam.ac.uk/conts/retrieving.html. The crystallography data of the TSUA cis crystals were not resolved because of difficulty in obtaining single TSUA cis crystals.

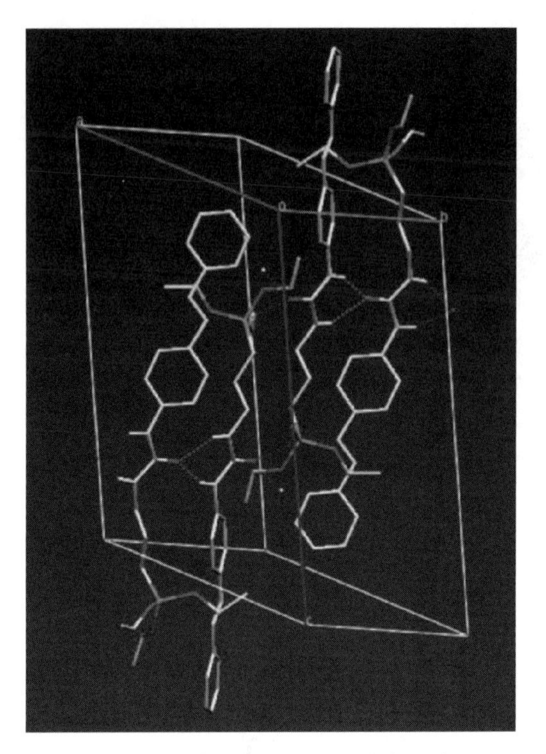

Figure 13.3. Unit cell structure of TSUA. Dotted bonds are H-bonds between two urea groups in the unit cell. *Source*: From Liu et al., 2003b. Reprinted with permission.

13.2.3. Self-Directed Self-Assembly of the BSUA Compound

The as-synthesized pure BSUA is a yellow needlelike powder with a melting point of 225°C and a decomposition temperature of 228°C. Thin films were prepared by dissolution of the as-synthesized BSUA in THF followed by casting on a silicon substrate. Slower evaporation of this same solution produced a shiny yellow-faceted solid. These samples showed some interesting properties. The ability of BSUA to both photoisomerize and self-assemble allows it to adopt various molecular-level mesoscale and macroscale structures. At the molecular level, exposure to UV light transforms the trans isomer to the cis isomer, whereas exposure to longer wavelength light or heat causes the reverse cis$\rightarrow$trans isomerization. On the mesoscale level, hydrogen bonding between the bisurea groups and $\pi-\pi$ stacking interactions between azobenzene bridges cause supramolecular assembly into well-defined lamellar structures. On the macroscale, the supramolecular structures exhibit multiple morphologies, sheetlike, twisted rope-like, or ribbonlike structures (Fig. 13.4).

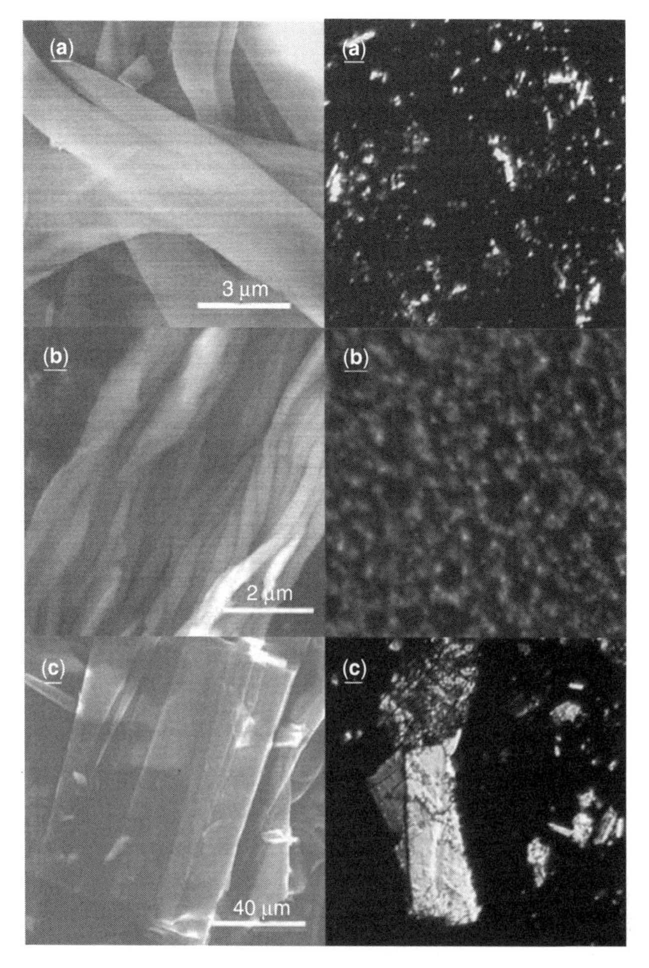

Figure 13.4. SEM images (*left column*) and polarized light microscopy (PLM) images (*right column*) of BSUA solids formatted under different conditions.

In dilute solution under ambient illumination, the azo group of BSUA is found mainly in a trans form, consistent with the greater stability of the trans isomer known for other azobenzene derivatives. Photo and thermal isomerization are reversible. However, after precipitation from THF, the azo group mainly adopts its cis form. Brinker and colleagues attributed this to hydrogen bonding through the urea groups and $\pi-\pi$ interactions between adjacent benzene groups that freeze-in the more stable cis conformation in the solid product, and prevent further photoisomerization (Liu et al., 2002). UV irradiation before and during evaporation, however, allows the ratio of cis to trans in solid samples to be controlled.

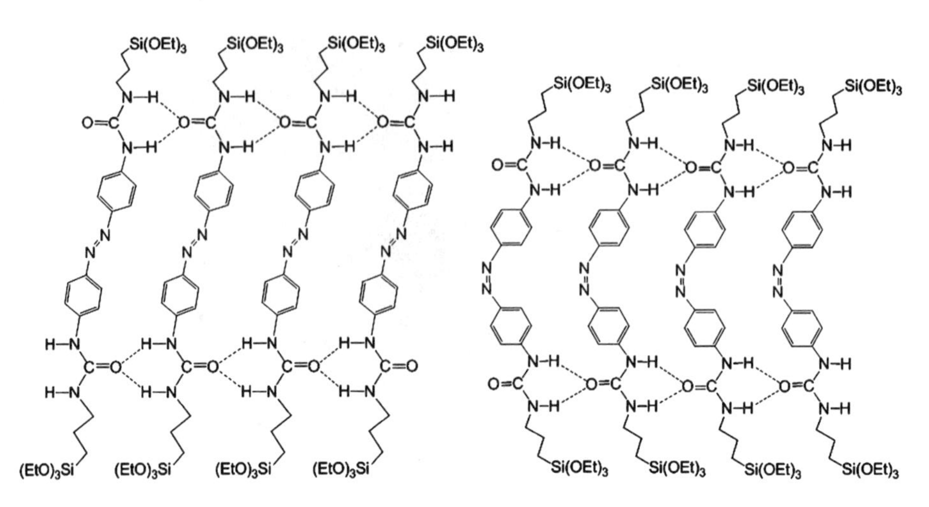

Figure 13.5. Supramolecular structures formed by self-directed assembly in trans (*left*) or cis (*right*) forms.

Brinker and coworkers proposed the structures of the trans and cis supramolecular assemblies (Fig. 13.5) (Liu et al., 2002). The two urea groups in a BSUA molecule form multiple intermolecular hydrogen-bonding which, in combination with the $\pi-\pi$ stacking of the azobenzene group, drives the self-assembly of BSUA into a highly ordered lamellar structure. As detected using XRD, the *d*-spacings of the cis assembly and the trans assembly are ca. 1.9 and 2.2 nm, respectively. Because of the extensive hydrogen-bonding interactions, the trans$\leftrightarrow$cis isomerization is severely inhibited in the supramolecular solids. This result indicates that realization of the full potential of transduction of optical excitation into work for applications in nanomechanical devices will require BSUA to be sufficiently dilute or chemically modified to prevent extensive hydrogen-bonding.

Polarized light microscopy images (Fig. 13.4, right column) shows that BSUA solids are bright and colorful, demonstrating that they have the birefringence that is very common in crystals. The anisotropic property of the BSUA compound is indicative of an ordered molecular assembly in which the light propagation in the direction parallel to the azobenzene axis is different from that perpendicular to the azobenzene axis.

13.3. PHOTORESPONSIVE MESOPOROUS MATERIALS

In 1992, Kresge and coworkers at Mobil invented the so-called MCM-41 mesoporous materials that have a hexagonal arrangement of monosized mesopores with pore size adjustable from 1.5 nm to 10 nm (Kresge et al., 1992).

This novel molecular sieve was synthesized by hydrothermal reactions of capatal alumina, tetramethylammonium silicate, and sodium silicate in the presence of ionic surfactants. It opened a new era of mesoporous materials because of their high surface area ($>700\,m^2\,g^{-1}$), controllable pore size in a wide range (1.5–10 nm), and versatile pore surface chemistries. Following this pioneering work, material scientists have synthesized mesoporous inorganic networks with different mesostructures (lamellar, two-dimensional (2-D) hexagonal, three-dimensional (3-D) hexagonal, face-centered cubic (FCC), body-centered cubic (BCC), and primitive cubic), using a wide range of surfactants including cationic surfactants (CTAB), anionic surfactants (SDS), nonionic surfactants (Brij 56), and amphiphilic block copolymers (P123), of interest in potential applications in catalysis, membrane separations, adsorption, and chemical sensing.

Evaporation-Induced Self-Assembly. The hydrothermal process designed for MCM-41 is time consuming, and more importantly, the mesoporous products are usually ill-defined powders that are not transparent, precluding their general applications in thin films and optical applications. Brinker and colleagues developed a new technique denoted as evaporation-induced self-assembly (EISA) to improve the MCM-41 type materials for the fast formation of mesostructured nanoparticles (Lu et al., 1999) and thin films (Lu et al., 1997). In the EISA process for film formation during dip-coating, a substrate is dipped into a homogeneous solution of soluble silica, surfactant, water, and acid in ethanol solvent, with an initial surfactant concentration far below the critical micelle concentration (CMC), and withdrawn from the sol solution at a constant speed. Preferential evaporation of ethanol concentrates the surfactant, water, acid, and silica species in the depositing film. Further evaporation of ethanol concentrates the surfactant to exceed CMC and causes the formation of micelles. Beyond this point, progressively increasing surfactant concentration drives self-assembly of silica-surfactant micelles and their organization into highly ordered liquid crystalline mesophases (cubic, hexagonal, or lamellar) that depend on the

Figure 13.6. Representative TEM images of mesoporous silica thin films. Insets are the corresponding electron diffraction patterns.

composition of surfactant, silica, water, acid, and solvent. After surfactant removal by calcination or solvent extraction, mesoporous silica materials are formed (Fig. 13.6). EISA provides a highly controllable approach to fabricate nanoporous films and particles. The fast process, easy integration of nanoporous films into devices, and controllable properties (pore sizes, pore surface chemistries, pore arrangements, and porosity) are especially suitable for the application of nanotechnology in industry.

Concept of Stimuli-Responsive Nanocomposite Materials. To expand the range of applications as well as create new properties, various organic functional groups have been covalently incorporated onto the pore surfaces of mesoporous materials, including phenyl, octyl, aminoalkyl, cyanoalkyl, thioloalkyl, epoxyl, and vinyl groups (Nicole et al., 2005). However, these modifications have provided mainly "passive" functionalities, such as controlled wetting properties, reduced dielectric constants, or enhanced adsorption of metal ions. By comparison, materials with "stimuli-responsive" functionalities would enable properties to be dynamically controlled by external stimuli, such as pH, temperature, or light.

The concept of stimuli-responsive nanocomposite materials was proposed by Brinker and coworkers (Liu et al., 2003a; Fig. 13.7). Incorporation of covalently bonded, "switchable" ligands as pendant groups (as opposed to bridging groups) within a porous (rather than a layered) framework is expected to result in nanocomposites in which the pore size, surface area, and adsorptive properties are externally controllable. In such synergistic nanocomposites, the 3-D organization of the responsive ligands allows transduction of photo, chemical, or thermal energy into a useful mechanical response of interest for molecular valves or gates. In addition, the rigid framework could serve to enhance the mechanical, chemical, and thermal stability of the responsive moieties, fostering their integration into devices.

As described earlier, switchable azobenzene derivatives are promising candidate molecules for the synthesis of photoresponsive nanocomposite materials. To best utilize photoisomerization to accurately control pore size, azobenzene ligands need to be anchored to the pore surfaces of a uniform mesostructure with monosized pores. This allows the rigid inorganic framework to precisely position azobenzene ligands in 3-D configurations where switching results in a well-defined change in pore size.

13.3.1. Synthesis and Characterization of Photoresponsive Nanoporous Materials

Two synthetic methods have been exploited to incorporate azobenzene moieties into mesoporous silica frameworks. The postgrafting method involves functionalizing mesoporous sol–gel materials with reactive azobenzene derivatives to tether azobenzene moieties to sol–gel matrices. The one-pot synthesis method requires

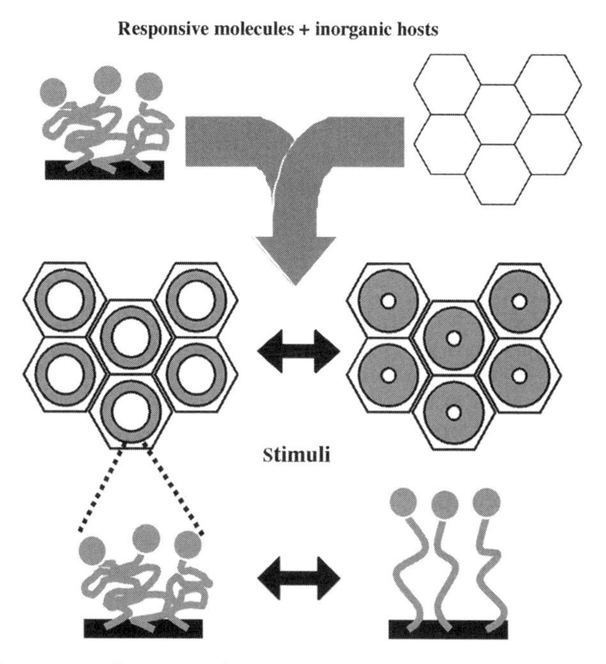

Figure 13.7. Concept of responsive nanocomposites. Organization of stimuli responsive ligands within a 3-D porous framework imparts new molecular-level functionality, useful for opening and closing a nanoscale valve. *Source*: From Liu et al., 2003a. Reprinted with permission.

cohydrolysis and cocondensation of azobenzene-containing organosilanes with other silica precursors, such as tetraethyl orthosilicate (TEOS).

13.3.1.1. Preparation of Photoresponsive Mesoporous Materials. Brinker and coworkers prepared photoresponsive thin film nanocomposites via an EISA procedure using the nonionic surfactant Brij 56 or the triblock copolymer surfactant P123 as structure-directing agents (Liu, 2004; Liu et al., 2003a). In the EISA process, TSUA is mixed with TEOS in a homogeneous ethanol–water solution. The amphiphilic nature of the hydrolyzed TSUA molecule positions the hydrophobic propylureidoazobenzene groups in the hydrophobic micellar cores and the silicic acid groups coorganize with hydrolyzed TEOS moieties at the hydrophilic micellar exteriors. In this fashion, TSUA is ultimately incorporated on the pore surfaces with the azobenzene ligands disposed toward the pore interiors. Subsequent solvent extraction results in a mesoporous silica framework modified with azobenzene ligands, which are isomerizable by light and thermal stimuli (Fig. 13.8). The dimensional change of the propylureidoazobenzene ligand associated with this isomerization mechanism is estimated to be 3.4 Å on the basis of molecular modeling with Chem3D™ 5.5 software (Cambridgesoft).

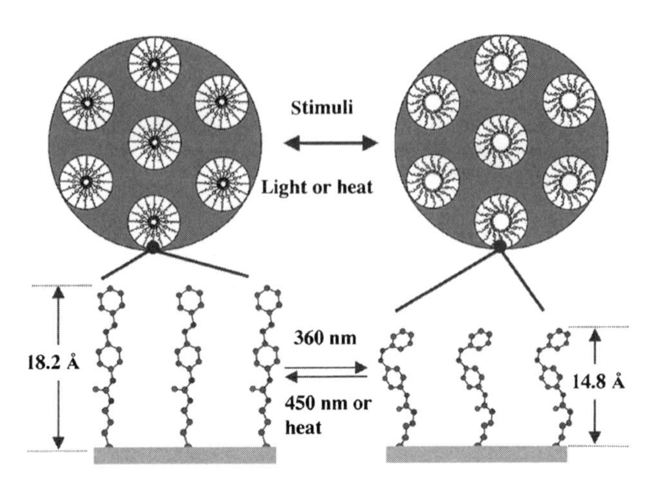

Figure 13.8. Photoresponsive nanocomposites prepared by EISA. The trans or cis conformation of azobenzene unit was calculated using Chem3D Pro™ (Cambridgesoft) molecular modeling analysis software. *Source*: From Liu et al., 2003a. Reprinted with permission.

Brinker and coworkers also prepared photoresponsive nanoporous silica particles using aerosol-assisted assembly, EISA, and surfactant-directed self-assembly (SDSA) techniques (Liu, 2004). The as-prepared photoresponsive nanocomposite particles were prepared so that the azobenzene ligands pointed toward the hydrophobic micellar interiors. After surfactant removal by solvent extraction, nanoporous particles with azobenzene ligands positioned on the pore surfaces were formed (Fig. 13.8).

Corriu and coworkers prepared several kinds of hybrid materials using BSPA as the azobenzene-containing precursor, including a xerogel monolith and thin films with (or without) structure-directing agents (Besson et al., 2005). The xerogel monolith was prepared by hydrolysis and cocondensation of BSPA with TEOS (19 equivalent) at pH = 1.5 using a nonionic triblock polymer P123 surfactant as the template. Thin film samples were deposited on silicon wafers by dip-coating a sol-containing BSPA, TEOS (19 equivalent), HCl, H_2O, P123, and EtOH in EISA process.

The postgrafting method was also utilized to functionalize mesoporous silica materials with azobenzene moieties. Zink and coworkers prepared a series of azobenzene-containing dendrons from G0 to G3 (Fig. 13.9) and tethered them to the nanoporous silica materials using a three-step synthetic procedure (Sierocki et al., 2006). First, nanostructured silica materials were synthesized according to the well-developed SDSA approach (Kresge et al., 1992) and the EISA technique (Brinker et al., 1999). Then, the pore surfaces of the nanostructured silica were modified with isocyanopropyltriethoxysilane (ICPES) through a vapor deposition

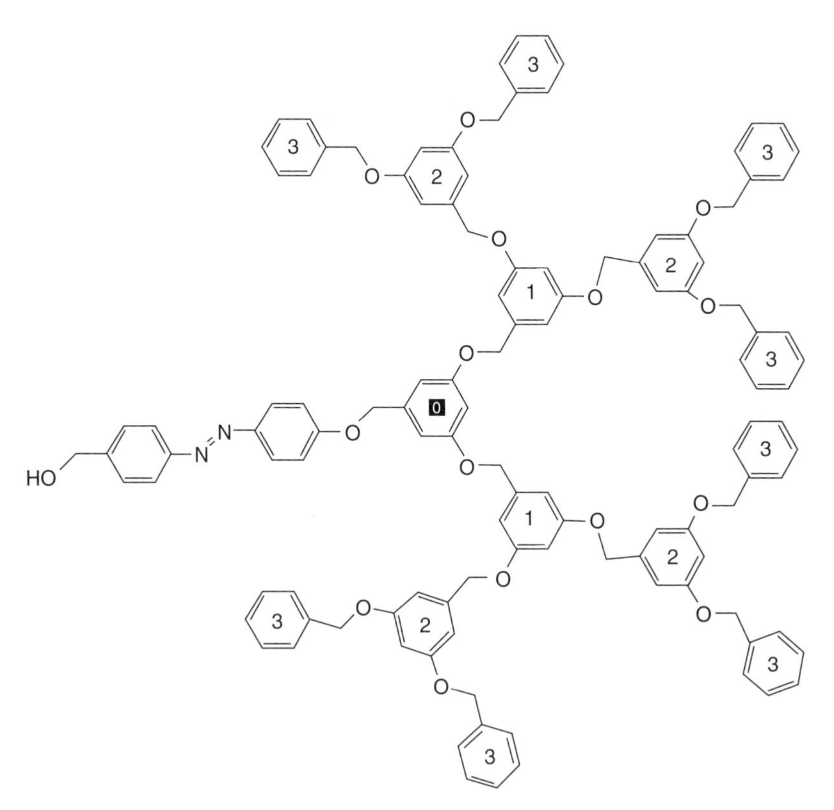

Figure 13.9. Chemical structures of the azobenzene-containing dendrons. The number 0, 1, 2, and 3 notations within the phenyl rings indicate the number of generations (AzoG0–AzoG3).

procedure (Chia et al., 2001). Finally the azobenzene-containing dendrons were tethered to the –N=C=O groups on the pore surfaces by the coupling reaction between the –N=C=O and –OH groups at the end of the azobenzene branches.

Maeda et al. (2006) also grafted alkyl-derivatized azobenzene moieties to the pore surfaces of nanoporous silica materials. Three steps were utilized to synthesize the azobenzene-tethered mesoporous silica materials: synthesis of 2-D-hexagonal mesoporous silica (MSU-H) using the SDSA technique, modification of the pore surfaces with aminopropylsilyl groups based on the reactions between the surface silanol (–OH) groups and organosilane, and tethering the azobenzene ligands to the amine groups using azobenzene derivative bromides under basic reaction conditions. Combining the nitrogen sorption isotherms and the CHN elementary analysis data, the density of amine groups and azobenzene containing pendant groups on pore surfaces were estimated to be $1.15/nm^2$ and $0.54/nm^2$, respectively.

13.3.1.2. Mesostructures of Azobenzene-Modified Mesoporous Materials. Highly ordered mesostructures of azobenzene-modified mesoporous materials have been studied using XRD, transmission electron microscopy (TEM), and grazing incidence, small angle X-ray-scattering (GISAXS) techniques. To date, 2-D hexagonal, wormlike, and BCC nanostructured silica powders and thin films have been reported.

Brinker and coworkers performed a comprehensive study over the azobenzene-modified mesoporous silica films and nanoparticles employing different surfactants as the structure-directing agents. Using EISA and SDSA techniques and photoresponsive TSUA and BSUA precursors, they synthesized Brij 56– and P123-templated thin films and CTAB-templated nanoparticles and thin films (Liu, 2004; Liu et al., 2003a). Figure 13.10a shows a representative TEM cross-sectional image of a Brij 56–templated TSUA-modified nanocomposite film, which exhibits a highly ordered cubic mesostructure. The average center-to-center pore spacing is 5.6 nm. After surfactant removal, one-dimensional (1-D) XRD of the nanocomposite film shows a strong reflection at 2.9 nm and a weaker reflection at 1.4 nm. The electron diffraction and 2-D GISAXS data (Fig. 13.10b,c) confirm that the pores are arranged in a BCC mesostructure ($Im\,\bar{3}m$ space group) with a unit cell dimension $a = 5.7$ nm ($d_{200} = 2.85$ nm, averaged between 2.9 nm from 2-D GISAXS and 2.8 nm from electron diffraction data), allowing the two XRD peaks to be assigned to the d_{200} and d_{400} reflections, respectively. The 2-D GISAXS data indicate that the (100) plane of the BCC mesostructure is oriented parallel to the silicon substrate. The slight distortion of the BCC mesostructure is due to the greater shrinkage in the direction normal to the substrate than parallel to the substrate resulting from siloxane condensation. Indeed, the thickness of the Brij 56–templated nanocomposite film measured using an ellipsometer changed from 312 nm (before surfactant removal) to 260 nm (after surfactant removal).

A representative TEM image of CTAB-templated TSUA-modified nanoparticles is shown in Fig. 13.10d. The particles are spherical with a broad size distribution (from 40 to 1600 nm) and exhibit an onionlike structure with an average spacing of 3.3 nm. The powder XRD of the nanocomposite particles exhibit a strong reflection at 3.4 nm and a weak hump centered at 1.7 nm. The combined TEM and XRD results suggest that the pores are arranged into a lamellar structure, allowing the two reflections in the XRD pattern to be indexed as d_{100} and d_{200}, respectively.

Mesoporous hybrid BSUA–silica thin films were formed using Brij 56 and CTAB as the structure-directing agents. The percentage of BSUA was <5 mol% because of its low solubility in EtOH. The XRD data show one strong reflection and one weaker reflection, indicating long-range mesostructural order of the prepared materials. For Brij 56– and CTAB-templated films, the first strong reflections are at 4.9 and 4.0 nm, respectively. The TEM images (Fig. 13.10e,f) show a highly ordered structure, which is consistent with (100) cubic symmetry. The average neighboring pore distances are 3.6 nm for Brij 56–templated films and 4 nm for CTAB-templated films. On the basis of these results, it is most plausible that for Brij 56–templated films, the pores are arranged in a BCC structure ($Im\,\bar{3}m$

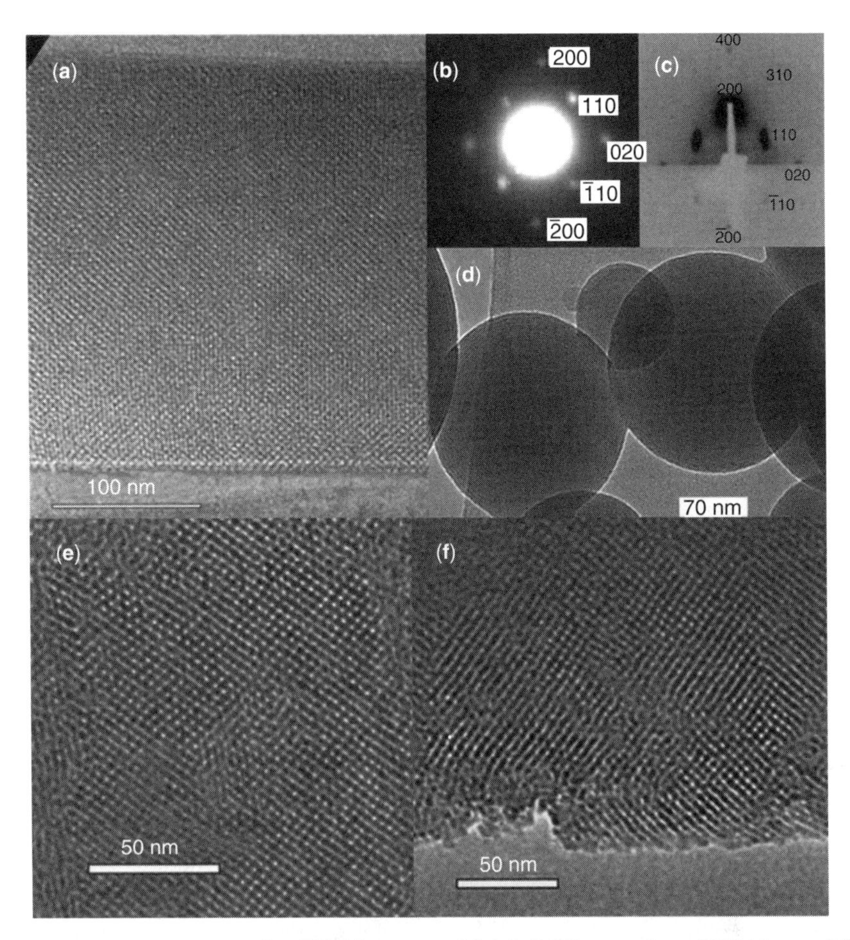

Figure 13.10. Representative TEM images of (**a**) Brij 56–templated TSUA-modified mesoporous films, (**b**) electron diffraction pattern of the cross-sectional TEM image (**a**), (**c**) 2-D GISAXS data of a Brij 56–templated photoresponsive nanocomposite thin film, (**d**) CTAB-templated TSUA-modified mesoporous nanoparticles, (**e**) CTAB-templated BSUA-modified mesoporous films, and (**f**) Brij 56–templated BSUA-modified mesoporous films.

space group) with a lattice parameter $a = 7.0\,\text{nm}$, allowing the two reflections in the XRD pattern to be indexed as d_{110} and d_{220}, respectively. For CTAB-templated films, the pores are arranged in a FCC (*Fm3m* space group) structure with the unit cell parameter $a = 8.0\,\text{nm}$, allowing the two reflections in the XRD pattern to be indexed as d_{200} and d_{400}, respectively.

13.3.1.3. Pore size and Distribution. Nitrogen sorption isotherms were widely used to estimate the pore size and distribution of mesoporous materials.

For Brij 56–templated TSUA-modified mesoporous films, the analysis of surface acoustic wave (SAW)-based nitrogen sorption isotherms exhibited type IV characteristics and gave a pore diameter of 1.3 nm using the classical Barrett–Joyner–Halenda (BJH) model, and a Brunauer–Emmett–Teller (BET) surface area of $355\,m^2\,g^{-1}$ (Liu et al., 2003a). For Brij 56–templated TSUA-modified nanoparticles, the nitrogen adsorption–desorption isotherms exhibit a progressive increase of nitrogen molecules loaded in the particles to ca. $290\,mL\,g^{-1}$ at the relative pressure near 1. This is a type II isotherm, which is consistent with the particles' onionlike mesostructure (Liu, 2004). The pore size is 2.5 nm with a broad-size distribution using the BJH model. The BET surface area was calculated to be $427\,m^2\,g^{-1}$. The N_2 adsorption–desorption isotherms of P123-templated BSPA-derivatized, nanoporous silica materials (Besson et al., 2005) are of type IV, characteristic of a mesoporous material. The specific BET surface area was large ($700\,m^2\,g^{-1}$). The mean pore diameter is 5.1 nm and the total pore volume amounts to $0.92\,cm^3\,g^{-1}$.

13.3.1.4. Environmentally Sensitive Color Change of the BSUA-Modified Mesoporous Films. The pure polysilsesquioxane gel (cast on a petri dish) made from BSUA has two colors: purple in acid and yellow in H_2O or base. The purple gel changes to yellow when it is soaked in a basic solution or H_2O, whereas the yellow gel changes to purple when it is soaked in acidic solution. The color change may result from the protonation reaction of the azo- and urea groups, which causes the conjugality change and correspondingly the absorption change. However, these processes were slow because of the microporous nature of the polysilsesquioxane gels. It takes time for water molecules to diffuse into the material.

Similar to the phenomenon described earlier, the surfactant-templated BSUA–silica nanocomposite thin films also exhibited color changes in different environments (Liu, 2004). As shown in Fig. 13.11, the prepared film was yellow after washing with water and EtOH. The color changed to purple when the film was soaked in acidic solution. Furthermore, the color change is reversible when the film is soaked in a basic solution or water. In a detailed study, it was found that the BSUA–silica nanocomposite material changed color around $pH = 2$ ($pH < 2$, purple; $pH > 2$, yellow). Compared with the pure polysilsesquioxane gels made from BSUA, the color change of the mesoporous nanocomposite film occurs more rapidly. This film changed colors immediately because of the mesoporous nature of the material, suggesting potential applications in environmental sensing.

Figure 13.11. Color changes of mesoporous BSUA/silica nanocomposite thin films.

13.3.2. Photoisomerization of Azobenzene Ligands in Mesoporous Materials

The azobenzene ligands incorporated into the mesoporous silica materials undergo photoisomerization upon light irradiation. Compared with the photoisomerization of the azobenzene ligands in solution, it exhibited some difference as described later.

It is widely known that the thermal cis→trans isomerization of azobenzene derivatives in solution follows first-order kinetics ($\ln([\text{cis}]_0/[\text{cis}]_t) = kt$, $[\text{cis}]_0$ and $[\text{cis}]_t$ denote the concentration of the cis isomer at time 0 and t, respectively, and k is the rate constant) (Dillow et al., 1998; Rau, 1990a,b; Asano and Okada, 1984). However, for azobenzene ligands contained in sol–gel matrices or polymers, deviation from first-order kinetics was observed. For example, Ueda et al. (1994, 1993, 1992) investigated the photoisomerization of MTAB and DTAB encapsulated in sol–gel films. They found that the rate constants of the cis→trans isomerization of MTAB and DTAB decreased with time at room temperature. At a higher temperature of 60°C, the cis→trans isomerization rates of MTAB and DTAB were initially faster than those in solution and gradually decreased with time. Ueda and colleagues attributed the early rate enhancement generated by UV light irradiation to the residual strain in the cis isomer within the confined silica matrix. The lower isomerization rate was ascribed to the interaction between the azobenzene ligands and the silica sol–gel matrix.

13.3.2.1. Photoisomerization of Photoresponsive Mesoporous Films in Air and Aqueous Solution. Brinker and coworkers investigated the photo and thermal responses of the incorporated azobenzene ligands in the mesoporous silica films (Liu, 2004; Liu et al., 2003a). Figure 13.12a shows the UV/visible spectra (reflection mode) of Brij 56–templated TSUA-modified thin films deposited on silicon after exposure to varying conditions of UV irradiation (8 W, $\lambda = 302$ nm), room light, or heat. The absorption band at 350 nm is attributed to the π–π^* transition of the trans isomer and the band at 460 nm to the n–π^* transition of the cis isomer. UV irradiation causes a progressive trans→cis isomerization as noted by the decrease of the 350-nm band and increase of the 460-nm band, with a photostationary state reached after 30 min (spectrum b). Exposure to room light or heat caused the reverse cis→trans isomerization. For example, from the photostationary state (spectrum b), room light exposure increased the intensity of the 350-nm absorption band and decreased the 460-nm absorption band progressively. Prolonged exposure (12 h) to room light increased the intensity of the 350-nm absorption band, close to that of the as-prepared film. Alternatively, heat treatment (100°C, for 5 min in the dark) caused the intensity of the 350-nm absorption band to exceed that of the as-prepared nanocomposite film (spectrum c), indicating that a certain amount of azobenzene ligands was in the cis conformation in the as-prepared films. These results, which are completely reversible, demonstrate the facile photoisomerization characteristics of the pendant azobenzene ligands in the ordered porous nanocomposite. Using a molar

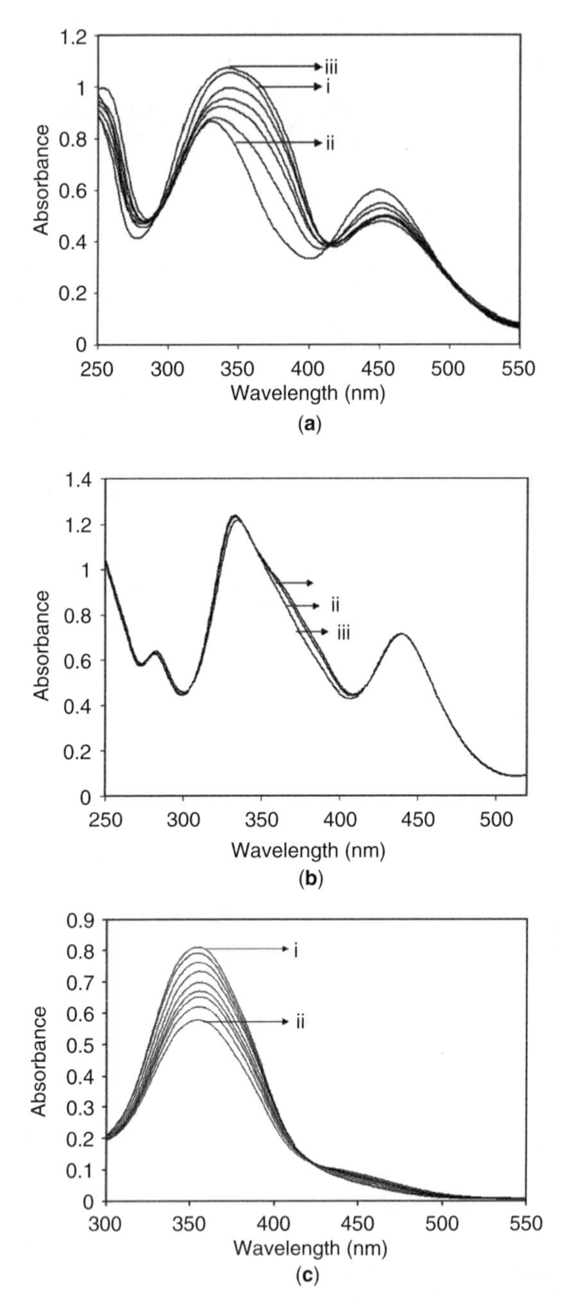

Figure 13.12. UV/visible spectra of Brij 56–templated TSUA-modified nanoporous films. (**a**) After surfactant removal, reflection mode. (i) As-prepared, (ii) after UV irradiation of sample (i) for 30 min, and (iii) after heating sample (ii) at 100°C for 5 min. (**b**) Before surfactant removal, reflection mode. (i) As-prepared, (ii) after UV exposure of sample (i) for 20 min, and (iii) after room light exposure of sample (ii) for 60 min. (**c**) After surfactant removal, transmission mode. (i) Before UV exposure, (ii) UV exposure for 5 min.

absorption coefficient of $2.59 \times 10^4\,M^{-1}\,cm^{-1}$ for the trans isomer, the trans isomer concentration in the films was estimated to be 1.55, 1.52, and 1.12 M, respectively, for samples iii, i, and ii in Fig. 13.12a. The molar concentrations of azobenzene ligands in the self-assembled films greatly exceed those attainable by dye adsorption in mesoporous films (Ogawa et al., 2000).

As a control experiment, thin films were prepared with the same sol, but without the Brij 56 surfactant. In this case, the azobenzene ligands were randomly incorporated in a microporous silica matrix and exhibited no detectable photoisomerization. Similarly, no significant photoisomerization was observed for the ordered self-assembled films before solvent extraction of the surfactant templates (Fig. 13.12b). The pore volume required for photoisomerization was created only upon surfactant removal (estimated as $127\,\text{Å}^3$ (Victor and Torkelson, 1987)). These results unambiguously locate the photosensitive azobenzene ligands, along with surfactant, within the uniform nanopores of the self-assembled films and emphasize the need to accommodate the steric demands of the photoisomerization process by engineering the pore size (as noted previously for adsorbed azobenzene dyes (Ogawa et al., 2000)) and positioning the photoactive species on the pore surfaces. Correspondingly, nanocomposite films prepared using the triblock copolymer surfactant P123, which had larger pores (ca. 6.5 nm in diameter), exhibited a faster and more complete trans$\rightarrow$cis isomerization, close to that of TSUA in EtOH solution, presumably because the azobenzene ligands were less sterically constrained.

It is noteworthy that there exists a difference between the UV/visible spectra of Brij 56–templated nanocomposite films before (Fig. 13.12a) and after (Fig. 13.12b) the surfactant removal. For the azobenzene contained in the nanocomposite film before the surfactant removal, the $\pi-\pi^*$ transition of the trans isomer is at 333 nm and the $n-\pi^*$ transition of the cis isomer is at 440 nm. After the surfactant removal, the $\pi-\pi^*$ transition of the trans isomer and the $n-\pi^*$ transition of the cis isomer are shifted to 350 and 460 nm, respectively. In both situations, the absorbance of the $n-\pi^*$ transition is quite high, indicating that the molar absorption coefficient (ε) of the $n-\pi^*$ transition is larger than that in solution owing to the *out-of-plane* position of the azobenzene phenyl rings. This effect may be the result of steric constraint and hydrogen-bonding interactions between the azo- and silanol groups on the silica wall. In comparison to the azobenzene contained in the nanocomposite film after the surfactant removal, the appearance of the $\pi-\pi^*$ and the $n-\pi^*$ transitions at lower wavelength (blueshift) before the surfactant removal indicates that the azobenzene ligands are *out of plane* to a larger extent. The absorbance at the $n-\pi^*$ transition of azobenzene ligands contained in P123-templated film is lower than that in the Brij 56–templated film, indicating a less sterically constrained environment because of the larger pore sizes of P123-templated film.

The preceding UV/visible data were measured using a reflection mode. For TSUA-modified nanocomposite thin films coated on indium tin oxide (ITO)/glass substrates, a transmission mode was adopted to measure the absorbance of the films. In the experiment, the film was soaked in a Tris buffer solution containing

50 mM KCl in a cubic cell. Fig. 13.12c shows the UV/visible spectra data of a Brij 56–templated TSUA-modified, nanocomposite film coated on an ITO/glass substrate. The film exhibits an excellent reversible photoisomerization phenomenon. UV irradiation causes gradual decreases in the π–π^* transition band (355 nm) and increases in the n–π^* transition band (456 nm). The reverse cis→ trans isomerization can be achieved by a visible light exposure ($\lambda = 435$ nm). A comparison between the absorbance in air (Fig. 13.12a) and absorbance in buffer solution (Fig. 13.12c) shows the aqueous environment causing a decrease in the absorbance in the n–π^* transition band. This indicates weaker hydrogen-bonding interactions between the azo and silanol groups on the silica wall than those present in air.

Brinker and coworkers also monitored the absorbance (A) at 355 nm (π–π^* transition) in situ to calculate the kinetics of the cis→trans isomerization of the azobenzene ligands contained in the nanocomposite film deposited on the ITO/glass substrate in buffer solution (Liu, 2004). The first-order plot of $\ln((A_\infty - A_0)/(A_\infty - A_t))$ vs. t was not perfectly linear in the entire region. The slope (rate constant, k) gradually decreases ($t < 250$ min) and remains constant ($t > 250$ min). The deviation from first-order kinetics is common for azobenzene ligands confined in sol–gel matrices or polymers (Böhm et al., 1996; Ueda et al., 1992). These data imply that the kinetics is composed of two parts, a fast one (k_1) and a slow one (k_2). When t is small, the fast process is predominant. On the contrary, the slow process becomes predominant as time (t) increases. A double exponential equation was used to fit the data and evaluate k_1 and k_2.

$$A_t = A_\infty - a_1 e^{-k_1 t} - a_2 e^{-k_2 t}$$

Here A_t and A_∞ are the absorbance at time t and infinite time. k_1 and k_2 are the rate constants of the fast and the slow processes. a_1 and a_2 represent the relative contributions of the fast and slow processes to the kinetics of the cis→ trans isomerization. The rate constants of the fast (k_1) and the slow (k_2) processes at 20°C are $1.7 \times 10^{-4}\,\mathrm{s}^{-1}$ and $1.7 \times 10^{-5}\,\mathrm{s}^{-1}$, respectively. The rate constant of the thermal cis→trans isomerization of azobenzene derivatives in solution is usually on the order of $10^{-5}\,\mathrm{s}^{-1}$. Compared with this value, the rate of cis→trans isomerization of azobenzene ligands incorporated into the nanocomposite films is faster for k_1 and is in the same order for k_2, which is good for applications requiring fast switching between the two isomerization states.

For P123-templated nanocomposite films, a similar analysis was performed to evaluate the rate constants of the fast and the slow processes. The rate constants of the fast (k_1) and the slow (k_2) processes are $5.7 \times 10^{-4}\,\mathrm{s}^{-1}$ and $6.5 \times 10^{-5}\,\mathrm{s}^{-1}$, respectively. Similar to those of Brij 56–templated nanocomposite films, the cis→ trans isomerization of P123-templated films is faster for k_1 and is in the same order for k_2.

Brinker and coworkers proposed a physical explanation to the foregoing experimental results (Liu, 2004). In the photoresponsive nanocomposites, a portion of the azobenzene ligands positioned on the mesopore surfaces has sufficient free volume to isomerize (cis → trans) via a rotation mechanism at a rate (k_1) comparable to that in solution, whereas a second portion of azobenzene ligands anchored at the constrictions between the mesopores (inherent to the cubic symmetry) is constrained and isomerizes (cis → trans) faster via an inversion mechanism (k_2). This is reasonable because the activation energy and extra volume needed for isomerization via the inversion mechanism are less than those needed for the rotation mechanism. The fast isomerization fractions ($a_1/(a_1 + a_2)$) of Brij 56– and P123-templated nanocomposite films are 38% and 29%, respectively, and both are larger than that of the azobenzene ligands contained in hybrid polymers (14%) (Böhm et al., 1996; Ueda et al., 1992).

13.3.2.2. Photoisomerization of Azobenzene-Containing Dendrons in Nanostructured Silica. Zink and coworkers studied the photoisomerization of azobenzene-containing dendrons in 2-D hexagonal mesoporous silica materials (Sierocki et al., 2006). Figure 13.13 shows a representative absorption spectra evolution of AzoG1 (see Fig. 13.9) in dichloromethane in the dark after

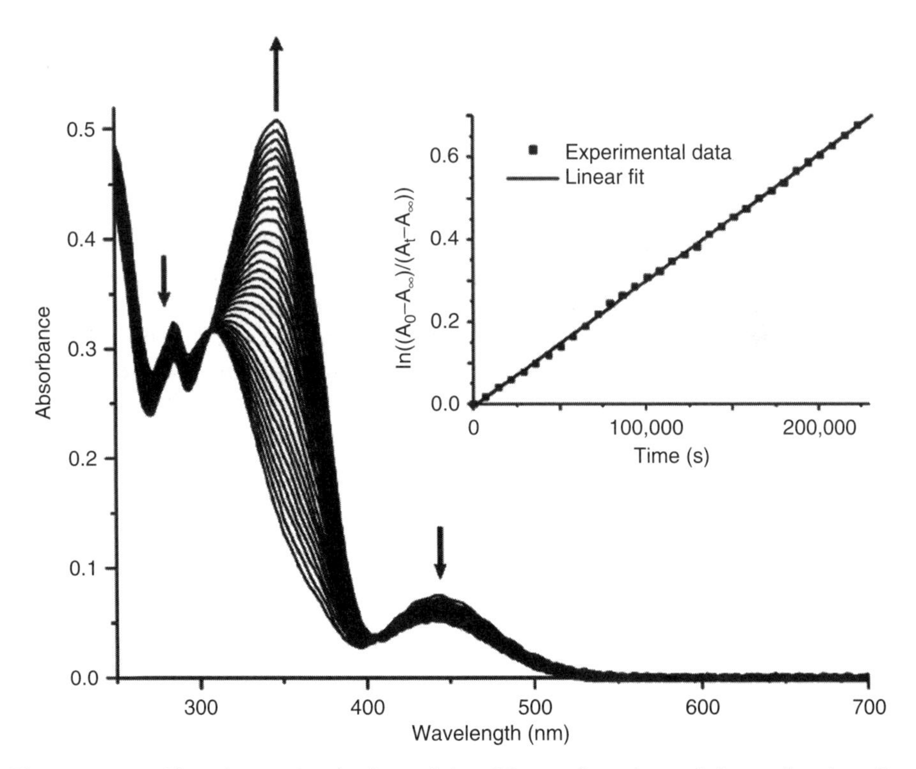

Figure 13.13. Absorbance evolution of AzoG1 as a function of time after irradiation with 344-nm light to reach the photostationary state. Inset is the plot of the first-order kinetics. *Source*: Sierocki et al., 2006. Reprinted with permission.

irradiation with 344 nm to reach the trans→cis photostationary state. The absorption at the 351-nm band progressively increased with time and the absorption at the 450-nm band decreased with time, indicating the cis→trans thermal isomerization. The rate constants were determined according to the first-order kinetics that fits the experimental data very well. There was a small change in the rate constants, that is, from AzoG0 to AzoG3 the rate constants of the cis→ trans thermal isomerization decreased from $3.3 \times 10^{-6} s^{-1}$ to $2.6 \times 10^{-6} s^{-1}$.

After the azodendrimers were tethered to the pore walls of the MCM-41 materials, the rate constants of the cis→trans back isomerization were investigated using a similar method. The size of the dendrons showed a significant effect on the rate constants. The AzoG0 dendron with the smallest size exhibited the fastest reverse isomerization ($10.2 \times 10^{-6} s^{-1}$), whereas the bulkier AzoG3 dendron had the slowest reaction ($2.6 \times 10^{-6} s^{-1}$). This result is not surprising because the smaller AzoG0 has more freedom to isomerize within the confined space of the MCM-41 materials. The azobenzene derivatives require an extra volume of at least 127 $Å^3$ to undergo complete isomerization (Victor and Torkelson, 1987). For the bulky AzoG3 dendron, the confined space hinders the conformational change of the azobenzene ligands.

The thermal cis→trans isomerization of the azobenzene ligands confined in the nanopores has a constant rate constant and exhibits faster isomerization than in solution with the exception of the bulky AzoG3 dendrimer. This result is consistent with Brinker and coworkers' observation of azobenzene-modified nanoporous silica films and supports their two-rate-constant physical model. The nanoporous silica materials prepared by Brinker and coworkers have a cubic (BCC) pore structure. The azobenzene ligands positioned on the pore connections have different local environments from those positioned on the spherical pore surfaces. Thus the azobenzene ligands isomerize at two different rates—fast and slow. The MCM-41 nanoporous materials prepared by Zink and coworkers have a hexagonal array pore structure in which all the azobenzene ligands positioned on the channel surfaces have the same local environment. Thus simple one-rate-constant first-order kinetics is sufficient to describe the isomerization process.

13.3.2.3. Photoisomerization of Bridged Azobenzene Ligands in Mesoporous Silica Materials. Corriu and coworkers found that for thin film samples using P123 surfactant templates, BSPA incorporated into a silica matrix exhibited similar reversible photoisomerization to that in solution (Besson et al., 2005). However, for BSPA–TEOS co-gel thin film samples without P123 template and pure BSPA gel films, the azobenzene moieties showed no change in the UV/visible spectra under UV light irradiation, indicating hindered photoisomerization. This observation is consistent with Brinker and coworkers' findings on the hybrid silica materials prepared using azobenzene-bridged organosilane (BSUA). In Brinker's case (Liu et al., 2002), the hydrogen-bonding interaction between the urea groups not only self-assembles BSUA into highly ordered lamellar structures but also locks in the structures and hinders the photoisomerization of the azobenzene ligands. Only after dilution of BSUA with TEOS and

HMDS to prevent the close packing of the BSUA molecules did the azobenzene ligands exhibit reversible photoisomerization. In Corriu's case, the BSPA/TEOS co-gel films and the pure BSPA gel films did not exhibit photoisomerization because of the lack of sufficient space for isomerization or the close packing of azobenzene ligands in the film. The P123-templated nanoporous films prepared using BSPA and TEOS precursors exhibited similar photoisomerization to those in solution because the removal of the surfactant molecules provided large free volume for the isomerization of the azobenzene ligands positioned on the pore surfaces.

13.3.3. Photoswitched Azobenzene Nanovalves

Nanovalves are devices that can regulate mass transport at the nanoscale by means of light, pH, heat, or other external stimuli. Mesoporous materials provide an excellent platform for constructing the nanovalves because of their size-controllable nanoscale pores and versatile surface chemistries. In an effort to make such devices, Mal et al. (2003a,b) synthesized coumarin-modified MCM-41 mesoporous silica powders that showed reversible control of release of guest molecules (cholestane) using UV light ($\lambda = 250$ nm). A team led by Stoddart and Zink built nanovalves that can regulate the controlled-release of guest molecules from nanoporous silica particles using redox-controllable bistable[2]rotaxane molecules (Nguyen et al., 2005; Hernandez et al., 2004) or pH-controllable supermolecules (Leung et al., 2006; Nguyen et al., 2006) to switch the opening and closing of nanopores. Azobenzene derivatives have also been used to control mass transport through lipid bilayer membranes.

13.3.3.1. Photocontrolled Transport Phenomenon in Lipid Bilayer Membranes. Photocontrolled ion transport across lipid bilayer membranes using photoresponsive compounds such as azobenzene derivatives has been of great interest for potential applications in optoelectronic devices and optical transducers. Most research has exploited membrane capacitance change because of the disruption of membrane structures resulting from photoisomerization of azobenzene-containing compounds incorporated into the lipid bilayers. Others have used the volumetric change of azobenzene moieties associated with photoisomerization.

In the 1990s, Yonezawa and coworkers pioneered the studies on the photoresponses of planar black lipid membranes containing an amphiphilic azobenzene derivative, 4-octyl-4′-(5-carbox-pentamethylene-oxy) azobenzene (8A5). (Tanaka and Yonezawa, 1997; Tanaka et al., 1996a,b, 1995; Fujiwara and Yonezawa, 1991) After 8A5 was inserted into the lipid membrane, transient current pulses were generated under alternating irradiation of UV and visible light. This method was chosen because of the lipid membrane's capacity change induced by azobenzene trans$\leftrightarrow$cis photoisomerization (Fig. 13.14a). The corresponding cis-8A5 concentration change (Fig. 13.14b) during alternating light irradiation correlated very well with the current change, demonstrating that the conformational change of the azobenzene ligands controls the K^+ flux across the

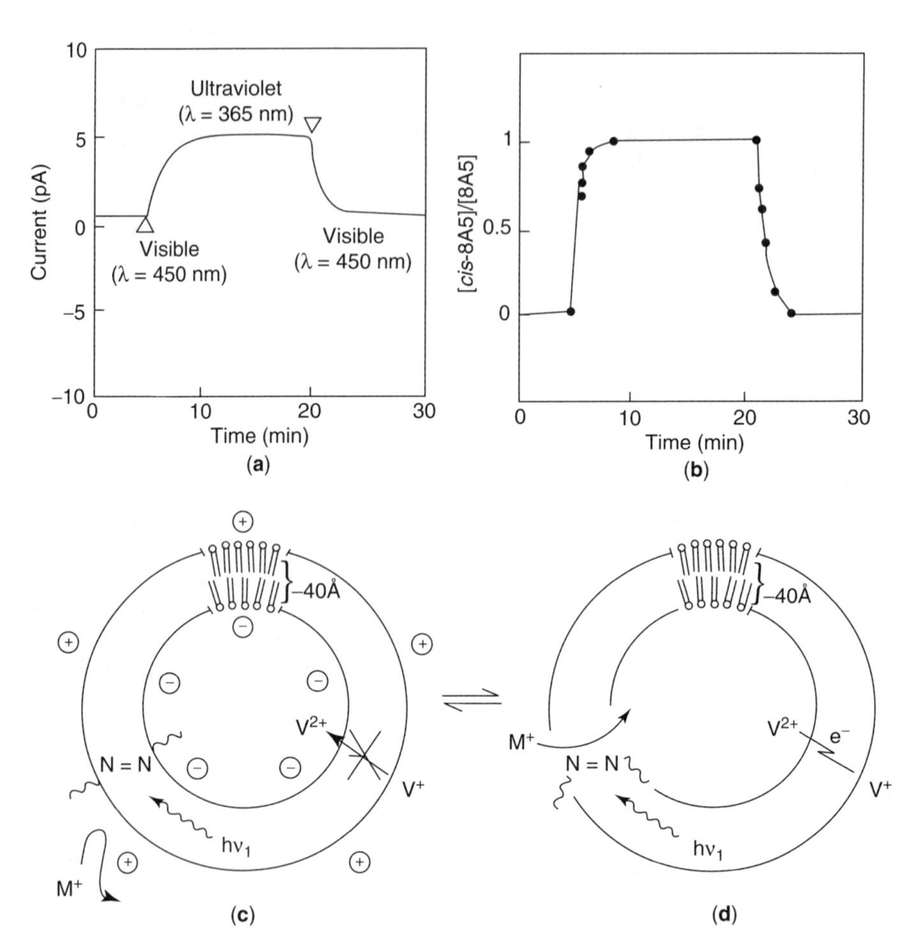

Figure 13.14. (**a**) The current response of the 8A5-doped planar lipid membrane under alternating UV (365 nm) and visible (450 nm) light irradiation. (**b**) The corresponding *cis*-8A5 concentration change during alternating light irradiation. (**c**, **d**) Concept of a bifunctional vesicle proposed by Hurst et al. (1999). In state **c**, reduction of internal viologen (V^{2+}) is blocked by membrane polarization; in state **d**, photoisomerization of the membrane-localized dye increase the permeation and the corresponding transmembrane redox reaction. *Source*: From Tanaka and Yonezawa, 1997 (parts **a** and **b**); and Lei and Hurst, 1999 (parts **c** and **b**). Reprinted with permission.

membrane. In addition to this photoelectric response, photoregulated K^+ transport through the lipid membrane and liposome membrane was also reported.

As an extension of Yonezawa's work, Hurst and coworkers reported photocontrolled K^+ permeation through a dihexadecyl phosphate (DHP) bilayer containing azobenzene amphiphilic molecules (Lei and Hurst, 1999). The concept

(Fig. 13.14c,d) was to build a vesicle containing two functional elements, a photoisomerizable molecule (dye) to control the ion permeation and an electrogenic transmembrane redox system to control the membrane polarization. The conformational change of the dye molecule disrupts the side-chain packing of the membrane surfactants, and, hence, changes the permeation rates. In the ideal circumstance, the dye molecule functions as an ion gate with an on/off switch and the membrane is on (ion permeable) in one conformation but off (ion impermeable) in the other. Photoisomerizable azobenzene derivatives are good candidates for the ion gates.

Hurst and coworkers synthesized an amphiphilic azobenzene derivative, 4-dodecy-4′-(3-phosphate-trimethyleneoxy)azobenzene (DPMA) and incorporated it into DHP vesicles. K^+ ions were entrapped in the vesicles during the vesicle formation. Photocontrolled release of K^+ ions across the vesicle membrane was conducted by monitoring the K^+ ion concentration released into the solution under UV and visible light irradiation conditions. UV/visible spectroscopic studies confirmed that DPMA molecules were dispersed into the vesicle membrane without aggregation and isomerized reversibly in the vesicle membrane under UV and visible light irradiation.

Diffusion of entrapped K^+ from DHP vesicles was extremely slow but photocontrollable. The transmembrane K^+ diffusion rate at room temperature was only slightly enhanced by the trans$\rightarrow$cis photoisomerization of the azobenzene ligands. At higher temperatures (close to the phase transition temperature), the release rate of K^+ from DHP vesicles was increased several fold (up to fivefold). The reversible switching of azobenzene between trans and cis conformations leads to alternately increasing and decreasing K^+ diffusion rates under alternating UV and visible light irradiation. Permeability coefficients at 41°C were estimated to be $4 \times 10^{12}\,\text{cm}\,\text{s}^{-1}$ and $2 \times 10^{11}\,\text{cm}\,\text{s}^{-1}$ for vesicles containing the trans and cis forms of azobenzene, respectively. As a comparison, the K^+ release rate from DHP vesicles containing predominantly *cis*-stilbene, which has a similar molecular structure to that of azobenzene, was only enhanced to ca. twofold of that from the *trans*-stilbene.

Hurst and coworkers attempted to use viologens for constructing an artificial ion-gated electron transport system because electrogenic transmembrane redox pathways can be selected by appropriate structural confirmations (Lymar and Hurst, 1994, 1992; Patterson and Hurst, 1993). However, the system was not very successful because the reducing potential of the viologen radical cations is sufficiently high to reduce the azobenzene species doped in the DHP vesicle membrane. Thus, the vesicles lost the capacity to photoregulate ionic flux across the membrane. Further investigation is needed to develop electrogenic transmembrane pathways, using less reducing electron carriers and strong oxidizing electron acceptors within the inner vesicle core.

Recently Jin (2007) synthesized an azobenzene-modified calix[4]arene, which acts as a photoresponsive ion carrier for the control of Na^+ flux across lipid bilayer membranes. The photoresponsive carrier was synthesized in two steps: first reacting hydrazine with calix[4]arene to form monohydrozide calix[4]arene and then grafting 4-dimethylaminoazobenzene to the hydrazide group through a reaction between the hydrazide and the sulfonyl chloride groups. (Fig. 13.15a) The azobenzene-modified calix[4]arene has two kinds of functionalities: the ether derivative of calix[4]arene acts as the Na^+ carrier (Jin et al., 1998, 1996) and the azobenzene provides photoresponses. Upon combination of these two functionalities, the azobenzene-modified calix[4]arene can regulate the Na^+ flux across the lipid bilayer membrane after it is incorporated into the membrane.

The azobenzene-modified calix[4]arene is one of the photochromic compounds that undergoes trans→cis isomerization upon visible light irradiation ($\pi-\pi^*$ transition at 440 nm). It can be easily incorporated into planar soybean phospholipid membranes by using the folding method (Montal and Mueller, 1972). An amplifier was used to record the trans-membrane currents and to control the voltages across the bilayer membrane. In the experiments, the ionic flux current was measured using a voltage-clamping method. It was found that the trans-membrane

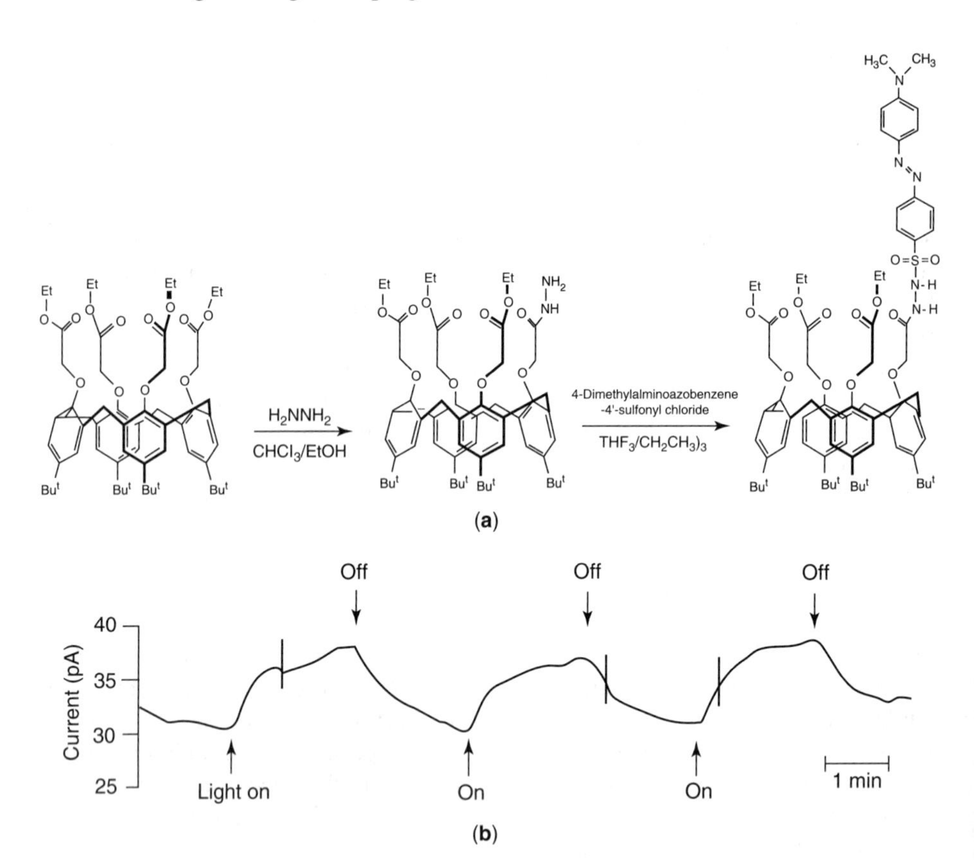

Figure 13.15. **(a)** Synthetic scheme of the azobenzene-modified calix[4]arene. **(b)** Photoresponses of the membrane currents at 50 mV in the presence of azobenzene-modified calix[4]arene. *Source*: From Jin, 2007. Reprinted with permission.

current linearly increased with increasing concentration of the azobenzene-modified calix[4]arene, indicating that the azobenzene-modified calix[4]arene complexed with Na^+ in a 1:1 ratio. The ion transport selectivity of Na^+ ion over Li^+ and K^+ was ~22 and fourfold, respectively. Fig. 13.15b showed the photoresponses of the Na^+ flux current across the lipid bilayer membrane doped with azobenzene-modified calix[4]arene. The trans-membrane current was progressively increased by ca. 30% under visible light (>400 nm) irradiation and decreased to the current level before the visible light exposure within several minutes after the visible light was turned off. This repeatable photoresponse correlated closely to the absorption spectra of the azobenzene-modified calix[4]arene incorporated into the lipid bilayer membrane. The control experiment using a pure lipid bilayer membrane did not show the ion flux current, and the membrane current did not change upon visible light irradiation. These results demonstrated that the azobenzene-modified calix[4]arene caused the ion flux current change when the visible light was switched on and off. The trans form of azobenzene-modified calix[4]arene is more stable than the cis form. When the visible light was on, the azobenzene ligands were excited and isomerized to their cis form. The *cis*-azobenzene-modified calix[4]arene is smaller than the *trans*-azobenzene-modified calix[4]arene, resulting in faster diffusion through the membrane. When the visible light was switched off, the azobenzene ligands thermally relaxed back to the more stable trans form. Thus the diffusion rate of the carrier-complexed Na^+ decreased because of the larger size of the *trans*-azobenzene-modified calix[4]arene. In short, the conformation (trans or cis) of the azobenzene ligands tethered to the calix[4]arene carrier determines the diffusion rate of the complexed ions and the corresponding ion flux current across the lipid bilayer membrane.

As described earlier, most of the studies on photocontrolled transport phenomenon focused on azobenzene-doped organic platforms such as planar lipid membranes and spherical vesicles, which have an intrinsic disadvantage. Those lipid membranes and vesicles are delicate and unstable, thus limiting their practical applications. Incorporation of the azobenzene moieties into a robust inorganic matrix greatly enhances the system stability and facilitates the device fabrication. For example, azobenzene moieties precisely positioned onto the pore surfaces of mesoporous silica membranes enable novel photocontrolled transport in the resulting composite materials (Liu et al., 2004).

13.3.3.2. Photoregulation of Mass Transport through Azobenzene-Modified, Mesoporous Nanocomposite Thin Films. Brinker and coworkers created azobenzene nanovalves using TSUA compound and the EISA and SDSA techniques. To demonstrate optical control of mass transport, a chronoamperometry experiment was performed using an azobenzene-functionalized, nanocomposite thin film–modified electrode as the working electrode in an electrochemical cell. The chronoamperometry experiment used ferrocenedimethanol (FDM) and its derivatives (ferrocene dimethanol diethylene glycol, FDMDG, and ferrocene dimethanol tetraethylene glycol, FDMTG) as molecular probes. This experiment provided a measurement of mass transport through the

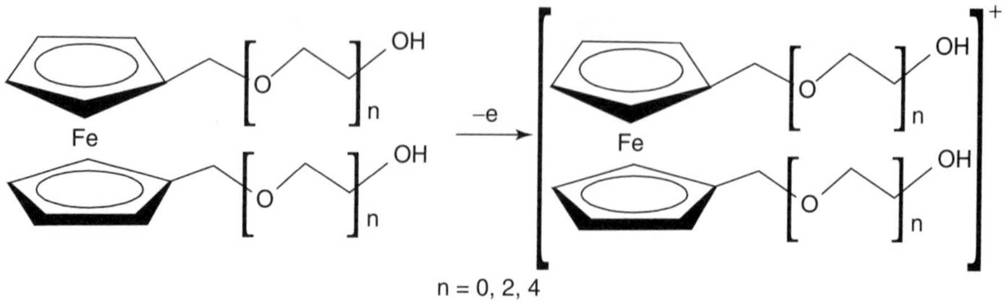

Figure 13.16. Electrochemical reactions taking place on the ITO working electrode surface.

nanocomposite film by monitoring the steady-state oxidative current at constant potential for the reactions (Fig. 13.16) taking place on the working electrode surface (ITO). By performing chronoamperometry under UV or visible light exposure, changes in the transport due to photoisomerization are reflected as changes in current.

The concept of this experiment is illustrated in Fig. 13.17. An ITO electrode is coated with a photoresponsive nanoporous film. At constant potential, the effective pore size limits the diffusion rates of the probe molecules (FDM, FDMDG, and FDMTG) to the electrode surface during electrolysis (as the probes are uncharged molecules, the electromigration effect can be ignored). Under dark conditions, the azobenzene moieties are predominately in their extended trans form. Upon UV irradiation ($\lambda = 360$ nm), the azobenzene moieties isomerize to the more compact cis form, which should increase the diffusion rate and correspondingly the oxidative current. Exposure to visible light ($\lambda = 435$ nm) triggers the cis→trans isomerization of the azobenzene moieties, decreasing the current to the pre-UV exposure level.

Figure 13.18 shows a typical *current–time* photoresponse of the azobenzene-modified nanocomposite film using FDM as the molecular probe. After initial decay due to electrode charging and the formation of a depletion layer of FDM through the film, the oxidative current of FDM reaches steady state after ~360 s. Upon UV ($\lambda = 360$ nm) irradiation, the current increases progressively because of the trans→cis isomerization, which increases the pore size before reaching a new plateau after 270 s, corresponding to the photostationary state of the trans→cis isomerization. Visible light ($\lambda = 435$ nm) exposure decreases the oxidative current owing to the cis→trans isomerization, which decreases the pore size. After 200 s of this light exposure, the current is reduced to its pre-UV exposure level, corresponding to the photostationary state of the cis→trans isomerization. Three cycles of alternating UV/visible light exposure were performed to show the reversibility and repeatability of this process. In the last cycle, a much weaker

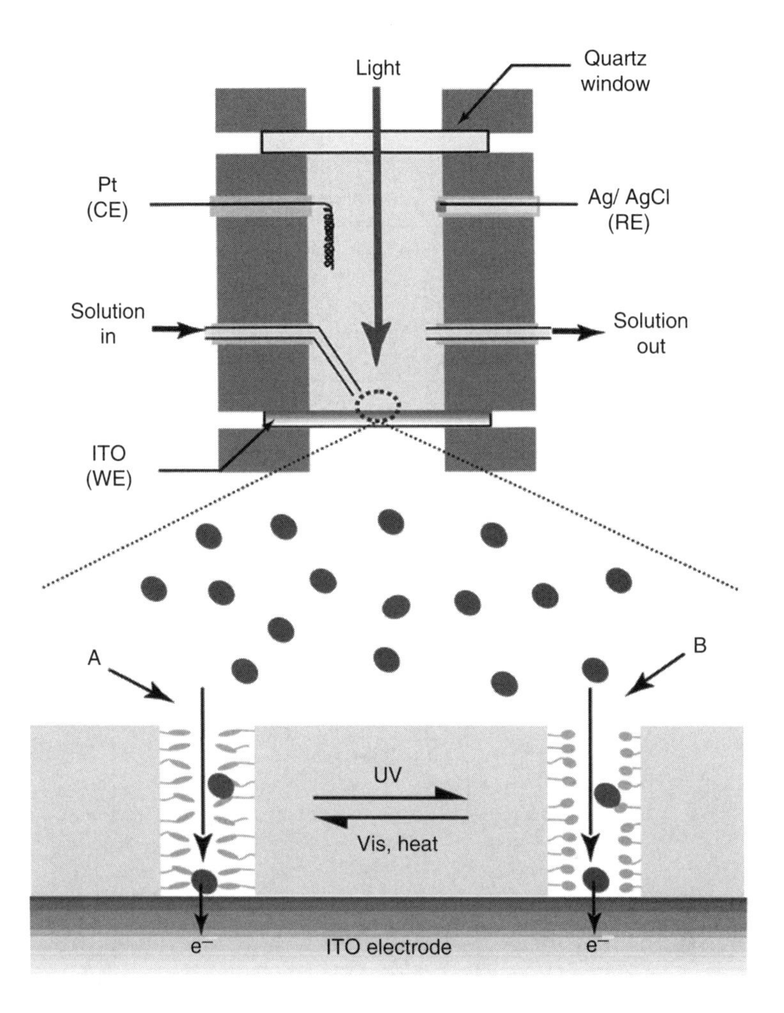

Figure 13.17. Schematic drawing of the electrochemical cell (*top*) and mass transport of probe molecules through the photoresponsive nanocomposite membrane integrated on an ITO electrode (*bottom*). **A** shows slower diffusion through smaller pores with azobenzene ligands in their trans configuration. **B** shows rapid diffusion through larger pores with azobenzene ligands in their cis configuration. *Source*: Liu et al., 2004. Reprinted with permission.

visible light (0.5 mW) was used instead of the stronger visible light (5.6 mW). We observed a much slower decrease in current, demonstrating the slower response under lower illumination levels. Control experiments were performed on a bare ITO electrode and a mesoporous silica-coated (2.5-nm pore diameter, without azobenzene) ITO electrode. However, no photoresponse was observed in either of these systems, demonstrating that the photoresponse in the present electrode/film

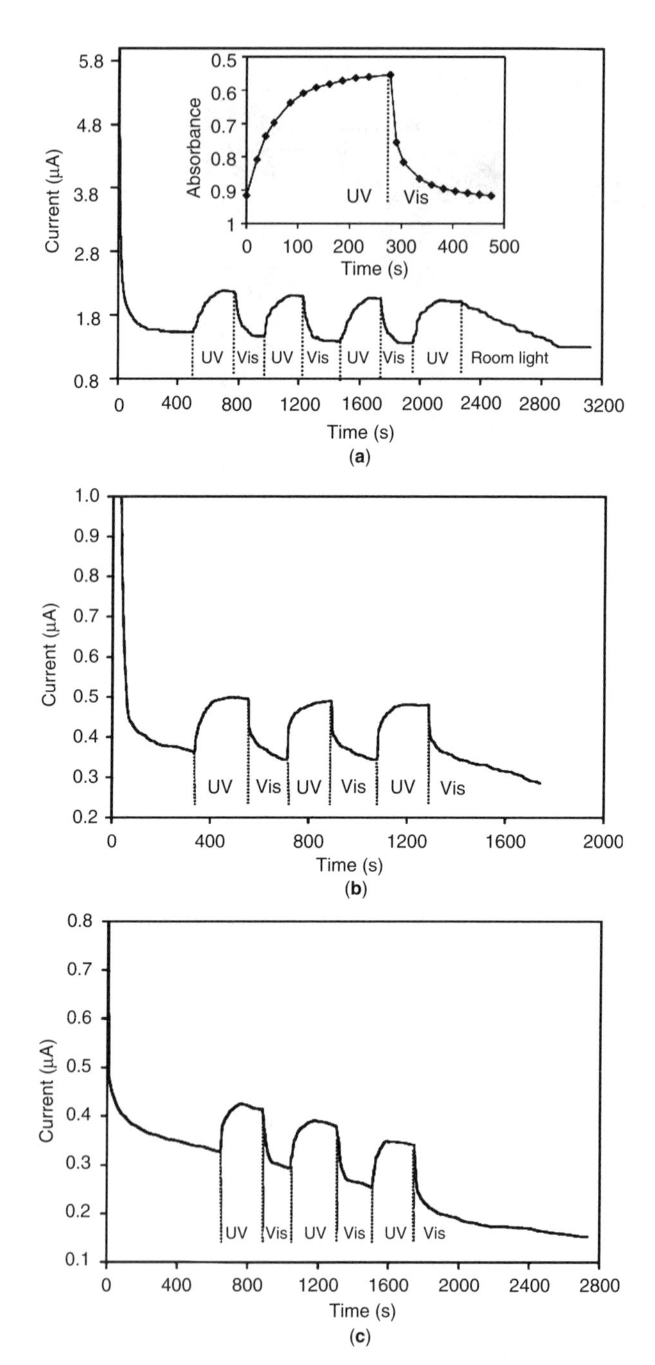

Figure 13.18. The *I–t* behavior of a Brij 56–templated photoresponsive nanocomposite film under alternate exposure to UV (360 nm) and visible light (435 nm) using various molecular probes. **(a)** FDM. Inset is the absorbance of the same film at 356 nm attributed to the $\pi-\pi^*$ transition of the trans isomer. The time scale of the UV/visible data corresponds to that of the first cycle in *I–t* response curve. **(b)** FDMDG. **(c)** FDMTG.

architecture is not an artifact of the ITO electrode or mesoporous silica film but a true effect due to the isomerization of azobenzene moieties attached to the nanopore surfaces.

To correlate the observed current changes with the actual isomerization state, a UV/visible spectroscopy study was performed, where the nanocomposite film used for chronoamperometry was immersed in the electrolyte solution and illuminated in the same manner as the first I–t cycle in the chronoamperometry experiment. As shown in the inset, Fig. 13.18a, the absorbance of the nanocomposite film at 355 nm (λ_{max} of π–π^* transition of azobenzene in trans form) decreases progressively under UV (λ = 360 nm) irradiation and reaches a plateau, which corresponds to the photostationary state of the trans→cis isomerization. The following visible light (λ = 435 nm) exposure causes the reverse cis→trans isomerization, increasing the absorbance gradually before reaching the pre-UV state. These data exactly correlate to the conformational changes of azobenzene moieties in the nanocomposite film with the oxidative current changes measured in the chronoamperometry experiment. This demonstrates for the first time optical control of mass transport through a nanocomposite thin film.

To investigate the influence of the size of the probing molecules on mass transport, FDMDG and FDMTG were synthesized according to a previously reported procedure (Lindner et al., 2001) and chronoamperometry experiments were performed using FDMDG and FDMTG as the molecular probes. The I–t curves (Fig. 13.18b,c) of the photoresponsive nanocomposite films show similar photoresponses to those of the FDM probe, that is, the oxidative current increases under UV light (360 nm) irradiation and decreases under visible light (435 nm) exposure. However, the oxidative current cannot reach steady state but decreases gradually to a very low value ($<$0.05 μA) because of the adsorption of the probe molecules and their oxidative species. The strong hydrogen-bonding interaction between the ethylene glycol groups in the probe molecules and the silanol (–OH), urea (–NH–CO–NH–), and azo groups (–N=N–) on the silica wall and the electrostatic interaction between the positively charged oxidative species of the probe molecules and negatively charged silica wall lead to the adsorption of the probe molecules and accumulation of the oxidative species on the ITO electrode surface and, correspondingly, a decrease in the current. In comparison to the hydrogen-bonding interaction, the electrostatic interaction is not predominant because of the low charge density on the silica wall in the azobenzene-modified nanocomposite membranes. As a support, steady-state oxidative current can be reached with an FDM probe that has no ethoxy groups because of less hydrogen-bonding interaction.

The Connolly solvent–excluded volumes (the volume contained within the contact molecular surface) of the probe molecules are 269 Å^3, 392 Å^3, and 515 Å^3 for FDM, FDMDG, and FDMTG, respectively. It is clear that increasing the volume of the probe molecule lowers the overall mass transport, thereby decreasing the oxidative current, but increases the selectivity as evidenced by the greater normalized current ratio ($\Delta I/I_{ss}$).

In Fig. 13.18, the normalized current ratio of Brij 56–templated TSUA-modified nanocomposite films is ~40%. However, in another experiment using P123-templated films that have a larger pore size (ca. 6.5 nm), the normalized current ratio is only ~1.2%. Because the molecular length of azobenzene ligands located on the pore surfaces is ~1.8 and 1.5 nm in the trans and cis forms, respectively, the optically triggered restriction in pore size is expected to have a diminished effect on transport for P123-templated films compared with the smaller pore size Brij 56–templated films.

As described earlier, azobenzene-functionalized nanoporous silica films exhibit dynamic photocontrol of their pore size, which in turn enables photoregulation of mass transport through the film. Their potential applications in nanofluidic devices, nanovalves, nanogates, smart gas masks, membrane separation, and controlled release can be anticipated.

13.3.3.3. Effective Diffusion Coefficient of Ferrocenedimethanol in the Nanopores. To evaluate the effective diffusion coefficient of FDM in the azobenzene-modified nanocomposite films, the chronoamperometry experiments were performed as a function of the concentration of the FDM probe. At low concentrations of FDM, the oxidative current reached the steady state within 2 min, whereas high concentrations of FDM took longer to stabilize. The steady-state current at 10 min was chosen to evaluate the effective diffusion coefficient.

On the basis of Fick's and Faraday's laws for diffusion-limited current, the following formula was used to estimate the effective diffusivity of probe molecules in the nanoporous membranes.

$$I = \frac{nFAD_{\text{eff}}}{H} C$$

where n is the number of electrons involved in the electrode reaction; in this case, $n = 1$; F is the Faraday constant, 96485 C/mol; D is the effective diffusion coefficient; A is the O-ring (7/16 in. in diameter) area exposed to the buffer solution; C is the concentration of FDM in the nanocomposite film; H is the film thickness.

As shown in Fig. 13.19, the steady-state current vs. the concentration of FDM curve is not linear. It tends to saturate at high concentrations of FDM. However, in the dilute solution region, it is linear. The slopes of the two curves without UV exposure are 5.9×10^{-4} A/M and $4.1 \times 10^{-4} A/M$, respectively. The effective diffusion coefficients (D_{eff}) of the FDM probing molecules across the photoresponsive nanocomposite film under UV light irradiation and room light exposure are estimated to be $1.6 \times 10^{-10}\,\text{cm}^2\,\text{s}^{-1}$ and $1.1 \times 10^{-10}\,\text{cm}^2\,\text{s}^{-1}$, respectively. These values are comparable with that of the transport of rohdamine 6G dye ($8 \times 10^{-11}\,\text{cm}^2\,\text{s}^{-1}$) in the mesoporous PNIPAM-modified silica particles at 50°C as reported by López and colleagues recently (Fu et al., 2003). These values are much smaller than the diffusion of free molecules in solution ($\sim 10^{-5}\,\text{cm}^2\,\text{s}^{-1}$). However, they are reasonable because the tortuous nature of the nanoporous film greatly reduces the diffusion coefficient.

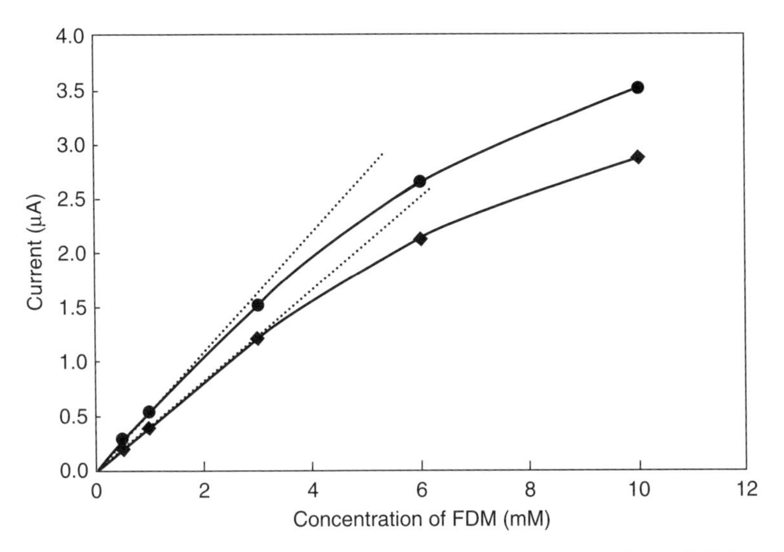

Figure 13.19. Plot of the steady-state current vs. concentration of FDM. Circles represent the current measured under UV exposure whereas rectangles represent the current measured under room light exposure.

13.3.4. Photocontrolled Release of Dye Molecules from Azobenzene-Modified Nanocomposite Particles

Brinker and coworkers prepared TSUA-modified mesoporous silica particles using the aerosol-assisted self-assembly route and demonstrated photocontrolled release of dye molecules from the nanocomposite particles (Liu, 2004). A dye molecule, oil blue N (OBN), was used to probe the effective pore size change of the azobenzene-modified nanocomposite particles in response to light stimuli. Uptake of OBN molecules into the pores of the particles was carried out by stirring the suspended particles in an OBN–EtOH solution (5 mmol/L) for 48 h under UV irradiation. As described previously, the UV irradiation causes trans→cis isomerization of the azobenzene ligands located on the pore surfaces of the nanocomposite material, and correspondingly, an enlargement of the effective pore size because of the shorter effective molecular chain length of the azobenzene ligands in the cis conformation. In this case, the pores were filled completely. After the uptake process, the particles were centrifuged, washed with ethanol (in the dark) to remove the OBN molecules adsorbed on the particle surface, and dried in a vacuum.

In the release process, the dye-filled particles were dispersed in 5 mL EtOH. One sample was stirred under UV light irradiation and the other was stirred in the dark. After filtration, the concentration of the OBN molecules released into EtOH was determined by the absorbance (A) of the solution at λ_{max} (644 nm) using UV/visible spectroscopy. The release data of the two samples are shown in Fig. 13.20. There are two factors influencing the release rate: the effective pore size

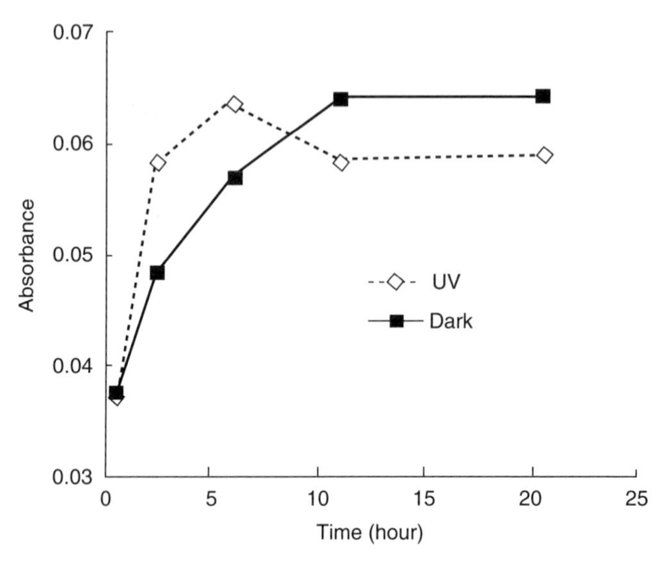

Figure 13.20. Release of the dye molecules from the azobenzene-modified nanocomposite particles under UV exposure and in the dark.

and the difference of the concentration of the probe molecules (OBN) between the nanocomposite particles and the EtOH solution. In comparison to the nanocomposite particles in the dark, the release rate of the nanocomposite particles under UV light irradiation is faster when $t < 5$ h because of the larger effective pore size. As we know, the total amounts of OBN molecules taken into the nanocomposite particles in the two samples are the same. With time, the difference in concentration of the OBN molecules between the nanocomposite particles and the EtOH solution under UV light irradiation is less than that in the dark, thereby decreasing the release rate. In the dark at $t \approx 11$ h, the concentration of OBN molecules in EtOH solution levels off, indicating a complete release of OBN molecules from the nanocomposite particles. Under UV light irradiation, the absorbance reaches the plateau value of the complete release in the dark at $t = 6$ h and decreases thereafter. The decrease in the absorbance occurs because the UV light irradiation bleaches the OBN dye molecules, as demonstrated by a control experiment. The absorbance at 644 nm of OBN in ethanol solution under UV light irradiation decreases gradually at the beginning and decreases dramatically after $t > 5$ h.

A simple model was developed to evaluate the effective diffusivity of OBN in the photoresponsive nanocomposite particles. In this model, it was assumed that the particles had uniform size and each particle was surrounded symmetrically by an equal volume of solution (volume of solution/number of particles). Although the particles had a wide size distribution, it is reasonable to use the average particle size to evaluate the effective diffusivity.

Using an average particle diameter of 2 μm and a D_s (diffusion coefficient of OBN in solution) value of $10^{-9}\,m^2\,s^{-1}$, the effective diffusion coefficients in the dark

($D_{eff,dark}$) and under UV irradiation ($D_{eff,UV}$) were estimated to be $9.5 \times 10^{-17}\,m^2\,s^{-1}$ and $1.7 \times 10^{-16}\,m^2\,s^{-1}$, respectively. These values are much smaller than the effective diffusivity of FDM in Brij 56–templated photoresponsive nanocomposite membranes ($D_{eff,dark} = 1.1 \times 10^{-14}\,m^2\,s^{-1}$, $D_{eff,UV} = 1.6 \times 10^{-14}\,m^2\,s^{-1}$). As the pore size and probing molecule size are comparable in the two cases, the pore accessibility plays a major role. It was demonstrated that the photoresponsive nanocomposite membrane had a BCC mesostructure whereas the photoresponsive nanocomposite particles had an onionlike lamellar structure. Only the defects located on the particle surfaces could provide the pathway for uptake and release of OBN probe molecules, which lowered the effective diffusivity considerably.

13.3.5. Reversible Photoswitching Liquid-Adsorption of Azobenzene-Modified Mesoporous Silica Materials

Seki et al. (1993) developed a "command surface" concept, which uses the azobenzene derivatives bound on the Langmuir–Blodgett (LB) films to control the orientation of the liquid crystal molecules on the film surfaces by the photoswitching of the azobenzene moieties. Maeda et al. (2006) extended this concept to control the adsorption properties of mesoporous silica materials by reversible photoswitching of the azobenzene moieties anchored on the pore surfaces.

A UV/visible spectra study confirmed that the azobenzene moieties isomerized from trans to cis conformation under UV ($\lambda = 365\,nm$) light irradiation at 10°C and the reverse isomerization (cis to trans) occurred under visible ($\lambda = 436\,nm$) light irradiation. Both isomerization processes saturated after 3–5 min. Although the azobenzene moieties isomerized from cis to trans in the dark because of the lower energy of the trans form, especially at elevated temperatures, it was verified that up to 85% cis conformation was retained while standing at 0°C for 24 h after UV light irradiation. It should be possible to measure the adsorption–desorption isotherms of the azobenzene-modified mesoporous materials and compare the difference of the sorption isotherms between trans and cis conformations. However, neither Brinker's nor Maeda's groups observed a significant difference in the nitrogen sorption isotherms at 77 K. One possible reason is that the probe molecules may be too small to discriminate the slight difference between the two conformations.

Instead of the chronoamperometry method employed by Brinker and coworkers to demonstrate the photocontrolled mass transport through the azobenzene-modified mesoporous silica membranes, Maeda and coworkers designed a liquid-phase adsorption experiment using 1,4,8,11,15,18,22,25-octabutoxy-29H-phthalocyanine (OBPc, ~2.3 nm in diameter) as the molecular probe to demonstrate the photoswitching of adsorption of OBPc on the azobenzene-modified pore surfaces. In this experiment, the azobenzene-modified MSU-H nanoparticles were suspended in an organic solvent containing specific amounts of OBPc molecular probes under alternating irradiation of UV and visible light. It was confirmed that the concentration of the unadsorbed OBPc molecules increased under UV light irradiation and decreased under visible light irradiation

(Fig. 13.21), an indication that the cis form azobenzene adsorbed less OBPc than the trans form azobenzene. The nearly constant OBPc concentration of each solution after repeated UV and visible light irradiation indicates that the change in the adsorption is reversible and reproducible. Fig. 13.21 shows the adsorption isotherms with clear difference in the adsorption amount of OBPc between the two forms (the trans form adsorbs more OBPc than the cis form). The head alkyl group affects adsorption amount of OBPc. Longer alkyl chain results in lower adsorption. A significant difference in adsorption between the trans form and the cis form was reported. A maximum value of $4.2\,\mathrm{mg\,g^{-1}}$ was observed when the alkyl group had four carbons, and a minimum value of $0.1\,\mathrm{mg\,g^{-1}}$ was measured when there was no alkyl head group.

These results conflict with the nanovalve–molecular gate mechanism (Liu et al., 2004), which suggests that the cis form should have a higher OBPc adsorption amount than the trans form. Maeda proposed a mechanism to explain the results. As in the case of the command surface on a flat surface, the direction of the head alkyl groups control the orientation of the OBPc molecules adsorbed on the channel surfaces through the van der Waals interaction between the butoxy groups and the alkyl groups of azobenzene. When the azobenzene groups are in cis form, the OBPc molecules adsorbed on the pore surfaces are more randomly

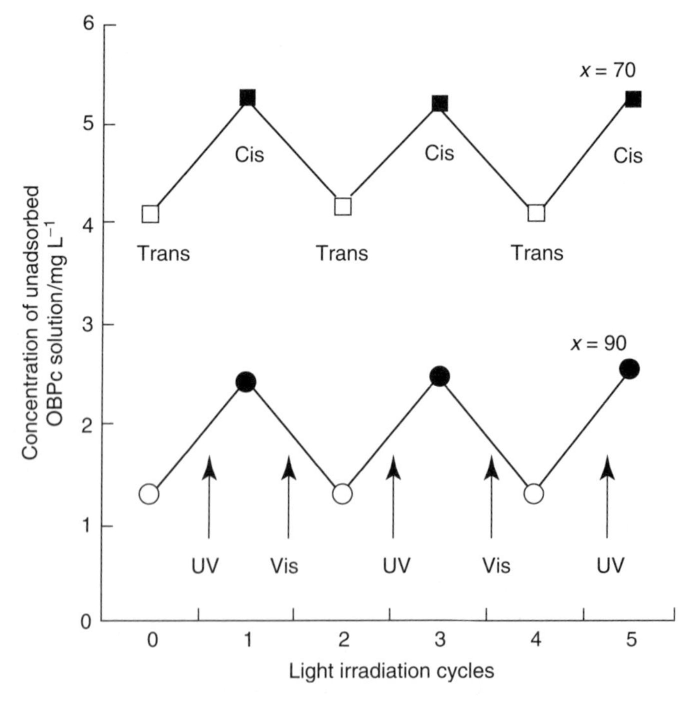

Figure 13.21. Variation in concentration of unadsorbed OBPc solution under UV and visible light irradiation. *Source*: From Maeda et al., 2006. Reprinted with permission.

oriented, resulting in lower adsorption. Therefore, it is reasonable to conclude that variation of OBPc packing on the channel surfaces contributes to photoresponsive adsorption behavior.

13.4. PHOTORESPONSIVE POLYSILSESQUIOXANE GELS

Hybrid polysilsesquioxane gels can be formed using conventional sol–gel hydrolysis and condensation chemistries in the absence of surfactant templates. These highly disordered xerogels and aerogels synthesized using silsesquioxane precursors can be fabricated into various forms that range from nonporous films to microporous monoliths in a single step, with organic functional groups dispersed in the gel at the molecular level. These features imply their potential applications in separations, catalysis, optics, sensors, and low dielectric coatings. Azobenzene-containing polysilsesquioxane gels are of particular interest for photocontrolled properties such as refractive index (Kudo et al., 2006) and stress (Liu, 2004).

13.4.1. Azobenzene-Modified Polysilsesquioxanes for Photocontrol of Refractive Index

In an effort to make novel materials that accommodate large changes in refractive index, Nishikubo and coworkers prepared a polysilsesquioxane (M_n = 10000, M_w/M_n = 2.89) by hydrolysis and condensation of 4-chloromethylphenyltrimethoxysilane (CPTS) in toluene at 60°C for 12 h and modified the polysilsesquioxane side chains with 4-*tert*-butylphenylazophenol or 4-phenylazophenol (Kudo et al., 2006). The resulting polysilsesquioxanes with azobenzene-containing side chains exhibited facile photoisomerization in thin films. Upon UV light irradiation, the absorption band located at ~334 nm ($\pi-\pi^*$ transition) progressively decreased to a photostationary state in ca. 25 s. Two isosbestic points were also observed at ~290 and 410 nm, indicating the trans→cis photoisomerization of the azobenzene moieties. The photochemical reaction followed the first-order kinetics with rate constants of 0.109 s^{-1} and 0.131 s^{-1} for 4-*tert*-butylphenylazophenol and 4-phenylazophenol modified polysilsesquioxanes, respectively. The rate constants are several orders of magnitude larger than those of other azobenzene-containing sol–gel materials (Sierocki et al., 2006; Liu et al., 2003b), probably owing to the very high power of UV light (500 W) used in the experiments, which is ca. 2-order of magnitude higher than the UV light sources used by others.

The refractive indices of the azobenzene-modified polysilsesquioxane films in trans and cis conformations were measured. The differences between the refractive indices ($\Delta n_D = n_{D,trans} - n_{D,cis}$) are 0.007 and 0.009 for 4-*tert*-butylphenylazophenol and 4-phenylazophenol modified polysilsesquioxanes, respectively. We can expect that the photocontrolled refractive indices of the azobenzene-modified polysilsesquioxanes will have novel applications in optoelectronics, switching devices, optical waveguides, and holographic image records.

13.4.2. Azobenzene-Modified Polysilsesquioxane Gels for Optomechanical Devices

Photomechanical properties of azobenzene-containing materials were widely studied because of the promising applications in photocontrolled switches and optomechanical actuators. Upon UV or visible light irradiation, the azobenzene ligands change conformation and corresponding molecular dimension. This molecular level conformational change can be transformed into microscopic and even macroscopic deformation of devices (Kim et al., 2005a,b,c).

13.4.2.1. Photomechanical Responses of Azobenzene-Containing Materials. Ji et al. (2004) reported that a self-assembled monolayer (SAM) of azobenzene molecules on a silicon microcantilever created a photon-driven switch. Commercial silicon microcantilevers were electron-beam coated on one side with 3 nm of primer layer (chromium), followed by 20 nm of gold. An azobenzene derivative with a decanthiol substituent on one ring was used to functionalize the gold surface. The thiol (-SH) end attached to the gold surface and the other azobenzene end disposed outward.

Before UV light irradiation, the azobenzene moieties adopted the more stable trans form. UV irradiation of the *trans*-azobenzene-functionalized surface caused the downward deflection of the microcantilever with a maximum bending at $\sim$210 nm (Fig. 13.22). When the UV light was switched off, the position of the microcantilever relaxed back to the level before the UV irradiation. This microcantilever deformation under UV irradiation was not observed in control experiments (Fig. 13.22) for silicon microcantilevers coated with a bare gold surface and a SAM of 1-dodecanethiol on the gold surface. This effect was demonstrated to be reversible through numerous cycles (Fig. 13.22 inset), suggesting potential use in photomechanical devices.

Ji and coworkers attributed the photoresponses of the azobenzene-functionalized microcantilever to the size change accompanying the trans$\leftrightarrow$cis isomerizations. Before UV irradiation, the azobenzene moieties were close-packed as a SAM on the gold surface. Upon UV irradiation, the azobenzene moieties changed to the cis conformation, which is larger than the trans conformation, leading to the repulsion of molecules in the SAM and, consequently, the downward deflection of the microcantilever. Switching off the UV irradiation caused thermal cis$\rightarrow$trans isomerization of the azobenzene moieties and incurred the complete relaxation of the repulsion force within the SAM, returning the cantilever to its original position.

Endo and coworkers also investigated the photomechanical behavior of a semi-interpenetrating network (IPN) and a polymer blend that were composed of poly(vinyl ether) containing azobenzene moieties in the side chain and a polycarbonate matrix (Kim et al., 2005a,b,c). Under UV (or UV plus visible) light irradiation, the azobenzene moieties contained in the semi-IPN film and the polymer blend undergo a trans$\rightarrow$cis conformational change that immediately decreases its elasticity and correspondingly increases its strain.

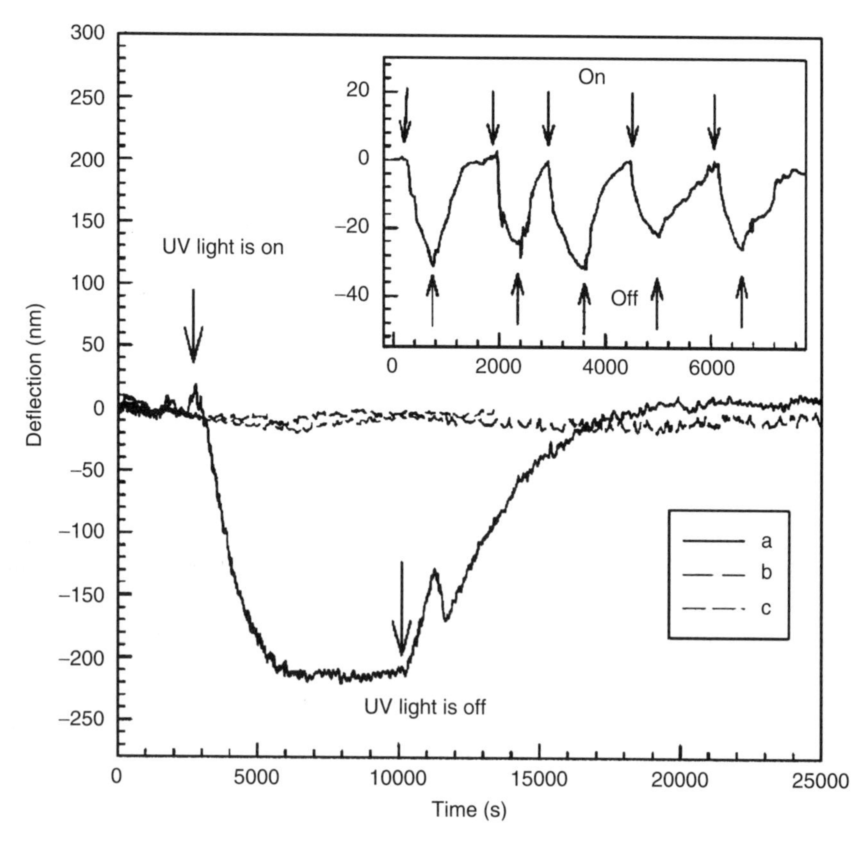

Figure 13.22. Bending responses as a function of time for silicon microcantilevers coated with (**a**) a monolayer of the *trans*-azobenzene compound on the gold surface, (**b**) a bare gold surface, (**c**) a monolayer of 1-dodecanethiol on the gold surface after exposure to a 365 nm UV light. *Source*: From Ji et al., 2004. Reprinted with permission.

Switching off the UV (or UV plus visible) light irradiation decreases the strain immediately.

A thermal mechanical analyzer (TMA) was used to measure the strain response of the semi-IPN film under a constant tensile stress loading sufficient to cause the elastic deformation of the film. When irradiated with a UV light ($\lambda = 365$ nm), elongation of the film was observed (Fig. 13.23). The strain reached a maximum value (0.01%–0.1%) in 2–4 s and stayed at the same level with small fluctuations during the UV exposure. Upon turning off the UV irradiation, the film contracted immediately and recovered its original length. To show the reversibility and reproducibility, 4–5 cycles were repeated. The same experiment was repeated using a visible light source ($\lambda > 440$ nm) instead of the UV light. A similar strain response was observed when the visible light was switched on (Fig. 13.23c), that is, the strain reached the maximum value in a very short time.

But when the visible light was turned off, the strain exhibited a different behavior, that is, the strain recovered its original length in two steps, a rapid contraction process followed by a slow contraction process. Unfortunately, the mechanism explaining this phenomenon is not clear. Since both UV and visible light caused the elongation of the semi-IPN film, an unfiltered light source with a broad wavelength (200–600 nm, $P = 24\,\mathrm{mW/cm^2}$) was used to control the deformation of the film. It was found that the maximum strain was 0.02% (Fig. 13.23c), which was higher than that obtained using a single light source (UV or visible light). This difference is probably a result of the higher illumination intensity from the broad band of wavelength.

It is surprising that the direction of the strain induced by the photomechanical response associated with the trans$\leftrightarrow$cis isomerization of the azobenzene moieties in the semi-IPN film is independent of the wavelength of the light. However, this property distinguishes the system from other azobenzene-based photoresponsive systems. Almost every photoresponse (contact angle, stress, pore size, etc.) (Liu et al., 2004; Lei and Hurst, 1999; Patterson and Hurst, 1993) in other systems under visible light exposure changes in the opposite direction to that induced by UV light. It suggests that the photomechanical response might be a result of the heat evolution in the film during light exposure. However, in a control experiment using a pure polycarbonate film, no deformation was observed during UV or visible exposure. The temperature time curve in the deformation process shown in Fig. 13.23a and c exhibited an increase of $\sim 0.2°\mathrm{C}$ and $1°\mathrm{C}$ in one cycle with respect to UV and visible light irradiation, respectively, which contradicted the similar photomechanical response (strain $= \sim 0.01\%$) of the film. Furthermore, the temperature change upon switching the light off and on was slow and could not explain the rapid photomechanical response of the film. On the basis of these results, the photomechanical response of the semi-IPN film is a true phenomenon associated with the azobenzene conformational change upon UV or visible light exposure.

The modulus of the semi-IPN film under UV and visible light exposure was investigated using TMA. Compared with the modulus value of the original film in the dark, it was decreased by light irradiation in the order: UV + visible < UV < visible. The elasticity change of the film caused the deformation. To investigate the relationship between the azobenzene conformational change and the elasticity change, the isomerization of the azobenzene moieties was monitored using UV/visible spectroscopy under light irradiation. It was evident that UV light isomerized the azobenzene ligands more completely from the trans to cis conformation than UV plus visible light. There was no significant change in the azobenzene conformation upon visible light irradiation. The photoisomerization of the azobenzene moieties is a slow process over 5 min, which cannot explain the rapid deformation of the film. Nevertheless, under visible light irradiation, the film did not show significant change in the trans:cis ratio as indicated in the UV/visible spectra data. It showed the deformation change immediately. From these results, it is impossible to correlate the photoinduced modulus change of the film with the trans:cis ratio of the azobenzene moieties in an equilibrium state during the irradiation.

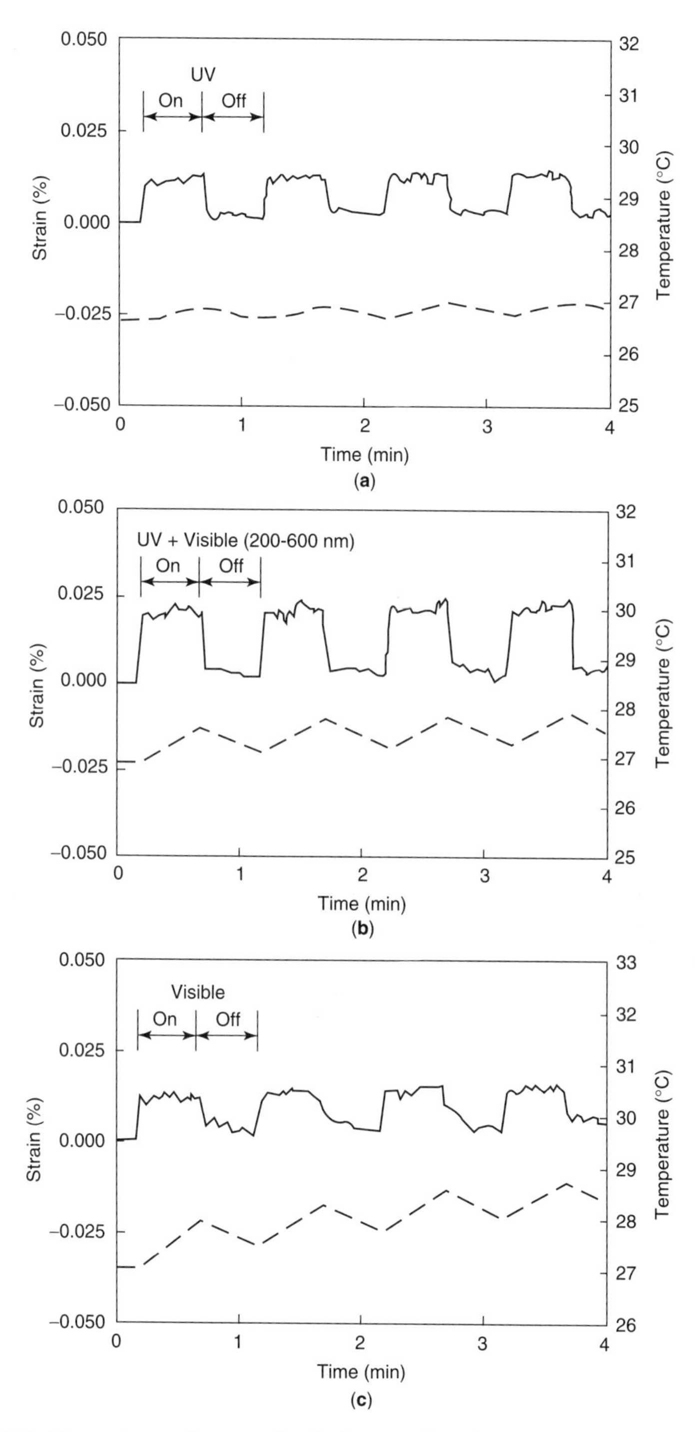

Figure 13.23. Time dependency of relative strain of semi-IPN films under alternating switching on and off of (**a**) UV, (**b**) UV+visible light, and (**c**) visible. *Source*: Kim et al., 2005a,b. Reprinted with permission.

The aggregation–dissociation mechanism (Kim et al., 2005a,c) tentatively proposed to explain the modulus change of the film under the alternating switching on and off of the UV irradiation is inadequate to account for the systems driven by the visible light. Under visible light irradiation, the film exhibited a behavior similar to that under UV irradiation. Endo and coworkers proposed a dynamic mechanism (Kim et al., 2005b) to explain this phenomenon. The azobenzene ligands under light irradiation change their conformation (trans or cis) dynamically at any moment. In other words, azobenzene moieties resonate to the incident light and the energy from irradiation is transformed into the vibratory motion of the excited azobenzene. The *trans*-azobenzene can be excited by UV light and the *cis*-azobenzene can be excited by visible light. Thus both types of irradiation can trigger the vibratory motion, which disturbs the local packing of the polymer chain and creates local free volume, resulting in a decrease in the modulus and elongation of the film. When the light is switched off, the azobenzene moieties cannot harvest photo energy for subsequent resonating. Thus the local free volume in the matrix disappears and the initial modulus is recovered completely. This mechanism is evidenced by several other studies. One example is that the frequent photoisomerization of azobenzene moieties could induce the softening of polymer films because of the light-induced free volume formation under UV and visible irradiation (Mechau et al., 2005, 2002).

Macroscopic deformation is a collective behavior of the numerous azobenzene ligands incorporated into the material. It is much easier to detect experimentally than molecular-level deformation. Despite the difficulty in device fabrication and microscopic force measurement, rigorous studies of single-molecule optomechanical transduction using an individual polymer chain containing photoresponsive azobenzene ligands in the backbone were reported.

Gaub and coworkers demonstrated for the first time the optomechanical conversion in a single-molecule device using atomic force microscope (AFM) techniques (Hugel et al., 2002). One end of a single polyazopeptide molecule with elasticity of $E = 20000$ pN was attached to the cantilever of the AFM and the other end to a total internal reflection surface. UV (365-nm) pulses shortened the molecule by 2.8 nm and 420-nm light pulses lengthened the molecule by 1.4 nm. Alternating UV (365 nm) and 420-nm pulses caused reversible contraction and extension of the molecule. As shown in Fig. 13.24a, at the first step, 420-nm pulses lengthened the polymer with a force of 80 pN (I), then expanded mechanically to a restoring force of 200pN (II). UV pulses caused a contraction of the polymer chain (III) and the force on the polymer was reduced to 85 pN (IV). Finally, 420-nm pulses changed the system back to its original state. From point II to III, the contraction of the polymer chain lifted the weight and thereby did mechanical work. Fig. 13.24b describes the corresponding force extension. The slope of the optical branches of the cycle are determined by the stiffness of the cantilever. The work output of the system (shadow area in Fig. 13.24b) is the mechanical energy related to the contraction of the polymer. The energy input needed to perform this mechanical work is light energy. However, the efficiency of energy conversion

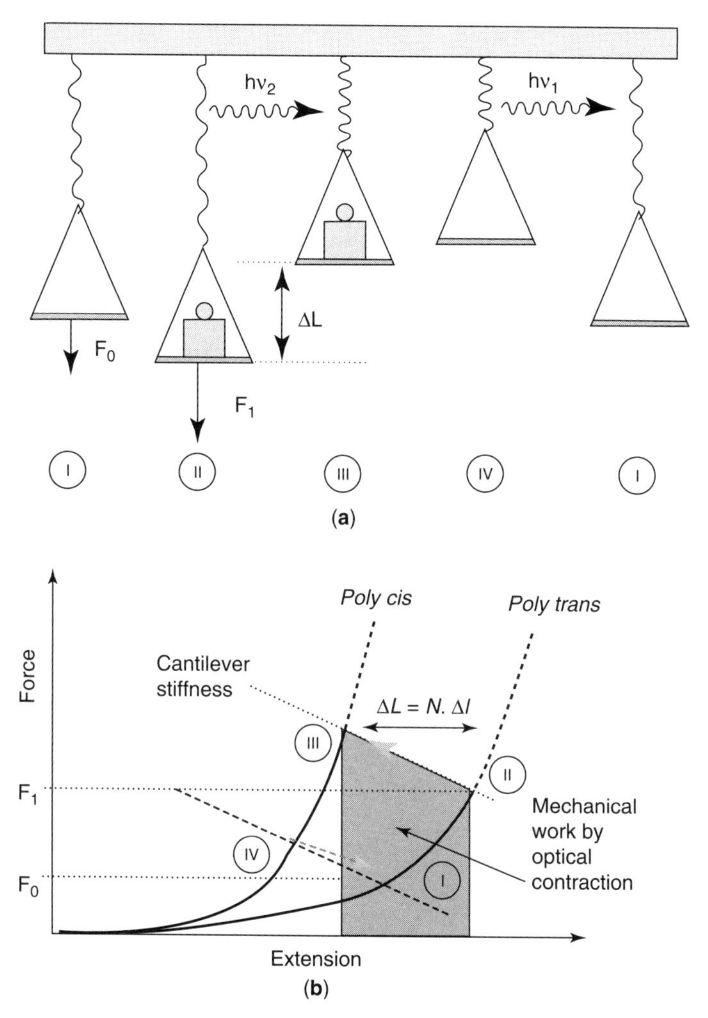

Figure 13.24. Schematics of the single-molecule operating cycle. Points I–IV in (**a**) represent the four states of an operating cycle. (**b**) The corresponding force extension plot. *Source*: Hugel et al., 2002. Reprinted with permission.

($\eta = W_{out}/W_{in}$) is estimated to be 0.1 in this single-molecule device, which is relatively low for optomechanical applications.

Other than the SAMs and azobenzene-containing organic polymers, azobenzene-modified hybrid polysilsesquioxane gels were also studied for their photomechanical responses.

13.4.2.2. Preparation of Azobenzene-Modified Polysilsesquioxane Gels. The BSUA molecules in highly concentrated solid states are very difficult to isomerize under UV irradiation because of the intermolecular hydrogen-bonding and π–π stacking interactions. To avoid these interactions, the BSUA molecules could be diluted in an aerogel network, providing the extra volume (~127 Å^3) and flexible framework needed for the isomerization of the azobenzene ligands. In the aerogel formation process, 15.3 mol% BSUA was diluted with 20.7 mol% bis(triethoxysilyl)ethane (BTE) and 64 mol% TEOS. BTE molecules with $-CH_2CH_2-$ bridges were used to increase the flexibility of the gel frameworks. The gel was finally derivitized with hexamethyldisilazine (HMDS) to promote stability and flexibility. Thin films were deposited on the substrates by the casting or spin-coating of the gel dispersed in EtOH.

The prepared thin films exhibited reversible photo and thermal isomerization of azobenzene ligands as detected using UV/visible spectroscopy. As shown in Fig. 13.25, the absorbance of the π–π* transition band of the trans isomer at 365 nm decreases under UV irradiation, and the corresponding absorbance of the n–π* transition band of the cis isomer at 480 nm increases (Fig. 13.25 curve b), indicating the trans → cis isomerization of azobenzene ligands contained in the film. Compared with BSUA in EtOH solvent, the n–π* transition of the cis isomer in the gel film exhibits a much higher absorbance, indicating a higher concentration of the cis isomer and a higher molar absorption coefficient (ε), which is due to the steric effect. The reverse cis → trans isomerization was achieved when the film was exposed to room light or heat (Fig. 13.25 curves c,d). The heat treatment at 100°C for 10 min even caused the 365-nm absorption band to exceed that of the as-prepared film, indicating that a certain amount of azobenzene ligands were in the

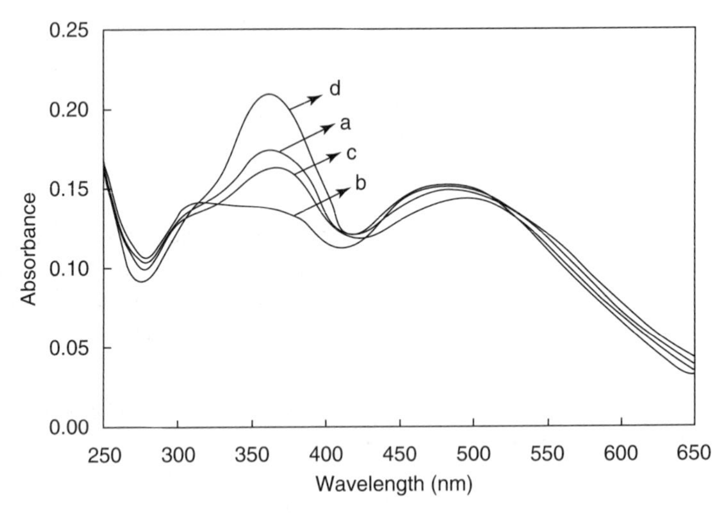

Figure 13.25. Photo and thermal isomerization of azobenzene ligands in the photoresponsive aerogel. (**a**) As-prepared, (**b**) after UV irradiation of sample (**a**) for 40 min, (**c**) after room light exposure of sample (**b**) for 60 min, (**d**) after heating sample (**b**) to 100°C for 10 min.

cis conformation in the as-prepared film. Using a molar absorption coefficient (ε) of $2.59 \times 10^4\,M^{-1}\,cm^{-1}$ for the trans isomer of TSUA, the concentration of the trans isomer in the film was estimated to be 0.27, 0.22, 0.20, and 0.17 *M* for film samples in Fig. 13.25d, a, c and b, respectively. It is difficult to estimate the concentration of the cis isomer in the film because the ε value of the n–π^* transition of the cis isomer is unknown. It is reasonable that a portion of the azobenzene ligands are in the cis conformation because the gel shrinkage during the film formation process caused by ethanol evaporation leads to a portion of the azobenzene ligands adopting the cis conformation with shorter molecular length. Only upon powerful stimulation (heat) to overcome the force (~200 pN/azobenzene ligand [Hugel et al., 2002]) involved in the conformational change does the cis isomer change to the trans isomer that has a longer molecular length.

13.4.2.3. Stress Development and Relaxation of Azobenzene-Modified Polysilsesquioxane Gels. The dimensional change associated with the trans ↔ cis isomerization of the azobenzene ligands contained in the polysilsesquioxane gel films causes the gel network to shrink or expand in response to light stimuli, which involves stress development and relaxation of the film. Brinker and coworkers used a homemade beam bending setup to observe the stress behavior in response to UV light irradiation.

The in-plane stress during alternating UV light exposure and dark exposure is calculated from Stoney's equation as follows:

$$\sigma = \frac{E_s d_s^2}{3L^2(1 - V_s)d_f}\delta$$

where σ is the in plane biaxial stress, E_s is Young's modulus of the silicon substrate (1.48×10^{12} dynes/cm^2), d_s is the substrate thickness (75 μm), L is the substrate length, V_s is Poisson's ratio of the silicon substrate (0.18), d_f is the film thickness, and δ is the measured deflection of the silicon substrate (Fig. 13.26). Tensile stress corresponds to a positive deflection (δ), which is in an upward direction. Negative δ generates contraction stress. Stoney's equation is valid, provided $d_s \gg d_f$, $d_s \gg \delta$, L is greater than twice the width of the substrate, and the elastic modulus of the substrate is much larger than that of the film.

The photoresponsive gel films were deposited on the silicon substrates using the casting or dip-coating methods. Figure 13.27 shows a typical stress response of the photoresponsive gel films (225 nm in thickness) *in situ* during UV light exposure. Before the UV light ($\lambda = 320$ nm) was turned on, the beam-bending system was zeroed by stabilizing it for half an hour. After the UV light was turned on, the tensile stress (positive direction) that comes from the trans → cis isomerization of the azobenzene ligands increases rapidly to a plateau value of ca. 0.58 MPa in 2 min. The extended UV light irradiation for ca. 26 min did not change the plateau value, which was the photostationary state of the trans → cis isomerization. Then the UV light was turned off and the whole system was kept in the dark. The corresponding cis → trans thermal isomerization caused the stress

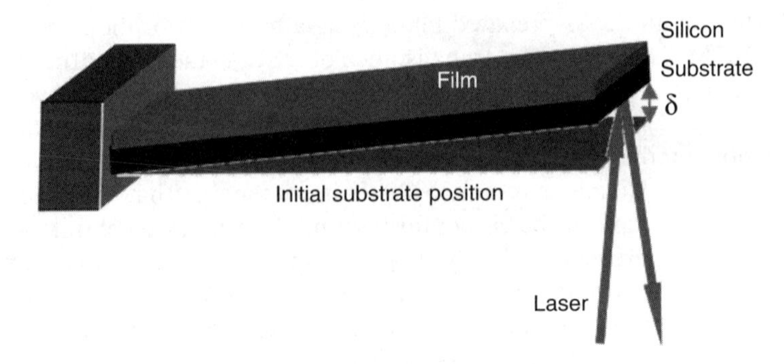

Figure 13.26. Schematic of the silicon substrate deflection under stress.

relaxation. The stress decreased gradually to a new plateau value of 0.20 MPa in ca. 70 min, which was related to the equilibrium state of the cis→trans thermal isomerization. This stress development and relaxation triggered by UV light was repeated for three cycles. In the latter two cycles, the stress curves were almost the same, demonstrating that the stress development and relaxation in response to UV light are reversible and reproducible.

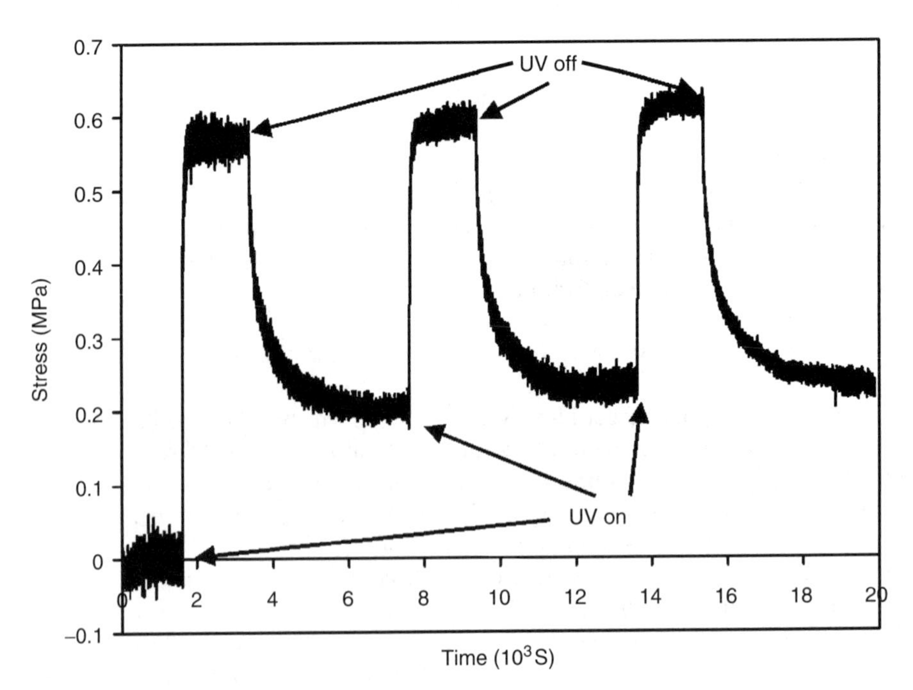

Figure 13.27. Stress development and relaxation of the photoresponsive gel films (thickness: 225 nm) in response to alternating UV light and dark exposure.

However, in the first cycle, there was a residual stress of 0.20 MPa that was not recovered completely by the cis → trans thermal isomerization in the dark. UV light can cause the condensation of sol–gel materials (Clark et al., 2000), and in this case, the UV light irradiation in the first cycle caused the further condensation of the photoresponsive gel framework, which induced tensile stress in the film. Such stress from the chemical change is not reversible and not recoverable. However, the cis → trans thermal isomerization of azobenzene derivatives at room temperature is reversible though slow ($k \approx 10^{-5}\,s^{-1}$, $t_{1/2} \approx 10^4\,s$), requiring a very long time for a complete recovery of the cis → trans thermal isomerization. In the experiment, the interval of the UV light off is ca. 70 min, not long enough for the complete recovery of the trans → cis conformational change of the azobenzene ligands.

In another experiment on a thicker film (376 nm in thickness) prepared using the same gel solution, a similar phenomenon was observed. The maximum tensile stress under UV light irradiation is 1 Mpa, and the residual stress in the first cycle is 0.32 MPa; that is, 32% of the stress was not relaxed, which is comparable with that of the former sample (34%) with a thinner thickness.

The photoresponsive stress change of the BSUA-derivatized gel films can be explained by the dimensional change accompanying the photoisomerization of the azobenzene ligand. Under UV irradiation, the trans → cis isomerization of the azobenzene ligands causes the shrinkage of the flexible gel network because of the shorter molecular length of the cis conformation than the trans conformation, which bends the silicon substrate upwardly and generates a tensile stress on the substrate. At the photostationary state, the ratio between the two isomers (trans:cis) reaches a maximum, thus the stress stays at the maximum level. When the UV light is switched off, the thermal cis → trans isomerization of the azobenzene moieties relieves the tensile stress.

13.5. FUTURE WORK

The azobenzene-modified mesoporous silica materials are ideal platforms to study photocontrolled transport behavior because of their easy fabrication, easy integration into nanodevices, and well-controlled pore connection, pore dimension, and surface chemistries. To date, all the related work reported in the literature has dealt with numerous nanochannels. It will be a milestone if we can study the photocontrolled transport behavior of a few nanochannels or even a single nanochannel with azobenzene-modified pore surfaces. However, significant engineering problems exist in the fabrication of such a single-channel device.

On the one hand, an azobenzene-modified single channel device mimics both the structure and chemistry of nature ion channels, which have intricate applications in stochastic sensing, molecular recognition, DNA sequencing, and chemical and bioweapon detection because of the ultrasensitive detection of matter passage through the channel. On the other hand, the photoisomerization of the azobenzene moieties can regulate the diffusion rate of molecules and ions through the channel at molecular scale, which opens a new era for sensitive reactions requiring minute amounts of reactants.

Other than the fabrication of single-channel devices, new azobenzene-containing organosilanes must be synthesized to make the photoresponsive nanoporous materials more suitable for applications in nanofluidic devices, molecular gates, and controlled releases. The reported azobenzene-containing organosilanes, that is, TSUA, have flexible "spacer" groups between the trialkoxysilyl and rigid azobenzene groups. After the azobenzene groups are anchored on the pore surfaces, they float around because of the flexibility of the "spacer" groups, and correspondingly compromise the extent of the pore size change triggered by light stimuli. Thus, it is of particular interest to synthesize new azobenzene-containing organosilanes with rigid pendant groups, such as organosilane with an azobenzene pendant group directly bonded to a triethoxysilyl group. Furthermore, the organosilanes with multiple azobenzene groups in series are capable of a larger change in the molecular dimension and effective pore size after the assembly on the pore surfaces, enabling better control of mass transport through the corresponding photoresponsive nanocomposite films.

REFERENCES

Asano T, Okada T. 1984. Thermal 2-E isomerization of azobenzenes. The pressure, solvent, and substituent effects. J Org Chem 49:4387–4391.

Besson E, Mehdi A, Lerner DA, Reye C, Corriu RJP. 2005. Photoresponsive ordered hybrid materials containing a bridged azobenzene group. J Mater Chem 15:803–809.

Böhm N, Materny A, Kiefer W, Steins H, Müller MM, Schottner G. 1996. Spectroscopic investigation of the thermal cis–trans isomerization of disperse red 1 in hybrid polymers. Macromolecule. 29:2599–2604.

Brinker CJ, Lu Y, Sellinger A, Fan H. 1999. Evaporation-induced self-assembly: Nanostructures made easy. Adv Mater 11:579–585.

Chia S, Cao J, Stoddart JF, Zink JI. 2001. Working supramolecular machines trapped in glass and mounted on a film surface Angew Chem Int Ed 40:2447–2451.

Clark T, Ruiz JD, Fan H, Brinker CJ, Swanson BJ, Parikh AN. 2000. A new application of uv-ozone treatment in the preparation of substrate-supported, mesoporous thin films. Chem Mater 12:3879–3884.

Dillow AK, Brown JS, Liotta CL, Eckert CA. 1998. Supercritical fluid tuning of reactions rates: the cis-trans isomerization of 4-4′-disubstituted azobenzenes. J Phys Chem A 102:7609–7617.

Fu Q, Rao GVR, Ista LK, Wu Y, Andrzejewski BP, Aklar LA, Ward TL, López GP. 2003. Control of molecular transport through stimuli–responsive ordered mesoporous materials. Adv Mater 15:1262–1266.

Fujiwara H, Yonezawa Y. 1991. Photoelectric response of a black lipid membrane containing an amphiphilic azobenzene derivative. Nature 351:724–726.

Hernandez R, Tseng HR, Wong JW, Stoddart JF, Zink JI. 2004. An operational supramolecular nanovalve. J Am Chem Soc 126:3370–3371.

Hugel T, Holland NB, Cattani A, Moroder L, Seitz M, Gaub HE. 2002. Single-molecule optomechanical cycle. Science 296:1103–1106.

Ji HF, Feng Y, Xu X, Purushotham V, Thundat T, Brown GM. 2004. Photon-driven nanomechanical cyclic motion. Chem Commun 22:2532–2533.

Jin T. 2007. Calixarene-based photoresponsive ion carrier for the control of Na^+ flux across a lipid bilayer membrane by visible light. Mater Lett 61:805–808.

Jin T, Kinjo K, Koyama T, Kobayashi Y, Hirata H. 1996. Selective Na^+ transport through phospholipid bilayer membrane by a synthetic calix[4]arene carrier. Langmuir 12: 2684.

Jin T, Kinjo M, Kobayashi Y, Hirata H. 1998. Ion transport activity of calix[n]arene (n = 4, 5, 6, 7, 8) esters toward alkali-metal cations in a phospholipid bilayer membrane. J Chem Soc Faraday Trans 94:3135.

Kim HK, Wang XS, Fujita Y, Sudo A, Nishida H, Fujii M, Endo T. 2005a. A rapid photomechanical switching polymer blend system composed of azobenzene-carrying poly(vinylether) and poly(carbonate). Polymer 46:5879–5883.

Kim HK, Wang XS, Fujita Y, Sudo A, Nishida H, Fujii M, Endo T. 2005b. Reversible photo-mechanical switching behavior of azobenzene-containing semi-interpenetrating network under UV and visible light irradiation. Macromol Chem Phys 206:2106–2111.

Kim HK, Wang XS, Fujita Y, Sudo A, Nishida H, Fujii M, Endo T. 2005c. Photo-mechanical switching behavior of semi-interpenetrating polymer network consisting of azobenzene-carrying crosslinked poly(vinyl ether) and polycarbonate. Macromol Rapid Commun 26:1032.

Kresge CT, Leonowicz ME, Roth WJ, Vartuli JC, Beck JS. 1992. Ordered mesoporous molecular-sieves synthesized by a liquid-crystal template mechanism. Nature 359:710–712.

Kudo H, Yamamoto M, Nishikubo T, Moriya O. 2006. Novel materials for large change in refractive index: synthesis and photochemical reaction of the ladderlike poly(silsesquioxane) containing norbornadiene, azobenzene, and anthracene groups in the side chains. Macromolecule 39:1759–1765.

Kumar GS, Neckers DC. 1989. Photochemistry of azobenzene-containing polymers. Chem Rev 89:1915–1925.

Lei YB, Hurst JK. 1999. Photoregulated potassium ion permeation through dihexadecyl phosphate bilayers containing azobenzene and stilbene surfactants. Langmuir 15:3424–3429.

Leung KCF, Nguyen TD, Stoddart JF, Zink JI. 2006. Supramolecular nanovalves controlled by proton abstraction and competitive binding. Chem Mater 18:5919–5928.

Lindner E, Kehrer U, Steimann M, Ströbele M. 2001. Preparation, properties, and reactions of metal-containing heterocycles part cv. Synthesis and structure of polyoxadiphosphaplatinaferrocenophanes. J Organomet Chem 630:266–274.

Liu NG 2004. Photoresponsive nanocomposite materials. Chemical Engineering [dissertation]. Albuquerqe (NM): the University of New Mexico.

Liu NG, Yu K, Smarsly B, Dunphy DR, Jiang YB, Brinker CJ. 2002. Self-directed assembly of photo-active hybrid silicates derived from an azobenzene-bridged silsesquioxane. J Am Chem Soc 124:14540–14541.

Liu NG, Chen Z, Dunphy Dr, Jiang YB, Assink Ra, Brinker CJ. 2003a. Photoresponsive nanocomposite formed by self-assembly of an azobenzene-modified silane. Angew Chem Int Ed 42:1731–1734.

Liu NG, Dunphy DR, Rodriguez MA, Singer S, Brinker CJ. 2003b. Synthesis and crystallographic structures of a novel photoresponsive azobenzene-containing organosilane. Chem Commun 10:1144–1145.

Liu NG, Dunphy DR, Atanassov P, Bunge SD, Chen Z, López GP, Boyle TJ, Brinker CJ. 2004. Photoregulation of mass transport through a photoresponsive azobenzene–modified nanoporous membrane. Nano Lett 4:551–554.

Lu Y, Ganguli R, Drewien CA, Anderson MT, Brinker CJ, Gong WL, Guo YX, Soyez H, Dunn B, Huang MH, Zink JI. 1997. Continuous formation of supported cubic and hexagonal mesoporous films by sol-gel dip-coating. Nature 389:364–368.

Lu Y, Fan H, Stump A, Ward TL, Reiker T, Brinker CJ. 1999. Aerosol-assisted self-assembly of spherical, silica nanoparticles exhibiting hexagonal, cubic, and vesicular mesophases. Nature 398:223–226.

Lymar SV, Hurst JK. 1992. Mechanisms of viologen-mediated charge separation across bilayer membranes deduced from mediator permeabilities. J Am Chem Soc 114: 9498.

Lymar SV, Hurst JK. 1994. Electrogenic and electroneutral pathways for methyl viologen-mediated transmembrane oxidation-reduction across dihexadecylphosphate vesicular membranes. J Phys Chem 98: 989.

Maclachlan MJ, Asefa T, Ozin GA. 2000. Writing on the wall with a new synthetic quill. Chem Eur J 6:2507–2511.

Maeda K, Nishiyama T, Yamazaki T, Suzuki T, Seki T. 2006. Reversible photoswitching liquid-phase adsorption on azobenzene derivative-grafted mesoporous silica. Chem lett 35:736–737.

Mal NK, Fujiwara M, Tanaka Y. 2003a. Photocontrolled reversible release of guest molecules from coumarinmodified mesoporous silica. Nature 421:350–353.

Mal NK, Fujiwara M, Tanaka Y, Taguchi T, Matsukata M. 2003b. Photo-switched storage and release of guest molecules in the pore void of coumarin-modified mcm-41. Chem Mater 15:3385–3394.

Mechau N, Neher D, Borger V, Menzel H, Urayama K. 2002. Appl Phys Lett 81: 4175.

Mechau N, Saphiannikiva M, Neher D. 2005. Dielectric and mechanical properties of azobenzene polymer layers under visible and ultraviolet irradiation Macromolecule 38:3894.

Montal M, Mueller P. 1972. Formation of bimolecular membranes from lipid monolayers and a study of their electrical properties. Proc Natl Acad Sci USA 69:3561.

Natansohn A. 1999. Azobenzene-containing materials. Weinheim: Wiley. p. 165.

Nguyen TD, Tseng HR, Celestre PC, Flood AH, Liu Y, Stoddart JF, Zink JI. 2005. A reversible molecular valve. Proc Natl Acad Sci USA 102:10029–10034.

Nguyen TD, Leung KCF, Liong M, Pentecost CD, Stoddart JF, Zink JI. 2006. Construction of a ph-driven supramolecular nanovalve. Org Lett 8:3363–3366.

Nicole L, Boissiere C, Grosso D, Quach A, Sanchez C. 2005. Mesostructured hybrid organic-inorganic thin films. J Mater Chem 15:3598–3627.

Ogawa M, Kurodacd K, Moric JI. 2000. Aluminium-containing mesoporous silica films as nano-vessels for organic photochemical reactions. Chem Commun:2441–2442.

Patterson BC, Hurst JK. 1993. Pathways of viologen-mediated oxidation-reduction reactions across dihexadecyl phosphate bilayer membranes. J Phys Chem 97: 454.

Rau H. 1990a. Photochemistry and photophysics. Rabek J, editor. Boca Raton (FL): CRC Press Inc. Vol II; chap. 4.

Rau H. 1990b. Azo compounds. In Studies in Organic Chemistry, Photochromism, Molecules and Systems. Dürr H, Bouas–Lauran H, editors. Amsterdam: Elsevier. pp. 165–192.

Seki T, Sakuragi M, Kawanishi Y, Tamaki T, Fukuda R, Ichimura K, Suzuki Y. 1993. "Command surfaces" of Langmuir-Blodgett films. Photoregulations of liquid crystal alignment by molecularly tailored surface azobenzene layers. Langmuir 9:211–218.

Sierocki P, Maas H, Dragut P, Richardt G, Vogtle F, Cola L, Brouwer FAM, Zink JI. 2006. Photoisomerization of azobenzene derivatives in nanostructured silica. J Phys Chem B 110:24390–24398.

Tanaka M, Yonezawa Y. 1997. Photochemical regulation of ion transport through 'quasi-channels' embedded in black lipid membrane. Mater Sci Eng C 4(4):297–301.

Tanaka M, Sato T, Yonezawa Y. 1995. Permeability enhancement in phospholipid bilayer containing azobenzene derivative around the phase transition temperature. Langmuir 11:2834.

Tanaka M, Sato T, Yonezawa Y. 1996a. Specific photoresponse of dmpc liposomes doped with azobenzene derivative around the phase transition temperature. Thin Solid Films 284/285:829–832.

Tanaka M, Yonezawa Y, Sato T. 1996b. Photochemical control of DC current across planar bilayer lipid membranes doped with azobenzene derivative. Thin Solid Films 284/285:833–835.

Ueda M, Kim HB, Ikeda T, Ichimura K. 1992. Photoisomerization of an azobenzene in sol-gel glass films. Chem Mater 4:1229–1233.

Ueda M, Kim Hb, Ikeda T, Ichimura K. 1993. Photoisomerizability of an azobenzene covalently attached to silica gel matrix. J Non-Crystal Solids 163:125–132.

Ueda M, Kim HB, Ichimura K. 1994. Photochemical and thermal isomerization of azobenzene derivatives in sol-gel bulk materials. Chem Mater 6:1771–1775.

Victor JG, Torkelson JM. 1987. On measuring the distribution of local free-volume in glassy polymers by photochromic and fluorescence techniques. Macromolecule 20:2241–2250.

INDEX

Smart Light-Responsive Materials. Edited by Yue Zhao and Tomiki Ikeda